Springer Complexity

Springer Complexity is an interdisciplinary program publishing the best research and academic-level teaching on both fundamental and applied aspects of complex systems – cutting across all traditional disciplines of the natural and life sciences, engineering, economics, medicine, neuroscience, social and computer science.

Complex Systems are systems that comprise many interacting parts with the ability to generate a new quality of macroscopic collective behavior the manifestations of which are the spontaneous formation of distinctive temporal, spatial or functional structures. Models of such systems can be successfully mapped onto quite diverse "real-life" situations like the climate, the coherent emission of light from lasers, chemical reaction-diffusion systems, biological cellular networks, the dynamics of stock markets and of the internet, earthquake statistics and prediction, freeway traffic, the human brain, or the formation of opinions in social systems, to name just some of the popular applications.

Although their scope and methodologies overlap somewhat, one can distinguish the following main concepts and tools: self-organization, nonlinear dynamics, synergetics, turbulence, dynamical systems, catastrophes, instabilities, stochastic processes, chaos, graphs and networks, cellular automata, adaptive systems, genetic algorithms and computational intelligence.

The three major book publication platforms of the Springer Complexity program are the monograph series "Understanding Complex Systems" focusing on the various applications of complexity, the "Springer Series in Synergetics", which is devoted to the quantitative theoretical and methodological foundations, and the "SpringerBriefs in Complexity" which are concise and topical working reports, case-studies, surveys, essays and lecture notes of relevance to the field. In addition to the books in these two core series, the program also incorporates individual titles ranging from textbooks to major reference works.

Springer Series in Synergetics

Founding Editor: H. Haken

The Springer Series in Synergetics was founded by Herman Haken in 1977. Since then, the series has evolved into a substantial reference library for the quantitative, theoretical and methodological foundations of the science of complex systems.

Through many enduring classic texts, such as Haken's *Synergetics and Information and Self-Organization*, Gardiner's *Handbook of Stochastic Methods*, Risken's *The Fokker Planck-Equation* or Haake's *Quantum Signatures of Chaos*, the series has made, and continues to make, important contributions to shaping the foundations of the field.

The series publishes monographs and graduate-level textbooks of broad and general interest, with a pronounced emphasis on the physico-mathematical approach.

More information about this series at http://www.springer.com/series/712

Peter Schuster

Stochasticity in Processes

Fundamentals and Applications to Chemistry and Biology

Springer

Peter Schuster
Institut für Theoretische Chemie
Universität Wien
Wien, Austria

ISSN 0172-7389 ISSN 2198-333X (electronic)
Springer Series in Synergetics
ISBN 978-3-319-39500-5 ISBN 978-3-319-39502-9 (eBook)
DOI 10.1007/978-3-319-39502-9

Library of Congress Control Number: 2016940829

Printed on acid-free paper

This Springer imprint is published by Springer Nature
The registered company is Springer International Publishing AG Switzerland

Dedicated to my wife Inge

Preface

The theory of probability and stochastic processes is often neglected in the education of chemists and biologists, although modern experimental techniques allow for investigations of small sample sizes down to single molecules and provide experimental data that are sufficiently accurate for direct detection of fluctuations. Progress in the development of new techniques and improvement in the resolution of conventional experiments have been enormous over the last 50 years. Indeed, molecular spectroscopy has provided hitherto unimaginable insights into processes at atomic resolution down to time ranges of a hundred attoseconds, whence observations of single particles have become routine, and as a consequence current theory in physics, chemistry, and the life sciences cannot be successful without a deeper understanding of fluctuations and their origins. Sampling of data and reproduction of processes are doomed to produce interpretation artifacts unless the observer has a solid background in the mathematics of probabilities. As a matter of fact, stochastic processes are much closer to observation than deterministic descriptions in modern science, as indeed they are in everyday life, and presently available computer facilities provide new tools that can bring us closer to applications by supplementing analytical work on stochastic phenomena with simulations.

The relevance of fluctuations in the description of real-world phenomena ranges, of course, from unimportant to dominant. The motions of planets and moons as described by celestial mechanics marked the beginning of modeling by means of differential equations. Fluctuations in these cases are so small that they cannot be detected, not even by the most accurate measurements: sunrise, sunset, and solar eclipses are predictable with almost no scatter. Processes in the life sciences are entirely different. A famous and typical historical example is Mendel's laws of inheritance: regularities are detectable only in sufficiently large samples of individual observations, and the influence of stochasticity is ubiquitous. Processes in chemistry lie between the two extremes: the deterministic approach in conventional chemical reaction kinetics has not become less applicable, nor have the results become less reliable in the light of modern experiments. What has increased dramatically are the accessible resolutions in amounts of materials, space, and time. Deeper insights into mechanisms provide new access to information regarding molecular properties for theory and practice.

Biology is currently in a state of transition: the molecular connections with chemistry have revolutionized the sources of biological data, and this sets the stage for a new theoretical biology. Historically, biology was based almost exclusively on observation and theory in biology engaged only in the interpretation of observed regularities. The development of biochemistry at the end of the nineteenth and the first half of the twentieth century introduced quantitative thinking concerning chemical kinetics into some biological subdisciplines. Biochemistry also brought a new dimension to experiments in biology in the form of *in vitro* studies on isolated and purified biomolecules. A second influx of mathematics into biology came from population genetics, first developed in the 1920s as a new theoretical discipline uniting Darwin's natural selection and Mendelian genetics. This became part of the theoretical approach more than 20 years before evolutionary biologists completed the so-called *synthetic theory*, achieving the same goal.

Then, in the second half of the twentieth century, molecular biology started to build a solid bridge from chemistry to biology, and the enormous progress in experimental techniques created a previously unknown situation in biology. Indeed, the volume of information soon went well beyond the capacities of the human mind, and new procedures were required for data handling, analysis, and interpretation. Today, biological cells and whole organisms have become accessible to complete description at the molecular level. The overwhelming amount of information required for a deeper understanding of biological objects is a consequence of two factors: (i) the complexity of biological entities and (ii) the lack of a universal theoretical biology.

Primarily, apart from elaborate computer techniques, the current flood of results from molecular genetics and genomics to systems biology and synthetic biology requires suitable statistical methods and tools for verification and evaluation of data. However, analysis, interpretation, and understanding of experimental results are impossible without proper modeling tools. In the past, these tools were primarily based on differential equations, but it has been realized within the last two decades that an extension of the available methodological repertoire by stochastic methods and techniques from other mathematical disciplines is inevitable. Moreover, the enormous complexity of the genetic and metabolic networks in the cell calls for radically new methods of modeling that resemble the mesoscopic level of description in solid state physics. In mesoscopic models, the overwhelming and for many purposes dispensable wealth of detailed molecular information is cast into a partially probabilistic description in the spirit of *dissipative particle dynamics* [358, 401], for example, and such a description cannot be successful without a solid mathematical background.

The field of stochastic processes has not been bypassed by the digital revolution. Numerical calculation and computer simulation play a decisive role in present-day stochastic modeling in physics, chemistry, and biology. Speed of computation and digital storage capacities have been growing exponentially since the 1960s, with a doubling time of about 18 months, a fact commonly referred to as Moore's law

[409]. It is not so well known, however, that the spectacular exponential growth in computer power has been overshadowed by progress in numerical methods, as attested by an enormous increase in the efficiency of algorithms. To give just one example, reported by Martin Grötschel from the Konrad Zuse-Zentrum in Berlin [260, p. 71]:

> The solution of a benchmark production planning model by linear programming would have taken – extrapolated – 82 years CPU time in 1988, using the computers and the linear programming algorithms of the day. In 2003 – fifteen years later – the same model could be solved in one minute and this means an improvement by a factor of about 43 million. Out of this, a factor of roughly 1 000 resulted from the increase in processor speed whereas a factor of 43 000 was due to improvement in the algorithms.

There are many other examples of similar progress in the design of algorithms. However, the analysis and design of high-performance numerical methods require a firm background in mathematics. The availability of cheap computing power has also changed the attitude toward exact results in terms of complicated functions: it does not take much more computer time to compute a sophisticated hypergeometric function than to evaluate an ordinary trigonometric expression for an arbitrary argument, and operations on confusingly complicated equations are enormously facilitated by symbolic computation. In this way, present-day computational facilities can have a significant impact on analytical work, too.

In the past, biologists often had mixed feelings about mathematics and reservations about using too much theory. The new developments, however, have changed this situation, if only because the enormous amount of data collected using the new techniques can neither be inspected by human eyes nor comprehended by human brains. Sophisticated software is required for handling and analysis, and modern biologists have come to rely on it [483]. The biologist Sydney Brenner, an early pioneer of molecular life sciences, makes the following point [64]:

> But of course we see the most clear-cut dichotomy between hunters and gatherers in the practice of modern biological research. I was taught in the pregenomic era to be a hunter. I learnt how to identify the wild beasts and how to go out, hunt them down and kill them. We are now, however, being urged to be gatherers, to collect everything lying about and put it into storehouses. Someday, it is assumed, someone will come and sort through the storehouses, discard all the junk and keep the rare finds. The only difficulty is how to recognize them.

The recent developments in molecular biology, genomics, and organismic biology, however, seem to initiate this change in biological thinking, since there is practically no way of shaping modern life sciences without mathematics, computer science, and theory. Brenner advocates the development of a comprehensive theory that would provide a proper framework for modern biology [63]. He and others are calling for a *new theoretical biology* capable of handling the enormous biological complexity. Manfred Eigen stated very clearly what can be expected from such a theory [112, p. xii]:

> Theory cannot remove complexity but it can show what kind of 'regular' behavior can be expected and what experiments have to be done to get a grasp on the irregularities.

Among other things, the new theoretical biology will have to find an appropriate way to combine randomness and deterministic behavior in modeling, and it is safe to predict that it will need a strong anchor in mathematics in order to be successful.

In this monograph, an attempt is made to bring together the mathematical background material that would be needed to understand stochastic processes and their applications in chemistry and biology. In the sense of the version of Occam's razor attributed to Albert Einstein [70, pp. 384–385; p. 475], viz., "everything should be made as simple as possible, but not simpler," dispensable refinements of higher mathematics have been avoided. In particular, an attempt has been made to keep mathematical requirements at the level of an undergraduate mathematics course for scientists, and the monograph is designed to be as self-contained as possible. A reader with sufficient background should be able to find most of the desired explanations in the book itself. Nevertheless, a substantial set of references is given for further reading. Derivations of key equations are given wherever this can be done without unreasonable mathematical effort. The derivations of analytical solutions for selected examples are given in full detail, because readers interested in applying the theory of stochastic processes in a practical context should be in a position to derive new solutions on their own. Some sections that are not required if one is primarily interested in applications are marked by a star ($\star$) for skipping by readers who are willing to accept the basic results without explanations.

The book is divided into five chapters. The first provides an introduction to probability theory and follows in part the introduction to probability theory by Kai Lai Chung [84], while Chap. 2 deals with the link between abstract probabilities and measurable quantities through statistics. Chapter 3 describes stochastic processes and their analysis and has been partly inspired by Crispin Gardiner's handbook [194]. Chapters 4 and 5 present selected applications of stochastic processes to problem-solving in chemistry and biology. Throughout the book, the focus is on stochastic methods, and the scientific origin of the various equations is never discussed, apart from one exception: chemical kinetics. In this case, we present two sections on the theory and empirical determination of reaction rate parameters, because for this example it is possible to show how Ariadne's red thread can guide us from first principles in theoretical physics to the equations of stochastic chemical kinetics. We have refrained from preparing a separate section with exercises, but case studies which may serve as good examples of calculations done by the reader himself are indicated throughout the book. Among others, useful textbooks would be [84, 140, 160, 161, 194, 201, 214, 222, 258, 290, 364, 437, 536, 573]. For a brief and concise introduction, we recommend [277]. Standard textbooks in mathematics used for our courses were [21, 57, 383, 467]. For dynamical systems theory, the monographs [225, 253, 496, 513] are recommended.

This book is derived from the manuscript of a course in stochastic chemical kinetics for graduate students of chemistry and biology given in the years 1999, 2006, 2011, and 2013. Comments by the students of all four courses were very helpful in the preparation of this text and are gratefully acknowledged. All figures in this monograph were drawn with the *COREL* software and numerical computations were done with *Mathematica 9*. *Wikipedia, the free encyclopedia*, has been used extensively by the author in the preparation of the text, and the indirect help by the numerous contributors submitting entries to Wiki is thankfully acknowledged.

Several colleagues gave important advice and made critical readings of the manuscript, among them Edem Arslan, Reinhard Bürger, Christoph Flamm, Thomas Hoffmann-Ostenhof, Christian Höner zu Siederissen, Ian Laurenzi, Stephen Lyle, Eric Mjolsness, Eberhard Neumann, Paul E. Phillipson, Christian Reidys, Bruce E. Shapiro, Karl Sigmund, and Peter F. Stadler. Many thanks go to all of them.

Wien, Austria
April 2016

Peter Schuster

Contents

1 Probability ... 1
1.1 Fluctuations and Precision Limits ... 2
1.2 A History of Probabilistic Thinking ... 6
1.3 Interpretations of Probability ... 11
1.4 Sets and Sample Spaces ... 16
1.5 Probability Measure on Countable Sample Spaces ... 20
1.5.1 Probability Measure ... 21
1.5.2 Probability Weights ... 24
1.6 Discrete Random Variables and Distributions ... 27
1.6.1 Distributions and Expectation Values ... 27
1.6.2 Random Variables and Continuity ... 29
1.6.3 Discrete Probability Distributions ... 34
1.6.4 Conditional Probabilities and Independence ... 38
1.7 ⋆ Probability Measure on Uncountable Sample Spaces ... 44
1.7.1 ⋆ Existence of Non-measurable Sets ... 46
1.7.2 ⋆ Borel σ-Algebra and Lebesgue Measure ... 49
1.8 Limits and Integrals ... 55
1.8.1 Limits of Series of Random Variables ... 55
1.8.2 Riemann and Stieltjes Integration ... 59
1.8.3 Lebesgue Integration ... 63
1.9 Continuous Random Variables and Distributions ... 70
1.9.1 Densities and Distributions ... 71
1.9.2 Expectation Values and Variances ... 76
1.9.3 Continuous Variables and Independence ... 77
1.9.4 Probabilities of Discrete and Continuous Variables ... 78

2 Distributions, Moments, and Statistics ... 83
2.1 Expectation Values and Higher Moments ... 83
2.1.1 First and Second Moments ... 84
2.1.2 Higher Moments ... 91
2.1.3 ⋆ Information Entropy ... 95

2.2 Generating Functions ... 101
2.2.1 Probability Generating Functions ... 101
2.2.2 Moment Generating Functions ... 103
2.2.3 Characteristic Functions ... 105
2.3 Common Probability Distributions ... 107
2.3.1 The Poisson Distribution ... 109
2.3.2 The Binomial Distribution ... 111
2.3.3 The Normal Distribution ... 115
2.3.4 Multivariate Normal Distributions ... 120
2.4 Regularities for Large Numbers ... 124
2.4.1 Binomial and Normal Distributions ... 125
2.4.2 Central Limit Theorem ... 130
2.4.3 Law of Large Numbers ... 133
2.4.4 Law of the Iterated Logarithm ... 135
2.5 Further Probability Distributions ... 137
2.5.1 The Log-Normal Distribution ... 137
2.5.2 The χ^2-Distribution ... 140
2.5.3 Student's t-Distribution ... 143
2.5.4 The Exponential and the Geometric Distribution ... 147
2.5.5 The Pareto Distribution ... 151
2.5.6 The Logistic Distribution ... 154
2.5.7 The Cauchy–Lorentz Distribution ... 156
2.5.8 The Lévy Distribution ... 159
2.5.9 The Stable Distribution ... 161
2.5.10 Bimodal Distributions ... 166
2.6 Mathematical Statistics ... 168
2.6.1 Sample Moments ... 169
2.6.2 Pearson's Chi-Squared Test ... 173
2.6.3 Fisher's Exact Test ... 180
2.6.4 The Maximum Likelihood Method ... 182
2.6.5 Bayesian Inference ... 190

3 Stochastic Processes ... 199
3.1 Modeling Stochastic Processes ... 203
3.1.1 Trajectories and Processes ... 203
3.1.2 Notation for Probabilistic Processes ... 208
3.1.3 Memory in Stochastic Processes ... 209
3.1.4 Stationarity ... 214
3.1.5 Continuity in Stochastic Processes ... 216
3.1.6 Autocorrelation Functions and Spectra ... 220
3.2 Chapman–Kolmogorov Forward Equations ... 224
3.2.1 Differential Chapman–Kolmogorov Forward Equation ... 225
3.2.2 Examples of Stochastic Processes ... 235
3.2.3 Master Equations ... 260

3.2.4 Continuous Time Random Walks 273
3.2.5 Lévy Processes and Anomalous Diffusion 284
3.3 Chapman–Kolmogorov Backward Equations 303
3.3.1 Differential Chapman–Kolmogorov Backward Equation ... 305
3.3.2 Backward Master Equations 307
3.3.3 Backward Poisson Process 310
3.3.4 Boundaries and Mean First Passage Times 313
3.4 Stochastic Differential Equations 319
3.4.1 Mathematics of Stochastic Differential Equations 321
3.4.2 Stochastic Integrals 323
3.4.3 Integration of Stochastic Differential Equations 337

4 Applications in Chemistry 347
4.1 A Glance at Chemical Reaction Kinetics 350
4.1.1 Elementary Steps of Chemical Reactions 351
4.1.2 Michaelis–Menten Kinetics 358
4.1.3 Reaction Network Theory 372
4.1.4 Theory of Reaction Rate Parameters 388
4.1.5 Empirical Rate Parameters 407
4.2 Stochasticity in Chemical Reactions 415
4.2.1 Sampling of Trajectories 416
4.2.2 The Chemical Master Equation 418
4.2.3 Stochastic Chemical Reaction Networks 425
4.2.4 The Chemical Langevin Equation 432
4.3 Examples of Chemical Reactions 435
4.3.1 The Flow Reactor 436
4.3.2 Monomolecular Chemical Reactions 441
4.3.3 Bimolecular Chemical Reactions 450
4.3.4 Laplace Transform of Master Equations 459
4.3.5 Autocatalytic Reaction 477
4.3.6 Stochastic Enzyme Kinetics 485
4.4 Fluctuations and Single Molecule Investigations 490
4.4.1 Single Molecule Enzymology 491
4.4.2 Fluorescence Correlation Spectroscopy 500
4.5 Scaling and Size Expansions 509
4.5.1 Kramers–Moyal Expansion 509
4.5.2 Small Noise Expansion 512
4.5.3 Size Expansion of the Master Equation 514
4.5.4 From Master to Fokker–Planck Equations 521
4.6 Numerical Simulation of Chemical Master Equations 526
4.6.1 Basic Assumptions 527
4.6.2 Tau-Leaping and Higher-Level Approaches 531
4.6.3 The Simulation Algorithm 533
4.6.4 Examples of Simulations 542

5 Applications in Biology 569
5.1 Autocatalysis and Growth 572
5.1.1 Autocatalysis in Closed Systems 572
5.1.2 Autocatalysis in Open Systems 575
5.1.3 Unlimited Growth 580
5.1.4 Logistic Equation and Selection 583
5.2 Stochastic Models in Biology 585
5.2.1 Master Equations and Growth Processes 585
5.2.2 Birth-and-Death Processes 589
5.2.3 Fokker–Planck Equation and Neutral Evolution 605
5.2.4 Logistic Birth-and-Death and Epidemiology 611
5.2.5 Branching Processes 631
5.3 Stochastic Models of Evolution 649
5.3.1 The Wright–Fisher and the Moran Process 651
5.3.2 Master Equation of the Moran Process 658
5.3.3 Models of Mutation 665
5.4 Coalescent Theory and Phylogenetic Reconstruction 673

Notation 679

References 683

Author Index 707

Index 711

Chapter 1
Probability

The man that's over-cautious will achieve little.
Wer gar zu viel bedenkt, wird wenig leisten.
Friedrich Schiller, Wilhelm Tell, III

Abstract Probabilistic thinking originated historically when people began to analyze the chances of success in gambling, and its mathematical foundations were laid down together with the development of statistics in the seventeenth century. Since the beginning of the twentieth century statistics has been an indispensable tool for bridging the gap between molecular motions and macroscopic observations. The classical notion of probability is based on counting and dealing with finite numbers of observations. Extrapolation to limiting values for hypothetical infinite numbers of observations is the basis of the frequentist interpretation, while more recently a *subjective* approach derived from the early works of Bayes has become useful for modeling and analyzing complex biological systems. The Bayesian interpretation of probability accounts explicitly for the incomplete but improvable knowledge of the experimenter. In the twentieth century, set theory became the ultimate basis of mathematics, thus constituting also the foundation of current probability theory, based on Kolmogorov's axiomatization of 1933. The modern approach allows one to handle and compare finite, countably infinite, and also uncountable sets, the most important class, which underlie the proper consideration of continuous variables in set theory. In order to define probabilities for uncountable sets such as subsets of real numbers, we define Borel fields, families of subsets of sample space. The notion of random variables is central to the analysis of probabilities and applications to problem solving. Random variables are elements of discrete and countable or continuous and uncountable probability spaces. They are conventionally characterized by their distributions.

Classical probability theory, in essence, can handle all cases that are modeled by discrete quantities. It is based on counting and accordingly runs into problems when it is applied to uncountable sets. Uncountable sets occur with continuous variables and are therefore indispensable for modeling processes in space as well as for handling large particle numbers, which are described as continuous concentrations in chemical kinetics. Current probability theory is based on set theory and can handle variables on discrete—hence countable—as well as continuous—hence

P. Schuster, *Stochasticity in Processes*, Springer Series in Synergetics,
DOI 10.1007/978-3-319-39502-9_1

uncountable—sets. After a general introduction, we present a history of probability theory through examples. Different notions of probability are compared, and we then provide a short account of probabilities which are derived axiomatically from set theoretical operations. Separate sections deal with countable and uncountable sample spaces. Random variables are characterized in terms of probability distributions and those properties required for applications to stochastic processes are introduced and analyzed.

1.1 Fluctuations and Precision Limits

When a scientist reproduces an experiment, what does he expect to observe? If he were a physicist of the early nineteenth century, he would expect the same results within the precision limits of the apparatus he is using for the measurement. Uncertainty in observations was considered to be merely a consequence of technical imperfection. Celestial mechanics comes close to this ideal and many of us, for example, were witness to the outstanding accuracy of astronomical predictions in the precise timing of the eclipse of the sun in Europe on August 11, 1999. Terrestrial reality, however, tells that there are limits to reproducibility that have nothing to do with lack of experimental perfection. Uncontrollable variations in initial and environmental conditions on the one hand and the broad intrinsic diversity of individuals in a population on the other hand are daily problems in biology. Predictive limitations are commonplace in complex systems: we witness them every day when we observe the failures of various forecasts for the weather or the stock market. Another no less important source of randomness comes from the irregular thermal motions of atoms and molecules that are commonly characterized as thermal fluctuations. The importance of fluctuations in the description of ensembles depends on population size: they are—apart from exceptions—of moderate importance in chemical reaction kinetics, but highly relevant for the evolution of populations in biology.

Conventional chemical kinetics handles molecular ensembles involving large numbers of particles,[1] $N \approx 10^{20}$ and more. Under the majority of common conditions, for example, at or near chemical equilibrium or stable stationary states, and in the absence of autocatalytic self-enhancement, random fluctuations in particle numbers are proportional to $\sqrt{N}$. This so-called $\sqrt{N}$ *law* is introduced here as a kind of heuristic, but we shall derive it rigorously for the Poisson distribution in Sect. 2.3.1 and we shall see many specific examples where it holds to a good approximation. Typical experiments in chemical laboratories deal with amounts of

[1] In this monograph we shall use the notion of particle number as a generic term for discrete population variables. Particle numbers may be numbers of molecules or atoms in a chemical system, numbers of individuals in a population, numbers of heads in sequences of coin tosses, or numbers of dice throws yielding the same number of pips.

substance of about 10^{-4} mol—of the order of $N = 10^{20}$ particles—so these give rise to natural fluctuations which typically involve $\sqrt{N} = 10^{10}$ particles, i.e., in the range of $\pm 10^{-10}N$. Under such conditions the detection of fluctuations would require an accuracy of the order of $1{:}10^{10}$, which is (almost always) impossible to achieve in direct measurements, since most techniques in analytical chemistry encounter serious difficulties when concentration accuracies of $1{:}10^{-6}$ or higher are required.

Exceptions are new techniques for observing single molecules (Sect. 4.4). In general, the chemist uses concentrations rather than particle numbers, i.e., $c = N/(N_L V)$, where $N_L = 6.022 \times 10^{23}\,\mathrm{mol}^{-1}$ and V are Avogadro's constant[2] and the volume in dm^3 or liters. Conventional chemical kinetics considers concentrations as continuous variables and applies deterministic methods, in essence differential equations, for analysis and modeling. It is thereby implicitly assumed that particle numbers are sufficiently large to ensure that the limit of infinite particle numbers is essentially correct and fluctuations can be neglected. This scenario is commonly not justified in biology, where particle numbers are much smaller than in chemistry and uncontrollable environmental effects introduce additional uncertainties.

Nonlinearities in chemical kinetics may amplify fluctuations through autocatalysis in such a way that the random component becomes much more important than the $\sqrt{N}$ law suggests. This is already the case with simple autocatalytic reactions, as discussed in Sects. 4.3.5, 4.6.4, and 5.1, and becomes a dominant effect, for example, with processes exhibiting oscillations or deterministic chaos. Some processes in physics, chemistry, and biology have no deterministic component at all. The most famous is *Brownian motion*, which can be understood as a visualized form of microscopic diffusion. In biology, other forms of entirely random processes are encountered, in which fluctuations are the only or the major driving force of change. An important example is random drift of populations in the space of genotypes, leading to fixation of mutants in the absence of any differences in fitness. In evolution, after all, particle numbers are sometimes very small: every new molecular species starts out from a single variant.

In 1827, the British botanist Robert Brown detected and analyzed irregular motions of particles in aqueous suspensions. These motions turned out to be independent of the nature of the suspended materials—pollen grains or fine particles of glass or minerals served equally well [69]. Although Brown himself had already

---

[2]The amount of a chemical compound A is commonly specified by the number N_A of molecules in the reaction volume V, via the number density $C_A = N_A/V$, or by the concentration $c_A = N_A/N_L V$, which is the number of moles in one liter of solution, where N_L is Avogadro's constant $N_L = 6.02214179 \times 10^{23}\,\mathrm{mol}^{-1}$, i.e., the number of atoms or molecules in one mole of substance. Loschmidt's constant $n_0 = 2.6867774 \times 10^{25}\,\mathrm{m}^{-3}$ is closely related to Avogadro's constant and counts the number of particles in one liter of ideal gas at standard temperature and pressure, which are 0° and 1 atm = 101.325 kPa. Both quantities have physical dimensions and are not *numbers*, a point often ignored in the literature. In order to avoid ambiguity errors we shall refer to Avogadro's constant as N_L, because N_A is needed for the number of particles A (for units used in this monograph see appendix Notation).

demonstrated that the motion was not caused by any (mysterious) biological effect, its origin remained something of a riddle until Albert Einstein [133], and independently Marian von Smoluchowski [559], published satisfactory explanations in 1905 and 1906, respectively.[3] These revealed two main points:

(i) The motion is caused by highly frequent collisions between the pollen grain and the steadily moving molecules in the liquid in which the particles are suspended, and
(ii) the motion of the molecules in the liquid is so complicated and irregular that its effect on the pollen grain can only be described probabilistically in terms of frequent, statistically independent impacts.

In order to model Brownian motion, Einstein considered the number of particles per unit volume as a function of space[4] and time, viz., $f(x,t) = N(x,t)/V$, and derived the equation

$$\frac{\partial f}{\partial t} = D\frac{\partial^2 f}{\partial x^2}\,, \qquad \text{with solution } f(x,t) = \frac{C}{\sqrt{4\pi D}}\frac{\exp(-x^2/4Dt)}{\sqrt{t}}\,,$$

where $C = N/V = \int f(x,t)\,\mathrm{d}x$ is the number density, the total number of particles per unit volume, and D is a parameter called the *diffusion coefficient*. Einstein showed that his equation for $f(x,t)$ was identical to the differential equation of diffusion already known as Fick's second law [165], which had been derived 50 years earlier by the German physiologist Adolf Fick. Einstein's original treatment was based on small discrete time steps $\Delta t = \tau$ and thus contains a—well justified—approximation that can be avoided by application of the modern theory of stochastic processes (Sect. 3.2.2.2). Nevertheless, Einstein's publication [133] represents the first analysis based on a probabilistic concept that is actually comparable to current theories, and Einstein's paper is correctly considered as the beginning of stochastic modeling. Later Einstein wrote four more papers on diffusion with different derivations of the diffusion equation [134]. It is worth mentioning that 3 years after the publication of Einstein's first paper, Paul Langevin presented an alternative mathematical treatment of random motion [325] that we shall discuss at length in the form of the Langevin equation in Sect. 3.4. Since the days of Brown's discovery, interest in Brownian motion has never ceased and publications on recent theoretical and experimental advances document this fact nicely—two interesting recent examples are [344, 491].

[3] The first mathematical model of Brownian motion was conceived as early as 1880, by Thorvald Thiele [330, 528]. Later, in 1900, a process involving random fluctuations of the Brownian motion type was used by Louis Bachelier [31] to describe the stock market at the Paris stock exchange. He gets the credit for having been the first to write down an equation that was later named after Paul Langevin (Sect. 3.4). For a recent and detailed monograph on Brownian motion and the mathematics of normal diffusion, we recommend [214].

[4] For the sake of simplicity we consider only motion in one spatial direction x.

From the solution of the diffusion equation, Einstein computed the diffusion parameter D and showed that it is linked to the mean square displacement $\langle \Delta x^2 \rangle$ of the particle in the x-direction:

$$D = \frac{\langle \Delta x^2 \rangle}{2t} , \qquad \text{or } \lambda_x = \sqrt{\langle \Delta x^2 \rangle} = \sqrt{2Dt} .$$

Here Δx is the net distance the particle travels during the time interval t. Extension to three-dimensional space is straightforward and results only in a different numerical factor: $D = \langle \Delta x^2 \rangle / 6t$. Both quantities, the diffusion parameter D and the mean displacement λ_x, are measurable, and Einstein concluded correctly that a comparison of the two quantities should allow for an experimental determination of Avogadro's constant [450].

Brownian motion was indeed the first completely random process that became accessible to a description within the frame of classical physics. Although James Clerk Maxwell and Ludwig Boltzmann had identified thermal motion as the driving force causing irregular collisions of molecules in gases, physicists in the second half of the nineteenth century were not interested in the details of molecular motion unless they were required in order to describe systems in the thermodynamic limit. In statistical mechanics the measurable macroscopic functions were, and still are, derived by means of global averaging techniques. By the first half of the twentieth century, thermal motion was no longer the only uncontrollable source of random natural fluctuations, having been supplemented by quantum mechanical uncertainty as another limitation to achievable precision.

The occurrence of complex dynamics in physics and chemistry has been known since the beginning of the twentieth century through the groundbreaking theoretical work of the French mathematician Henri Poincaré and the experiments of the German chemist Wilhelm Ostwald, who explored chemical systems with periodicities in space and time. Systematic studies of dynamical complexity, however, required the help of electronic computers and the new field of research on complex dynamical systems was not initiated until the 1960s. The first pioneer of this discipline was Edward Lorenz [354] who used numerical integration of differential equations to demonstrate what is nowadays called *deterministic chaos*. What was new in the second half of the twentieth century were not so much the concepts of complex dynamics but the tools to study it. Easy access to previously unimagined computer power and the development of highly efficient algorithms made numerical computation an indispensable technique for scientific investigation, to the extent that it is now almost on a par with theory and experiment.

Computer simulations have shown that a large class of dynamical systems modeled by nonlinear differential equations exhibit irregular, i.e., nonperiodic, behavior for certain ranges of parameter values. Hand in hand with complex dynamics go limitations on predictability, a point of great practical importance: although the differential equations used to describe and analyze chaos are still deterministic, initial conditions of an accuracy that could never be achieved in reality would be required for correct long-time predictions. Sensitivity to small changes makes a stochastic treatment indispensable, and solutions were indeed found to be extremely sensitive to small changes in initial and boundary conditions in these *chaotic regimes*. Solution curves that are almost identical at the beginning can deviate exponentially from each other and appear completely different after sufficiently long times. Deterministic chaos gives rise to a third kind of uncertainty, because initial conditions cannot be controlled with greater precision than the experimental setup allows. It is no accident that Lorenz first discovered chaotic dynamics in the equations for atmospheric motions, which are indeed so complex that forecasts are limited to the short or mid-term at best.

In this monograph we shall focus on the mathematical handling of processes that are irregular and often simultaneously sensitive to small changes in initial and environmental conditions, but we shall not be concerned with the physical origin of these irregularities.

1.2 A History of Probabilistic Thinking

The concept of probability originated much earlier than its applications in physics and resulted from the desire to analyze by rigorous mathematical methods the chances of winning when gambling. An early study that has remained largely unnoticed, due to the sixteenth century Italian mathematician Gerolamo Cardano, already contained the basic ideas of probability. However, the beginning of classical probability theory is commonly associated with the encounter between the French mathematician Blaise Pascal and a professional gambler, the Chevalier de Méré, which took place in France a 100 years after Cardano. This tale provides such a nice illustration of a pitfall in probabilistic thinking that we repeat it here as our first example of conventional probability theory, despite the fact that it can be found in almost every textbook on statistics or probability.

On July 29, 1654, Blaise Pascal addressed a letter to the French mathematician Pierre de Fermat, reporting a careful observation by the professional gambler Chevalier de Méré. The latter had noted that obtaining at least one *six* with one die in 4 throws is successful in more than 50 % of cases, whereas obtaining at least one *double six* with two dice in 24 throws comes out in fewer than 50 % of cases. He considered this paradoxical, because he had calculated naïvely and erroneously that the chances should be the same:

$$\text{4 throws with one die yields} \quad 4 \times \frac{1}{6} = \frac{2}{3} \,,$$

$$\text{24 throws with two dice yields} \quad 24 \times \frac{1}{36} = \frac{2}{3} \,.$$

Blaise Pascal became interested in the problem and correctly calculated the probability as we would do it now in classical probability theory, by careful counting of events:

$$\text{probability} = P = \frac{\text{number of favorable events}}{\text{total number of events}} . \tag{1.1}$$

According to (1.1), the probability is always a positive quantity between zero and one, i.e., $0 \le P \le 1$. The sum of the probabilities that a given event has either occurred or not occurred is always one. Sometimes, as in Pascal's example, it is easier to calculate the probability q of the unfavorable case and to obtain the desired probability by computing $p = 1 - q$. In the one-die example, the probability of not throwing a *six* is $5/6$, while in the two-die case, the probability of not obtaining a *double six* is $35/36$. Provided the events are independent, their probabilities are multiplied[5] and we finally obtain for 4 and 24 trials, respectively:

$$q(1) = \left(\frac{5}{6}\right)^4 \quad \text{and} \quad p(1) = 1 - \left(\frac{5}{6}\right)^4 = 0.51775 ,$$

$$q(2) = \left(\frac{35}{36}\right)^{24} \quad \text{and} \quad p(2) = 1 - \left(\frac{35}{36}\right)^{24} = 0.49140 .$$

It is remarkable that Chevalier de Méré was able to observe this rather small difference in the probability of success—indeed, he must have watched the game very often!

In order to see where the Chevalier made a mistake, and as an exercise in deriving correct probabilities, we calculate the first case—the probability of obtaining at least one *six* in four throws—by a more direct route than the one used above. We are throwing the die four times and the favorable events are: 1 time *six*, 2 times *six*, 3 times *six*, and 4 times *six*. There are four possibilities for 1 *six*—the *six* appearing in the first, the second, the third, or the fourth throw, six possibilities for 2 *sixes*, four possibilities for 3 *sixes*, and one possibility for 4 *sixes*. With the probabilities $1/6$ for obtaining a *six* and $5/6$ for any other number of pips, we get finally

$$\binom{4}{1}\times\frac{1}{6}\left(\frac{5}{6}\right)^3+\binom{4}{2}\times\left(\frac{1}{6}\right)^2\left(\frac{5}{6}\right)^2+\binom{4}{3}\times\left(\frac{1}{6}\right)^3\frac{5}{6}+\binom{4}{4}\times\left(\frac{1}{6}\right)^4=\frac{671}{1296} . \qquad \square$$

For those who want to become champion probability calculators, we suggest calculating $p(2)$ directly as well.

[5] We shall come back to a precise definition of independent events later, when we introduce modern probability theory in Sect. 1.6.4.

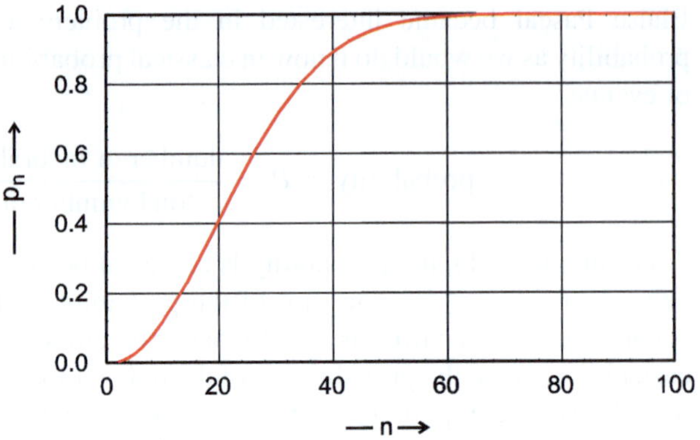

Fig. 1.1 The birthday problem. The *curve* shows the probability p_n that two people in a group of n people celebrate their birthday on the same day of the year

The second example presented here is the *birthday problem*.[6] It can be used to demonstrate the common human inability to estimate probabilities:

> Let your friends guess – without calculating – how many people you need in a group so that there is a fifty percent chance that at least two of them celebrate their birthday on the same day. You will be surprised by some of the answers!

With our knowledge of the gambling problem, this probability is easy to calculate. First we compute the negative event, that is, when everyone celebrates their birthday on a different day of the year, assuming that it is not a leap year, so that there are 365 days. For n people in the group, we find[7]

$$q = \frac{365}{365} \cdot \frac{364}{365} \cdot \frac{363}{365} \cdot \ldots \cdot \frac{365-(n-1)}{365} \quad \text{and} \quad p = 1 - q\,.$$

The function $p(n)$ is shown in Fig. 1.1. For the above-mentioned 50 % chance, we need only 27 people. With 41 people, we already have more than 90 % chance that two of them will celebrate their birthday on the same day, while 57 would yield a probability above 99 %, and 70 a probability above 99.9 %. An implicit assumption in this calculation has been that births are uniformly distributed over the year, i.e., the probability that somebody has their birthday on some particular day does not depend on that particular day. In mathematical statistics, such an assumption may be subjected to test and then it is called a *null hypothesis* (see [177] and Sect. 2.6.2).

Laws in classical physics are considered to be deterministic, in the sense that a single measurement is expected to yield a precise result. Deviations from this result

[6]The birthday problem was invented in 1939 by Richard von Mises [557] and it has fascinated mathematicians ever since. It has been discussed and extended in many papers, such as [3, 89, 255, 430], and even found its way into textbooks on probability theory [160, pp. 31–33].

[7]The expression is obtained by the following argument. The first person's birthday can be chosen freely. The second person's must not be chosen on the same day, so there are 364 possible choices. For the third, there remain 363 choices, and so on until finally, for the nth person, there are $365 - (n-1)$ possibilities.

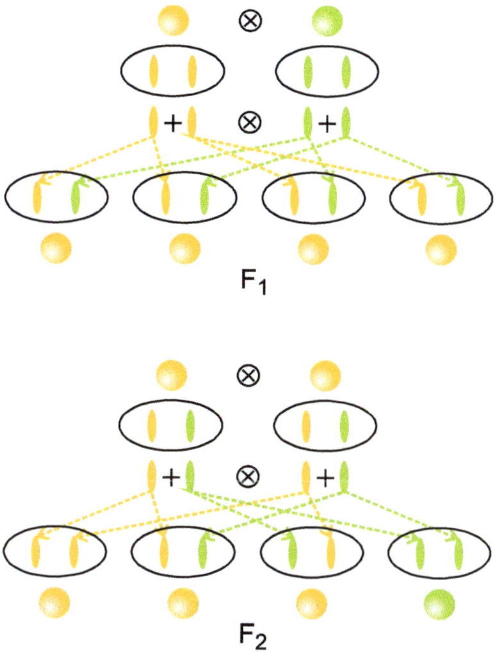

Fig. 1.2 Mendel's laws of inheritance. The sketch illustrates Mendel's laws of inheritance: (i) the law of segregation and (ii) the law of independent assortment. Every (diploid) organism carries two copies of each gene, which are separated during the process of reproduction. Every offspring receives one randomly chosen copy of the gene from each parent. *Encircled* are the genotypes formed from two alleles, yellow or green, and above or below the genotypes are the phenotypes expressed as the colors of seeds of the garden pea (*pisum sativum*). The *upper part* of the figure shows the first generation (F_1) of progeny of two homozygous parents—parents who carry two identical alleles. All genotypes are heterozygous and carry one copy of each allele. The yellow allele is dominant and hence the phenotype expresses yellow color. Crossing two F_1 individuals (*lower part* of the figure) leads to two homozygous and two heterozygous offspring. Dominance causes the two heterozygous genotypes and one homozygote to develop the dominant phenotype and accordingly the observable ratio of the two phenotypes in the F_2 generation is 3:1 on the average, as observed by Gregor Mendel in his statistics of fertilization experiments (see Table 1.1)

are then interpreted as due to a lack of precision in the equipment used. When it is observed, random scatter is thought to be caused by variations in experimental conditions that are not sufficiently well controlled. Apart from deterministic laws, other regularities are observed in nature, which become evident only when sample sizes are made sufficiently large through repetition of experiments. It is appropriate to call such regularities *statistical laws*. Statistical results regarding the biology of plant inheritance were pioneered by the Augustinian monk Gregor Mendel, who discovered regularities in the progeny of the *garden pea* in controlled fertilization experiments [392] (Fig. 1.2).

As a third and final example, we consider some of Mendel's data in order to exemplify a statistical law. Table 1.1 shows the results of two typical experiments

Table 1.1 Statistics of Gregor Mendel's experiments with the garden pea (*pisum sativum*)

	Form of seed			Color of seed		
Plant	Round	Wrinkled	Ratio	Yellow	Green	Ratio
1	45	12	3.75	25	11	2.27
2	27	8	3.38	32	7	4.57
3	24	7	3.43	14	5	2.80
4	19	10	1.90	70	27	2.59
5	32	11	2.91	24	13	1.85
6	26	6	4.33	20	6	3.33
7	88	24	3.67	32	13	2.46
8	22	10	2.20	44	9	4.89
9	28	6	4.67	50	14	3.57
10	25	7	3.57	44	18	2.44
Total	336	101	**3.33**	355	123	**2.89**

In total, Mendel analyzed 7324 seeds from 253 hybrid plants in the second trial year. Of these, 5474 were round or roundish and 1850 angular and wrinkled, yielding a ratio 2.96:1. The color was recorded for 8023 seeds from 258 plants, out of which 6022 were yellow and 2001 were green, with a ratio of 3.01:1. The results of two typical experiments with ten plants, which deviate more strongly because of the smaller sample size, are shown in the table

distinguishing roundish or wrinkled seeds with yellow or green color. The ratios observed with single plants exhibit a broad scatter. The mean values for ten plants presented in the table show that some averaging has occurred in the sample, but the deviations from the ideal values are still substantial. Mendel carefully investigated several hundred plants, whence the statistical law of inheritance demanding a ratio of 3:1 subsequently became evident [392].[8] In a somewhat controversial publication [176], Ronald Fisher reanalyzed Mendel's experiments, questioning his statistics and accusing him of intentionally manipulating his data, because the results were too close to the ideal ratio. Fisher's publication initiated a long-lasting debate in which many scientists spoke up in favor of Mendel [427, 428], but there were also critical voices saying that most likely Mendel had unconsciously or consciously eliminated outliers [127]. In 2008, one book declared *the end of the Mendel–Fisher controversy* [186]. In Sect. 2.6.2, we shall discuss statistical laws and Mendel's experiments in the light of present day mathematical statistics, applying the so-called χ^2 test.

Probability theory in its classical form is more than 300 years old. It is no accident that the concept arose in the context of gambling, originally considered to be a domain of chance in stark opposition to the rigours of science. Indeed it was rather a long time before the concept of probability finally entered the realms

[8] According to modern genetics this ratio, like other ratios between distinct inherited phenotypes, are idealized values that are found only for completely independent genes [221], i.e., lying either on different chromosomes or sufficiently far apart on the same chromosome.

of scientific thought in the nineteenth century. The main obstacle to the acceptance of probabilities in physics was the strong belief in determinism that held sway until the advent of quantum theory. Probabilistic concepts in nineteenth century physics were still based on deterministic thinking, although the details of individual events at the microscopic level were considered to be too numerous to be accessible to calculation. It is worth mentioning that probabilistic thinking entered physics and biology almost at the same time, in the second half of the nineteenth century. In physics, James Clerk Maxwell pioneered statistical mechanics with his dynamical theory of gases in 1860 [375–377]. In biology, we may mention the considerations of pedigree in 1875 by Sir Francis Galton and Reverend Henry William Watson [191, 562] (see Sect. 5.2.5), or indeed Gregor Mendel's work on the genetics of inheritance in 1866, as discussed above. The reason for the early considerations of statistics in the life sciences lies in the very nature of biology: sample sizes are typically small, while most of the regularities are probabilistic and become observable only through the application of probability theory. Ironically, Mendel's investigations and papers did not attract a broad scientific audience until they were rediscovered at the beginning of the twentieth century. In the second half of the nineteenth century, the scientific community was simply unprepared for quantitative and indeed probabilistic concepts in biology.

Classical probability theory can successfully handle a number of concepts like conditional probabilities, probability distributions, moments, and so on. These will be presented in the next section using set theoretic concepts that can provide a much deeper insight into the structure of probability theory than mere counting. In addition, the more elaborate notion of probability derived from set theory is absolutely necessary for extrapolation to countably infinite and uncountable sample sizes. Uncountability is an unavoidable attribute of sets derived from continuous variables, and the set theoretic approach provides a way to define probability measures on certain sets of real numbers $\mathbf{x} \in \mathbb{R}^n$. From now on we shall use only the set theoretic concept, because it can be introduced straightforwardly for countable sets and discrete variables and, in addition, it can be straightforwardly extended to probability measures for continuous variables.

1.3 Interpretations of Probability

Before introducing the current standard theory of probability we make a brief digression into the dominant philosophical interpretations:

(i) the classical interpretations that we have adopted in Sect. 1.2,
(ii) the frequency-based interpretation that stand in the background for the rest of the book, and
(iii) the Bayesian or *subjective* interpretation.

The *classical interpretation of probability* goes back to the concepts laid out in the works of the Swiss mathematician Jakob Bernoulli and the French mathematician and physicist Pierre-Simon Laplace. The latter was the first to present a clear definition of probability [328, pp. 6–7]:

> The theory of chance consists in reducing all the events of the same kind to a certain number of equally possible cases, that is to say, to such as we may be equally undecided about in regard of their existence, and in determining the number of cases favorable to the event whose probability is sought. The ratio of this number to that of all possible cases is the measure of this probability, which is thus simply a fraction whose numerator is the number of favorable cases and whose denominator is the number of all possible cases.

Clearly, this definition is tantamount to (1.1) and the explicitly stated assumption of equal probabilities is now called the *principle of indifference*. This classical definition of probability was questioned during the nineteenth century by the two British logicians and philosophers George Boole [58] and John Venn [549], among others, initiating a paradigm shift from the classical view to the modern frequency interpretations of probabilities.

Modern interpretations of the concept of probability fall essentially into two categories that can be characterized as *physical probabilities* and *evidential probabilities* [228]. Physical probabilities are often called *objective* or frequency-based probabilities, and their advocates are referred to as *frequentists*. Besides the pioneer John Venn, influential proponents of the frequency-based probability theory were the Polish–American mathematician Jerzy Neyman, the British statistician Egon Pearson, the British statistician and theoretical biologist Ronald Fisher, the Austro-Hungarian–American mathematician and scientist Richard von Mises, and the German–American philosopher of science Hans Reichenbach. Physical probabilities are derived from some real process like radioactive decay, a chemical reaction, the turn of a roulette wheel, or rolling dice. In all such systems the notion of probability makes sense only when it refers to some well defined experiment with a random component.

Frequentism comes in two versions: (i) *finite frequentism* and (ii) *hypothetical frequentism*. Finite frequentism replaces the notion of the *total number of events* in (1.1) by the *actually recorded number of events*, and is thus congenial to philosophers with empiricist scruples. Philosophers have a number of problems with finite frequentism. For example, we may mention problems arising due to small samples: one can never speak about probability for a single experiment and there are cases of unrepeated or unrepeatable experiments. A coin that is tossed exactly once yields a relative frequency of heads being either zero or one, no matter what its bias really is. Another famous example is the spontaneous radioactive decay of an atom, where the probabilities of decaying follow a continuous exponential law, but according to finite frequentism it decays with probability one only once, namely at its actual decay time. The evolution of the universe or the origin of life can serve as cases of unrepeatable experiments, but people like to speak about the probability that the development has been such or such. Personally, I think it would do no harm to replace *probability* by *plausibility* in such estimates dealing with unrepeatable single events.

Hypothetical frequentism complements the empiricism of finite frequentism by the admission of infinite sequences of trials. Let N be the total number of repetitions of an experiment and n_A the number of trials when the event A has been observed. Then the relative frequency of recording the event A is an *approximation of the probability* for the occurrence of A :

$$\text{probability}\,(A) = P(A) \approx \frac{n_A}{N}\,.$$

This equation is essentially the same as (1.1), but the claim of the hypothetical frequentists' interpretation is that there exists a *true frequency* or *true probability* to which the relative frequency would converge if we could repeat the experiment an infinite number of times[9]:

$$P(A) = \lim_{N\to\infty} \frac{n_A}{N} = \frac{|A|}{|\Omega|}\,, \quad \text{with } A \in \Omega\,. \tag{1.2}$$

The probability of an event A relative to a sample space Ω is then defined as the limiting frequency of A in Ω. As N goes to infinity, $|\Omega|$ becomes infinitely large and, depending on whether $|A|$ is finite or infinite, $P(A)$ is either zero or may be a nonzero limiting value. This is based on two a priori assumptions that have the character of axioms:

(i) *Convergence*. For any event A, there exists a limiting relative frequency, the probability $P(A)$, satisfying $0 \le P(A) \le 1$.
(ii) *Randomness*. The limiting relative frequency of each event in a set Ω is the same for any *typical* infinite subsequence of Ω.

A typical sequence is *sufficiently random*[10] in order to avoid results biased by predetermined order. As a negative example of an acceptable sequence, consider *heads, heads, heads, heads, ...* recorded by tossing a coin. If it was obtained with a fair coin—not a coin with two heads—$|A|$ is 1 and $P(A) = 1/|\Omega| = 0$, and we may say that this particular event has measure zero and the sequence is not typical. The sequence *heads, tails, heads, tails, ...* is not typical either, despite the fact that it yields the same probabilities for the average number of *heads* and *tails* as a fair coin. We should be aware that the extension to infinite series of experiments leaves the realm of empiricism, leading purist philosophers to reject the claim that the interpretation of probabilities by hypothetical frequentism is more *objective* than others.

Nevertheless, the frequentist probability theory is not in conflict with the mathematical axiomatization of probability theory and it provides straightforward

---

[9]The absolute value symbol $|A|$ means here the size or cardinality of A, i.e., the number of elements in A (Sect. 1.4).

[10]Sequences are sufficiently random when they are obtained through recordings of random events. *Random sequences* are approximated by the sequential outputs of *pseudorandom number generators*. 'Pseudorandom' implies here that the approximately random sequence is created by some deterministic, i.e., nonrandom, algorithm.

guidance in applications to real-world problems. The pragmatic view that prefigures the dominant concept in current probability theory has been nicely put by William Feller, the Croatian–American mathematician and author of the two-volume classic introduction to probability theory [160, 161, Vol. I, pp. 4–5]:

> The success of the modern mathematical theory of probability is bought at a price: the theory is limited to one particular aspect of 'chance'. (...) we are not concerned with modes of inductive reasoning but with something that might be called physical or statistical probability.

He also expresses clearly his attitude towards pedantic scruples of philosophic purists:

> (...) in analyzing the coin tossing game we are not concerned with the accidental circumstances of an actual experiment, the object of our theory is sequences or arrangements of symbols such as 'head, head, tail, head, ...'. There is no place in our system for speculations concerning the probability that the sun will rise tomorrow. Before speaking of it we should have to agree on an idealized model which would presumably run along the lines 'out of infinitely many worlds one is selected at random ...'. Little imagination is required to construct such a model, but it appears both uninteresting and meaningless.

We shall adopt the frequentist interpretation throughout this monograph, but give brief mention here briefly to two more interpretations of probability in order to show that it is not the only reasonable probability theory.

The *propensity interpretation* of probability was proposed by the American philosopher Charles Peirce in 1910 [448] and reinvented by Karl Popper [455, pp. 65–70] (see also [456]) more than 40 years later [228, 398]. *Propensity* is a tendency to do or achieve something. In relation to probability, the propensity interpretation means that it makes sense to talk about the probabilities of single events. As an example, we can talk about the probability—or propensity—of a radioactive atom to decay within the next 1000 years, and thereby conclude from the behavior of an ensemble to that of a single member of the ensemble. Likewise, we might say that there is a probability of 1/2 of getting 'heads' when a fair coin is tossed, and precisely expressed, we should say that the coin has a propensity to yield a sequence of outcomes in which the limiting frequency of scoring 'heads' is 1/2. The single case propensity is accompanied by, but distinguished from, the *long-run* propensity [215]:

> A long-run propensity theory is one in which propensities are associated with repeatable conditions, and are regarded as propensities to produce in a long series of repetitions of these conditions frequencies, which are approximately equal to the probabilities.

In these theories, a long run is still distinct from an infinitely long run, in order to avoid basic philosophical problems. Clearly, the use of *propensities* rather than *frequencies* provides a somewhat more careful language than the frequentist interpretation, making it more acceptable in philosophy.

Finally, we sketch the most popular example of a theory based on *evidential probabilities*: Bayesian statistics, named after the eighteenth century British mathematician and Presbyterian minister Thomas Bayes. In contrast to the frequentist view, probabilities are *subjective* and exist only in the human mind. From a

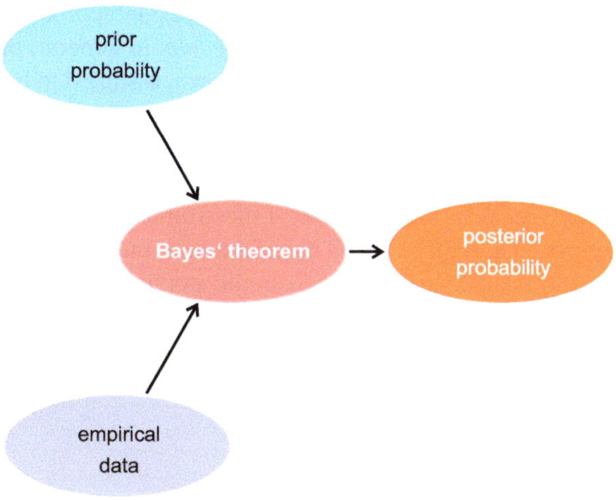

Fig. 1.3 A sketch of the Bayesian method. Prior information on probabilities is confronted with empirical data and converted by means of Bayes' theorem into a new distribution of probabilities called *posterior* probability [120, 507]

practitioner's point of view, one major advantage of the Bayesian approach is that it gives a direct insight into the way we improve our knowledge of a given subject of investigation. In order to understand Bayes' theorem, we need the notion of conditional probability, presented in Sect. 1.6.4. We thus postpone a precise formulation of the Bayesian approach to Sect. 2.6.5. Here we sketch only the basic principle of the method in a narrative manner.[11]

In physics and chemistry, we common deal with well established theories and models that are assumed to be essentially correct. Experimental data have to be fitted to the model and this is done by adjusting unknown model parameters using fitting techniques like the maximum-likelihood method (Sect. 2.6.4). This popular statistical technique is commonly attributed to Ronald Fisher, although it has been known for much longer [8, 509]. Researchers in biology, economics, social sciences, and other disciplines, however, are often confronted with situations where no commonly accepted models exist, so they cannot be content with parameter estimates. The model must then be tested and the basic formalisms improved.

Figure 1.3 shows schematically how Bayes' theorem works: the inputs of the method are (i) a preliminary or *prior* probability distribution derived from the initial model and (ii) a set of empirical data. Bayes theorem converts the inputs into a *posterior* probability distribution, which encapsulates the improvement of the model in the light of the data sample.[12] What is missing here is a precise probabilistic formulation of the process shown in Fig. 1.3, but this will be added in Sect. 2.6.5.

[11]In this context it is worth mentioning the contribution of the great French mathematician and astronomer the Marquis de Laplace, who gave an interpretation of statistical inference that can be considered equivalent to Bayes' theorem [508].

[12]It is worth comparing the Bayesian approach with conventional data fitting: the inputs are the same, a model and data, but the nature of the probability distribution is kept constant in data fitting methods, whereas it is conceived as flexible in the Bayes method.

Accordingly, the advantage of the Bayesian approach is that a *change of opinion* in the light of new data is *part of the game*. In general, parameters are input quantities of frequentist statistics and, if unknown, assumed to be available through data fitting or consecutive repetition of experiments, whereas they are understood as random variables in the Bayesian approach. In practice, direct application of the Bayesian theorem involves quite elaborate computations that were not possible in real world examples before the advent of electronic computers. An example of the Bayesian approach and the relevant calculations is presented in Sect. 2.6.5.

Bayesian statistics has become popular in disciplines where model building is a major issue. Examples are bioinformatics, molecular genetics, modeling of ecosystems, and forensics, among others. Bayesian statistics is described in many monographs, e.g., [92, 199, 281, 333]. For a brief introduction, we recommend [510].

1.4 Sets and Sample Spaces

Conventional probability theory is based on several axioms rooted in set theory. These will be introduced and illustrated in this section. The development of set theory in the 1870s was initiated by Georg Cantor and Richard Dedekind. Among many other things, it made it possible to put the concept of probability on a firm basis, allowing for an extension to certain families of uncountable samples of the kind that arise when we are dealing with continuous variables. Present day probability theory can thus be understood as a convenient extension of the classical concept by means of set and measure theory. We begin by stating a few indispensable notions of set theory.

Sets are collections of objects with two restrictions: (i) each object belongs to one set and cannot be a member of two or more sets, and (ii) a member of a set must not appear twice or more often. In other words, objects are assigned to sets unambiguously. In the application to probability theory we shall denote the *elementary objects* by the lower case Greek letter ω, if necessary with various sub- and superscripts, and call them *sample points* or *individual results*. The collection of all objects ω under consideration, the *sample space*, is denoted by the upper case Greek letter Ω, so $\omega \in \Omega$. *Events* A are subsets of sample points that satisfy some condition[13]

$$A = \{\boldsymbol{\omega}, \omega_k \in \Omega : f(\boldsymbol{\omega}) = c\} , \tag{1.3}$$

[13]The meaning of such a *condition* will become clearer later on. For the moment it suffices to understand a condition as a restriction specified by a function $f(\boldsymbol{\omega})$, which implies that not all subsets of sample points belong to A. Such a condition, for example, is a score 6 when rolling two dice, which comprises the five sample points: $A = \{1+5, 2+4, 3+3, 4+2, 5+1\}$.

where $\boldsymbol{\omega} = (\omega_1, \omega_2, \ldots)$ is the set of individual results which satisfy the condition $f(\boldsymbol{\omega}) = c$. When dealing with stochastic processes, we shall characterize the sample space as a *state space*,

$$\Omega = \{\ldots , \Sigma_{-n} , \ldots , \Sigma_{-1} , \Sigma_0 , \Sigma_1, , \ldots , \Sigma_n , \ldots\} , \tag{1.4}$$

where Σ_k is a particular state and completeness is indicated by the index running from $-\infty$ to $+\infty$.[14]

Next we consider the basic logical operations with sets. Any partial collection of points $\omega_k \in \Omega$ is a *subset* of Ω. We shall be dealing with fixed Ω and, for simplicity, often just refer to these subsets of Ω as sets. There are two extreme cases, the entire sample space Ω and the *empty set* $\emptyset$. The number of points in a set S is called its size or *cardinality*, written $|S|$, whence $|S|$ is a nonnegative integer or infinity. In particular, the size of the empty set is $|\emptyset| = 0$. The unambiguous assignment of points to sets can be expressed by[15]

$$\omega \in S \quad \text{exclusive or} \quad \omega \notin S \, .$$

Consider two sets A and B. If every point of A belongs to B, then A is contained in B. In this case, A is a subset of B and B is a superset of A:

$$A \subset B \quad \text{and} \quad B \supset A \, .$$

Two sets are identical if they contain exactly the same points, and then we write $A = B$. In other words, $A = B$ iff[16] $A \subset B$ and $B \subset A$.

Some basic operations with sets are illustrated in Fig. 1.4. We repeat them briefly here:

Complement The complement of the set A is denoted by A^{c} and consists of all points not belonging to A[17]:

$$A^{\mathrm{c}} = \{\omega | \omega \notin A\} \, . \tag{1.5}$$

There are three obvious relations which are easily checked: $(A^{\mathrm{c}})^{\mathrm{c}} = A$, $\Omega^{\mathrm{c}} = \emptyset$, and $\emptyset^{\mathrm{c}} = \Omega$.

---

[14] Strictly speaking, sample space Ω and state space Σ are related by a mapping $\mathcal{Z} : \Omega \to \Sigma$, where Σ is the state space and the (measurable) function $\mathcal{Z}$ is a random variable (Sect. 1.6.2).

[15] In order to be unambiguously clear we shall write *or* for *and/or* and *exclusive or* for *or* in the strict sense.

[16] The word *iff* stands for *if and only if*.

[17] Since we are considering only fixed sample sets Ω, these points are uniquely defined.

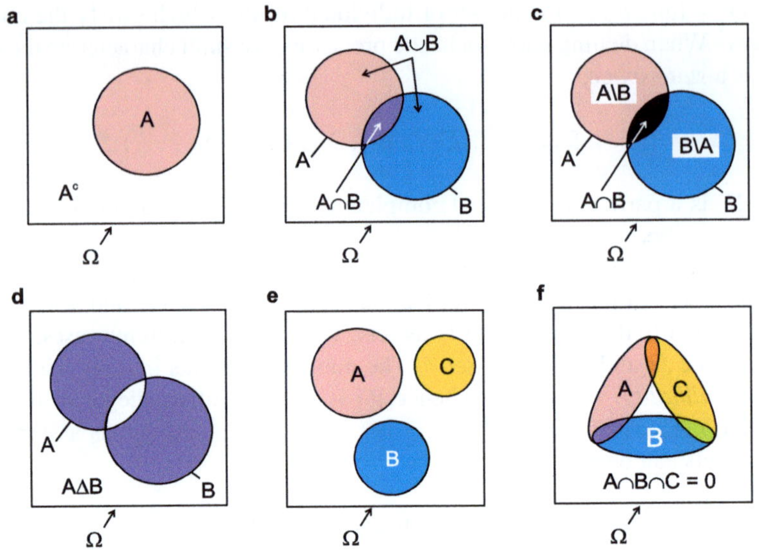

Fig. 1.4 Some definitions and examples from set theory. (**a**) The complement A^c of a set A in the sample space Ω. (**b**) The two basic operations union and intersection, $A \cup B$ and $A \cap B$, respectively. (**c**) and (**d**) Set-theoretic difference $A \setminus B$ and $B \setminus A$, and the symmetric difference, $A \Delta B$. (**e**) and (**f**) Demonstration that a vanishing intersection of three sets does not imply pairwise disjoint sets. The illustrations use *Venn diagrams* [223, 224, 547, 548]

Union The union $A \cup B$ of the two sets A and B is the set of points which belong to at least one of the two sets:

$$A \cup B = \{\omega | \omega \in A \text{ or } \omega \in B\} . \tag{1.6}$$

Intersection The intersection $A \cap B$ of the two sets A and B is the set of points which belong to both sets[18]:

$$A \cap B = AB = \{\omega | \omega \in A \text{ and } \omega \in B\} . \tag{1.7}$$

Unions and intersections can be executed in sequence and are also defined for more than two sets, or even for a countably infinite number of sets:

$$\bigcup_{n=1,\dots} A_n = A_1 \cup A_2 \cup \dots = \{\omega | \omega \in A_n \text{ for at least one value of } n\} ,$$

$$\bigcap_{n=1,\dots} A_n = A_1 \cap A_2 \cap \dots = \{\omega | \omega \in A_n \text{ for all values of } n\} .$$

[18]For short, $A \cap B$ is often written simply as AB.

The proof of these relations is straightforward, because the commutative and the associative laws are fulfilled by both operations, intersection and union:

$$A \cup B = B \cup A \,, \qquad A \cap B = B \cap A \,,$$
$$(A \cup B) \cup C = A \cup (B \cup C) \,, \qquad (A \cap B) \cap C = A \cap (B \cap C) \,.$$

Difference The set theoretic difference $A \setminus B$ is the set of points which belong to A but not to B:

$$A \setminus B = A \cap B^{\mathrm{c}} = \{\omega | \omega \in A \text{ and } \omega \notin B\} \,. \tag{1.8}$$

When $A \supset B$, we write $A - B$ for $A \setminus B$, whence $A \setminus B = A - (A \cap B)$ and $A^{\mathrm{c}} = \Omega - A$.

Symmetric Difference The symmetric difference $A \triangle B$ is the set of points which belong to exactly one of the two sets A and B. It is used in advanced set theory and is symmetric, since it satisfies the commutativity condition $A \triangle B = B \triangle A$:

$$A \triangle B = (A \cap B^{\mathrm{c}}) \cup (A^{\mathrm{c}} \cap B) = (A \setminus B) \cup (B \setminus A) \,. \tag{1.9}$$

Disjoint Sets Disjoint sets A and B have no points in common, so their intersection $A \cap B$ is empty. They fulfill the following relations:

$$A \cap B = \emptyset \,, \quad A \subset B^{\mathrm{c}} \text{ and } B \subset A^{\mathrm{c}} \,. \tag{1.10}$$

Several sets are disjoint only if they are pairwise disjoint. For three sets, A, B, and C, this requires $A \cap B = \emptyset$, $B \cap C = \emptyset$, and $C \cap A = \emptyset$. When two sets are disjoint the addition symbol is (sometimes) used for the union, i.e., we write $A + B$ for $A \cup B$. Clearly, we always have the *decomposition* $\Omega = A + A^{\mathrm{c}}$.

Sample spaces may contain finite or infinite numbers of sample points. As shown in Fig. 1.5, it is important to distinguish further between different classes of infinity[19]: *countable* and *uncountable* numbers of points. The set of *rational numbers* $\mathbb{Q}$, for example, is countably infinite since these numbers can be labeled and assigned each to a different positive integer or *natural number* $\mathbb{N}_{>0}$: $1 < 2 < 3 < \ldots < n < \ldots$. The set of *real numbers* $\mathbb{R}$ cannot be assigned in this way, and so is uncountable. (The notations used for number systems are summarized in appendix at the end of the book.)

[19]Georg Cantor attributed the cardinality $\aleph_0$ to countably infinite sets and characterized uncountable sets by the sizes $\aleph_1$, $\aleph_2$, etc. Important relations between infinite cardinalities are: $\aleph_0 + \aleph_0 = \aleph_0$, $\aleph_0 \cdot \aleph_0 = \aleph_0$ but $2^{\aleph_k} = \aleph_{k+1}$. In particular we have $2^{\aleph_0} = \aleph_1$, the exponential function of a countable infinite set leads to an uncountable infinite set.

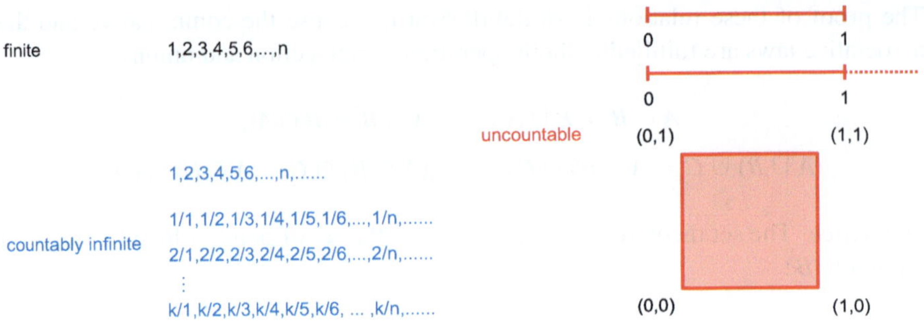

Fig. 1.5 Sizes of sample sets and countability. Finite (*black*), countably infinite (*blue*), and uncountable sets (*red*) are distinguished. We show examples of every class. A set is countably infinite when its elements can be assigned uniquely to the natural numbers ($\mathbb{N}_{>0}$=1,2,3,..., n, ...). This is possible for the rational numbers $\mathbb{Q}$, but not for the positive real numbers $\mathbb{R}_{>0}$ (see, for example, [517])

1.5 Probability Measure on Countable Sample Spaces

For countable sets it is straightforward and almost trivial to measure the size of the set by counting the numbers of sample points they contain. The ratio

$$P(A) = \frac{|A|}{|\Omega|} \tag{1.11}$$

gives the probability for the occurrence of event A and the expression is, of course, identical with the one in (1.1) defining the classical probability. For another event, for example B, one has $P(B) = |B| \;/\; |\Omega|$. Calculating the sum of the two probabilities, $P(A) + P(B)$, requires some care, since Fig. 1.4 suggests that there will only be an inequality (see previous Sect. 1.4):

$$|A| + |B| \;\geq\; |A \cup B| \;.$$

The excess of $|A| + |B|$ over the size of the union $|A \cup B|$ is precisely the size of the intersection $|A \cap B|$, and thus we find

$$|A| + |B| = |A \cup B| + |A \cap B| \;.$$

Dividing by the size of sample space Ω, we obtain

$$\begin{aligned} P(A) + P(B) &= P(A \cup B) + P(A \cap B) \\ \text{or} \;\; P(A \cup B) &= P(A) + P(B) - P(A \cap B) \;. \end{aligned} \tag{1.12}$$

Only when the intersection is empty, i.e., $A \cap B = \emptyset$, are the two sets disjoint and their probabilities additive, so that $|A \cup B| = |A| + |B|$. Hence,

$$P(A + B) = P(A) + P(B) \text{ iff } A \cap B = \emptyset \,. \tag{1.13}$$

It is important to memorize this condition for later use, because it is implicitly assumed when computing probabilities.

1.5.1 Probability Measure

We are now in a position to define a probability measure by means of the basic axioms of probability theory. We present the three axioms as they were first formulated by Andrey Kolmogorov [311]:

A *probability measure* on the sample space Ω is a function of subsets of Ω, $P : S \mapsto P(S)$, which is defined by the following three axioms:

(i) For every set $A \subset \Omega$, the value of the probability measure is a nonnegative number $P(A) \geq 0$ for all A.
(ii) The probability measure of the entire sample set—as a subset—is equal to one, $P(\Omega) = 1$.
(iii) For any two disjoint subsets A and B, the value of the probability measure for the union, $A \cup B = A + B$, is equal to the sum of its values for A and B:

$$P(A \cup B) = P(A + B) = P(A) + P(B) \text{ provided } P(A \cap B) = \emptyset \,.$$

Condition (iii) implies that for any countable—possibly infinite—collection of disjoint or non-overlapping sets, A_i, $i = 1, 2, 3, \ldots$, with $A_i \cap A_j = \emptyset$ for all $i \neq j$, the following *σ-additivity* or *countable additivity* relation holds:

$$P\left(\bigcup_i A_i\right) = \sum_i P(A_i) \,, \quad \text{or } P\left(\sum_{i=1}^{\infty} A_i\right) = \sum_{i=1}^{\infty} P(A_i) \,. \tag{1.14}$$

In other words, the probabilities associated with disjoint sets are additive. Clearly, we also have $P(A^c) = 1 - P(A)$, $P(A) = 1 - P(A^c) \leq 1$, and $P(\emptyset) = 0$. For any two sets $A \subset B$, we find $P(A) \leq P(B)$ and $P(B - A) = P(B) - P(A)$, and for any two

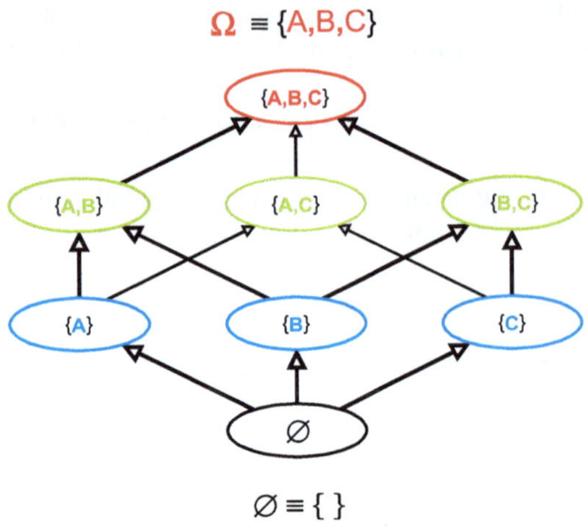

Fig. 1.6 The powerset. The powerset $\Pi(\Omega)$ is a set containing all subsets of Ω, including the empty set $\emptyset$ (*black*) and Ω itself (*red*). The figure shows the construction of the powerset for a sample space of three events A, B, and C (single events in *blue* and double events in *green*). The relation between sets and sample points is also illustrated in a set level diagram (see the *black* and *red levels* in Fig. 1.15)

arbitrary sets A and B, we can write the union as a sum of two disjoint sets:

$$A \cup B = A + A^c \cap B ,$$

$$P(A \cup B) = P(A) + P(A^c \cap B) .$$

Since $B \subset A^c \cap B$, we obtain $P(A \cup B) \leq P(A) + P(B)$.

The set of all subsets of Ω is called the *powerset* $\Pi(\Omega)$ (Fig. 1.6). It contains the empty set $\emptyset$, the entire sample space Ω, and all other subsets of Ω, and this includes the results of all set theoretic operations that were listed in the previous Sect. 1.4. Cantor's theorem named after the mathematician Georg Cantor states that, for any set A, the cardinality of the powerset $\Pi(A)$ is strictly greater than the cardinality $|A|$ [518]. For the example shown in Fig. 1.6, we have $|A| = 3$ and $|\Pi(A)| = 2^3 = 8$. Cantor's theorem is particularly important for countably infinite sample sets [517] like the set of the natural numbers $\mathbb{N}$: $|\Omega| = \aleph_0$ and $|\Pi(\Omega)| = 2^{\aleph_0} = \aleph_1$, the power set of the natural numbers is uncountable.

We illustrate the relationship between the sample point ω, an event A, the sample space Ω, and the powerset $\Pi(\Omega)$ by means of an example, the repeated coin toss, which we shall analyze as a Bernoulli process in Sect. 3.1.3. Flipping a coin has two outcomes: '0' for *heads* and '1' for *tails*. One particular coin toss experiment might give the sequence $(0, 1, 1, 1, 0, \ldots, 1, 0, 0)$. Thus the sample points ω for flipping the coin n times are binary n-tuples or strings,[20] $\omega = (\omega_1, \omega_2, \ldots, \omega_n)$ with $\omega_i \in \sigma = \{0, 1\}$. Then the sample space Ω is the space of all binary strings of length n, commonly denoted by σ^n, and it has the cardinality $|\sigma^n| = 2^n$. The

[20]There is a trivial but important distinction between strings and sets: in a string, the position of an element matters, whereas in a set it does not. The following three sets are identical: $\{1, 2, 3\} = \{3, 1, 2\} = \{1, 2, 2, 3\}$. In order to avoid ambiguities strings are written in round brackets and sets in curly brackets.

extension to the set of all strings of any finite length is straightforward:

$$\sigma^* = \bigcup_{i \in \mathbb{N}} \sigma^i = \{\varepsilon\} \cup \sigma^1 \cup \sigma^2 \cup \sigma^3 \ldots . \qquad (1.15)$$

This set is called the *Kleene star*, after the American mathematician Stephen Kleene. Here $\sigma^0 = \{\varepsilon\}$, where ε denotes the unique string over σ^0, called the *empty string*, while $\sigma^1 = \{0, 1\}$, $\sigma^2 = \{00, 01, 10, 11\}$, etc. The importance of the Kleene star is the *closure* property[21] under concatenation of the sets σ^i:

$$\sigma^m \sigma^n = \sigma^{m+n} = \{wv | w \in \sigma^m \text{ and } v \in \sigma^n\} \text{ with } m, n > 0 . \qquad (1.16)$$

Concatenation of strings is the operation

$$w = (0001) , \quad v = (101) \quad \Longrightarrow \quad wv = (0001101) ,$$

which can be extended to concatenation of sets in the sense of (1.16):

$$\sigma^1 \sigma^2 = \{0, 1\}\{00, 01, 10, 11\}$$
$$= \{000, 001, 010, 011, 100, 101, 110, 111\} = \sigma^3 .$$

The Kleene star set σ^* is the smallest superset of σ, which contains the empty string ε and which is closed under the string concatenation operation. Although all individual strings in σ^* have finite length, the set σ^* itself is countably infinite.

We end this brief excursion into strings and string operations by considering infinite numbers of repeats, i.e., we consider the space σ^n of strings of length n in the limit $n \to \infty$, yielding strings like $\omega = (\omega_1, \omega_2, \ldots) = (\omega_i)_{i \in \mathbb{N}}$ with $\omega_i \in \{0, 1\}$. In this limit, the space $\Omega = \sigma^n = \{0, 1\}^{\mathbb{N}}$ becomes the sample space of all infinitely long binary strings. Whereas the natural numbers are countable, $|\mathbb{N}| = \aleph_0$, binary strings of infinite length are not as follows from a simple argument: Every real number, rational or irrational, can be encoded in binary representation provided the number of digits is infinite, and hence $|\mathbb{R}| = |\{0, 1\}^{\mathbb{N}}| = \aleph_1$ (see also Sect. 1.7.1).

A subset of Ω will be called an *event* A when a probability measure derived from axioms (i), (ii), and (iii) has been assigned. Often one is not interested in a probabilistic result in all its detail, and events can be formed simply by lumping together sample points. This can be illustrated in statistical physics by the microstates in the partition function, which are lumped together according to some macroscopic property. Here, we ask, for example, for the probability A that n coin

[21] Closure under a given operation is an important property of a set that we shall need later on. For example, the natural numbers $\mathbb{N}$ are closed under addition and the integers $\mathbb{Z}$ are closed under addition and substraction.

flips show *tails* at least s times or, in other words, yield a score $k \geq s$:

$$A = \left\{ \omega = (\omega_1, \omega_2, \ldots, \omega_n) \in \Omega : \sum_{i=1}^{n} \omega_i = k \geq s \right\} ,$$

where the sample space is $\Omega = \{0, 1\}^n$. The task is now to find a system of events Ξ that allows for a consistent assignment of a probability $P(A)$ to all possible events A. For countable sample spaces Ω, the powerset $\Pi(\Omega)$ represents such a system Ξ: we characterize $P(A)$ as a probability measure on $\big(\Omega, \Pi(\Omega)\big)$, and the further handling of probabilities is straightforward, following the procedure outlined below. For uncountable sample spaces Ω, the powerset $\Pi(\Omega)$ will turn out to be too large and a more sophisticated procedure will be required (Sect. 1.7).

Among all possible collections of subsets of Ω, a class called σ-algebras plays a special role in measure theory, and their properties will be important for handling uncountable sets:

A *σ-algebra* Σ on some set Ξ is a subset $\Sigma \subset \Pi(\Xi)$ of its powerset satisfying the following three conditions:

(i) $\Xi \in \Sigma$.
(ii) Σ is closed under complements, i.e., if $A \in \Sigma$ then $A^c = \Xi \backslash A \in \Sigma$.
(iii) Σ is closed under countable unions, i.e., if $A_1 \in \Sigma, A_2 \in \Sigma, \ldots$, then $A_1 \cup A_2 \cup \ldots \in \Sigma$.

Closure under countable unions also implies closure under countable intersections by De Morgan's laws [437, pp. 18–19]. From (ii), it follows that every σ-algebra necessarily contains the empty set $\emptyset$, and accordingly the smallest possible σ-algebra is $\{\emptyset, \Xi\}$. If a σ-algebra contains an event A, then the complement A^c is also contained in it, so $\{\emptyset, A, A^c, \Xi\}$ is a σ-algebra.

1.5.2 *Probability Weights*

So far we have constructed, compared, and analyzed sets but have not yet introduced weights or numbers for application to real world situations. In order to construct a probability measure that can be adapted to calculations on countable sample space $\Omega = \{\omega_1, \omega_2, \ldots, \omega_n, \ldots\}$, we have to assign a weight ϱ_n to every sample point ω_n and it must satisfy the conditions

$$\forall n : \ \varrho_n \geq 0 \ \text{and} \ \sum_n \varrho_n = 1 . \tag{1.17}$$

Then for $P(\{\omega_n\}) = \varrho_n \; \forall \, n$ the two equations

$$\begin{aligned} P(A) &= \sum_{\omega \in A} \varrho(\omega) \;\; \text{for} \;\; A \in \Pi(\Omega) \,, \\ \varrho(\omega) &= P(\{\omega\}) \;\; \text{for} \;\; \omega \in \Omega \end{aligned} \tag{1.18}$$

represent a bijective relation between the probability measure P on $\big(\Omega, \Pi(\Omega)\big)$ and the sequences $\varrho = \big(\varrho(\omega)\big)_{\omega \in \Omega}$ in [0,1] with $\sum_{\omega \in \Omega} \varrho(\omega) = 1$. Such a sequence is called a discrete probability density.

The function $\varrho(\omega_n) = \varrho_n$ has to be prescribed by some null hypothesis, estimated or determined empirically, because it is the result of factors lying outside mathematics or probability theory. The uniform distribution is commonly adopted as null hypothesis in gambling, as well as for many other purposes: the discrete *uniform distribution* $\mathcal{U}_\Omega$ assumes that all elementary results $\omega \in \Omega$ appear with equal probability,[22] whence $\varrho(\omega) = 1/|\Omega|$. What is meant here by 'elementary' will become clear when we come to discuss applications. Throwing more than one die at a time, for example, can be reduced to throwing one die more often.

In science, particularly in physics, chemistry, or biology, the correct assignment of probabilities has to meet the conditions of the experimental setup. A simple example from *scientific gambling* will make this point clear: the question as to whether a die is fair and shows all its six faces with equal probability, whether it is imperfect, or whether it has been manipulated and shows, for example, the 'six' more frequently then the other faces, is a matter of physics, not mathematics. Empirical information—for example, a calibration curve of the faces determined by carrying out and recording a few thousand die-rolling experiments—replaces the principle of indifference, and assumptions like the null hypothesis of a uniform distribution become obsolete.

Although the application of a probability measure in the discrete case is rather straightforward, we illustrate by means of a simple example. With the assumption of a uniform distribution $\mathcal{U}_\Omega$, we can measure the size of sets by counting sample points, as illustrated by considering the scores from throwing dice. For one die, the sample space is $\Omega = \{1, 2, 3, 4, 5, 6\}$, and for the *fair* die we make the assumption

$$P(\{k\}) = \frac{1}{6} \,, \quad k = 1, 2, 3, 4, 5, 6 \,,$$

[22] The assignment of equal probabilities $1/n$ to n mutually exclusive and collectively exhaustive events, which are indistinguishable except for their tags, is known as the *principle of insufficient reason* or the *principle of indifference*, as it was called by the British economist John Maynard Keynes [299, Chap. IV, pp. 44–70]. The equivalent in Bayesian probability theory, the *a priori* assignment of equal probabilities, is characterized as the simplest *non-informative prior* (see Sect. 1.3).

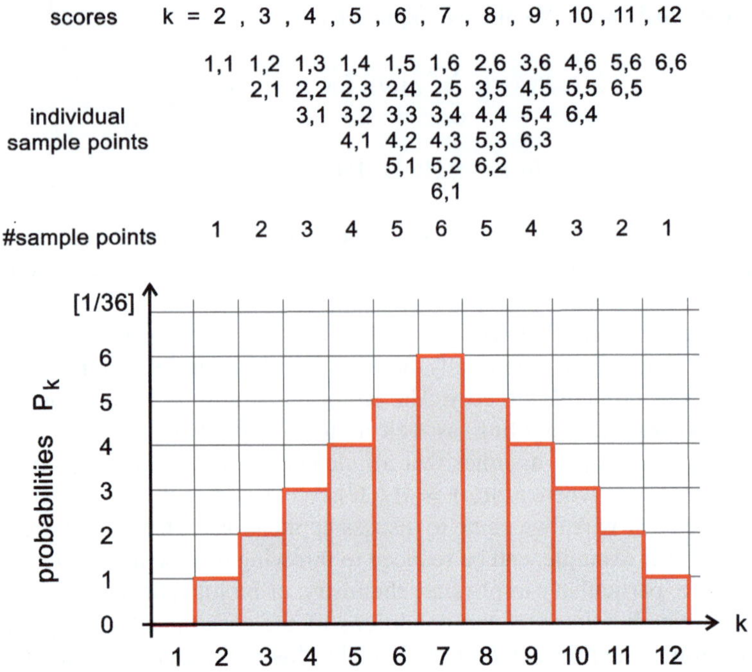

Fig. 1.7 Histogram of probabilities when throwing two dice. The probabilities of obtaining scores of 2–12 when throwing two perfect or fair dice are based on the equal probability assumption for obtaining the individual faces of a single die. The probability $P(N)$ rises linearly for scores from 2 to 7 and then decreases linearly between 7 and 12: $P(N)$ is a discretized tent map with the additivity or normalization condition $\sum_{k=2}^{12} P(N = k) = 1$. The histogram is equivalent to the probability mass function (pmf) of a random variable $\mathcal{Z}: f_{\mathcal{Z}}(x)$ as shown in Fig. 1.11

that all six outcomes corresponding to the different faces of the die are equally likely. Assuming $\mathcal{U}_\Omega$, we obtain the probabilities for the outcome of two simultaneously rolled fair dice (Fig. 1.7). There are $6^2 = 36$ possible outcomes with scores in the range $k = 2, 3, \ldots, 12$, and the most likely outcome is a count of $k = 7$ points because it has the highest multiplicity: $\{(1,6),(2,5),(3,4),(4,3),(5,2),(6,1)\}$. The probability distribution is shown here as a histogram, an illustration introduced into statistics by Karl Pearson [443]. It has the shape of a discretized tent function and is equivalent to the *probability mass function* (pmf) shown in Fig. 1.11. A generalization to simultaneously rolling n dice is presented in Sect. 1.9.1 and Fig. 1.23.

1.6 Discrete Random Variables and Distributions

Conventional deterministic variables are not suitable for describing processes with limited reproducibility. In probability theory and statistics we shall make use of *random* or *stochastic variables*, $\mathcal{X}, \mathcal{Y}, \mathcal{Z}, \ldots$, which were invented especially for dealing with random scatter and fluctuations. Even if an experiment is repeated under precisely the same conditions, the random variable will commonly assume a different value. The probabilistic nature of random variables is expressed by an equation, which is particularly useful for the definition of probability distribution functions[23]:

$$P_k = P\big(\mathcal{Z} = k\big) \quad \text{with } k \in \mathbb{N} \,. \tag{1.19a}$$

A deterministic variable $z(t)$ is defined by a function that returns a unique value for a given argument $z(t) = z_t$.[24] For a random variable $\mathcal{Z}(t)$, the single value of the conventional variable has to be replaced by a series of probabilities $P_k(t)$. This series could be visualized, for example, by means of an L_1 normalized probability vector[25] with the probabilities P_k as components, i.e., $\mathbf{P} = \big(P_0, P_1, \ldots\big)$, with $\|\mathbf{P}\|_1 = \sum_k P_k = 1$.

1.6.1 Distributions and Expectation Values

In probability theory, a random variable is characterized by a probability distribution function rather than a vector, because these functions can be applied with minor modifications to both the discrete and the continuous case. Two probability functions are particularly important and in general use (see Sect. 1.6.3): the *probability mass function* or pmf (see Fig. 1.11)

$$f_{\mathcal{Z}}(x) = \begin{cases} P(\mathcal{Z} = k) = P_k \,, & \forall\, x = k \in \mathbb{N} \,, \\ 0 \,, & \text{anywhere else} \,. \end{cases} \tag{1.19b}$$

and the *cumulative distribution function* or cdf (see Fig. 1.12)

$$F_{\mathcal{Z}}(x) = P(\mathcal{Z} \le k) = \sum_{i \le k} P_i \,. \tag{1.19c}$$

[23]Whenever possible we shall use k, l, m, n for discrete counts, $k \in \mathbb{N}$, and t, x, y, z for continuous variables, $x \in \mathbb{R}^1$ (see appendix on notation at the back of the book).

[24]We use here t as independent variable of the function but do not necessarily imply that t is always time.

[25]The notation for vectors and matrices as used in this book is described in appendix at the back of the book.

The probability mass function $f_{\mathcal{Z}}(x)$ is not a function in the usual sense, because it has the value zero almost everywhere. In fact, it is only nonzero at points where x is a natural number, $x = k \in \mathbb{N}$. In this respect it is related to the Dirac delta function (Sect. 1.6.3). Two properties of the cumulative distribution function follow directly from the properties of probabilities:

$$\lim_{k\to-\infty} F_{\mathcal{Z}}(k) = 0\,, \qquad \lim_{k\to+\infty} F_{\mathcal{Z}}(k) = 1\,.$$

The limit at low k values is chosen in analogy to definitions that will be applied later on. Taking $-\infty$ instead of zero as lower limit makes no difference, because $f_{\mathcal{Z}}(-|k|) = P_{-|k|} = 0$ ($k \in \mathbb{N}$), i.e., negative particle numbers have zero probability. Simple examples of the two probability functions are shown in Figs. 1.11 and 1.12.

All measurable quantities, such as expectation values and variances, can be computed equally well from either of the probability functions:

$$\mathrm{E}(\mathcal{Z}) = \sum_{k=-\infty}^{+\infty} k f_{\mathcal{Z}}(k) = \sum_{k=0}^{+\infty}\big(1 - F_{\mathcal{Z}}(k)\big)\,, \tag{1.20a}$$

$$\begin{aligned}\mathrm{var}(\mathcal{Z}) &= \sum_{k=-\infty}^{k=+\infty} k^2 f_{\mathcal{Z}}(k) - \mathrm{E}(\mathcal{Z})^2 \\ &= 2\sum_{k=0}^{+\infty} k\big(1 - F_{\mathcal{Z}}(k)\big) - \mathrm{E}(\mathcal{Z})^2\,.\end{aligned} \tag{1.20b}$$

In both equations the expressions calculated directly from the cumulative distribution function are valid only for exclusively nonnegative random variables $\mathcal{Z} \in \mathbb{N}$.

To exemplify the use of the cumulative distribution function, we present a proof for the computation of the expectation values for positive random variables: $\mathrm{E}(\mathcal{Z}) = \sum_{k=0}^{\infty}\big(1 - F_{\mathcal{Z}}(k)\big)$.[26] We show the validity of the expression $\mathrm{E}(\mathcal{Z}) = \sum_{k=1}^{\infty} P(\mathcal{Z} \geq k)$ with $k \in \mathbb{N}$ by first expanding the '$\geq$' relation and interchanging the order of summation:

$$\begin{aligned}\sum_{k=1}^{\infty} P(\mathcal{Z} \geq k) &= \sum_{k=1}^{\infty}\sum_{j=k}^{\infty} P(\mathcal{Z} = j) = \sum_{j=1}^{\infty}\sum_{k=1}^{j} P(\mathcal{Z} = j) \\ &= \sum_{j=1}^{\infty}\sum_{k=1}^{j} P_j = \sum_{j=1}^{\infty} jP_j = \mathrm{E}(\mathcal{Z})\,.\end{aligned}$$

[26] The proof is taken from en.wikipedia.org/wiki/Expected_value as of 16 March 2014.

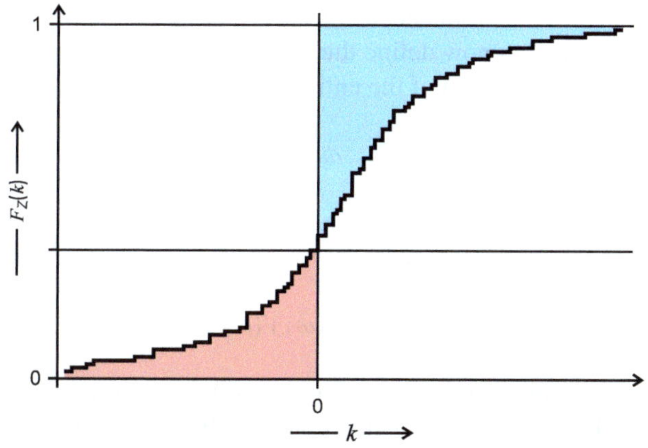

Fig. 1.8 Construction for the calculation of expectation values from cumulative distribution functions. The expectation value is obtained from the cumulative distribution function of a discrete variable as the difference between two contributions: $\sum_{k=0}^{\infty}\big(1-F_{\mathcal{Z}}(k)\big)$ (*blue*) and $\sum_{k=-\infty}^{0} F_{\mathcal{Z}}(k)$ (*red*)

We then introduce the cumulative distribution function:

$$F_{\mathcal{Z}}(k) = P(\mathcal{Z} \le k) = 1 - P(\mathcal{Z} > k) \,,$$

$$F_{\mathcal{Z}}(k-1) = P(\mathcal{Z} \le k-1) = 1 - P(\mathcal{Z} > k-1) = 1 - P(\mathcal{Z} \ge k) \,,$$

$$\mathrm{E}(\mathcal{Z}) = \sum_{k=0}^{\infty}\big(1 - F_{\mathcal{Z}}(k)\big) \,. \qquad \square$$

The generalization to the entire range of integers is possible but requires two summations. For the expectation value, we get

$$\mathrm{E}(\mathcal{Z}) = \sum_{k=0}^{+\infty}\big(1 - F_{\mathcal{Z}}(k)\big) - \sum_{k=-\infty}^{0} F_{\mathcal{Z}}(k) \,. \tag{1.20c}$$

The partitioning of $\mathrm{E}(\mathcal{Z})$ into positive and negative parts is visualized in Fig. 1.8. The expression will be derived for the continuous case in Sect. 1.9.1.

1.6.2 Random Variables and Continuity

Random variables on countable sample spaces require a *probability triple* $(\Omega, \Pi(\Omega), P)$ for a precise definition: Ω contains the sample points or individual results, the powerset $\Pi(\Omega)$ provides the events A as subsets, and P represents a probability measure as introduced in (1.18). Such a probability triple defines a *probability space*. We can now define the *random variable* as a numerically valued function $\mathcal{Z}$ of ω on the domain of the entire sample space Ω :

$$\omega \in \Omega : \omega \mapsto \mathcal{Z}(\omega) . \tag{1.21}$$

Random variables $\mathcal{X}(\omega)$ and $\mathcal{Y}(\omega)$ can be manipulated by conventional operations to yield other random variables, such as

$$\mathcal{X}(\omega) + \mathcal{Y}(\omega) , \quad \mathcal{X}(\omega) - \mathcal{Y}(\omega) , \quad \mathcal{X}(\omega)\mathcal{Y}(\omega) , \quad \mathcal{X}(\omega)/\mathcal{Y}(\omega) \ (\mathcal{Y}(\omega) \neq 0) .$$

In particular, any linear combination $\alpha\mathcal{X}(\omega) + \beta\mathcal{Y}(\omega)$ of random variables is also a random variable. Likewise, as a function of a function is still a function, so a function of a random variable is a random variable:

$$\omega \in \Omega : \ \omega \mapsto \varphi\big(\mathcal{X}(\omega), \mathcal{Y}(\omega)\big) = \varphi(\mathcal{X}, \mathcal{Y}) .$$

Particularly important cases of derived quantities are the partial sums of variables[27]:

$$\mathcal{S}_n(\omega) = \mathcal{Z}_1(\omega) + \cdots + \mathcal{Z}_n(\omega) = \sum_{k=1}^{n} \mathcal{Z}_k(\omega) . \tag{1.22}$$

Such a partial sum $\mathcal{S}_n$ could, for example, be the cumulative outcome of n successive throws of a die. The series could in principle be extended to infinity, thereby covering the entire sample space, in which case the probability conservation relation $\mathcal{S}_n = \sum_{k=1}^{\infty} \mathcal{Z}_k = 1$ must be satisfied. The terms in the sum can be arbitrarily permuted since no ordering criterion has been introduced so far. Most frequently, and in particular in the context of stochastic processes, events will be ordered according to their time of occurrence t (see Chap. 3). An ordered series of events where the current cumulative outcome is given by the sum $\mathcal{S}_n(t) = \sum_{k=1}^{n} \mathcal{Z}_k(t)$ is shown in Fig. 1.9: the plot of the random variable $\mathcal{S}(t)$ is a multi-step function over a continuous time axis t.

Continuity

Steps are inherent discontinuities, and without some further convention we do not know how the value at the step is handled by various step functions. In order to avoid ambiguities, which concern not only the value of the function but also the problem of partial continuity or discontinuity, we must first decide upon a convention that makes expressions like (1.21) or (1.22) precise. The Heaviside step or Θ function is defined by:

---

[27] The use of *partial* in this context expresses the fact that the sum need not cover the entire sample space, at least not for the moment. Dice-rolling series, for example, could be continued in the future.

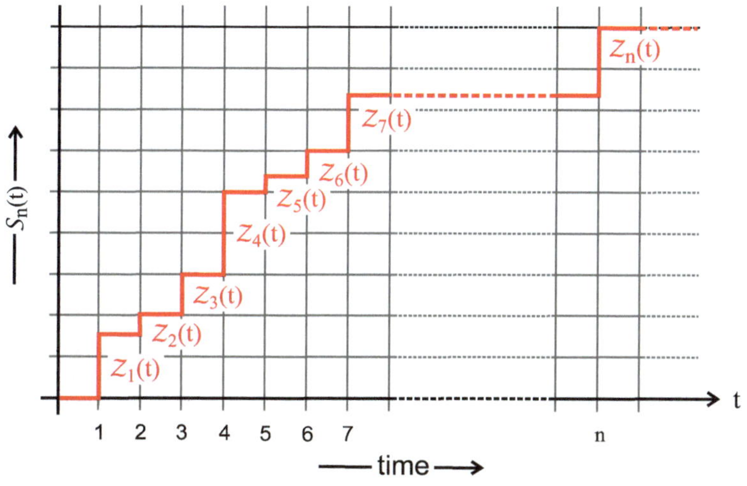

Fig. 1.9 Ordered partial sum of random variables. The sum of random variables, $\mathcal{S}_n(t) = \sum_{k=1}^{n} \mathcal{Z}_k(t)$, represents the cumulative outcome of a series of events described by a class of random variables $\mathcal{Z}_k$. The series can be extended to $+\infty$, and such cases will be encountered, for example, with probability distributions. The ordering criterion specified in this sketch is *time* t, and we are dealing with a stochastic process, here a jump process. The time intervals need not be equal as shown here. The ordering criterion could equally well be a *spatial coordinate* x, y, or z

$$\mathrm{H}(x) = \begin{cases} 0\,, & \text{if } x < 0\,, \\ \text{undefined}\,, & \text{if } x = 0\,, \\ 1\,, & \text{if } x > 0\,. \end{cases} \tag{1.23}$$

It has a discontinuity at the origin $x = 0$ and is undefined there. The Heaviside step function can be interpreted as the integral of the Dirac delta function, viz.,

$$\mathrm{H}(x) = \int_{-\infty}^{x} \delta(\xi)\, \mathrm{d}\xi\,,$$

and this expression becomes ambiguous or meaningless for $x = 0$ as well. The ambiguity can be removed by specifying the value at the origin

$$\mathrm{H}_\gamma(x) = \begin{cases} 0\,, & \text{if } x < 0\,, \\ \gamma \in [0,1]\,, & \text{if } x = 0\,, \\ 1\,, & \text{if } x > 0\,. \end{cases} \tag{1.24}$$

In particular, the three definitions shown in Fig. 1.10 for the value of the function at the step are commonly encountered.

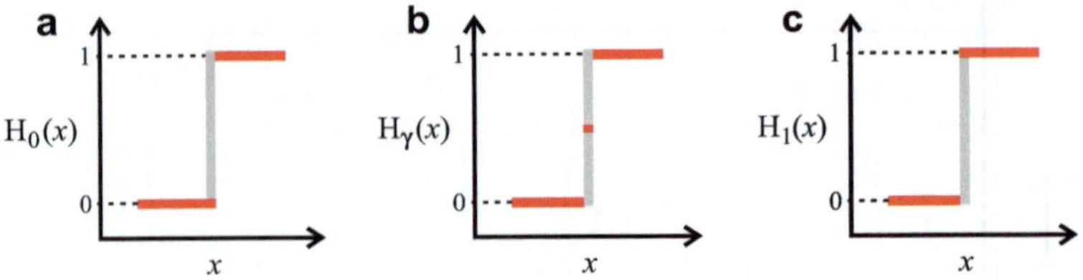

Fig. 1.10 Continuity in probability theory and step processes. Three possible choices of partial continuity or no continuity are shown for the step of the Heaviside function $H_\gamma(x)$: (**a**) $\gamma = 0$ with left-hand continuity, (**b**) $\gamma \notin \{0, 1\}$ implying no continuity, and (**c**) $\gamma = 1$ with right-hand continuity. The step function in (**a**) is left-hand semi-differentiable, the step function in (**c**) is right-hand semi-differentiable, and the step function in (**b**) is neither right-hand nor left-hand semi-differentiable. Choice (**b**) with $\gamma = 1/2$ allows one to exploit the inherent symmetry of the Heaviside function. Choice (**c**) is the standard assumption in Lebesgue–Stieltjes integration, probability theory, and stochastic processes. It is also known as the *càdlàg*-property (Sect. 3.1.3)

For a general step function $F(x)$ with the step at x_0—discrete cumulative probability distributions $F_{\mathcal{Z}}(x)$ may serve as examples—the three possible definitions of the discontinuity at x_0 are expressed in terms of the values (immediately) below and immediately above the step, which we denote by f_{low} and f_{high}, respectively:

(i) Figure 1.10a: $\lim_{\epsilon \to 0} F(x_0 - \epsilon) = f_{\text{low}}$ and $\lim_{\epsilon \to \delta > 0} F(x_0 + \epsilon) = f_{\text{high}}$, with $\epsilon > \delta$ and δ arbitrarily small. The value f_{low} at $x = x_0$ for the function $F(x)$ implies left-hand continuity and the function is semi-differentiable to the left, that is towards decreasing values of x.

(ii) Figure 1.10b: $\lim_{\epsilon \to \delta > 0} F(x_0 - \epsilon) = f_{\text{low}}$ and $\lim_{\epsilon \to \delta > 0} F(x_0 + \epsilon) = f_{\text{high}}$, with $\epsilon > \delta$ and δ arbitrarily small, and the value of the step function at $x = x_0$ is neither f_{low} nor f_{high}. Accordingly, $F(x)$ is not differentiable at $x = x_0$. A special definition is chosen if we wish to emphasize the inherent inversion symmetry of a step function: $F(x_0) = \big(f_{\text{low}} + f_{\text{high}}\big)/2$ (see the sign function below).

(iii) Figure 1.10c: $\lim_{\epsilon \to \delta > 0} F(x_0 - \epsilon) = f_{\text{low}}$, with $\epsilon > \delta$ and δ arbitrarily small and $\lim_{\epsilon \to 0} F(x_0 + \epsilon) = f_{\text{high}}$. The value $F(x_0) = f_{\text{high}}$ results in right-hand continuity and semi-differentiability to the right as expressed by *càdlàg*, which is an acronym from French for *'continue à droite, limites à gauche'*. Right-hand continuity is the standard assumption in the theory of stochastic processes. The cumulative distribution functions $F_{\mathcal{Z}}(x)$, for example, are semi-differentiable to the right, that is towards increasing values of x.

A frequently used example of the second case (Fig. 1.10b) is the sign function or signum function, $\mathrm{sgn}(x) = 2\,\mathrm{H}_{1/2}(x) - 1$:

$$\mathrm{sgn}(x) = \begin{cases} -1\,, & \text{if } x < 0\,, \\ 0\,, & \text{if } x = 0\,, \\ 1\,, & \text{if } x > 0\,, \end{cases} \tag{1.25}$$

which has inversion symmetry at the origin $x_0 = 0$. The sign function is also used in combination with the Heaviside Theta function in order to specify real parts and absolute values in unified analytical expressions.[28]

The value 1 at $x = x_0 = 0$ in $\mathrm{H}_1(x)$ implies right-hand continuity. As mentioned, this convention is adopted in probability theory. In particular, the cumulative distribution functions, $F_{\mathcal{Z}}(x)$ are defined to be right-hand continuous, as are the integrator functions $h(x)$ in Lebesgue–Stieltjes integration (Sect. 1.8). This leads to semi-differentiability to the right. Right-hand continuity is applied in conventional handling of stochastic processes. An example are semimartigales (Sect. 3.1.3), for which the *càdlàg* property is basic.

The behavior of step functions is easily expressed in terms of *indicator functions*, which we discuss here as another class of step function. The indicator function of the event A in Ξ is a mapping of Ξ onto 0 and 1, $\mathbf{1}_A : \Xi \to \{0, 1\}$, with the properties

$$\mathbf{1}_A(x) = \begin{cases} 1\,, & \text{if } x \in A\,, \\ 0\,, & \text{if } x \notin A\,. \end{cases} \tag{1.26a}$$

Accordingly, $\mathbf{1}_A(x)$ extracts the point of the subset $A \in \Xi$ from a set Ξ that might be the entire sample set $\Xi \equiv \Omega$. For a probability space characterized by the triple $(\Omega, \Xi, \mathcal{P})$ with $\Xi \in \Pi(\Omega)$, we define an indicator random variable $\mathbf{1}_A : \Omega \to \{0, 1\}$ with the properties $\mathbf{1}_A(\omega) = 1$ if $\omega \in A$, otherwise $\mathbf{1}_A(\omega) = 0$, and this yields the expectation value

$$\mathrm{E}\big(\mathbf{1}_A(\omega)\big) = \int_{\Xi} \mathbf{1}_A(x)\,\mathrm{d}\mathcal{P}(x) = \int_A \mathrm{d}\mathcal{P}(x) = P(A)\,, \tag{1.26b}$$

[28]Program packages for computer-assisted calculations commonly contain several differently defined step functions. For example, *Mathematica* uses a *Heaviside Theta* function with the definition (1.23), i.e., H(0) is undefined but $\mathrm{H}(0) - \mathrm{H}(0) = 0$ and $\mathrm{H}(0)/\mathrm{H}(0) = 1$, a *Unit Step* function with right-hand continuity, which is defined as $\mathrm{H}_1(x)$, and a *Sign* function specified by (1.25).

and the variance and covariance

$$\begin{aligned} \mathrm{var}\big(\mathbf{1}_A(\omega)\big) &= P(A)\,\big(1 - P(A)\big)\,, \\ \mathrm{cov}\big(\mathbf{1}_A(\omega), \mathbf{1}_B(\omega)\big) &= P(A \cap B) - P(A)P(B)\,. \end{aligned} \tag{1.26c}$$

We shall use indicator functions in the forthcoming sections for the calculation of Lebesgue integrals (Sect. 1.8.3) and for convenient solutions of principal value integrals by partitioning the domain of integration (Sect. 3.2.5).

1.6.3 Discrete Probability Distributions

Discrete random variables are fully characterized by either of the two probability distributions, the probability mass function (pmf) or the cumulative distribution function (cdf). Both functions have been mentioned already and were illustrated in Figs. 1.7 and 1.9, respectively. They are equivalent in the sense that essentially all observable properties can be calculated from either of them. Because of their general importance, we summarize the most important properties of discrete probability distributions.

Making use of our knowledge of the probability space, the probability mass function (pmf) can be formulated as a mapping from the sample space into the real numbers, delivering the probability that a discrete random variable $\mathcal{Z}(\omega)$ attains exactly some value $x = x_k$. Let $\mathcal{Z}(\omega) : \Omega \to \mathbb{R}$ be a discrete random variable on the sample space Ω. Then the probability mass function is a mapping onto the unit interval, i.e., $f_{\mathcal{Z}} : \mathbb{R} \to [0, 1]$, such that

$$f_{\mathcal{Z}}(x_k) = P\big(\{\omega \in \Omega \mid \mathcal{Z}(\omega) = x_k\}\big)\,, \quad \text{with} \ \sum_{k=1}^{\infty} f_{\mathcal{Z}}(x_k) = 1\,, \tag{1.27}$$

where the probability could also be more simply expressed by $P(\mathcal{Z} = x_k)$. Sometimes it is useful to be able to treat a discrete probability distribution as if it were continuous. In this case, the function $f_{\mathcal{Z}}(x)$ is defined for all real numbers $x \in \mathbb{R}$, including those outside the sample set. We then have $f_{\mathcal{Z}}(x) = 0\,,\ \forall\, x \notin \mathcal{Z}(\Omega)$. A simple but straightforward representation of the probability mass function makes use of the Dirac delta-function.[29] The nonzero scores are assumed to lie exactly at

[29]The delta-function is not a proper function, but a generalized function or *distribution*. It was introduced by Paul Dirac in quantum mechanics. For more detail see, for example, [481, pp. 585–590] and [469, pp. 38–42].

the positions x_k with $k \in \mathbb{N}_{>0}$ and $p_k = P(\mathcal{Z} = x_k)$:

$$f_{\mathcal{Z}}(x) = \sum_{k=1}^{\infty} P(\mathcal{Z} = x_k)\,\delta(x - x_k) = \sum_{k=1}^{\infty} p_k \delta(x - x_k)\,. \tag{1.27$'$}$$

In this form, the probability density function is suitable for deriving probabilities by integration (1.28′).

The cumulative distribution function (cdf) of a discrete probability distribution is a step function and contains, in essence, the same information as the probability mass function. Once again, it is a mapping $F_{\mathcal{Z}} : \mathbb{R} \to [0, 1]$ from the sample space into the real numbers on the unit interval, defined by

$$F_{\mathcal{Z}}(x) = P(\mathcal{Z} \le x)\,, \quad \text{with} \lim_{x\to-\infty} F_{\mathcal{Z}}(x) = 0 \text{ and } \lim_{x\to+\infty} F_{\mathcal{Z}}(x) = 1\,. \tag{1.28}$$

By definition the cumulative distribution functions are continuous and differentiable on the right-hand side of the steps. They cannot be integrated by conventional Riemann integration, but they are Riemann–Stieltjes or Lebesgue integrable (see Sect. 1.8). Since the integral of the Dirac delta-function is the Heaviside function, we may also write

$$F_{\mathcal{Z}}(x) = \int_{-\infty}^{x} f_{\mathcal{Z}}(s)\,\mathrm{d}s = \sum_{x_k \le x} p_k\,. \tag{1.28$'$}$$

This integral expression is convenient because it holds for both discrete and continuous probability distributions.

Special cases of importance in physics and chemistry are integer-valued positive random variables $\mathcal{Z} \in \mathbb{N}$, corresponding to a countably infinite sample space, which is the set of nonnegative integers, i.e., $\Omega = \mathbb{N}$, with

$$p_k = P(\mathcal{Z} = k)\,, \quad k \in \mathbb{N} \quad \text{and } F_{\mathcal{Z}}(x) = \sum_{0\le k\le x} p_k\,. \tag{1.29}$$

Such integer-valued random variables will be used, for example, in master equations for modeling particle numbers or other discrete quantities in stochastic processes.

For the purpose of illustration we consider dice throwing again (see Figs. 1.11 and 1.12). If we throw one die with s faces, the pmf consists of s isolated peaks, $f_{1\mathrm{d}}(x_k) = 1/s$ at $x_k = 1, 2, \ldots, s$, and has the value $f_{\mathcal{Z}}(x) = 0$ everywhere else ($x \ne 1, 2, \ldots, s$). Rolling two dice leads to a pmf in the form of a tent function, as shown in Fig. 1.11:

$$f_{2\mathrm{d}}(x_k) = \begin{cases} \dfrac{1}{s^2}(k-1)\,, & \text{for } k = 1, 2, \ldots, s\,, \\[2ex] \dfrac{1}{s^2}(2s+1-k)\,, & \text{for } k = s+1, s+2, \ldots, 2s\,. \end{cases}$$

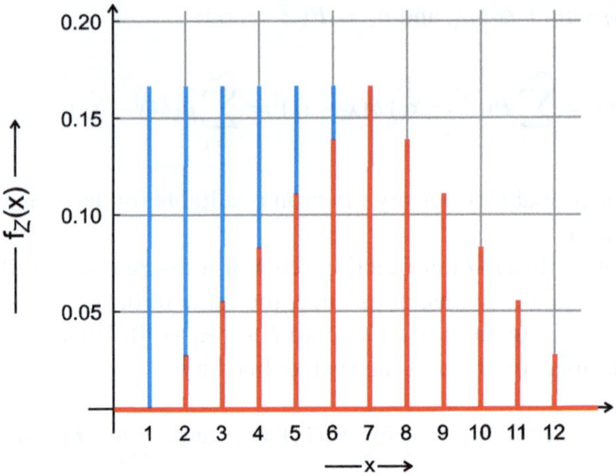

Fig. 1.11 Probability mass function for fair dice. The figure shows the probability mass function (pmf) $f_{\mathcal{Z}}(x_k)$ when rolling one die or two dice simultaneously. The scores x_k are plotted as *abscissa*. The pmf is zero everywhere on the x-axis except at a set of points $x_k \in \{1, 2, 3, 4, 5, 6\}$ for one die and $x_k \in \{2, 3, 4, 5, 6, 7, 8, 9, 10, 11, 12\}$ for two dice, corresponding to the possible scores, with $f_{\mathcal{Z}}(x_k) = (1/6, 1/6, 1/6, 1/6, 1/6, 1/6)$ for one die (*blue*) and $f_{\mathcal{Z}}(x_k) = (1/36, 1/18, 1/12, 1/9, 5/36, 1/6, 5/36, 1/9, 1/12, 1/18, 1/36)$ (*red*) for two dice, respectively. In the latter case the maximal probability value is obtained for the score $x = 7$ [see also (1.27′) and Fig. 1.7]

Here k is the score and s the number of faces of the die, which is six for the most commonly used dice. The cumulative probability distribution function (cdf) is an example of an ordered sum of random variables. The scores when rolling one die or two dice simultaneously are the events. The cumulative probability distribution is simply given by the sum of the scores (Fig. 1.12):

$$F_{2\mathrm{d}}(k) = \sum_{i=2}^{k} f_{2\mathrm{d}}(i)\,, \quad k = 2, 3, \ldots, 2s\,.$$

A generalization to rolling n dice will be presented in Chap. 2.6 when we come to discuss the *central limit theorem*.

Finally, we generalize to sets that define the domain of a random variable on the closed interval[30] $[a, b]$. This is tantamount to restricting the sample set to these

[30]The notation we are applying here uses square brackets [,] for closed intervals, reversed square brackets] , [for open intervals, and] ,] and [, [for intervals open at the left and right ends, respectively. An alternative notation uses round brackets instead of reversed square brackets, e.g., (,) instead of] , [, and so on.

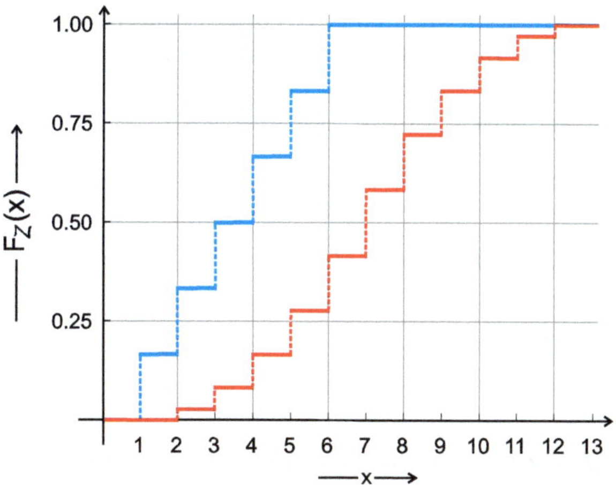

Fig. 1.12 The cumulative distribution function for rolling fair dice. The cumulative probability distribution function (cdf) is a mapping from the sample space Ω onto the unit interval $[0, 1]$ of $\mathbb{R}$. It corresponds to the ordered partial sum with ordering parameter the score given by the stochastic variable. The example considers the case of fair dice: the distribution for one die (*blue*) consists of six steps of equal height $p_k = 1/6$ at the scores $x_k = 1, 2, \ldots, 6$. The second curve (*red*) is the probability that a simultaneous throw of two dice will yield the scores $x_k = 2, 3, \ldots, 12$, where the weights for the individual scores are $p_k = (1/36, 1/18, 1/12, 1/9, 5/36, 1/6, 5/36, 1/9, 1/12, 1/18, 1/36)$. The two limits of any cdf are $\lim_{x\to-\infty} F_{\mathcal{Z}}(x) = 0$ and $\lim_{x\to+\infty} F_{\mathcal{Z}}(x) = 1$

sample points, which give rise to values of the random variable on the interval:

$$\{a \le \mathcal{Z} \le b\} = \{\omega |\, a \le \mathcal{Z}(\omega) \le b\}\,,$$

and defining their probabilities by $P(a \le \mathcal{Z} \le b)$. Naturally, the set of sample points for event A need not be a closed interval: it may be open, half-open, infinite, or even a single point x. In the latter case, it is called a *singleton* $\{x\}$ with $P(\mathcal{Z} = x) = P(\mathcal{Z} \in \{x\})$.

For any countable sample space Ω, i.e., finite or countably infinite, the exact range of $\mathcal{Z}$ is just the set of real numbers w_i:

$$W_{\mathcal{Z}} = \bigcup_{\omega\in\Omega} \{\mathcal{Z}(\omega)\} = \big\{w_1, w_2, \ldots, w_n, \ldots\big\}\,, \quad p_k = P(\mathcal{Z} = w_k)\,, \quad w_k \in W_{\mathcal{Z}}\,.$$

As with the probability mass function (1.27′), we have $P(\mathcal{Z} = x) = 0$ if $x \notin W_{\mathcal{Z}}$. Knowledge of all p_k values is tantamount to having full information on all probabilities derivable for the random variable $\mathcal{Z}$:

$$P(a \le \mathcal{Z} \le b) = \sum_{a\le w_k\le b} p_k\,, \quad \text{or in general,}\ \ P(\mathcal{Z} \in A) = \sum_{w_k\in A} p_k\,. \tag{1.30}$$

The cumulative distribution function (1.28) of $\mathcal{Z}$ is the special case for which A is the infinite interval $]-\infty, x]$. It satisfies several properties on intervals, viz.,

$$F_{\mathcal{Z}}(a) - F_{\mathcal{Z}}(b) = P(\mathcal{Z} \le b) - P(\mathcal{Z} \le a) = P(a < \mathcal{Z} \le b) ,$$

$$P(\mathcal{Z} = x) = \lim_{\epsilon \to 0} \big(F_{\mathcal{Z}}(x + \epsilon) - F_{\mathcal{Z}}(x - \epsilon)\big) ,$$

$$P(a < \mathcal{Z} < b) = \lim_{\epsilon \to 0} \big(F_{\mathcal{Z}}(b - \epsilon) - F_{\mathcal{Z}}(a + \epsilon)\big) ,$$

which are easily verified.

1.6.4 Conditional Probabilities and Independence

Probabilities of events A have been defined so far in relation to the entire sample space Ω by $P(A) = |A|/|\Omega| = \sum_{\omega \in A} P(\omega) \,/\, \sum_{\omega \in \Omega} P(\omega)$. Now we want to know the probability of an event A relative to some subset S of the sample space Ω. This means that we wish to calculate the proportional weight of the part of the subset A in S, as expressed by the intersection $A \cap S$, relative to the weight of the set S. This yields

$$\sum_{\omega \in A \cap S} P(\omega) \Big/ \sum_{\omega \in S} P(\omega) .$$

In other words, we switch from Ω to S as the new universe and the sets to be weighted are sets of sample points belonging to both, i.e., to both A and to S. It is often helpful to call the event S a *hypothesis*, reducing the sample space from Ω to S for the definition of conditional probabilities.

The *conditional probability* measures the probability of A *relative to* S:

$$P(A|S) = \frac{P(A \cap S)}{P(S)} = \frac{P(AS)}{P(S)} , \qquad (1.31)$$

provided $P(S) \neq 0$. The conditional probability $P(A|S)$ is undefined for a hypothesis of zero probability, such as $S = \emptyset$. Clearly, the conditional probability vanishes when the intersection is empty, that is,[31] $P(A|S) = 0$ if $A \cap S = AS = \emptyset$, and $P(AS) = 0$. When S is a true subset of A, $AS = S$, we have $P(A|S) = 1$ (Fig. 1.13).

The definition of the conditional probability implies that all general theorems on probabilities hold by the same token for conditional probabilities. For example,

[31] From here on we shall use the short notation $AS \equiv A \cap S$ for the intersection.

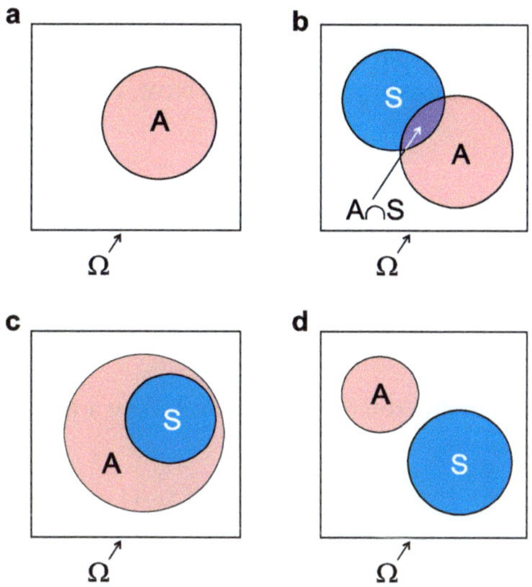

Fig. 1.13 Conditional probabilities. Conditional probabilities measure the intersection $A \cap S$ of the sets for two events relative to the set S: $P(A|S) = |AS|/|S|$. In essence, this is the same kind of weighting that defines the probabilities in sample space: $P(A) = |A|/|\Omega|$. (**a**) shows $A \subset \Omega$ and (**b**) shows $A \cap S \subset S$. The two extremes are $A \cap S = S$ and $P(A|S) = 1$ (**c**) and $A \cap S = 0$ and $P(A|S) = 0$ (**d**)

(1.12) implies that

$$P(A \cup B|S) = P(A|S) + P(B|S) - P(AB|S) . \tag{1.12$'$}$$

Additivity of conditional probabilities, for example, requires an empty intersection $AB = \emptyset$.

Equation (1.31) is particularly useful when written in the slightly different form

$$P(AS) = P(A|S)P(S) . \tag{1.31$'$}$$

This is known as the *theorem of compound probabilities* and is easily generalized to more events. For three events, we derive [160, Chap.V]

$$P(ABC) = P(A|BC)P(B|C)P(C)$$

by applying (1.31′) twice—first by setting $BC \equiv S$ and then by setting $BC \equiv AS$. For n arbitrary events A_i, $i = 1, \ldots, n$, this leads to

$$P(A_1A_2 \ldots A_n) = P(A_1|A_2A_3 \ldots A_n)P(A_2|A_3 \ldots A_n) \ldots P(A_{n-1}|A_n)P(A_n),$$

provided that $P(A_2A_3 \ldots A_n) > 0$. If the intersection $A_2 \ldots A_n$ does not vanish, all conditional probabilities are well defined, since

$$P(A_n) \geq P(A_{n-1}A_n) \geq \ldots \geq P(A_2A_3 \ldots A_n) > 0 .$$

Next we derive an equation that we shall need in Chap. 3 to model stochastic processes. We assume that the sample space Ω is partitioned into n disjoint sets, viz., $\Omega = \sum_n S_n$. Then we have, for any set A,

$$A = AS_1 \cup AS_2 \cup \ldots \cup AS_n ,$$

and from (1.31′) we get

$$P(A) = \sum_n P(A|S_n)P(S_n) . \tag{1.32}$$

From this relation it is straightforward to derive the conditional probability

$$P(S_j|A) = \frac{P(S_j)P(A|S_j)}{\sum_n P(S_n)P(A|S_n)} ,$$

provided that $P(A) > 0$.

The conditional probability can also be interpreted as information about the occurrence of an event S as reflected by the probability of A. Independence of the events—implying that considering $P(A)$ does not allow for any inference on whether or not S has occurred—is easily formulated in terms of conditional probabilities: it implies that S has no influence on A, so $P(A|S) = P(A)$ defines *stochastic independence*. Making use of (1.31′), we define

$$P(AS) = P(A)P(S) , \tag{1.33}$$

and thereby observe an important symmetry of stochastic independence: A is independent of S implies S is independent of A. We may account for this symmetry in defining independence by stating that A and S are independent if (1.33) holds. We remark that the definition (1.33) is also acceptable when $P(S) = 0$, even though $P(A|S)$ is undefined [160, p. 125].

The case of more than two events needs some care. We take three events A, B, and C as an example. So far we have been dealing only with pairwise independence, and accordingly we have

$$P(AB) = P(A)P(B) , \quad P(BC) = P(B)P(C) , \quad P(CA) = P(C)P(A) . \tag{1.34a}$$

Pairwise independence, however, does not necessarily imply that

$$P(ABC) = P(A)P(B)P(C) . \tag{1.34b}$$

Moreover, examples can be constructed in which the last equation is satisfied but the sets are not in fact pairwise independent [200].

Independence or lack of independence of three events is easily visualized using weighted Venn diagrams. In Fig. 1.14 and Table 1.2 (row a), we show a case where

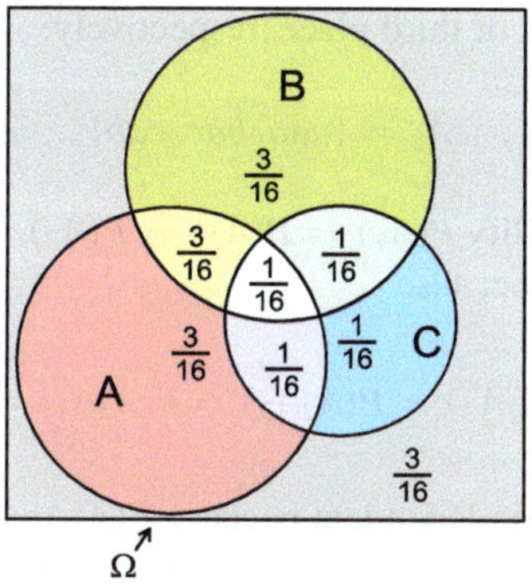

Fig. 1.14 Testing for stochastic independence of three events. The case shown here is a example for independence of three events and corresponds to example (a) in Table 1.2. The numbers in the sketch satisfy (1.34a) and (1.34b). The probability of the union of all three sets is given by the relation

$$P(A \cup B \cup C) = P(A) + P(B) + P(C) - P(AB) - P(BC) - P(AC) + P(ABC),$$

and by addition of the remainder, one checks that $P(\Omega) = 1$

Table 1.2 Testing for stochastic independence of three events

	Probability P						
	Singles			Pairs			Triple
	A	B	C	AB	BC	CA	ABC
Case a	1/2	1/2	1/4	1/4	1/8	1/16	1/16
Case b	1/2	1/2	1/4	1/4	1/8	1/8	**1/10**
Case c	1/5	2/5	1/2	**1/10**	**6/25**	**7/50**	1/25

We show three examples: case (**a**) satisfies (1.34a) and (1.34b), and represents a case of mutual independence (Fig. 1.14). Case (**b**) satisfies only (1.34a) and not (1.34b), and is an example of pairwise independent but not mutually independent events. Case (**c**) is a specially constructed example satisfying (1.34b) with three sets that are not pairwise independent. Deviations from (1.34a) and (1.34b) are indicated in boldface

independence of the three sets A, B, and C is easily tested. Although situations with three pairwise independent events but lacking mutual independence are not particularly common, they can nevertheless be found: the situation illustrated in Fig. 1.4f allows for straightforward construction of examples with lack of pairwise independence, but $P(ABC) = 0$. Let us also consider the opposite situation, namely, pairwise independence but non-vanishing triple dependence $P(ABC) \neq 0$, using an example attributed to Sergei Bernstein [160, p. 127]. The six permutations of the three letters a, b and c together with the three triples (aaa), (bbb), and (ccc) constitute the sample space and a probability $P = 1/9$ is attributed to each sample point. We now define three events A_1, A_2, and A_3 according to the appearance of the letter a at the first, second, or third place, respectively:

$$A_1 = \{aaa, abc, acb\}, \quad A_2 = \{aaa, bac, cab\}, \quad A_3 = \{aaa, bca, cba\}.$$

Every event has a probability $P(A_1) = P(A_2) = P(A_3) = 1/3$ and the three events are pairwise independent because

$$P(A_1A_2) = P(A_2A_3) = P(A_3A_1) = \frac{1}{9},$$

but they are not mutually independent because $P(A_1A_2A_3) = 1/9$ instead of $1/27$, as required by (1.34b). In this case it is easy to detect the cause of the mutual dependence: the occurrence of two events implies the occurrence of the third and therefore we have $P(A_1A_2) = P(A_2A_3) = P(A_3A_1) = P(A_1A_2A_3)$. Table 1.2 presents numerical examples for all three cases.

Generalization to n events is straightforward [160, p. 128]. The events $A_1, A_2, \ldots, A_n$ are mutually independent if the multiplication rules apply for all combinations $1 \le i < j < k < \ldots \le n$, whence we have the following $2^n - n - 1$ conditions[32]:

$$\begin{aligned} P(A_iA_j) &= P(A_i)P(A_j), \\ P(A_iA_jA_k) &= P(A_i)P(A_j)P(A_k), \\ &\vdots \\ P(A_1A_2\ldots A_n) &= P(A_1)\cdot P(A_2)\cdot\ldots\cdot P(A_n). \end{aligned} \tag{1.35}$$

Two or More Random Variables

Two variables,[33] for example $\mathcal{X}$ and $\mathcal{Y}$, can be subsumed in a *random vector* $\boldsymbol{\mathcal{V}} = (\mathcal{X}, \mathcal{Y})$, which is expressed by the *joint probability*

$$P(\mathcal{X} = x_i, \mathcal{Y} = x_j) = p(x_i, y_j). \tag{1.36}$$

[32]These conditions consist of $\binom{n}{2}$ equations in the first line, $\binom{n}{3}$ equations in the second line, and so on, down to $\binom{n}{n} = 1$ equations in the last line. Summing yields $\sum_{i=2}^{n}\binom{n}{i} = (1+1)^n - \binom{n}{1} - \binom{n}{0} = 2^n - n - 1$.

[33]For simplicity, we restrict ourselves to the two-variable case here. The extension to any finite number of variables is straightforward.

The random vector $\mathcal{V}$ is fully determined by the *joint probability mass function*

$$\begin{aligned} f_{\mathcal{V}}(x,y) &= P(\mathcal{X} = x, \mathcal{Y} = y) = P(\mathcal{X} = x \ \wedge \ \mathcal{Y} = y) \\ &= P(\mathcal{Y} = y|\mathcal{X} = x)P(\mathcal{X} = x) \\ &= P(\mathcal{X} = x|\mathcal{Y} = y)P(\mathcal{Y} = y) \ . \end{aligned} \tag{1.37}$$

This density constitutes the probabilistic basis of the random vector $\mathcal{V}$. It is straightforward to define a *cumulative probability distribution* in analogy to the single variable case:

$$F_{\mathcal{V}}(x,y) = P(\mathcal{X} \le x, \mathcal{Y} \le y) \ . \tag{1.38}$$

In principle, both of these probability functions contain full information about both variables, but depending on the specific situation, either the pmf or the cdf may be more efficient.

Often no detailed information is required regarding one particular random variable. Then, summing over one variable of the vector $\mathcal{V}$, we obtain the probabilities for the corresponding *marginal distribution*:

$$\begin{aligned} P(\mathcal{X} = x_i) &= \sum_{y_j} p\,(x_i, y_j) = p\,(x_i, *) \ , \\ P(\mathcal{Y} = y_j) &= \sum_{x_i} p\,(x_i, y_j) = p\,(*, y_j) \ , \end{aligned} \tag{1.39}$$

of $\mathcal{X}$ and $\mathcal{Y}$, respectively.

Independence of random variables will be a highly relevant problem in the forthcoming chapters. Countably-valued random variables $\mathcal{X}_1, \ldots, \mathcal{X}_n$ are defined to be *independent* if and only if, for any combination $x_1, \ldots, x_n$ of real numbers, the joint probabilities can be factorized:

$$P(\mathcal{X}_1 = x_1, \ldots, \mathcal{X}_n = x_n) = P(\mathcal{X}_1 = x_1) \cdots P(\mathcal{X}_n = x_n) \ . \tag{1.40}$$

An extension of (1.40) replaces the single values x_i by arbitrary sets S_i :

$$P(\mathcal{X}_1 \in S_1, \ldots, \mathcal{X}_n \in S_n) = P(\mathcal{X}_1 \in S_1) \cdots P(\mathcal{X}_n \in S_n) \ .$$

In order to justify this extension, we sum over all points belonging to the sets $S_1, \ldots, S_n$:

$$\sum_{x_1 \in S_1} \cdots \sum_{x_n \in S_n} P(\mathcal{X}_1 = x_1, \ldots, \mathcal{X}_n = x_n)$$

$$= \sum_{x_1 \in S_1} \cdots \sum_{x_n \in S_n} P(\mathcal{X}_1 \in S_1) \cdots P(\mathcal{X}_n \in S_n)$$

$$= \left(\sum_{x_1 \in S_1} P(\mathcal{X}_1 \in S_1) \right) \cdots \left(\sum_{x_n \in S_n} P(\mathcal{X}_n \in S_n) \right) ,$$

which is equal to the right-hand side of the equation we wish to justify. □

Since the factorization is fulfilled for arbitrary sets $S_1, \ldots S_n$, it holds also for all subsets of $(\mathcal{X}_1 \ldots \mathcal{X}_n)$, and accordingly the events

$$\{\mathcal{X}_1 \in S_1\}, \ \ldots, \ \{\mathcal{X}_n \in S_n\}$$

are also independent. It can also be checked that, for arbitrary real-valued functions $\varphi_1, \ldots, \varphi_n$ on $]-\infty, +\infty[$, the random variables $\varphi_1(\mathcal{X}_1), \ldots, \varphi_n(\mathcal{X}_n)$ are independent, too.

Independence can also be extended in straightforward manner to the joint distribution function of the random vector $\mathcal{V} = (\mathcal{X}_1, \ldots, \mathcal{X}_n)$

$$F_{\mathcal{V}}(x_1, \ldots, x_n) = F_{\mathcal{X}_1}(x_1) \cdots F_{\mathcal{X}_n}(x_n) ,$$

where the $F_{\mathcal{X}_j}$ are the marginal distributions of the $\mathcal{X}_j$, $1 \le j \le n$. Thus, the marginal distributions completely determine the joint distribution when the random variables are independent.

1.7 ⋆ Probability Measure on Uncountable Sample Spaces

In the previous sections we dealt with countably finite or countably infinite sample spaces where classical probability theory would have worked as well as the set theoretic approach. A new situation arises when the sample space Ω is uncountable (see, e.g., Fig. 1.5) and this is the case, for example, for continuous variables defined on nonzero, open, half open, or closed segments of the real line, viz., $]a, b[$, $]a, b]$, $[a, b[$, or $[a, b]$ for $a < b$. We must now ask *how we can assign a measure on an uncountable sample space*.

The most straightforward way to demonstrate the existence of such measures is the assignment of length (m), area (m^2), volume (m^3), or generalized volume (m^n)

to uncountable sets. In order to illustrate the problem we may ask a very natural question: does every proper subset of the real line $-\infty < x < +\infty$ have a length? It seems natural to assign length 1 to the interval $[0, 1]$ and length $b - a$ to the interval $[a, b]$ with $a \leq b$, but here we have to analyze such an assignment using set theory in order to check that it is consistent.

Sometimes the weight of a homogeneous object is easier to determine than the length or volume and we assign mass to sets in the sense of homogeneous bars with uniform density. For example, we attribute to $[0, 1]$ a bar of length 1 that has mass 1, and accordingly, to the stretch $[a, b]$, a bar of mass $b-a$. Taken together, two bars corresponding to the set $[0, 2] \cup [6, 9]$ have mass 5, with $\cup$ symbolizing σ-additivity. More ambitiously, we might ask for the mass of the set of rational numbers $\mathbb{Q}$, given that the mass of the interval $[0, 1]$ is one? Since the rational numbers are *dense* in the real numbers,[34] any nonnegative value for the mass of the rational numbers appears to be acceptable *a priori*. The real numbers $\mathbb{R}$ are uncountable and so are the irrational numbers $\mathbb{R}\backslash\mathbb{Q}$. Assigning mass $b - a$ to $[a, b]$ leaves no room for the rational numbers, and indeed the rational numbers $\mathbb{Q}$ have measure zero, like any other set of countably many objects.

Now we have to be more precise and introduce a measure called *Lebesgue measure*, which measures generalized volume.[35] As argued above the rational numbers should be attributed Lebesgue measure zero, i.e., $\lambda(\mathbb{Q}) = 0$. In the following, we shall show that the Lebesgue measure does indeed assign precisely the values to the intervals on the real axis that we have suggested above, i.e., $\lambda([0, 1]) = 1$, $\lambda([a, b]) = b - a$, etc. Before discussing the definition and the properties of Lebesgue measures, we repeat the conditions for measurability and consider first a simpler measure called Borel measure μ, which follows directly from σ-additivity of disjoint sets as expressed in (1.14).

For countable sample spaces Ω, the powerset $\Pi(\Omega)$ represents the set of all subsets, including the results of all set theoretic operations of Sect. 1.4, and is the appropriate reference for measures since all subsets $A \in \Pi(\Omega)$ have a defined probability, $P(A) = |A|/|\Omega|$ (1.11) and are measurable. Although it would seem natural to proceed in the same way for countable and uncountable sample spaces Ω, it turns out that the powerset of uncountable sample spaces Ω is too large, because equation (1.11) may be undefined for some sets V. Then, no probability exists and V is not measurable (Sect. 1.7.1). Recalling Cantor's theorem the cardinality of the powerset $\Pi(\Omega)$ is $\aleph_2$ if $|\Omega| = \aleph_1$. What we have to search for is an *event system* Ξ with $A \in \Xi$, which is a subset of the powerset Π, and which allows to define a probability measure (Fig. 1.15).

---

[34] A subset D of real numbers is said to be *dense* in $\mathbb{R}$ if every arbitrarily small interval $]a, b[$ with $a < b$ contains at least one element of D. Accordingly, the set of rational numbers $\mathbb{Q}$ and the set of irrational numbers $\mathbb{R}\backslash\mathbb{Q}$ are both dense in $\mathbb{R}$.

[35] Generalized volume is understood as a line segment in $\mathbb{R}^1$, an area in $\mathbb{R}^2$, a volume in $\mathbb{R}^3$, etc.

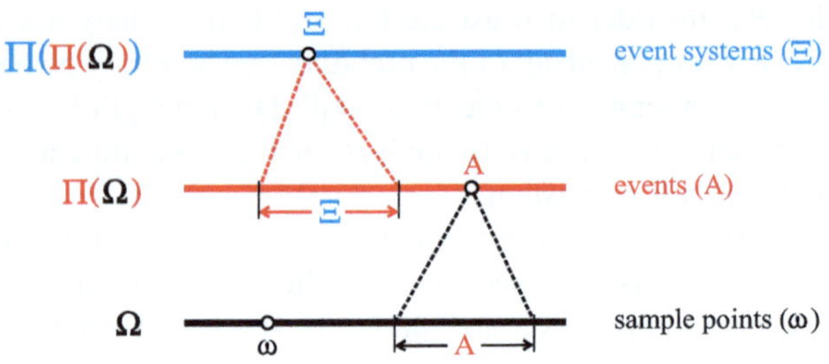

Fig. 1.15 Conceptual levels of sets in probability theory. The lowest level is the sample space Ω (*black*), which contains the sample points or individual results ω as elements. Events A are subsets of Ω: $\omega \in \Omega$ and $A \subset \Omega$. The next higher level is the powerset $\Pi(\Omega)$ (*red*). Events A are elements of the powerset and event systems Ξ constitute subsets of the powerset: $A \in \Pi(\Omega)$ and $\Xi \subset \Pi(\Omega)$. The highest level is the power powerset $\Pi\big(\Pi(\Omega)\big)$, which contains event systems Ξ as elements: $\Xi \in \Pi\big(\Pi(\Omega)\big)$ (*blue*). Adapted from [201, p. 11]

Three properties of probability measures μ are indispensable and have to be fulfilled by all measurable collections Ξ of events A on uncountable sample spaces like $\Omega = [0, \infty[$:

(i) Nonnegativity: $\mu(A) \geq 0\,, \ \forall A \in \Xi\,.$
(ii) Normalization: $P(\Omega) = 1\,.$
(iii) Additivity: $\mu(A) + \mu(B) = \mu(A \cup B)$, whenever $P(A \cap B) = \emptyset\,.$

In essence, the task is now to find measures for uncountable sets that are derived from event systems Ξ, which are collections of subsets of the powerset. Problems concerning measurability arise from the impossibility of assigning a probability to every subset of Ω; in other words, there may be sets to which no measure—no length, no mass, etc.—can be assigned. The rigorous derivation of the concept of measurable sets is highly demanding and requires advanced mathematical techniques, in particular a sufficient knowledge of measure theory [51, 523, 527]. For the probability concept we are using here, however, the simplest bridge from countability to uncountability is sufficient and we need only derive a measure for a certain family of sets, the *Borel sets* $\mathcal{B} \subset \Omega$. For this goal, the introduction of σ-additivity (1.14) and Lebesgue measure $\lambda(A)$ is sufficient. Still unanswered so far, however, is the question of whether there are in fact non-measurable sets (Sect. 1.7.1).

1.7.1 ⋆ Existence of Non-measurable Sets

A general description of non-measurable sets is difficult. However, Giuseppe Vitali [552, 553] provided a proof of existence by contradiction. For a given example, the infinitely repeated coin flip on $\Omega = \{0, 1\}^{\mathbb{N}}$, there exists no mapping $P : \Pi(\Omega) \to$

$[0,1]$ which satisfies the indispensable properties for probabilities (see, e.g., [201, p. 9, 10]):

(N) Normalization: $P(\Omega) = 1$.
(A) σ-additivity: for pairwise disjoint events $A_1, A_2, \ldots \subset \Omega$,

$$P\left(\bigcup_{i \geq 1} A_i\right) = \sum_{i \geq 1} P(A_i) \ .$$

(I) Invariance: for all $A \subset \Omega$ and $k \geq 1$, we have $P(T_k A) = P(A)$, where T_k is an operator that reverses the outcome of the k th toss.

The sample points of Ω are infinitely long strings $\omega = (\omega_1, \omega_2, \ldots)$, the operators T_k are defined by

$$T_k : \omega = (\omega_1, \ldots, \omega_{k-1}, \omega_k, \omega_{k+1}, \ldots) \longrightarrow (\omega_1, \ldots, \omega_{k-1}, 1 - \omega_k, \omega_{k+1}, \ldots) \,,$$

and $T_k A = \{T_k(\omega) : \omega \in A\}$ is the image of A under the operation T_k, which defines a mapping of Ω onto itself. The first two conditions, (N) and (A), are the criteria for probability measures, and the invariance condition (I) is specific for coin flipping and encapsulates the properties derived from the uniform distribution $\mathcal{U}_\Omega$: $P(\omega_k) = P(1 - \omega_k) = 1/2$ for the single coin toss.

Proof In order to prove the conjecture of incompatibility with all three conditions, we define an equivalence relation $\sim$ in Ω by saying that $\omega \sim \omega'$ iff $\omega_k = \omega'_k$ for all sufficiently large k. In other words the sequences in a given equivalence class are the same in their infinitely long tails. The elements of an equivalence class are sequences, which have the same digits from some position on. The *axiom of choice*,[36] states the existence of a set $A \subset \Omega$, which contains exactly one element of each equivalence class.

Next we define $\mathcal{S} = \{S \subset \mathbb{N} : |S| < \infty\}$ to be the set of all finite subsets of $\mathbb{N}$. Since $\mathcal{S}$ is the union of a countable number of finite sets $\{S \subset \mathbb{N} : \max S = m\}$ with $m \in \mathbb{N}$, $\mathcal{S}$ is countable too. For $S = \{k_1, \ldots, k_n\} \in \mathcal{S}$, we define $T_S = \prod_{k_i \in S} T_{k_i} = T_{k_1} \circ \cdots \circ T_{k_n}$, the simultaneous reversal of all elements ω_{k_i} corresponding to the integers in S. Then we have:

(i) $\Omega = \bigcup_{S \in \mathcal{S}} T_S A$, since for every sequence $\omega \in \Omega$, there exists an $\omega' \in A$ with $\omega \sim \omega'$, and accordingly an $S \in \mathcal{S}$ such that $\omega = T_S\, \omega' \in T_S A$.
(ii) The sets $(T_S A)_{S \in \mathcal{S}}$ are pairwise disjoint: if $T_S A \cup T_{S'} A \neq \emptyset$ were true for $S, S' \in \mathcal{S}$, then there would exist $\omega, \omega' \in A$ with $T_S \omega = T_{S'}\, \omega'$ and accordingly $\omega \sim T_S \omega = T_S \omega \sim \omega'$. By definition of A, we would have $\omega = \omega'$ and hence $S = S'$.

[36]The axiom of choice is as follows. Suppose that $A_\theta : \theta \in \Theta$ is a decomposition of Ω into nonempty sets. The axiom of choice guarantees that there exists at least one set C which contains exactly one point from each A_θ, so that $C \cap A_\theta$ is a singleton for each θ in Θ (see [51, p. 572] and [117]).

Applying the properties (N), (A), and (I) of the probability P, we find

$$1 = P(\Omega) = \sum_{S \in \mathcal{S}} P(T_S A) = \sum_{S \in \mathcal{S}} P(A) . \tag{1.41}$$

Equation (1.41) cannot be satisfied for infinitely long series of coin tosses, since all values $P(A)$ or $P(T_S A)$ are the same, and infinite summation by σ-additivity (A) is tantamount to an infinite sum of the same number, which yields either 0 or ∞, but never 1 as required to satisfy (N). □

It is straightforward to show that the set of all binary strings with countably infinite length, viz., $B = \{0, 1\}^{\mathbb{N}}$, is bijective[37] with the unit interval $[0, 1]$. A more or less explicit bijection $f : B \leftrightarrow [0, 1]$ can be obtained by defining an auxiliary function

$$g(s) \doteq \sum_{k=1}^{\infty} \frac{s_k}{2^k} .$$

This interprets a binary string $s = (s_1, s_2, \ldots) \in B$ as an infinite binary fraction

$$\frac{s_1}{2} + \frac{s_2}{4} + \cdots .$$

The function $g(s)$ maps B only *almost* bijectively onto $[0, 1]$, because each dyadic rational in $]0, 1[$ has two preimages,[38] e.g.,

$$g(1, 0, 0, 0, \ldots) = \frac{1}{2} = \frac{1}{4} + \frac{1}{8} + \frac{1}{16} + \ldots = g(0, 1, 1, 1, \ldots) .$$

In order to fix this problem we reorder the dyadic rationals:

$$\left(q_n\right)_{n \geq 1} = \left(\frac{1}{2}, \frac{1}{4}, \frac{3}{4}, \frac{1}{8}, \frac{3}{8}, \frac{5}{8}, \frac{7}{8}, \frac{1}{16}, \ldots\right) ,$$

and take for the bijection

$$f(s) \doteq \begin{cases} q_{2n-1} , & \text{if } g(s) = q_n , \text{ and } s_k = 1 \text{ for almost all } k , \\ q_{2n} , & \text{if } g(s) = q_n , \text{ and } s_k = 0 \text{ for almost all } k , \\ g(s) , & \text{otherwise} . \end{cases} \tag{1.42}$$

---

[37] A *bijection* or bijective function specifies a one-to-one correspondence between the elements of two sets.

[38] Suppose a function $f : X \to Y$ with $(X, Y) \in \Omega$. Then the image of a subset $A \subseteq X$ is the subset $f(A) \subseteq Y$ defined by $f(A) = \{y \in Y | \, y = f(x) \text{ for some } x \in A\}$, and the preimage or inverse image of a set $B \subseteq Y$ is $f^{-1}(B) = \{x \in X | f(x) \in B\} \subseteq X$.

Hence Vitali's theorem applies equally well to the unit interval [0, 1], where we are also dealing with an uncountable number of non-measurable sets. For other more detailed proofs of Vitali's theorem, see, e.g., [51, p. 47].

The proof of Vitali's theorem shows the existence of non-measurable subsets called Vitali sets within the real numbers by contradiction. More precisely, it provides evidence for subsets of the real numbers that are not Lebesgue measurable (see Sect. 1.7.2). The problem to be solved now is a rigorous reduction of the powerset to an event system Ξ such that the subsets causing the lack of countability can be left aside (Fig. 1.15).

1.7.2 * Borel σ-Algebra and Lebesgue Measure

In Fig. 1.15, we consider the three levels of sets in set theory that are relevant for our construction of an event system Ξ. The objects on the lowest level are the sample points $\omega \in \Omega$ corresponding to individual results. The next higher level is the powerset $\Pi(\Omega)$, containing the events $A \in \Pi(\Omega)$. The elements of the powerset are subsets $A \subset \Omega$ of the sample space. To illustrate the role of event systems Ξ, we need a still higher level, the powerset $\Pi\big(\Pi(\Omega)\big)$ of the powerset: event systems Ξ are elements of the power powerset, i.e., $\Xi \in \Pi\big(\Pi(\Omega)\big)$ and subsets $\Xi \subset \Pi(\Omega)$ of the powerset.[39]

The minimal requirements for an *event system* Ξ are summarized in the following definition of a *σ-algebra* on Ω with $\Omega \neq \emptyset$ and $\Xi \subset \Pi(\Omega)$:

Condition (1): $\Omega \in \Xi\,,$
Condition (2): $A \in \Xi \implies A^{c} \doteq \Omega \backslash A \in \Xi\,,$
Condition (3): $A_1, A_2, \ldots \in \Xi \implies \bigcup_{i \geq 1} A_i \in \Xi\,.$

Condition (2) requires the existence of a complement A^{c} for every subset $A \in \Xi$ and defines the logical negation as expressed by the difference between the entire sample space and the event A. Condition (3) represents the logical *or* operation as required for σ-additivity. The pair (Ω, Ξ) is called an *event space* and represents here a *measurable space*. Other properties follow from the three properties (1) to (3). The intersection, for example, is the complement of the union of the complements $A \cap B = (A^{c} \cup B^{c})^{c} \in \Xi$, and the argument is easily extended to the intersection of a countable number of subsets of Ξ, so such countable intersections must also belong to Ξ as well. As already mentioned in Sect. 1.5.1, a σ-algebra is closed

[39] Recalling the situation in the countable case, we chose the entire powerset $\Pi(\Omega)$ as reference instead of a smaller event system Ξ.

under the operations of complement, union, and intersection. Trivial examples of σ-algebras are $\{\emptyset, \Omega\}$, $\{\emptyset, A, A^c, \Omega\}$, or the family of all subsets. The Borel σ-algebra on $\Omega = \mathbb{R}$ is the smallest σ-algebra which contains all open sets, or equivalently, all closed sets of $\mathbb{R}$.

Completeness of Measure Spaces
We consider a probability space defined by the measure triple $(\Omega, \mathcal{B}, \mu)$, sometimes also called a *measure space*, where $\mathcal{B}$ is a measurable collection of sets and the measure is a function $\mu : \mathcal{B} \to [0, \infty)$ that returns a value $\mu(A)$ for every set $A \in \mathcal{B}$. The real line, $\Omega = \mathbb{R}$, allows for the definition of a Borel measure that assigns $\mu([a, b]) = b - a$ to the interval $[a, b]$. The Borel measure is defined on the σ-algebra (see Sects. 1.5.1 and 1.7.2)[40] of the Borel sets $\mathcal{B}(\mathbb{R})$ and it is the smallest σ-algebra that contains all open—or equivalently all closed—intervals on $\mathbb{R}$. The Borel set $\mathcal{B}$ is formed from open or from closed sets through the operations of (countable) unions, (countable) intersections, and complements. It is important to note that the numbers of unions or the number of intersections have to be countable, even though the intervals $[a, b]$ contain uncountably many elements.

In practice the Borel measure μ is not the most useful measure defined on the σ-algebra of Borel sets since it is not a *complete measure*. Completeness of a measure space (Ω, Ξ, μ) requires that every subset S of every *null set* N is measurable and has measure zero:

$$S \subseteq N \in \Xi \ \text{ and } \ \mu(N) = 0 \quad \Longrightarrow \quad S \in \Xi \ \text{ and } \ \mu(S) = 0 \,.$$

Completeness is not a mere question of esthetics. It is needed for the construction of higher dimensional spaces using the Cartesian product, e.g., $\mathbb{R}^n = \mathbb{R} \times \mathbb{R} \times \cdots \times \mathbb{R}$. Otherwise unmeasurable sets may sneak in and corrupt the measurability of the product space. Complete measures can be constructed from incomplete measure spaces (Ω, Ξ, μ) through a minimal extension: Z is the set of all subsets z of Ω that have measure $\mu(z) = 0$ and intuitively the elements of Z that are not yet in Ξ are those that prevent the measure μ from being complete. The σ-algebra generated by Ξ and Z, the smallest σ-algebra containing every element of Ξ and every element of Z, is denoted by Ξ_0. The unique extension of μ to μ_0 completes the measure space by adding the elements of Z to Ξ in order to yield Ξ_0. It is given by the infimum:

$$\mu_0(C) \doteq \inf\{\mu(D) \,|\, C \subseteq D \in \Xi_0\} \,.$$

[40]For our purposes here it is sufficient to remember that a σ-algebra on a set Ξ is a collection Σ of subsets $A \in \Xi$ which have certain properties, including σ-additivity (see Sect. 1.5.1).

Accordingly, the space (Ω, Ξ_0, μ_0) is the completion of (Ω, Ξ, μ). In particular, every member of Ξ_0 is of the form $A \cup B$ with $A \in \Xi$ and $B \in Z$ and $\mu_0(A \cup B) = \mu(A)$. The Borel measure if completed in this way becomes the Lebesgue measure λ on $\mathbb{R}$. Every Borel-measurable set A is also a Lebesgue-measurable set, and the two measures coincide on Borel sets A: $\lambda(A) = \mu(A)$. As an illustration of the incompleteness of the Borel measure space, we consider the Cantor set,[41] named after Georg Cantor. The set of all Borel sets over $\mathbb{R}$ has the same cardinality as $\mathbb{R}$. The Cantor set is a Borel set and has measure zero. By Cantor's theorem, its powerset has a cardinality strictly greater than that of the real numbers and hence there must be a subset of the Cantor set that is not contained in the Borel sets. Therefore, the Borel measure cannot be complete.

Construction of σ-Algebras

A construction principle for σ-algebras starts out from some event system $\mathcal{G} \subset \Pi(\Omega)$ (for $\Omega \neq \emptyset$) that is sufficiently small and otherwise arbitrary. Then, there exists exactly one smallest σ-algebra $\Xi = \Sigma(\mathcal{G})$ in Ω with $\Xi \supset \mathcal{G}$, and we call Ξ the σ-algebra induced by $\mathcal{G}$. In other words, $\mathcal{G}$ is the generator of Ξ. In probability theory, we deal with three cases: (i) countable sample spaces Ω, (ii) the uncountable space of real numbers $\Omega = \mathbb{R}$, and (iii) the Cartesian product spaces $\Omega = \mathbb{R}^n$ of vectors with real components in n dimensions. Case (i) has already been discussed in Sect. 1.5.

The Borel σ-algebra for case (ii) is constructed with the help of a generator representing the set of all compact intervals in one-dimensional Cartesian space $\Omega = \mathbb{R}$ which have rational endpoints, viz.,

$$\mathcal{G} = \{[a, b] : a < b\, ,\ (a, b) \in \mathbb{Q}\}\ , \tag{1.43a}$$

where $\mathbb{Q}$ is the set of all rational numbers. The restriction to rational endpoints is the trick that makes the event system Ξ tractable in comparison to the powerset, which as we have shown is too large for the definition of a Lebesgue measure. The σ-algebra induced by this generator is known as the Borel σ-algebra $\mathcal{B} \doteq \Sigma(\mathcal{G})$ on $\mathbb{R}$, and each $A \in \mathcal{B}$ is a Borel set.

---

[41] The Cantor set is generated from the interval [0, 1] by consecutively taking out the open middle third:

$$[0,1] \to \left[0, \frac{1}{3}\right] \cup \left[\frac{2}{3}, 1\right] \to \left[0, \frac{1}{9}\right] \cup \left[\frac{2}{9}, \frac{1}{3}\right] \cup \left[\frac{2}{3}, \frac{7}{9}\right] \cup \left[\frac{8}{9}, 1\right] \to \dots .$$

An explicit formula for the set is

$$C = [0,1] \backslash \bigcup_{m=1}^{\infty} \bigcup_{k=0}^{(3^{m-1}-1)} \left]\frac{3k+1}{3^m}, \frac{3k+2}{3^m}\right[.$$

The extension to n dimensions as required in case (iii) is straightforward if one recalls that a *product measure* $\mu = \mu_1 \times \mu_2$ is defined for a product measurable space $(\mathbb{X}_1 \times \mathbb{X}_2, \Xi_1 \otimes \Xi_2, \mu_1 \times \mu_2)$ when $(\mathbb{X}_1, \Xi_1, \mu_1)$ and $(\mathbb{X}_2, \Xi_2, \mu_2)$ are two measurable spaces. The generator $\mathcal{G}_n$ is the set of all compact cuboids in n-dimensional Cartesian space $\Omega = \mathbb{R}^n$ which have rational corners:

$$\mathcal{G}_n = \left\{ \prod_{k=1}^{n} [a_k, b_k] : a_k < b_k \,,\ (a_k, b_k) \in \mathbb{Q} \right\} \,. \tag{1.43b}$$

The σ-algebra induced by this generator is called a Borel σ-algebra in n dimensions, $\mathcal{B}^{(n)} \doteq \Sigma(\mathcal{G}_n)$ on $\mathbb{R}^n$. Each $A \in \mathcal{B}^{(n)}$ is a Borel set. Then, $\mathcal{B}_k$ is a Borel σ-algebra on the subspace $\mathbb{E}_k$ with $\Xi_k : \Omega \to \mathbb{E}_k$ the projection onto the kth coordinate. The generator

$$\mathcal{G}_k = \left\{ \Xi_k^{-1} A_k : k \in \mathcal{I} \,,\ A_k \in \mathcal{B}_k \right\} \,, \quad \text{with } \mathcal{I} \text{ as index set} \,,$$

is the system of all sets in Ω that are determined by an event on coordinate k, $\Xi^{-1} A_k$ is the preimage of A_k in $(\mathbb{R}^n)_k$, and $\mathcal{B}^{(n)} = \bigotimes_{k \in I} \mathcal{B}_k \doteq \Sigma(\mathcal{G}_n)$ is the product σ-algebra of the sets $\mathcal{B}_k$ on Ω. In the important case of equivalent Cartesian coordinates, $\mathbb{E}_k = \mathbb{E}$ and $\mathcal{B}_k = \mathcal{B}$ for all $k \in \mathcal{I}$, we have that the Borel σ-algebra $\mathcal{B}^{(n)} = \mathcal{B}^n$ on $\mathbb{R}^n$ is represented by the n-dimensional product σ-algebra of the Borel σ-algebra $\mathcal{B}$ on $\mathbb{R}$.[42]

A Borel σ-algebra is characterized by five properties, which are helpful for visualizing its enormous size:

(i) Each open interval $] \, [= A \subset \mathbb{R}^n$ is Borel. Every $\omega \in A$ has a neighborhood $Q \in \mathcal{G}$ such that $Q \subset A$ and Q has rational endpoints. We thus have

$$A = \bigcup_{Q \in \mathcal{G},\, Q \subset A} Q \,,$$

which is a union of countably many sets in $\mathcal{B}^n$. This follows from condition (3) for σ-algebras.

(ii) Each closed set $[\,] = A \subset \mathbb{R}^n$ is Borel, since A^c is open and Borel, according to item (i).

(continued)

[42]For $n = 1$, one commonly writes $\mathcal{B}$ instead of $\mathcal{B}^1$, or $\mathcal{B}^n = \mathcal{B}^{\otimes n}$.

(iii) The σ-algebra $\mathcal{B}^n$ cannot be described in a constructive way, because it consists of much more than the union of cuboids and their complements. In order to create $\mathcal{B}^n$, the operation of adding complements and countable unions has to be repeated as often as there are countable ordinal numbers (and this involves an uncountable number of operations [50, pp. 24, 29]). For practical purposes, it is sufficient to remember that $\mathcal{B}^n$ covers almost all sets in $\mathbb{R}^n$, but not all of them.
(iv) The Borel σ-algebra $\mathcal{B}$ on $\mathbb{R}$ is generated not only by the system of compact sets (1.43), but also by the system of intervals that are unbounded on the left and closed on the right:

$$\tilde{\mathcal{G}} = \{]-\infty, c] : c \in \mathbb{R}\} . \tag{1.44}$$

By analogy, $\mathcal{B}$ is also generated by all open left-unbounded intervals, by all closed intervals, and by all open right-unbounded intervals.
(v) The event system $\mathcal{B}^n_\Omega = \{A \cap \Omega : A \in \mathcal{B}^n\}$ on $\Omega \subset \mathbb{R}^n$, $\Omega \neq \emptyset$, is a σ-algebra on Ω called the Borel σ-algebra on Ω.

Item (iv) follows from condition (2), which requires $\tilde{\mathcal{G}} \subset \mathcal{B}$ and, because of minimality of $\sigma(\tilde{\mathcal{G}})$, also $\sigma(\tilde{\mathcal{G}}) \subset \mathcal{B}$. Alternatively, $\sigma(\tilde{\mathcal{G}})$ contains all left-open intervals, since $]a, b] =]-\infty, b] \setminus]-\infty, a]$, and also all compact or closed intervals, since $[a, b] = \bigcap_{n\geq 1}]a - 1/n, b]$, and hence also the σ-algebra $\mathcal{B}$ generated by these intervals (1.43a). All intervals discussed in items (i)–(iv) are Lebesgue measurable, while certain other sets such as the Vitali sets are not.

The Lebesgue measure is the conventional way of assigning lengths, areas, and volumes to subsets of three-dimensional Euclidean space and to objects with higher dimensional volumes in formal Cartesian spaces. Sets to which generalized volumes can be assigned are called Lebesgue measurable and the measure or the volume of such a set A is denoted by $\lambda(A)$. The Lebesgue measure on $\mathbb{R}^n$ has the following properties:

(1) If A is a Lebesgue measurable set, then $\lambda(A) \geq 0$.
(2) If A is a Cartesian product of intervals, $I_1 \times I_2 \times \ldots \times I_n$, then A is Lebesgue measurable and $\lambda(A) = |I_1||I_2| \ldots |I_n|$.
(3) If A is Lebesgue measurable, its complement A^c is measurable, too.
(4) If A is a disjoint union of countably many disjoint Lebesgue measurable sets, $A = \bigcup_k A_k$, then A is Lebesgue measurable and $\lambda(A) = \sum_k \lambda(A_k)$.
(5) If A and B are Lebesgue measurable and $A \subset B$, then $\lambda(A) \leq \lambda(B)$.
(6) Countable unions and countable intersections of Lebesgue measurable sets are Lebesgue measurable.[43]

[43] This is not a consequence of items (3) and (4): a family of sets, which is closed under complements and countable disjoint unions, need not be closed under countable non-disjoint

(7) If A is an open or closed subset or Borel set of $\mathbb{R}^n$, then A is Lebesgue measurable.
(8) The Lebesgue measure is strictly positive on non-empty open sets, and its domain is the entire $\mathbb{R}^n$.
(9) If A is a Lebesgue measurable set with $\lambda(A) = 0$, called a null set, then every subset of A is also a null set, and every subset of A is measurable.
(10) If A is Lebesgue measurable and $\mathbf{r}$ is an element of $\mathbb{R}^n$, then the translation of A by $\mathbf{r}$, defined by $A + \mathbf{r} = \{\mathbf{a} + \mathbf{r} | \mathbf{a} \in A\}$, is also Lebesgue measurable and has the same measure as A.
(11) If A is Lebesgue measurable and $\delta > 0$, then the dilation of A by δ, defined by $\delta A = \{\delta\mathbf{r} | \mathbf{r} \in A\}$, is also Lebesgue measurable and has measure $\delta^n \lambda(A)$.
(12) Generalizing items (10) and (11), if L is a linear transformation and A is a measurable subset of $\mathbb{R}^n$, then $T(A)$ is also measurable and has measure $\lambda = |\det(T)|\, \lambda(A)$.

All 12 items listed above can be summarized succinctly in one lemma:

The Lebesgue measurable sets form a σ-algebra on $\mathbb{R}^n$ containing all products of intervals, and λ is the unique complete translation-invariant measure on that σ-algebra with

$$\lambda\big([0,1] \otimes [0,1] \otimes \ldots \otimes [0,1]\big) = 1 \, .$$

We conclude this section on Borel σ-algebras and Lebesgue measure by mentioning a few characteristic and illustrative examples:

(i) Any closed interval $[a, b]$ of real numbers is Lebesgue measurable, and its Lebesgue measure is the length $b - a$. The open interval $]a, b[$ has the same measure, since the difference between the two sets consists only of the two endpoint a and b and has measure zero.
(ii) Any Cartesian product of intervals $[a, b]$ and $[c, d]$ is Lebesgue measurable and its Lebesgue measure is $(b-a)(d-c)$, the area of the corresponding rectangle.
(iii) The Lebesgue measure of the set of rational numbers in an interval of the line is zero, although this set is dense in the interval.

unions. Consider, for example, the set

$$\{\emptyset, \{1,2\}, \{1,3\}, \{2,4\}, \{3,4\}, \{1,2,3,4\}\} \, .$$

(iv) The Cantor set is an example of an uncountable set that has Lebesgue measure zero.
(v) Vitali sets are examples of sets that are not measurable with respect to the Lebesgue measure.

In the forthcoming sections, we shall make implicit use of the fact that the continuous sets on the real axes become countable and Lebesgue measurable if rational numbers are chosen as beginnings and end points of intervals. For all practical purposes, we can work with real numbers with almost no restriction.

1.8 Limits and Integrals

A few technicalities concerning the definition of limits will facilitate the discussion of continuous random variables and their distributions. Precisely defined limits of sequences are required for problems of convergence and for approximating random variables. Taking limits of stochastic variables often needs some care and problems may arise when there are ambiguities, although they can be removed by a sufficiently rigorous approach.

In previous sections we encountered functions of discrete random variables like the probability mass function (pmf) and the cumulative probability distribution function (cdf), which contain peaks and steps that cannot be subjected to conventional Riemannian integration. Here, we shall present a brief introduction to generalizations of the conventional integration scheme that can be used in the case of functions with discontinuities.

1.8.1 Limits of Series of Random Variables

A sequence of random variables, $\mathcal{X}_n$, is defined on a probability space Ω and is assumed to have the limit

$$\mathcal{X} = \lim_{n\to\infty} \mathcal{X}_n \,. \tag{1.45}$$

We assume now that the probability space Ω has elements ω with probability density $p(\omega)$. Four different definitions of the stochastic limit are common in probability theory [194, pp. 40, 41].

Almost Certain Limit The series $\mathcal{X}_n$ converges *almost certainly* to $\mathcal{X}$ if, for all ω except a set of probability zero, we have

$$\mathcal{X}(\omega) = \lim_{n\to\infty} \mathcal{X}_n(\omega) , \tag{1.46}$$

and each realization of $\mathcal{X}_n$ converges to $\mathcal{X}$.

Limit in the Mean The limit in the mean or the mean square limit of a series requires that the mean square deviation of $\mathcal{X}_n(\omega)$ from $\mathcal{X}(\omega)$ vanishes in the limit. The condition is

$$\lim_{n\to\infty} \int_{\Omega} \mathrm{d}\omega\, p(\omega)\Big(\mathcal{X}_n(\omega) - \mathcal{X}(\omega)\Big)^2 \equiv \lim_{n\to\infty} \big\langle(\mathcal{X}_n - \mathcal{X})^2\big\rangle = 0 . \tag{1.47}$$

The mean square limit is the standard limit in Hilbert space theory and it is commonly used in quantum mechanics.

Stochastic Limit A limit in probability is called the stochastic limit $\mathcal{X}$ if it fulfils the condition

$$\lim_{n\to\infty} P\big(|\mathcal{X}_n - \mathcal{X}| > \varepsilon\big) = 0 , \tag{1.48a}$$

for any $\varepsilon > 0$. The approach to the stochastic limit is sometimes characterized as convergence in probability:

$$\lim_{n\to\infty} \mathcal{X}_n \stackrel{\mathrm{P}}{\longrightarrow} \mathcal{X} , \tag{1.48b}$$

where $\stackrel{\mathrm{P}}{\longrightarrow}$ stands for *convergence in probability* (see also Sect. 2.4.3).

Limit in Distribution Probability theory also uses a weaker form of convergence than the previous three limits, known as the limit in distribution. This requires that, for a sequence of random variables $\mathcal{X}_1, \mathcal{X}_2, \ldots$, the sequence $f_1(x), f_2(x), \ldots$, should satisfy

$$\lim_{n\to\infty} f_n(x) \stackrel{\mathrm{d}}{\longrightarrow} f(x) ,\ \ \forall\, x \in \mathbb{R} , \tag{1.49}$$

where $\stackrel{\mathrm{d}}{\longrightarrow}$ stands for *convergence in distribution*. The functions $f_n(x)$ are quite general, but they may for instance be probability mass functions or cumulative probability distributions $F_n(x)$. This limit is particularly useful for characteristic functions $\phi_n(s) = \int_{-\infty}^{\infty} \exp(\mathrm{i}xs) f_n(x)\, \mathrm{d}x$ (see Sect. 2.2.3): if the characteristic functions $\phi_n(s)$ approach $\phi(s)$, the probability density of $\mathcal{X}_n$ converges to that of $\mathcal{X}$.

As an example for convergence in distribution we present here the probability mass function of the scores for rolling n dice. A collection of n dice is thrown

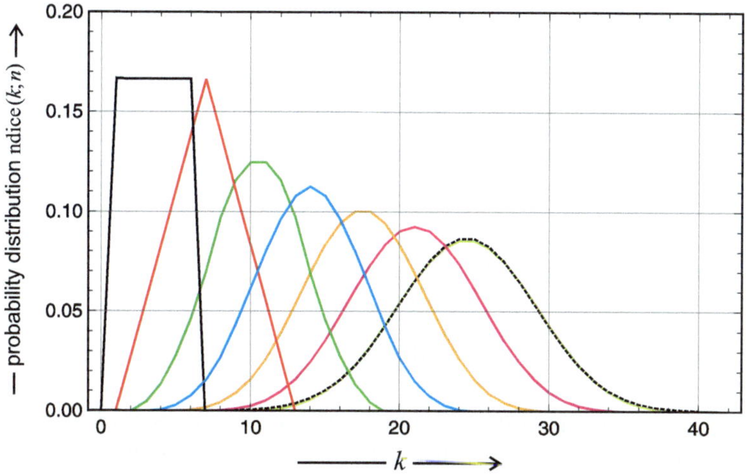

Fig. 1.16 Convergence to the normal density of the probability mass function for rolling n dice. The probability mass functions $f_{6,n}(k)$ of (1.50) for rolling n conventional dice are used here to illustrate convergence in distribution. We begin with a pulse function $f_{6,1}(k) = 1/6$ for $i = 1, \ldots, 6$ ($n = 1$). Next there is a tent function ($n = 2$), and then follows a gradual approach towards the normal distribution for $n = 3, 4, \ldots$. For $n = 7$, we show the fitted normal distribution (*broken black curve*), coinciding almost perfectly with $f_{6,7}(k)$. Choice of parameters: $s = 6$ and $n = 1$ (*black*), 2 (*red*), 3 (*green*), 4 (*blue*), 5 (*yellow*), 6 (*magenta*), and 7 (*chartreuse*)

simultaneously and the total score of all the dice together is recorded (Fig. 1.16). We are already familiar with the cases $n = 1$ and 2 (Figs. 1.11 and 1.12) and the extension to arbitrary cases is straightforward. The general probability of a total score of k points obtained when rolling n dice with s faces is obtained combinatorically as

$$f_{s,n}(k) = \frac{1}{s^n} \sum_{i=0}^{\lfloor (k-n)/s \rfloor} (-1)^i \binom{n}{i} \binom{k - si - 1}{n - 1} . \tag{1.50}$$

The results for small values of n and ordinary dice ($s = 6$) are illustrated in Fig. 1.16. The convergence to a continuous probability density is nicely illustrated. For $n = 7$, the deviation from the Gaussian curve of the normal distribution is barely visible. We shall come back to convergence to the normal distribution in Fig. 1.23 and in Sect. 2.4.2.

Finally, we mention stringent conditions for the convergence of functions that are important for probability distributions as well. We distinguish *pointwise convergence* and *uniform convergence*. Consider a series of functions $f_0(x), f_1(x), f_2(x), \ldots$, defined on some interval $I \in \mathbb{R}$. The series converges pointwise to the function $f(x)$

if the limit holds for every point x:

$$\lim_{n\to\infty} f_n(x) = f(x)\,, \quad \forall\, x \in I\,. \tag{1.51}$$

It is readily checked that a series of functions can be written as a sum of functions whose convergence is to be tested:

$$f(x) = \lim_{n\to\infty} f_n(x) = \lim_{n\to\infty} \sum_{i=1}^{n} g_i(x)\,,$$
$$g_i(x) = \varphi_{i-1}(x) - \varphi_i(x)\,, \quad \text{and hence } f_n(x) = \varphi_0(x) - \varphi_n(x)\,, \tag{1.52}$$

because $\sum_{i=1}^{n} g_i(x)$ expressed in terms of the functions φ_i is a *telescopic sum*. An example of a series of curves with $\varphi_n(x) = (1 + nx^2)^{-1}$ and hence $f_n(x) = nx^2 \,/\, (1+nx^2)$ exhibiting pointwise convergence is shown in Fig. 1.17. It is easily checked that the limit takes the form

$$f(x) = \lim_{n\to\infty} \frac{nx^2}{1+nx^2} = \begin{cases} 1\,, & \text{for } x \neq 0\,, \\ 0\,, & \text{for } x = 0\,. \end{cases}$$

All the functions $f_n(x)$ are continuous on the interval $]-\infty, \infty[$, but the limit $f(x)$ is discontinuous at $x = 0$. An interesting historical detail is worth mentioning. In 1821 the famous mathematician Augustin Louis Cauchy gave the wrong answer to the question of whether or not infinite sums of continuous functions were necessarily continuous, and his obvious error was only corrected 30 years later. It is not hard to imagine that pointwise convergence is compatible with discontinuities in the convergence limit (Fig. 1.17), since the convergent series may have very different limits at two neighboring points. There are many examples of series of functions which have a discontinuous infinite limit. Two further cases that we shall need later on are $f_n(x) = x^n$ with $I = [0, 1] \in \mathbb{R}$ and $f_n(x) = \cos(\pi x)^{2n}$ on $I =]-\infty, \infty[\in \mathbb{R}$.

Uniform convergence is a stronger condition. Among other things, it guarantees that the limit of a series of continuous functions is *continuous*. It can be defined in terms of (1.52): the sum $f_n(x) \sum_{i=1}^{n} g_i(x)$ with $\lim_{n\to\infty} f_n(x) = f(x)$ and $x \in I$ is uniformly convergent in the interval $x \in I$ for every given positive error bound ϵ if there exists a value $\nu \in \mathbb{N}$ such that, for any $\nu \geq n$, the relation $|f(x) - f_\nu(x)| < \epsilon$ holds for all $x \in I$. In compact form, this convergence condition may be expressed by

$$\lim_{n\to\infty} \sup\left\{|f_n(x) - f(x)|\right\} = 0 \;\; \forall\, x \in I\,. \tag{1.53}$$

A simple illustration is given by the power series $f(x) = \lim_{n\to\infty} x^n$ with $x \in [0, 1]$, which converges pointwise to the discontinuous function $f(x) = 1$ for $x = 1$ and 0 otherwise. A slight modification to $f(x) = \lim_{n\to\infty} x^n/n$ leads to a uniformly converging series, because $f(x) = 0$ is now valid for the entire domain $[0, 1]$ (including the point $x = 1$).

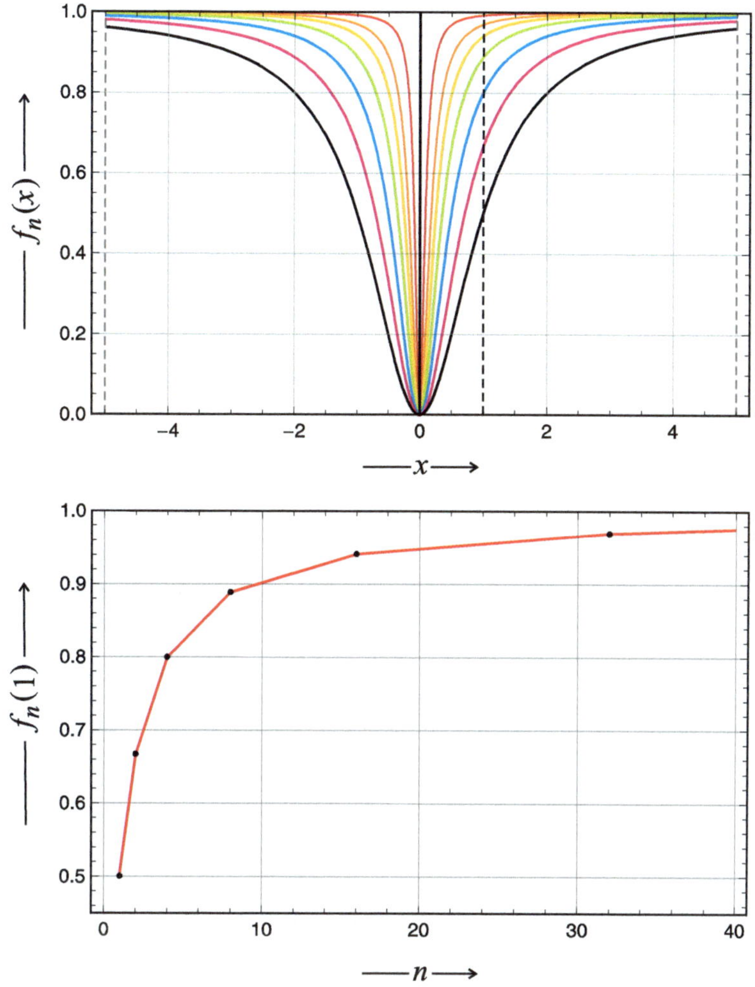

Fig. 1.17 Pointwise convergence. *Upper*: Convergence of the series of functions $f_n(x) = nx^2/(1+nx^2)$ to the limit $\lim_{n\to\infty} f_n(x) = f(x)$ on the real axis $]-\infty, \infty[$. *Lower*: Convergence as a function of n at the point $x = 1$. Color code of the upper plot: $n = 1$ *black*, $n = 2$ *violet*, $n = 4$ *blue*, $n = 8$ *chartreuse*, $n = 16$ *yellow*, $n = 32$ *orange*, and $n = 128$ *red*

1.8.2 Riemann and Stieltjes Integration

Although the reader is assumed to be familiar with Riemann integration, we briefly summarize the conditions for the existence of a Riemann integral (Fig. 1.18). For

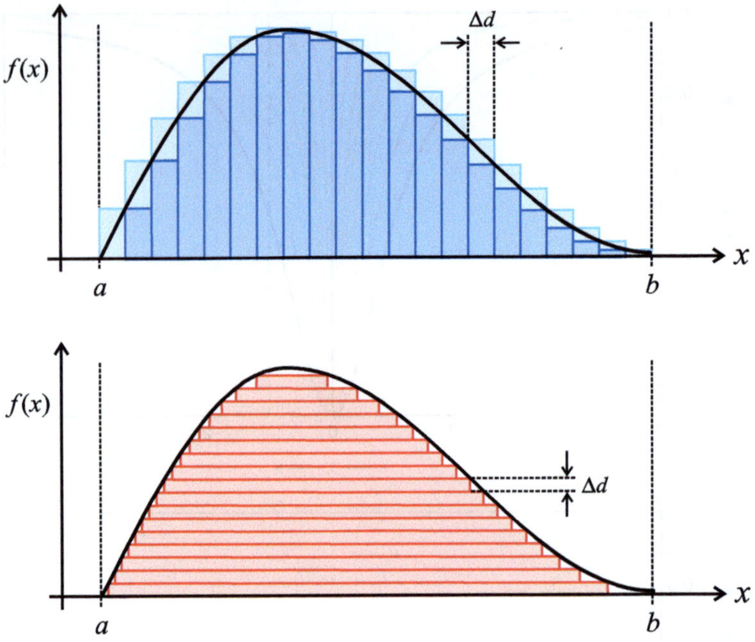

Fig. 1.18 Comparison of Riemann and Lebesgue integrals. In the conventional Riemann–Darboux integration, the integrand is embedded between an upper sum (*light blue*) and a lower sum (*dark blue*) of rectangles. The integral exists iff the upper sum and the lower sum converge to the integrand in the limit $\Delta d \to 0$. The Lebesgue integral can be visualized as an approach to calculating the area enclosed by the x-axis and the integrand by partitioning it into horizontal stripes (*red*) and considering the limit $\Delta d \to 0$. The definite integral $\int_a^b f(x)\,\mathrm{d}x$ confines integration to a closed interval $[a, b]$ or $a \le x \le b$

this purpose, we define the *Darboux sum*[44] as follows. A function $f : \mathcal{D} \to \mathbb{R}$ is considered on a closed interval $I = [a, b] \in \mathcal{D}$, which is partitioned by $n - 1$ additional points

$$a = x_0^{(n)} < x_1^{(n)} < \ldots < x_{n-1}^{(n)} < x_n^{(n)} = b$$

into n intervals[45]:

$$S_n = \left([x_0^{(n)}, x_1^{(n)}], [x_1^{(n)}, x_2^{(n)}], \ldots, [x_{n-1}^{(n)}, x_n^{(n)}] \right), \qquad \Delta x_i = x_i - x_{i-1} \ .$$

[44]The idea of representing an integral by the convergence of two sums is due to the French mathematician Gaston Darboux. A function is Darboux integrable iff it is Riemann integrable, and the values of the Riemann and the Darboux integral are equal whenever they exist.

[45]The intervals $|x_{k+1}^{(n)} - x_k^{(n)}| > 0$ can be assumed to be equal, although this is not essential.

The Darboux sum is defined by

$$\Sigma_{[a,b]}(S) = \sum_{i=1}^{n} f(\hat{x}_i)\Delta x_i = \sum_{i=1}^{n} \hat{f}_i \Delta x_i , \quad \text{for } x_{i-1} \leq \hat{x}_i \leq x_i , \tag{1.54}$$

where $\hat{x}$ is any point on the corresponding interval. Two particular choices of $\hat{x}$ are important for Riemann integration: (i) the *upper Riemann sum* $\Sigma^{(\text{high})}_{[a,b]}(S)$ with $\hat{f}_i = \sup\{f(x), x \in [x_{i-1}, x_i]\}$ and (ii) the *lower Riemann sum* $\Sigma^{(\text{low})}_{[a,b]}(S)$ with $\hat{f}_i = \inf\{f(x), x \in [x_{i-1}, x_i]\}$. Then the definition of the Riemann integral is given by taking the limit $n \to \infty$, which implies $\Delta x_i \to 0\,,\ \forall\, i$:

$$\int_a^b f(x)\,\mathrm{d}x = \lim_{n\to\infty} \Sigma^{(\text{high})}_{[a,b]}(S) = \lim_{n\to\infty} \Sigma^{(\text{low})}_{[a,b]}(S) . \tag{1.55}$$

If $\lim_{n\to\infty} \Sigma^{(\text{high})}_{[a,b]}(S) \neq \lim_{n\to\infty} \Sigma^{(\text{low})}_{[a,b]}(S)$, the Riemann integral does not exist.

Some generalizations of the conventional Riemann integral which are important in probability theory are introduced briefly here. Figure 1.18 presents a sketch that compares Riemann's and the Lebesgue's approaches to integration. Stieltjes integration is a generalization of Riemann or Lebesgue integration which allows one to calculate integrals over step functions, of the kind that occur, for example, when properties are derived from cumulative probability distributions. The *Stieltjes integral* is commonly written in the form

$$\int_a^b g(x)\,\mathrm{d}h(x) . \tag{1.56}$$

Here $g(x)$ is the integrand, $h(x)$ is the integrator, and the conventional Riemann integral is recovered for $h(x) = x$. The integrator is best visualized as a weighting function for the integrand. When $g(x)$ and $h(x)$ are continuous and continuously differentiable, the Stieltjes integral can be resolved by partial integration:

$$\begin{aligned}
\int_a^b g(x)\,\mathrm{d}h(x) &= \int_a^b g(x)\frac{\mathrm{d}h(x)}{\mathrm{d}x}\,\mathrm{d}x \\
&= \Big(g(x)h(x)\Big)\Big|_{x=a}^{b} - \int_a^b \frac{\mathrm{d}g(x)}{\mathrm{d}x} h(x)\,\mathrm{d}x \\
&= g(b)h(b) - g(a)h(a) - \int_a^b \frac{\mathrm{d}g(x)}{\mathrm{d}x} h(x)\,\mathrm{d}x .
\end{aligned}$$

However, the integrator $h(x)$ need not be continuous. It may well be a step function $F(x)$, e.g., a cumulative probability distribution. When $g(x)$ is continuous and $F(x)$ makes jumps at the points $x_1, \ldots, x_n \in\,]a, b[$ with heights $\Delta F_1, \ldots, \Delta F_n \in \mathbb{R}$,

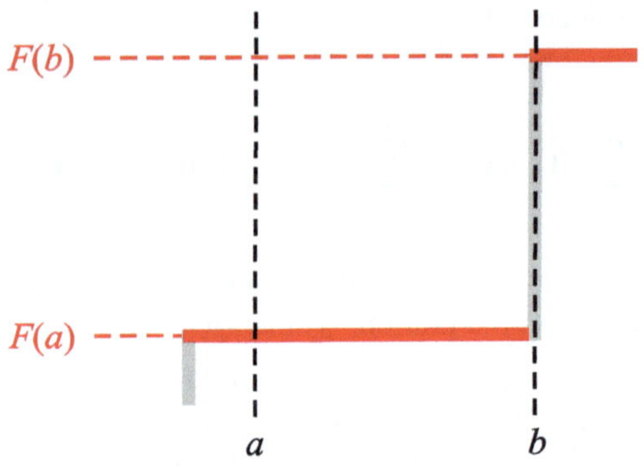

Fig. 1.19 Stieltjes integration of step functions. Stieltjes integral of a step function according to the definition of right-hand continuity applied in probability theory (Fig. 1.10): $\int_a^b \mathrm{d}F(x) = F(b) - F(a) = \Delta F\big|_{x=b}$. The figure also illustrates the Lebesgue–Stieltjes measure $\lambda_F\big((a,b]\big) = F(b) - F(a)$ in (1.63)

respectively, and $\sum_{i=1}^{n} \Delta F_n \leq 1$, the Stieltjes integral has the form

$$\int_a^b g(x)\,\mathrm{d}F(x) = \sum_{i=1}^{n} g(x_i)\Delta F_i\,, \tag{1.57}$$

where the constraint on $\sum_i \Delta F_i$ is the normalization of probabilities. With $g(x) = 1$, $b = x$ and in the limit $\lim_{a\to-\infty}$ the integral becomes identical with the (discrete) cumulative probability distribution function (cdf). Figure 1.19 illustrates the influence of the definition of continuity in probability theory (Fig. 1.10) on the Stieltjes integral.

Riemann–Stieltjes integration is used in probability theory for the computation of functions of random variables, for example, for the computation of moments of probability densities (Sect. 2.1). If $F(x)$ is the cumulative probability distribution of a random variable $\mathcal{X}$ for the discrete case, the expected value (see Sect. 2.1) for any function $g(\mathcal{X})$ is obtained from

$$E\big(g(\mathcal{X})\big) = \int_{-\infty}^{\infty} g(x)\,\mathrm{d}F(x) = \sum_i g(x_i)\Delta F_i\,.$$

If the random variable $\mathcal{X}$ has a probability density $f(x) = \mathrm{d}F(x)/\mathrm{d}x$ with respect to the Lebesgue measure, continuous integration can be used:

$$E\big(g(\mathcal{X})\big) = \int_{-\infty}^{\infty} g(x)f(x)\,\mathrm{d}x\,.$$

Important special cases are the moments $E(\mathcal{X}^n) = \int_{-\infty}^{\infty} x^n\,\mathrm{d}F(x)$.

1.8.3 Lebesgue Integration

Lebesgue integration differs from conventional integration in two respects: (i) the basis of Lebesgue integration is set theory and measure theory and (ii) the integrand is partitioned in horizontal segments, whereas Riemannian integration makes use of vertical slices. For nonnegative functions like probability functions, an important difference between the two integration methods can be visualized in three-dimensional space: in Riemannian integration the volume below a surface given by the function $f(x, y)$ is measured by summing the volumes of cuboids with square cross-sections of edge Δd, whereas the Lebesgue integral sums the volumes of layers with thickness Δd between constant level sets. Every continuous bounded function $f \in \mathcal{C}(a, b)$ on a compact finite interval $[a, b]$ is Riemann integrable and also Lebesgue integrable, and the Riemann and Lebesgue integrals coincide.

The Lebesgue integral is a generalization of the Riemann integral in the sense that certain functions may be Lebesgue integrable in cases where the Riemann integral does not exist. The opposite situation may occur with improper Riemann integrals:[46] Partial sums with alternating signs may converge for the improper Riemann integral whereas Lebesgue integration leads to divergence, as illustrated by the alternating harmonic series. The Lebesgue integral can be generalized by the Stieltjes integration technique using integrators $h(x)$, very much in the same way as we showed it for the Riemann integral.

Lebesgue integration theory assumes the existence of a probability space defined by the triple (Ω, Ξ, μ), which represents the sample space Ω, a σ-algebra Ξ of subsets $A \in \Omega$, and a probability measure $\mu \geq 0$ satisfying $\mu(\Omega) = 1$. The construction of the Lebesgue integral is similar to the construction of the Riemann integral: the shrinking rectangles (or cuboids in higher dimensions) of Riemannian integration are replaced by horizontal strips of shrinking height that can be represented by *simple functions* (see below). Lebesgue integrals over nonnegative functions on A, viz.,

$$\int_{\Omega} f \, \mathrm{d}\mu \,, \quad \text{with } f : (\Omega, \Xi, \mu) \to (\mathbb{R}_{\geq 0}, \mathcal{B}, \lambda) \,, \tag{1.58}$$

---

[46]An improper integral is the limit of a definite integral in a series in which the endpoint of the interval of integration either approaches a finite number b at which the integrand diverges or becomes $\pm\infty$:

$$\int_a^b f(x) \, \mathrm{d}x = \lim_{\varepsilon \to +0} \int_a^{b-\varepsilon} f(x) \, \mathrm{d}x \,, \quad \text{with } f(b) = \pm\infty \,,$$

or

$$\lim_{b \to \infty} \int_a^b f(x) \, \mathrm{d}x \quad \text{and} \quad \lim_{a \to -\infty} \int_a^b f(x) \, \mathrm{d}x \,.$$

are defined for measurable functions f satisfying

$$f^{-1}\big([a,b]\big) \in \Omega \quad \text{for all } \ a < b \,. \tag{1.59}$$

This condition is equivalent to the requirement that the preimage of any Borel subset $[a, b]$ of $\mathbb{R}$ is an element of the event system $\mathcal{B}$. The set of measurable functions is closed under algebraic operations and also closed under certain pointwise sequential limits like

$$\sup_{k\in\mathbb{N}} f_k \,, \quad \liminf_{k\in\mathbb{N}} f_k \,, \quad \limsup_{k\in\mathbb{N}} f_k \,,$$

which are measurable if the sequence of functions $(f_k)_{k\in\mathbb{N}}$ contains only measurable functions.

An integral $\int_\Omega f \, \mathrm{d}\mu = \int_\Omega f(x) \, \mu(\mathrm{d}x)$ is constructed in steps. We first apply the indicator function (1.26):

$$\mathbf{1}_A(x) = \begin{cases} 1 \,, & \text{iff } x \in A \,, \\ 0 \,, & \text{otherwise} \,, \end{cases} \tag{1.26a$'$}$$

to define the integral over $A \in \mathcal{B}$ by

$$\int_A f(x) \, \mathrm{d}x \doteq \int \mathbf{1}_A(x) f(x) \, \mathrm{d}x \,.$$

The indicator function $\mathbf{1}_A$ assigns a volume to Lebesgue measurable sets A by setting $f \equiv 1$:

$$\int \mathbf{1}_A \, \mathrm{d}\mu = \mu(A) \,.$$

This is the Lebesgue measure $\mu(A) = \lambda(A)$ for a mapping $\lambda\colon \mathcal{B} \to \mathbb{R}$. It is often useful to consider the expectation value and the variance of the indicator function (1.26):

$$\mathrm{E}\big(\mathbf{1}_A(\omega)\big) = \frac{A}{\Omega} = P(A) \,, \qquad \mathrm{var}\big(\mathbf{1}_A(\omega)\big) = P(A)\big(1 - P(A)\big) \,.$$

We shall make use of this property of the indicator function in Sect. 1.9.2.

Next we define *simple functions*, which are understood as finite linear combinations of indicator functions $g = \sum_j \alpha_j \mathbf{1}_{A_j}$. They are measurable if the coefficients α_j are real numbers and the sets A_j are measurable subsets of Ω. For nonnegative coefficients α_j, the linearity property of the integral leads to a measure for nonnegative simple functions:

$$\int \left(\sum_j \alpha_j \mathbf{1}_{A_j} \right) \mathrm{d}\mu = \sum_j \alpha_j \int \mathbf{1}_{A_j} \, \mathrm{d}\mu = \sum_j \alpha_j \mu(A_j) \, .$$

Often a simple function can be written in several ways as a linear combination of indicator functions, but the value of the integral will necessarily be the same.[47]

An arbitrary nonnegative function $g : (\Omega, \Xi, \mu) \to (\mathbb{R}_{\geq 0}, \mathcal{B}, \lambda)$ is measurable iff there exists a sequence of simple functions $(g_k)_{k \in \mathbb{N}}$ that converges *pointwise* and approaches g, i.e., $g = \lim_{k \to \infty} g_k$ monotonically. The Lebesgue integral of a nonnegative and measurable function g is defined by

$$\int_\Omega g \, \mathrm{d}\mu = \lim_{k \to \infty} \int_\Omega g_k \, \mathrm{d}\mu \, , \tag{1.60}$$

where g_k are simple functions which converge pointwise and monotonically towards g, as described. The limit is independent of the particular choice of the functions g_k. Such a sequence of simple functions is easily visualized, for example, by the bands below the function $g(x)$ in Fig. 1.18: the band width Δd decreases and converges to zero as the index increases $k \to \infty$.

The extension to general functions with positive and negative value domains is straightforward. As shown in Fig. 1.20, the function to be integrated, $f(x) : [a, b] \to \mathbb{R}$, is split into two regions that may consist of disjoint domains:

$$f_+(x) \doteq \max\{0, f(x)\} \, , \qquad f_-(x) \doteq \max\{0, -f(x)\} \, .$$

These are considered separately. The function is Lebesgue integrable on the entire domain $[a, b]$ iff both $f_+(x)$ and $f_-(x)$ are Lebesgue integrable, and then we have

$$\int_a^b f(x) \, \mathrm{d}x = \int_a^b f_+(x) \, \mathrm{d}x - \int_a^b f_-(x) \, \mathrm{d}x \, . \tag{1.61}$$

This yields precisely the same result as obtained for the Riemann integral. Lebesgue integration readily yields the value for the integral of the absolute value of the function:

$$\int_a^b |f(x)| \, \mathrm{d}x = \int_a^b f_+(x) \, \mathrm{d}x + \int_a^b f_-(x) \, \mathrm{d}x \, . \tag{1.62}$$

[47]Care is sometimes needed for the construction of a real-valued simple function $g = \sum_j \alpha_j \mathbf{1}_{A_j}$, in order to avoid undefined expressions of the kind $\infty - \infty$. Choosing $\alpha_i = 0$ implies that $\alpha_i \mu(A_i) = 0$ always holds, because $0 \times \infty = 0$ by convention in measure theory.

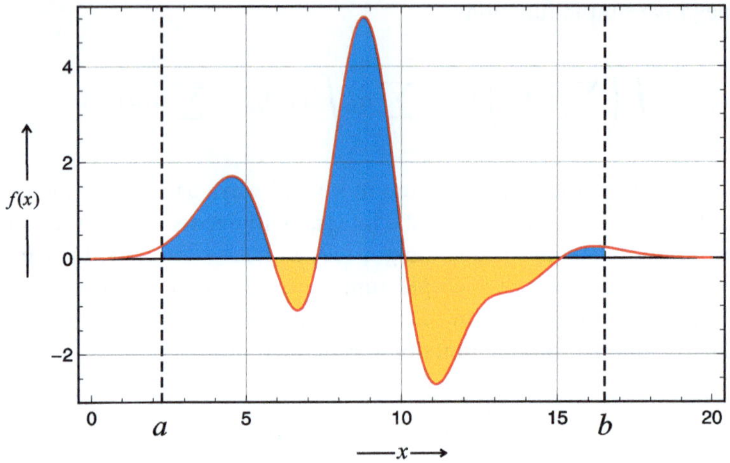

Fig. 1.20 Lebesgue integration of general functions. Lebesgue integration of general functions, i.e., functions with positive and negative regions, is performed in three steps: (i) the integral $I = \int_a^b f \, d\mu$ is split into two parts, viz., $I_+ = \int_a^b f_+(x) \, d\mu$ (*blue*) and $I_- = \int_a^b f_-(x) \, d\mu$ (*yellow*), (ii) the positive part $f_+(x) \doteq \max\{0, f(x)\}$ is Lebesgue integrated like a nonnegative function yielding $I_+ = \int_a^b f_+(x) \, d\mu$ and the negative part $f_-(x) \doteq \max\{0, -f(x)\}$ is first reflected through the x-axis and then Lebesgue integrated like a nonnegative function yielding $I_- = \int_a^b f_-(x) \, d\mu$, and (iii) the value of the integral is obtained as $I = I_+ - I_-$

Whenever the Riemann integral exists, it is identical with the Lebesgue integral, and for practical purposes the calculation by the conventional technique of Riemann integration is to be preferred, since much more experience is available.

For the purpose of illustration, we consider cases where Riemann and Lebesgue integration yield different results. For $\Omega = \mathbb{R}$ and the Lebesgue measure λ, functions which are Riemann integrable on a compact and finite interval $[a, b]$ are Lebesgue integrable, too, and the values of the two integrals are the same. However, the converse is not true: not every Lebesgue integrable function is Riemann integrable. As an example, we consider the Dirichlet step function $D(x)$, which is the characteristic function of the rational numbers, assuming the value 1 for rationals and the value 0 for irrationals[48]:

$$D(x) = \begin{cases} 1\,, & \text{if } x \in \mathbb{Q}\,, \\ 0\,, & \text{otherwise}\,, \end{cases} \qquad \text{or } D(x) = \lim_{k \to \infty} \left(\lim_{n \to \infty} \cos^{2n}(k!\, \pi\, x) \right).$$

[48]It is worth noting that the highly irregular, nowhere continuous Dirichlet function $D(x)$ can be formulated as the (double) pointwise convergence limit, $\lim_{k \to \infty}$ and $\lim_{n \to \infty}$, of a trigonometric function.

$D(x)$ has no Riemann integral, but it does have a Lebesgue integral. The proof is straightforward.

Proof $D(x)$ fails Riemann integrability for every arbitrarily small interval: each partitioning S of the integration domain $[a, b]$ into intervals $[x_{k-1}, x_k]$ leads to parts that necessarily contain at least one rational and one irrational number. Hence the lower Darboux sum vanishes, viz.,

$$\Sigma_{[a,b]}^{(\text{low})}(S) = \sum_{k=1}^{n}(x_k - x_{k-1}) \times \inf_{x_{k-1}<x<x_k} D(x) = 0 ,$$

because the infimum is always zero, while the upper Darboux sum, viz.,

$$\Sigma_{[a,b]}^{(\text{high})}(S) = \sum_{k=1}^{n}(x_k - x_{k-1}) \times \sup_{x_{k-1}<x<x_k} D(x) = b - a ,$$

is the length $b-a = \sum_k (x_k - x_{k-1})$ of the integration interval, because the supremum is always one and the sum runs over all partial intervals. Riemann integrability requires

$$\sup_S \Sigma_{[a,b]}^{(\text{low})}(S) = \int_a^b f(x)\mathrm{d}x = \inf_S \Sigma_{[a,b]}^{(\text{high})}(S) ,$$

whence $D(x)$ cannot be Riemann integrated. The Dirichlet function $D(x)$, on the other hand, has a Lebesgue integral for every interval: $D(x)$ is a nonnegative simple function, so we can write the Lebesgue integral over an interval S by sorting into irrational and rational numbers:

$$\int_S D\,\mathrm{d}\lambda = 0 \times \lambda(S \cap \mathbb{R}\backslash\mathbb{Q}) + 1 \times \lambda(S \cap \mathbb{Q}) ,$$

with λ the Lebesgue measure. The evaluation of the integral is straightforward. The first term vanishes since multiplication by zero yields zero no matter how large $\lambda(S \cap \mathbb{R}\backslash\mathbb{Q})$ may be—recall that $0 \times \infty$ is zero by convention in measure theory—and the second term $\lambda(S \cap \mathbb{Q})$ is also zero since the set of rational numbers $\mathbb{Q}$ is countable. Hence we have $\int_S D\,\mathrm{d}\lambda = 0$. □

Another difference between Riemann and Lebesgue integration can, however, occur when the integration is extended to infinity in an improper Riemann integral. Then, the positive and negative contributions may cancel locally in the Riemann summation, whereas divergence may occur in both $f_+(x)$ and $f_-(x)$, since all positive parts and all negative parts are added first in the Lebesgue integral. An example is the improper Riemann integral $\int_0^\infty \sin x/x\,\mathrm{d}x$, which has the value $\pi/2$, whereas the corresponding Lebesgue integral does not exist, because $f_+(x)$ and $f_-(x)$ diverge.

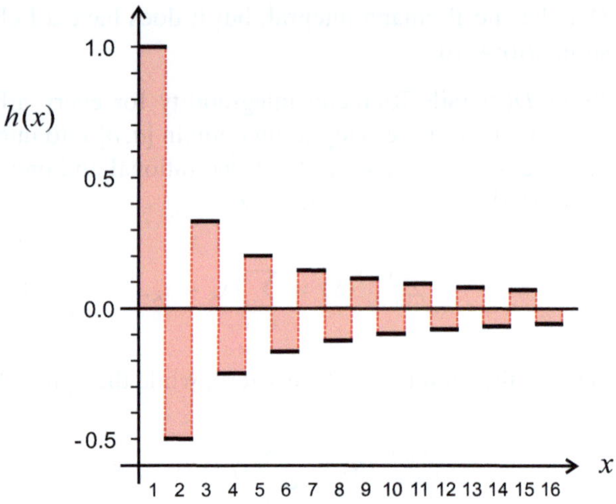

Fig. 1.21 The alternating harmonic series. The alternating harmonic step function, $h(x) = n_k = (-1)^{k+1}/k$ with $(k-1) \leq x < k$ and $n_k \in \mathbb{N}$, has an improper Riemann integral since $\sum_{k=1}^{\infty} n_k = \ln 2$. It is not Lebesgue integrable because the series $\sum_{k=1}^{\infty} |n_k|$ diverges

A typical example of a function that has an improper Riemann integral but is not Lebesgue integrable is the step function $h(x) = (-1)^{k+1}/k$ with $k - 1 \leq x < k$ and $k \in \mathbb{N}$ shown in Fig. 1.21. Under Riemann integration, this function yields a series of contributions of alternating sign that has a finite infinite sum

$$\int_0^{\infty} h(x)\,\mathrm{d}x = 1 - \frac{1}{2} + \frac{1}{3} - \cdots = \ln 2\,.$$

However, Lebesgue integrability of h requires $\int_{\mathbb{R}_{\geq 0}} |h|\,\mathrm{d}\lambda < \infty$ and this does not hold: both f_+ and f_- diverge. The proof is straightforward if one uses Leonhard Euler's result that the series of reciprocal prime numbers diverges:

$$\sum_{p \text{ prime}} \frac{1}{p} = \frac{1}{2} + \frac{1}{3} + \frac{1}{5} + \frac{1}{7} + \frac{1}{11} + \frac{1}{13} + \cdots = \infty\,,$$

$$\sum_{o \text{ odd}} \frac{1}{o} = 1 + \frac{1}{3} + \frac{1}{5} + \frac{1}{7} + \frac{1}{9} + \frac{1}{11} + \frac{1}{13} + \cdots > \sum_{p \text{ prime}} \frac{1}{p}\,,$$

$$1 + \sum_{e \text{ even}} \frac{1}{e} = 1 + \frac{1}{2} + \frac{1}{4} + \frac{1}{6} + \frac{1}{8} + \frac{1}{10} + \frac{1}{12} + \cdots > \sum_{o \text{ odd}} \frac{1}{o}\,.$$

Since $\infty - 1 = \infty$, both partial sums $\sum_{o \text{ odd}} 1/o$ and $\sum_{e \text{ even}} 1/e$ diverge. □

The first case discussed here—no Riemann integral but Lebesgue integrability—is the more important issue, since it provides a proof that the set of rational numbers $\mathbb{Q}$ has Lebesgue measure zero.

Finally, we introduce the Lebesgue–Stieltjes integral in a way that will allow us to summarize the most important results of this section. For each right-hand continuous and monotonically increasing function $F : \mathbb{R} \to \mathbb{R}$, there exists a uniquely determined Lebesgue–Stieltjes measure λ_F satisfying

$$\lambda_F\big((a,b]\big) = F(b) - F(a)\,, \quad \text{for all } (a,b] \subset \mathbb{R}\,. \tag{1.63}$$

Right-hand continuous and monotonically increasing functions $F : \mathbb{R} \to \mathbb{R}$ are said to be *measure generating*. The Lebesgue integral of a λ_F integrable function f is called a Lebesgue–Stieltjes integral:

$$\int_A f\,\mathrm{d}\lambda_F\,, \quad \text{with } A \in \mathcal{B}\,, \tag{1.64}$$

and it is Borel measurable. Let F be the identity function[49] on $\mathbb{R}$:

$$F = \mathrm{id} : \mathbb{R} \to \mathbb{R}\,, \quad \mathrm{id}(x) = x\,.$$

Then the corresponding Lebesgue–Stieltjes measure is the Lebesgue measure itself: $\lambda_F = \lambda_{\mathrm{id}} = \lambda$. For proper Riemann integrable functions f, we have stated that the Lebesgue integral is identical with the Riemann integral:

$$\int_{[a,b]} f\,\mathrm{d}\lambda = \int_a^b f(x)\,\mathrm{d}x\,.$$

The interval $[a,b] = a \le x \le b$ is partitioned into a sequence

$$S_n = \left(x_0^{(n)} = a, x_1^{(n)}, \ldots, x_n^{(n)} = b\right),$$

where the superscript (n) indicates a Riemann sum converging to the integral in the limit $|S| = \Delta x_i \to 0\,,\ \forall\, i$, and the Riemann integral on the right-hand side is replaced by the limit of the Riemann summation:

$$\begin{aligned}\int_{[a,b]} f\,\mathrm{d}\lambda &= \lim_{n\to\infty} \sum_{k=1}^{n} f(x_{k-1}^{(n)}) \left(x_k^{(n)} - x_{k-1}^{(n)}\right) \\ &= \lim_{n\to\infty} \sum_{k=1}^{n} f(x_{k-1}^{(n)}) \left(\mathrm{id}(x_k^{(n)}) - \mathrm{id}(x_{k-1}^{(n)})\right).\end{aligned}$$

The Lebesgue measure λ was introduced above for the special case $F = \mathrm{id}$ and therefore the general Stieltjes–Lebesgue integral is obtained by replacing λ by λ_F

[49]The identity function $\mathrm{id}(x) \doteq x$ maps a domain like $[a,b]$ point by point onto itself.

and id by F:

$$\int_{[a,b]} f \, \mathrm{d}\lambda_F = \lim_{n\to\infty} \sum_{k=1}^{n} f(x_{k-1}^{(n)}) \left(F(x_k^{(n)}) - F(x_{k-1}^{(n)}) \right) .$$

The details of the derivation can be found in [77, 390].

In summary, we define a Stieltjes–Lebesgue integral by $(F,f)\colon \mathbb{R} \to \mathbb{R}$, where the two functions F and f are partitioned on the interval $[a, b]$ by the sequence $S_n = (a = x_0, x_1, \ldots, x_n = b)$:

$$\sum_{S_n} f \, \mathrm{d}F \doteq \sum_{k=1}^{n} f(x_{k-1}) \big(F(x_k) - F(x_{k-1}) \big) .$$

The function f is F-integrable on $[a, b]$ if

$$\int_a^b f \, \mathrm{d}F = \lim_{|S|\to 0} \sum_S f \, \mathrm{d}F \tag{1.65}$$

exists in $\mathbb{R}$. Then $\int_a^b f \, \mathrm{d}F$ is called the Stieltjes–Lebesgue integral or sometimes also the F-integral of f. In the theory of stochastic processes, the Stieltjes–Lebesgue integral is required for the formulation of the Itō integral, which is used in Itō calculus applied to the integration of stochastic differential equations or SDEs (Sect. 3.4) [272, 273].

1.9 Continuous Random Variables and Distributions

Random variables on uncountable sets are completely characterized by a *probability triple* (Ω, Ξ, P). The *triple* is essentially the same as in the case of discrete variables (Sect. 1.6.3), except that the powerset $\Pi(\Omega)$ has been replaced by the event system $\Xi \subset \Pi(\Omega)$. We recall that the powerset $\Pi(\Omega)$ is too large to define probabilities since it contains uncountably many subsets or events A (Fig. 1.15). The sets in Ξ are the Borel σ-algebras. They are measurable and they *alone* have probabilities. Accordingly, we are now in a position to handle probabilities on uncountable sets:

$$\{\omega | \mathcal{X}(\omega) \leq x\} \in \Xi \quad \text{and} \quad P(\mathcal{X} \leq x) = \frac{|\{\mathcal{X}(\omega) \leq x\}|}{|\Omega|} , \tag{1.66a}$$

$$\{a < \mathcal{X} \leq b\} = \{\mathcal{X} \leq b\} - \{\mathcal{X} \leq a\} \in \Xi , \quad \text{with } a < b , \tag{1.66b}$$

$$P(a < \mathcal{X} \leq b) = \frac{|\{a < \mathcal{X} \leq b\}|}{|\Omega|} = F_{\mathcal{X}}(b) - F_{\mathcal{X}}(a) \,. \tag{1.66c}$$

Equation (1.66a) contains the definition of a real-valued function $\mathcal{X}$ which is called a random variable iff it satisfies $P(\mathcal{X} \leq x)$ for any real number x, (1.66b) is valid since Ξ is closed under difference, and finally, (1.66c) provides the basis for defining and handling probabilities on uncountable sets. The three equations (1.66) together constitute the basis of the probability concept on uncountable sample spaces that will be applied throughout this book.

1.9.1 Densities and Distributions

Random variables on uncountable sets Ω are commonly characterized by *probability density functions* (pdf). The probability density function—or density for short—is the continuous analogue of the probability mass function (pmf). A density is a function f on $\mathbb{R} =]-\infty, +\infty[$, $u \mapsto f(u)$, which satisfies the two conditions[50]:

$$\text{(i)} \quad \forall u\,, \quad f(u) \geq 0\,,$$
$$\text{(ii)} \quad \int_{-\infty}^{\infty} f(u)\,\mathrm{d}u = 1\,. \tag{1.67}$$

Now we can define a class of continuous random variables[51] on general sample spaces: $\mathcal{X}$ is a function on $\Omega : \omega \to \mathcal{X}(\omega)$ whose probabilities are prescribed by means of a density function $f(u)$. For any interval $[a, b]$, the probability is given by

$$P(a \leq \mathcal{X} \leq b) = \int_a^b f(u)\,\mathrm{d}u \,. \tag{1.68}$$

If A is the union of not necessarily disjoint intervals, some of which may even be infinite, the probability can be derived in general from the density

$$P(\mathcal{X} \in A) = \int_A f(u)\,\mathrm{d}u \,.$$

[50]From here on we shall omit the random variable as subscript and simply write $f(x)$ or $F(x)$, unless a nontrivial specification is required.

[51]Random variables with a density are often called *continuous* random variables, in order to distinguish them from *discrete* random variables, defined on countable sample spaces.

In particular, A can be split into disjoint intervals, i.e., $A = \bigcup_{j=1}^{k}[a_j, b_j]$, and the integral can then be rewritten as

$$\int_A f(u)\,\mathrm{d}u = \sum_{j=1}^{k}\int_{a_j}^{b_j} f(u)\,\mathrm{d}u\ .$$

For the interval $A =]-\infty, x]$, we define the *cumulative probability distribution function* (cdf) $F(x)$ of the continuous random variable $\mathcal{X}$ to be

$$F(x) = P(\mathcal{X} \le x) = \int_{-\infty}^{x} f(u)\,\mathrm{d}u\ .$$

An easy to verify and useful relation defines the complementary cumulative distribution function (ccdf):

$$\tilde{F}(x) = P(\mathcal{X} > x) = 1 - F(x)\ . \tag{1.69}$$

If f is continuous, then it is the derivative of F, as follows from the fundamental theorem of calculus:

$$F'(x) = \frac{\mathrm{d}F(x)}{\mathrm{d}x} = f(x)\ .$$

If the density f is not continuous everywhere, the relation is still true for every x at which f is continuous.

If the random variable $\mathcal{X}$ has a density, then by setting $a = b = x$, we find

$$P(\mathcal{X} = x) = \int_{x}^{x} f(u)\,\mathrm{d}u = 0\ ,$$

reflecting the trivial geometric result that every line segment has zero area. It seems somewhat paradoxical that $\mathcal{X}(\omega)$ must be some number for every ω, whereas any given number has probability zero. The paradox is resolved by looking at countable and uncountable sets in more depth, as we did in Sects. 1.5 and 1.7.

To exemplify continuous probability functions, we present here the *normal distribution* (Fig. 1.22), which is of primary importance in probability theory for several reasons: (i) it is mathematically simple and well behaved, (ii) it is exceedingly *smooth*, since it can be differentiated an infinite number of times, and (iii) all distributions converge to the normal distribution in the limit of large sample numbers, a result known as the *central limit theorem* (Sect. 2.4.2). The density of the normal distribution is a Gaussian function named after the German mathematician

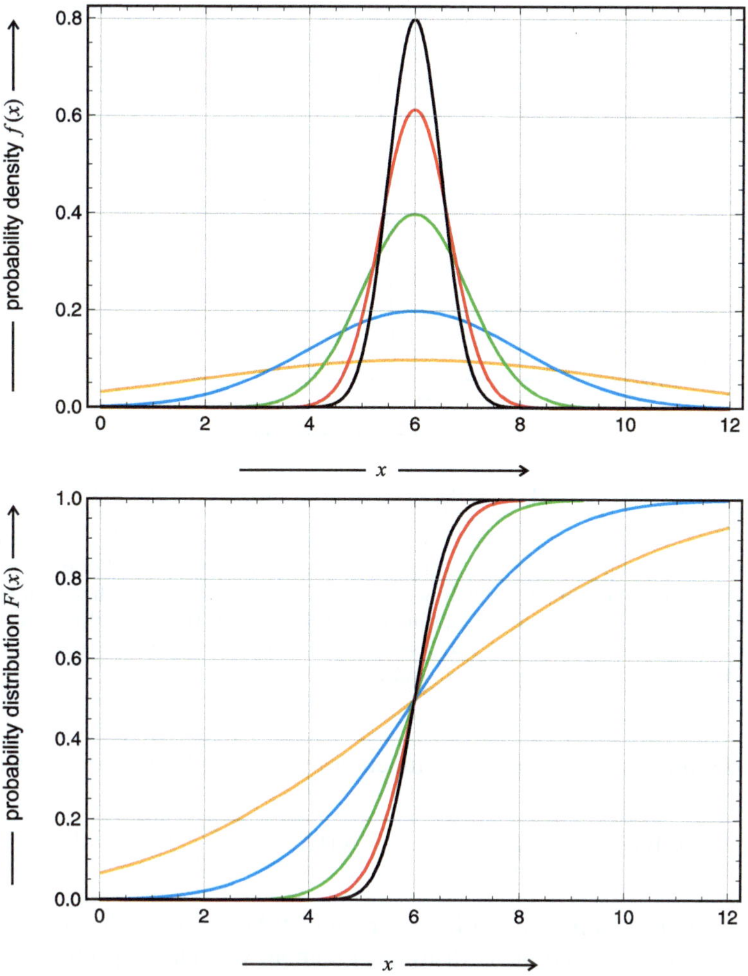

Fig. 1.22 Normal density and distribution. The normal distribution $\mathcal{N}(\mu, \sigma)$ is shown in the form of the probability density $f(x) = \exp\big(-(x-\mu)^2/2\sigma^2\big) \big/ \sqrt{2\pi}\sigma$ and the probability distribution $F(x) = \Big(1+\mathrm{erf}\big((x-\mu)/\sqrt{2\sigma^2}\big)\Big)/2$, where erf is the error function. Choice of parameters: $\mu = 6$ and $\sigma = 0.5$ (*black*), 0.65 (*red*), 1 (*green*), 2 (*blue*) and 4 (*yellow*)

Carl Friedrich Gauss and also sometimes called the symmetric bell curve:

$$\mathcal{N}(x; \mu, \sigma^2): \; f(x) = \frac{1}{\sqrt{2\pi\sigma^2}} \exp\left(-\frac{(x-\mu)^2}{2\sigma^2}\right), \tag{1.70}$$

$$F(x) = \frac{1}{2}\left(1 + \mathrm{erf}\left(\frac{x-\mu}{\sqrt{2\sigma^2}}\right)\right). \tag{1.71}$$

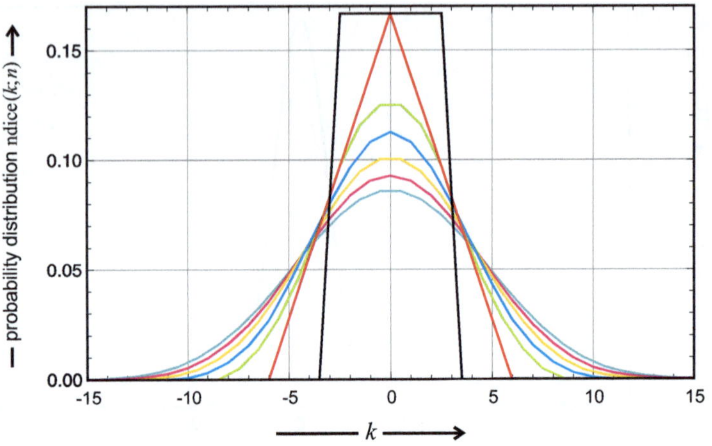

Fig. 1.23 Convergence to the normal density. The series of probability mass functions for rolling n conventional dice, $f_{s,n}(k)$ with $s = 6$ and $n = 1, 2, \ldots$, begins with a pulse function $f_{6,1}(k) = 1/6$ for $k = 1, \ldots, 6$ ($n = 1$), followed by a tent function ($n = 2$), and then a gradual approach towards the normal distribution ($n = 3, 4, \ldots$). For $n = 7$, we show the fitted normal distribution (*broken black curve*) coinciding almost perfectly with $f_{7d}(k)$. The series of densities has been used as an example for convergence in distribution (Fig. 1.16 in Sect. 1.8.1). The probability mass functions are centered around the mean values $\mu_{s,n} = n(s-1)/2$. Color code: $n = 1$ (*black*), 2 (*red*), 3 (*green*), 4 (*blue*), 5 (*yellow*), 6 (*magenta*), and 7 (*sea green*)

Here erf is the error function.[52] This function and its complement erfc are defined by

$$\mathrm{erf}(x) = \frac{2}{\sqrt{\pi}} \int_0^x \mathrm{e}^{-z^2} \mathrm{d}z \,, \qquad \mathrm{erfc}(x) = \frac{2}{\sqrt{\pi}} \int_x^\infty \mathrm{e}^{-z^2} \mathrm{d}z \,.$$

The two parameters μ and σ^2 of the normal distribution are the expectation value and the variance of a normally distributed random variable, respectively, and σ is called the standard deviation.

The central limit theorem will be discussed separately in Sect. 2.4.2, but here we present an example of the convergence of a probability distribution towards the normal distribution with which we are already familiar: the dice-rolling problem extended to n dice. A collection of n dice is thrown simultaneously and the total score of all the dice together is recorded (Fig. 1.23). The probability of obtaining a total score of k points by rolling n dice with s faces can be calculated by means of

[52] We remark that $\mathrm{erf}(x)$ and $\mathrm{erfc}(x)$ are not normalized in the same way as the normal density, since we have $\mathrm{erf}(x) + \mathrm{erfc}(x) = (2/\sqrt{\pi}) \int_0^\infty \exp(-t^2)\,\mathrm{d}t = 1$, but $\int_0^\infty f(x)\,\mathrm{d}x = (1/2) \int_{-\infty}^\infty f(x)\,\mathrm{d}x = 1/2$.

combinatorics:

$$f_{s,n}(k) = \frac{1}{s^n} \sum_{i=0}^{\lfloor (k-n)/s \rfloor} (-1)^i \binom{n}{i} \binom{k-si-1}{n-1} . \tag{1.50'}$$

The results for small values of n and ordinary dice ($s = 6$) are illustrated in Fig. 1.23. The convergence to a continuous probability density is nicely illustrated. For $n = 7$ the deviation from the Gaussian curve of the normal distribution is hardly recognizable (see Fig. 1.16).

It is sometimes useful to discretize a density function in order to yield a set of elementary probabilities. The x-axis is divided up into m pieces (Fig. 1.24), not necessarily equal and not necessarily small, and we denote the piece of the integral on the interval $\Delta_k = x_{k+1} - x_k$, i.e., between the values $u(x_k)$ and $u(x_{k+1})$ of the variable u, by

$$p_k = \int_{x_k}^{x_{k+1}} f(u)\,\mathrm{d}u , \quad 0 \le k \le m-1 , \tag{1.72}$$

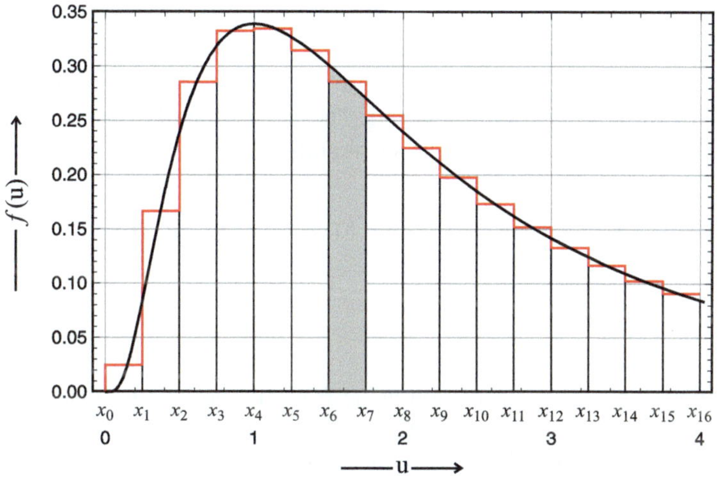

Fig. 1.24 Discretization of a probability density. A segment $[x_0, x_m]$ on the u-axis is divided up into m not necessarily equal intervals, and elementary probabilities are obtained by integration. The curve shown here is the density of the lognormal distribution $\ln \mathcal{N}(\nu, \sigma^2)$:

$$f(u) = \frac{1}{u\sqrt{2\pi\sigma^2}} \mathrm{e}^{-(\ln u - \nu)^2/2\sigma^2} .$$

The *red step function* represents the discretized density. The *hatched area* is the probability $p_6 = \int_{x_6}^{x_7} f(u)\,\mathrm{d}u$ with the parameters $\nu = \ln 2$ and $\sigma = \sqrt{\ln 2}$

where the p_k values satisfy

$$\forall\, k\,,\; p_k \geq 0\,,\quad \sum_{k=0}^{m-1} p_k = 1\;.$$

If we choose $x_0 = -\infty$ and $x_m = +\infty$, we are dealing with a partition that is not finite but countable, provided we label the intervals suitably, e.g., $\ldots, p_{-2}, p_{-1}, p_0, p_1, p_2, \ldots$. Now we consider a random variable $\mathcal{Y}$ such that

$$P(\mathcal{Y} = x_k) = p_k\;, \tag{1.72$'$}$$

where we may replace x_k by any value of x in the subinterval $[x_k, x_{k+1}]$. The random variable $\mathcal{Y}$ can be interpreted as the discrete analogue of the continuous random variable $\mathcal{X}$. Making the intervals Δ_k smaller increases the accuracy of the discretization approximation and this procedure has a lot in common with Riemann integration.

1.9.2 Expectation Values and Variances

Although we shall treat expectation values and other moments of probability distributions extensively in Chap. 2, we make here a short digression to present examples of various integration concepts. The calculation of expectation values and variances from continuous densities is straightforward:

$$\mathrm{E}(\mathcal{X}) = \int_{-\infty}^{\infty} x f(x)\,\mathrm{d}x = \int_0^{\infty} \big(1 - F(x)\big)\,\mathrm{d}x - \int_{-\infty}^{0} F(x)\,\mathrm{d}x\;, \tag{1.73a}$$

$$\mathrm{var}(\mathcal{X}) = \int_{-\infty}^{\infty} x^2 f(x)\,\mathrm{d}x - \mathrm{E}(\mathcal{X})^2\;. \tag{1.73b}$$

The computation of the expectation value from the probability distribution is the analogue of the discrete case (1.20a). We present the derivation of the expression here as an exercise in handling probabilities and integrals [229]. As in a Lebesgue integral, we decompose $\mathcal{X}$ into positive and negative parts: $\mathcal{X} = \mathcal{X}^+ - \mathcal{X}^-$ with $\mathcal{X}^+ = \max\{\mathcal{X}, 0\}$ and $\mathcal{X}^- = \max\{-|\mathcal{X}|, 0\}$. Then, we express both parts by means of indicator functions:

$$\mathcal{X}^+ = \int_0^{\infty} \mathbf{1}_{\mathcal{X}>\vartheta}\,\mathrm{d}\vartheta\;, \qquad \mathcal{X}^- = \int_{-\infty}^{0} \mathbf{1}_{\mathcal{X}\leq\vartheta}\,\mathrm{d}\vartheta\;.$$

By applying Fubini's theorem named after the Italian mathematician Guido Fubini [189] we reverse the order of taking the expectation value and integration, make use of (1.26b) and (1.69), and find

$$\begin{aligned}
\mathrm{E}(\mathcal{X}) &= \mathrm{E}(\mathcal{X}^+ - \mathcal{X}^-) = \mathrm{E}(\mathcal{X}^+) - \mathrm{E}(\mathcal{X}^-) \\
&= \mathrm{E}\left(\int_0^\infty \mathbf{1}_{\mathcal{X}>\vartheta}\,\mathrm{d}\vartheta\right) - \mathrm{E}\left(\int_{-\infty}^0 \mathbf{1}_{\mathcal{X}\leq\vartheta}\,\mathrm{d}\vartheta\right) \\
&= \int_0^\infty \mathrm{E}\left(\mathbf{1}_{\mathcal{X}>\vartheta}\right)\mathrm{d}\vartheta - \int_{-\infty}^0 \mathrm{E}\left(\mathbf{1}_{\mathcal{X}\leq\vartheta}\right)\mathrm{d}\vartheta \\
&= \int_0^\infty P(\mathcal{X}>\vartheta)\,\mathrm{d}\vartheta - \int_{-\infty}^0 P(\mathcal{X}\leq\vartheta)\,\mathrm{d}\vartheta \\
&= \int_0^\infty \big(1-F(\vartheta)\big)\,\mathrm{d}\vartheta - \int_{-\infty}^0 F(\vartheta)\,\mathrm{d}\vartheta\,. \qquad \square
\end{aligned}$$

The calculation of expectation values directly from the cumulative distribution function has the advantage of being applicable to cases where densities do not exist or where they are hard to handle.

1.9.3 Continuous Variables and Independence

In the joint distribution function of the random vector $\mathcal{V} = (\mathcal{X}_1, \ldots, \mathcal{X}_n)$, the property of independence of variables is tantamount to factorizability into (marginal) distributions, i.e.,

$$F(x_1, \ldots, x_n) = F_1(x_1) \cdots F_n(x_n)\,,$$

where the F_j are the marginal distributions of the random variables, the $\mathcal{X}_j$ $(1 \leq j \leq n)$. As in the discrete case, the marginal distributions are sufficient to calculate joint distributions of independent random variables.

For the continuous case, we can formulate the definition of independence for sets $S_1, \ldots, S_n$ forming a Borel family. In particular, when there is a joint density function $f(u_1, \ldots, u_n)$, we have

$$\begin{aligned}
P(\mathcal{X}_1 \in S_1, \ldots, \mathcal{X}_n \in S_n) &= \int_{S_1} \cdots \int_{S_n} f(u_1, \ldots, u_n)\,\mathrm{d}u_1 \ldots \mathrm{d}u_n \\
&= \int_{S_1} \cdots \int_{S_n} f_1(u_1) \ldots f_n(u_n)\,\mathrm{d}u_1 \ldots \mathrm{d}u_n \\
&= \left(\int_{S_1} f_1(u_1)\,\mathrm{d}u_1\right) \cdots \left(\int_{S_n} f_n(u_n)\,\mathrm{d}u_n\right)\,,
\end{aligned}$$

where $f_1, \ldots, f_n$ are the marginal densities, e.g.,

$$f_1(u_1) = \int_{S_2} \cdots \int_{S_n} f(u_1, \ldots, u_n)\, \mathrm{d}u_2 \ldots \mathrm{d}u_n \,. \tag{1.74}$$

Eventually, we find for the density case:

$$f(u_1, \ldots, u_n) = f_1(u_1) \cdots f_n(u_n) \,. \tag{1.75}$$

As we have seen here, stochastic independence is the basis for factorization of joint probabilities, distributions, densities, and other functions. Independence is a stronger criterion than lack of correlation, as we shall show in Sect. 2.3.4.

1.9.4 Probabilities of Discrete and Continuous Variables

We close this chapter with a comparison of the formalisms of probability theory on countable and uncountable sample spaces. To this end, we repeat and compare in Table 1.3 the basic features of discrete and continuous probability distributions as they have been discussed in Sects. 1.6.3 and 1.9.1, respectively.

Discrete probability distributions are defined on countable sample spaces and their random variables are discrete sets of events $\omega \in \Omega$, e.g., sample points on a closed interval $[a, b]$:

$$\{a \le \mathcal{X} \le b\} = \{\omega \,|\, a \le \mathcal{X} \le b\} \,.$$

If the sample space Ω is finite or countably infinite, the exact range of $\mathcal{X}$ is a set of real numbers w_i:

$$W_{\mathcal{X}} = \{w_1, w_2, \ldots, w_n, \ldots\} \,, \quad \text{with } w_k \in \Omega \,,\ \forall\, k \,.$$

Introducing probabilities for individual events, $p_n = P\big(\mathcal{X} = w_n | \, w_n \in W_{\mathcal{X}}\big)$ and $P\big(\mathcal{X}(x) = 0 | x \notin W_{\mathcal{X}}\big)$, yields

$$P(\mathcal{X} \in A) = \sum_{w_n \in A} p_n \,, \quad \text{with } A \in \Omega \,,$$

or, in particular,

$$P(a \le \mathcal{X} \le b) = \sum_{a \le w_n \le b} p_n \,. \tag{1.30}$$

Table 1.3 Comparison of the formalism of probability theory on countable and uncountable sample spaces

	Expression	Countable case		Uncountable case		
Domain, full	$A \in \Omega$	$w_n,\ n = \dots, -2, -1, 0, 1, 2, \dots$	$w_n \in \mathbb{Z}$	$-\infty < u < \infty$	$]-\infty, \infty[$	$u \in \mathbb{R}$
nonnegative		$w_n,\ n = 0, 1, 2, \dots$	$w_n \in \mathbb{N}$	$0 \le u < \infty$	$[0, \infty[$	$u \in \mathbb{R}_{\ge 0}$
positive		$w_n,\ n = 1, 2, \dots$	$w_n \in \mathbb{N}_{>0}$	$0 < u < \infty$	$]0, \infty[$	$u \in \mathbb{R}_{>0}$
Probability	$P(\mathcal{X} \in A) : a \in \Omega$	p_n		$dF(u) = f(u)\,du$		
Interval	$P(a \le \mathcal{X} \le b)$	$\sum_{a \le w_n \le b} p_n$		$\int_a^b f(u)\,du$		
Density, pmf or pdf	$f(x) = P(\mathcal{X} = x)$	$\begin{cases} p_n & \text{if } x \in W_{\mathcal{X}} = \{w_1, \dots, w_n, \dots\} \\ 0 & \text{if } x \notin W_{\mathcal{X}} = \{w_1, \dots\} \end{cases}$		$f(u)\,du$		
Distribution, cdf	$F(x) = P(\mathcal{X} \le x)$	$\sum_{w_n \le x} p_n$		$F(u) = \int_{-\infty}^{x} f(u)\,du$		
Expectation value	$\mathrm{E}(\mathcal{X}) = \sum_n n f_{\mathcal{X}}(n)$	$\sum_n p_n w_n$	$\sum_n p_n \lvert w_n \rvert < \infty$	$\int_{-\infty}^{\infty} u f(u)\,du$		$\int_{-\infty}^{\infty} \lvert u \rvert f(u)\,du < \infty$
		$\mathrm{E}(\mathcal{X}) = \sum_n \big(1 - F(n)\big)$	$n \in \mathbb{N}$	$\int_0^{\infty} \big(1 - F(u)\big)\,du$		$u \in \mathbb{R}_{\ge 0}$
Variance	$\mathrm{var}(\mathcal{X}) = \sum_n n^2 f_{\mathcal{X}}(n) - \mathrm{E}(\mathcal{X})^2$	$\sum_n p_n w_n^2 - \mathrm{E}(\mathcal{X})^2$	$\sum_n p_n w_n^2 < \infty$	$\int_{-\infty}^{\infty} u^2 f(u)\,du - \mathrm{E}(\mathcal{X})^2$		$\int_{-\infty}^{\infty} u^2 f(u)\,du < \infty$
		$2 \sum_n n\big(1 - F(n)\big) - \mathrm{E}(\mathcal{X})^2$	$n \in \mathbb{N}$	$2 \int_0^{\infty} u\big(1 - F(u)\big)\,du - \mathrm{E}(\mathcal{X})^2$		$u \in \mathbb{R}_{\ge 0}$

The table shows the basic formulas for discrete and continuous random variables

Two probability functions are in common use: the probability mass function (pmf)

$$f_{\mathcal{X}}(x) = P(\mathcal{X} = x) = \begin{cases} p_n \,, & \text{if } x = w_n \in W_{\mathcal{X}} \,, \\ 0 \,, & \text{if } x \neq W_{\mathcal{X}} \,, \end{cases}$$

and the cumulative distribution function (cdf)

$$F_{\mathcal{X}}(x) = P(\mathcal{X} \leq x) = \sum_{w_n \leq x} p_n \,,$$

with two properties following form those of the probabilities:

$$\lim_{x \to -\infty} F_{\mathcal{X}}(x) = 0 \,, \qquad \lim_{x \to +\infty} F_{\mathcal{X}}(x) = 1 \,.$$

Continuous probability distributions are defined on uncountable Borel measurable sample spaces, and their random variables $\mathcal{X}$ have densities. A *probability density function* (pdf) is a mapping

$$f : \mathbb{R} \to \mathbb{R}_{\geq 0}$$

which satisfies the two conditions:

$$\text{(i)} \quad f(u) \geq 0 \,, \ \forall\, u \in \mathbb{R} \,,$$

$$\text{(ii)} \quad \int_{-\infty}^{\infty} f(u)\, \mathrm{d}u = 1 \,. \tag{1.76}$$

Random variables $\mathcal{X}$ are functions on Ω: $\omega \mapsto \mathcal{X}(\omega)$ whose probabilities are derived from density functions $f(u)$:

$$P(a \leq \mathcal{X} \leq b) = \int_a^b f(u)\, \mathrm{d}u \,. \tag{1.68}$$

As in the discrete case, the probability functions come in two forms: (i) the probability density function (pdf) defined above, viz.,

$$f(u)\, \mathrm{d}u = \mathrm{d}F(u) \,,$$

and (ii) the cumulative distribution function (cdf), viz.,

$$F(x) = P(\mathcal{X} \leq x) = \int_{-\infty}^{x} f(u)\, \mathrm{d}u \,, \quad \text{with} \quad \frac{\mathrm{d}F(x)}{\mathrm{d}x} = f(x) \,,$$

provided that the function $f(x)$ is continuous.

Conventional thinking in terms of probabilities has been extended in two important ways in the last two sections. Firstly, the handling of the uncountable sets that are important in probability theory has allowed us to define and calculate with probabilities when comparison by counting is not possible, and secondly, Lebesgue–Stieltjes integration has provided an extension of calculus to the step functions encountered with discrete probabilities.

Chapter 2
Distributions, Moments, and Statistics

Everything should be made as simple as possible, but not simpler.

Attributed to Albert Einstein 1950

Abstract The moments of probability distributions represent the link between theory and observations since they are readily accessible to measurement. Rather abstract-looking generating functions have become important as highly versatile concepts and tools for solving specific problems. The probability distributions which are most important in applications are reviewed. Then the central limit theorem and the law of large numbers are presented. The chapter is closed by a brief digression into mathematical statistics and shows how to handle real world samples that cover a part, sometimes only a small part, of sample space.

In this chapter we make an attempt to bring probability theory closer to applications. Random variables are accessible to analysis via their probability distributions. Full information is derived from ensembles defined on the entire sample space Ω. Complete coverage of sample space, however, is an ideal that can rarely be achieved in reality. Samples obtained in experimental observations are almost always far from being exhaustive collections. We begin here with a theoretical discussion and introduce mathematical statistics afterwards.

Probability distributions and densities are used to calculate measurable quantities like expectation values, variances, and higher moments. The moments provide relevant partial information on probability distributions since full information would require a series expansion up to infinite order.

2.1 Expectation Values and Higher Moments

Distributions can be characterized by moments that are powers of variables averaged over the entire sample space. Most important are the first two moments, which have a straightforward interpretation: the expectation value $\mathrm{E}(\mathcal{X})$ is the average value of a distribution, and the variance $\mathrm{var}(\mathcal{X})$ or $\sigma^2(\mathcal{X})$ is a measure of the width of a distribution.

P. Schuster, *Stochasticity in Processes*, Springer Series in Synergetics,
DOI 10.1007/978-3-319-39502-9_2

2.1.1 First and Second Moments

The most natural and important ensemble property is the *expectation value* or *average*, written $\mathrm{E}(\mathcal{X})$ or $\langle\mathcal{X}\rangle$ as preferred in physics. We begin with a countable sample space Ω:

$$\mathrm{E}(\mathcal{X}) = \sum_{\omega\in\Omega} \mathcal{X}(\omega)P(\omega) = \sum_{n} w_n p_n \,. \tag{2.1}$$

In the most common special case of a random variable $\mathcal{X}$ on $\mathbb{N}$, we have $w_n = n$ and find

$$\mathrm{E}(\mathcal{X}) = \sum_{n=0}^{\infty} n p_n = \sum_{n=1}^{\infty} n p_n \,.$$

The expectation value (2.1) of a distribution exists when the series in the sum converges in absolute values: $\sum_{\omega\in\Omega} |\mathcal{X}(\omega)|P(\omega) < \infty$. Whenever the random variable $\mathcal{X}$ is bounded, which means that there exists a number m such that $|\mathcal{X}(\omega)| \le m$ for all $\omega \in \Omega$, then it is summable and in fact

$$\mathrm{E}(|\mathcal{X}|) = \sum_{\omega} |\mathcal{X}(\omega)|\, P(\omega) \le m \sum_{\omega} P(\omega) = m \,.$$

It is straightforward to show that the sum $\mathcal{X}+\mathcal{Y}$ of two summable random variables is summable, and the expectation value of the sum is the sum of the expectation values:

$$\mathrm{E}(\mathcal{X}+\mathcal{Y}) = \mathrm{E}(\mathcal{X}) + \mathrm{E}(\mathcal{Y}) \,.$$

This relation can be extended to an arbitrary countable number of random variables:

$$\mathrm{E}\left(\sum_{k=1}^{n} \mathcal{X}_k\right) = \sum_{k=1}^{n} \mathrm{E}(\mathcal{X}_k) \,.$$

In addition, the expectation values satisfy $\mathrm{E}(a) = a$ and $\mathrm{E}(a\mathcal{X}) = a\mathrm{E}(\mathcal{X})$, which can be combined in

$$\mathrm{E}\left(\sum_{k=1}^{n} a_k\mathcal{X}_k\right) = \sum_{k=1}^{n} a_k\mathrm{E}(\mathcal{X}_k) \,. \tag{2.2}$$

Accordingly, $\mathrm{E}(\cdot)$ fulfils all conditions required for a *linear operator*.

For a random variable $\mathcal{X}$ on an arbitrary sample space Ω the expectation value may be written as an abstract integral on Ω or as an integral over $\mathbb{R}$, provided the density $f(u)$ exists:

$$\mathrm{E}(\mathcal{X}) = \int_{\Omega} \mathcal{X}(\omega)\,\mathrm{d}\omega = \int_{-\infty}^{+\infty} uf(u)\,\mathrm{d}u\,. \tag{2.3}$$

In this context it is worth reconsidering the discretization of a continuous density (Fig. 1.24): the discrete expression for the expectation value is based upon $p_n = P(\mathcal{Y} = x_n)$ as outlined in (1.72) and (1.72′),

$$\mathrm{E}(\mathcal{Y}) = \sum_n x_n p_n \approx \mathrm{E}(\mathcal{X}) = \int_{-\infty}^{+\infty} uf(u)\,\mathrm{d}u\,,$$

and approximates the exact value similarly, just as the Darboux sum does in case of a Riemann integral.

For two or more variables, for example, $\boldsymbol{\mathcal{V}} = (\mathcal{X}, \mathcal{Y})$, described by a joint density $f(u, v)$, we have

$$\mathrm{E}(\mathcal{X}) = \int_{-\infty}^{+\infty} uf(u, *)\,\mathrm{d}u\,, \qquad \mathrm{E}(\mathcal{Y}) = \int_{-\infty}^{+\infty} vf(*, v)\,\mathrm{d}v\,,$$

where $f(u, *) = \int_{-\infty}^{+\infty} f(u, v)\,\mathrm{d}v$ and $f(*, v) = \int_{-\infty}^{+\infty} f(u, v)\,\mathrm{d}u$ are the marginal densities.

The expectation value of the sum $\mathcal{X} + \mathcal{Y}$ of the variables can be evaluated by iterated integration:

$$\begin{aligned}
\mathrm{E}(\mathcal{X} + \mathcal{Y}) &= \int_{-\infty}^{+\infty}\int_{-\infty}^{+\infty} (u + v) f(u, v)\,\mathrm{d}u\,\mathrm{d}v \\
&= \int_{-\infty}^{+\infty} u\,\mathrm{d}u \left(\int_{-\infty}^{+\infty} f(u, v)\,\mathrm{d}v\right) + \int_{-\infty}^{+\infty} v\,\mathrm{d}v \left(\int_{-\infty}^{+\infty} f(u, v)\,\mathrm{d}u\right) \\
&= \int_{-\infty}^{+\infty} uf(u, *)\,\mathrm{d}u + \int_{-\infty}^{+\infty} vf(*, v)\,\mathrm{d}v \\
&= \mathrm{E}(\mathcal{X}) + \mathrm{E}(\mathcal{Y})\,,
\end{aligned}$$

which yields the same expression as previously derived in the discrete case.

The *multiplication theorem* of probability theory requires that the two variables $\mathcal{X}$ and $\mathcal{Y}$ be independent and summable, and this implies for the discrete and the continuous case[1]:

$$\mathrm{E}(\mathcal{X} \cdot \mathcal{Y}) = \mathrm{E}(\mathcal{X}) \cdot \mathrm{E}(\mathcal{Y}) \, , \tag{2.4a}$$

$$\begin{aligned} \mathrm{E}(\mathcal{X} \cdot \mathcal{Y}) &= \int_{-\infty}^{+\infty} \int_{-\infty}^{+\infty} uvf(u,v)\, \mathrm{d}u\, \mathrm{d}v \\ &= \int_{-\infty}^{+\infty} uf(u,*)\, \mathrm{d}u \int_{-\infty}^{+\infty} vf(*,v)\, \mathrm{d}v \qquad (2.4b) \\ &= \mathrm{E}(\mathcal{X}) \cdot \mathrm{E}(\mathcal{Y}) \, , \qquad (2.4c) \end{aligned}$$

respectively. The multiplication theorem is easily extended to any finite number of independent and summable random variables:

$$\mathrm{E}(\mathcal{X}_1, \ldots, \mathcal{X}_n) = \mathrm{E}(\mathcal{X}_1) \cdots \mathrm{E}(\mathcal{X}_n) \, . \tag{2.4c}$$

Next we consider the expectation values of functions of random variables and start with the expectation values of their powers $\mathcal{X}^r$, which give rise to the *raw moments* of the probability distribution: $\hat{\mu}_r = E(\mathcal{X}^r)$, $r = 1, 2, \ldots$.[2] In general, moments are defined about some point a according to a shifted random variable

$$\mathcal{X}^{(a)} = \mathcal{X} - a \, .$$

For $a = 0$ we obtain the raw moments

$$\hat{\mu}_r(\mathcal{X}) = \mathrm{E}(\mathcal{X}^r) \, . \tag{2.5a}$$

For the *centered moments* the random variable is centered around the expectation value $a = \mathrm{E}(\mathcal{X})$,

$$\tilde{\mathcal{X}} = \mathcal{X} - \mathrm{E}(\mathcal{X}) \, ,$$

and this leads to the following expression for the moments:

$$\mu_r(\mathcal{X}) = \mathrm{E}\big(\mathcal{X}^r - \mathrm{E}(\mathcal{X})\big) \, . \tag{2.5b}$$

[1] A proof is given in [84, pp. 164–166].

[2] Since the moments centered around the expectation value will be used more frequently than the raw moments, we denote them by μ_r and reserve $\hat{\mu}_r$ for the raw moments. The first centered moment vanishes and since confusion is unlikely, we shall write the expectation value μ instead of $\hat{\mu}_1$. The rth moment of a distribution is also called the moment of order r.

The first raw moment is the expectation value, $\mathrm{E}(\mathcal{X}) \equiv \mu(\mathcal{X}) \equiv \langle\mathcal{X}\rangle$, the first centered moment vanishes, $\mathrm{E}(\tilde{\mathcal{X}}) \equiv \mu_1(\mathcal{X}) = 0$, and the second centered moment is the variance $\mathrm{var}(\mathcal{X}) \equiv \mu_2(\mathcal{X}) \equiv \sigma^2(\mathcal{X})$. The positive square root of the variance, $\sigma(\mathcal{X}) = \sqrt{\mathrm{var}(\mathcal{X})}$ is called the *standard deviation*.

In the case of continuous random variables the expressions for the rth raw and centered moments are obtained from the densities $f(u)$ by integration:

$$\mathrm{E}(\mathcal{X}^r) = \hat{\mu}_r(\mathcal{X}) = \int_{-\infty}^{+\infty} u^r f(u)\,\mathrm{d}u\,, \tag{2.6a}$$

$$\mathrm{E}(\tilde{\mathcal{X}}^r) = \mu_r(\mathcal{X}) = \int_{-\infty}^{+\infty} (u-\mu)^r f(u)\,\mathrm{d}u\,. \tag{2.6b}$$

As in the discrete case the second centered moment is called the *variance*, $\mathrm{var}(\mathcal{X})$ or $\sigma^2(\mathcal{X})$, and its positive square root is the *standard deviation* $\sigma(\mathcal{X})$.

Several properties of the moments are valid independently of whether the random variable is discrete or continuous:

(i) The variance is always a nonnegative quantity as can be easily shown:

$$\begin{aligned}\mathrm{var}(\mathcal{X}) &= \mathrm{E}(\tilde{\mathcal{X}}^2) = \mathrm{E}\left(\left(\mathcal{X} - \mathrm{E}(\mathcal{X})\right)^2\right)\\ &= \mathrm{E}\left(\mathcal{X}^2 - 2\mathcal{X}\,\mathrm{E}(\mathcal{X}) + \mathrm{E}(\mathcal{X})^2\right)\\ &= \mathrm{E}(\mathcal{X}^2) - 2\mathrm{E}(\mathcal{X})\,\mathrm{E}(\mathcal{X}) + \mathrm{E}(\mathcal{X})^2\\ &= \mathrm{E}(\mathcal{X}^2) - \mathrm{E}(\mathcal{X})^2\,.\end{aligned} \tag{2.7}$$

The variance is an expectation value of squares $\left(\mathcal{X} - E(\mathcal{X})\right)^2$, which are nonnegative by the law of multiplication, whence the variance is necessarily a nonnegative quantity, $\mathrm{var}(\mathcal{X}) \geq 0$, and the standard deviation is always real.

(ii) If $\mathcal{X}$ and $\mathcal{Y}$ are independent and have finite variances, then we obtain

$$\mathrm{var}(\mathcal{X} + \mathcal{Y}) = \mathrm{var}(\mathcal{X}) + \mathrm{var}(\mathcal{Y})\,,$$

as follows readily by simple calculation:

$$\begin{aligned}\mathrm{E}\left((\tilde{\mathcal{X}} + \tilde{\mathcal{Y}})^2\right) &= \mathrm{E}\left(\tilde{\mathcal{X}}^2 + 2\tilde{\mathcal{X}}\tilde{\mathcal{Y}} + \tilde{\mathcal{Y}}^2\right)\\ &= \mathrm{E}(\tilde{\mathcal{X}}^2) + 2\mathrm{E}(\tilde{\mathcal{X}})\mathrm{E}(\tilde{\mathcal{Y}}) + \mathrm{E}(\tilde{\mathcal{Y}}^2) = \mathrm{E}(\tilde{\mathcal{X}}^2) + \mathrm{E}(\tilde{\mathcal{Y}}^2)\,,\end{aligned}$$

where we use the fact that the first centered moments vanish, viz., $\mathrm{E}(\tilde{\mathcal{X}}) = \mathrm{E}(\tilde{\mathcal{Y}}) = 0$.

(iii) For two general, not necessarily independent, random variables $\mathcal{X}$ and $\mathcal{Y}$, the Cauchy–Schwarz inequality holds for the mixed expectation value:

$$\mathrm{E}(\mathcal{X}\mathcal{Y})^2 \le \mathrm{E}(\mathcal{X}^2)\mathrm{E}(\mathcal{Y}^2) . \tag{2.8}$$

If both random variables have finite variances, the *covariance* is defined by

$$\begin{aligned} \mathrm{cov}(\mathcal{X},\mathcal{Y}) = \mathrm{var}(\mathcal{X},\mathcal{Y}) &= \mathrm{E}\Big(\big(\mathcal{X}-\mathrm{E}(\mathcal{X})\big)\big(\mathcal{Y}-\mathrm{E}(\mathcal{Y})\big)\Big) \\ &= \mathrm{E}\big(\mathcal{X}\mathcal{Y}-\mathcal{X}\,\mathrm{E}(\mathcal{Y})-\mathrm{E}(\mathcal{X})\,\mathcal{Y}+\mathrm{E}(\mathcal{X})\,\mathrm{E}(\mathcal{Y})\big) \\ &= \mathrm{E}(\mathcal{X}\mathcal{Y})-\mathrm{E}(\mathcal{X})\,\mathrm{E}(\mathcal{Y}) . \end{aligned} \tag{2.9}$$

The covariance $\mathrm{cov}(\mathcal{X},\mathcal{Y})$ and the *coefficient of correlation* $\rho(\mathcal{X},\mathcal{Y})$,

$$\mathrm{cov}(\mathcal{X},\mathcal{Y}) = \mathrm{E}(\mathcal{X}\mathcal{Y}) - \mathrm{E}(\mathcal{X})\mathrm{E}(\mathcal{Y}) , \qquad \rho(\mathcal{X},\mathcal{Y}) = \frac{\mathrm{cov}(\mathcal{X},\mathcal{Y})}{\sigma(\mathcal{X})\sigma(\mathcal{Y})} , \tag{2.9$'$}$$

are measures of the correlation between the two variables. As a consequence of the Cauchy–Schwarz inequality, we have $-1 \le \rho(\mathcal{X},\mathcal{Y}) \le 1$. If the covariance and correlation coefficient are equal to zero, the two random variables $\mathcal{X}$ and $\mathcal{Y}$ are *uncorrelated*. Independence implies lack of correlation but is in general the stronger property (Sect. 2.3.4).

Two more quantities are used to characterize the center of probability distributions in addition to the expectation value (Fig. 2.1):

(i) The *median* $\bar{\mu}$ is the value at which the number of points or the cumulative probability distribution at lower values exactly matches the number of points or the distribution at higher values as expressed in terms of two inequalities:

$$\begin{gathered} P(\mathcal{X} \le \bar{\mu}) \ge \frac{1}{2} , \qquad P(\mathcal{X} \ge \bar{\mu}) \ge \frac{1}{2} , \\ \text{or} \\ \int_{-\infty}^{\bar{\mu}} \mathrm{d}F(x) \ge \frac{1}{2} , \qquad \int_{\bar{\mu}}^{+\infty} \mathrm{d}F(x) \ge \frac{1}{2} , \end{gathered} \tag{2.10}$$

where Lebesgue–Stieltjes integration is applied. In the case of an absolutely continuous distribution, the condition simplifies to

$$P(\mathcal{X} \le \bar{\mu}) = P(\mathcal{X} \ge \bar{\mu}) = \int_{-\infty}^{\bar{\mu}} f(x)\,\mathrm{d}x = \int_{\bar{\mu}}^{+\infty} f(x)\,\mathrm{d}x = \frac{1}{2} . \tag{2.10$'$}$$

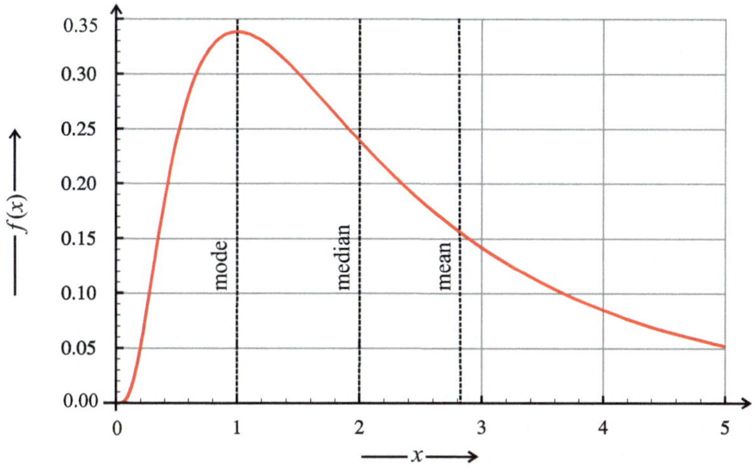

Fig. 2.1 Probability densities and moments. As an example of an asymmetric distribution with very different values for the mode, median, and mean, the log-normal density

$$f(x) = \frac{1}{\sigma x \sqrt{2\pi}} \exp\bigl(-(\ln x - \nu)^2/(2\sigma^2)\bigr)$$

is shown. Parameter values: $\sigma = \sqrt{\ln 2}$, $\nu = \ln 2$ yielding $\tilde{\mu} = \exp(\nu - \sigma^2/2) = 1$ for the mode, $\bar{\mu} = \exp \nu = 2$ for the median and $\mu = \exp(\nu + \sigma^2/2) = 2\sqrt{2}$ for the mean, respectively. The ordering mode < median < mean is characteristic for distributions with positive skewness, whereas the opposite ordering mean < median < mode is found in cases of negative skewness (see also Fig. 2.3)

(ii) The *mode* $\tilde{\mu}$ of a distribution is the most frequent value—the value that is most likely obtained through sampling—and it coincides with the maximum of the probability mass function for discrete distributions or the maximum of the probability density in the continuous case. An illustrative example for the discrete case is the probability mass function of the scores for throwing two dice, where the mode is $\tilde{\mu} = 7$ (Fig. 1.11). A probability distribution may have more than one mode. Bimodal distributions occur occasionally and then the two modes provide much more information on the expected outcomes than the mean or the median (Sect. 2.5.10).

The median and the mean are related by an inequality, which says that the difference between them is bounded by one standard deviation [365, 394]:

$$\begin{aligned} |\mu - \bar{\mu}| &= |\mathrm{E}(\mathcal{X} - \bar{\mu})| \;\le\; \mathrm{E}(|\mathcal{X} - \bar{\mu}|) \\ &\le \mathrm{E}(|\mathcal{X} - \mu|) \le \sqrt{\mathrm{E}\bigl((\mathcal{X} - \mu)^2\bigr)} = \sigma \,. \end{aligned} \tag{2.11}$$

The absolute difference between the mean and the median cannot be greater than one standard deviation of the distribution.

For many purposes a generalization of the median from two to n equally sized data sets is useful. The *quantiles* are points taken at regular intervals from the cumulative distribution function $F(x)$ of a random variable $\mathcal{X}$. Ordered data are divided into n essentially equal-sized subsets, and accordingly $(n-1)$ points on the x-axis separate the subsets. Then, the k th n-quantile is defined by $P(\mathcal{X} < x) \le k/n = p$ (Fig. 2.2), or equivalently,

$$F^{-1}(p) \doteq \inf\left\{x \in \mathbb{R} : F(x) \ge p\right\}, \qquad p = \int_{-\infty}^{x} \mathrm{d}F(u)\,. \tag{2.12}$$

When the random variable has a probability density, the integral simplifies to $p = \int_{-\infty}^{x} f(u)\,\mathrm{d}u$. The median is simply the value of x for $p = 1/2$. For partitioning into four parts we have the first or lower quartile at $p = 1/4$, the second quartile or median at $p = 1/2$, and the third or upper quartile at $p = 3/4$. The lower quartile contains 25 % of the data, the median 50%, and the upper quartile eventually 75 % of the data.

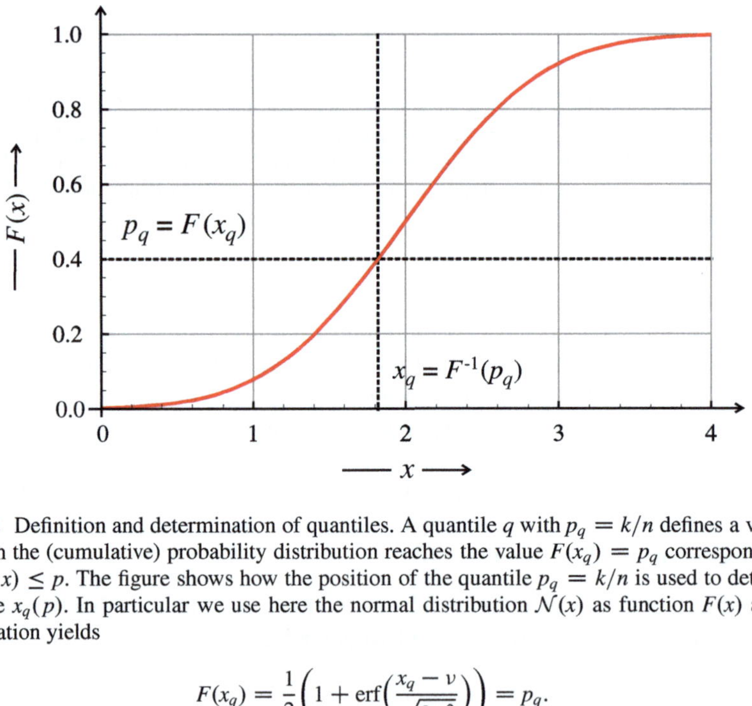

Fig. 2.2 Definition and determination of quantiles. A quantile q with $p_q = k/n$ defines a value x_q at which the (cumulative) probability distribution reaches the value $F(x_q) = p_q$ corresponding to $P(\mathcal{X} < x) \le p$. The figure shows how the position of the quantile $p_q = k/n$ is used to determine its value $x_q(p)$. In particular we use here the normal distribution $\mathcal{N}(x)$ as function $F(x)$ and the computation yields

$$F(x_q) = \frac{1}{2}\left(1 + \mathrm{erf}\left(\frac{x_q - \nu}{\sqrt{2\sigma^2}}\right)\right) = p_q.$$

Parameter choice: $\nu = 2$, $\sigma^2 = 1/2$, and for the quantile $(n = 5, k = 2)$, yielding $p_q = 2/5$ and $x_q = 1.8209$

2.1.2 Higher Moments

Two other quantities related to higher moments are frequently used for a more detailed characterization of probability distributions[3]:

(i) The *skewness*, which describes properties determined by the moments of third order:

$$\gamma_1 = \frac{\mu_3}{\mu_2^{3/2}} = \frac{\mu_3}{\sigma^3} = \frac{\mathrm{E}\left(\left(\mathcal{X} - \mathrm{E}(\mathcal{X})\right)^3\right)}{\left(\mathrm{E}\left(\left(\mathcal{X} - \mathrm{E}(\mathcal{X})\right)^2\right)\right)^{3/2}} . \tag{2.13}$$

(ii) The *kurtosis*, which is either defined as the fourth standardized moment β_2 or as *excess kurtosis* γ_2 in terms of the cumulants κ_2 and κ_4:

$$\beta_2 = \frac{\mu_4}{\mu_2^2} = \frac{\mu_4}{\sigma^4} = \frac{\mathrm{E}\left(\left(\mathcal{X} - \mathrm{E}(\mathcal{X})\right)^4\right)}{\left(\mathrm{E}\left(\left(\mathcal{X} - \mathrm{E}(\mathcal{X})\right)^2\right)\right)^{2}} ,$$

$$\gamma_2 = \frac{\kappa_4}{\kappa_2^2} = \frac{\mu_4}{\sigma^4} - 3 = \beta_2 - 3 . \tag{2.14}$$

Skewness is a measure of the asymmetry of the probability density: curves that are symmetric about the mean have zero skew, while *negative skew* implies a longer left tail of the distribution caused by more low values, and *positive skew* is characteristic for a distribution with a longer right tail. Positive skew is quite common with empirical data (see, for example the log-normal distribution in Sect. 2.5.1).

Kurtosis characterizes the degree of *peakedness* of a distribution. High kurtosis implies a sharper peak and flat tails, while low kurtosis characterizes flat or round peaks and thin tails. Distributions are said to be *leptokurtic* if they have a positive excess kurtosis and therefore a sharper peak and a thicker tail than the *normal distribution* (Sect. 2.3.3), which is taken as a reference with zero kurtosis, or they are characterized as *platykurtic* when the excess kurtosis is negative in the sense of a broader peak and thinner tails. Figure 2.3 compares the following seven distributions, standardized to $\mu = 0$ and $\sigma^2 = 1$, with respect to kurtosis:

(i) Laplace distribution: $f(x) = \frac{1}{2b} \exp\left(-\frac{|x-\mu|}{b}\right),\ b = \frac{1}{\sqrt{2}}$.

(ii) Hyperbolic secant distribution: $f(x) = \frac{1}{2} \mathrm{sech} \frac{\pi x}{2}$.

[3]In contrast to expectation value, variance and standard deviation, skewness and kurtosis are not uniquely defined and it is necessary therefore to check the author's definitions carefully when reading the literature.

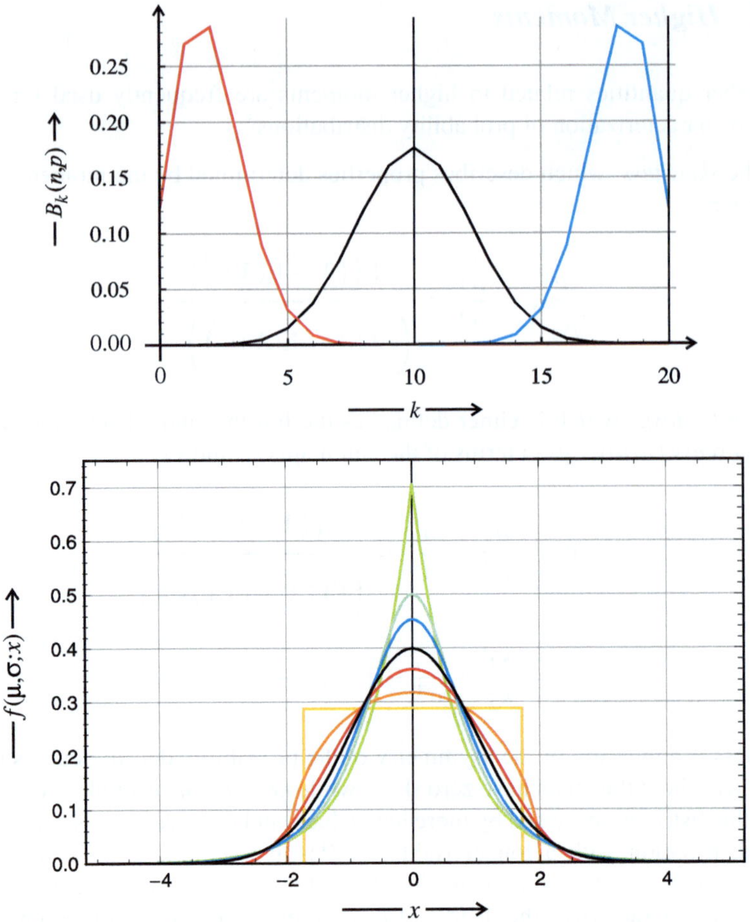

Fig. 2.3 Skewness and kurtosis. The *upper part* of the figure illustrates the sign of skewness with asymmetric density functions. The examples are taken form the binomial distribution $B_k(n,p)$: $\gamma_1 = (1-2p)/\sqrt{np\,(1-p)}$ with $p = 0.1$ (*red*), 0.5 (*black*, symmetric), and 0.9 (*blue*) with the values $\gamma_1 = 0.596, 0, -0.596$. Densities with different kurtosis are compared in the *lower part* of the figure. The Laplace distribution (*chartreuse*), the hyperbolic secant distribution (*green*), and the logistic distribution (*blue*) are leptokurtic with excess kurtosis values 3, 2, and 1.2, respectively. The normal distribution is the reference curve with zero excess kurtosis (*black*). The raised cosine distribution (*red*), the Wigner semicircle distribution (*orange*), and the uniform distribution (yellow) are platykurtic with excess kurtosis values of -0.593762, -1, and -1.2, respectively. All densities are calibrated such that $\mu = 0$ and $\sigma^2 = 1$. Recalculated and redrawn from http://en.wikipedia.org/wiki/Kurtosis, March 30, 2011

(iii) Logistic distribution: $f(x) = \dfrac{e^{-(x-\mu)/s}}{s(1+e^{-(x-\mu)/s})^2}$, $s = \sqrt{3}/\pi$.

(iv) Normal distribution: $f(x) = \dfrac{1}{\sqrt{2\pi\sigma^2}} e^{-(x-\mu)^2/2\sigma^2}$.

(v) Raised cosine distribution: $f(x) = \dfrac{1}{2s}\left(1 + \cos\dfrac{\pi(x-\mu)}{s}\right)$, $s = \dfrac{1}{\sqrt{\dfrac{1}{3} - \dfrac{2}{\pi^2}}}$.

(vi) Wigner's semicircle: $f(x) = \dfrac{2}{\pi r^2}\sqrt{r^2 - x^2}$, $r = 2$.

(vii) Uniform distribution: $f(x) = \dfrac{1}{b-a}$, $b - a = 2\sqrt{3}$.

These seven functions span the whole range of maxima from a sharp peak to a completely flat plateau, with the normal distribution chosen as the reference function (Fig. 2.3) with excess kurtosis $\gamma_2 = 0$. Distributions (i), (ii), and (iii) are leptokurtic whereas (v), (vi), and (vii) are platykurtic. It is important to note one property of skewness and kurtosis that follows from the definition: the expectation value, the standard deviation, and the variance are quantities with dimensions, whereas skewness and kurtosis are defined as dimensionless numbers.

The cumulants κ_n provide another way to expand probability distributions that has certain advantages because of its relation to generating functions discussed in Sect. 2.2. The first five cumulants κ_n $(n = 1, \ldots, 5)$ expressed in terms of the expectation value μ and the central moments μ_n $(\mu_1 = 0)$ are:

$$\kappa_1 = \mu\,, \quad \kappa_2 = \mu_2\,, \quad \kappa_3 = \mu_3\,, \quad \kappa_4 = \mu_4 - 3\mu_2^2\,, \quad \kappa_5 = \mu_5 - 10\mu_2\mu_3\,. \tag{2.15}$$

The relationships between the cumulants and the moment generating function (2.29) and the characteristic function (2.32), which is the Fourier transform of the probability density function $f(x)$, are:

$$\begin{aligned} k(s) &= \ln E\left(e^{\mathcal{X}s}\right) = \sum_{i=1}^{\infty} \kappa_n \frac{s^n}{n!}\,, \\ h(s) &= \ln \phi(s) = \sum_{n=1}^{\infty} \kappa_n \frac{(is)^n}{n!}\,, \quad \text{with } \phi(s) = \int_{-\infty}^{+\infty} \exp(isx) f(x)\,dx\,. \end{aligned} \tag{2.16}$$

The two series expansions are also called the real and the complex expansion of cumulants. We shall come back to the use of cumulants κ_n in Sects. 2.3 and 2.5 when we compare frequently used individual probability densities and in Sect. 2.6 when we apply k-statistics in order to compute empirical moments from incomplete data sets.

Finally, we mention another example of composite raw moments, the factorial moments, which will turn out to be useful in the context of probability generating functions (Sect. 2.2.1):

$$\mathrm{E}\big((\mathcal{X})_r\big) = \mathrm{E}\big(\mathcal{X}(\mathcal{X}-1)(\mathcal{X}-2)\cdots(\mathcal{X}-r+1)\big)\,, \tag{2.17}$$

where $(x)_r = x(x-1)(x-2)\cdots(x-r+1)$ is the Pochhammer symbol abbreviating the falling factorial named after the German mathematician Leo August Pochhammer.[4] If the factorial moments are known, the raw moments of the random variable $\mathcal{X}$ can be obtained from

$$\mathrm{E}(\mathcal{X}^n) = \sum_{r=0}^{n} \left\{ {n \atop r} \right\} \mathrm{E}\big((\mathcal{X})_r\big)\,, \tag{2.18}$$

where the Stirling numbers of the second kind, named after the Scottish mathematician James Stirling, are denoted by

$$S(n,k) = \left\{ {n \atop k} \right\} = \frac{1}{k!}\sum_{i=0}^{k}(-1)^{k-i}\binom{k}{i} i^n\,. \tag{2.19}$$

The factorial moments of certain distributions assume very simple expressions and can be very useful. The moments of the Poisson distribution (Sect. 2.3.1), for example, satisfy $\mathrm{E}\big((\mathcal{X})_r\big) = \alpha^r$ where α is a parameter.

---

[4]The definition of the Pochhammer symbol is ambiguous [308, p. 414]. In combinatorics, the Pochhammer symbol $(x)_n$ is used for the *falling factorial*,

$$(x)_n = x(x-1)(x-2)\cdots(x-n+1) = \frac{\Gamma(x+1)}{\Gamma(x-n+1)}\,,$$

whereas the *rising factorial* is

$$x^{(n)} = x(x+1)(x+2)\cdots(x+n-1) = \frac{\Gamma(x+n)}{\Gamma(x)}\,.$$

We also mention a useful identity between the partial factorials

$$(-x)^{(n)} = (-1)^n\,(x)_n\,.$$

In the theory of special functions in physics and chemistry, in particular in the context of the hypergeometric functions, however, $(x)_n$ is used for the rising factorial. Here, we shall use the unambiguous symbols from combinatorics and we shall say whether we mean the rising or the falling factorial. Clearly, expressions in terms of Gamma functions are unambiguous.

2.1.3 ⋆ *Information Entropy*

Information theory was developed during World War Two as the theory of communication of secret messages. No wonder that the theory was conceived and worked out at Bell Labs, and the leading figure in this area was an American cryptographer, electronic engineer, and computer scientist, Claude Elwood Shannon [497, 498]. One of the central issues of information theory is *self-information* or the *content of information*

$$I(\omega) = \mathrm{ld}\left(\frac{1}{P(\omega)}\right) = -\mathrm{ld}\, P(\omega) \tag{2.20}$$

that can be encoded, for example, in a sequence of given length. Commonly one thinks about binary sequences and therefore the information is measured in *binary digits* or *bits*.[5] The rationale behind this expression is the definition of a measure of information that is positive and additive for independent events. From (1.33), we have

$$P(AB) = P(A)P(B) \implies I(A \cap B) = I(AB) = I(A) + I(B)\,,$$

and this relation is satisfied by the logarithm. Since $P(\omega) \leq 1$ by definition, the negative logarithm is a positive quantity. Equation (2.20) yields zero information for an event taking place with certainty, i.e., $P(\omega) = 1$. The outcome of the fair coin toss with $P(\omega) = 1/2$ provides 1 bit of information, and rolling two sixes with two dice in one throw has a probability $P(\omega) = 1/36$ and yields 5.17 bits. For a modern treatise on information theory and entropy, see [220].

Countable Sample Space

In order to measure the information content of a probability distribution, Claude Shannon introduced the information entropy, which is simply the expectation value of the information content, represented by a function that resembles the expression for the thermodynamic entropy in statistical mechanics. We consider first the discrete case of a probability mass function $p_k = P(\mathcal{X} = x_k)$, $k \in \mathbb{N}_{>0}$, $k \leq n$:

$$H\big(f(p)\big) = H\big(\{p_k\}\big) = -\sum_{k=1}^{n} p_k \log p_k\,, \quad \text{with } p_k \geq 0\,, \quad \sum_{k=1}^{n} p_k = 1\,. \tag{2.21}$$

---

[5]The logarithm is taken to the base 2 and it is commonly called *binary logarithm* or *logarithmus dualis*, $\log_2 \equiv \mathrm{lb} \equiv \mathrm{ld}$, with the dimensionless unit 1 binary digit (bit). The conventional unit of information in informatics is the *byte*: 1 byte (B) = 8 bits being tantamount to the coding capacity of an eight digit binary sequence. Although there is little chance of confusion, one should be aware that in the International System of Units, B is the abbreviation for the acoustical unit 'bel', which is the unit for measuring the signal strength of sound.

For short we also write $H(p)$, where p stands for the pmf of the distribution. Thus, the entropy can be visualized as the expectation value of the negative logarithm of the probabilities, viz.,

$$H(p) = \mathrm{E}(-\log p_k) = \mathrm{E}\left(\log \frac{1}{p_k}\right),$$

where the term $\log(1/p_k)$ can be viewed as the number of bits to be assigned to the point x_k, provided the binary logarithm $\log = \log_2 \equiv \mathrm{ld}$ is used.

The functional relationship $H(x) = -x \log x$ on the interval $0 \le x \le 1$ underlying the information entropy is a concave function (Fig. 2.4). It is easily seen that the entropy of a discrete probability distribution is always nonnegative. This conjecture can be checked, for example, by considering the two extreme cases:

(i) There is almost certainly only one outcome, $p_1 = P(\mathcal{X} = x_1) = 1$ and $p_j = P(\mathcal{X} = x_j) = 0 \ \forall \ j \in \mathbb{N}_{>0}, j \neq 1$, and then the information entropy fulfils $H = 0$ in this completely determined case.
(ii) All events have the same probability, whence we are dealing with the uniform distribution $p_k = P(\mathcal{X} = x_k) = 1/n$, or a case of the principle of indifference. The entropy is then positive and takes on its maximum value $H(p) = \log n$.

The entropies of all other discrete distributions lie in-between:

$$0 \le H(p) \le \log n . \tag{2.22}$$

The value of the entropy is a measure of the lack of information on the distribution. Case (i) is deterministic and we have full information on the outcome *a priori*,

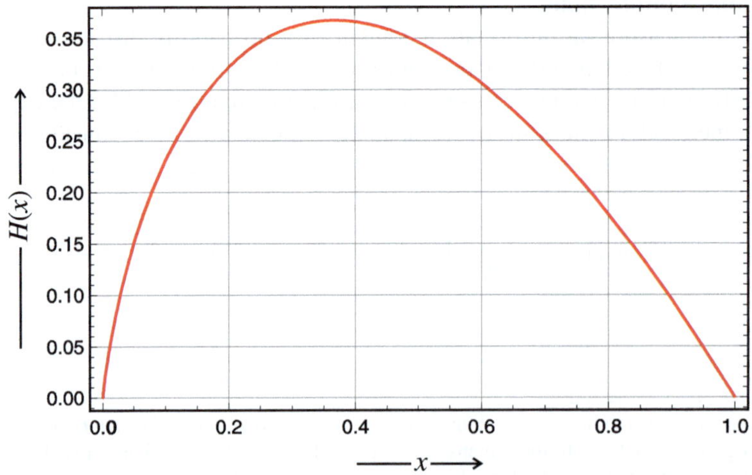

Fig. 2.4 The functional relation of information entropy. The plot shows the function $H = -x \ln x$ in the range $0 \le x \le 1$. For $x = 0$, we apply the probability theory convention $-0 \ln 0 = 0 \times \infty = 0$

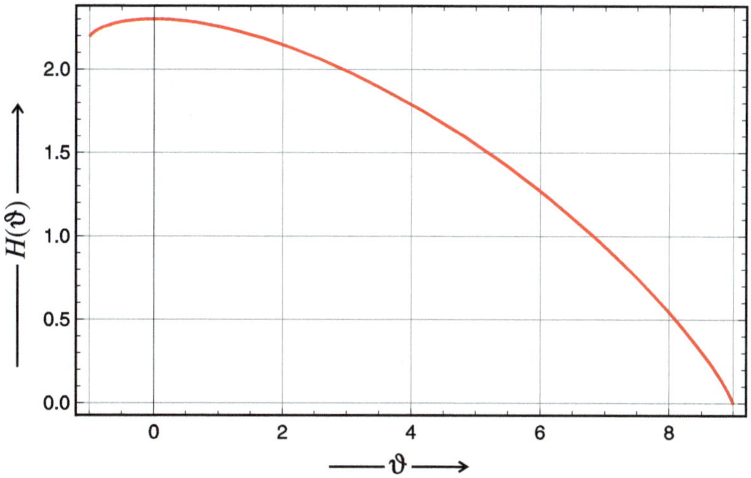

Fig. 2.5 Maximum information entropy. The discrete probability distribution with maximal information entropy is the uniform distribution $\mathcal{U}_p$. The entropy of the probability distribution $p_1 = \frac{1+\vartheta}{n}$ and $p_j = \frac{1}{n}\left(1 - \frac{\vartheta}{n-1}\right)$, $\forall j = 2, 3, \ldots, n$ with $n = 10$ is plotted against the parameter ϑ. All probabilities p_k are defined and the entropy $H(\vartheta)$ is real and nonnegative on the interval $-1 \leq \vartheta \leq 9$ and has a maximum at $\vartheta = 0$

whereas case (ii) provides maximal uncertainty because all outcomes have the same probability. A rigorous proof that the uniform distribution has maximum information entropy among all discrete distributions can be found in the literature [86, 90]. We dispense from reproducing the proof here but illustrate by means of Fig. 2.5. The starting point is the uniform distribution of n events with a probability of $p = 1/n$ for each one, and then we attribute a different probability to a single event:

$$p_1 = \frac{1+\vartheta}{n}\,, \qquad p_j = \frac{1}{n}\left(1 - \frac{\vartheta}{n-1}\right), \quad j = 2, 3, \ldots, n\,.$$

The entropy of the distribution is considered as a function of ϑ, and indeed the maximum occurs at $\vartheta = 0$.

Uncountable Measurable Sample Space

The information entropy of a continuous probability density $p(x)$ with $x \in \mathbb{R}$ is calculated by integration:

$$H\big(f(x)\big) = -\int_{-\infty}^{+\infty} p(x) \log p(x)\, \mathrm{d}x\,, \quad p_k \geq 0\,, \quad \int_{-\infty}^{+\infty} p(x)\, \mathrm{d}x = 1\,. \qquad (2.21')$$

As in the discrete case we can write the entropy as an expectation value of $\log(1/p)$:

$$H(p) = \mathrm{E}\big(-\log p(x)\big) = \mathrm{E}\left(\log \frac{1}{p(x)}\right) .$$

We consider two specific examples representing distributions with maximum entropy: (i) the exponential distribution (Sect. 2.5.4) on $\Omega = \mathbb{R}_{\geq 0}$ with the density

$$f_{\mathrm{exp}}(x) = \frac{1}{\mu} \mathrm{e}^{-x/\mu} ,$$

the mean μ, and the variance μ^2, and (ii) the normal distribution (Sect. 2.3.3) on $\Omega = \mathbb{R}$ with the density

$$f_{\mathcal{N}}(x) = \frac{1}{\sqrt{2\pi\sigma^2}} \mathrm{e}^{-(x-\mu)^2/2\sigma^2} ,$$

the mean μ, and the variance σ^2.

In the discrete case we made a seemingly unconstrained search for the distribution of maximum entropy, although the discrete uniform distribution contained the number of sample points n as input restriction and n does indeed appear as parameter in the analytical expression for the entropy (Table 2.1). Now, in the continuous case the constraints become more evident, since we shall use fixed mean (μ) or fixed mean and variance (μ, σ^2) as the basis of comparison in the search for distributions with maximum entropy.

The entropy of the exponential density on the sample space $\Omega = \mathbb{R}_{\geq 0}$ with mean μ and variance μ^2 is calculated to be

$$H(f_{\mathrm{exp}}) = -\int_0^\infty \frac{1}{\mu} \mathrm{e}^{-x/\mu} \left(-\log \mu - \frac{x}{\mu}\right) \mathrm{d}x = 1 + \log \mu . \tag{2.23}$$

In contrast to the discrete case the entropy of the exponential probability density can become negative for small μ values, as can be easily visualized by considering

Table 2.1 Probability distributions with maximum information entropy

Distribution	Space Ω	Density	Mean	Var	Entropy
Uniform	$\mathbb{N}_{>0}$	$\frac{1}{n}, \ \forall\, k = 1, \ldots, n$	$\frac{n+1}{2}$	$\frac{n^2-1}{12}$	$\log n$
Exponential	$\mathbb{R}_{\geq 0}$	$\frac{1}{\mu} \mathrm{e}^{-x/\mu}$	μ	μ^2	$1 + \log \mu$
Normal	$\mathbb{R}$	$\frac{1}{\sqrt{2\pi\sigma^2}} \mathrm{e}^{-(x-\mu^2)/2\sigma^2}$	μ	σ^2	$\frac{1}{2}(1 + \log(2\pi\sigma^2))$

The table compares three probability distributions with maximum entropy: (i) the discrete uniform distribution on the support $\Omega = \{1 \leq k \leq n, k \in \mathbb{N}\}$, (ii) the exponential distribution on $\Omega = \mathbb{R}_{\geq 0}$, and (iii) the normal distribution on $\Omega = \mathbb{R}$

the shape of the density. Since $\lim_{x \to 0} f_{\exp}(x) = 1/\mu$, an appreciable fraction of the density function adopts values $f_{\exp}(x) > 1$ for sufficiently small μ and then $-p \log p < 0$ is negative. Among all continuous probability distributions with mean $\mu > 0$ on the support $\mathbb{R}_{\geq 0} = [0, \infty[$, the exponential distribution has the maximum entropy. Proofs for this conjecture are available in the literature [86, 90, 438].

For the normal density, (2.21′) implies

$$H(f_{\mathcal{N}}) = -\int_{-\infty}^{+\infty} \frac{1}{\sqrt{2\pi\sigma^2}} \mathrm{e}^{-(x-\mu)^2/2\sigma^2} \left(-\log(\sqrt{2\pi\sigma^2}) - \frac{1}{2}\left(\frac{x-\mu}{\sigma}\right)^2 \right) \mathrm{d}x$$
$$= \frac{1}{2}\Big(1 + \log(2\pi\sigma^2)\Big) . \tag{2.24}$$

It is not unexpected that the information entropy of the normal distribution should be independent of the mean μ, which causes nothing but a shift of the whole distribution along the x-axis: all Gaussian densities with the same variance σ^2 have the same entropy. Once again we see that the entropy of the normal probability density can become negative for sufficiently small values of σ^2. The normal distribution is distinguished among all continuous distributions on $\Omega = \mathbb{R}$ with given variance σ^2 since it is the distribution with maximum entropy. Several proofs of this theorem have been devised. We refer again to the literature [86, 90, 438]. The three distributions with maximum entropy are compared in Table 2.1.

Principle of Maximum Entropy

The information entropy can be interpreted as the required amount of information we would need in order to fully describe the system. Equations (2.21) and (2.21′) are the basis of a search for probability distribution with maximum entropy under certain constraints, e.g., constant mean μ or constant variance σ^2. The maximum entropy principle was introduced by the American physicist Edwin Thompson Jaynes as a method of statistical inference [279, 280]. He suggested using those probability distributions which satisfy the prescribed constraints and have the largest entropy. The rationale for this choice is to use a probability distribution that reflects our knowledge and does not contain any unwarranted information. The predictions made on the basis of a probability distribution with maximum entropy should be least surprising. If we chose a distribution with smaller entropy, this distribution would contain more information than justified by our *a priori* understanding of the problem. It is useful to illustrate a typical strategy [86]:

> [. . .] the principle of maximum entropy guides us to the best probability distribution that reflects our current knowledge and it tells us what to do if experimental data do not agree with predictions coming from our chosen distribution: understand why the phenomenon being studied behaves in an unexpected way, find a previously unseen constraint, and maximize the entropy over the distributions that satisfy all constraints we are now aware of, including the new one.

We realize a different way of thinking about probability that becomes even more evident in Bayesian statistics, which is sketched in Sects. 1.3 and 2.6.5.

The choice of the word *entropy* for the expected information content of a distribution is not accidental. Ludwig Boltzmann's statistical formula is[6]

$$S = k_B \ln W \,, \quad \text{with } W = \frac{N!}{N_1!N_2!\cdots N_m!} \,, \tag{2.25}$$

where W is the so-called thermodynamic probability, k_B is Boltzmann's constant, $k_B = 1.38065 \times 10^{-23}\,\mathrm{J\,K^{-1}}$, and $N = \sum_{j=1}^{m} N_j$ is the total number of particles, distributed over m states with the frequencies $p_k = N_k/N$ and $\sum_{j=1}^{m} p_j = 1$. The number of particles N is commonly very large and we can apply Stirling's formula $n! \approx n \ln n$, named after the Scottish mathematician James Stirling. This leads to

$$\begin{aligned} S = k_B \left(N \ln N - \sum_{i=1}^{m} N_i \ln N_i \right) &= -k_B N \left(-\ln N + \sum_{i=1}^{m} \frac{N_i}{N} \ln N_i \right) \\ &= -k_B N \sum_{i=1}^{m} p_i \ln p_i \,. \end{aligned}$$

For the entropy per particle we obtain

$$s = \frac{S}{N} = -k_B \sum_{i=1}^{m} p_i \ln p_i \,, \tag{2.25'}$$

which is identical with Shannon's formula (2.21), except for the factor containing the universal constant k_B.

Eventually, we shall point out important differences between thermodynamic entropy and information entropy that should be kept in mind when discussing analogies between them. The thermodynamic principle of maximum entropy is a physical law known as the second law of thermodynamics: the entropy of an isolated system[7] is non-decreasing in general and increasing whenever processes are taking place, in which case it approaches a maximum. The principle of maximum entropy in statistics is a rule for appropriate design of distribution functions and should be considered as a guideline and not a natural law. Thermodynamic entropy is an *extensive property* and this means that it increases with the size of the system. Information entropy, on the other hand, is an *intensive property* and insensitive

[6]Two remarks are worth noting: (2.25) is Max Planck's expression for the entropy in statistical mechanics, although it has been carved on Boltzmann's tombstone, and W is called a probability despite the fact that it is not normalized, i.e., $W \geq 1$.

[7]An isolated system exchanges neither matter nor energy with its environment. For isolated, closed, and open systems, see also Sect. 4.3.

to size. The difference has been exemplified by the Russian biophysicist Mikhail Vladimirovich Volkenshtein [554]: considering the process of flipping a coin in reality and calculating all contributions to the process shows that the information entropy constitutes only a minute contribution to the thermodynamic entropy. The change in the total thermodynamic entropy that results from the coin-flipping process is dominated by far by the metabolic contributions of the flipping individual, involving muscle contractions and joint rotations, and by heat production on the surface where the coin lands, etc. Imagine the thermodynamic entropy production if you flip a coin two meters in diameter—the gain in information is still one bit, just as it would be for a small coin!

2.2 Generating Functions

In this section we introduce auxiliary functions, which are compact representations of probability distributions and which provide convenient tools for handling functions of probabilities. The generating functions commonly contain one or more auxiliary variables—here denoted by s—that have no direct physical meaning but enable straightforward calculation of functions of random variables at certain values of s. In particular we shall introduce the probability generating functions $g(s)$, the moment generating functions $M(s)$, and the characteristic functions $\phi(s)$. A characteristic function $\phi(s)$ exists for every distribution, but we shall encounter cases where no probability or moment generating functions exist (see, for example, the Cauchy–Lorentz distribution in Sect. 2.5.7). In addition to these three generating functions several other generating functions are also used. One example is the *cumulant generating function*, which lacks a uniform definition. It is either the logarithm of the moment generating function or the logarithm of the characteristic function—we shall mention both.

2.2.1 *Probability Generating Functions*

Let $\mathcal{X}$ be a random variable taking only nonnegative integer values with a probability distribution given by

$$P(\mathcal{X} = j) = a_j \, , \quad j = 0, 1, 2, \ldots \, . \tag{2.26}$$

An auxiliary variable s is introduced and the *probability generating function* is expressed by the infinite power series

$$g(s) = a_0 + a_1 s + a_2 s^2 + \cdots = \sum_{j=0}^{\infty} a_j s^j = \mathrm{E}(s^{\mathcal{X}}) \, . \tag{2.27}$$

As we shall show later, the full information on the probability distribution is encapsulated in the coefficients a_j ($j \in \mathbb{N}$). Intuitively, this is no surprise since the coefficients a_j are the individual probabilities of a probability mass function in (1.27′): $a_j = p_j$. The expression for the probability generation function as an expectation value is useful in the comparison with other generating functions.

In most cases, s is a real-valued variable, although it can be of advantage to consider also complex s. Recalling $\sum_j a_j = 1$ from (2.26), we can easily check that the power series (2.27) converges for $|s| \le 1$:

$$|g(s)| \le \sum_{j=0}^{\infty} |a_j|\,|s|^j \le \sum_{j=0}^{\infty} a_j = 1\,, \quad \text{for } |s| \le 1\,.$$

The *radius of convergence* of the series (2.27) determines the meaningful range of the auxiliary variable: $0 \le |s| \le 1$.

For $|s| \le 1$, we can differentiate[8] the series term by term in order to calculate the derivatives of the generating function $g(s)$:

$$\frac{\mathrm{d}g}{\mathrm{d}s} = g'(s) = a_1 + 2a_2 s + 3a_3 s^2 + \cdots = \sum_{n=1}^{\infty} n a_n s^{n-1}\,,$$

$$\frac{\mathrm{d}^2 g}{\mathrm{d}s^2} = g''(s) = 2a_2 + 6a_3 s + \cdots = \sum_{n=2}^{\infty} n(n-1) a_n s^{n-2}\,,$$

and, in general, we have

$$\begin{aligned}\frac{\mathrm{d}^j g}{\mathrm{d}s^j} = g^{(j)}(s) &= \sum_{n=j}^{\infty} n(n-1)\cdots(n-j+1) a_n s^{n-j} \\ &= \sum_{n=j}^{\infty} (n)_j a_n s^{n-j} = \sum_{n=j}^{\infty} \binom{n}{j} j!\, a_n s^{n-j}\,,\end{aligned}$$

where $(x)_n \equiv (x-n+1)^{(n)}$ stands for the falling Pochhammer symbol. Setting $s = 0$, all terms vanish except the constant term:

$$\left.\frac{\mathrm{d}^j g}{\mathrm{d}s^j}\right|_{s=0} = g^{(j)}(0) = j!\, a_j\,, \quad \text{or } a_j = \frac{1}{j!} g^{(j)}(0)\,.$$

[8]Since we shall often need the derivatives in this section, we shall use the shorthand notations $\mathrm{d}g(s)/\mathrm{d}s = g'(s)$, $\mathrm{d}^2 g(s)/\mathrm{d}s^2 = g''(s)$, and $\mathrm{d}^j g(s)/\mathrm{d}s^j = g^{(j)}(s)$, and for simplicity also $(\mathrm{d}g/\mathrm{d}s)|_{s=k} = g'(k)$ and $(\mathrm{d}^2 g/\mathrm{d}s^2)|_{s=k} = g''(k)$ ($k \in \mathbb{N}$).

In this way all the a_j may be obtained by consecutive differentiation from the generating function, and alternatively the generating function can be determined from the known probability distribution.

Setting $s = 1$ in $g'(s)$ and $g''(s)$, we can compute the first and second moments of the distribution of $\mathcal{X}$:

$$g'(1) = \sum_{n=0}^{\infty} n a_n = \mathrm{E}(\mathcal{X}) \,,$$

$$g''(1) = \sum_{n=0}^{\infty} n^2 a_n - \sum_{n=0}^{\infty} n a_n = \mathrm{E}(\mathcal{X}^2) - \mathrm{E}(\mathcal{X}) \,, \tag{2.28}$$

$$\mathrm{E}(\mathcal{X}) = g'(1) \,,$$

$$\mathrm{E}(\mathcal{X}^2) = g'(1) + g''(1) \,, \quad \mathrm{var}(\mathcal{X}) = g'(1) + g''(1) - g'(1)^2 \,.$$

To sum up, the probability distribution of a nonnegative integer-valued random variable can be converted into a generating function without losing information. The generating function is uniquely determined by the distribution and *vice versa*.

2.2.2 *Moment Generating Functions*

The basis of the moment generating function is the series expansion of the exponential function of the random variable $\mathcal{X}$:

$$\mathrm{e}^{\mathcal{X}s} = 1 + \mathcal{X}s + \frac{\mathcal{X}^2}{2!}s^2 + \frac{\mathcal{X}^3}{3!}s^3 + \cdots \,.$$

The *moment generating function* (mgf) allows for direct computation of the moments of a probability distribution as defined in (2.26), since we have

$$M_{\mathcal{X}}(s) = \mathrm{E}(\mathrm{e}^{\mathcal{X}s}) = 1 + \hat{\mu}_1 s + \frac{\hat{\mu}_2}{2!}s^2 + \frac{\hat{\mu}_3}{3!}s^3 + \cdots = 1 + \sum_{n=1}^{\infty} \hat{\mu}_n \frac{s^n}{n!} \,, \tag{2.29}$$

where $\hat{\mu}_i$ is the i th raw moment. The moments can be obtained by differentiating $M_{\mathcal{X}}(s)$ with respect to s and then setting $s = 0$. From the n th derivative, we obtain

$$\mathrm{E}(\mathcal{X}^n) = \hat{\mu}_n = M_{\mathcal{X}}^{(n)} = \left.\frac{\mathrm{d}^n M_{\mathcal{X}}}{\mathrm{d}s^n}\right|_{s=0} \,.$$

A probability distribution thus has (at least) as many moments as the number of times that the moment generating function can be continuously differentiated (see also the characteristic function in Sect. 2.2.3). If two distributions have the same moment generating functions, they are identical at all points:

$$M_{\mathcal{X}}(s) = M_{\mathcal{Y}}(s) \iff F_{\mathcal{X}}(x) = F_{\mathcal{Y}}(x) .$$

However, this statement does not imply that two distributions are identical when they have the same moments, because in some cases the moments exist but the moment generating function does not, since the limit $\lim_{n\to\infty}\sum_{k=0}^{n}\hat{\mu}_k s^k/k!$ diverges, as with the log-normal distribution.

Cumulant Generating Function

The real cumulant generating function is the formal logarithm of the moment generating function that can be expanded in a power series

$$\begin{aligned} k(s) = \ln \mathrm{E}\big(\mathrm{e}^{\mathcal{X}s}\big) &= -\sum_{n=1}^{\infty}\frac{1}{n}\Big(1-\mathrm{E}\big(\mathrm{e}^{\mathcal{X}s}\big)\Big)^n \\ &= -\sum_{n=1}^{\infty}\frac{1}{n}\left(-\sum_{m=1}^{\infty}\hat{\mu}_m\frac{s^m}{m!}\right)^n \\ &= \hat{\mu}_1 s + \big(\hat{\mu}_2-\hat{\mu}_1^2\big)\frac{s^2}{2!} + \big(\hat{\mu}_3 - 3\hat{\mu}_2\hat{\mu}_1 + 2\hat{\mu}_1^3\big)\frac{s^3}{3!} + \cdots . \end{aligned} \tag{2.30}$$

The cumulants κ_n are obtained from the cumulant generating function by differentiating $k(s)$ a total of n times and calculating the derivative at $s = 0$:

$$\begin{aligned} \kappa_1 &= \frac{\partial k(s)}{\partial s}\bigg|_{s=0} = \hat{\mu}_1 = \mu , \\ \kappa_2 &= \frac{\partial^2 k(s)}{\partial s^2}\bigg|_{s=0} = \hat{\mu}_2 - \mu^2 = \sigma^2 , \\ \kappa_3 &= \frac{\partial^3 k(s)}{\partial s^3}\bigg|_{s=0} = \hat{\mu}_3 - 3\hat{\mu}_2\mu + 2\mu^3 = \mu_3 , \\ &\ \vdots \\ \kappa_n &= \frac{\partial^n k(s)}{\partial s^n}\bigg|_{s=0} , \\ &\ \vdots \end{aligned} \tag{2.15'}$$

As shown in (2.15), the first three cumulants coincide with the centered moments μ_1, μ_2, and μ_3. All higher cumulants are polynomials of two or more centered moments.

In probability theory, the Laplace transform[9]

$$\hat{f}(s) = \int_0^{\infty} \mathrm{e}^{-sx} f_{\mathcal{X}}(x)\, \mathrm{d}x = \mathcal{L}\big(f_{\mathcal{X}}(x)\big)(s) \tag{2.31}$$

can be visualized as an expectation value that is closely related to the moment generating function: $\mathcal{L}\big(f_{\mathcal{X}}(x)\big)(s) = \mathrm{E}\big(\mathrm{e}^{-s\mathcal{X}}\big)$, where $f_{\mathcal{X}}(x)$ is the probability density. The cumulative distribution function $F_{\mathcal{X}}(x)$ can be recovered by means of the inverse Laplace transform:

$$F_{\mathcal{X}}(x) = \mathcal{L}_s^{-1}\left(\frac{\mathrm{E}\big(\mathrm{e}^{-s\mathcal{X}}\big)}{s}\right)(x) = \mathcal{L}_s^{-1}\left(\frac{\mathcal{L}\big(f_{\mathcal{X}}(x)\big)(s)}{s}\right)(x)\,.$$

We shall not use the Laplace transform here as a pendant to the moment generating function, but we shall apply it in Sect. 4.3.4 to the solution of chemical master equations, where the inverse Laplace transform is also discussed.

2.2.3 Characteristic Functions

Like the moment generating function the *characteristic function* (cf) of a random variable $\mathcal{X}$, denoted by $\phi(s)$, completely describes the cumulative probability distribution $F(x)$. It is defined by

$$\phi(s) = \int_{-\infty}^{+\infty} \exp(\mathrm{i}sx)\, \mathrm{d}F(x) = \int_{-\infty}^{+\infty} \exp(\mathrm{i}sx) f(x)\, \mathrm{d}x\,, \tag{2.32}$$

where the integral over $\mathrm{d}F(x)$ is of Riemann–Stieltjes type. When a probability density $f(x)$ exists for the random variable $\mathcal{X}$, the characteristic function is (almost)

[9]We remark that the same symbol s is used for the Laplace transformed variable and the dummy variable of probability generating functions (Sect. 2.2) in order to be consistent with the literature. We shall point out the difference wherever confusion is possible.

the Fourier transform of the density[10]:

$$\mathcal{F}\big(f(x)\big) = \tilde{f}(k) = \frac{1}{\sqrt{2\pi}} \int_{-\infty}^{+\infty} f(x) \mathrm{e}^{\mathrm{i}kx} \, \mathrm{d}x \,. \tag{2.33}$$

Equation (2.32) implies the following useful expression for the expansion in the discrete case:

$$\phi(s) = \mathrm{E}\big(\mathrm{e}^{\mathrm{i}s\mathcal{X}}\big) = \sum_{n=-\infty}^{\infty} P_n \mathrm{e}^{\mathrm{i}ns} \,, \tag{2.32$'$}$$

which we shall use, for example, to solve master equations for stochastic processes (Chaps. 3 and 4). For more details on characteristic functions, see, e.g., [359, 360].

The characteristic function exists for all random variables since it is an integral of a bounded continuous function over a space of finite measure. There is a bijection between distribution functions and characteristic functions:

$$\phi_{\mathcal{X}}(s) = \phi_{\mathcal{Y}}(s) \iff F_{\mathcal{X}}(x) = F_{\mathcal{Y}}(x) \,.$$

If a random variable $\mathcal{X}$ has moments up to k th order, then the characteristic function $\phi(x)$ is k times continuously differentiable on the entire real line, and vice versa, if a characteristic function $\phi(x)$ has a k th derivative at zero, then the random variable $\mathcal{X}$ has all moments up to k if k is even and up to $k-1$ if k is odd:

$$\mathrm{E}(\mathcal{X}^k) = (-\mathrm{i})^k \frac{\mathrm{d}^k\phi(s)}{\mathrm{d}s^k}\bigg|_{s=0} \quad \text{and} \quad \frac{\mathrm{d}^k\phi(s)}{\mathrm{d}s^k}\bigg|_{s=0} = \mathrm{i}^k\mathrm{E}(\mathcal{X}^k) \,. \tag{2.34}$$

An interesting example is the Cauchy distribution (see Sect. 2.5.7) with $\phi(s) = \exp|s|$: it is not differentiable at $s = 0$ and the distribution has no moments, not even the expectation value.

The moment generating function is related to the probability generating function $g(s)$ (Sect. 2.2.1) and the characteristic function $\phi(s)$ (Sect. 2.2.3) by

$$g\left(\mathrm{e}^s\right) = \mathrm{E}\left(\mathrm{e}^{\mathcal{X}s}\right) = M_{\mathcal{X}}(s) \ \text{ and } \ \phi(s) = M_{\mathrm{i}\mathcal{X}}(s) = M_{\mathcal{X}}(\mathrm{i}s) \,.$$

---

[10]The difference between the Fourier transform $\tilde{f}(k)$ and the characteristic function $\phi(s)$ of a function $f(x)$, viz.,

$$\tilde{f}(k) = \frac{1}{\sqrt{2\pi}} \int_{-\infty}^{+\infty} f(x) \exp(+\mathrm{i}kx) \, \mathrm{d}x \quad \text{and} \quad \phi(s) = \int_{-\infty}^{\infty} f(x) \exp(\mathrm{i}sx) \, \mathrm{d}x \,,$$

is only a matter of the factor $(\sqrt{2\pi})^{-1}$. The Fourier convention used here is the same as the one in modern physics. For other conventions, see, e.g., [568] and Sect. 3.1.6.

The three generating functions are closely related, as seen by comparing the expressions as expectation values:

$$g(s) = \mathrm{E}(s^{\mathcal{X}}) \,, \quad M_{\mathcal{X}} = \mathrm{E}(\mathrm{e}^{s\mathcal{X}}) \,, \quad \text{and } \phi(s) = \mathrm{E}(\mathrm{e}^{\mathrm{i}s\mathcal{X}}) \,,$$

but it may happen that not all three actually exist. As mentioned, characteristic functions exist for all probability distributions.

The cumulant generating function was formulated as the logarithm of the moment generating function in the last section. It can be written equally well as the logarithm of the characteristic function [514, p. 84 ff]:

$$h(s) = \ln \phi(s) = \sum_{n=1}^{\infty} \kappa_n \frac{(\mathrm{i}s)^n}{n!} \,. \tag{2.16$'$}$$

It might seem a certain advantage that $E\big(\mathrm{e}^{\mathrm{i}s\mathcal{X}}\big)$ is well defined for all values of s, even when $E\big(\mathrm{e}^{s\mathcal{X}}\big)$ is not. Although $h(s)$ is well defined, the MacLaurin series[11] need not exist for higher orders in the argument s. The Cauchy distribution (Sect. 2.5.7) is an example where not even the linear term exists.

2.3 Common Probability Distributions

After a comparative overview of the important characteristics of the most frequently used distributions in Table 2.2, we enter the discussion of individual probability distributions. We begin in this section by analyzing Poisson, binomial, and normal distributions, along with the transformations between them. The central limit theorem and the law of large numbers are presented in separate sections, following the analysis of multivariate normal distributions. In Sect. 2.5, we have also listed several less common but nevertheless frequently used probability distributions, which are of importance for special purposes. We shall make use of them in Chaps. 3, 4, and 5, which deal with stochastic processes and applications.

Table 2.2 compares probability mass functions or densities, cumulative distributions, moments up to order four, and the moment generating functions and characteristic functions for several common probability distributions. The Poisson

[11]The Taylor series $f(s) = \sum_{n=0}^{\infty} \frac{f^{(n)}(a)}{n!}(s-a)^n$ is named after the English mathematician Brook Taylor who invented the calculus of finite differences in 1715. Earlier series expansions were already in use in the seventeenth century. The MacLaurin series, in particular, is a Taylor expansion centered around the origin $a = 0$, named after the eighteenth century Scottish mathematician Colin MacLaurin.

Table 2.2 Comparison of several common probability densities

Name	Parameters	Support	pmf/pdf	cdf	Mean	Median	Mode	Variance	Skewness	Kurtosis	mgf	cf
Poisson $\pi(\alpha)$	$\alpha > 0 \in \mathbb{R}$	$k \in \mathbb{N}$	$\frac{\alpha^k}{k!}e^{-\alpha}$	$Q(k+1,\alpha) = \frac{\Gamma(k+1,\alpha)}{k!}$	α	$\alpha - \ln 2 \leq \bar{\mu} \leq \alpha + \frac{1}{3}$	$\lceil\alpha\rceil - 1$	α	$\frac{1}{\sqrt{\alpha}}$	$\frac{1}{\alpha}$	$\exp\big(\alpha(e^s - 1)\big)$	$\exp\big(\alpha(e^{is} - 1)\big)$
Binomial $B(n,p)$	$n \in \mathbb{N}$, $p \in [0,1]$	$k \in \mathbb{N}$, $p \in [0,1]$	$\binom{n}{k}p^k(1-p)^{n-k}$	$I_{1-p} = (n-k, 1+k)$	np	$\lfloor np \rfloor \leq \bar{\mu} \leq \lceil np \rceil$	$\lfloor (n+1)p \rfloor$ or $\lfloor (n+1)p - 1 \rfloor$	$np(1-p)$	$\frac{1-2p}{\sqrt{np(1-p)}}$	$\frac{1-6p(1-p)}{np(1-p)}$	$(1-p+p^s)^n$	$(1-p+p^{is})^n$
Normal $\varphi(\nu,\sigma)$	$\nu \in \mathbb{R}$, $\sigma^2 \in \mathbb{R}_{>0}$	$x \in \mathbb{R}$	$\frac{1}{\sqrt{2\pi\sigma^2}}e^{-\frac{(x-\nu)^2}{2\sigma^2}}$	$\frac{1}{2}\left(1+\mathrm{erf}\left(\frac{x-\nu}{\sqrt{2\sigma^2}}\right)\right)$	ν	ν	ν	σ^2	0	0	$\exp\left(\nu s + \frac{1}{2}\sigma^2 s^2\right)$	$\exp\left(i\nu s - \frac{1}{2}\sigma^2 s^2\right)$
Chi-square $\chi^2(k)$	$k \in \mathbb{N}$	$x \in [0, \infty[$	$\frac{x^{\frac{k}{2}-1}e^{-\frac{x}{2}}}{2^{\frac{k}{2}}\Gamma\left(\frac{k}{2}\right)}$	$\frac{\gamma\left(\frac{k}{2},\frac{x}{2}\right)}{\Gamma\left(\frac{k}{2}\right)}$	k	$\bar{\mu} \approx k\left(1-\frac{2}{9k}\right)^3$	$\max\{k-2, 0\}$	$2k$	$\sqrt{\frac{8}{k}}$	$\frac{12}{k}$	$(1-2s)^{-\frac{k}{2}}$ for $s < \frac{1}{2}$	$(1-2is)^{-\frac{k}{2}}$
Logistic	$a \in \mathbb{R}, b > 0$	$x \in \mathbb{R}$	$\frac{\mathrm{sech}^2\left((x-a)/2b\right)}{4b}$	$\frac{1}{1+\exp\left(-(x-a)/b\right)}$	a	a	a	$\pi^2 b^2/3$	0	4.2	$\frac{\pi b s e^a s}{\sin(\pi b s)}$	$\frac{i\pi b s e^a s}{\sin(i\pi b s)}$
Laplace	$\nu \in \mathbb{R}$, $b > 0$	$x \in \mathbb{R}$	$\frac{1}{2b}e^{-\frac{\lvert x-\nu\rvert}{b}}$	$\begin{cases} \frac{1}{2}e^{-\frac{\nu-x}{b}}, & x < a \\ 1-\frac{1}{2}e^{-\frac{x-\nu}{b}}, & x \geq a \end{cases}$	ν	ν	ν	$2b^2$	0	3	$\frac{\exp(\nu s)}{1-b^2 s^2}$ for $\lvert s\rvert < \frac{1}{b}$	$\frac{\exp(i\nu s)}{1-b^2 s^2}$
Uniform $\mathcal{U}(a,b)$	$a < b$, $a, b \in \mathbb{R}$	$x \in [a,b]$	$\begin{cases} \frac{1}{b-a}, & x \in [a,b] \\ 0 & \text{otherwise} \end{cases}$	$\begin{cases} 0, & x < a \\ \frac{x-a}{b-a}, & x \in [a,b] \\ 1, & x \geq b \end{cases}$	$\frac{a+b}{2}$	$\frac{a+b}{2}$	$\tilde{m} \in [a,b]$	$\frac{(b-a)^2}{12}$	0	$-\frac{6}{5}$	$\frac{e^{bs}-e^{as}}{(b-a)s}$	$\frac{e^{ibs}-e^{ias}}{i(b-a)s}$
Cauchy	$x_0 \in \mathbb{R}$, $\gamma \in \mathbb{R}_{>0}$	$x \in \mathbb{R}$	$\frac{1}{\pi\gamma\left(1+\left(\frac{x-x_0}{\gamma}\right)^2\right)}$	$\frac{1}{\pi}\arctan\left(\frac{x-x_0}{\gamma}\right)$	–	x_0	x_0	–	–	–	–	$\exp(ix_0 s - \gamma\lvert s\rvert)$

Abbreviations and notations used in the table are as follows: $\Gamma(r,x) = \int_x^\infty s^{r-1}e^{-s}ds$ and $\gamma(r,x) = \int_0^x s^{r-1}e^{-s}ds$ are the upper and lower incomplete gamma functions, respectively, while $I_x(a,b) = B(x;a,b)/B(1;a,b)$ is the regularized incomplete beta function with $B(x;a,b) = \int_0^x s^{a-1}(1-s)^{b-1}ds$. For more details, see [142] distribution is discrete, has only one parameter α, which is the expectation value that coincides with the variance and approaches the normal distribution for large values of α. The Poisson distribution has positive skewness $\gamma_1 = 1/\sqrt{\alpha}$, and becomes symmetric as it converges to the normal distribution, i.e., $\gamma_1 \to 0$ as $\alpha \to \infty$. The binomial distribution is symmetric for $p = 1/2$. Discrete probability distributions—the Poisson and the binomial distribution in the table—need some care, because median and mode are more tricky to define in the case of tie modes occurring when the pmf has the same maximal value at two neighboring points. All continuous distributions in the table except the chi-square distribution are symmetric with zero skewness. The Cauchy distribution is of special interest since it has a perfect shape, well defined pdf, cdf, and characteristic function, while no moments exist. For further details, see the forthcoming discussion on the individual distributions.

2.3.1 The Poisson Distribution

The Poisson distribution, named after the French physicist and mathematician Siméon Denis Poisson, is a discrete probability distribution expressing the probability of occurrence of independent events within a given interval. A popular example deals with the arrivals of phone calls, emails, and other independent events within a fixed time interval Δt. The expected number of events α occurring per unit time is the only parameter of the distribution $\pi_k(\alpha)$, which returns the probability that k events are recorded during time Δt. In physics and chemistry, the Poisson process is the stochastic basis of first order processes, radioactive decay, or irreversible first order chemical reactions, for example. In general, the Poisson distribution is the probability distribution underlying the time course of particle numbers, atoms, or molecules, satisfying the deterministic rate law $\mathrm{d}N(t) = -\alpha N(t)\,\mathrm{d}t$. The events to be counted need not be on the time axis. The interval can also be defined as a given distance, area, or volume.

Despite its major importance in physics and biology, the Poisson distribution with probability mass function (pmf) $\pi_k(\alpha)$ is a fairly simple mathematical object. As mentioned, it contains a single parameter only, the real-valued positive number α:

$$P(\mathcal{X} = k) = \pi_k(\alpha) = \frac{\mathrm{e}^{-\alpha}}{k!}\alpha^k\,, \quad k \in \mathbb{N}\,, \tag{2.35}$$

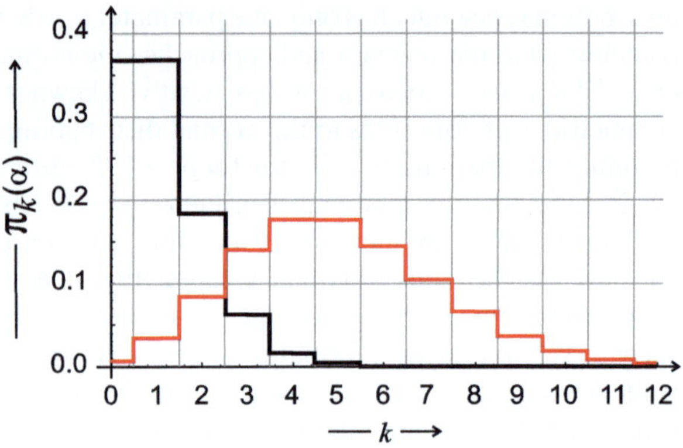

Fig. 2.6 The Poisson probability density. Two examples of Poisson distributions $\pi_k(\alpha) = \alpha^k \mathrm{e}^{-\alpha}/k!$ are shown, with $\alpha = 1$ (*black*) and $\alpha = 5$ (*red*). The distribution with the larger α has mode shifted further to the right and a thicker tail

where $\mathcal{X}$ is a random variable with Poissonian density. As an exercise we leave it to the reader to check the following properties[12]:

$$\sum_{k=0}^{\infty} \pi_k = 1 \,, \quad \mu = \sum_{k=0}^{\infty} k\pi_k = \alpha \,, \quad \hat{\mu}_2 = \sum_{k=0}^{\infty} k^2\pi_k = \alpha + \alpha^2 \,.$$

Examples of Poisson distributions with two different parameter values, $\alpha = 1$ and 5, are shown in Fig. 2.6. The cumulative distribution function (cdf) is obtained by summation:

$$P(\mathcal{X} \leq k) = \exp(-\alpha) \sum_{j=0}^{k} \frac{\alpha^j}{j!} = \frac{\Gamma(k+1, \alpha)}{k!} = Q(k+1, \alpha) \,, \tag{2.36}$$

where $\Gamma(a, z)$ is the incomplete and $Q(a, z)$ the regularized Γ-function.

By means of a Taylor series expansion we can find the generating function of the Poisson distribution:

$$g(s) = \mathrm{e}^{\alpha(s-1)} \,. \tag{2.37}$$

[12] In order to be able to solve the problems, note the following basic infinite series:

$$\mathrm{e} = \sum_{n=0}^{\infty} \frac{1}{n!} \,, \quad \mathrm{e}^x = \sum_{n=0}^{\infty} \frac{x^n}{n!} \,, \text{ for } |x| < \infty \,,$$

$$\mathrm{e} = \lim_{n\to\infty} \left(1 + \frac{1}{n}\right)^n \,, \quad \mathrm{e}^{-\alpha} = \lim_{n\to\infty} \left(1 - \frac{\alpha}{n}\right)^n \,.$$

From the generating function, we calculate

$$g'(s) = \alpha e^{\alpha(s-1)} \,, \qquad g''(s) = \alpha^2 e^{\alpha(s-1)} \,.$$

The expectation value and second moment follow straightforwardly from the derivatives and (2.28):

$$E(\mathcal{X}) = g'(1) = \alpha \,, \tag{2.37a}$$

$$E(\mathcal{X}^2) = g'(1) + g''(1) = \alpha + \alpha^2 \,, \tag{2.37b}$$

$$\mathrm{var}(\mathcal{X}) = \alpha \,. \tag{2.37c}$$

Both the expectation value and the variance are equal to the parameter α, whence the standard deviation amounts to $\sigma(\mathcal{X}) = \sqrt{\alpha}$. Accordingly, the Poisson distribution is the discrete prototype of a distribution satisfying a $\sqrt{N}$-law. This remarkable property of the Poisson distribution is not limited to the second moment. The *factorial moments* (2.17) satisfy

$$E\big((\mathcal{X})_r\big) = E\big(\mathcal{X}(\mathcal{X}-1)\cdots(\mathcal{X}-r+1)\big) = \alpha^r \,, \tag{2.37d}$$

which is easily checked by direct calculation.

The characteristic function and the moment generating function of the Poisson distribution are obtained straightforwardly:

$$\phi_{\mathcal{X}}(s) = \exp\big(\alpha(e^{is} - 1)\big) \,, \tag{2.38}$$

$$M_{\mathcal{X}}(s) = \exp\big(\alpha(e^{s} - 1)\big) \,. \tag{2.39}$$

The characteristic function will be used for characterization and analysis of the Poisson process (Sects. 3.2.2.4 and 3.2.5).

2.3.2 The Binomial Distribution

The binomial distribution $B(n, p)$ expresses the cumulative scores of n independent trials with two-valued outcomes, for example, yes/no decisions or successive coin tosses, as discussed already in Sects. 1.2 and 1.5:

$$\mathcal{S}_n = \sum_{i=1}^{n} \mathcal{X}_i \,, \quad i \in \mathbb{N}_{>0} \,, \ n \in \mathbb{N}_{>0} \,. \tag{1.22$'$}$$

In general, we assume that heads is obtained with probability p and tails with probability $q = 1 - p$. The $\mathcal{X}_i$ are called Bernoulli random variables, named after the Swiss mathematician Jakob Bernoulli, and the sequence of events $\mathcal{S}_n$ is called a Bernoulli process (Sect. 3.1.3). The corresponding random variable is said to have a Bernoulli or binomial distribution:

$$P(\mathcal{S}_n = k) = B_k(n,p) = \binom{n}{k} p^k q^{n-k} \,, \quad q = 1 - p \,, \; k, n \in \mathbb{N} \,, \; k \leq n \,. \tag{2.40}$$

Two examples of binomial distributions are shown in Fig. 2.7. The distribution with $p = q = 1/2$ is symmetric with respect to $k = n/2$. The symmetric binomial distribution corresponding to fair coin tosses $p = q = 1/2$ is, of course, also obtained from the probability distribution of n independent generalized dice throws in (1.50) by choosing $s = 2$.

The generating function for the single trial is $g(s) = q + ps$. Since we have n independent trials the complete generating function is

$$g(s) = (q + ps)^n = \sum_{k=0}^{n} \binom{n}{k} q^{n-k} p^k s^k \,. \tag{2.41}$$

From the derivatives of the generating function, viz.,

$$g'(s) = np(q + ps)^{n-1} \,, \qquad g''(s) = n(n-1)p^2(q + ps)^{n-2} \,,$$

we readily compute the expectation value and variance:

$$\mathrm{E}(\mathcal{S}_n) = g'(1) = np \,, \tag{2.41a}$$

$$\mathrm{E}(\mathcal{S}_n^2) = g'(1) + g''(1) = np + n^2p^2 - np^2 = npq + n^2p^2 \,, \tag{2.41b}$$

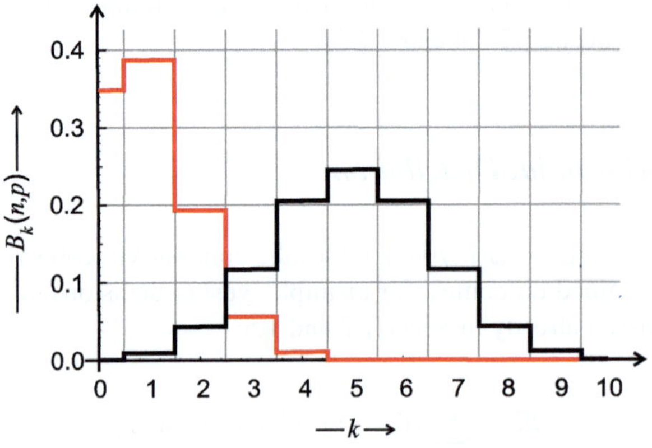

Fig. 2.7 The binomial probability density. Two examples of binomial distributions $B_k(n,p) = \binom{n}{k}p^k(1-p)^{n-k}$, with $n = 10$, $p = 0.5$ (*black*) and $p = 0.1$ (*red*) are shown. The former distribution is symmetric with respect to the expectation value $\mathrm{E}(B_k) = n/2$, and accordingly has zero skewness. The latter case is asymmetric with positive skewness (see Fig. 2.3)

$$\mathrm{var}(\mathcal{S}_n) = npq\,, \tag{2.41c}$$

$$\sigma(\mathcal{S}_n) = \sqrt{npq}\,. \tag{2.41d}$$

For the *symmetric binomial distribution*, the case of the unbiased coin with $p = 1/2$, the first and second moments are $\mathrm{E}(\mathcal{S}_n) = n/2$, $\mathrm{var}(\mathcal{S}_n) = n/4$, and $\sigma(\mathcal{S}_n) = \sqrt{n}/2$. We note that the expectation value is proportional to the number of trials n, and the standard deviation is proportional to its square root $\sqrt{n}$.

Relation Between Binomial and Poisson Distribution

The binomial distribution $B(n,p)$ can be transformed into a Poisson distribution $\pi(\alpha)$ in the limit $n \to \infty$. In order to show this we start from

$$B_k(n,p) = \binom{n}{k} p^k (1-p)^{n-k}\,, \quad k, n \in \mathbb{N}\,,\ k \le n\,.$$

The symmetry parameter p is assumed to vary with n according to the relation $p(n) = \alpha/n$ for $n \in \mathbb{N}_{>0}$, and thus we have

$$B_k\left(n, \frac{\alpha}{n}\right) = \binom{n}{k} \left(\frac{\alpha}{n}\right)^k \left(1 - \frac{\alpha}{n}\right)^{n-k}\,, \quad k, n \in \mathbb{N}\,,\ k \le n\,.$$

We let n go to infinity for fixed k and start with $B_0(n,p)$:

$$\lim_{n\to\infty} B_0\left(n, \frac{\alpha}{n}\right) = \lim_{n\to\infty} \left(1 - \frac{\alpha}{n}\right)^n = \mathrm{e}^{-\alpha}\,.$$

Now we compute the ratio B_{k+1}/B_k of two consecutive terms, viz.,

$$\frac{B_{k+1}\left(n, \frac{\alpha}{n}\right)}{B_k\left(n, \frac{\alpha}{n}\right)} = \frac{n-k}{k+1}\left(\frac{\alpha}{n}\right)\left(1 - \frac{\alpha}{n}\right)^{-1} = \frac{\alpha}{k+1}\left(\left(\frac{n-k}{n}\right)\left(1 - \frac{\alpha}{n}\right)^{-1}\right)\,.$$

Both terms in the outer brackets converge to one as $n \to \infty$, and hence we find:

$$\lim_{n\to\infty} \frac{B_{k+1}\left(n, \frac{\alpha}{n}\right)}{B_k\left(n, \frac{\alpha}{n}\right)} = \frac{\alpha}{k+1}\,.$$

Starting from the limit of B_0, we compute all terms by iteration, i.e.,

$$\lim_{n\to\infty} B_0 = \exp(-\alpha) ,$$
$$\lim_{n\to\infty} B_1 = \alpha \exp(-\alpha) ,$$
$$\lim_{n\to\infty} B_2 = \alpha^2 \exp(-\alpha)/2! ,$$

and so on until eventually,

$$\lim_{n\to\infty} B_k = \frac{\alpha^k}{k!} \exp(-\alpha) . \qquad \square$$

Accordingly, we have shown Poisson's limit law:

$$\lim_{n\to\infty} B_k \left(n, \frac{\alpha}{n}\right) = \pi_k(\alpha) , \quad k \in \mathbb{N} . \tag{2.42}$$

It is worth keeping in mind that the limit was performed in a rather peculiar way, since the symmetry parameter $p(n) = \alpha/n$ was shrinking with increasing n, and as a matter of fact vanished in the limit of $n \to \infty$.

Multinomial Distribution

The multinomial distribution of m random variables, $\mathcal{X}_i$, $i = 1, 2, \ldots, m$, is an important generalization of the binomial distribution. It is defined on a finite domain of integers, $\mathcal{X}_i \leq n$, $\mathcal{X}_i \in \mathbb{N}$, with $\sum_{i=1}^{m} \mathcal{X}_i = \sum_{i=1}^{m} n_i = n$. The parameters for the individual event probabilities are p_i, $i = 1, 2, \ldots, m$, with $p_i \in [0, 1]\ \forall\, i$ and $\sum_{i=1}^{m} p_i = 1$, and the probability mass function (pmf) of the multinomial distribution has the form

$$M_{n_1,\ldots,n_m}(n, p_1, \ldots, p_m) = \frac{n!}{n_1!\, n_2! \cdots n_m!} p_1^{n_1} p_2^{n_2} \cdots p_m^{n_m} . \tag{2.43}$$

For the first and second moments, we find

$$\begin{gathered} \mathrm{E}(\mathcal{X}_i) = np_i , \quad \mathrm{var}(\mathcal{X}_i) = np_i(1 - p_i) , \\ \mathrm{cov}(\mathcal{X}_i, \mathcal{X}_j) = -np_i p_j . \end{gathered} \tag{2.44}$$

We shall encounter multinomial distributions as solutions for the probability densities of chemical reactions in closed systems (Sects. 4.2.3 and 4.3.2).

2.3.3 *The Normal Distribution*

The normal or Gaussian distribution is of central importance in probability theory. Indeed most distributions converge to it in the limit of large numbers since the central limit theorem (CLT) states that under mild conditions the sums of large numbers of random variables follow approximately a normal distribution (Sect. 2.4.2). The normal distribution is a special case of the *stable distribution* (Sect. 2.5.9) and this fact is not unrelated to the central limit theorem. Historically the normal distribution is attributed to the French mathematician Marquis de Laplace [326, 327] and the German mathematician Carl Friedrich Gauss [197]. Although the Laplace's research in the eighteenth century came earlier than Gauss's contributions, the latter is commonly considered to have provided the more significant contribution, so the probability distribution is now named after him (but see also [508]). The famous English statistician Karl Pearson [446] comments on the priority discussion:

> Many years ago I called the Laplace–Gaussian curve the normal curve, which name, while it avoids an international question of priority, has the disadvantage of leading people to believe that all other distributions of frequency are in one sense or another 'abnormal'.

The normal distribution has several advantageous technical features. It is the only absolutely continuous distribution that has only zero cumulants except for the first two, i.e., the expectation value and the variance, which have the straightforward meaning of the position and the width of the distribution. In other words a normal distribution is completely determined by the mean and variance.

For given variance, the normal distribution has the largest information entropy of all distributions on $\Omega = \mathbb{R}$ (Sect. 2.1.3). As a matter of fact, the mean μ does not enter the expression for the entropy of the normal distribution (Table 2.1):

$$H(\sigma) = \frac{1}{2}\left(1 + \log\left(2\pi\sigma^2\right)\right) . \tag{2.24'}$$

In other words, shifting the normal distribution along the x-axis does not change the information entropy of the distribution.

The normal distribution is fundamental for estimating statistical errors, so we shall discuss it in some detail. Because of this, the normal distribution is extremely popular in statistics and experts sometimes claim that it is 'overapplied'. Empirical samples are often not symmetrically distributed but skewed to the right, and yet they are analyzed by means of normal distributions. The log-normal distribution [346] or the Pareto distribution, for example, might do better in such cases. Statistics based on normal distribution is not robust in the presence of outliers where a description by more heavy-tailed distributions like Student's t-distribution is superior. Whether or not the tails have more weight in the distribution is easily checked by means of the excess kurtosis. Student's distribution has an excess kurtosis of

$$\gamma_2 = \begin{cases} \dfrac{6}{\nu - 4} , & \text{for } \nu > 4 , \\ \infty , & \text{for } 2 < \nu \le 4 , \\ \text{undefined} , & \text{otherwise} , \end{cases}$$

which is always positive, whereas the excess kurtosis of the normal distribution is zero.

The density of the normal distribution is[13]

$$f_{\mathcal{N}}(x) = \frac{1}{\sigma\sqrt{2\pi}} \mathrm{e}^{-(x-\mu)^2/2\sigma^2} , \qquad \int_{-\infty}^{+\infty} f(x)\,\mathrm{d}x = 1 . \tag{2.45}$$

The corresponding random variable $\mathcal{X}$ has moments $\mathrm{E}(\mathcal{X}) = \mu$, $\mathrm{var}(\mathcal{X}) = \sigma^2$, and $\sigma(\mathcal{X}) = \sigma$. For many purposes it is convenient to use the normal density in centered and normalized form, i.e., $\widetilde{\mathcal{X}} = (\mathcal{X} - \mu)/\sigma$, $\tilde{\mu} = 0$, and $\tilde{\sigma}^2 = 1$, which is called the standard normal distribution or the Gaussian bell-shaped curve:

$$f_{\mathcal{N}}(x; 0, 1) = \varphi(x) = \frac{1}{\sqrt{2\pi}} \mathrm{e}^{-x^2/2} , \qquad \int_{-\infty}^{+\infty} \varphi(x)\,\mathrm{d}x = 1 . \tag{2.45$'$}$$

In this form we clearly have $\mathrm{E}(\widetilde{\mathcal{X}}) = 0$, $\mathrm{var}(\widetilde{\mathcal{X}}) = 1$, and $\sigma(\widetilde{\mathcal{X}}) = 1$.

Integration of the density yields the cumulative distribution function

$$\begin{aligned} P(\mathcal{X} \le x) = F_{\mathcal{N}}(x) &= \frac{1}{\sigma\sqrt{2\pi}} \int_{-\infty}^{x} \mathrm{e}^{-(u-\mu)^2/2\sigma^2}\,\mathrm{d}u \\ &= \frac{1}{2}\left(1 + \mathrm{erf}\left(\frac{x-\mu}{\sigma\sqrt{2}}\right)\right) . \end{aligned} \tag{2.46}$$

The function $F_{\mathcal{N}}(x)$ is not available in analytical form, but it can be easily formulated in terms of a special function, the error function $\mathrm{erf}(x)$. This function and its complement $\mathrm{erfc}(x)$ are defined by

$$\mathrm{erf}(x) = \frac{2}{\sqrt{\pi}} \int_0^x \mathrm{e}^{-u^2}\,\mathrm{d}u , \qquad \mathrm{erfc}(x) = \frac{2}{\sqrt{\pi}} \int_x^{\infty} \mathrm{e}^{-u^2}\,\mathrm{d}u ,$$

[13]The notation applied here for the normal distribution is as follows: $\mathcal{N}(\mu, \sigma)$ in general, $F_{\mathcal{N}}(x; \mu, \sigma)$ for the cumulative distribution, and $f_{\mathcal{N}}(x; \mu, \sigma)$ for the density. Commonly, the parameters (μ, σ) are omitted, when no misinterpretation is possible. For standard stable distributions (Sect. 2.5.9), a variance $\gamma^2 = \sigma^2/2$ is applied.

and are available in tables and in standard mathematical packages.[14] Examples of the normal density $f_{\mathcal{N}}(x)$ and the integrated distribution $F_{\mathcal{N}}(x)$ with different values of the standard deviation σ are shown in Fig. 1.22. The normal distribution is also used in statistics to define confidence intervals: 68.2 % of the data points lie within an interval $\mu \pm \sigma$, 95.4 % within an interval $\mu \pm 2\sigma$, and 99.7 % within an interval $\mu \pm 3\sigma$.

The normal density function $f_{\mathcal{N}}(x)$ has, among other remarkable properties, derivatives of all orders. Each derivative can be written as product of $f_{\mathcal{N}}(x)$ by a polynomial, of the order of the derivative, known as a Hermite polynomial. The function $f_{\mathcal{N}}(x)$ decreases to zero very rapidly as $|x| \to \infty$. The existence of all derivatives makes the *bell-shaped* Gaussian curve $x \to f(x)$ particularly smooth, and the moment generating function of the normal distribution is especially attractive (see Sect. 2.2.2) since $M(s)$ can be obtained directly by integration:

$$\begin{aligned} M(s) &= \int_{-\infty}^{+\infty} \mathrm{e}^{xs} f(x)\, \mathrm{d}x = \int_{-\infty}^{+\infty} \exp\left(xs - \frac{x^2}{2}\right) \mathrm{d}x \\ &= \int_{-\infty}^{+\infty} \exp\left(\frac{s^2}{2} - \frac{(x-s)^2}{2}\right) \mathrm{d}x = \mathrm{e}^{s^2/2} \int_{-\infty}^{+\infty} f(x-s)\, \mathrm{d}x \\ &= \mathrm{e}^{s^2/2} . \end{aligned} \tag{2.47}$$

All raw moments of the normal distribution are defined by the integrals

$$\hat{\mu}_n = \int_{-\infty}^{+\infty} x^n f(x)\, \mathrm{d}x . \tag{2.48}$$

They can be obtained, for example, by successive differentiation of $M(s)$ with respect to s (Sect. 2.2.2). In order to obtain the moments more efficiently, we expand the first and the last expression in (2.47) in a power series of s:

$$\begin{aligned} &\int_{-\infty}^{+\infty} \left(1 + xs + \frac{(xs)^2}{2!} + \cdots + \frac{(xs)^n}{n!} + \cdots\right) f(x)\, \mathrm{d}x \\ &\qquad = 1 + \frac{s^2}{2} + \frac{1}{2!}\left(\frac{s^2}{2}\right)^2 + \cdots + \frac{1}{n!}\left(\frac{s^2}{2}\right)^n + \cdots , \end{aligned}$$

[14] We remark that erf(x) and erfc(x) are not normalized in the same way as the normal density:

$$\lim_{x\to\infty} \mathrm{erf}(x) = \frac{2}{\sqrt{\pi}} \int_0^{\infty} \exp(-u^2)\, \mathrm{d}u = 1 , \quad \int_0^{\infty} \varphi(u)\, \mathrm{d}u = \frac{1}{2} \int_{-\infty}^{+\infty} \varphi(u)\, \mathrm{d}u = \frac{1}{2} .$$

or express it in terms of the moments $\hat{\mu}_n$:

$$\sum_{n=0}^{\infty} \frac{\hat{\mu}_n}{n!} s^n = \sum_{n=0}^{\infty} \frac{1}{2^n n!} s^{2n} ,$$

from which we compute the moments of $\varphi(x)$ by equating the coefficients of equal powers of s on each side of the expansion. For $n \geq 1$, we find[15]:

$$\hat{\mu}_{2n-1} = 0 , \qquad \hat{\mu}_{2n} = \frac{(2n)!}{2^n n!} . \tag{2.49}$$

All odd moments vanish due to symmetry. In the case of the fourth moment, kurtosis, it is common to apply a kind of standardization which assigns zero excess kurtosis, viz., $\gamma_2 = 0$, to the normal distribution. In other words, excess kurtosis monitors peak shape with respect to the normal distribution: positive excess kurtosis implies peaks that are sharper than the normal density, while negative excess kurtosis implies peaks that are broader than the normal density (Fig. 2.3).

As already mentioned, all cumulants (2.15) of the normal distribution except $\kappa_1 = \mu$ and $\kappa_2 = \sigma^2$ are zero, since the moment generating function of the general normal distribution with mean μ and variance σ^2 is of the form

$$M_{\mathcal{N}}(s) = \exp\Big(\mu s + \frac{1}{2}\sigma^2 s^2\Big) . \tag{2.50}$$

The expression for the standardized Gaussian distribution is the special case with $\mu = 0$ and $\sigma^2 = 1$.

Finally, we give the characteristic function of the normal distribution:

$$\phi_{\mathcal{N}}(s) = \exp\Big(\mathrm{i}\mu s - \frac{1}{2}\sigma^2 s^2\Big) . \tag{2.51}$$

This will be used, for example, in the derivation of the central limit theorem (Sect. 2.4.2).

A Poisson density with sufficiently large values of α resembles a normal density (see Fig. 2.8) and it can indeed be shown that the two curves become more and more

---

[15] The definite integrals are:

$$\int_{-\infty}^{+\infty} x^n \exp(-x^2)\,\mathrm{d}x = \begin{cases} \sqrt{\pi} , & n = 0 , \\ 0 , & n \geq 1 , \; n \text{ odd} , \\ \dfrac{(n-1)!!}{2^{n/2}} \sqrt{\pi} , & n \geq 2 , \; n \text{ even} , \end{cases}$$

where $(n-1)!! = 1 \times 3 \times \cdots \times (n-1)$ is the *double factorial*.

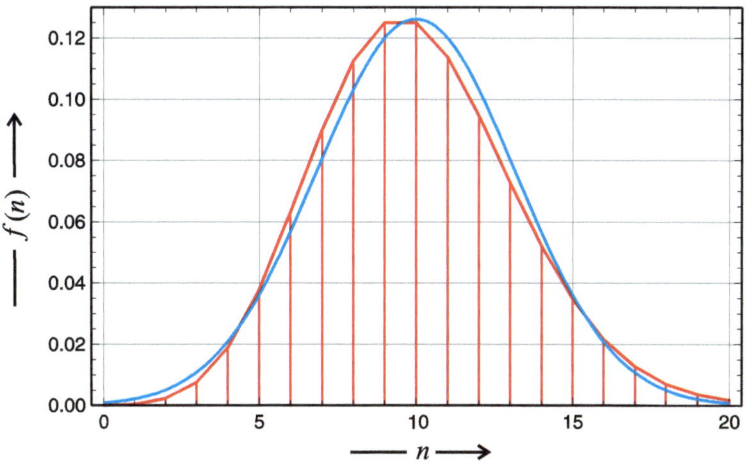

Fig. 2.8 Comparison between Poisson and normal density. The figure compares the pmf of the Poisson distribution with the parameter α (*red*) and a best fit normal distribution with mean $\mu = \alpha$ and standard deviation $\sigma = \sqrt{\alpha}$ (*blue*) according to (2.52). Parameter choice $\alpha = 10$

alike with increasing α[16]:

$$\pi_k(\alpha) = \frac{\alpha^k}{k!}\mathrm{e}^{-\alpha} \approx \frac{1}{\sqrt{2\pi\alpha}} \exp\left(-\frac{(k-\alpha)^2}{2\,\alpha}\right) , \quad \text{for } \alpha \gg 1 . \tag{2.52}$$

We present a short proof based on the moment generating functions for the approximation of the standardized Poisson distribution by a standard normal distribution. The Poisson variable $\mathcal{X}_\alpha$ with $P(\mathcal{X}_\alpha = k) = \pi_k(\alpha)$ is standardized to $\mathcal{Y}_\alpha = (\mathcal{X}_\alpha - \alpha)/\sqrt{\alpha}$ and we obtain for the moment generating functions:

$$M_{\mathcal{X}_\alpha}(s) = \mathrm{E}\left(\mathrm{e}^{\mathcal{X}_\alpha s}\right) = \exp\big(\alpha(\mathrm{e}^s - 1)\big) \implies M_{\mathcal{Y}_\alpha}(s) = \mathrm{E}\left(\exp\left(\frac{\mathcal{X}_\alpha - \alpha}{\sqrt{\alpha}} s\right)\right) .$$

We now take the limit $n \to \infty$, expand the exponential function, and truncate after the first non-vanishing term [334]:

$$\begin{aligned}
\lim_{\alpha\to\infty} M_{\mathcal{Y}_\alpha}(s) &= \lim_{\alpha\to\infty} \mathrm{E}\left(\exp\left(\frac{\mathcal{X}_\alpha - \alpha}{\sqrt{\alpha}} s\right)\right) = \lim_{\alpha\to\infty} \mathrm{e}^{-\sqrt{\alpha} s}\mathrm{E}\left(\exp\left(\frac{\mathcal{X}_\alpha s}{\sqrt{\alpha}}\right)\right) \\
&= \lim_{\alpha\to\infty} \mathrm{e}^{-\sqrt{\alpha} s} \exp\big(\alpha(\mathrm{e}^{s/\sqrt{\alpha}} - 1)\big) \\
&= \lim_{\alpha\to\infty} \exp\left(\frac{s^2}{2} + \frac{s^3}{6\sqrt{\alpha}} + \cdots\right) = \exp(s^2/2) . \qquad \square
\end{aligned}$$

[16]It is important to remember that k is a discrete variable on the left-hand side, whereas it is continuous on the right-hand side of (2.52).

In the limit of large α, we do indeed obtain the moment generating function of the standardized normal distribution $\mathcal{N}(0,1)$. The result is an example of the central limit theorem, which will be presented and analyzed in Sect. 2.4.2. We shall require this approximation of the Poissonian distribution by a normal distribution in Sects. 3.4.3 and 4.2.4 for the derivation of a chemical Langevin equation.

2.3.4 Multivariate Normal Distributions

In applications to real world problems it is often necessary to consider probability distributions in multiple dimensions. Then, a random vector $\boldsymbol{\mathcal{X}} = (\mathcal{X}_1, \ldots, \mathcal{X}_n)$ with the joint probability distribution

$$P(\mathcal{X}_1 = x_1, \ldots, \mathcal{X}_n = x_n) = p(x_1, \ldots, x_n) = p(\mathbf{x})$$

replaces the random variable $\mathcal{X}$. This multivariate normal probability density can be written as

$$f(\mathbf{x}) = \frac{1}{\sqrt{(2\pi)^n |\Sigma|}} \exp\left(-\frac{1}{2}(\mathbf{x}-\boldsymbol{\mu})^{\mathrm{t}}\, \Sigma^{-1}(\mathbf{x}-\boldsymbol{\mu})\right) .$$

The vector $\boldsymbol{\mu}$ consists of the (raw) first moments along the different coordinates, viz., $\boldsymbol{\mu} = (\mu_1, \ldots, \mu_n)$, and the variance–covariance matrix Σ contains the n variances in the diagonal while the covariances are represented by the off-diagonal elements:

$$\Sigma = \begin{pmatrix} \mathrm{var}(\mathcal{X}_1) & \mathrm{cov}(\mathcal{X}_1,\mathcal{X}_2) & \ldots & \mathrm{cov}(\mathcal{X}_1,\mathcal{X}_n) \\ \mathrm{cov}(\mathcal{X}_2,\mathcal{X}_1) & \mathrm{var}(\mathcal{X}_2) & \ldots & \mathrm{cov}(\mathcal{X}_2,\mathcal{X}_n) \\ \vdots & \vdots & \ddots & \vdots \\ \mathrm{cov}(\mathcal{X}_n,\mathcal{X}_1) & \mathrm{cov}(\mathcal{X}_n,\mathcal{X}_2) & \ldots & \mathrm{var}(\mathcal{X}_n) \end{pmatrix} = \begin{pmatrix} \sigma_{11} & \sigma_{12} & \ldots & \sigma_{1n} \\ \sigma_{12} & \sigma_{22} & \ldots & \sigma_{2n} \\ \vdots & \vdots & \ddots & \vdots \\ \sigma_{1n} & \sigma_{2n} & \ldots & \sigma_{nn} \end{pmatrix} .$$

The matrix Σ is symmetric, $\sigma_{ij} = \mathrm{cov}(\mathcal{X}_i, \mathcal{X}_j) = \mathrm{cov}(\mathcal{X}_j, \mathcal{X}_i) = \sigma_{ji}$, by the definition of covariances, and $\sigma_{ii} = \sigma_i^2$.

The mean and variance are given by $\hat{\boldsymbol{\mu}} = \boldsymbol{\mu}$ and the variance–covariance matrix Σ, and expressed by the dummy vector variable $\mathbf{s} = (s_1, \ldots, s_n)$. The moment generating function is of the form

$$M(\mathbf{s}) = \exp\left(\boldsymbol{\mu}^{\mathrm{t}}\mathbf{s}\right) \exp\left(\frac{1}{2}\mathbf{s}^{\mathrm{t}}\Sigma\,\mathbf{s}\right) .$$

Finally, the characteristic function is given by

$$\phi(\mathbf{s}) = \exp\left(\mathrm{i}\boldsymbol{\mu}^{\mathrm{t}}\mathbf{s}\right) \exp\left(-\frac{1}{2}\mathbf{s}^{\mathrm{t}}\Sigma\,\mathbf{s}\right) .$$

Without showing the details, we remark that this particularly simple characteristic function implies that all moments higher than order two can be expressed in terms of first and second moments, in particular expectation values, variances, and covariances. To give an example that we shall require in Sect. 3.4.2, the fourth order moments can be derived from

$$\begin{aligned} \mathrm{E}(\mathcal{X}_i^4) &= 3\sigma_{ii}^2 , \\ \mathrm{E}(\mathcal{X}_i^3 \mathcal{X}_j) &= 3\sigma_{ii}\sigma_{ij} , \\ \mathrm{E}(\mathcal{X}_i^2 \mathcal{X}_j^2) &= \sigma_{ii}\sigma_{jj} + 2\sigma_{ij}^2 , \\ \mathrm{E}(\mathcal{X}_i^2 \mathcal{X}_j \mathcal{X}_k) &= \sigma_{ii}\sigma_{jk} + 2\sigma_{ij}\sigma_{ik} , \\ \mathrm{E}(\mathcal{X}_i \mathcal{X}_j \mathcal{X}_k \mathcal{X}_l) &= \sigma_{ij}\sigma_{kl} + \sigma_{li}\sigma_{jk} + \sigma_{ik}\sigma_{jl} , \end{aligned} \tag{2.53}$$

with $i, j, k, l \in \{1, 2, 3, 4\}$.

The entropy of the multivariate normal distribution is readily calculated and appears as a straightforward extension of (2.24) to higher dimensions:

$$\begin{aligned} H(f) &= -\int_{-\infty}^{+\infty}\int_{-\infty}^{+\infty}\cdots\int_{-\infty}^{+\infty} f(\mathbf{x}) \log f(\mathbf{x})\, \mathrm{d}\mathbf{x} \\ &= \frac{1}{2}\Big(n + \log\big((2\pi)^n |\Sigma|\big)\Big) , \end{aligned} \tag{2.54}$$

where $|\Sigma|$ is the determinant of the variance–covariance matrix.

The marginal distributions of a multivariate normal distribution are obtained straightforwardly by simply dropping the marginalized variables. If $\boldsymbol{\mathcal{X}} = (\mathcal{X}_i, \mathcal{X}_j, \mathcal{X}_k)$ is a multivariate normally distributed variable with the mean vector $\boldsymbol{\mu} = (\mu_i, \mu_j, \mu_k)$ and variance–covariance matrix Σ, then after elimination of $\mathcal{X}_j$, the marginal joint distribution of the vector $\widetilde{\boldsymbol{\mathcal{X}}} = (\mathcal{X}_i, \mathcal{X}_k)$ is multivariate normal with mean vector $\tilde{\boldsymbol{\mu}} = (\mu_i, \mu_k)$ and variance–covariance matrix

$$\widetilde{\Sigma} = \begin{pmatrix} \Sigma_{ii} & \Sigma_{ik} \\ \Sigma_{ki} & \Sigma_{kk} \end{pmatrix} = \begin{pmatrix} \mathrm{var}(\mathcal{X}_i) & \mathrm{cov}(\mathcal{X}_i, \mathcal{X}_k) \\ \mathrm{cov}(\mathcal{X}_k, \mathcal{X}_i) & \mathrm{var}(\mathcal{X}_k) \end{pmatrix} .$$

It is worth noting that non-normal bivariate distributions have been constructed which have normal marginal distributions [317].

Uncorrelatedness Versus Independence

The multivariate normal distribution presents an excellent example for discussing the difference between uncorrelatedness and independence. Two random variables are independent if

$$f_{\mathcal{X}\mathcal{Y}}(x,y) = f_{\mathcal{X}}(x) f_{\mathcal{Y}}(y) \,, \quad \forall\, x, y \,,$$

whereas uncorrelatedness of two random variables requires

$$\begin{aligned} \sigma_{\mathcal{X}\mathcal{Y}} = \mathrm{cov}(\mathcal{X}, \mathcal{Y}) = 0 = \mathrm{E}(\mathcal{X}\mathcal{Y}) - \mathrm{E}(\mathcal{X})\mathrm{E}(\mathcal{Y}) \,, \\ \mathrm{E}(\mathcal{X}\mathcal{Y}) = \mathrm{E}(\mathcal{X})\mathrm{E}(\mathcal{Y}) \,, \end{aligned}$$

which implies only factorizability of the joint expectation value. The covariance between two independent random variables vanishes, and hence,

$$\begin{aligned} \mathrm{E}(\mathcal{X}\mathcal{Y}) &= \int_{-\infty}^{+\infty} \int_{-\infty}^{+\infty} xy f_{\mathcal{X},\mathcal{Y}}(x,y) \, \mathrm{d}x \, \mathrm{d}y \\ &= \int_{-\infty}^{+\infty} \int_{-\infty}^{+\infty} xy f_{\mathcal{X}}(x) f_{\mathcal{Y}}(y) \, \mathrm{d}x \, \mathrm{d}y \\ &= \int_{-\infty}^{+\infty} x f_{\mathcal{X}}(x) \mathrm{d}x \int_{-\infty}^{+\infty} y f_{\mathcal{Y}}(y) \, \mathrm{d}y = \mathrm{E}(\mathcal{X})\mathrm{E}(\mathcal{Y}) \,. \qquad \square \end{aligned}$$

Note that we nowhere made use of the fact that the variables are normally distributed, and the statement that *independent variables are uncorrelated* holds in full generality. The converse, however, is not true as has been shown by means of specific examples [391]. Indeed, uncorrelated random variables $\mathcal{X}_1$ and $\mathcal{X}_2$ which have the same (marginal) normal distribution need not be independent. A counterexample can be constructed from a two-dimensional random vector $\boldsymbol{\mathcal{X}} = (\mathcal{X}_1, \mathcal{X}_2)^{\mathrm{t}}$ with a bivariate normal distribution with mean $\boldsymbol{\mu} = (0,0)^{\mathrm{t}}$, variance $\sigma_1^2 = \sigma_2^2 = 1$, and covariance $\mathrm{cov}(\mathcal{X}_1, \mathcal{X}_2) = 0$:

$$\begin{aligned} f(x_1, x_2) &= \frac{1}{2\pi} \exp\left(-\frac{1}{2} (x_1, x_2) \begin{pmatrix} 1 & 0 \\ 0 & 1 \end{pmatrix} \begin{pmatrix} x_1 \\ x_2 \end{pmatrix} \right) \\ &= \frac{1}{2\pi} \mathrm{e}^{-(x_1^2 + x_2^2)/2} = \frac{1}{\sqrt{2\pi}} \mathrm{e}^{-x_1^2/2} \frac{1}{\sqrt{2\pi}} \mathrm{e}^{-x_2^2/2} = f(x_1) f(x_2) \,. \end{aligned}$$

The two random variables are independent. Next we introduce a modification in one of the two random variables: $\mathcal{X}_1$ remains unchanged and has the density $f(x_1) = \frac{1}{\sqrt{2\pi}} \exp(-x_1^2/2)$, whereas the second random variable is modulated by an ideal coin flip $\mathcal{W}$ with the density

$$f(w) = \frac{1}{2} \big(\delta(w+1) + \delta(w-1) \big) \,.$$

In other words, we have $\mathcal{X}_2 = \mathcal{W}\mathcal{X}_1 = \pm\mathcal{X}_1$ with equal weights for both signs, and accordingly the density function is

$$f(x_2) = \frac{1}{2}f(x_1) + \frac{1}{2}f(-x_1) = f(x_1) ,$$

since the normal distribution with zero mean $\mathrm{E}(\mathcal{X}_1) = 0$ is symmetric, i.e., $f(x_1) = f(-x_1)$. Equality of the two distribution functions with the same normal distribution can also be derived directly:

$$\begin{aligned} P(\mathcal{X}_2 \le x) &= \mathrm{E}\big(P(\mathcal{X}_2 \le x|\mathcal{W})\big) \\ &= P(\mathcal{X}_1 \le x)P(\mathcal{W} = 1) + P(-\mathcal{X}_1 \le x)P(\mathcal{W} = -1) \\ &= F_{\mathcal{N}}(x) \times \frac{1}{2} + F_{\mathcal{N}}(x) \times \frac{1}{2} = F_{\mathcal{N}}(x) = P(\mathcal{X}_1 \le x) . \end{aligned}$$

The covariance of $\mathcal{X}_1$ and $\mathcal{X}_2$ is readily calculated:

$$\begin{aligned} \mathrm{cov}(\mathcal{X}_1\mathcal{X}_2) &= \mathrm{E}(\mathcal{X}_1\mathcal{X}_2) - \mathrm{E}(\mathcal{X}_1)\mathrm{E}(\mathcal{X}_2) = \mathrm{E}(\mathcal{X}_1\mathcal{X}_2) - 0 \\ &= \mathrm{E}\big(\mathrm{E}(\mathcal{X}_1\mathcal{X}_2)|\mathcal{W}\big) = \mathrm{E}(\mathcal{X}_1^2)P(\mathcal{W} = 1) + \mathrm{E}(-\mathcal{X}_1^2)P(\mathcal{W} = -1) \\ &= 1 \times \frac{1}{2} + (-1) \times \frac{1}{2} = 0 , \end{aligned}$$

whence $\mathcal{X}_1$ and $\mathcal{X}_2$ are uncorrelated. The two random variables, however, are not independent because

$$\begin{aligned} p(x_1, x_2) &= P(\mathcal{X}_1 = x_1, \mathcal{X}_2 = x_2) \\ &= \frac{1}{2}P(\mathcal{X}_1 = x_1, \mathcal{X}_2 = x_1) + \frac{1}{2}P(\mathcal{X}_1 = x_1, \mathcal{X}_2 = -x_1) \\ &= \frac{1}{2}p(x_1) + \frac{1}{2}p(x_1) = p(x_1) , \\ f(x_1, x_2) &= f(x_1) \neq f(x_1) \cdot f(x_2) , \end{aligned}$$

since $f(x_1) = f(x_2)$. Lack of independence can also be shown simply by considering $|\mathcal{X}_1| = |\mathcal{X}_2|$. Two random variables that have the same absolute value cannot be independent.

The example is illustrated in Fig. 2.9. The fact that marginal distributions are identical does not imply that the joint distribution is also the same! The statement

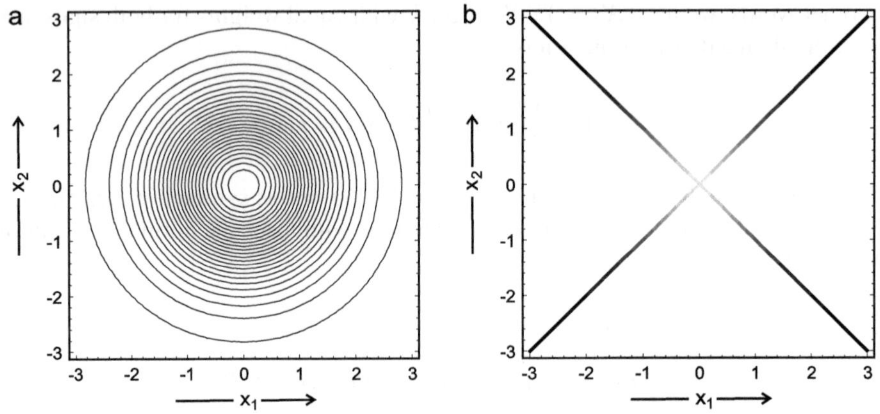

Fig. 2.9 Uncorrelated but not independent normal distributions. The figure compares two different joint densities which have identical marginal densities. The contour plot in (**a**) shows the joint distribution $f(x_1, x_2) = \frac{1}{2\pi} e^{-(x_1^2+x_2^2)/2}$. The contour lines are circles equidistant in f and plotted for $f = 0.03, 0.09, \ldots, 0.153$. The marginal distributions of this joint distribution are standard normal distributions in x_1 or x_2. The density in (**b**) is derived from one random variable $\mathcal{X}_1$ with standard normal density $f(x_1) = \frac{1}{\sqrt{2\pi}} e^{-x_1^2/2}$ and a second random variable that is modulated by a perfect coin flip: $\mathcal{X}_2 = \mathcal{X}_1 \mathcal{W}$ with $\mathcal{W} = \pm 1$. The two variables $\mathcal{X}_1$ and $\mathcal{X}_2$ are uncorrelated but not independent

about independence, however, can be made stronger and then it turns out to be true [391]:

> If random variables have a multivariate normal distribution and are pairwise uncorrelated, then the random variables are always independent.

2.4 Regularities for Large Numbers

The expression *normal distribution* actually originated from the fact that many distributions can be transformed in a natural way to yield the probability density $f_{\mathcal{N}}(x)$ for large numbers n. In Sects. 1.9.1 and 2.3.3, we demonstrated convergence to the normal distribution for specific probabilities derived from samples with large numbers of trials, and this raises the question as to whether or not a more general regularity lies behind the special cases. Therefore we consider a sum of independent random variables resulting from a sequence of Bernoulli trials according to (1.22′). The partial sums follow a binomial distribution with the expectation value

$$\overline{\mathcal{X}} = \frac{1}{n} \mathcal{S}_n = \frac{1}{n}(\mathcal{X}_1 + \mathcal{X}_2 + \cdots + \mathcal{X}_n) .$$

First we shall prove here that the binomial distribution converges to the normal distribution in the limit $n \to \infty$. Then follows the generalization to sequences of independent variables with arbitrary but identical distributions in the form of the central limit theorem (CLT). As an extension of CLT in the simplest manifestation, we show convergence of sums of random variables no matter whether they are identically distributed or not: sufficient conditions are only a finite expectation value $\mathrm{E}(\mathcal{X}_j) = \mu_j$ and a finite variance $\mathrm{var}(\mathcal{X}_j) = \sigma_j^2$ for each random variable $\mathcal{X}_j$.

Two other regularities concern the first and second moments of $\mathcal{S}_n$: the *law of large numbers* guarantees convergence of the sum $\mathcal{S}_n$ to the expectation value in strong and weak form, viz.,

$$\lim_{n\to\infty} \mathcal{S}_n = n\mu ,$$

and the *law of the iterated logarithm* bounds the fluctuations, viz.,

$$\limsup_{n\to\infty} (\mathcal{S}_n - n\mu) = +\sigma\sqrt{n}\sqrt{2\ln(\ln n)} ,$$

$$\liminf_{n\to\infty} (\mathcal{S}_n - n\mu) = -\sigma\sqrt{n}\sqrt{2\ln(\ln n)} .$$

For larger values of n the iterated logarithm $\ln(\ln n)$ is a very slowly increasing function of n, so the upper and lower bounds on the stochastic variable are not too different from $\sqrt{n}$ (Fig. 2.13). The law of the iterated logarithm is the rigorous final answer to the conjectured $\sqrt{n}$-law for fluctuations that we have mentioned several times already.

2.4.1 Binomial and Normal Distributions

Here we prove convergence of the binomial distribution to the normal distribution, which is the case where it appears most natural (Fig. 2.11). A binomial density,

$$B_k(n,p) = \binom{n}{k} p^k(1-p)^{n-k} , \quad k,n \in \mathbb{N} , \; 0 \le k \le n ,$$

becomes a normal density through extrapolation[17] to large values of n at constant p. The transformation from the binomial distribution to the normal distribution is properly done in two steps: (i) standardization and (ii) taking the limit $n \to \infty$ (see also [84, pp. 210–217]).

[17]This differs from the extrapolation performed in Sect. 2.3.2 because the limit $\lim_{n\to\infty} B_k(n, \alpha/n) = \pi_k(\alpha)$ leading to the Poisson distribution was performed for vanishing $p = \alpha/n$.

First we make the binomial distribution comparable to the standard normal density $\varphi(x) = \mathrm{e}^{-x^2/2}/\sqrt{2\pi}$ by shifting the maximum towards $x = 0$ and adjusting the width (Fig. 2.12). For $0 < p < 1$ and $q = 1 - p$, the discrete variable k is replaced by a new variable ξ :

$$\xi = \frac{k - np}{\sqrt{npq}} \,, \qquad 0 \le k \le n \,.$$

Note that the new variable ξ depends on k and n, but for short we dispense with subscripts. Instead of the variables $\mathcal{X}_k$ and $\mathcal{S}_k$ in (1.22′), we introduce new random variables $\mathcal{X}_k^*$ and $\mathcal{S}_n^* = \sum_{k=1}^{n} \mathcal{X}_k^*$ which account for centering around $x = 0$ and adjustment to the width of a standard Gaussian $\varphi(x)$ by making use of the expectation value $\mathrm{E}(\mathcal{S}_n) = np$ and the standard deviation $\sigma(\mathcal{S}_n) = \sqrt{npq}$ of the binomial distribution.

The Theorem of de Moivre and Laplace

The theorem states that for large values of n and k values in a neighborhood of $k = np$ with $|\xi| = |k - np| \le c$ and c being an arbitrarily small and fixed positive constant, the approximation

$$\binom{n}{k} p^k q^{n-k} \approx \frac{1}{\sqrt{2\pi npq}} \mathrm{e}^{-\xi^2/2} \,, \quad p + q = 1 \,, \ p > 0 \,, \ q > 0 \,, \tag{2.55′}$$

becomes exact in the sense that the ratio of the left-hand side to the right-hand side converges to one as $n \to \infty$ [160, Sect. VII.3]. The convergence is uniform with respect to k in the range specified above. A short and elegant proof of this convergence provides a nice exercise in performing properly the limits of large numbers [84, pp. 214–215]. Here we reproduce the proof in a slightly different and more straightforward way.

First we transform the left-hand side by making use of Stirling's approximation to the factorial, viz., $n! \approx n^n \mathrm{e}^{-n}\sqrt{2\pi n}$ as $n \to \infty$:

$$\binom{n}{k} p^k q^{n-k} = \frac{n!}{k!(n-k)!} p^k q^{n-k} \approx \sqrt{\frac{n}{2\pi k(n-k)}} \left(\left(\frac{k}{np}\right)^{-k} \left(\frac{n-k}{nq}\right)^{-(n-k)} \right) .$$

Next we introduce the variable $\xi = (k - np)/\sqrt{np\,q} = \big(-(n-k) + nq\big)/\sqrt{npq}$, and find

$$k = np + \xi\sqrt{npq} \,, \qquad n - k = nq - \xi\sqrt{npq} \,.$$

Neglecting $\sqrt{n}$ with respect to n in the limit $n \to \infty$, $k \approx np$ and $n - k \approx nq$, and we get

$$\sqrt{\frac{n}{2\pi k(n-k)}} \approx \frac{1}{\sqrt{2\pi npq}} .$$

A transformation to the exponential function yields

$$\begin{aligned}\binom{n}{k} p^k q^{n-k} &\approx \frac{1}{\sqrt{2\pi npq}} \left(\left(1 + \xi\sqrt{\frac{q}{np}}\right)^{-k} \left(1 - \xi\sqrt{\frac{p}{nq}}\right)^{-(n-k)} \right) \\ &= \frac{1}{\sqrt{2\pi npq}} \mathrm{e}^{\ln\left(\left(1+\xi\sqrt{q/np}\right)^{-k}\left(1-\xi\sqrt{p/nq}\right)^{-(n-k)}\right)} .\end{aligned}$$

Then the evaluation of the logarithm yields

$$\begin{aligned}&\ln\left(\left(1 + \xi\sqrt{\frac{q}{np}}\right)^{-k} \left(1 - \xi\sqrt{\frac{p}{nq}}\right)^{-(n-k)} \right) \\ &\qquad = -k \ln\left(1 + \xi\sqrt{\frac{q}{np}}\right) - (n-k)\ln\left(1 - \xi\sqrt{\frac{p}{nq}}\right) .\end{aligned}$$

Making use of the series expansion $\ln(1 \pm \gamma) \approx \pm\gamma - \gamma^2/2 \pm \gamma^3/3 - \cdots$, truncation after the second term yields

$$\begin{aligned}&-(np + \xi\sqrt{npq})\left(\xi\sqrt{\frac{q}{np}} - \frac{\xi^2}{2}\frac{q}{np} + \cdots\right) \\ &\qquad -(nq - \xi\sqrt{npq})\left(-\xi\sqrt{\frac{p}{nq}} - \frac{\xi^2}{2}\frac{p}{nq} + \cdots\right) .\end{aligned}$$

The linear terms cancel and the sum of the quadratic terms has the first non-vanishing coefficient. Evaluation of the expressions eventually yields

$$\ln\left(\left(1 + \xi\sqrt{\frac{q}{np}}\right)^{-k} \left(1 - \xi\sqrt{\frac{p}{nq}}\right)^{-(n-k)} \right) = -\frac{\xi^2}{2} + o(\xi^3) ,$$

and

$$\binom{n}{k} p^k q^{n-k} \approx \frac{1}{\sqrt{2\pi npq}} \mathrm{e}^{-\xi^2/2} ,$$

which proves the conjecture (2.55′). □

Comparing Figs. 2.10, 2.11, and 2.12, we see that the convergence of the binomial distribution to the normal distribution is particularly effective in the symmetric case $p = q = 0.5$. A value of $n = 20$ is sufficient to make the difference hardly recognizable with the unaided eye. Figure 2.12 also shows the effect of standardization on the binomial distribution. The difference is somewhat greater for the asymmetric case $p = 0.1$: in Fig. 2.11, we went up to the case $n = 500$, where the binomial and the normal density are almost indistinguishable.

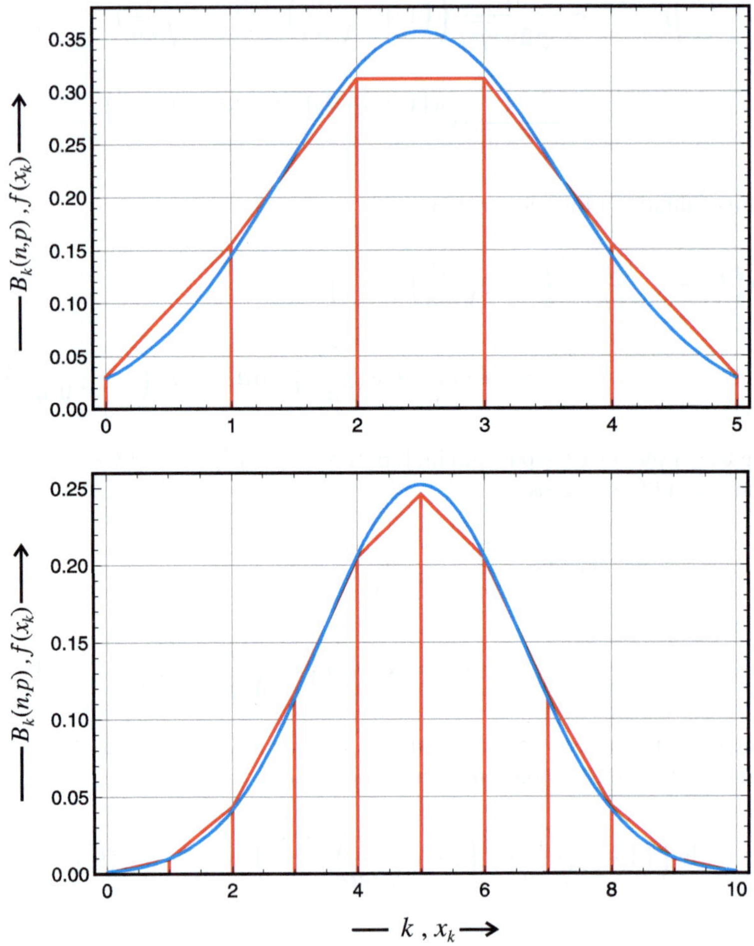

Fig. 2.10 Fit of the normal distribution to symmetric binomial distributions. The *curves* represent two examples of normal densities (*blue*) that were fitted to the points of the binomial distribution (*red*). Parameter choices for the binomial distributions: ($n = 5, p = 0.5$) and ($n = 10, p = 0.5$), for the *upper* and *lower plots*, respectively. The normal densities are determined by $\mu = np$ and $\sigma = \sqrt{np(1-p)}$

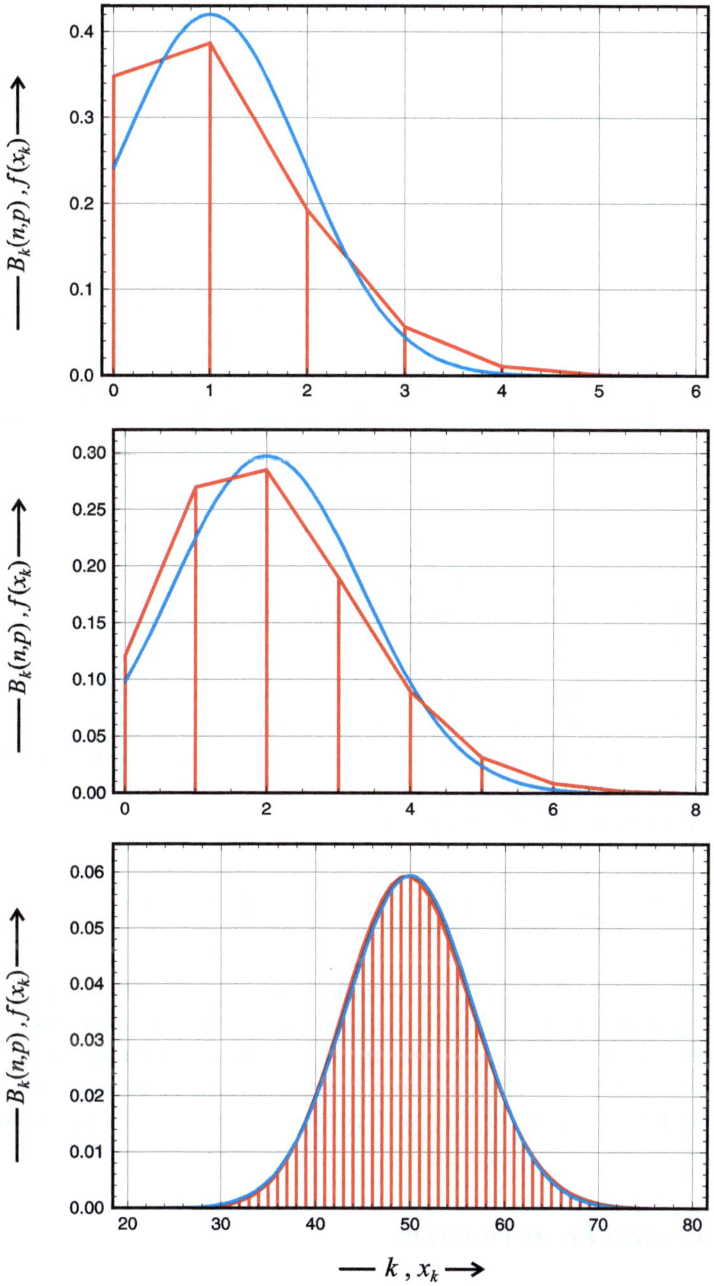

Fig. 2.11 Fit of the normal distribution to asymmetric binomial distributions. The *curves* represent three examples of normal densities (*blue*) that were fitted to the points of the binomial distribution (*red*). Parameter choices for the binomial distributions: $(n = 10, p = 0.1)$, $(n = 20, p = 0.1)$, and $(n = 500, p = 0.1)$, for the *upper*, *middle*, and *lower plots*, respectively. The normal densities are determined by $\mu = np$ and $\sigma = \sqrt{np(1-p)}$

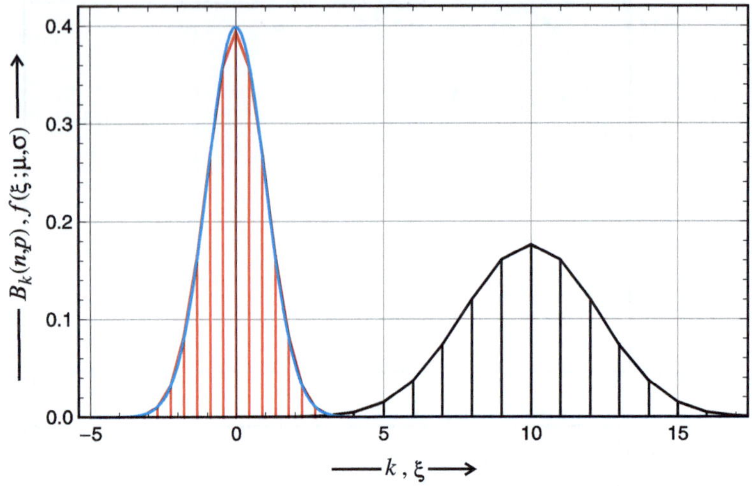

Fig. 2.12 Standardization of the binomial distribution. The figure shows a symmetric binomial distribution $B(20, 1/2)$ which is centered around $\mu = 10$ (*black*). The transformation yields a binomial distribution centered around the origin with unit variance: $\sigma = \sigma^2 = 1$ (*red*). The *blue curve* is a standardized normal density $\varphi(x)$ ($\mu = 0, \sigma^2 = 1$)

In the context of the central limit theorem (Sect. 2.4.2), it is appropriate to formulate the theorem of de Moivre and Laplace in a slightly different way: the distribution of the standardized random variable $\mathcal{S}_n^*$ with a binomial distribution converges in the limit of large numbers n to the normal distribution $\varphi(x)$ on any finite constant interval $]a, b]$ with $a < b$:

$$\lim_{n\to\infty} P\left(\left(\frac{\mathcal{S}_n - np}{\sqrt{npq}}\right) \in]a, b]\right) = \frac{1}{\sqrt{2\pi}} \int_a^b \mathrm{e}^{-x^2/2}\, \mathrm{d}x\,. \tag{2.55}$$

In the proof [84, pp. 215–217], the definite integral $\int_a^b \varphi(x)\, \mathrm{d}x$ is partitioned into n small segments just as in the Riemann integral, where the segments still reflect the discrete distribution. In the limit $n \to \infty$, the partition becomes finer and eventually converges to the continuous function described by the integral. In the sense of Sect. 1.8.1, we are dealing with convergence to a limit in distribution.

2.4.2 *Central Limit Theorem*

In addition to the transformation of the binomial distribution into the normal distribution analyzed in Sect. 2.4.1, we have already encountered two cases where other probability distributions approach the normal distribution in the limit of large numbers n: (i) the distribution of scores for rolling n dice simultaneously

(Sect. 1.9.1) and (ii) the Poisson distribution (Sect. 2.3.3). Therefore it is reasonable to conjecture a more general role for the normal distribution in the limit of large numbers. The Russian mathematician Aleksandr Lyapunov pioneered the formulation and derivation of the generalization known as the *central limit theorem* (CLT) [361, 362]. Research on CLT continued and was completed at least for practical purposes through extensive studies during the twentieth century [6, 493]. The *central limit theorem* comes in various stronger and weaker forms. We mention three of them here:

(i) The so-called *classical central limit theorem* is commonly associated with the names of the Finnish mathematician Jarl Waldemar Lindeberg [349] and the French mathematician Paul Pierre Lévy [339]. It is the most common version used in practice. In essence, the Lindeberg–Lévy central limit theorem is nothing but the generalization of the de Moivre–Laplace theorem (2.55) that was used in Sect. 2.4.1 to prove the transition from the binomial to the normal distribution in the limit $n \to \infty$. The generalization proceeds from Bernoulli variables to *independent and identically distributed* (iid) random variables $\mathcal{X}_i$. The distribution is arbitrary, i.e., it need not be specified, and the only requirements are a finite expectation value and a finite variance: $\mathrm{E}(\mathcal{X}_i) = \mu < \infty$ and $\mathrm{var}(\mathcal{X}_i) = \sigma^2 < \infty$. Again we consider the sum $\mathcal{S}_n = \sum_{i=1}^{n} \mathcal{X}_i$ of n random variables, standardize to yield $\mathcal{X}_i^*$ and $\mathcal{S}_n^*$, and instead of (2.55), obtain

$$\lim_{n\to\infty} P\left(\frac{\mathcal{S}_n - n\mu}{\sqrt{n}\sigma} \in]a,b]\right) = \frac{1}{\sqrt{2\pi}} \int_a^b \mathrm{e}^{-x^2/2}\,\mathrm{d}x\,. \tag{2.56}$$

For every segment $a < b$, the arbitrary initial distribution converges to the normal distribution in the limit $n \to \infty$. Although this is already a remarkable extension of the validity in the limit of the normal distribution, the results can be made more general.

(ii) Lyapunov's earlier version of the central limit theorem [361, 362] requires only independent and not necessarily identically distributed variables $\mathcal{X}_i$ with finite expectation values μ_i and variances σ_i^2, provided a criterion called the *Lyapunov condition* is satisfied by the sum $s_n^2 = \sum_{i=1}^{n} \sigma_i^2$ of the variances:

$$\lim_{n\to\infty} \frac{1}{s^{2+\delta}} \sum_{i=1}^{n} \mathrm{E}\left(|\mathcal{X}_i - \mu_i|^{2+\delta}\right) = 0\,. \tag{2.57}$$

Then the sum $\sum_{i=1}^{n} (\mathcal{X}_i - \mu_i)/s_n$ converges in distribution in the limit $n \to \infty$ to the standard normal distribution:

$$\frac{1}{s_n} \sum_{i=1}^{n} (\mathcal{X}_i - \mu_i) \xrightarrow{\mathrm{d}} \mathcal{N}(0,1)\,. \tag{2.58}$$

In practice, whether or not a given sequence of random variables satisfies the Lyapunov condition is commonly checked by setting $\delta = 1$.

(iii) Lindeberg showed in 1922 [350] that a weaker condition than Lyapunov's was sufficient to guarantee convergence in distribution to the standard normal distribution:

$$\lim_{n\to\infty} \frac{1}{s_n^2} \sum_{i=1}^{n} \mathrm{E}\Big((X_i - \mu_i)^2 \mathbf{1}_{|\mathcal{X}_i - \mu_i| > \epsilon s_n} \Big) = 0 \,, \tag{2.59}$$

where $\mathbf{1}_{|\mathcal{X}_i - \mu_i| > \epsilon s_n}$ is the indicator function (1.26a) identifying the sample space

$$\big\{ |\mathcal{X}_i - \mu_i| > \epsilon\, s_n \big\} \doteq \big\{ \omega \in \Omega : |\mathcal{X}_i(\omega) - \mu_i| > \epsilon s_n \big\} \,.$$

If a sequence of random variables satisfies Lyapunov's condition, it also satisfies Lindeberg's condition, but the converse does not hold in general. Lindeberg's condition is sufficient but not necessary in general, and the condition for necessity is

$$\max_{i=1,\ldots,n} \frac{\sigma_i^2}{s_n^2} \longrightarrow 0 \,, \quad \text{as } n \to \infty \,.$$

In other words, the Lindeberg condition is satisfied if and only if the central limit theorem holds.

The three versions of the central limit theorem are related to each other: Lindeberg's condition (iii) is the most general form, and hence both the classical CLT (i) and the Lyapunov CLT (ii) can be derived as special cases from (iii). It is worth noting, however, that (i) does not necessarily follow from (ii), because (i) requires a finite second moment whereas the condition for (ii) is a finite moment of order $(2+\delta)$.

In summary the central limit theorem for a sequence of independent random variables $\mathcal{S}_n = \sum_{i=1}^{n} \mathcal{X}_i$ with finite means, $\mathrm{E}(\mathcal{X}_i) = \mu_i < \infty$, and variances, $\mathrm{var}(\mathcal{X}_i) = \sigma_i^2 < \infty$, states that the sum $\mathcal{S}_n$ converges in distribution to a standardized normal density $\mathcal{N}(0,1)$ without any further restriction on the densities of the variables. The literature on the central limit theorem is enormous and several proofs with many variants have been derived (see, for example, [83] or [84, pp. 222–224]). We dispense here with a repetition of this elegant proof that makes use of the characteristic function, and present only the key equation for the convergence where the number n approaches infinity with s fixed:

$$\lim_{n\to\infty} \mathrm{E}\big(\mathrm{e}^{\mathrm{i}s\,\mathcal{S}_n^*}\big) = \lim_{n\to\infty} \left(1 - \frac{s^2}{2n} \left(1 + \varepsilon \frac{s}{\sqrt{n}} \right) \right)^n = \mathrm{e}^{-s^2/2} \,, \tag{2.60}$$

with ε being any small positive constant.

For practical applications used in the *statistics of large samples*, the central limit theorem as encapsulated in (2.60) is turned into the rough approximation

$$P(\sigma\sqrt{n}x_1 < \mathcal{S}_n - n\mu < \sigma\sqrt{n}x_2) \approx F_{\mathcal{N}}(x_2) - F_{\mathcal{N}}(x_1)\ . \tag{2.61}$$

The spread around the mean μ is obtained by setting $x = x_1 = -x_2$:

$$P(|\mathcal{S}_n - n\mu| < \sigma\sqrt{n}x) \approx 2F_{\mathcal{N}}(x) - 1\ . \tag{2.61$'$}$$

In pre-computer days, (2.61) was used extensively with the aid of tabulations of the functions $F_{\mathcal{N}}(x)$ and $F_{\mathcal{N}}^{-1}(x)$, which are still found in most textbooks of statistics.

2.4.3 Law of Large Numbers

The *law of large numbers* states that in the limit of infinitely large samples the sum of random variables converges to the expectation value:

$$\frac{1}{n}\mathcal{S}_n = \frac{1}{n}(\mathcal{X}_1 + \mathcal{X}_2 + \cdots + \mathcal{X}_n) \longrightarrow \mu\ , \quad \text{for } n \to \infty\ .$$

In its strong form the law can be expressed as

$$P\left(\lim_{n\to\infty}\frac{1}{n}\mathcal{S}_n = \mu\right) = 1\ . \tag{2.62a}$$

In other words, the sample average converges almost certainly to the expectation value.

The weaker form of the law of large numbers is written as

$$P\left(\lim_{n\to\infty}\left|\frac{1}{n}\mathcal{S}_n - \mu\right| > \varepsilon\right) = 0\ , \tag{2.62b}$$

and implies convergence in probability: $\mathcal{S}_n/n \xrightarrow{\text{P}} \mu$. The weak law states that, for any sufficiently large sample, there exists a zone $\mu \pm \varepsilon$ around the expectation value, no matter how small ε is, such that the average of the observed quantity will come so close to the expectation value that it lies within this zone.

It is also instructive to visualize the difference between the strong and the weak law from a dynamical perspective. The weak law says that the average $\mathcal{S}_n/n$ will be near μ, provided n is sufficiently large. The sample, however, may rarely but infinitely often leave the zone and satisfy $|\mathcal{S}_n/n - \mu| > \varepsilon$, and the frequency with which this happens is of measure zero. The strong law asserts that such excursions will almost certainly never happen and the inequality $|\mathcal{S}_n/n - \mu| < \varepsilon$ holds for all large enough n.

The law of large numbers can be derived as a straightforward consequence of the central limit theorem (2.56) [84, pp. 227–233]. For any fixed but arbitrary constant $\varepsilon > 0$, we have

$$\lim_{n\to\infty} P\left(\left|\frac{\mathcal{S}_n}{n} - \mu\right| < \varepsilon\right) = 1 \,. \tag{2.63}$$

The constant ε is fixed and therefore we can define a positive constant ℓ that satisfies $\ell < \varepsilon\sqrt{n}/\sigma$ and for which

$$\left(\left|\frac{\mathcal{S}_n - n\mu}{\sqrt{n}\sigma}\right| < \ell\right) \implies \left(\left|\frac{\mathcal{S}_n - n\mu}{n}\right| < \varepsilon\right) ,$$

and hence,

$$P\left(\left|\frac{\mathcal{S}_n - n\mu}{\sqrt{n}\sigma}\right| < \ell\right) \le P\left(\left|\frac{\mathcal{S}_n - n\mu}{n}\right| < \varepsilon\right) ,$$

provided n is sufficiently large. Now we go back to (2.56) and choose a symmetric interval $a = -\ell$ and $b = +\ell$ for the integral. Then the left-hand side of the inequality converges to $\int_{-l}^{+l} \exp(-x^2/2)\mathrm{d}x/\sqrt{2\pi}$ in the limit $n \to \infty$. For any $\delta > 0$, we can choose ℓ so large that the value of the integral exceeds $1 - \delta$, and for sufficiently large values of n, we get

$$P\left(\left|\frac{\mathcal{S}_n}{n} - \mu\right| < \varepsilon\right) = 1 - \delta \,. \tag{2.64}$$

This proves that the law of large numbers (2.63) is a corollary of (2.56). □

Related to and a consequence of (2.63) is Chebyshev's inequality for random variables $\mathcal{X}$ that have a finite second moment, which is named after the Russian mathematician Pafnuty Lvovich Chebyshev:

$$P(|\mathcal{X}| \ge c) \le \frac{\mathrm{E}(\mathcal{X}^2)}{c^2} \,, \tag{2.65}$$

and which is true for any constant $c > 0$. We dispense here with a proof, which can be found in [84, pp. 228–233]. Using Chebyshev's inequality, the law of large numbers (2.63) can be extended to a sequence of independent random variables $\mathcal{X}_j$ with different expectation values and variances, $\mathrm{E}(\mathcal{X}_j) = \mu_j$ and $\mathrm{var}(\mathcal{X}_j) = \sigma_j^2$, with the restriction that there exists a constant $\Sigma^2 < \infty$ such that $\sigma_j^2 \le \Sigma^2$ is satisfied for all $\mathcal{X}_j$. Then we have, for each $c > 0$,

$$\lim_{n\to\infty} P\left(\left|\frac{\mathcal{X}_1 + \cdots + \mathcal{X}_n}{n} - \frac{\mu^{(1)} + \cdots + \mu^{(n)}}{n}\right| < c\right) = 1 \,. \tag{2.66}$$

The main message of the law of large numbers is that, for a sufficiently large number of independent events, the statistical errors in the sum will vanish and the mean will converge to the exact expectation value. Hence, the law of large numbers provides the basis for the assumption of convergence in mathematical statistics (Sect. 2.6).

2.4.4 Law of the Iterated Logarithm

The *law of the iterated logarithm* consists of two asymptotic regularities derived for sums of random variables, which are related to the central limit theorem and the law of large numbers, and in an important way complete the predictions of both. The name of the law arises due to the appearance of the function log log in the forthcoming expressions—it does not refer to the notion of the *iterated logarithm* in computer science[18] – and the derivation is attributed to the two Russian scholars of mathematics Aleksandr Khinchin [300] and Andrey Kolmogorov [309]. To the degree of generality used here, the proof was provided later [157, 242]. The law of the iterated logarithm provides upper and lower bounds for the values of sums of random variables, and in this ways confines the size of fluctuations.

For a sum of n independent and identically distributed (iid) random variables with expectation value $\mathrm{E}(\mathcal{X}_i) = \mu$ and finite variance $\mathrm{var}(\mathcal{X}) = \sigma^2 < \infty$, viz.,

$$\mathcal{S}_n = \mathcal{X}_1 + \mathcal{X}_1 + \cdots + \mathcal{X}_n \,,$$

the following two limits are satisfied with probability one:

$$\limsup_{n\to\infty} \frac{\mathcal{S}_n - n\mu}{\sqrt{2n\ln(\ln n)}} = +|\sigma| \,, \tag{2.67a}$$

$$\liminf_{n\to\infty} \frac{\mathcal{S}_n - n\mu}{\sqrt{2n\ln(\ln n)}} = -|\sigma| \,. \tag{2.67b}$$

The two theorems (2.67) are equivalent and this follows directly from the symmetry of the standardized normal distribution $\mathcal{N}(0,1)$. We dispense here with the presentation of a proof for the law of the iterated logarithm. This can be found,

---

[18] In computer science, the iterated logarithm of n is commonly written $\log^* n$ and represents the number of times the logarithmic function must be iteratively applied before the result is less than or equal to one:

$$\log^* \doteq \begin{cases} 0\,, & \text{if } n \le 1\,, \\ 1 + \log^*(\log n)\,, & \text{if } n > 1\,. \end{cases}$$

The iterated logarithm is well defined for base e, for base 2, and in general for any base greater than $\mathrm{e}^{1/\mathrm{e}} = 1.444667\ldots$.

for example, in the monograph by Henry McKean [380] or in the publication by William Feller [157]. For the purpose of illustration, we compare with the already mentioned heuristic $\sqrt{n}$-law (see Sect. 1.1), which is based on the properties of the symmetric standardized binomial distribution $B(n, p)$ with $p = 1/2$. Accordingly, we have $2\sigma/n = 1/\sqrt{n}$ and consequently most values of $\mathcal{S}_n - n\mu$ lie in the interval $-|\sigma| \leq \mathcal{S}_n \leq +|\sigma|$. The corresponding result from the law of the iterated logarithm is

$$-\sqrt{\frac{2\ln(\ln n)}{n}} \leq \mathcal{S}_n \leq +\sqrt{\frac{2\ln(\ln n)}{n}}$$

with probability one. One particular case of iterated Bernoulli trials—tosses of a fair coin—is shown in Fig. 2.13, where the envelope of the sum $\mathcal{S}_n$ of the cumulative score of n trials, $\pm\sqrt{2\ln(\ln n)/n}$ is compared with the results of the naïve square root n law, $\mu \pm \sigma = \pm\sqrt{1/n}$. We remark that the sum quite frequently takes on values close to the envelopes. The special importance of the law of the iterated logarithm for the Wiener process will be discussed in Sect. 3.2.2.2.

In essence, we may summarize the results of this section in three statements, which are part of *large sample theory*. For independent and identically distributed

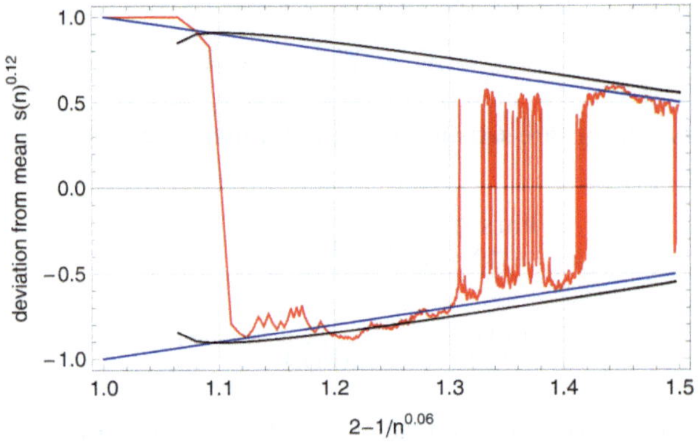

Fig. 2.13 Illustration of the law of the iterated logarithm. The picture shows the sum of the score of a sequence of Bernoulli trials with the outcome $\mathcal{X}_i = \pm 1$ and $\mathcal{S}_n = \sum_{i=1}^{n} \mathcal{X}_i$. The standardized sum, $\mathcal{S}(n)/n - \mu = s(n) - \mu = s(n)$ since $\mu = 0$, is shown as a function of n. In order to make the plot illustrative, we adopt the scaling of the axes proposed by Dean Foster [184] which yields a straight line for the function $\sigma(n) = 1/\sqrt{n}$. On the x-axis, we plot $x(n) = 2 - 1/n^{0.06}$, and this results in the following pairs of values: $(x, n) = (1, 1)$, $(1.129, 10)$, $(1.241, 100)$, $(1.339, 1000)$, $(1.564, 10^6)$, $(1.810, 10^{12})$, and $(2, \infty)$. The y-axis is split into two halves corresponding to positive and negative values of $s(n)$. In the positive half we plot $s(n)^{0.12}$ and in the negative half $-|s(n)|^{0.12}$ in order to yield symmetry between the positive and the negative zones. The two *blue curves* provide an envelope $\mu \pm \sigma = \mu \pm \sqrt{1/n}$, and the two *black curves* present the results of the law of the iterated logarithm, $\mu \pm \sqrt{2\ln(\ln n)/n}$. Note that the function $\ln(\ln n)$ assumes negative values for $1 < x < 1.05824$ $(1 < n < 2.71828)$

(iid) random variables $\mathcal{X}_i$ and $\mathcal{S}_n = \sum_{i=1}^n \mathcal{X}_i$, with $\mathrm{E}(\mathcal{X}_i) = \mathrm{E}(\mathcal{X}) = \mu$ and finite variance $\mathrm{var}(\mathcal{X}_i) = \sigma < \infty$, we have the three large sample results:

(i) The *law of large numbers*: $\mathcal{S}_n \longrightarrow n\mathrm{E}(\mathcal{X}) = n\mu$.
(ii) The *law of the iterated logarithm*:

$$\limsup \frac{(\mathcal{S}_n - n\mu)}{\sqrt{2n\ln(\ln n)}} \longrightarrow +|\sigma| \,, \qquad \liminf \frac{(\mathcal{S}_n - n\mu)}{\sqrt{2n\ln(\ln n)}} \longrightarrow -|\sigma| \,.$$

(iii) The *central limit theorem*: $\frac{1}{\sqrt{n}}\big[\mathcal{S}_n - n\mathrm{E}(\mathcal{X})\big] \longrightarrow \mathcal{N}(0,1)$.

Theorem (1) defines the limit of the sample average, while theorem (2) determines the size of fluctuations, and theorem (3) refers to the limiting probability density, which turns out to be the normal distribution. All three theorems can be extended in their range of validity to independent random variables with arbitrary distributions, provided that the mean and variance are finite.

2.5 Further Probability Distributions

In Sect. 2.3, we presented the three most important probability distributions: (i) the Poisson distribution is highly relevant, because it describes the distribution of occurrence of independent events, (ii) the binomial distribution deals with the most frequently used simple model of randomness, independent trials with two outcomes, and (iii) the normal distribution is the limiting distribution of large numbers of individual events, irrespective of the statistics of single events. In this section we shall discuss ten more or less arbitrarily selected distributions which play an important role in science and/or in statistics. The presentation here is inevitably rather brief, and for a more detailed treatment, we refer to [284, 285]. Other probability distributions will be mentioned together with the problems to which they are applied, e.g., the *Erlang distribution* in the discussion of the Poisson process (Sect. 3.2.2.4) and the *Maxwell–Boltzmann distribution* in the derivation of the chemical rate parameter from molecular collisions (Sect. 4.1.4).

2.5.1 The Log-Normal Distribution

The *log-normal* distribution is a continuous probability distribution of a random variable $\mathcal{Y}$ with a normally distributed logarithm. In other words, if $\mathcal{X} = \ln \mathcal{Y}$ is normally distributed, then $\mathcal{Y} = \exp(\mathcal{X})$ has a log-normal distribution. Accordingly, $\mathcal{Y}$ can assume only positive real values. Historically, this distribution had several other names, the most popular of them being *Galton's distribution*, named after the pioneer of statistics in England, Francis Galton, or *McAlister's distribution*, named after the statistician Donald McAlister [284, chap. 14, pp. 207–258].

The log-normal distribution meets the need for modeling empirical data that show frequently observed deviations from the conventional normal distribution: (i) meaningful data are nonnegative, (ii) positive skew implying that there are more values above than below the maximum of the probability density function (pdf), and (iii) a more obvious meaning attributed to the geometric rather than the arithmetic mean [191, 378]. Despite its obvious usefulness and applicability to problems in science, economics, and sociology the log-normal distribution is not popular among non-statisticians [346].

The log-normal distribution contains two parameters, $\ln\mathcal{N}(\mu,\sigma^2)$ with $\mu \in \mathbb{R}$ and $\sigma^2 \in \mathbb{R}_{>0}$, and is defined on the domain $x \in]0,\infty[$. The density function (pdf) and the cumulative distribution (cdf) are given by (Fig. 2.14):

$$
\begin{aligned}
f_{\ln\mathcal{N}}(x) &= \frac{1}{x\sqrt{2\pi\sigma^2}}\exp\left(-\frac{(\ln x-\mu)^2}{2\sigma^2}\right) \qquad \text{(pdf)}\,,\\
F_{\ln\mathcal{N}}(x) &= \frac{1}{2}\left(1+\operatorname{erf}\left(\frac{\ln x-\mu}{\sqrt{2\sigma^2}}\right)\right) \qquad \text{(cdf)}\,.
\end{aligned}
\tag{2.68}
$$

By definition, the logarithm of the variable $\mathcal{X}$ is normally distributed, and this implies

$$\mathcal{X} = \mathrm{e}^{\mu+\sigma\mathcal{N}}\,,$$

where $\mathcal{N}$ stands for a standard normal variable. The moments of the log-normal distribution are readily calculated[19]:

$$
\begin{array}{ll}
\text{Mean} & \mathrm{e}^{\mu+\sigma^2/2}\\
\text{Median} & \mathrm{e}^{\mu}\\
\text{Mode} & \mathrm{e}^{\mu-\sigma^2}\\
\text{Variance} & (\mathrm{e}^{\sigma^2}-1)\,\mathrm{e}^{2\mu+\sigma^2}\\
\text{Skewness} & (\mathrm{e}^{\sigma^2}+2)\,\sqrt{\mathrm{e}^{\sigma^2}-1}\\
\text{Kurtosis} & \mathrm{e}^{4\sigma^2}+2\mathrm{e}^{3\sigma^2}+3\mathrm{e}^{2\sigma^2}-6
\end{array}
\tag{2.69}
$$

The skewness γ_1 is always positive and so is the (excess) kurtosis, since $\sigma^2 = 0$ yields $\gamma_2 = 0$, and $\sigma^2 > 0$ implies $\gamma_2 > 0$.

[19]Here and in the following listings for other distributions, 'kurtosis' stands for excess kurtosis $\gamma_2 = \beta_2 - 3 = \mu_4/\sigma^4$.

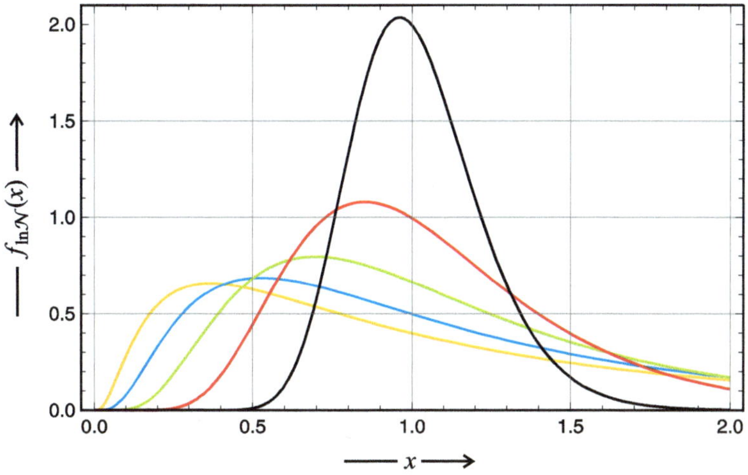

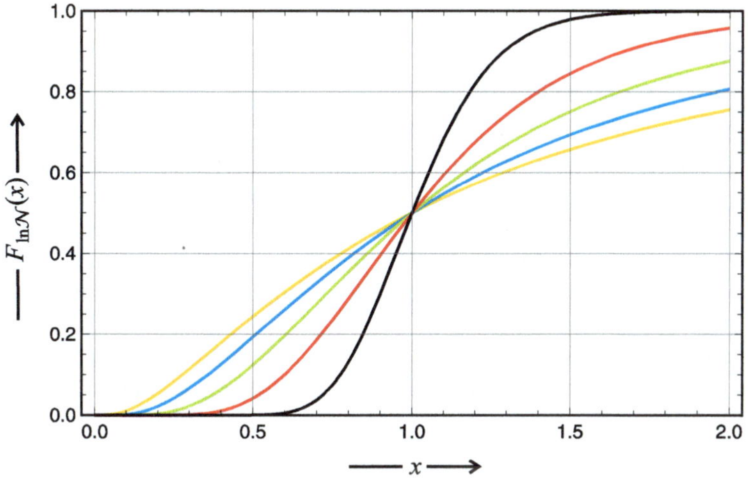

Fig. 2.14 The log-normal distribution. The log-normal distribution $\ln\mathcal{N}(\mu, \sigma)$ is defined on the positive real axis $x \in]0, \infty[$ and has the probability density (pdf)

$$f_{\ln\mathcal{N}}(x) = \frac{\exp\big(-(\ln x - \mu)^2/2\sigma^2\big)}{x\sqrt{2\pi\sigma^2}}$$

and the cumulative distribution function (cdf)

$$F_{\ln\mathcal{N}}(x) = \frac{1}{2}\Big(1 + \mathrm{erf}\big((\ln x - \mu)/\sqrt{2\sigma^2}\big)\Big)\,.$$

The two parameters are restricted by the relations $\mu \in \mathbb{R}$ and $\sigma^2 > 0$. Parameter choice and color code: $\mu = 0$, $\sigma =$0.2 (*black*), 0.4 (*red*), 0.6 (*green*), 0.8 (*blue*), and 1.0 (*yellow*)

The entropy of the log-normal distribution is

$$H(f_{\ln \mathcal{N}}) = \frac{1}{2}\Big(1 + \ln(2\pi\sigma^2) + 2\mu\Big) . \qquad (2.70)$$

As the normal distribution has the maximum entropy of all distributions defined on the real axis $x \in \mathbb{R}$, the log-normal distribution is the maximum entropy probability distribution for a random variable $\mathcal{X}$ for which the mean and variance of $\ln \mathcal{X}$ are fixed.

Finally, we mention that the log-normal distribution can be well approximated by a distribution [519]

$$F(x; \mu\sigma) = \left(\left(\frac{e^{\mu}}{x}\right)^{\pi/\sigma\sqrt{3}} + 1\right)^{-1} ,$$

which has integrals that can be expressed in terms of elementary functions.

2.5.2 The χ^2-Distribution

The χ^2-distribution, also written chi-squared distribution, is one of the most frequently used distributions in inferential statistics for hypothesis testing and construction of confidence intervals.[20] In particular, the χ^2 distribution is applied in the common χ^2-test for the quality of the fit of an empirically determined distribution to a theoretical one (Sect. 2.6.2). Many other statistical tests are based on the χ^2-distribution as well.

The chi-squared distribution χ_k^2 is the distribution of a random variable $\mathcal{Q}$ which is given by the sum of the squares of k independent, standard normal variables with distribution $\mathcal{N}(0, 1)$:

$$\mathcal{Q} = \sum_{i=1}^{k} \mathcal{X}_i^2 . \qquad (2.71)$$

The only parameter of the distribution, namely k, is called the number of *degrees of freedom*. It is tantamount to the number of independent variables $\mathcal{X}_i$. $\mathcal{Q}$ is defined on the positive real axis (including zero) $x \in [0, \infty[$ and has the following density

[20] The chi-squared distribution is sometimes written $\chi^2(k)$, but we prefer the subscript since the number of degrees of freedom, the parameter k, specifies the distribution. Often the random variables $\mathcal{X}_i$ satisfy a conservation relation and then the number of independent variables is reduced to $k-1$, and we have χ_{k-1}^2 (Sect. 2.6.2).

function and cumulative distribution (Fig. 2.15):

$$f_{\chi_k^2}(x) = \frac{x^{k/2-1}e^{-x/2}}{2^{k/2}\Gamma(k/2)} \,, \quad x \in \mathbb{R}_{\geq 0} \qquad \text{(pdf)} \,,$$
$$F_{\chi_k^2}(x) = \frac{\gamma(k/2, x/2)}{\Gamma(k/2)} = Q\left(\frac{k}{2}, \frac{x}{2}\right) \qquad \text{(cdf)} \,, \tag{2.72}$$

where $\gamma(k, z)$ is the lower incomplete Gamma function and $Q(k, z)$ is the regularized Gamma function. The special case with $k = 2$ has the particularly simple form $F_{\chi_2^2}(x) = 1 - e^{-x/2}$.

The conventional χ^2-distribution is sometimes referred to as the *central* χ^2-distribution in order to distinguish it from the *noncentral* χ^2-distribution, which is derived from k independent and normally distributed variables with means μ_i and variances σ_i^2. The random variable

$$\mathcal{Q} = \sum_{i=1}^{k} \left(\frac{\mathcal{X}_i}{\sigma_i}\right)^2$$

is distributed according to the noncentral χ^2-distribution $\chi_k^2(\lambda)$ with two parameters, k and λ, where $\lambda = \sum_{i=1}^{k}(\mu_i/\sigma_i)^2$ is the *noncentrality* parameter.

The moments of the central χ_k^2-distribution are readily calculated:

Mean	k	
Median	$\approx k\left(1 - \frac{2}{9k}\right)^3$	
Mode	$\max\{k-2, 0\}$	
Variance	$2k$	(2.73)
Skewness	$\sqrt{8/k}$	
Kurtosis	$12/k$	

The skewness γ_1 is always positive and so is the excess kurtosis γ_2. The raw moments $\hat{\mu}_n = E(\mathcal{Q}^n)$ and the cumulants of the χ_k^2-distribution have particularly simple expressions:

$$E(\mathcal{Q}^n) = \hat{\mu}_n = k(k+2)(k+4)\cdots(k+2n-2) = 2^n \frac{\Gamma(n+k/2)}{\Gamma(k/2)} \,, \tag{2.74}$$

$$\kappa_n = 2^{n-1}(n-1)!\,k \,. \tag{2.75}$$

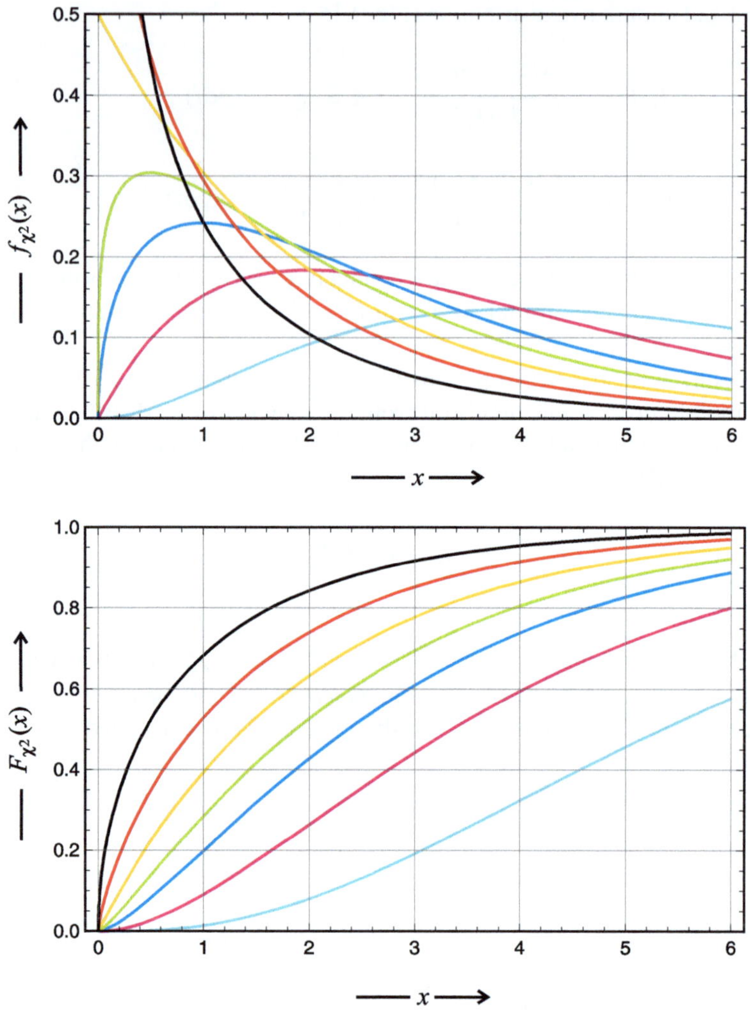

Fig. 2.15 The χ^2 distribution. The chi-squared distribution χ_k^2, $k \in \mathbb{N}$, is defined on the positive real axis $x \in [0, \infty[$. The parameter k, called the number of degrees of freedom, has the probability density (pdf)

$$f_{\chi_k^2}(x) = \frac{x^{k/2-1}\mathrm{e}^{-x/2}}{2^{k/2}\Gamma(k/2)}$$

and the cumulative distribution function (cdf)

$$F_{\chi_k^2}(x) = \frac{\gamma\,(k/2, x/2)}{\Gamma\,(k/2)} \ .$$

Parameter choice and color code: k =1 (*black*), 1.5 (*red*), 2 (*yellow*), 2.5 (*green*), 3 (*blue*), 4 (*magenta*), and 6 (*cyan*). Although k, the number of degrees of freedom, is commonly restricted to integer values, we also show here the curves for two intermediate values (k=1.5, 2.5)

The entropy of the χ_k^2-distribution is readily calculated by integration:

$$H(f_{\chi^2}) = \frac{k}{2} + \ln\left(2\Gamma\left(\frac{k}{2}\right)\right) + \left(1 - \frac{k}{2}\right)\psi\left(\frac{k}{2}\right), \tag{2.76}$$

where $\psi(x) = \frac{\mathrm{d}}{\mathrm{d}x} \ln \Gamma(x)$ is the digamma function.

The χ_k^2-distribution has the simple characteristic function

$$\phi_{\chi^2}(s) = (1 - 2\mathrm{i}s)^{-k/2}. \tag{2.77}$$

The moment generating function is only defined for $s < 1/2$:

$$M_{\chi^2}(s) = (1 - 2s)^{-k/2}, \quad \text{for } s < 1/2. \tag{2.78}$$

Because of its central importance in significance tests, numerical tables of the χ^2-distribution are found in almost every textbook of mathematical statistics.

2.5.3 *Student's t-Distribution*

Student's t-distribution has a remarkable history. It was discovered by the famous English statistician William Sealy Gosset, who published his works under the pen name 'Student' [441]. Gosset was working at the brewery of Arthur Guinness in Dublin, Ireland, where it was forbidden to publish any paper, regardless of the subject matter, because Guinness was afraid that trade secrets and other confidential information might be disclosed. Almost all of Gosset's papers, including the one describing the t-distribution, were published under the pseudonym 'Student' [516]. Gosset's work was known to and supported by Karl Pearson, but it was Ronald Fisher who recognized and appreciated the importance of Gosset's work on small samples and made it popular [171].

Student's t-distribution is a family of continuous, normal-like probability distributions that apply to situations where the sample size is small, the variance is unknown, and one wants to derive a reliable estimate of the mean. Student's distribution plays a role in a number of commonly used tests for analyzing statistical data. An example is Student's test for assessing the significance of differences between two sample means—for example to find out whether or not a difference in mean body height between basketball players and soccer players is significant—or the construction of confidence intervals for the difference between population means. In a way, Student's t-distribution is required for *higher order statistics* in the sense of a statistics of statistics, for example, to estimate how likely it is to find the true mean within a given range around the finite sample mean (Sect. 2.6). In other words, n samples are taken from a population with a normal distribution having fixed but unknown mean and variance, the sample mean and the sample variance are computed from these n points, and the t-distribution is the distribution of the location of the true mean relative to the sample mean, calibrated by the sample standard deviation.

To make the meaning of Student's t-distribution precise, we assume n independent random variables $\mathcal{X}_i$, $i = 1, \ldots, n$, drawn from the same population, which is normally distributed with mean value $\mathrm{E}(\mathcal{X}_i) = \mu$ and variance $\mathrm{var}(\mathcal{X}_i) = \sigma^2$. Then the sample mean and the unbiased sample variance are the random variables

$$\overline{\mathcal{X}}_n = \frac{1}{n}\sum_{i=1}^{n} \mathcal{X}_i \,, \qquad \mathcal{S}_n^2 = \frac{1}{n-1}\sum_{i=1}^{n}(\mathcal{X}_i - \overline{\mathcal{X}}_n)^2 \,.$$

According to Cochran's theorem [85], the random variable $\mathcal{V} = (n-1)\mathcal{S}_n^2/\sigma^2$ follows a χ^2-distribution with $k = r = n-1$ degrees of freedom. The deviation of the sample mean from the population mean is properly expressed by the variable

$$\mathcal{Z} = (\overline{\mathcal{X}}_n - \mu)\frac{\sqrt{n}}{\sigma} \,, \tag{2.79}$$

which is the basis for the calculation of z-scores.[21] The variable $\mathcal{Z}$ is normally distributed with mean zero and variance one, as follows from the fact that the sample mean $\overline{\mathcal{X}}_n$ obeys a normal distribution with mean μ and variance σ^2/n. In addition, the two random variables $\mathcal{Z}$ and $\mathcal{V}$ are independent, and the *pivotal quantity*[22]

$$\mathcal{T} \doteq \frac{\mathcal{Z}}{\sqrt{\mathcal{V}/(n-1)}} = (\overline{\mathcal{X}}_n - \mu)\frac{\sqrt{n}}{\mathcal{S}_n} \tag{2.80}$$

follows a Student's t-distribution, which depends on the degrees of freedom $r = n-1$, but on neither μ nor σ.

Student's distribution is a one-parameter distribution with r the number of sample points or the so-called degree of freedom. It is symmetric and bell-shaped like the normal distribution, but the tails are heavier in the sense that more values fall further away from the mean. Student's distribution is defined on the real axis $x \in]-\infty, +\infty[$

[21] In mathematical statistics (Sect. 2.6), the quality of measured data is often characterized by scores. The z-score of a sample corresponds to the random variable $\mathcal{Z}$ (2.79) and it is measured in standard deviations from the population mean as units.

[22] A pivotal quantity or pivot is a function of measurable and unmeasurable parameters whose probability distribution does not depend on the unknown parameters.

and has the following density function and cumulative distribution (Fig. 2.16):

$$
\begin{aligned}
f_{\mathrm{stud}}(x) &= \frac{\Gamma\big((r+1)/2\big)}{\sqrt{\pi r}\,\Gamma\,(r/2)}\left(1+\frac{x^2}{r}\right)^{-(r+1)/2}, \quad x\in\mathbb{R} \qquad \text{(pdf)}\,,\\
F_{\mathrm{stud}}(x) &= \frac{1}{2}+x\Gamma\left(\frac{r+1}{2}\right)\frac{{}_2F_1\left(\frac{1}{2},\frac{r+1}{2},\frac{3}{2},-\frac{x^2}{r}\right)}{\sqrt{\pi r}\,\Gamma\,(r/2)} \qquad \text{(cdf)}\,,
\end{aligned}
\tag{2.81}
$$

where ${}_2F_1$ is the hypergeometric function. The t-distribution has simple expressions for several special cases:

(i) $r = 1$, Cauchy-distribution:

$$
f(x) = \frac{1}{\pi(1+x^2)}\,, \quad F(x) = \frac{1}{2}+\frac{1}{\pi}\arctan(x)\,,
$$

(ii) $r = 2$: $\quad f(x) = \dfrac{1}{(2+x^2)^{3/2}}\,, \quad F(x) = \dfrac{1}{2}\left(1+\dfrac{x}{\sqrt{2+x^2}}\right)\,,$

(iii) $r = 3$: $\quad f(x) = \dfrac{6\sqrt{3}}{\pi(3+x^2)^2}\,, \quad F(x) = \dfrac{1}{2}+\dfrac{\sqrt{3}x}{\pi(3+x^2)}+\dfrac{1}{\pi}\arctan\left(\dfrac{x}{\sqrt{3}}\right)\,,$

(iv) $r = \infty$, normal distribution:

$$
f(x) = \varphi(x) = \frac{1}{\sqrt{2\pi}}\mathrm{e}^{-x^2/2}\,, \quad F(x) = F_{\mathcal{N}}(x)\,.
$$

Formally the t-distribution represents an interpolation between the Cauchy–Lorentz distribution (Sect. 2.5.7) and the normal distribution, both standardized to mean zero and variance one. In this sense it has a lower maximum and heavier tails than the normal distribution and a higher maximum and less heavy tails than the Cauchy–Lorentz distribution.

The moments of Student's distribution are readily calculated:

Mean	$0\,,$ for $r > 1$, otherwise undefined
Median	0
Mode	0
Variance	$\begin{cases} \infty\,, & \text{for } 1 < r \le 2 \\ \dfrac{r}{r-2}\,, & \text{for } r > 2 \\ \text{undefined}\,, & \text{otherwise} \end{cases}$ (2.82)
Skewness	$0\,,$ for $r > 3$, otherwise undefined

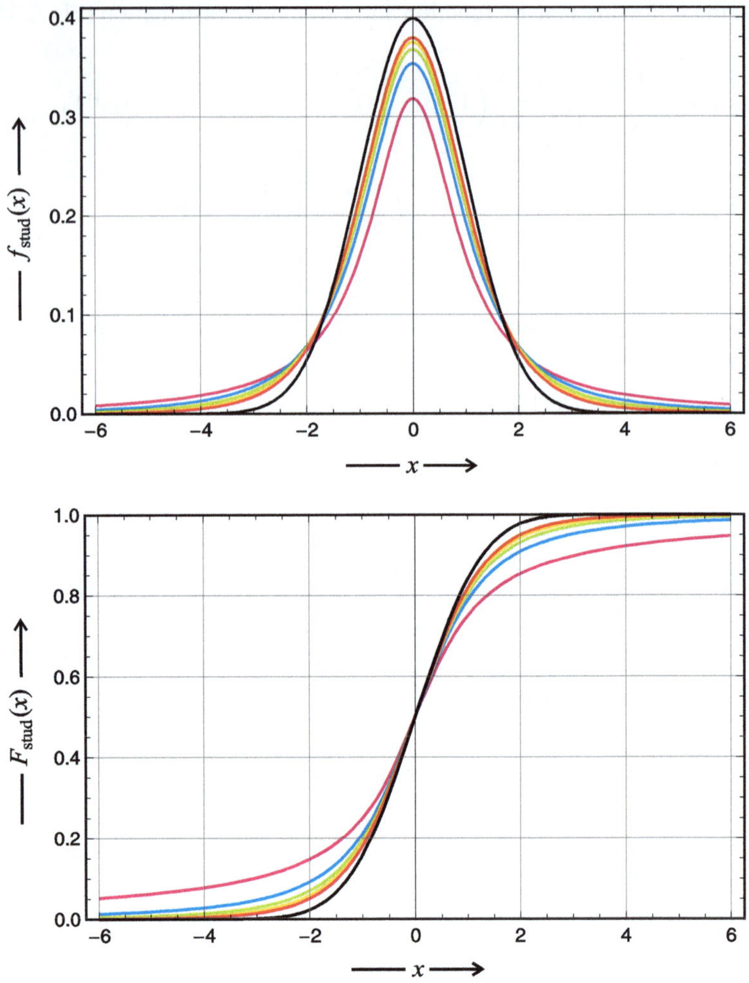

Fig. 2.16 Student's t-distribution. Student's distribution is defined on the real axis $x \in]-\infty, +\infty[$. The parameter $r \in \mathbb{N}_{>0}$ is called the number of degrees of freedom. This distribution has the probability density (pdf)

$$f_{\text{stud}}(x) = \frac{\Gamma\left((r+1)/2\right)}{\sqrt{\pi r}\,\Gamma\left(r/2\right)} \left(1 + \frac{x^2}{r}\right)^{(r+1)/2}$$

and the cumulative distribution function (cdf)

$$F_{\text{stud}}(x) = \frac{1}{2} + x\Gamma\left(\frac{r+1}{2}\right) \frac{{}_2F_1\left(\frac{1}{2}, \frac{r+1}{2}, \frac{3}{2}, -\frac{x^2}{r}\right)}{\sqrt{\pi r}\,\Gamma\,(r/2)}\,.$$

The *first curve* (*magenta*, $r = 1$) represents the density of the Cauchy–Lorentz distribution (Fig. 2.20). Parameter choice and color code: $r = 1$ (*magenta*), 2 (*blue*), 3 (*green*), 4 (*yellow*), 5 (*red*) and $+\infty$ (*black*). The *black curve* representing the limit $r \to \infty$ of Student's distribution is the standard normal distribution

$$\text{Kurtosis} \qquad \begin{cases} \infty\,, & \text{for } 2 < r \leq 4 \\ \dfrac{6}{r-4}\,, & \text{for } r > 4 \\ \text{undefined}\,, & \text{otherwise} \end{cases}$$

If it is defined, the variance of the Student t-distribution is greater than the variance of the standard normal distribution ($\sigma^2 = 1$). In the limit of infinite degrees of freedom, Student's distribution converges to the standard normal distribution and so does the variance: $\sigma^2 = \lim_{r\to\infty} \frac{r}{r-2} = 1$. Student's distribution is symmetric and hence the skewness γ_1 is either zero or undefined, and the (excess) kurtosis γ_2 is undefined or positive and converges to zero in the limit $r \to \infty$.

The raw moments $\hat{\mu}_n = \mathrm{E}(\mathcal{T}^n)$ of the t-distribution have fairly simple expressions:

$$\mathrm{E}(\mathcal{T}^k) = \begin{cases} 0\,, & k \text{ odd}\,,\ 0 < k < r\,, \\ \dfrac{1}{\sqrt{\pi}\,\Gamma\,(r/2)} r^{k/2} \Gamma\left(\dfrac{k+1}{2}\right) \Gamma\left(\dfrac{r-k}{2}\right)\,, & k \text{ even}\,,\ 0 < k < r\,, \\ \text{undefined}\,, & k \text{ odd}\,,\ 0 < r \leq k\,, \\ \infty\,, & k \text{ even}\,,\ 0 < r \leq k\,. \end{cases} \tag{2.83}$$

The entropy of Student's t-distribution is readily calculated by integration:

$$H(f_{\text{stud}}) = \frac{k+1}{2}\left(\psi\left(\frac{1+r}{2}\right) - \psi\left(\frac{r}{2}\right)\right) + \ln\left(\sqrt{r}B\left(\frac{r}{2}, \frac{1}{2}\right)\right)\,, \tag{2.84}$$

where $\psi(x) = \frac{\mathrm{d}}{\mathrm{d}x} \ln \Gamma(x)$ and $B(x, y) = \int_0^1 t^{x-1}(1-t)^{y-1}\,\mathrm{d}t$ are the digamma function and the beta function, respectively. Student's-distribution has the characteristic function

$$\phi_{\text{stud}}(s) = \frac{(\sqrt{r}|s|)^{r/2} K_{r/2}(\sqrt{r}|s|)}{2^{r/2-1}\,\Gamma\,(r/2)}\,, \quad \text{for } r > 0\,, \tag{2.85}$$

where $K_\alpha(x)$ is a modified Bessel function.

2.5.4 The Exponential and the Geometric Distribution

The *exponential distribution* is a continuous probability distribution which describes the distribution of the time intervals between events in a Poisson process

(Sect. 3.2.2.4).[23] A Poisson process is one where the number of events within any time interval is distributed according to a Poissonian. The Poisson process is a process where events occur independently of each other and at a constant average rate $\lambda \in \mathbb{R}_{>0}$, which is the only parameter of the exponential distribution and the Poisson process as well.

The exponential distribution has widespread applications in science and sociology. It describes the decay time of radioactive atoms, the time to reaction events in irreversible first order processes in chemistry and biology, the waiting times in queues of independently acting customers, the time to failure of components with constant failure rates and other instances.

The exponential distribution is defined on the positive real axis, $x \in [0, \infty[$, with a positive rate parameter $\lambda \in]0, \infty[$. The density function and cumulative distribution are of the form (Fig. 2.17)

$$\begin{aligned} f_{\text{exp}}(x) &= \lambda \exp(-\lambda x) \,, \quad x \in \mathbb{R}_{>0} \qquad \text{(pdf)} \,, \\ F_{\text{exp}}(x) &= 1 - \exp(-\lambda x) \,, \quad x \in \mathbb{R}_{>0} \qquad \text{(cdf)} \,. \end{aligned} \tag{2.86}$$

The moments of exponential distribution are readily calculated

$$\begin{array}{ll} \text{Mean} & \lambda^{-1} = \mu \\ \text{Median} & \lambda^{-1} \ln 2 \\ \text{Mode} & 0 \\ \text{Variance} & \lambda^{-2} \\ \text{Skewness} & 2 \\ \text{Kurtosis} & 6 \end{array} \tag{2.87}$$

A commonly used alternative parametrization makes use of a *survival parameter* $\beta = \lambda^{-1} = \mu$ instead of the rate parameter, and survival is often measured in terms of *half-life*, which is the expectation value of the time when one half of the events will have taken place—for example, 50 % of the atoms have decayed—and is in fact just another name for the median: $\bar{\mu} = \beta \ln 2 = \ln 2/\lambda$. The exponential

[23] It is important to distinguish the exponential distribution and the class of *exponential families of distributions*, which comprises a number of distributions like the normal distribution, the Poisson distribution, the binomial distribution, the exponential distribution and others [142, pp. 82–84]. The common form of the exponential family in the pdf is:

$$f_\vartheta(x) = \exp\big(A(\vartheta) \cdot B(x) + C(x) + D(\vartheta)\big),$$

where the parameter ϑ can be a scalar or a vector.

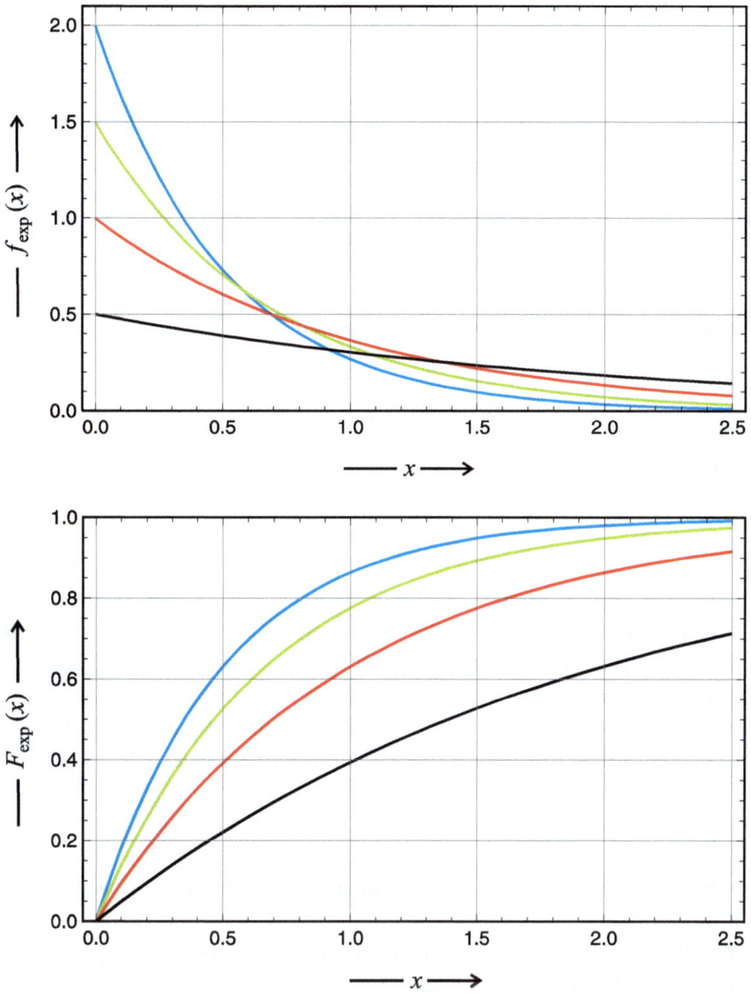

Fig. 2.17 The exponential distribution. The exponential distribution is defined on the real axis including zero $x \in [0, +\infty[$, with a parameter $\lambda \in \mathbb{R}_{>0}$ called the rate parameter. It has the probability density (pdf)

$$f_{\mathrm{exp}}(x) = \lambda \exp(-\lambda x)$$

and the cumulative distribution function (cdf)

$$F_{\mathrm{exp}}(x) = 1 - \exp(-\lambda x) .$$

Parameter choice and color code: $\lambda = 0.5$ (*black*), 2 (*red*), 3 (*green*), and 4 (*blue*)

distribution provides an easy to verify test case for the median–mean inequality:

$$\left|\mathrm{E}(\mathcal{X}) - \bar{\mu}\right| = \frac{1 - \ln 2}{\lambda} < \frac{1}{\lambda} = \sigma .$$

The raw moments of the exponential distribution are given simply by

$$\mathrm{E}(\mathcal{X}^n) = \hat{\mu}_n = \frac{n!}{\lambda^n} . \tag{2.88}$$

Among all probability distributions with the support $[0, \infty[$ and mean μ, the exponential distribution with $\lambda = 1/\mu$ has the largest entropy (Sect. 2.1.3):

$$H(f_{\exp}) = 1 - \log\lambda = 1 + \log\mu . \tag{2.23$'$}$$

The moment generating function of the exponential distribution is

$$M_{\exp}(s) = \left(1 - \frac{s}{\lambda}\right)^{-1} , \tag{2.89}$$

and the characteristic function is

$$\phi_{\exp}(s) = \left(1 - \frac{\mathrm{i}s}{\lambda}\right)^{-1} . \tag{2.90}$$

Finally, we mention a property of the exponential distribution that makes it unique among all continuous probability distributions: it is *memoryless*. Memorylessness can be encapsulated in an example called the hitchhiker's dilemma: waiting for hours on a lonely road does not increase the probability of arrival of the next car. Cast into probabilities, this means that for a random variable $\mathcal{T}$,

$$P(\mathcal{T} > s + t \,|\, \mathcal{T} > s) = P(\mathcal{T} > t) , \quad \forall\, s, t \geq 0 . \tag{2.91}$$

In other words, the probability of arrival does not change, no matter how many events have happened.[24]

In the context of the exponential distribution, we mention the Laplace distribution named after the Marquis de Laplace, which is an exponential distribution doubled by mirroring in the line $x = \mu$, with the density $f_{\mathcal{L}}(x) = \lambda \exp(-\lambda|x - \mu|)/2$. Sometimes it is also called the *double exponential distribution*. Knowing the results for the exponential distribution, it is a simple exercise to calculate the various properties of the Laplace distribution.

The discrete analogue of the exponential distribution is the geometric distribution. We consider a sequence of independent Bernoulli trials with p the probability of success and the only parameter of the distribution: $0 < p \leq 1$. The random variable $\mathcal{X} \in \mathbb{N}$ is the number of trials before the first success.

[24]We remark that memorylessness is not tantamount to independence. Independence requires $P(\mathcal{T} > s + t \,|\, \mathcal{T} > s) = P(\mathcal{T} > s + t)$.

The probability mass function and the cumulative distribution function of the geometric distribution are:

$$f_{k;p}^{\text{geom}} = p(1-p)^k\ , \quad k \in \mathbb{N} \qquad \text{(pdf)}\ ,$$
$$F_{k;p}^{\text{geom}} = 1-(1-p)^{k+1}\ , \quad k \in \mathbb{N} \qquad \text{(cdf)}\ . \tag{2.92}$$

The moments of the geometric distribution are readily calculated:

$$\begin{array}{ll} \text{Mean} & \dfrac{1-p}{p} \\ \text{Median} & p^{-1}\ln 2 \\ \text{Mode} & 0 \\ \text{Variance} & \dfrac{1-p}{p^2} \\ \text{Skewness} & \dfrac{2-p}{\sqrt{1-p}} \\ \text{Kurtosis} & 6+\dfrac{p^2}{1-p} \end{array} \tag{2.93}$$

Like the exponential distribution the geometric distribution lacks memory in the sense of (2.91). The information entropy has the form

$$H(f_{k;p}^{\text{geom}}) = -\frac{1}{p}\Big((1-p)\log(1-p) + p\log p\Big)\ . \tag{2.94}$$

Finally, we present the moment generating function and the characteristic function of the geometric distribution:

$$M_{\text{geom}}(s) = \frac{p}{1-(1-p)\exp(s)}\ , \tag{2.95}$$

$$\phi_{\text{geom}}(s) = \frac{p}{1-(1-p)\exp(\mathrm{i}s)}\ , \tag{2.96}$$

respectively.

2.5.5 *The Pareto Distribution*

As already mentioned, the Pareto distribution $\mathcal{P}(\tilde{\mu}, \alpha)$ is named after the Italian civil engineer and economist Vilfredo Pareto and represents a power law distribution with widespread applications from social sciences to physics. A definition is most easily visualized in terms of the complement of the cumulative distribution function, $\overline{F}(x) = 1 - F(x)$,

$$\overline{F}_{\mathcal{P}}(x) = P(\mathcal{X} > x) = \begin{cases} (\tilde{\mu}/x)^{\alpha} \,, & \text{for } x \geq \tilde{\mu} \,, \\ 1 \,, & \text{for } x < \tilde{\mu} \,. \end{cases} \tag{2.97}$$

The mode $\tilde{\mu}$ is the necessarily smallest relevant value of $\mathcal{X}$, and by the same token $f_{\mathcal{P}}(\tilde{\mu})$ is the maximum value of the density. The parameter $\tilde{\mu}$ is often referred to as the scale parameter of the distribution, and in the same spirit α is called the shape parameter. Other names for α are the *Pareto index* in economics and the *tail index* in probability theory.

The Pareto distribution is defined on the real axis with values above the mode, $x \in [\tilde{\mu}, \infty[$, with two real and positive parameters $\tilde{\mu} \in \mathbb{R}_{>0}$ and $\alpha \in \mathbb{R}_{>0}$. The density function and cumulative distribution are of the form:

$$\begin{aligned} f_{\mathcal{P}}(x) &= \frac{\alpha\, \tilde{\mu}^{\alpha}}{x^{\alpha-1}} \,, & x \in [\tilde{\mu}\,, \infty[\quad & \text{(pdf)} \,, \\ F_{\mathcal{P}}(x) &= 1 - \left(\frac{\tilde{\mu}}{x}\right)^{\alpha} \,, & x \in [\tilde{\mu}\,, \infty[\quad & \text{(cdf)} \,. \end{aligned} \tag{2.98}$$

The moments of Pareto distribution are readily calculated:

$$\begin{array}{lll} \text{Mean} & \begin{cases} \infty \,, & \text{for } \alpha \leq 1 \\ \dfrac{\alpha\tilde{\mu}}{\alpha - 1} \,, & \text{for } \alpha > 1 \end{cases} \\ \text{Median} & \tilde{\mu}\, 2^{\alpha/2} \\ \text{Mode} & \tilde{\mu} \\ \text{Variance} & \begin{cases} \infty \,, & \text{for } \alpha \in]1, 2] \\ \dfrac{\alpha\tilde{\mu}^2}{(\alpha-1)^2(\alpha-2)} \,, & \text{for } \alpha > 2 \end{cases} \\ \text{Skewness} & \dfrac{2(\alpha+1)}{\alpha-3}\sqrt{\dfrac{\alpha-2}{\alpha}} \,, \quad \text{for } \alpha > 3 \\ \text{Kurtosis} & \dfrac{\alpha^3 + \alpha^2 - 6\alpha - 2}{\alpha(\alpha-3)(\alpha-4)} \,, \quad \text{for } \alpha > 4 \end{array} \tag{2.99}$$

The shapes of the distributions for different values of the parameter α are shown in Fig. 2.18.

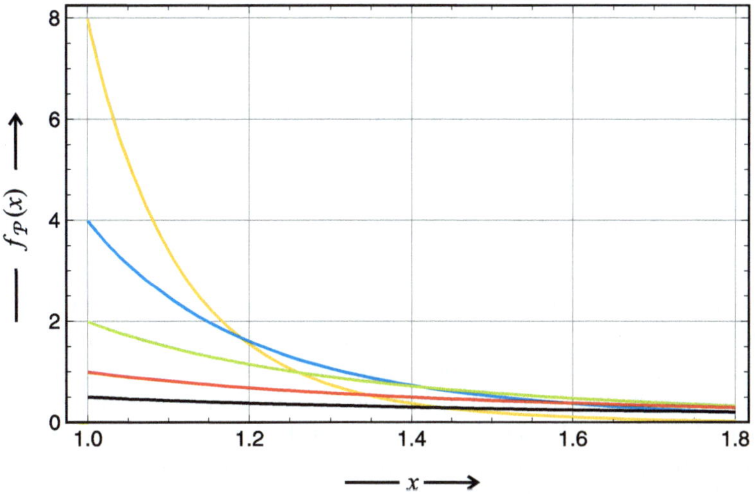

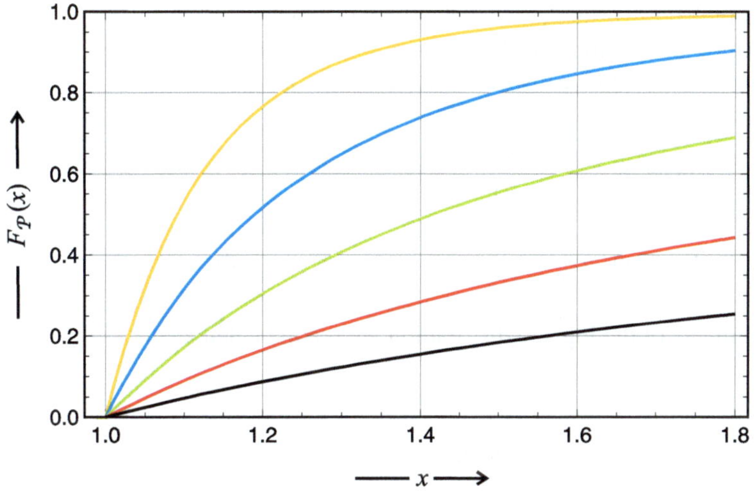

Fig. 2.18 The Pareto distribution. The Pareto distribution $\mathcal{P}(\tilde{\mu}, \alpha)$ is defined on the positive real axis $x \in]\tilde{\mu}, \infty[$. It has the density (pdf) $f_{\mathcal{P}}(x) = \alpha \tilde{\mu}^{\alpha} / x^{\alpha - 1}$ and the cumulative distribution function (cdf) $F_{\mathcal{P}}(x) = 1 - (\tilde{\mu}/x)^{\alpha}$. The two parameters are restricted by the relations $\tilde{\mu}, \alpha > \in \mathbb{R}_{\geq 0}$. Parameter choice and color code: $\tilde{\mu} = 1, \alpha = 1/2$ (*black*), 1 (*red*), 2 (*green*), 4 (*blue*), and 8 (*yellow*)

The relation between a Pareto distributed random variable $\mathcal{X}$ and an exponentially distributed variable $\mathcal{Y}$ is obtained straightforwardly:

$$\mathcal{Y} = \log\left(\frac{\mathcal{X}}{\tilde{\mu}}\right), \qquad \mathcal{X} = \tilde{\mu}\, e^{\mathcal{Y}},$$

where the Pareto index or shape parameter α corresponds to the rate parameter of the exponential distribution.

Finally, we mention that the Pareto distribution comes in different types and that type I was described here. The various types differ mainly with respect to the definitions of the parameters and the location of the mode [142]. We shall come back to the Pareto distribution when we discuss Pareto processes (Sect. 3.2.5).

2.5.6 The Logistic Distribution

The logistic distribution is commonly used as a model for growth with limited resources. It is applied in economics, for example, to model the market penetration of a new product, in biology for population growth in an ecosystem, and in agriculture for the expansion of agricultural production or weight gain in animal fattening. It is a continuous probability distribution with two parameters, the position of the mean μ and the scale b. The cumulative distribution function of the logistic distribution is the *logistic function*.

The logistic distribution is defined on the real axis $x \in]-\infty, \infty[$, with two parameters, the position of the mean $\mu \in \mathbb{R}$ and the scale $b \in \mathbb{R}_{>0}$. The density function and cumulative distribution are of the form (Fig. 2.19):

$$\begin{aligned} f_{\text{logist}}(x) &= \frac{\mathrm{e}^{-(x-\mu)b}}{b\big(1+\mathrm{e}^{-(x-\mu)/b}\big)^2}\,, \quad x \in \mathbb{R} \qquad \text{(pdf)}\,, \\ F_{\text{logist}}(x) &= \frac{1}{1+\mathrm{e}^{-(x-\mu)/b}}\,, \quad x \in \mathbb{R} \qquad \text{(cdf)}\,, \end{aligned} \tag{2.100}$$

The moments of the logistic distribution are readily calculated:

Mean	μ
Median	μ
Mode	μ
Variance	$\pi^2 b^2/3$
Skewness	0
Kurtosis	$6/5$

(2.101)

A frequently used alternative parametrization uses the variance as parameter, $\sigma = \pi b/\sqrt{3}$ or $b = \sqrt{3}\sigma/\pi$. The density and the cumulative distribution can also be expressed in terms of hyperbolic functions:

$$f_{\text{logist}}(x) = \frac{1}{4b}\operatorname{sech}^2\frac{x-\mu}{2b}\,, \qquad F_{\text{logist}}(x) = \frac{1}{2} + \frac{1}{2}\tanh\frac{x-\mu}{2b}\,.$$

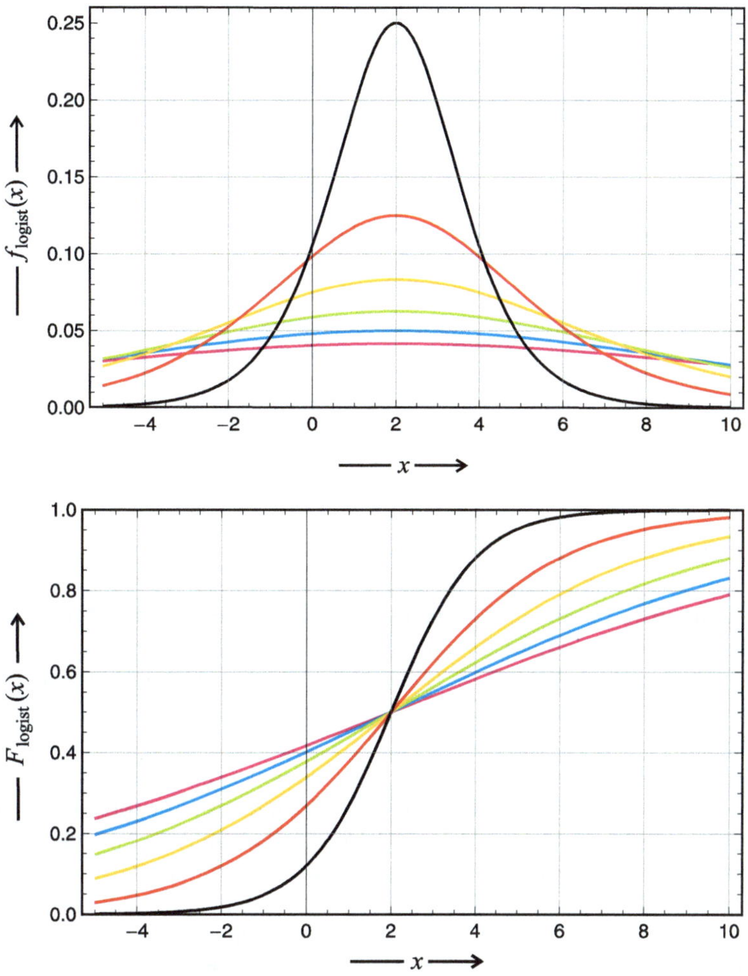

Fig. 2.19 The logistic distribution. The logistic distribution is defined on the real axis, $x \in]-\infty, +\infty[$, with two parameters, the location $\mu \in \mathbb{R}$ and the scale $b \in \mathbb{R}_{>0}$. It has the probability density (pdf)

$$f_{\text{logist}}(x) = \frac{e^{-(x-\mu)b}}{b\left(1 + e^{-(x-\mu)/b}\right)^2}$$

and the cumulative distribution function (cdf)

$$F_{\text{logist}}(x) = \frac{1}{1 + e^{-(x-\mu)/b}} \ .$$

Parameter choice and color code: $\mu = 2$, $b = 1$ (*black*), 2 (*red*), 3 (*yellow*), 4 (*green*), 5 (*blue*), and 6 (*magenta*)

The logistic distribution resembles the normal distribution, and like Student's distribution the logistic distribution has heavier tails and a lower maximum than the normal distribution. The entropy takes on the simple form

$$H(f_{\text{logist}}) = \log b + 2 \,. \tag{2.102}$$

The moment generating function of the logistic distribution is

$$M_{\text{logist}}(s) = \exp(\mu s)\, B(1 - bs, 1 + bs) \,, \tag{2.103}$$

for $|bs| < 1$, where $B(x, y)$ is the beta function. The characteristic function of the logistic distribution is

$$\phi_{\text{logist}}(s) = \frac{\pi bs \exp(\mathrm{i}\mu s)}{\sinh(\pi bs)} \,. \tag{2.104}$$

2.5.7 The Cauchy–Lorentz Distribution

The Cauchy–Lorentz distribution $\mathcal{C}(\gamma, \vartheta)$ is a continuous distribution with two parameters, the position ϑ and the scale γ. It is named after the French mathematician Augustin Louis Cauchy and the Dutch physicist Hendrik Antoon Lorentz. In order to facilitate comparison with the other distributions one might be tempted to rename the parameters, $\vartheta = \mu$ and $\gamma = \sigma^2$, but we shall refrain from changing the notation because the first and second moments are undefined for the Cauchy distribution.

The Cauchy distribution is important in mathematics, and in particular in physics, where it occurs as the solution to the differential equation for forced resonance. In spectroscopy, the Lorentz curve is used for the description of spectral lines that are homogeneously broadened. The Cauchy distribution is a typical heavy-tailed distribution in the sense that larger values of the random variable are more likely to occur in the two tails than in the tails of the normal distribution. Heavy-tailed distributions need not have *two* heavy tails like the Cauchy distribution, and then we speak of heavy right tails or heavy left tails. As we shall see in Sects. 2.5.9 and 3.2.5, the Cauchy distribution belongs to the class of stable distributions and hence can be partitioned into a linear combination of other Cauchy distributions.

The Cauchy probability density function and the cumulative probability distribution are of the form (Fig. 2.20)

$$\begin{aligned} f_{\mathcal{C}}(x) &= \frac{1}{\pi\gamma} \frac{1}{1 + \left(\dfrac{x - \vartheta}{\gamma}\right)^2} \\ &= \frac{1}{\pi} \frac{\gamma}{(x - \vartheta)^2 + \gamma^2} \,, \quad x \in \mathbb{R} \qquad \text{(pdf)} \,, \\ F_{\mathcal{C}}(x) &= \frac{1}{2} + \frac{1}{\pi} \arctan \frac{x - \vartheta}{\gamma} \qquad \text{(cdf)} \,. \end{aligned} \tag{2.105}$$

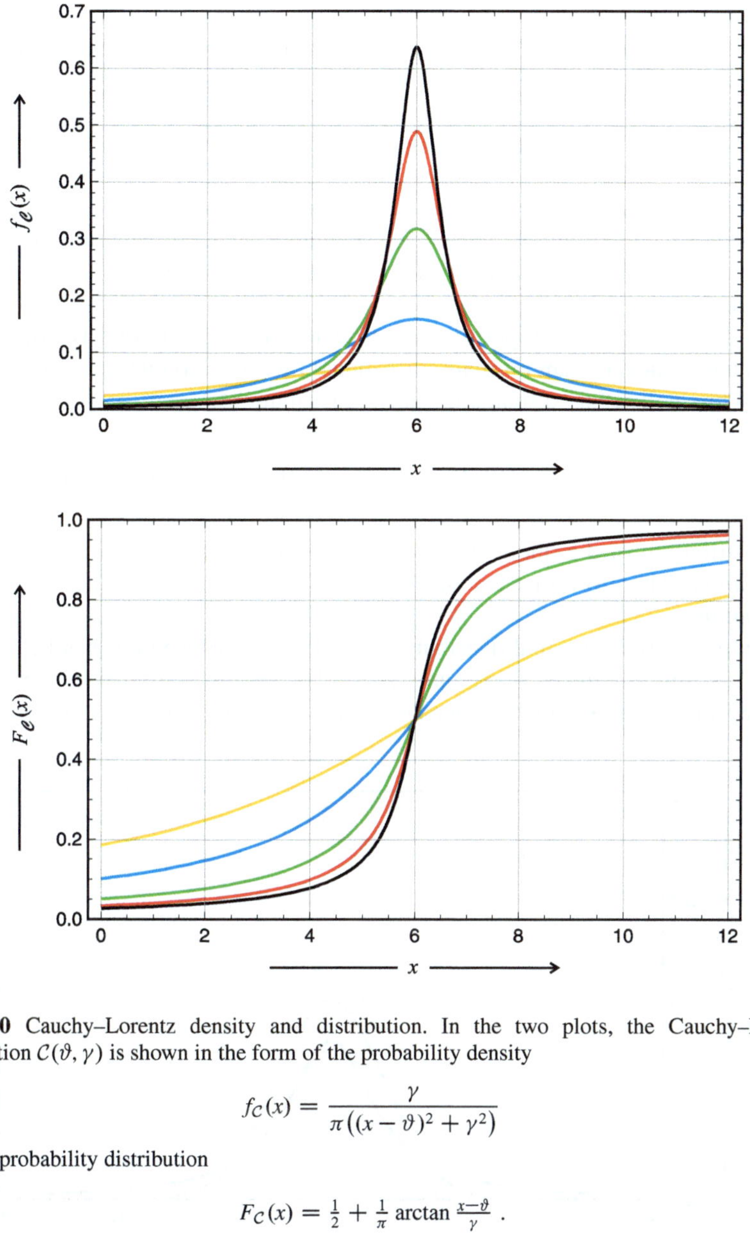

Fig. 2.20 Cauchy–Lorentz density and distribution. In the two plots, the Cauchy–Lorentz distribution $\mathcal{C}(\vartheta, \gamma)$ is shown in the form of the probability density

$$f_{\mathcal{C}}(x) = \frac{\gamma}{\pi\big((x-\vartheta)^2 + \gamma^2\big)}$$

and the probability distribution

$$F_{\mathcal{C}}(x) = \tfrac{1}{2} + \tfrac{1}{\pi} \arctan \tfrac{x-\vartheta}{\gamma} \,.$$

Choice of parameters: $\vartheta = 6$ and $\gamma = 0.5$ (*black*), 0.65 (*red*), 1 (*green*), 2 (*blue*), and 4 (*yellow*)

The two parameters define the position of the peak ϑ and the width of the distribution γ (Fig. 2.20). The peak height or amplitude is $1/\pi\gamma$. The function $F_C(x)$ can be inverted to give

$$F_C^{-1}(p) = \vartheta + \gamma \tan\bigl(\pi\,(p - 1/2)\bigr) , \tag{2.105'}$$

and we obtain for the quartiles and the median the values $(\vartheta - \gamma, \vartheta, \vartheta + \gamma)$. As with the normal distribution, we define a standard Cauchy distribution $\mathcal{C}(\vartheta, \gamma)$ with $\vartheta = 0$ and $\gamma = 1$, which is identical to the Student t-distribution with one degree of freedom, $r = 1$ (Sect. 2.5.3).

Another remarkable property of the Cauchy distribution concerns the ratio $\mathcal{Z}$ between two independent normally distributed random variables $\mathcal{X}$ and $\mathcal{Y}$. It turns out that this will satisfy a standard Cauchy distribution:

$$\mathcal{Z} = \frac{\mathcal{X}}{\mathcal{Y}} , \quad F_{\mathcal{X}} = \mathcal{N}(0,1) , \quad F_{\mathcal{Y}} = \mathcal{N}(0,1) \implies F_{\mathcal{Z}} = \mathcal{C}(0,1) ,$$

The distribution of the quotient of two random variables is often called the *ratio distribution*. Therefore one can say the Cauchy distribution is the *normal ratio distribution*.

Compared to the normal distribution, the Cauchy distribution has heavier tails and accordingly a lower maximum (Fig. 2.21). In this case we cannot use the (excess) kurtosis as an indicator because all moments of the Cauchy distribution are

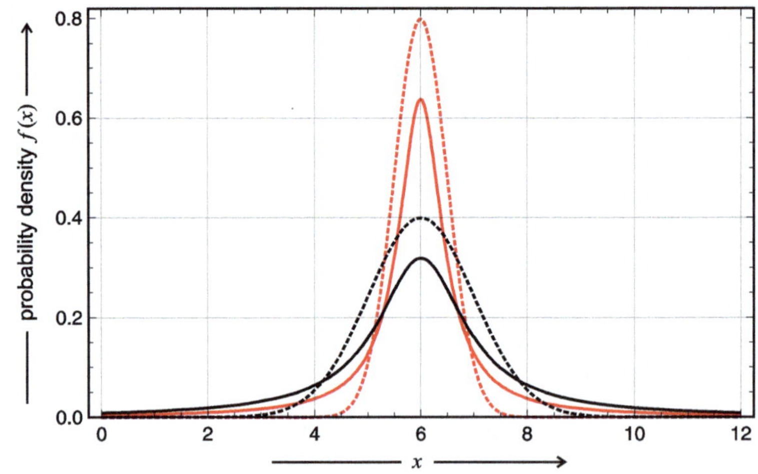

Fig. 2.21 Comparison of the Cauchy–Lorentz and normal densities. The plots compare the Cauchy–Lorentz density $\mathcal{C}(\vartheta, \gamma)$ (*full lines*) and the normal density $\mathcal{N}(\mu, \sigma^2)$ (*broken lines*). In the flanking regions, the normal density decays to zero much faster than the Cauchy–Lorentz density, and this is the cause of the abnormal behavior of the latter. Choice of parameters: $\vartheta = \mu = 6$ and $\gamma = \sigma^2 = 0.5$ (*black*) and $\gamma = \sigma^2 = 1$ (*red*)

undefined, but we can compute and compare the heights of the standard densities:

$$f_{\mathcal{C}}(x=\vartheta) = \frac{1}{\pi}\frac{1}{\gamma}\,, \qquad f_{\mathcal{N}}(x=\mu) = \frac{1}{\sqrt{2\pi}}\frac{1}{\sigma}\,,$$

which yields

$$f_{\mathcal{C}}(\vartheta) = \frac{1}{\pi}\,, \qquad f_{\mathcal{N}}(\mu) = \frac{1}{\sqrt{2\pi}}\,, \qquad \text{for } \gamma = \sigma = 1\,,$$

with $1/\pi < 1/\sqrt{2\pi}$. □

The Cauchy distribution nevertheless has a well defined median and mode, both of which coincide with the position of the maximum of the density function, $x = \vartheta$.

The entropy of the Cauchy density is $H(f_{\mathcal{C}(\vartheta,\gamma)}) = \log\gamma + \log 4$. It cannot be compared with the entropy of the normal distribution in the sense of the maximum entropy principle (Sect. 2.1.3), because this principle refers to distributions with variance σ^2, whereas the variance of the Cauchy distribution is undefined.

The Cauchy distribution has no moment generating function, but it does have a characteristic function:

$$\phi_{\mathcal{C}}(s) = \exp\left(\mathrm{i}\vartheta s - \gamma|s|\right)\,. \tag{2.106}$$

A consequence of the lack of defined moments is that the central limit theorem cannot be applied to a sequence of Cauchy variables. It is can be shown by means of the characteristic function that the mean $\mathcal{S} = \sum_{i=1}^{n}\mathcal{X}_i/n$ of a sequence of independent and identically distributed random variables with standard Cauchy distribution has the same standard Cauchy distribution and is not normally distributed as the central limit theorem would predict.

2.5.8 The Lévy Distribution

The Lévy distribution $\mathcal{L}(\gamma,\vartheta)$ is a continuous one-sided probability distribution which is defined for values of the variable x that are greater than or equal to a shift parameter ϑ , i.e., $x \in [\vartheta, \infty[$. It is a special case of the *inverse gamma distribution* and belongs, together with the normal and the Cauchy distribution, to the class of analytically accessible *stable distributions*.

The Lévy probability density function and the cumulative probability distribution are of the form (Fig. 2.22):

$$\begin{aligned} f_{\mathcal{L}}(x) &= \sqrt{\frac{\gamma}{2\pi}}\frac{1}{(x-\vartheta)^{3/2}}\exp\left(-\frac{\gamma}{2(x-\vartheta)}\right)\,, \quad x\in[\vartheta,\infty[\qquad \text{(pdf)}\,,\\ F_{\mathcal{L}}(x) &= \mathrm{erfc}\left(\sqrt{\frac{\gamma}{2(x-\vartheta)}}\right) \qquad \text{(cdf)}\,. \end{aligned} \tag{2.107}$$

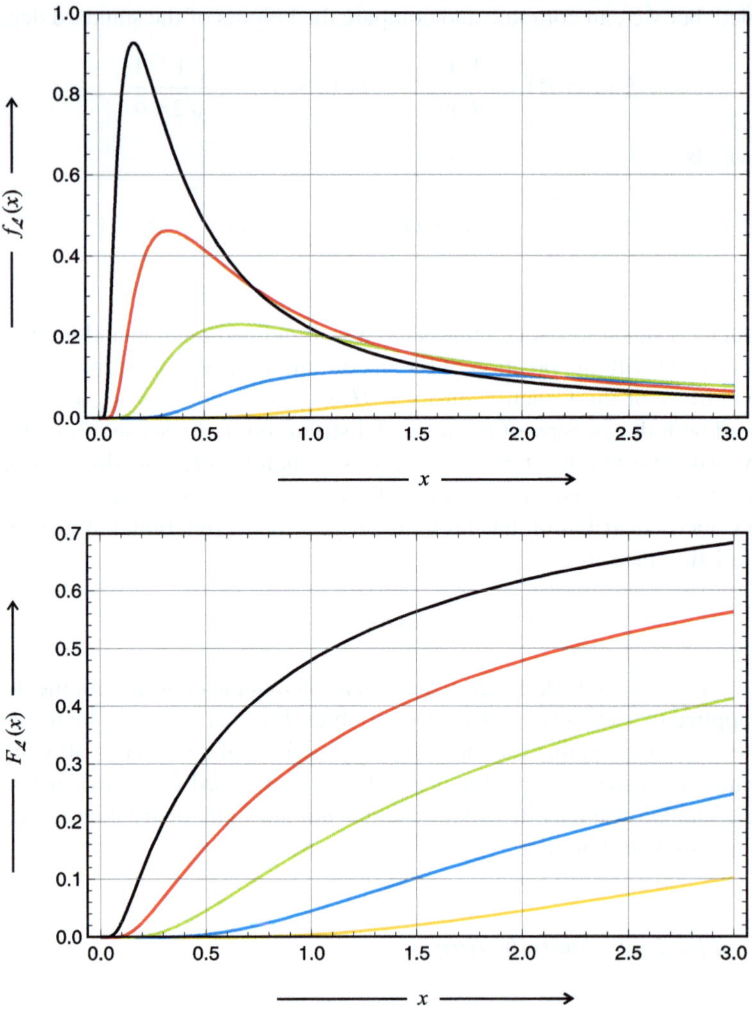

Fig. 2.22 Lévy density and distribution. In the two plots, the Lévy distribution $\mathcal{L}(\vartheta, \gamma)$ is shown in the form of the probability density

$$f_{\mathcal{L}}(x) = \sqrt{\frac{\gamma}{2\pi}} \frac{1}{(x-\vartheta)^{3/2}} \exp\left(-\frac{\gamma}{2(x-\vartheta)}\right)$$

and the probability distribution

$$F_{\mathcal{L}}(x) = \operatorname{erfc}\left(\sqrt{\frac{\gamma}{2(x-\vartheta)}}\right).$$

Choice of parameters: $\vartheta = 0$ and $\gamma = 0.5$ (*black*), 1 (*red*), 2 (*green*), 4 (*blue*) and 8 (*yellow*)

The two parameters $\vartheta \in \mathbb{R}$ and $\gamma \in \mathbb{R}_{>0}$ are the location of $f_{\mathcal{L}}(x) = 0$ and the scale parameter. The mean and variance of the Lévy distribution are infinite, while the skewness and kurtosis are undetermined. For $\vartheta = 0$, the mode of the distribution appears at $\tilde{\mu} = \gamma/3$ and the median takes on the value $\bar{\mu} = \gamma/2\big(\text{erfc}^{-1}(1/2)\big)^2$.

The entropy of the Lévy distribution is

$$H\big(f_{\mathcal{L}}(x)\big) = \frac{1 + 3C + \ln(16\pi\gamma^2)}{2} ,$$

where C is Euler's constant, and the characteristic function

$$\phi_{\mathcal{L}}(s) = \exp\big(\mathrm{i}\vartheta s - \sqrt{-2\mathrm{i}\gamma s}\big) \tag{2.108}$$

is the only defined generating function. We shall encounter the Lévy distribution when Lévy processes are discussed in Sect. 3.2.5.

2.5.9 *The Stable Distribution*

A whole family of distributions subsumed under the name *stable distribution* was first investigated in the 1920s by the French mathematician Paul Lévy. Compared to most of the probability distributions discussed earlier, stable distributions, with very few exceptions, have a number of unusual features like undefined moments or no analytical expressions for densities and cumulative distribution functions. On the other hand, they share several properties like infinite divisibility and shape stability, which will turn out to be important in the context of certain stochastic processes called Lévy processes (Sect. 3.2.5).

Shape Stability

Shape stability or stability for short comes in two flavors: stability in the broader sense and strict stability. For an explanation of stability we make the following definition: A random variable $\mathcal{X}$ has a stable distribution if any linear combination $\mathcal{X}_1$ and $\mathcal{X}_2$ of two independent copies of this variable satisfies the same distribution up to a shift in location and a change in the width as expressed by a scale parameter [423][25,26]:

$$a\mathcal{X}_1 + b\mathcal{X}_2 \overset{\mathrm{d}}{=} c\mathcal{X} + d , \tag{2.109}$$

[25] As mentioned for the Cauchy distribution (Sect. 2.5.7), the location parameter defines the center of the distribution ϑ and the scale parameter γ determines its width, even in cases where the corresponding moments μ and σ^2 do not exist.

[26] The symbol $\overset{\mathrm{d}}{=}$ means *equality in distribution.*

wherein a and b are positive constants, c is some positive number dependent on a, b, and the summation properties of $\mathcal{X}$, and $d \in \mathbb{R}$. *Strict stability* or *stability in the narrow sense* differs from *stability* or *stability in the broad sense* by satisfying the equality (2.109) with $d = 0$ for all choices of a and b. A random variable is said to be *symmetrically stable* if it is stable and symmetrically distributed around zero so that $\mathcal{X} \stackrel{\mathrm{d}}{=} -\mathcal{X}$.

Stability and strict stability of the normal distribution $\mathcal{N}(\mu, \sigma)$ are easily demonstrated by means of CLT:

$$\begin{aligned} \mathcal{S}_n &= \sum_{i=1}^{n} \mathcal{X}_i \,, \quad \text{with } \mathrm{E}(\mathcal{X}_i) = \mu \,, \ \mathrm{var}(\mathcal{X}_i) = \sigma^2 \,, \ \forall\, i = 1, \ldots, n \,, \\ \mathrm{E}(\mathcal{S}_n) &= n\mu \,, \quad \mathrm{var}(\mathcal{S}_n) = (n\sigma)^2 \,, \ \forall\, i = 1, \ldots, n \,. \end{aligned} \tag{2.110}$$

Equations (2.109) and (2.110) imply the conditions on the constants a, b, c, and d:

$$\mu(a\mathcal{X}) = a\mu(\mathcal{X}) \,, \quad \mu(b\mathcal{X}) = b\mu(\mathcal{X}) \,, \quad \mu(c\mathcal{X} + d) = c\mu(\mathcal{X}) + d$$
$$\Longrightarrow d = (a + b - c)\mu \,,$$

$$\mathrm{var}(a\mathcal{X}) = (a\sigma)^2 \,, \quad \mathrm{var}(b\mathcal{X}) = (b\sigma)^2 \,, \quad \mathrm{var}(c\mathcal{X} + d) = (c\sigma)^2$$
$$\Longrightarrow c^2 = a^2 + b^2 \,.$$

The two conditions $d = (a + b - c)\mu$ and $c = \sqrt{a^2 + b^2}$ with $d \neq 0$ are readily satisfied for pairs of arbitrary real constants $a, b \in \mathbb{R}$ and accordingly, the normal distribution $\mathcal{N}(\mu, \sigma)$ is stable. Strict stability, on the other hand, requires $d = 0$, and this can only be achieved by zero-centered normal distributions $\mathcal{N}(0, \sigma)$.

Infinite Divisibility

The property of *infinite divisibility* is defined for classes of random variables $\mathcal{S}_n$ with a density $f_{\mathcal{S}}(x)$ which can be partitioned into any arbitrary number $n \in \mathbb{N}_{>0}$ of independent and identically distributed (iid) random variables such that all individual variables $\mathcal{X}_k$, their sum $\mathcal{S}_n = \mathcal{X}_1 + \mathcal{X}_2 + \cdots + \mathcal{X}_n$, and all possible partial sums have the same probability density $f_{\mathcal{X}}(x)$.

In particular the probability density $f_{\mathcal{S}}(x)$ of a random variable $\mathcal{S}_n$ is infinitely divisible if there exists a series of independent and identically distributed (iid) random variables $\mathcal{X}_i$ such that for

$$\mathcal{S}_n \stackrel{\mathrm{d}}{=} \mathcal{X}_1 + \mathcal{X}_2 + \cdots + \mathcal{X}_n = \sum_{i=1}^{n} \mathcal{X}_i \,, \quad \text{with } n \in \mathbb{N}_{>0} \,, \tag{2.111a}$$

the density satisfies the convolution (see Sect. 3.1.6)

$$f_{\mathcal{S}}(x) = f_{\mathcal{X}_1}(x) * f_{\mathcal{X}_2}(x) * \cdots * f_{\mathcal{X}_n}(x) . \tag{2.111b}$$

In other words, infinite divisibility implies closure under convolution. The convolution theorem (3.27) allows one to convert the convolution into a product by applying a Fourier transform $\phi_{\mathcal{S}}(u) = \int_{\Omega} e^{iux} f_{\mathcal{S}}(x)\,dx$:

$$\phi_{\mathcal{S}}(u) = \left(\phi_{\mathcal{X}_i}(u)\right)^n . \tag{2.111c}$$

Infinite divisibility is closely related to shape stability: with the help of the central limit theorem (CLT) we can easily show that the shape stable standard normal distribution $\varphi(x)$ has the property of being infinitely divisible. All shape stable distributions are infinitely divisible, but there are infinitely divisible distributions which do not belong to the class of stable distributions. Examples are the Poisson distribution, the χ^2 distribution, and many others (Fig. 2.23).

Stable Distributions

A stable distribution $\mathcal{S}(\alpha, \beta, \gamma, \vartheta)$ is characterized by four parameters:

(i) a *stability parameter* $\alpha \in]0, 2]$,
(ii) a *skewness parameter* $\beta \in [-1, 1]$,
(iii) a *scale parameter* $\gamma \geq 0$, $\chi = \gamma^{\alpha}$,
(iv) a *location parameter* $\vartheta \in \mathbb{R}$.

Among other things, the stability parameter $0 < \alpha \leq 2$ determines the asymptotic behavior of the density and the distribution function (see the *Pareto distribution*). For stable distributions with $\alpha \leq 1$, the mean is undefined, and for stable distribution with $\alpha < 2$, the variance is undefined. The skewness parameter β determines the symmetry and skewness of the distribution: $\beta = 0$ implies a symmetric distribution, whereas $\beta > 0$ indicates more weight given to points on the right-hand side of the mode and $\beta < 0$ more weight to points on the left-hand side.[27] Accordingly, asymmetric stable distributions $\beta \neq 0$ have a light tail and a heavy tail. For $\beta > 0$, the heavy tail lies on the right-hand side, while for $\beta < 0$ it is on the left-hand side. For stability parameters $\alpha < 1$ and $|\beta| = 1$, the light tail is zero and the support of the distribution is only one of the two real half-lines, $x \in \mathbb{R}_{\geq 0}$ for $\beta = 1$ and $x \in \mathbb{R}_{\leq 0}$ for $\beta = -1$ (see, for example, the Levy distribution in Sect. 2.5.8). The parameters α and β together determine the shape of the distribution and are thus called *shape parameters* (Fig. 2.23). The scale parameter γ determines the width of the distribution, as the standard deviation σ would do if it existed. The location parameter ϑ generalizes the conventional mean μ when the latter does not exist.

[27] We remark that, for all stable distributions except the normal distribution, the conventional skewness (Sect. 2.1.2) is undefined.

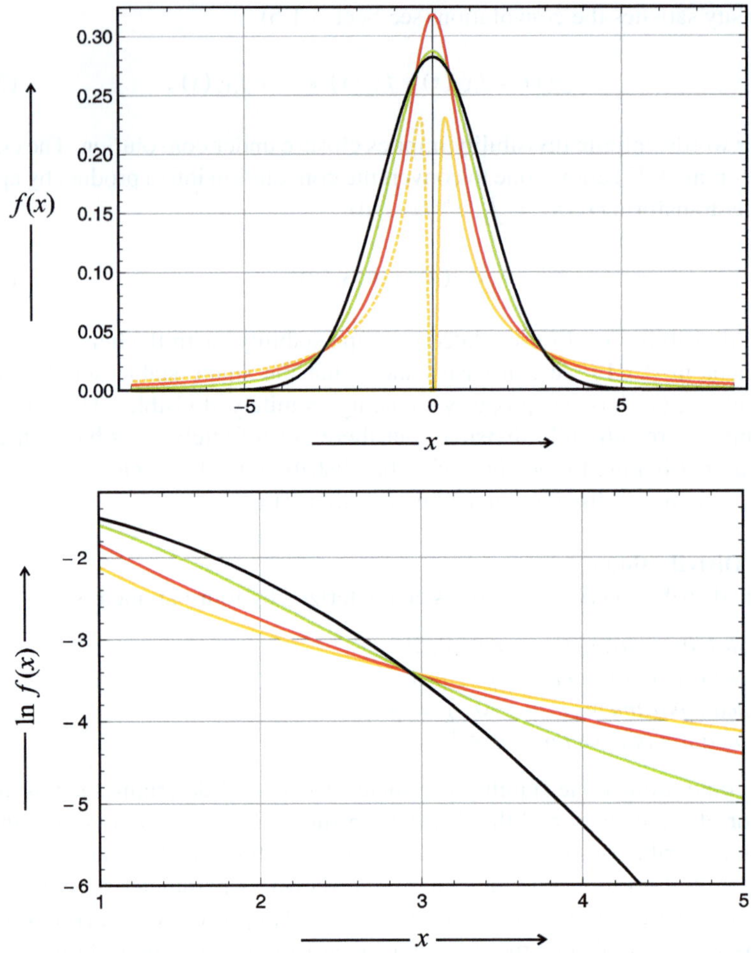

Fig. 2.23 A comparison of stable probability densities. *Upper*: Comparison between four different stable distributions with characteristic exponents $\alpha = 1/2$ (*yellow*), 1 (*red*), 3/2 (*green*), and 2 (*black*). For $\alpha < 1$, symmetric distributions ($\beta = 0$) are not stable and therefore we show the two extremal distributions with $\beta = \pm 1$ for the Lévy distribution ($\alpha = 1/2$). *Lower*: Log-linear plot of the densities against the position x. Within a small interval around $x = 2.9$, the curves for the individual probability densities cross and illustrate the increase in the probabilities for longer jumps

The parameters of the three already known stable distributions with analytical densities are as follows:

1. Normal distribution $\mathcal{N}(\mu, \sigma^2)$, with $\alpha = 2, \beta = 0, \gamma = \sigma/\sqrt{2}, \mu$.
2. Cauchy distribution $\mathcal{C}(\delta, \gamma)$, with $\alpha = 1, \beta = 0, \gamma, \vartheta$.
3. Lévy distribution $\mathcal{L}(\delta, \gamma)$, with $\alpha = 1/2, \beta = 1, \gamma, \vartheta$.

As for the normal distribution, we define standard stable distributions with only two parameters by setting $\gamma = 1$ and $\vartheta = 0$:

$$\mathcal{S}_{\alpha,\beta}(x) = \mathcal{S}_{\alpha,\beta,1,0}(x) = \mathcal{S}_{\alpha,\beta,1,0}\left(\frac{\xi - \vartheta}{\gamma}\right) = \gamma \mathcal{S}_{\alpha,\beta,\gamma,\vartheta}(\xi) \,.$$

All stable distributions except the normal distribution with $\alpha = 2$ are leptokurtic and have heavy tails. Furthermore, we stress that the central limit theorem in its conventional form is only valid for normal distributions. No other stable distributions satisfy CLT as follows directly from equation (2.109): linear combinations of a large number of Cauchy distributions, for example, form a Cauchy distribution and not a normal distribution, Lévy distributions form a Lévy distribution, and so on! The inapplicability of CLT follows immediately from the requirement of a finite variance $\text{var}(\mathcal{X})$, which is violated for all stable distributions with $\alpha < 2$.

There are no analytical expressions for the densities of stable distributions, with the exception of the Lévy, the Cauchy, and the normal distribution, and cumulative distributions can be given in analytical form only for the first two cases—the cumulative normal distribution is available only in the form of the error function. A general expression in closed form can be given, however, for the characteristic function:

$$\varphi_{\mathcal{S}}(s;\alpha,\beta,\gamma,\vartheta) = \exp\Big(\mathrm{i}s\vartheta - |\gamma s|^{\alpha}\big(1 - \mathrm{i}\beta\,\mathrm{sgn}(s)\big)\Phi\Big)\,,$$

$$\text{with} \quad \Phi = \begin{cases} \tan\dfrac{\pi\alpha}{2}\,, & \text{for } \alpha \neq 1\,, \\ -\dfrac{2}{\pi}\log|s|\,, & \text{for } \alpha = 1\,. \end{cases} \tag{2.112}$$

The characteristic function of symmetric stable distributions centered around the origin expressed by $\beta = 0$ and $\vartheta = 0$ takes on the simple real form $\varphi(s;\alpha,0,\gamma,0) = \exp(-\gamma^{\alpha}|s|^{\alpha})$. This equation is easily checked with the characteristic functions of the Cauchy distribution (2.106) and the normal distribution (2.51) with $\gamma = \sigma/\sqrt{2}$.

Asymptotic Densities of Stable Distributions

The characteristic exponent α is also called the *index of stability* since it determines the order of the singularity at $x = 0$ (Sect. 3.2.5), and at the same time it is basic for the long-distance scaling of the probability density [43, 81, 182, 454]. For $\alpha < 2$, we obtain

$$f_{\mathcal{S}}(x;\alpha,\beta,\gamma,0) \approx \frac{\gamma^{\alpha}\big(1 + \mathrm{sgn}(x)\beta\big)\sin(\alpha\pi/2)\Gamma(\alpha+1)/\pi}{|x|^{\alpha+1}}\,, \qquad \text{for } x \to \pm\infty\,.$$

For symmetric distributions, the asymptotic law simplifies to

$$f_S(x;\alpha,0,\gamma,0) \approx \frac{C(\alpha)}{|x|^{\alpha+1}}, \quad \text{for } x \to \pm\infty\,,$$

$$P(|\mathcal{X}| > |x|) \approx \frac{C(\alpha)}{|x|^{\alpha}}, \quad \text{for } x \to \pm\infty\,.$$

The asymptotic behavior of the normal distribution, where $\alpha = 2$, has been calculated, for example, by Feller [160, p. 193]:

$$P(|\mathcal{X}| > |x|) \approx \frac{\exp(-x^2/2)}{|x|\sqrt{2\pi}}, \quad \text{for } x \to \pm\infty\,.$$

We shall come back to the asymptotic densities in the discussion of anomalous diffusion (Sect. 3.2.5).

2.5.10 *Bimodal Distributions*

As the name of the bimodal distribution indicates the density function $f(x)$ has two maxima. It arises commonly as a mixture of two unimodal distribution in the sense that the bimodally distributed random variable $\mathcal{X}$ is defined as

$$P(\mathcal{X}) = \begin{cases} P(\mathcal{X} = \mathcal{Y}_1) = \alpha\,, \\ P(\mathcal{X} = \mathcal{Y}_2) = 1 - \alpha\,. \end{cases}$$

Bimodal distributions commonly arise from statistics of populations that are split into two subpopulations with sufficiently different properties. The sizes of weaver ants give rise to bimodal distributions because of the existence of two classes of worker [563]. If the differences are too small, as in the case of the combined distribution of body heights for men and women, monomodality is observed [478].

As an illustrative model we choose the superposition of two normal distributions with different means and variances (Fig. 2.24). The probability density for $\alpha = 1/2$ is then of the form

$$f(x) = \frac{1}{2\sqrt{2\pi}}\left(\frac{e^{-(x-\nu_1)^2/2\sigma_1^2}}{\sqrt{\sigma_1^2}} + \frac{e^{-(x-\nu_2)^2/2\sigma_2^2}}{\sqrt{\sigma_2^2}}\right). \tag{2.113}$$

The cumulative distribution function is readily obtained by integration. As in the case of the normal distribution, the result is not analytical, but formulated in terms

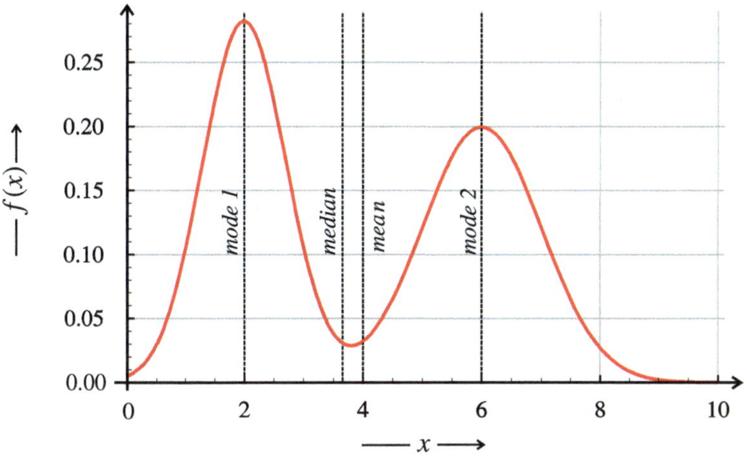

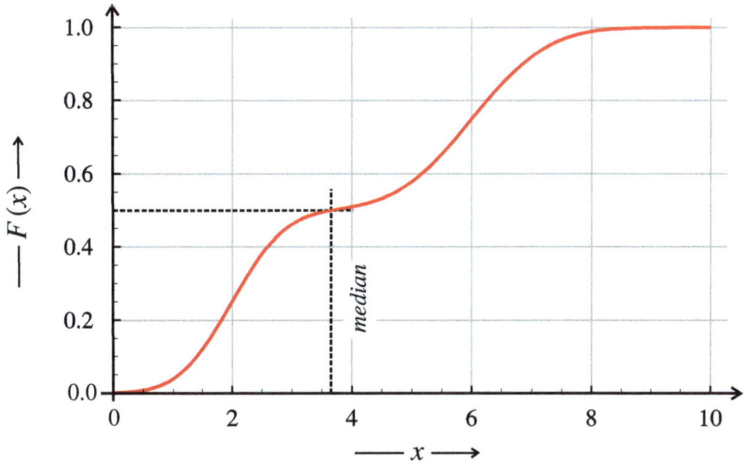

Fig. 2.24 A bimodal probability density. The figure illustrates a bimodal distribution modeled as a superposition of two normal distributions (2.113) with $\alpha = 1/2$ and different values for the mean and variance ($\nu_1 = 2, \sigma_1^2 = 1/2$) and ($\nu_2 = 6, \sigma_2^2 = 1$):

$$f(x) = \frac{\sqrt{2}e^{-(x-2)^2} + e^{-(x-6)^2/2}}{2\sqrt{2\pi}} .$$

Upper: Probability density corresponding to the two modes $\tilde{\mu}_1 = \nu_1 = 2$ and $\tilde{\mu}_2 = \nu_2 = 6$. The median $\bar{\mu} = 3.65685$ and mean $\mu = 4$ are situated near the density minimum between the two maxima. *Lower*: Cumulative probability distribution, viz.,

$$F(x) = \frac{1}{4}\left(2 + \text{erf}(x-2) + \text{erf}\left(\frac{x-6}{\sqrt{2}}\right)\right) ,$$

as well as the construction of the median. The variances in this example are $\hat{\mu}_2 = 20.75$ and $\mu_2 = 4.75$

of the error function, which is available only numerically through integration:

$$F(x) = \frac{1}{4}\left(2 + \mathrm{erf}\left(\frac{x-\nu_1}{\sqrt{2\sigma_1^2}}\right) + \mathrm{erf}\left(\frac{x-\nu_2}{\sqrt{2\sigma_2^2}}\right)\right) . \tag{2.114}$$

In the numerical example shown in Fig. 2.24, the distribution function shows two distinct steps corresponding to the maxima of the density $f(x)$.

The first and second moments of the bimodal distribution can be readily computed analytically as an exercise. The results are

$$\hat{\mu}_1 = \mu = \frac{1}{2}(\nu_1 + \nu_2) , \quad \mu_1 = 0$$

$$\hat{\mu}_2 = \frac{1}{2}(\nu_1^2 + \nu_2^2) + \frac{1}{2}(\sigma_1^2 + \sigma_2^2) , \quad \mu_2 = \frac{1}{4}(\nu_1 - \nu_2)^2 + \frac{1}{2}(\sigma_1^2 + \sigma_2^2) .$$

The centered second moment illustrates the contributions to the variance of the bimodal density. It is composed of the sum of the variances of the subpopulations and the square of the difference between the two means, viz., $(\nu_1 - \nu_2)^2$.

2.6 Mathematical Statistics

Mathematical statistics provides the bridge between probability theory and the analysis of real data, which is inevitably incomplete since samples are always finite. Nevertheless, it turns out to be very appropriate to use infinite samples as a reference (Sect. 1.3). Large sample theory, and in particular the law of large numbers (Sect. 2.4.2), deals with the asymptotic behavior of series of samples of increasing size. Although mathematical statistics is a discipline in its own right and would require a separate monograph for a comprehensive presentation, a brief account of the basic concepts will be included here, since they are of general importance for every scientist.[28]

First we shall be concerned with approximations to moments derived from finite samples. In practice, we cannot collect data for all points of the sample space Ω, except in very few exceptional cases. Otherwise exhaustive measurements are

[28]For the reader who is interested in more details on mathematical statistics, we recommend the classic textbook by the Polish mathematician Marek Fisz [179] and the comprehensive treatise by Stuart and Ord [514, 515], which is a new edition of Kendall's classic on statistics. An account that is useful as a not too elaborate introduction can be found in [257], while the monograph [88] is particularly addressed to experimentalists using statistics, and a wide variety of other, equally suitable texts are, of course, available in the rich literature on mathematical statistics.

impossible and we have to rely on limited samples as they are obtained in physics through experiments or in sociology through opinion polls. As an example, for the evaluation and justification of assumptions, we introduce Pearson's chi-squared test, present the ideas of the maximum likelihood method, and finally illustrate statistical inference by means of an example applying Bayes' theorem.

2.6.1 Sample Moments

As we did before for complete sample spaces, we evaluate functions Z from incomplete random samples $(\mathcal{X}_1, \ldots, \mathcal{X}_n)$ and obtain $\mathcal{Z} = Z(\mathcal{X}_1, \ldots, \mathcal{X}_n)$ as output random variables. Quantities calculated from incomplete samples are called *estimators* since they correspond to estimates of the values of the function computed from the entire sample space. Estimators of the moments of distributions are of primary importance and we shall compute sample expectation values, also called sample means, sample variances, and sample standard deviations from limited sets of data $\mathbf{x} = (x_1, x_2, \ldots, x_n)$. They are calculated as if the sample set covered the entire sample space. Using the same notations, but replacing μ by m, we obtain for the *sample mean*:

$$m = \hat{m}_1 = \frac{1}{n} \sum_{i=1}^{n} x_i \,. \tag{2.115}$$

For the *sample variance*, we calculate

$$m_2 = \frac{1}{n} \sum_{i=1}^{n} x_i^2 - \left(\frac{1}{n} \sum_{i=1}^{n} x_i \right)^2 , \tag{2.116}$$

and after some calculation, we find for the third and fourth moments:

$$m_3 = \frac{1}{n} \sum_{i=1}^{n} x_i^3 - \frac{3}{n^2} \left(\sum_{i=1}^{n} x_i \right) \left(\sum_{j=1}^{n} x_j^2 \right) + \frac{2}{n^3} \left(\sum_{i=1}^{n} x_i \right)^3 , \tag{2.117a}$$

$$m_4 = \frac{1}{n} \sum_{i=1}^{n} x_i^4 - \frac{4}{n^2} \left(\sum_{i=1}^{n} x_i \right) \left(\sum_{j=1}^{n} x_j^3 \right) + \frac{6}{n^3} \left(\sum_{i=1}^{n} x_i \right)^2 \left(\sum_{j=1}^{n} x_j^2 \right) - \frac{3}{n^4} \left(\sum_{i=1}^{n} x_i \right)^4 . \tag{2.117b}$$

These naïve estimators m_i ($i = 2, 3, 4, \ldots$) contain a bias because the exact expectation value μ around which the moments are centered is not known and has to be approximated by the sample mean m. For the variance, we illustrate the systematic deviation by calculating a correction factor known as Bessel's correction, named after the German astronomer, mathematician, and physicist Friedrich Bessel, although the correction would be more properly attributed to Carl Friedrich Gauss [295, Part 2, p. 161]. In order to obtain expectation values for the sample moments, we repeat the drawing of samples with n elements and denote their mean values by $\langle m_i \rangle$.[29] In particular, we have

$$\begin{aligned} m_2 &= \frac{1}{n}\sum_{i=1}^{n} x_i^2 - \left(\frac{1}{n}\sum_{i=1}^{n} x_i\right)^2 \\ &= \frac{1}{n}\sum_{i=1}^{n} x_i^2 - \frac{1}{n^2}\left(\sum_{i=1}^{n} x_i^2 + \sum_{i,j=1,\, i\neq j}^{n} x_i x_j\right) \\ &= \frac{n-1}{n^2}\sum_{i=1}^{n} x_i^2 - \frac{1}{n^2}\sum_{i,j=1,\, i\neq j}^{n} x_i x_j \,. \end{aligned}$$

The mean value is now of the form

$$\langle m_2 \rangle = \frac{n-1}{n}\left\langle \frac{1}{n}\sum_{i=1}^{n} x_i^2 \right\rangle - \frac{1}{n^2}\left\langle \sum_{i,j=1,\, i\neq j}^{n} x_i x_j \right\rangle \,.$$

Using $\langle x_i x_j \rangle = \langle x_i \rangle \langle x_j \rangle = \langle x_i \rangle^2$ for independent data, we find

$$\begin{aligned} \langle m_2 \rangle &= \frac{n-1}{n}\left\langle \frac{1}{n}\sum_{i=1}^{n} x_i^2 \right\rangle - \frac{n(n-1)}{n^2}\left\langle \sum_{i=1}^{n} x_i \right\rangle^2 \\ &= \frac{n-1}{n}\hat{\mu}_2 - \frac{n(n-1)}{n^2}\mu^2 = \frac{n-1}{n}(\hat{\mu}_2 - \mu^2) \,, \end{aligned}$$

[29]It is important to note that $\langle m_i \rangle$ is the expectation value of an average over a finite sample, whereas the genuine expectation value refers to the entire sample space. In particular, we find

$$\langle m \rangle = \left\langle \frac{1}{n}\sum_{i=1}^{n} x_i \right\rangle = \mu = \hat{\mu}_1 \,,$$

where μ is the first (raw) moment. For the higher moments, the situation is more complicated and requires some care (see text).

where $\hat{\mu}_2$ is the second raw moment. Using the identity $\hat{\mu}_2 = \mu_2 + \mu^2$, we find for the unbiased sample variance $\widetilde{\text{var}}$:

$$\langle m_2 \rangle = \frac{n-1}{n}\mu_2 \,, \qquad \widetilde{\text{var}}(\mathbf{x}) = \frac{1}{n-1}\sum_{i=1}^{n}(x_i - m)^2 \,. \tag{2.118}$$

The generalization of the bias to other estimators T yields

$$\mathcal{B}(T) = E(T) - \Theta = E(T - \Theta) \,, \tag{2.119}$$

and an unbiased estimator requires $\mathcal{B}(T) = 0 \,, \ \forall \, \Theta$. For the sample mean, we find

$$\mathcal{B}(m, \mu) = E(m) - \mu = E(m - \mu) = 0 \,.$$

For the sample variance we can make of use Bienaymé's formula, which gives $\text{var}(\sum_{i=1}^{n} x_i) = \sum_{i=1}^{n} \text{var}(x_i)$, to obtain directly for the bias

$$\mathcal{B}(m_2, \mu_2) = E(m_2) - \mu_2 = E(m_2 - \mu_2) = -\frac{1}{n}\mu_2 \,,$$

which is, of course, identical to (2.118). The bias, the biased mean value, and the *mean squared error* $\text{mse}(T) = \langle (T - \Theta)^2 \rangle$, are related by

$$\text{mse}(T) = \text{var}(T) + \big(\mathcal{B}(t, \Theta)\big)^2 \,.$$

The mean squared error and other issues of parameter optimization for probability distributions will be discussed in Sect. 2.6.4.

A useful expression for the first and second sample moments of a data series combining the data sets from two independent series of measurements, $\mathcal{S}_1 = \mathbf{x}_1 = \big(x_1^{(1)}, \ldots, x_{n_1}^{(1)}\big)$ and $\mathcal{S}_2 = \mathbf{x}_2 = \big(x_1^{(2)}, \ldots, x_{n_2}^{(2)}\big)$, is obtained as follows:

$$m_1 = \frac{1}{n_1}\sum_{i=1}^{n_1} x_i^{(1)} \,, \quad \widetilde{\text{var}}_1 = \frac{1}{n_1 - 1}\sum_{i=1}^{n_1}\big(x_i^{(1)} - m_1\big)^2 = \frac{n_1}{n_1 - 1}\Big(\text{E}\big(\mathbf{x}_1^2\big) - m_1^2\Big) \,,$$

$$m_2 = \frac{1}{n_2}\sum_{i=1}^{n_2} x_i^{(2)} \,, \quad \widetilde{\text{var}}_2 = \frac{1}{n_2 - 1}\sum_{i=1}^{n_2}\big(x_i^{(2)} - m_2\big)^2 = \frac{n_2}{n_2 - 1}\Big(\text{E}\big(\mathbf{x}_2^2\big) - m_2^2\Big) \,.$$

Combining the two data sets yields the set

$$\mathcal{S} = \mathbf{x} = \big(x_1^{(1)}, \ldots, x_{n_1}^{(1)}, x_1^{(2)}, \ldots, x_{n_2}^{(2)}\big)$$

with $n = n_1 + n_2$ entries. It is now straightforward to express the sample mean and the sample variance of the new set through the moments of $\mathcal{S}_1$ and $\mathcal{S}_2$:

$$\langle \mathbf{x} \rangle = m = \frac{n_1\, m_1 + n_2\, m_2}{n} \, ,$$
$$\widetilde{\text{var}} = \frac{1}{n-1} \left((n_i - 1)\widetilde{\text{var}}_i + (n_2 - 1)\widetilde{\text{var}}_2 + \frac{n_1 n_2}{n} (m_1 - m_2)^2 \right) . \tag{2.120}$$

Generalization to k independent data sets yields:

$$\langle \mathbf{x} \rangle = m = \frac{1}{n} \sum_{i=1}^{k} n_i m_i \, ,$$
$$\widetilde{\text{var}} = \frac{1}{n-1} \left(\sum_{i=1}^{k} (n_i - 1)\widetilde{\text{var}}_i + \sum_{i=1}^{k-1} \sum_{j=2, j\neq i}^{k} \frac{n_i n_j}{n} (m_i - m_j)^2 \right) . \tag{2.121}$$

The results for the biased samples are obtained in complete analogy and have the same form with the $n_{(i)} - 1$ terms replaced by $n_{(i)}$.

The measures of correlation between pairs of random variables can be calculated straightforwardly: the unbiased *sample covariance* is

$$\mathcal{M}_{\mathcal{X}\mathcal{Y}} = \frac{1}{n-1} \sum_{i=1}^{n} (x_i - m)\,(y_i - m) \, , \tag{2.122}$$

and the *sample correlation coefficient* is

$$\mathcal{R}_{\mathcal{X}\mathcal{Y}} = \frac{\sum_{i=1}^{n} (x_i - m)\,(y_i - m)}{\sqrt{\left(\sum_{i=1}^{n} (x_i - m)^2\right)\left(\sum_{i=1}^{n} (y_i - m)^2\right)}} \, . \tag{2.123}$$

For practical purposes, Bessel's correction is unimportant when the data sets are sufficiently large, but it is important to recognize the principle, in particular for more involved statistical properties than variances. Sometimes a problem is encountered in cases where the second moment μ_2 of a distribution diverges or does not exist. Then, computing variances from incomplete data sets is unstable and one may choose instead the *mean absolute deviation*, viz.,

$$\mathcal{D}(\mathcal{X}) = \frac{1}{n} \sum_{i=1}^{n} |\mathcal{X}_i - m| \, , \tag{2.124}$$

as a measure for the width of the distribution [458, pp. 455–459], because it is commonly more robust than the variance or the standard deviation.

Ronald Fisher conceived k-statistics in order to derive estimators for the moments of finite samples [173]. The cumulants of a probability distribution are derived as mean values k_i of finite set cumulants and are calculated in the same way as the analogues κ_i from a complete sample set [296, pp. 99–100]. The first four terms of k-statistics for n sample points are as follows:

$$\begin{aligned} k_1 &= m\,, \\ k_2 &= \frac{n}{n-1}\,m_2\,, \\ k_3 &= \frac{n^2}{(n-1)(n-2)}\,m_3\,, \\ k_4 &= \frac{n^2\Big((n+1)m_4 - 3(n-1)m_2^2\Big)}{(n-1)(n-2)(n-3)}\,, \end{aligned} \tag{2.125}$$

which can be derived by inversion of the following well known relationships:

$$\begin{aligned} \langle m\rangle &= \mu\,, \\ \langle m_2\rangle &= \frac{n-1}{n}\mu_2\,, \\ \langle m_3\rangle &= \frac{(n-1)(n-2)}{n^2}\mu_3\,, \\ \langle m_2^2\rangle &= \frac{(n-1)\Big((n-1)\mu_4 + (n^2-2n+3)\mu_2^2\Big)}{n^3}\,, \\ \langle m_4\rangle &= \frac{(n-1)\Big((n^2-3n+3)\mu_4 + 3(2n-3)\mu_2^2\Big)}{n^3}\,. \end{aligned} \tag{2.126}$$

The usefulness of these relations becomes evident in various applications.

The statistician computes moments and other functions from his empirical, non-exhaustive data sets, e.g., $\{x_1,\ldots,x_n\}$ or $\{(x_1,y_1),\ldots,(x_n,y_n)\}$ by means of (2.115) and (2.118) to (2.123). The underlying assumption is, of course, that the values of the empirical functions converge to the corresponding exact moments as the random sample increases. The theoretical basis for this assumption is provided by the law of large numbers.

2.6.2 *Pearson's Chi-Squared Test*

The main issue of mathematical statistics, however, is not so much to compute approximations to the moments but, as it has always been and still is, the development of independent tests that allow for the derivation of information on the appropriateness of models and the quality of data. Predictions about the reliability of computed values are made using a wide variety of tools. We dispense with the details, which are treated extensively in the literature [180, 514, 515], and present only the most frequently applied test as an example. In 1900 Karl Pearson conceived this test [445], which became popular under the name of the chi-squared test. It was used, for example, by Ronald Fisher when he analyzed Gregor Mendel's data on the genetics of the garden pea *Pisum sativum*, and we shall apply it here, for illustrative purposes, to the data given in Table 1.1.

The formula of Pearson's test can be made plausible by means of a simple example [258, pp. 407–414]. A random variable $\mathcal{Y}_1$ is binomially distributed according to $B_k(n, p_1)$ with expectation value $\mathrm{E}(\mathcal{Y}_1) = np_1$ and variance $\sigma_1^2 = np_1(1 - p_1)$ (Sect. 2.3.2). By the central limit theorem, the random variable

$$\mathcal{Z} = \frac{\mathcal{Y}_1 - np_1}{\sqrt{np_1(1 - p_1)}}$$

has a standardized binomial distribution which approaches $\mathcal{N}(0, 1)$ for sufficiently large n (Sect. 2.4.1). A second random variable is $\mathcal{Y}_2 = n - \mathcal{Y}_1$, which has expectation value $\mathrm{E}(\mathcal{Y}_2) = np_2$ and variance $\sigma_2^2 = \sigma_1^2 = np_2(1 - p_2) = np_1(1 - p_1)$, since $p_2 = 1 - p_1$. The sum $\mathcal{Z}^2 = \mathcal{Y}_1^2 + \mathcal{Y}_2^2$ is approximately χ^2-distributed:

$$\mathcal{Z}^2 = \frac{(\mathcal{Y}_1 - np_1)^2}{np_1(1 - p_1)} = \frac{(\mathcal{Y}_1 - np_1)^2}{np_1} + \frac{(\mathcal{Y}_2 - np_2)^2}{np_2} ,$$

since

$$(\mathcal{Y}_1 - np_1)^2 = \Big(n - \mathcal{Y}_1 - n(1 - p_1)\Big)^2 = (\mathcal{Y}_2 - np_2)^2 .$$

We can now rewrite the expression by introducing the expectation values

$$\mathcal{Q}_1 = \sum_{i=1}^{2} \frac{\big(\mathcal{Y}_i - \mathrm{E}(\mathcal{Y}_i)\big)^2}{\mathrm{E}(\mathcal{Y}_i)} ,$$

indicating the number of independent random variables as a subscript. Provided all products np_i are sufficiently large—a conservative estimate would be $np_i \geq 5 ,\ \forall\, i$—the quantity $\mathcal{Q}_1$ has an approximate chi-squared distribution with one degree of freedom χ_1^2.

The generalization to an experiment with k mutually exclusive and exhaustive outcomes $A_1, A_2, \ldots, A_k$ of the variables $\mathcal{X}_1, \mathcal{X}_2, \ldots, \mathcal{X}_k$, is straightforward. All variables $\mathcal{X}_i$ are assumed to have finite mean μ_i and finite variance σ_i^2 so that the central limit theorem applies and the distribution for large n converges to the normal distribution $\mathcal{N}(0, 1)$. We define the probability of obtaining the result A_i by $P(A_i) = p_i$. Due to conservation of probability, we have $\sum_{i=1}^{k} p_i = 1$, whence one variable lacks independence and we choose it to be $\mathcal{X}_k$:

$$\mathcal{X}_k = n - \sum_{i=1}^{k-1} \mathcal{X}_i \,. \tag{2.127}$$

The joint distribution of $k-1$ variables $\mathcal{X}_1, \mathcal{X}_2, \ldots, \mathcal{X}_{k-1}$ then has the joint probability mass function (pmf)

$$f(x_1, x_2, \ldots, x_{k-1}) = P(\mathcal{X}_1 = x_1, \mathcal{X}_2 = x_2, \ldots, \mathcal{X}_{k-1} = x_{k-1}) \,.$$

Next we consider n independent trials yielding x_1 times A_1, x_2 times A_2, ..., and x_k times A_k, where a particular outcome has the probability

$$P(\mathcal{X}_1 = x_1, \mathcal{X}_1 = x_1, \ldots, \mathcal{X}_{k-1} = x_{k-1}) = p_1^{x_1} p_2^{x_2} \cdots p_k^{x_k} \,,$$

with the frequency factor or statistical weight

$$g(x_1, x_2, \ldots, x_k) = \binom{n}{x_1, x_2, \ldots, x_k} = \frac{n!}{x_1! x_2! \cdots x_k!} \,,$$

and eventually we find for the pmf

$$\begin{aligned} f(x_1, x_2, \ldots, x_{k-1}) &= g(x_1, x_2, \ldots, x_k) P(\mathcal{X}_1 = x_1, \mathcal{X}_2 = x_2, \ldots, \mathcal{X}_{k-1} = x_{k-1}) \\ &= \frac{n!}{x_1! x_2! \cdots x_k!} p_1^{x_1} p_2^{x_2} \cdots p_k^{x_k} \,, \end{aligned} \tag{2.128}$$

with the two restrictions $x_k = n - \sum_{i=1}^{k-1} x_i$ and $p_k = 1 - \sum_{i=1}^{k-1} p_i$. Pearson's construction follows the lines we have shown before for the binomial distribution with $k = 2$. Considering (2.127), this yields

$$\mathcal{Q}_{k-1}(n) = X_{k-1}^2(n) = \sum_{i=1}^{k} \frac{\left(\mathcal{X}_i - \mathrm{E}(\mathcal{X}_i)\right)^2}{\mathrm{E}(\mathcal{X}_i)} \,. \tag{2.129}$$

The sum of squares $\mathcal{Q}_{k-1}(n)$ in (2.129) is called Pearson's *cumulative test statistic.* It has an approximate chi-squared distribution with $k-1$ degrees of freedom χ_{k-1}^2,[30] and again if n is sufficiently large to satisfy $np_i \geq 5\,,\ \forall\, i$, the distributions are close enough for most practical purposes.

In order to be able to test hypotheses we divide our sample space into k cells and record observations falling into individual cells (Fig. 2.25). In essence, these

[30]We indicate the expected converge in the sense of the central limit theorem by choosing the symbol X_{k-1}^2 for the finite n expression with $\lim_{n\to\infty} X_{k-1}^2(n) = \chi_{k-1}^2$.

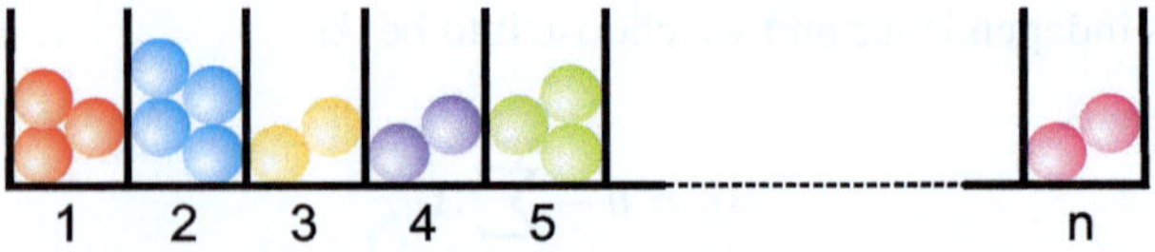

Fig. 2.25 Definition of cells for application of the χ^2-square test. The space of possible outcomes of recordings is partitioned into n cells, which correspond to features of classification. As an example, one could group animals into males and females, or scores according to the numbers on the top face of a rolled die. The characteristics of classification are visualized by different colors

cells $\mathcal{C}_i$ are tantamount to the outcomes A_i, but we can define them to be completely general, for example, collecting all instances that fall in a certain range. At the end of the registration period, the number of observations is n and the partitioning into the instances that were recorded in the cell $\mathcal{C}_i$ is ν_i with $\sum_{i=1}^{k} \nu_i = n$. Equation (2.129) is now applied to test a (null) hypothesis H_0 against empirically registered values for the different outcomes:

$$H_0 : \; E_i^{(0)}(\mathcal{X}_i) = \varepsilon_{i0} \, , \quad i = 1, \ldots, k \, . \tag{2.130}$$

In other words, the null hypothesis predicts the distribution of score values falling into the cells $\mathcal{C}_i$ to be ε_{i0} $(i = 1, \ldots, k)$ and this in the sense of expectation values $E_i^{(0)}$. If the null hypothesis were, for example, the uniform distribution, we would have $\varepsilon_{i0} = n/k, \; \forall\, i = 1, \ldots, k$. The cumulative test statistic $X^2(n)$ converges to the χ^2 distribution in the limit $n \to \infty$, just as the average value of a stochastic variable $\langle \mathcal{Z} \rangle = \sum_{i=1}^{n} z_i/n$ converges to the expectation value $\lim_{n\to\infty} \langle \mathcal{Z} \rangle = \mathrm{E}(\mathcal{Z})$. This implies that $X^2(n)$ is never exactly equal to χ^2, but the approximation will always become better when the sample size is increased. Empirical knowledge of statisticians defines a lower limit for the number of entries in the cells to be considered, which lies between 5 and 10.

If the null hypotheses H_0 were true, ν_i and ε_{i0} should be approximately equal. Thus we expect the deviation expressed by

$$X_{\mathrm{d}}^2 = \sum_{i=1}^{k} \frac{(\nu_i - \varepsilon_{i0})^2}{\varepsilon_{i0}} \approx \chi_{\mathrm{d}}^2 \tag{2.131}$$

to be small if H_0 is acceptable. If the deviation is too large, we shall reject H_0: $X_{\mathrm{d}}^2 \geq \chi_{\mathrm{d}}^2(\alpha)$, where α is the predefined level of significance for the test. Two basic quantities are still undefined: (i) the degree of freedom d and (ii) the significance level α.

First the number of degrees of freedom d of the theoretical distribution to which the data are fitted has to be determined. The number of cells k represents the maximal number of degrees of freedom, which is reduced by one because of the conservation relation $\sum_i \nu_i = n$ discussed above, so $d = k - 1$. The dimension d is reduced further when parameters are needed to fit the distribution of the null hypothesis. If the number of such parameters is s, we get $d = k - 1 - s$. Choosing the parameter-free uniform distribution $\mathcal{U}$ as null hypothesis we find, of course, $d = k - 1$.

The significance of the null hypothesis for a given set of data is commonly tested by means of the so-called p-value: for $p < \alpha$, the null hypothesis is rejected. More precisely, the p-value is the probability of obtaining a test statistic which is at least as extreme as the one actually observed under the assumption that the null hypothesis is true. We call a probability $P(A)$ *more extreme* than $P(B)$ if A is less likely to occur than B under the null hypothesis. As shown in Fig. 2.26, this probability is obtained as the integral below the probability density function from the calculated X_{d}^2-value to $+\infty$. For the χ_{d}^2 distribution, we have

$$p = \int_{X_{\mathrm{d}}^2}^{+\infty} \chi_{\mathrm{d}}^2(x)\,\mathrm{d}x = 1 - \int_0^{X_{\mathrm{d}}^2} \chi_{\mathrm{d}}^2(x)\,\mathrm{d}x = 1 - F_{\chi^2}(X^2; d)\,, \tag{2.132}$$

which involves the cumulative distribution function of the χ^2-distribution, viz., $F_{\chi^2}(x; d)$, defined in (2.72). Commonly, the null hypothesis is rejected when p is smaller than the significance level, i.e., $p < \alpha$, with the empirical choice $0.02 \le \alpha \le 0.05$ (Fig. 2.27). If the condition $p < \alpha$ is satisfied one says that *the null hypothesis is rejected by statistical significance*. In other words, the null hypothesis is statistically significant or statistically confirmed in the range $\alpha \le p \le 1$.

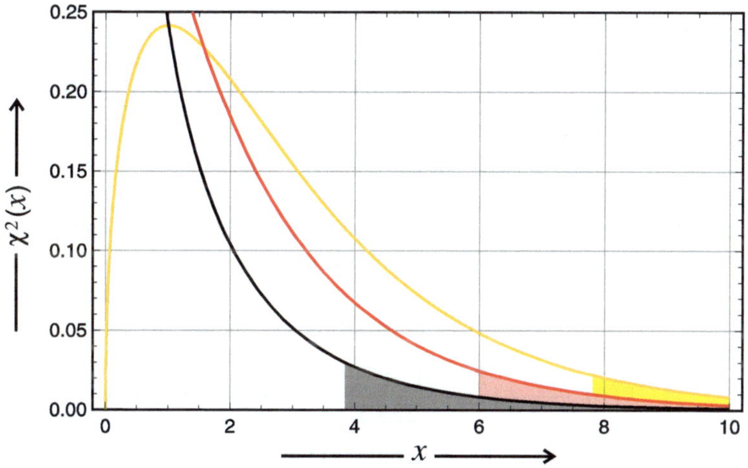

Fig. 2.26 Definition of the p-value in the significance test. The figure illustrates the definition of the p-value. The *three curves* represent the χ_k^2 probability densities with parameters $k = 1$ (*black*), 2 (*red*), and 3 (*yellow*). The three specific $x_k(\alpha)$-values are shown for the critical p-value with $\alpha = 0.05$: for $k = 1$ we find $x_1(0.05) = 3.84146$, for $k = 2$ we obtain $x_2(0.05) = 5.99146$, and for $k = 3$ we have $x_3(0.05) = 7.81473$. *Hatched areas* show the range of values of the random variable $\mathcal{Q}$ that are more extreme than the predefined critical p-value. The latter is defined as the cumulative probability within the indicated areas that were defined by $\alpha = 0.05$. If the p-value for an observed data set satisfies $p < \alpha$, the null hypothesis is rejected

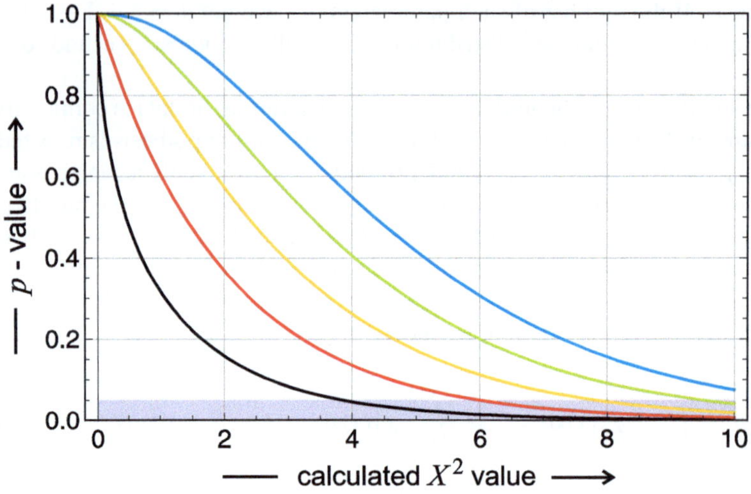

Fig. 2.27 The p-value in the significance test and rejection of the null hypothesis. The figure shows the p-values from (2.132) as a function of the calculated values of X_k^2 for k cells. Color code for the k-values: 1 (*black*), 2 (*red*), 3 (*yellow*), 4 (*green*), and 5 (*blue*). The *shaded area* at the bottom of the figure shows the range where the null hypothesis is rejected

A simple example can illustrate this. Two random samples of n animals are drawn from a population and it is found that ν_1 are males and ν_2 are females, with $\nu_1 + \nu_2 = n$. A first sample has

$$n = 322\,,\ \ \nu_1 = 170\,,\ \ \nu_2 = 152\,,\ \ X_1^2 = \frac{(170-161)^2 + (152-161)^2}{322} = 0.503\,,$$
$$p = 1 - F_{\chi^2}(0.503; 1) = 0.478\,,$$

which clearly supports the null hypothesis that males and females are equally frequent, since $p > \alpha \approx 0.05$. The second sample has

$$n = 467\,,\ \ \nu_1 = 207\,,\ \ \nu_2 = 260\,,$$
$$X_1^2 = \frac{(207-233.5)^2 + (260-233.5)^2}{233.5} = 6.015\,,$$
$$p = 1 - F_{\chi^2}(6.015; 1) = 0.0142\,,$$

and this leads to a p-value which is below the critical limit of significance, and hence to rejection of the null hypothesis. Then the hypothesis that the numbers of males and females are equal is statistically insignificant. In other words, there is very likely another reason for the difference, something other than random fluctuations.

As a second example we test here Gregor Mendel's experimental data on the garden pea *Pisum sativum*, as given in Table 1.1. Here the null hypothesis to be

Table 2.3 Pearson χ^2-test of Gregor Mendel's experiments with the garden pea

		Number of seeds		χ^2-statistics	
Property	Sample space	A/B	a/b	X_1^2	p
Shape (A,a)	Total	5474	1850	0.2629	0.6081
Color (B,b)	Total	6022	2001	0.0150	0.9025
Shape (A,a)	Plant 1	45	12	0.4737	0.4913
Color (B,b)	Plant 1	25	11	0.5926	0.4414
Shape (A,a)	Plant 5	32	11	0.00775	0.9298
Color (B,b)	Plant 5	24	13	2.0405	0.1532
Shape (A,a)	Plant 8	22	10	0.6667	0.4142
Color (B,b)	Plant 8	44	9	1.8176	0.1776

The total results as well as the data for three selected plants are analyzed using Karl Pearson's chi-squared statistics. Two characteristic features of the seeds are reported: the shape, roundish or angular (wrinkled), and the color, yellow or green. The phenotypes of the two dominant alleles are A = *round* and B = *yellow* and the recessive phenotypes are a = *wrinkled* and b = *green*. The data are taken from Table 1.1

tested is the ratio between different phenotypic features developed by the genotypes. We consider two features: (i) the shape of the seeds, roundish or wrinkled, and (ii) the color of the seeds, yellow or green, which are determined by two independent loci and two alleles each, viz., A and a or B and b, respectively. The two alleles form four diploid genotypes, AA, Aa, and aA, aa, or BB, Bb, and bB, bb, respectively. Since the alleles a and b are *recessive*, only the genotypes aa or bb develop the second phenotype, wrinkled and green, and based on the null hypothesis of a uniform distribution of genotypes, we expect a 3:1 ratio of phenotypes.

In Table 2.3, we apply Pearson's chi-squared hypothesis to the null hypothesis of 3:1 ratios for the phenotypes roundish and wrinkled or yellow and green. As examples we have chosen the total sample of Mendel's experiments as well as three plants (1, 5, and 8 in Table 1.1) which are typical (1) or show extreme ratios (5 having the best and the worst value for shape and color, respectively, and 8 showing the highest ratio, namely, 4.89). All p-values in this table are well above the critical limit and confirm the 3:1 ratio without the need for further discussion.[31]

The independence test is relevant for situations when an observer registers two outcomes and the null hypothesis is that these outcomes are statistically independent. Each observation is allocated to one cell of a two-dimensional array of cells called a *contingency table* (see Sect. 2.6.3). In the general case there are m rows and n columns in a table. Then, the theoretical frequency for a cell under the null hypothesis of independence is

$$\varepsilon_{ij} = \frac{\sum_{k=1}^{n} \nu_{ik} \sum_{k=1}^{m} \nu_{kj}}{N} , \tag{2.133}$$

[31]Recall the claim by Ronald Fisher and others to the effect that Mendel's data were too good to be true.

where N is the (grand) total sample size or the sum of all cells in the table. The value of the X^2 test statistic is

$$X^2 = \sum_{i=1}^{m} \sum_{j=1}^{n} \frac{(\nu_{ij} - \varepsilon_{ij})^2}{\varepsilon_{ij}} . \tag{2.134}$$

Fitting the model of independence reduces the number of degrees of freedom by $\pi = m+n-1$. Originally, the number of degrees of freedom is equal to the number of cells mn, and after reduction by π, we have $d = (m-1)(n-1)$ degrees of freedom for comparison with the χ^2 distribution. The p-value is again obtained by insertion into the cumulative distribution function (cdf), $p = 1 - F_{\chi^2}(X^2; d)$, and a value of p less than a predefined critical value, commonly $p < \alpha = 0.05$, is considered as justification for rejection of the null hypothesis, i.e., the conclusion that the row variables do not appear to be independent of the column variables.

2.6.3 *Fisher's Exact Test*

As a second example out of many statistical significance tests developed in mathematical statistics, we mention here *Fisher's exact test* for the analysis of contingency tables. In contrast to the χ^2-test, Fisher's test is valid for all sample sizes and not only for sufficiently large samples. We begin by defining a contingency table. This is an $m \times n$ matrix M where all possible outcomes of one variable x enter different columns in a row defined by a given outcome for y, while the distribution of outcomes of the second variable y for a specified outcome of x is contained in a column. The most common case, and the one that is most easily analyzed, is 2×2, i.e., two variables with two values each. Then the contingency table has the form

	x_1	x_2	Total
y_1	a	b	$a+b$
y_2	c	d	$c+d$
Total	$a+c$	$b+d$	N

where every variable x and y has two outcomes, and $N = a + b + c + d$ is the grand total number of trials. Fisher's contribution was to prove that the probability of obtaining the set of values (x_1, x_2, y_1, y_2) is given by the hypergeometric distribution with

$$\text{probability mass function} \quad f_{\mu,\nu}(k) = \frac{\binom{\mu}{k}\binom{N-\mu}{\nu-k}}{\binom{N}{\nu}},$$

$$\text{cumulative density function} \quad F_{\mu,\nu}(k) = \sum_{i=0}^{k} \frac{\binom{\mu}{k}\binom{N-\mu}{\nu-k}}{\binom{N}{\nu}}, \tag{2.135}$$

where $N \in \mathbb{N}_{>0}$, $\mu \in \{0, 1, \ldots, N\}$, $\nu \in \{1, 2, \ldots, N\}$, and

$$k \in \left\{ \max(0, \nu + \mu - N), \ldots, \min(\mu, \nu) \right\} .$$

Translating the contingency table into the notation of probability functions, we have $a \equiv k$, $b \equiv \mu - k$, $c \equiv \nu - k$, and $d \equiv N + k - (\mu + \nu)$, and hence Fisher's result for the p-value of the general 2×2 contingency table is

$$p = \frac{\binom{a+b}{a}\binom{c+d}{c}}{\binom{N}{a+c}} = \frac{(a+b)!(c+d)!(a+c)!(b+d)!}{a!\,b!\,c!\,d!\,N!}, \tag{2.136}$$

where the expression on the right-hand side shows beautifully the equivalence between rows and columns.

We present the right- or left-handedness of human males or females to illustrate Fisher's test. A sample consisting of 52 males and 48 females yields 9 left-handed males and 4 left-handed females. Is the difference statistically significant and does it allow us to conclude that left-handedness is more common among males than females? The contingency table in this case reads:

	x_m	x_f	Total
y_r	43	44	87
y_l	9	4	13
Total	52	48	100

The calculation yields $p \approx 0.10$, above the critical value $0.02 \leq \alpha \leq 0.05$, and $p > \alpha$ confirms the null hypothesis of men and women being equally likely to be left-handed. Therefore, the assumption that males are more likely to be left-handed can be rejected for this data sample.

2.6.4 The Maximum Likelihood Method

The maximum likelihood method (mle) is a widely used procedure for estimating unknown parameters in models with known functional relations. In probability theory the function is a probability density containing unknown parameters which are estimated by means of data sets. In Sect. 2.6.1 we carried out such tasks when we computed expressions for the moments of distributions derived from finite samples. Maximum likelihood searches for optimal estimates given fixed data sets (see also Sect. 4.1.5).

History of Maximum Likelihood
The maximum likelihood method has been around for a long time and many famous mathematicians have made contributions to it [509]. (For an extensive literature survey, see also [424, 425].) Examples are the French–Italian mathematician Joseph-Louis Lagrange and the Swiss mathematician Daniel Bernoulli in the second half of the eighteenth century, Carl Friedrich Gauss in his famous book [197], and Karl Pearson together with Louis Filon [447]. Ronald Fisher got interested in parameter optimization rather early on [169] and worked intensively on maximum likelihood. He published three proofs with the aim of showing that this approach is the most efficient strategy for parameter optimization [8, 170, 172, 175].

Maximum likelihood did indeed become the most used optimization strategy in practice and is still a preferred topic in estimation theory. The variance of estimators was shown to be bounded from below by the Cramér–Rao bound, named after Harald Cramér and Calyampudi Radhakrishna Rao [94, 463]. Unbiased estimators, which can achieve this lower bound are said to be fully efficient. At the present time, maximum likelihood is fairly well understood and most of its common failures and cases of inapplicability are known and documented [331], but care is needed in its application to complex problems, as pointed out by Stephen Stigler in the conclusion to his review [509]:

> We now understand the limitations of maximum likelihood better than Fisher did, but far from well enough to guarantee safety in its application in complex situations where it is most needed. Maximum likelihood remains a truly beautiful theory, even though tragedy may lurk around a corner.

Maximum Likelihood Estimation
Maximum likelihood estimation deals with a sample of n independent and identically distributed (iid) observations, $(\mathbf{x}_1, \ldots, \mathbf{x}_n)$, which follow a probability density f_0 with unknown parameters $\boldsymbol{\theta}_0$ from a parameter space that is characteristic for the family with distribution $f(\cdot|\boldsymbol{\theta} \in \boldsymbol{\Theta})$. The task is to find an estimator $\hat{\boldsymbol{\theta}}$ that comes as close as possible to the true parameter values $\boldsymbol{\theta}_0$. Both the observed data and the

parameters may be scalar quantities or vectors as indicated. For independent and identically distributed samples the density can be written as a product of n factors:

$$f(\mathbf{x}_1, \mathbf{x}_2, \ldots, \mathbf{x}_n|\boldsymbol{\theta}) = f(\mathbf{x}_1|\boldsymbol{\theta})f(\mathbf{x}_2|\boldsymbol{\theta})\cdots f(\mathbf{x}_n|\boldsymbol{\theta}) = \prod_{i=1}^{n} f(\mathbf{x}_i|\boldsymbol{\theta}) \,. \tag{2.137}$$

The probability density is expressed as a function of the sampled values $\mathbf{x}_i$ under the condition that $\boldsymbol{\theta}$ is the applied parameter set. For the purpose of optimization we look at (2.137) and turn around the interpretation:

$$\mathcal{L}(\boldsymbol{\theta}; \mathbf{x}_1, \mathbf{x}_2, \ldots, \mathbf{x}_n) = \prod_{i=1}^{n} f(\mathbf{x}_i|\boldsymbol{\theta}) \,, \tag{2.138}$$

where $\boldsymbol{\theta}$ is the variable and $(\mathbf{x}_1, \ldots, \mathbf{x}_n)$ is the fixed set of observations.[32] In general, it is simpler to operate on sums than on products and hence the logarithm $\log \mathcal{L}$ of the likelihood function is preferred over $\mathcal{L}$. The logarithm $\log \mathcal{L}$ is a strictly monotonically increasing function and therefore shows extremum values at exactly the same positions as the likelihood $\mathcal{L}$. Since we shall be interested only in the parameter values $\boldsymbol{\theta}_{\mathrm{mle}}$, it makes no difference whether we are using the function $\mathcal{L}$ or its logarithm $\log \mathcal{L}$. For a discussion of the behavior in the limit $n \to \infty$ of large sample numbers, it is advantageous to use the average log-likelihood function

$$\boldsymbol{\ell} = \frac{1}{n} \log \mathcal{L} \,. \tag{2.139}$$

The maximum likelihood estimate of $\boldsymbol{\theta}_0$ now consists in finding a value for $\boldsymbol{\theta}$ that maximizes the average log-likelihood, viz.,

$$\hat{\boldsymbol{\theta}}_{\mathrm{mle}} = \arg\max_{\boldsymbol{\theta} \in \boldsymbol{\Theta}} \boldsymbol{\ell}(\boldsymbol{\theta}; \mathbf{x}_1, \mathbf{x}_2, \ldots, \mathbf{x}_n) \,, \tag{2.140}$$

provided that such a maximum exists. There are, of course, situations where the approach might fail: (i) if no maximum occurs when the function increases above $\boldsymbol{\Theta}$ without adopting the supremum value, and (ii) if multiple equivalent maximum likelihood estimates are found.

Maximum likelihood represents an optimization technique maximizing average log-likelihood as the *objective function*:

$$\boldsymbol{\ell}(\boldsymbol{\theta}|\mathbf{x}_1, \mathbf{x}_2, \ldots, \mathbf{x}_n) = \frac{1}{n} \sum_{i=1}^{n} \log f(\mathbf{x}_i|\boldsymbol{\theta}) \,.$$

[32]Variables and parameters of a function are separated by a semicolon as in $f(\mathbf{x}; \mathbf{p})$.

This objective function is understood as the sample analogue of the expectation value of log-likelihood:

$$\boldsymbol{\ell}(\boldsymbol{\theta}_0) = \mathrm{E}\big(\log f(\mathbf{x}_i|\boldsymbol{\theta}_0)\big) = \lim_{n\to\infty} \frac{1}{n} \sum_{i=1}^{n} \log f(\mathbf{x}_i|\boldsymbol{\theta}) \ .$$

Maximum likelihood has a number of attractive properties in the limit as n approaches infinity:

(i) It is *consistent* in the sense that, with increasing sample size n, the sequence of estimators $\hat{\boldsymbol{\theta}}_{\mathrm{mle}}(n)$ converges in probability exactly to the value $\boldsymbol{\theta}_0$ that is estimated.

(ii) It has the property of *asymptotic normality*, since the distribution of the maximum likelihood estimator approaches a normal distribution with mean $\boldsymbol{\theta} \to \boldsymbol{\theta}_0$, and the covariance matrix is equal to the inverse Fisher information matrix as n goes to infinity.[33]

(iii) It is fully *efficient* since it reaches the Cramér–Rao lower bound when the sample size tends to infinity, expressing the fact that no estimator which does not approach this lower bound has a lower mean squared error (see below).

Two notions are relevant in the context of maximum likelihood estimates: Fisher information and sufficient statistic.

Fisher Information

The Fisher information is a way of measuring the mean information content in the parameter θ which is contained in a random variable $\mathcal{X}$ with probability density $f(\mathcal{X}|\theta)$. It is named after Ronald Fisher, who pointed out its importance for maximum likelihood estimators [170]. Prior to Fisher, similar ideas were pursued by Francis Edgeworth [122–124]. The Fisher information can be directly obtained from the *score function*, which is the derivative of the log-likelihood:

$$U(\mathcal{X};\theta) = \frac{\partial}{\partial\theta} \log f(\mathcal{X}|\theta) \ . \tag{2.141}$$

The expectation value of the score function is zero, i.e.,

$$\mathrm{E}\left(\frac{\partial}{\partial\theta} \log f(\mathcal{X}|\theta) \mid \theta\right) = 0 \ ,$$

[33]The prerequisite for asymptotic normality is, of course, that the central limit theorem should be applicable, requiring finite expectation value and finite variance of the distribution $f(\mathbf{x}|\boldsymbol{\theta})$.

and the second moment is the Fisher information[34]:

$$\mathcal{I}(\theta) = \mathrm{E}\left(\left(\frac{\partial}{\partial\theta} \log f(\mathcal{X}|\theta) \right)^2 \Bigg| \, \theta \right) . \tag{2.142}$$

Since the expectation value of the score function is zero, the Fisher information is also the variance of the score. Provided the density function is twice differentiable ($\mathcal{C}^2$), the expression for the Fisher information can be brought into a different form:

$$\begin{aligned} \frac{\partial^2}{\partial\theta^2} \log f(x,\theta) &= \frac{\partial}{\partial\theta} \left(\frac{1}{f(x,\theta)} \frac{\partial}{\partial\theta} f(x,\theta) \right) \\ &= \frac{\partial^2 f(x,\theta)/\partial\theta^2}{f(x,\theta)} - \left(\frac{\partial f(x,\theta)/\partial\theta}{f(x,\theta)} \right)^2 . \end{aligned}$$

Taking the expectation value shows that the first term vanishes:

$$E\left(\frac{\partial^2 f(x,\theta)/\partial\theta^2}{f(x,\theta)} \Bigg| \, \theta \right) = \int \frac{\partial f(x,\theta)/\partial\theta}{f(x,\theta)} f(x,\theta)\, \mathrm{d}x = \frac{\partial^2}{\partial\theta^2} \int f(x,\theta)\, \mathrm{d}x = 0 .$$

We thus obtain an alternative expression for the Fisher information:

$$\mathcal{I}(\theta) = -\mathrm{E}\left(\frac{\partial^2}{\partial\theta^2} \log f(\mathcal{X}|\theta) \Bigg| \, \theta \right) . \tag{2.142'}$$

According to (2.142), the Fisher information is non-negative, $0 \le \mathcal{I}(\theta) < \infty$. Equation (2.142′) allows for an illustrative interpretation. In essence, the Fisher information is the negative curvature of the log-likelihood function,[35] and a flat curve implies that the log-likelihood function contains little information about the parameter θ. Alternatively, a large absolute value of the curvature implies that the distribution $f(\mathcal{X}|\theta)$ varies strongly with changes in the parameter θ and carries plenty of information about it.

It is important to point out that the Fisher information is an expectation value and hence results from averaging over all possible values of the random variable $\mathcal{X}$

---

[34]The notation $\mathrm{E}(\ldots|\theta)$ stands for a *conditioned* expectation value. Here the average is taken over the random variable $\mathcal{X}$ for a given value of θ.

[35]The *signed curvature* of a function $y = f(x)$ is defined by

$$k(x) = \frac{\mathrm{d}^2 f(x)/\mathrm{d}x^2}{\left(1 + \left(\mathrm{d}f(x)/\mathrm{d}x\right)^2\right)^{3/2}} .$$

If the tangent $\mathrm{d}f(x)/\mathrm{d}x$ is small compared to unity, the curvature is determined by the second derivative $\mathrm{d}^2 f(x)/\mathrm{d}x^2$. Use of the function $\kappa(x) = |k(x)|$ as (unsigned) curvature is also common.

in the form of its probability density. The property before averaging is defined as *observed information*:

$$\mathcal{J}(\theta) = -\frac{\partial^2}{\partial\theta^2}\big(n\ell(\theta)\big) = -\frac{\partial^2}{\partial\theta^2}\left(\sum_{i=1}^{n}\log f(\mathcal{X}_i|\theta)\right), \tag{2.143}$$

which is related to the Fisher information by $\mathcal{I}(\theta) = \mathrm{E}\big(\mathcal{J}(\theta)\big)$.

In the case of multiple parameters $\boldsymbol{\theta} = (\theta_1, \theta_2, \ldots, \theta_n)^{\mathrm{t}}$, the Fisher information is expressed by means of a matrix with the elements

$$\begin{aligned}\big(\mathcal{I}(\boldsymbol{\theta})\big)_{i,j} &= \mathrm{E}\left(\left(\frac{\partial}{\partial\theta_i}\log f(\mathcal{X},\boldsymbol{\theta})\bigg|\,\boldsymbol{\theta}\right)\left(\frac{\partial}{\partial\theta_j}\log f(\mathcal{X},\boldsymbol{\theta})\bigg|\,\boldsymbol{\theta}\right)\right)\\ &= -\mathrm{E}\left(\frac{\partial^2}{\partial\theta_i\partial\theta_j}\log f(\mathcal{X},\boldsymbol{\theta})\bigg|\,\boldsymbol{\theta}\right).\end{aligned} \tag{2.144}$$

The second expression shows that the Fisher information is the expectation value of the Hessian matrix of the log-likelihood.

Sufficient Statistic

A statistic of a random sample $\boldsymbol{\mathcal{X}} = (\mathcal{X}_1, \mathcal{X}_2, \ldots, \mathcal{X}_n)$ is a function $T(\boldsymbol{\mathcal{X}}) = \varrho(\mathcal{X}_1, \mathcal{X}_2, \ldots, \mathcal{X}_n) = \varrho(\boldsymbol{\mathcal{X}})$. Examples of such functions are the sample moments, like sample means, sample variances and others, the minimum function $\min\{\boldsymbol{\mathcal{X}}\} = \mathcal{X}_{\mathrm{min}}$, the maximum function $\max\{\boldsymbol{\mathcal{X}}\} = \mathcal{X}_{\mathrm{max}}$, or the maximum likelihood function $\mathcal{L}(\boldsymbol{\theta};\boldsymbol{x})$. In the estimate of a parameter, many details of the sample do not matter in the sense that they have no influence on the result. In estimating the expectation value μ, for example, the samples $(5,2,4,7)$, $(1,4,6,7)$, and $(6,2,6,4)$ yield the same sample mean $m = 9/2$, and they are therefore equivalent for the statistics $T(\boldsymbol{\mathcal{X}}) = m(\boldsymbol{\mathcal{X}}) = \sum_{i=1}^{n} x_i/n$. The statistic m is sufficient for estimation of the expectation value μ. Generalizing, we say that, in the estimate of a parameter θ, it makes no difference for a statistician whether he has the full information consisting of all values of the random variable $\mathcal{X}$ or only the value of the statistic $\vartheta(\mathbf{x})$ with $\mathbf{x} = (x_1, \ldots, x_n)$, and accordingly we call ϑ a *sufficient statistic*.

In mathematical terms a statistic ϱ is sufficient if, for all $r = \varrho(\mathbf{x})$, the conditional distribution does not depend on the parameter θ:

$$f(\mathbf{x}|r,\theta) = f(\mathbf{x}|r), \quad \forall\, r. \tag{2.145}$$

This condition is met when the *factorization theorem* holds: the statistic T is sufficient if and only if the conditional density can be factorized according to

$$f(\mathbf{x}|\,\theta) = u(\mathbf{x})v\big(\varrho(\mathbf{x}),\theta\big). \tag{2.146}$$

The first factor $u(\mathbf{x})$ is independent of the unknown parameter θ, and the second factor v may depend on θ, but depends on the random sample exclusively through the statistic $\varrho(\boldsymbol{\mathcal{X}})$.

For the purpose of illustration, consider the family of normal distributions and assume that the variance $\widetilde{\mathrm{var}} = \sigma^2$ is known, but the expectation value $\mathrm{E} = \mu$ must be estimated from a random sample $\boldsymbol{\mathcal{X}}$. The joint density is of the form

$$f(\mathbf{x}|\mu) = \frac{1}{\left(\sqrt{2\pi\sigma^2}\right)^n} \mathrm{e}^{-\sum_{i=1}^n (x_i-\mu)^2/2\sigma^2}$$

$$= \frac{1}{\left(\sqrt{2\pi\sigma^2}\right)^n} \mathrm{e}^{-\sum_{i=1}^n x_i^2/2\sigma^2} \mathrm{e}^{\mu \sum_{i=1}^n x_i/\sigma^2} \mathrm{e}^{-n\mu^2/2\sigma^2} \, .$$

It is straightforward to choose

$$u(\mathbf{x}) = \frac{1}{\left(\sqrt{2\pi\sigma^2}\right)^n} \mathrm{e}^{-\sum_{i=1}^n x_i^2/2\sigma^2} \, ,$$

$$v\big(\varrho(\mathbf{x}), \mu\big) = \mathrm{e}^{(-n\mu^2+2\mu\varrho(\mathbf{x}))/2\sigma^2} \, , \qquad \varrho(\mathbf{x}) = \sum_{i=1}^n x_i \, .$$

Since the factorization theorem is satisfied, $T = \sum_{i=1}^n \mathcal{X}_i$ is a sufficient statistic and $m = \sum_{i=1}^n \mathcal{X}_i/n$ is a sufficient statistic as well.

It is straightforward to show that each of the following four statistics of normally distributed random variables with unknown variance $\mathcal{N}(0, \sigma^2)$ are sufficient: $T_1(\mathcal{X}) = (\mathcal{X}_1, \ldots, \mathcal{X}_n)$, $T_2(\mathcal{X}) = (\mathcal{X}_1^2, \ldots, \mathcal{X}_n^2)$, $T_3(\mathcal{X}) = \sum_{i=1}^n \mathcal{X}_n^2$, and $T_4(\mathcal{X}) = \sum_{i=1}^m \mathcal{X}_n^2 + \sum_{i=m+1}^n \mathcal{X}_n^2$, $\forall\, m = 1, 2, \ldots, n-1$.

As a second example we consider the uniform distribution $\mathcal{U}_\Omega(x)$ with $\Omega = [0, \theta]$, and a joint density

$$f(\mathbf{x}|\,\theta) = \theta^{-1} \mathbf{1}_{x_i \le \theta}(x) \, , \quad i = 1, 2, \ldots, n \, ,$$

where θ is unknown and $\mathbf{1}_A(x)$ is the indicator function (1.26). A necessary and sufficient condition for $x_1 \le \theta$, $\forall\, i$ is given by $\max\{x_1, \ldots, x_n\} \le \theta$. Applying the factorization theorem to

$$f(\mathbf{x}|\,\theta) = \theta^{-1} \mathbf{1}_{\max\{x_1,\ldots,x_n\} \le \theta}(x)$$

provides evidence that $T = \max\{\mathcal{X}_1, \ldots, \mathcal{X}_n\}$ is a sufficient statistic. It is instructive to demonstrate that the sample mean $m = \sum_{i=1}^n \mathcal{X}_i/n$ is not a sufficient statistic here, because it is impossible to write $\mathbf{1}_{\max\{x_1,\ldots,x_n\} \le \theta}(x)$ as a function of m and θ alone.

When estimating several independent parameters $\boldsymbol{\theta} = (\theta_1, \theta_2, \ldots)$, more statistics are required:

$$T_i = \varrho_i(\mathcal{X}_1, \ldots, \mathcal{X}_n), \quad i = 1, 2, \ldots, k,$$

and the condition for jointly sufficient statistics is

$$f(\mathbf{x}|\boldsymbol{\theta}) = u(\mathbf{x}) v\big(\varrho_1(\mathbf{x}), \ldots, \varrho_k(\mathbf{x}), \boldsymbol{\theta}\big) . \tag{2.147}$$

As before, u and v are non-negative functions and u may depend on the full random sample, but not on the parameters $\boldsymbol{\theta}$ that are to be estimated, whereas v may depend on $\boldsymbol{\theta}$, but the dependence on the sample $\mathbf{x}$ is restricted to the values of the statistics T_k.

On the basis of this generalization, it is straightforward to show that, for normally distributed random variables with unknown expectation value and variance $\boldsymbol{\theta}(\mu, \sigma^2)$, two jointly sufficient statistics are $T_1 = \sum_{i=1}^n \mathcal{X}_i$ and $T_2 = \sum_{i=1}^n \mathcal{X}_i^2$. Not surprisingly, another set of jointly sufficient statistics is the sample mean $m(\boldsymbol{\mathcal{X}}) = \sum_{i=1}^n \mathcal{X}_i/n$, and the sample variance $\widetilde{\text{var}}(\boldsymbol{\mathcal{X}}) = \sum_{i=1}^n (\mathcal{X}_i - m)^2/(n-1)$.

Examples of Maximum Likelihood

Two well known examples are presented here for the purpose of illustration. The first case deals with the arrival of phone calls in a call center with n operators at the switchboards. The n lines have the same average utilization and the arrival of calls follows a Poisson density $f(k_i|\alpha) = \pi_{k_i}(\alpha) = \alpha^{k_i} \mathrm{e}^{-\alpha}/k_i!$ $(i = 1, \ldots, n)$, where k_i is the number of phone calls put through to the switchboard of operator i and α is the unknown parameter which we want to determine by means of maximum likelihood. The likelihood function takes the form

$$\mathcal{L}(\alpha) = \prod_{i=1}^n \frac{\alpha^{k_i}}{k_i!} \mathrm{e}^{-\alpha} = \frac{\alpha^{nm}}{k_1! \cdots k_n!} \mathrm{e}^{-n\alpha}, \qquad m = \frac{1}{n} \sum_{i=1}^n k_i .$$

Calculating the logarithm and the extreme values of log-likelihood by differentiating and equating the result with zero yields

$$\ln \mathcal{L}(\alpha) = nm \ln \alpha - n\alpha - \ln(k_1! \cdots k_n!) ,$$

$$\frac{\mathrm{d}}{\mathrm{d}\alpha} \ln \mathcal{L}(\alpha) = n \left(\frac{m}{\alpha} - 1 \right) = 0 ,$$

$$\hat{\alpha}_{\text{mle}} = m = \frac{1}{n} \sum_{i=1}^n k_i .$$

By taking the second derivative, it is easy to check that the extremum is indeed a maximum. The maximum likelihood estimator of the parameter of the Poisson distribution is simply the sample mean of the incoming calls taken over all operators.

The second example concerns a set of n independent and identically distributed normal variables with unknown expectation value and variance $\boldsymbol{\theta} = (\mu, \sigma)$:

$$f(\mathbf{x}|\mu,\sigma) = \prod_{i=1}^{n} f(x_i|\mu,\sigma) = \left(\frac{1}{\sigma\sqrt{2\pi}}\right)^n \exp\left(-\frac{\sum_{i=1}^{n}(x_i-\mu)^2}{2\sigma^2}\right)$$

$$= \left(\frac{1}{\sigma\sqrt{2\pi}}\right)^n \exp\left(-\frac{\sum_{i=1}^{n}(x_i-m)^2 + n(m-\mu)^2}{2\sigma^2}\right) ,$$

where $m = \sum_{i=1}^{n} x_i/n$ is the sample mean.[36] The log-likelihood function

$$\ln \mathcal{L}(\mu,\sigma) = -\frac{n}{2}\ln(2\pi\sigma^2) - \frac{1}{2\sigma^2}\left(\sum_{i=1}^{n}(x_i-m)^2 + n(m-\mu)^2\right) ,$$

is searched for the existence of maximum values, which are determined by setting the two partial derivatives equal to zero:

$$\frac{\partial}{\partial\mu}\ln \mathcal{L}(\mu,\sigma) = \frac{2n(m-\mu)}{2\sigma^2} = 0 \implies \hat{\mu}_{\text{mle}} = m ,$$

$$\frac{\partial}{\partial\sigma}\ln \mathcal{L}(\mu,\sigma) = -\frac{n}{\sigma} + \frac{\sum_{i=1}^{n}(x_i-m)^2 + n(m-\mu)^2}{\sigma^3} = 0$$

$$\implies \hat{\sigma}^2 = \frac{1}{n}\sum_{i=1}^{n}(x_i-\mu)^2 .$$

In this particular case we were able to obtain the two estimators individually, but in general the results will be the solution of a system of two equations in two variables.

Considering the two maximum likelihood estimators of the normal distribution in detail, we see in the first case that the expectation value of the estimator $\hat{\mu}$ coincides with the parameter μ, viz., $\text{E}(\hat{\mu}) = \mu$, whence the maximum likelihood estimator in unbiased.

[36]The equivalence $\sum_{i=1}^{n}(x_i-\mu)^2 = \sum_{i=1}^{n}(x_i-m)^2 + n(m-\mu)^2$ is easy to check using the definition of the sample mean $m = \sum_{i=1}^{n} x_i/n$. We use it here because the dependence on the unknown parameter μ is reduced to a single term.

Insertion of the estimator $\hat{\mu}$ for the parameter value μ into the equation for $\hat{\sigma}^2$ yields

$$\hat{\sigma}^2 = \frac{1}{n}\sum_{i=1}^{n}(x_i - m)^2 = \frac{1}{n}\sum_{i=1}^{n} x_i^2 - \frac{1}{n^2}\sum_{i=1}^{n}\sum_{j=1}^{n} x_i x_j \, .$$

The introduction of new variables $\xi_i = x_i - \mu$ with zero expectation yields

$$\hat{\sigma}^2 = \frac{1}{n}\sum_{i=1}^{n}(\xi_i - \mu)^2 - \frac{1}{n^2}\sum_{i=1}^{n}\sum_{j=1}^{n}(\xi_i - \mu)(\xi_j - \mu) \, .$$

We now construct the expectation value using $\mathrm{E}(\xi) = 0$ and $\mathrm{E}(\xi^2) = \sigma^2$, perform some calculations similar to the derivation of (2.118), and find

$$\mathrm{E}(\hat{\sigma}^2) = \frac{n-1}{n}\sigma^2 \, .$$

Accordingly, the estimator $\hat{\sigma}^2$ is biased. As expected from consistency, we derived exactly the same results for the estimators of $\boldsymbol{\theta} = (\mu, \sigma)$ with the maximum likelihood method as we got from direct calculations of the expectation values (Sect. 2.6.1).

Finally, we mention without going into details that the normal log-likelihood at the maximum and the information entropy of the distribution are closely related functions of the variance σ^2, viz.,

$$\log \mathcal{L}(\hat{\mu}, \hat{\sigma}) = -\frac{n}{2}\big(\log(2\pi\hat{\sigma}^2) + 1)\big] \, , \qquad H\big(\mathcal{N}(\mu, \sigma)\big) = \frac{1}{2}\big(\log(2\pi\sigma^2) + 1\big) \, ,$$

and independent of the expectation value μ (Table 2.1).

2.6.5 *Bayesian Inference*

Finally, we sketch the most popular example of a theory based on *evidential probabilities*: Bayesian statistics, named after the eighteenth century English mathematician and Presbyterian minister Reverend Thomas Bayes.[37] Bayesian statistics has become popular in disciplines where model building is a major issue. Examples from biology include among others bioinformatics, molecular genetics, modeling of ecosystems, and forensics. In contrast to the frequentists' view, probabilities are

[37] Bayesian statistics is described in many monographs, for example, in references [92, 199, 281, 333]. As a brief introduction to Bayesian statistics, we recommend [510].

subjective and exist only in the human mind. From a practitioner's point of view, the major advantage of the Bayesian approach is a direct insight into the process of improving our knowledge of the subject of investigation.

The starting point of the Bayesian approach is the conditional probability

$$P(A|B) = \frac{P(AB)}{P(B)} , \tag{2.148}$$

which is the probability of simultaneous occurrence of events A and B divided by the probability of the occurrence of B alone. Conditional probabilities can be inverted straightforwardly in the sense that we can ask about the probability of B under the condition that event A has occurred:

$$P(B|A) = \frac{P(AB)}{P(A)} , \quad \text{since } P(AB) = P(BA) , \tag{2.148$'$}$$

which implies $P(A|B) \neq P(B|A)$ unless $P(A) = P(B)$. In other words $P(A|B)$ and $P(B|A)$ are on an equal footing in probability theory. Calculating $P(AB)$ from the two equations (2.148) and (2.148′) and setting the expressions equal yields

$$P(A|B)P(B) = P(AB) = P(B|A)P(A) \implies P(B|A) = P(A|B)\frac{P(B)}{P(A)} ,$$

which is already Bayes' theorem when properly interpreted.

Bayes' theorem provides a straightforward interpretation of conditional probabilities and their inversion in terms of models or hypotheses (H) and data (E). The conditional probability $P(E|H)$ corresponds to the conventional procedure in science: given a set of hypotheses cast into a model H, the task is to calculate the probabilities of the different outcomes E. In physics and chemistry, where we are dealing with well established theories and models, this is, in essence, the common situation. For a given model and a set of measured data the unknown parameters are calculated by means of a fitting technique, for example by the maximum-likelihood method (Sect. 2.6.4). Biology, economics, the social sciences, and other disciplines, however, are often confronted with situations where no confirmed models exist and then we want to test and improve the probability of a model. We need to invert the conditional probability since we are interested in testing the model in the light of the available data. In other words, the conditional probability $P(H|E)$ becomes important: what is the probability that a hypothesis H is justified given a set of measured data encapsulated in the evidence E? The Bayesian approach casts (2.148) and (2.148′) into Bayes' theorem,

$$P(H|E) = P(E|H)\frac{P(H)}{P(E)} = \frac{P(E|H)}{P(E)}P(H) , \tag{2.149}$$

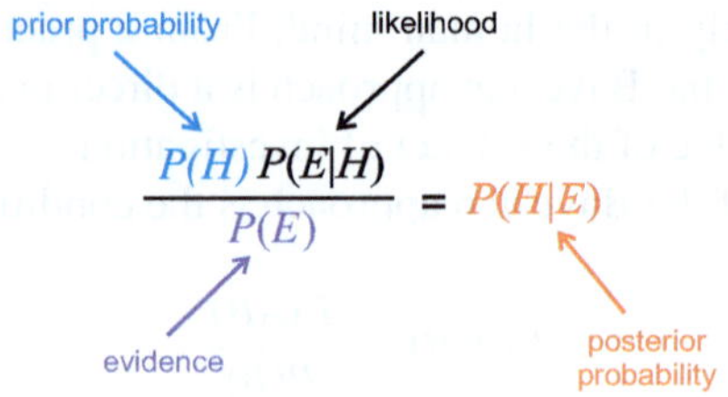

Fig. 2.28 A sketch of the Bayesian method. Prior information about probabilities is confronted with empirical data and converted by means of Bayes' theorem into a new distribution of probabilities called the *posterior probability* [120, 507]

and provides a hint on how to proceed, at least in principle (Fig. 2.28). A *prior probability* in the form of a hypothesis $P(H)$ is converted into evidence according to the likelihood principle $P(E|H)$. The basis of the prior is understood as *a priori* knowledge and comes form many sources: theory, previous experiments, gut feeling, etc. New empirical information is incorporated in the *inverse probability* computation from data to model, $P(H|E)$, thereby yielding the improved *posterior probability*. The advantage of the Bayesian approach is that a change of opinion in the light of new data is *part of the game*, so to speak. In general, parameters are input quantities of frequentist statistics and if unknown they are assumed to be available through repetition of experiments, whereas they are random variables in the Bayesian approach.

There is an interesting relation between the maximum likelihood estimation (Sect. 2.6.4) and the Bayesian approach that becomes evident when we rewrite Bayes theorem:

$$P(\boldsymbol{\theta}|\mathbf{x}) = \frac{f(\mathbf{x}|\boldsymbol{\theta})P(\boldsymbol{\theta})}{P(\mathbf{x})}\,, \tag{2.149$'$}$$

where $P(\mathbf{x})$ is the probability of the data set averaged over all parameters and $P(\boldsymbol{\theta})$ is the prior distribution of the parameters $\boldsymbol{\theta}$. The Bayesian estimator is obtained by maximizing the product $f(\mathbf{x}|\boldsymbol{\theta})P(\boldsymbol{\theta})$. For a uniform prior $P(\boldsymbol{\theta}) = \mathcal{U}(\boldsymbol{\theta})$, the Bayesian estimator is calculated from the maximum of $f(\mathbf{x}|\boldsymbol{\theta})$ and coincides with the maximum likelihood estimator.

In practice, direct application of the Bayesian theorem involves quite elaborate computations which were not possible for real world examples before the advent of electronic computers. Here we present a simple example of Bayesian statistics [120] which has been adapted from the original work of Thomas Bayes in the posthumous publication of 1763 [459]. It is called *table game* and allows for analytical calculations. Table game is played by two people, Alice (A) and Bob (B), along with a third person (C) who acts as game master and remains *neutral*. A (pseudo)random number generator is used to draw pseudorandom numbers from a uniform distribution in the range $0 \le \mathcal{R} < 1$. The pseudorandom number generator is operated by the game master and cannot be seen by the two players. In essence,

A and B are completely passive, they have no information about the game except knowledge of the basic setup of the game and the scores, which are $a(t)$ for A and $b(t)$ for B. The person who first reaches the predefined score value z has won. This simple game starts with the drawing of a pseudorandom number $\mathcal{R} = r_0$ by the game master. Consecutive drawings yielding numbers r_i assign points to A iff $0 \leq r_i < r_0$ is satisfied and B iff $r_0 \leq r_i < 1$. The game is continued until one person, A or B, reaches the final score z.

The problem is to compute fair odds of winning for A and B when the game is terminated prematurely and r_0 is unknown. Let us assume that the scores at the time of termination were $a(t) = a$ and $b(t) = b$ with $a < z$ and $b < z$, and to make the calculations easy, assume also that A is only one point away from winning so $a = z-1$ and $b < z-1$. If the parameter r_0 were known, the answer would be trivial. In the conventional approach we would make an assumption about the parameter r_0. Without further knowledge, we could make the null hypothesis $r_0 = \hat{r}_0 = 1/2$, and find simply

$$P_0(B) = P(B \text{ wins}) = (1-\hat{r}_0)^{z-b} = \left(\frac{1}{2}\right)^{z-b},$$

$$P_0(A) = P(A \text{ wins}) = 1-(1-\hat{r}_0)^{z-b} = 1-\left(\frac{1}{2}\right)^{z-b},$$

because the only way for B to win is to make $z-b$ scores in a row. Thus fair odds for A to win would be $(2^{z-b}-1) : 1$. An alternative approach is to make the *maximum likelihood* estimate of the unknown parameter $r_0 = \tilde{r}_0 = a/(a+b)$. Once again, we calculate the probabilities and find by the same token

$$P_{\text{ml}}(B) = P(B \text{ wins}) = (1-\tilde{r}_0)^{z-b} = \left(\frac{b}{a+b}\right)^{z-b},$$

$$P_{\text{ml}}(A) = P(A \text{ wins}) = 1-(1-\tilde{r}_0)^{z-b} = 1-\left(\frac{b}{a+b}\right)^{z-b},$$

while the odds in favor of A are

$$\left(\frac{a+b}{b}\right)^{z-b} - 1 .$$

The Bayesian solution considers $r_0 = p$ as an unknown but variable parameter about which no estimate is made. Instead the uncertainty is modeled rigorously by integrating over all possible values $0 \leq p \leq 1$. The expected probability for B to win is then

$$\mathrm{E}\big(P(B)\big) = \int_0^1 (1-p)^{z-b} P(p \,|\, a, b)\, \mathrm{d}p ,$$

where $(1-p)^{z-b}$ is the probability of B winning and $P(p\,|\,a,b)$ is the probability of a certain value of p provided the data a and b are obtained at the end of the game. The probability $P(p\,|\,a,b)$, written formally as $P(\text{model}\,|\,\text{data})$, is the inversion of the common problem $P(\text{data}\,|\,\text{model})$, i.e., given a certain model, what is the probability of finding a certain set of data? This is a so-called *inverse probability* problem. The solution of the problem is provided by Bayes' theorem, which is an almost trivial truism for two random variables $\mathcal{X}$ and $\mathcal{Y}$:

$$P(\mathcal{X}|\mathcal{Y}) = \frac{P(\mathcal{Y}|\mathcal{X})P(\mathcal{X})}{P(\mathcal{Y})} = \frac{P(\mathcal{Y}|\mathcal{X})P(\mathcal{X})}{\sum_{\mathcal{Z}} P(\mathcal{Y}|\mathcal{Z})P(\mathcal{Z})}, \tag{2.149''}$$

where the sum over the random variable $\mathcal{Z}$ covers the entire sample space. Equation (2.149′) yields in our example

$$P(p\,|\,a,b) = \frac{P(a,b\,|\,p)P(p)}{\int_0^1 P(a,b\,|\varrho)P(\varrho)\mathrm{d}\varrho}.$$

The interpretation of the equation is straightforward: the probability of a particular choice of p given the data (a,b), called the *posterior probability* (Fig. 2.28), is proportional to the probability of obtaining the observed data if p is true, i.e., the *likelihood* of p, multiplied by the *prior probability* of this particular value of p relative to all other values of p. The integral in the denominator takes care of the normalization of the probability and the summation is replaced by an integral, because p is a continuous variable, and $0 \le p \le 1$ in the entire domain of p.

The likelihood term is readily calculated from the binomial distribution

$$P(a,b\,|\,p) = \binom{a+b}{b} p^a(1-p)^b,$$

but the prior probability requires more care. By definition $P(p)$ is the probability of p before the data have been recorded. How can we estimate p before we have seen any data? We thus turn to the question of how r_0 is determined. We know it has been picked from the uniform distribution, so $P(p)$ is a constant that appears in the numerator and in the denominator and thus cancels in the equation (2.149′) for Bayes' theorem. After a little algebra, we eventually obtain for the probability of B winning:

$$\mathrm{E}\big(P(B)\big) = \frac{\int_0^1 p^a(1-p)^z\,\mathrm{d}p}{\int_0^1 p^a(1-p)^b\,\mathrm{d}p}.$$

Integration is straightforward, because the integrals are known as Euler integrals of the first kind, which have the beta-function as solution:

$$B(x,y) = \int_0^1 z^{x-1}(1-z)^{y-1}\,\mathrm{d}z = \frac{(x-1)!\,(y-1)!}{(x+y-1)!} = \frac{\Gamma(x)\Gamma(y)}{\Gamma(x+y)}\,. \tag{2.150}$$

Finally, we obtain the following expression for the probability of B winning:

$$\mathrm{E}\big(P(B)\big) = \frac{z!\,(a+b+1)!}{b!\,(a+z+1)!}\,,$$

while the Bayesian estimate for fair odds yields

$$\left(\frac{b!\,(a+z+1)!}{z!\,(a+b+1)!} - 1\right) : 1\,.$$

A specific numerical example is given in [120]: $a = 5$, $b = 3$, and $z = 6$. The null hypothesis of equal probabilities of winning for A and B, viz., $\hat{r}_0 = 0.5$, yields an advantage of 7:1 for A, the maximum likelihood approach with $\tilde{r}_0 = a/(a+b) = 5/8$ yields $\approx$ 18:1, and the Bayesian estimate yields 10:1. The large differences should not be surprising since the sample size is very small. The correct answer for the table game with these values for a, b, and z is indeed 10:1 as can be easily checked by numerical computation with a small computer program.

Finally, we show how the Bayesian approach operates on probability distributions (a simple but straightforward description can be found in [507]). According to (2.149″), the posterior probability $P(\mathcal{X}|\mathcal{Y})$ is obtained through multiplication of the prior probability $P(\mathcal{X})$ by the data likelihood function $P(\mathcal{Y}|\mathcal{X})$ and normalization. We illustrate the relation between the probability function by means of two normal distributions and their product (Fig. 2.29). For the prior probability and the data function, we assume

$$P(\mathcal{X}) = f_1(x) = \frac{1}{\sqrt{2\pi\sigma_1^2}}\mathrm{e}^{-(x-\mu_1)^2/2\sigma_1^2}\,,$$

$$P(\mathcal{X}|\mathcal{Y}) = f_2(x) = \frac{1}{\sqrt{2\pi\sigma_2^2}}\mathrm{e}^{-(x-\mu_2)^2/2\sigma_2^2}\,,$$

and obtain for the product, with normalization factor $\mathcal{N} = \mathcal{N}(\mu_1,\mu_2,\sigma_1,\sigma_2)$,

$$P(\mathcal{Y}|\mathcal{X}) = \mathcal{N} f_1(x) f_2(x) = \mathcal{N}\, g\, \mathrm{e}^{-(x-\bar{\mu})^2/2\bar{\sigma}^2}\,,$$

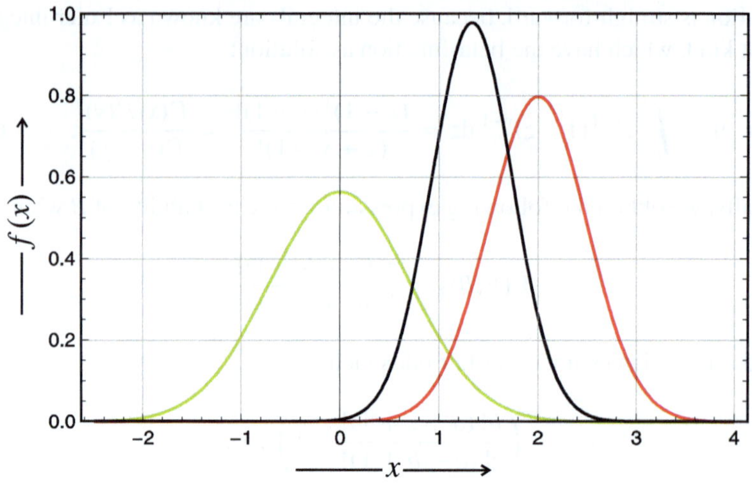

Fig. 2.29 The Bayesian method of inference. The figure outlines the Bayesian method by means of normal density functions. The sample data are given in the form of the likelihood function $P(\mathcal{Y}|\mathcal{X}) = \mathcal{N}(2, 1/2)$ (*red*) and additional external information on the parameters enters the analysis as the prior distribution $P(\mathcal{X}) = \mathcal{N}(0, 1/\sqrt{2})$ (*green*). The resulting posterior distribution $P(\mathcal{X}|\mathcal{Y}) = P(\mathcal{Y}|\mathcal{X})P(\mathcal{X})/P(\mathcal{Y})$ (*black*) is once again a normal distribution with mean $\bar{\mu} = (\mu_1\sigma_2^2 + \mu_2\sigma_1^2)/(\sigma_1^2 + \sigma_2^2)$ and variance $\bar{\sigma}^2 = \sigma_1^2\sigma_2^2/(\sigma_1^2 + \sigma_2^2)$. It is straightforward to show that the mean $\bar{\mu}$ lies between μ_1 and μ_2 and the variance has become smaller $\bar{\sigma} \leq \min(\sigma_1, \sigma_2)$ (see text)

with

$$\bar{\mu} = \frac{\mu_1\sigma_2^2 + \mu_2\sigma_1^2}{\sigma_1^2 + \sigma_2^2}\,, \quad \bar{\sigma}^2 = \frac{\sigma_1^2\sigma_2^2}{\sigma_1^2 + \sigma_2^2}\,, \quad g = \frac{1}{2\pi\sigma_1\sigma_2}\exp\left(-\frac{1}{2}\frac{(\mu_2-\mu_1)^2}{\sigma_1^2+\sigma_2^2}\right)\,,$$

and

$$\mathcal{N}g = \frac{\sqrt{\sigma_1^2+\sigma_2^2}}{\sqrt{2\pi}\sigma_1\sigma_2} = \frac{1}{\sqrt{2\pi\bar{\sigma}^2}}\,,$$

as required for normalization of the Gaussian curve.

Two properties of the posterior probability are easily tested by means of our example: (i) the averaged mean $\bar{\mu}$ lies always between μ_1 and μ_2, and (ii) the product distribution is sharper than the two factor distributions

$$\frac{\sigma_1^2\sigma_2^2}{\sigma_1^2+\sigma_2^2} \leq \min\{\sigma_1^2, \sigma_2^2\}\,,$$

with the equals sign requiring either $\sigma_1 = 0$ or $\sigma_2 = 0$. The improvement due to the Bayesian analysis thus reduces the difference in the mean values between expectation and model, and the distribution becomes narrower in the sense of reduced uncertainty.

Whereas the Bayesian approach does not seem to provide a lot more information in situations where the models are confirmed by many other independent applications, as, for example, in the majority of problems in physics and chemistry, the highly complex situations in modern biology, economics, or the social sciences require highly simplified and flexible models, and there is ample room for application of Bayesian statistics.

Chapter 3
Stochastic Processes

With four parameters I can fit an elephant
and with five I can make him wiggle his trunk.
Enrico Fermi quoting John von Neumann 1953 [119].

Abstract Stochastic processes are defined and grouped into different classes, their basic properties are listed and compared. The Chapman–Kolmogorov equation is introduced, transformed into a differential version, and used to classify the three major types of processes: (i) drift and (ii) diffusion with continuous sample paths, and (iii) jump processes which are essentially discontinuous. In pure form these prototypes are described by Liouville equations, stochastic diffusion equations, and master equations, respectively. The most popular and most frequently used continuous equation is the Fokker–Planck (FP) equation, which describes the evolution of a probability density by drift and diffusion. The pendant to FP equations on the discontinuous side are master equations which deal only with jump processes and represent the appropriate tool for modeling processes described by discrete variables. For technical reasons they are often difficult to handle unless population sizes are relatively small. Particular emphasis is laid on modeling conventional and anomalous diffusion processes. Stochastic differential equations (SDEs) model processes at the level of random variables by solving ordinary differential equations upon which a diffusion process, called a Wiener process, is superimposed. Ensembles of individual trajectories of SDEs are equivalent to time dependent probability densities described by Fokker–Planck equations.

Stochastic processes introduce time into probability theory and represent the most prominent way to combine dynamical phenomena and randomness resulting from incomplete information. In physics and chemistry the dominant source of randomness is thermal motion at the microscopic level, but in biology the overwhelming complexity of systems is commonly prohibitive for a complete description and then

P. Schuster, *Stochasticity in Processes*, Springer Series in Synergetics,
DOI 10.1007/978-3-319-39502-9_3

lack of information results also from unavoidable simplifications in the macroscopic model. In essence, there are two ways of dealing with stochasticity in processes:

(i) calculation or recording of stochastic variables as functions of time,
(ii) modeling of the temporal evolution of entire probability densities.

In the first case one particular computation or one experiment yields a single sample path or *trajectory*, and full information about the process is obtained by sampling trajectories from repetitions under identical conditions.[1] Sampling of trajectories leads to bundles of curves which can be evaluated in the spirit of mathematical statistics (Sect. 2.6) to yield time-dependent moments of time-dependent probability densities. For an illustrative example comparing superposition of trajectories and migration of the probability density, we refer to the Ornstein–Uhlenbeck process shown in Figs. 3.9 and 3.10.

For *linear* processes, the expectation value $\mathrm{E}\big(\mathcal{X}(t)\big)$ of the random variable as a function of time coincides with the deterministic solution $x(t)$ of the corresponding differential equation (Sect. 3.2.3). This is not the case in general, but the differences $\big|\mathrm{E}\big(\mathcal{X}(t)\big) - x(t)\big|$ will commonly be small unless we are dealing with very small numbers of molecules. For single-point initial conditions, the solution curves of ordinary or partial differential equations (ODEs or PDEs) consists of single trajectories as determined by the theorems of existence and uniqueness of solutions. As mentioned above, solutions of stochastic processes correspond to bundles of trajectories which differ in the sequence of random events and which as a rule surround the deterministic solution. Commonly, sharp initial conditions are chosen, and then the bundle of trajectories starts at a single point and diverges into the future as well as into the past, depending on whether the process is studied in the *forward* or the *backward* direction (see Fig. 3.21). The stochastic equations describing processes in the forward direction are different from those modeling backward processes. The typical symmetry of differential equations with respect to time reversal does not hold for stochastic processes, and the reason for symmetry breaking is the presence of a diffusion term [10, 135, 500]. Considering processes in the forward direction with sharp initial conditions, the variance $\mathrm{var}\big(\mathcal{X}(t)\big)$ increases with time and provides the basis for a useful distinction between different types of stochastic processes. In processes of type (i), the variance grows without limits. Clearly, such processes are idealizations and cannot occur in a finite world but they provide important insights into enhancement of fluctuations. Examples of type (i) processes are unlimited spatial diffusion and unlimited growth of biological populations. Type (ii) processes are confined by boundaries and can take place in reality. After some initial growth,

[1] Identical conditions means that all parameters are the same except for the random fluctuations. In computer simulations this is achieved by keeping everything precisely the same except the seeds for the pseudorandom number generator.

the variance settles down at some finite value. For the majority of such bounded processes, the long-time limit corresponds to a thermodynamic equilibrium state or a stationary state where the standard deviations satisfy an approximate $\sqrt{N}$-law. Type (iii) processes exhibit complex long-time behavior corresponding to oscillations or deterministic chaos in the deterministic system.

Figure 3.1 presents an overview of the most frequently used general model equations for stochastic processes,[2] which are introduced in this chapter, and it shows how they are interrelated [535, 536]. Two classes of equations are of central importance:

(i) the differential form of the Chapman–Kolmogorov equation (dCKE, see Sect. 3.2) describing the evolution of probability densities,
(ii) the stochastic differential equation (SDE, see Sect. 3.4) modeling stochastic trajectories.

The Fokker–Planck equation, named after the Dutch Physicist Adriaan Fokker and the German physicist Max Planck, and the master equation are derived from the differential Chapman–Kolmogorov equation by restriction to continuous processes or jump processes, respectively. The *chemical master equation* is a master equation adapted for modeling chemical reaction networks, where the jumps are changes in the integer particle numbers of chemical species (Sect. 4.2.2).

In this chapter we shall present a brief introduction to stochastic processes and the general formalisms for modeling them. The chapter is essentially based on three textbooks [91, 194, 543], and it uses in essence the notation introduced by Crispin Gardiner [193]. A few examples of stochastic processes of general importance will be discussed here in order to illustrate how the formalisms are used. In particular, we shall focus on random walks and diffusion. Other applications are presented in Chaps. 4 and 5. Mathematical analysis of stochastic processes is complemented by numerical simulations [213]. These have become more and more important over the years, essentially for two reasons:

(i) the accessibility of cheap and extensive computing power,
(ii) the need for stochastic treatment of complex reaction kinetics in chemistry and biology, in situations that escape analytical methods.

Numerical simulation methods will be presented in detail and applied in Chap. 4 (Sect. 4.6).

[2]By *general* we mean here methods that are widely applicable and not tailored specifically for deriving stochastic solutions for a single case or a small number of cases.

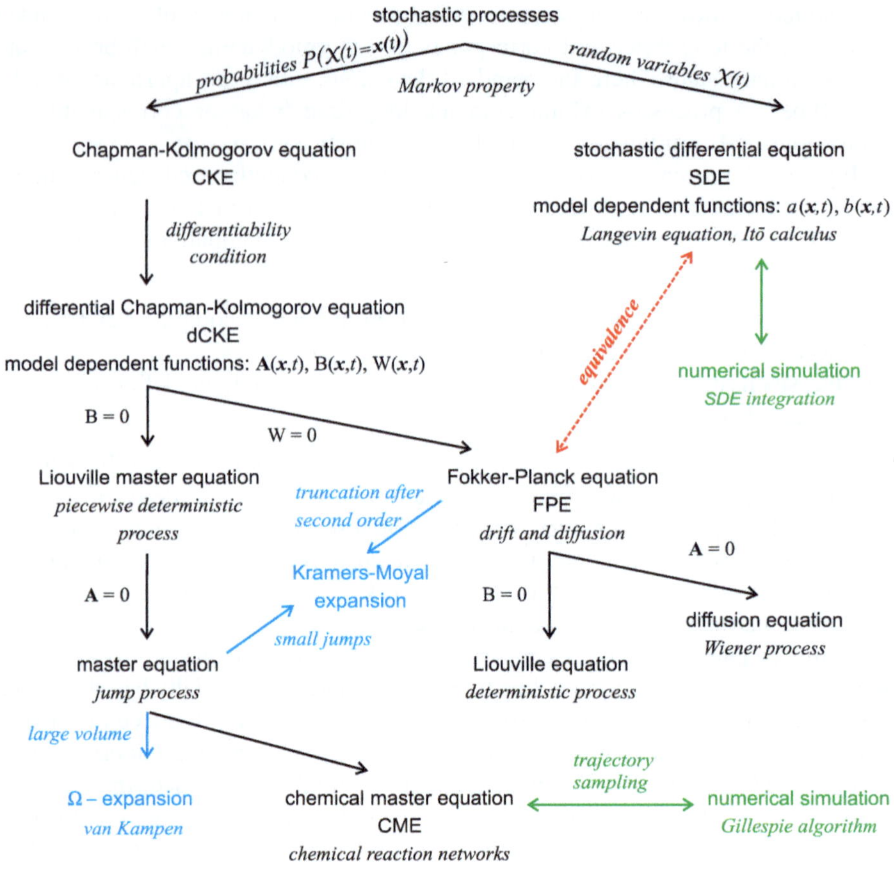

Fig. 3.1 Description of stochastic processes. The sketch presents a family tree of stochastic models [535]. Almost all stochastic models used in science are based on the Markov property of processes, which, in a nutshell, states that full information on the system at present is sufficient for predicting the future or past (Sect. 3.1.3). Models fall into two major classes depending on the objects they are dealing with: (1) random variables $\mathcal{X}(t)$ or (2) probability densities $P\big(\mathcal{X}(t) = x\big)$. In the center of stochastic modeling stands the Chapman–Kolmogorov equation (CKE), which introduces the Markov property into time series of probability densities. In differential form CKE contains three model dependent functions, viz., the vector $\mathbf{A}(\mathbf{x}, t)$ and the matrices $\mathrm{B}(\mathbf{x}, t)$ and $\mathrm{W}(\mathbf{x}, t)$, which determine the nature of the stochastic process. Different combinations of these functions yield the most important equations for stochastic modeling: the Fokker–Planck equation with $\mathrm{W} = 0$ ($\mathbf{A} \neq 0$ and $\mathrm{B} \neq 0$), the stochastic diffusion equation with $\mathrm{B} \neq 0$ ($\mathbf{A} = 0$ and $\mathrm{W} = 0$), and the master equation with $\mathrm{W} \neq 0$ ($\mathbf{A} = 0$ and $\mathrm{B} = 0$). For stochastic processes without jumps the solutions of the stochastic differential equation are trajectories, which when properly sampled describe the evolution of a probability density $P\big(\mathcal{X}(t) = x(t)\big)$ that is equivalent to the solution of a Fokker–Planck equation (*red arrow*). Common approximations by means of size expansions are shown in *blue*. *Green arrows* indicate where conventional numerical integration and simulation methods come into play. Adapted from [535, p. 252]

3.1 Modeling Stochastic Processes

The use of conventional differential equations for modeling dynamical systems implies determinism in the sense that full information about the system at a single time t_0, for example, is sufficient for exact computation of both future and past. In reality we encounter substantial limitations concerning prediction and reconstruction, especially in the case of deterministic chaos, because initial and boundary conditions are available only with finite accuracy, and even the smallest errors are amplified to arbitrary size after sufficiently long times. The theory of *stochastic processes* provides the tools for taking into account all possible sources of uncontrollable irregularities, and defines in a natural way the limits for predictions of the future as well as for reconstruction of the past. Different stochastic processes can be classified with respect to *memory effects*, making precise how the past acts on the future. Almost all stochastic models in science fall into the very broad class of *Markov processes*, named after the Russian mathematician Andrey Markov,[3] and which are characterized by lack of memory, in the sense that the future can be modeled and predicted probabilistically from knowledge of the present, and no information about historical events is required.

3.1.1 Trajectories and Processes

The probabilistic evolution of a system in time is described as a general stochastic process. We assume the existence of a time dependent random variable $\mathcal{X}(t)$ or random vector $\boldsymbol{\mathcal{X}}(t) = \big(\mathcal{X}_k(t); k = 1,\ldots,M; k \in \mathbb{N}_{>0}\big)$.[4] The random variable $\mathcal{X}$ and also the time t can be discrete or continuous, giving rise to four classes of stochastic models (Table 3.1). At first we shall assume discrete time because this case is easier to visualize, and as in the previous chapters, we shall distinguish the simpler case of discrete random variables, viz.,

$$P_n(t) = P\big(\mathcal{X}(t) = x_n\big) \,, \qquad n \in \mathbb{N} \,, \tag{3.1}$$

from the continuous or probability density case,

$$\mathrm{d}F(x,t) = f(x,t)\,\mathrm{d}x = P\big(x \le \mathcal{X}(t) \le x + \mathrm{d}x\big) \,, \qquad x \in \mathbb{R} \,. \tag{3.2}$$

---

[3]The Russian mathematician Andrey Markov (1856–1922) was one of the founders of Russian probability theory and pioneered the concept of memory-free processes, which are named after him. He expressed more precisely the assumptions that were made by Albert Einstein [133] and Marian von Smoluchowski [559] in their derivation of the diffusion process.

[4]For the moment we need not specify whether $\mathcal{X}(t)$ is a simple random variable or a random vector $\boldsymbol{\mathcal{X}}(t) = \big(\mathcal{X}_k(t); k = 1,\ldots,M\big)$, so we drop the index k determining the individual component. Later on, for example in chemical kinetics where the distinction between different (chemical) species becomes necessary, we shall make clear the sense in which $\mathcal{X}(t)$ is used, i.e., random variable or random vector.

Table 3.1 Notation used to model stochastic processes

Variables $\mathcal{X}$	Discrete time t	Continuous time t
Discrete	$P_{n,k} = P(\mathcal{X}_k = x_n)\,,\ \ k, n \in \mathbb{N}$	$P_n(t) = \big(\mathcal{X}(t) = x_n\big)\,,\ \ n \in \mathbb{N}\,,\ t \in \mathbb{R}$
Continuous	$p_k(x)\,\mathrm{d}x = P(x \le \mathcal{X}_k \le x + \mathrm{d}x) = f_k(x)\,\mathrm{d}x = \mathrm{d}F_k(x),\ \ k \in \mathbb{N},\ x \in \mathbb{R}$	$p(x,t)\,\mathrm{d}x = P(x \le \mathcal{X}_k \le x + \mathrm{d}x) = f(x,t)\,\mathrm{d}x = \mathrm{d}F(x,t),\ \ x, t \in \mathbb{R}$

Comparison between four different approaches to modeling stochastic processes by means of probability densities: (i) discrete values of the random variable $\mathcal{X}$ and discrete time, (ii) discrete values and continuous time, (iii) continuous values and discrete time, and (iv) continuous values and continuous time

A particular series of events—be it the result of a calculation or an experiment—constitutes a *sample path* or a *trajectory* in phase space.[5] The trajectory is a listing of the values of the random variable $\mathcal{X}(t)$ recorded at certain times and arranged in the form of pairs (x_i, t_i):

$$\mathcal{T} = \Big((x_1, t_1), (x_2, t_2), \ldots, (x_k, t_k), (x_{k+1}, t_{k+1}), \ldots, (x_n, t_n)\Big)\,. \tag{3.3}$$

For the sake of clarity, and although it is not essential for the application of probability theory, we shall always assume that the recorded values are time ordered, here with the earliest or oldest values in the rightmost position and the most recent values at the latest entry on the left-hand side. Assuming that the recorded series has started at some time t_n in the past with x_n, we have

$$t_1 \ge t_2 \ge t_3 \ge \ldots \ge t_k \ge t_{k+1} \ge \ldots \ge t_n\,.$$

Accordingly, a trajectory is a sequence of time ordered pairs (x, t).

It is worth noting that the conventional way of counting time in physics progresses in the opposite direction from some initial time $t = t_0$ to t_1, t_2, t_3 and so on up until t_n, the most recent instant, is reached (Fig. 3.2):

$$\mathcal{T} = \Big((x_n, t_n), (x_{n-1}, t_{n-1}), \ldots, (x_k, t_k), (x_{k-1}, t_{k-1}), \ldots, (x_0, t_0)\Big)\,, \tag{3.3$'$}$$

where we adopt the same notation as in (3.3) with the changed ordering

$$t_n \ge t_{n-1} \ge \ldots \ge t_k \ge t_{k-1} \ge \ldots \ge t_0\,.$$

[5]Here we shall use the notion of *phase space* in a loose way to mean an abstract space that is sufficient for the characterization of the system and for the description of its temporal development. For example, in a reaction involving n chemical species, the phase space will be a Cartesian space spanned by n axes for n independent concentrations. In classical mechanics and in statistical mechanics, the phase space is precisely defined as a—usually Cartesian—space spanned by the $3n$ spatial coordinates and the $3n$ coordinates of the linear momenta of an n-particle system.

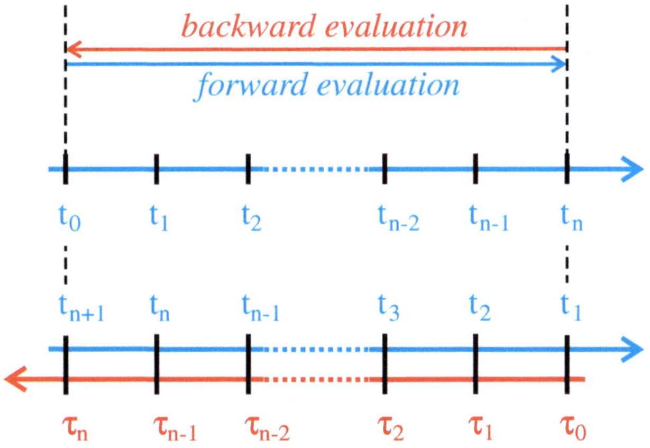

Fig. 3.2 Time order in modeling stochastic processes. Physical or real time goes from left to right and the most recent event is given by the rightmost recording. Conventional numbering of instances in physics starts at some time t_0 and ends at time t_n (*upper blue time axis*). In the theory of stochastic processes, an opposite ordering of times is often preferred, and then t_1 is the latest event of the series (*lower blue time axis*). The modeling of stochastic processes, for example by a Chapman–Kolmogorov equation, distinguishes two modes of description: (i) the forward equation, predicting the future from the past and present, and (ii) the backward equation that extrapolates back in time from present to past. Accordingly, we are dealing with two time scales, real time and computational time, which progresses in the same direction as real time in the forward evaluation (*blue*), but in the opposite direction for the backward evaluation (*red*)

In order to avoid confusion we shall always state explicitly when we are not using the convention shown in (3.3).[6]

Single trajectories are superimposed to yield bundles of trajectories in the sense of a summation of random variables, as in (1.22)[7]:

$$\begin{array}{cccc}
\mathcal{X}^{(1)}(t_0) & \mathcal{X}^{(1)}(t_1) & \dots & \mathcal{X}^{(1)}(t_n) \\
\mathcal{X}^{(2)}(t_0) & \mathcal{X}^{(2)}(t_1) & \dots & \mathcal{X}^{(2)}(t_n) \\
\vdots & \vdots & \ddots & \vdots \\
\mathcal{X}^{(N)}(t_0) & \mathcal{X}^{(N)}(t_1) & \dots & \mathcal{X}^{(N)}(t_n) \\
\hline
\mathcal{S}(t_0) & \mathcal{S}(t_1) & \dots & \mathcal{S}(t_n)
\end{array}$$

[6]The different numberings for the elements of trajectories should not be confused with forward and backward processes (Fig. 3.2), to be discussed in Sect. 3.3.

[7]In order to leave the subscript free to indicate discrete times or different chemical species, we use the somewhat clumsy superscript notation $\mathcal{X}^{(i)}$ or $x^{(i)}$ $(i = 1, \dots, N)$, to specify individual trajectories, and we use the physical numbering of times $t_0 \rightarrow t_n$.

and we obtain the summation random variable $\mathcal{S}(t)$ from the columns. The calculation of sample moments is straightforward and (2.115) and (2.118) imply the following:

$$m(t) = \tilde{\mu}(t) = \frac{1}{N}\mathcal{S}(t) = \frac{1}{N}\sum_{i=1}^{N} x^{(i)}(t) ,$$

$$\begin{aligned} m_2(t) = \widetilde{\mathrm{var}}(t) &= \frac{1}{N-1}\sum_{i=1}^{N}\left(x^{(i)}(t) - m(t)\right)^2 \\ &= \frac{1}{N-1}\left(\sum_{i=1}^{N} x^{(i)}(t)^2 - Nm(t)^2\right) . \end{aligned} \tag{3.4}$$

This is illustrated by a numerical example in Fig. 3.3.

So far almost all events and samples have been expressed as dimensionless numbers. Except in the discussion of particle numbers and concentrations, dimensions of quantities were more or less ignored and this was justified since the *scores* resulted from counting the outcomes of flipping coins or rolling dice, from counting incoming phone calls, seeds with specified colors or shapes, etc. Considering processes introduces time, and time has a dimension so we need to specify a unit in which the recorded data are measured, e.g., seconds, minutes, or hours. From now on, we shall in general specify which quantities the random variables $(\mathcal{A}, \mathcal{B}, \ldots, \mathcal{W}) \in \Omega$ describe, what exactly their realizations $(a, b, \ldots, w) \in \mathbb{R}$ are in some measurable space, and in which units they are measured. Some processes take place in three-dimensional physical space, where units for length, area, and volume are required. In applications, we shall be concerned with variables of other physical dimensions, for example, mass, viscosity, surface tension, electric charge, magnetic moments, electromagnetic radiation, etc. Wherever a quantity is introduced, we shall mention its dimension and the units commonly used in measurements.

Stochastic processes in chemistry and biology commonly model the time development of ensembles or populations. In spatially homogeneous chemical reaction systems, the variables are discrete particle numbers or continuous concentrations, $\mathcal{A}(t)$ or $a(t)$, and as a common notation we shall use $[\mathsf{A(t)}]$ and omit (t), whenever no misunderstanding is possible. Spatial heterogeneity, for example, is accounted for by explicit consideration of diffusion, and this leads to reaction–diffusion systems, where the solutions can be visualized as migrations of evolving probability densities in time and in three-dimensional space. Then, the variables are functions $\mathcal{A}(\mathbf{r}, t)$ or $a(\mathbf{r}, t)$ in 3D space and time, with $\mathbf{r} = (x, y, z) \in \mathbb{R}^3$ a vector in space. In biology, the variables are often numbers of individuals in populations, and then they depend on time, or in chemistry, on time and three-dimensional space when migration processes are considered. Sometimes it is an advantage to consider stochastic processes in formal spaces like the genotype or sequence space, which is a discrete

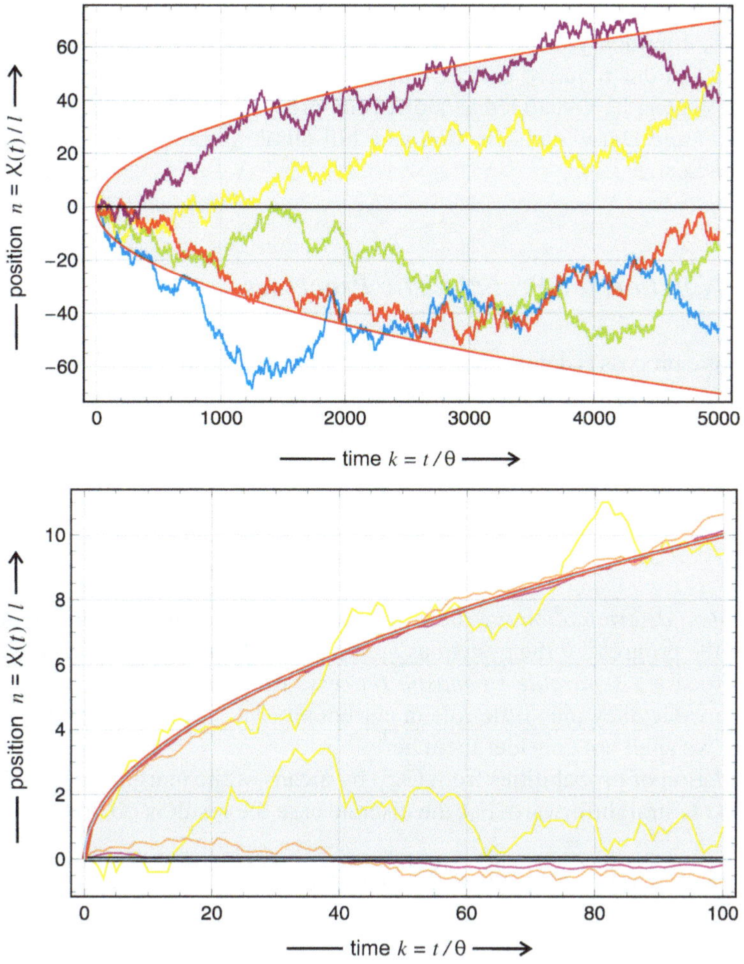

Fig. 3.3 The discrete time one-dimensional random walk. The random walk in one dimension on an infinite line $x \in \mathbb{R}$ is shown as an example of a martingale. The *upper part* shows five trajectories $\mathcal{X}(t)$ which were calculated with different seeds for the random number generator. The expectation value $\mathrm{E}\big(\mathcal{X}(t)\big) = x_0 = 0$ is constant (*black line*), the variance grows linearly with time $\mathrm{var}\big(\mathcal{X}(t)\big) = k = t/\theta$, and the standard deviation is $\sigma\big(\mathcal{X}(t)\big) = \sqrt{k}$. The *two red lines* correspond to the one standard deviation band $\mathrm{E}(t) \pm \sigma(t)$, while the *gray area* represents the confidence interval of 68.2 %. Choice of parameters: $\theta^{-1} = 1\,[\mathrm{t}]\ (= 2\vartheta)$; $l = 1\,[\mathrm{l}]$. Random number generator: Mersenne Twister with seeds: 491 (*yellow*), 919 (*blue*), 023 (*green*), 877 (*red*), 127 (*violet*). The *lower part* of the figure shows the convergence of the sample mean and the sample standard deviation according to (3.4) with increasing number N of sampled trajectories: $N = 10$ (*yellow*), 100 (*orange*), 1000 (*purple*), and 10^6 (*red and black*). The last curve is almost indistinguishable from the limit $N \to \infty$ (*ice blue line* on the *red and the black curves*). Parameters are the same as in the upper part. Mersenne Twister with seeds: 637

space where the points represent individual genotypes and the distance between genotypes, commonly called the Hamming distance, counts the minimal number of point mutations required to bridge the interval between them. Neutral evolution, for example, can be visualized as a diffusion process in genotype space [304] (see Sect. 5.2.3) and Darwinian selection as a hill-climbing process in genotype space [580] (see Sect. 5.3.2).

3.1.2 Notation for Probabilistic Processes

A stochastic process is determined by a set of joint probability densities, whose existence and analytical form are presupposed.[8] The probability density encapsulates the physical nature of the process and contains all parameters and data reflecting the internal dynamics and external conditions. It this way it completely determines the system under consideration:

$$p(x_1,t_1;x_2,t_2;x_3,t_3;\ldots;x_n,t_n;\ldots)\,. \tag{3.5}$$

By *complete determination* we mean that no additional information is required to describe the progress of the system as a time ordered series (3.3), and we shall call such a process a *separable stochastic process*. Although more general processes are conceivable, they play little role in current physics, chemistry, and biology, and therefore we shall not consider them here.

Calculation of probabilities from (3.5) by means of the marginal densities (1.39) and (1.74) is straightforward. For the discrete case the result is obvious:

$$P(\mathcal{X}=x_1)=p(x_1,*)=\sum_{x_k\neq x_1} p(x_1,t_1;x_2,t_2;x_3,t_3;\ldots;x_n,t_n;\ldots)\,.$$

The probability of recording the value x_1 for the random variable $\mathcal{X}$ at time t_1 is obtained through summation over all previous values $x_2, x_3, \ldots$. In the continuous case the summations are simply replaced by integrals:

$$\begin{aligned}P\big(\mathcal{X}_1 &= x_1 \in [a,b]\big)\\ &= \int_a^b \mathrm{d}x_1 \int_{-\infty}^{\infty} \mathrm{d}x_2 \int_{-\infty}^{\infty} \mathrm{d}x_3 \ldots \int_{-\infty}^{\infty} \mathrm{d}x_n \ldots p(x_1,t_1;x_2,t_2;x_3,t_3;\ldots;x_n,t_n;\ldots)\,.\end{aligned}$$

[8] The joint density p is defined as in (1.36) and in Sect. 1.9.3. We use it here with a slightly different notation, because in stochastic processes we are always dealing with pairs (x,t), which we separate by a semicolon: $\ldots;x_k,t_k;x_{k+1},t_{k+1};\ldots$.

Time ordering admits a formulation of the predictions of future values from the known past in terms of conditional probabilities:

$$p(x_1, t_1; x_2, t_2; \ldots \,|\, x_k, t_k; x_{k+1}, t_{k+1}, \ldots)$$
$$= \frac{p(x_1, t_1; x_2, t_2; \ldots ; x_k, t_k; x_{k+1}, t_{k+1}, \ldots)}{p(x_k, t_k; x_{k+1}, t_{k+1}, \ldots)} ,$$

with $t_1 \geq t_2 \geq \ldots \geq t_k \geq t_{k+1} \geq \ldots$. In other words, we may compute

$$\big\{(x_1, t_1), (x_2, t_2), \ldots\big\} \text{ from known } \big\{(x_k, t_k), (x_{k+1}, t_{k+1}), \ldots\big\} .$$

With respect to the temporal progress of the process we shall distinguish discrete and continuous time. A trajectory in discrete time is just a time ordered sequence $\mathcal{X}_1, \mathcal{X}_2, \ldots, \mathcal{X}_n$ of random variables, where time is implicitly included in the index of the variable in the sense that $\mathcal{X}_1$ is recorded at time t_1, $\mathcal{X}_2$ at time t_2, and so on. The discrete probability distribution is characterized by two indices, n for the integer values the random variable can adopt and k for time: $P_{n,k} = P(\mathcal{X}_k = x_n)$ with $n, k \in \mathbb{N}_{>0}$ (Table 3.1). The introduction of continuous time is straightforward, since we need only replace $k \in \mathbb{N}_{>0}$ by $t \in \mathbb{R}$. The random variable is still discrete and the probability mass function becomes a function of time, i.e., $P_{n,k} \Rightarrow P_n(t)$. The transition to a continuous sample space for the random variable is made in precisely the same way as for probability mass functions described in Sect. 1.9. For the discrete time case, we change the notation accordingly, to obtain $P_{n,k} \Rightarrow p_k(x)\,\mathrm{d}x = f_k(x)\,\mathrm{d}x = \mathrm{dF}_k(x)$, while for continuous time, we have $P_{n,k} \Rightarrow p(x,t)\,\mathrm{d}x = f(x,t)\,\mathrm{d}x = \mathrm{dF}(x,t)\,\mathrm{d}x$.

Before we derive a general concept that allows for flexible models of stochastic processes which are applicable to chemical kinetics and biological modeling, we introduce a few common classes of stochastic processes with certain characteristic properties that are meaningful in the context of applications. In addition we shall distinguish different behavior with respect to the past, present, and future as encapsulated in memory effects.

3.1.3 Memory in Stochastic Processes

Three simple stochastic processes with characteristic memory effects will be discussed here:

(i) The fully factorizable process with probability densities that are independent of other events, with the special case of the Bernoulli process, where the probability densities are also independent of time.

(ii) The martingale, where the (sharp) initial value of the stochastic variable is equal to the conditional mean value of the variable in the future.
(iii) The Markov process, where the future is completely determined by the present. This is the most common formalism for modeling dynamics stochasticity in science.

Independence and Bernoulli Processes
The simplest class of stochastic processes is characterized by *complete independence* of events. This allows for factorization of the density:

$$p(x_1, t_1; x_2, t_2; x_3, t_3; \ldots) = \prod_i p(x_i, t_i) . \tag{3.6}$$

Equation (3.6) implies that the current value $\mathcal{X}(t)$ is completely independent of its values in the past. A special case is the sequence of Bernoulli trials (see previous chapters, and in particular Sects. 1.5 and 2.3.2), where the probability densities are also independent of time: $p(x_i, t_i) = p(x_i)$. Then we have

$$p(x_1, t_1; x_2, t_2; x_3, t_3; \ldots) = \prod_i p(x_i) . \tag{3.6$'$}$$

Further simplification occurs, of course, when all trials are based on the same probability distribution, for example, if the same coin is tossed in Bernoulli trials or the same dice are thrown. The product can then be replaced by the power $p(x)^n$.

Martingales
The notion of *martingale* was introduced by the French mathematician Paul Pierre Lévy, and the development of the theory of martingales can be attributed to the American mathematician Joseph Leo Doob among others [367]. As appropriate, we distinguish discrete time and continuous time processes. A discrete-time martingale is a sequence of random variables, $\mathcal{X}_1, \mathcal{X}_2, \ldots$, which satisfy the conditions[9]

$$\mathrm{E}(\mathcal{X}_{n+1} | \mathcal{X}_n, \ldots, \mathcal{X}_1) = \mathcal{X}_n , \qquad \mathrm{E}(|\mathcal{X}_n|) < \infty . \tag{3.7}$$

Given all past values $\mathcal{X}_1, \ldots, \mathcal{X}_n$, the conditional expectation value for the next observation $\mathrm{E}(\mathcal{X}_{n+1})$ is equal to the last recorded value $\mathcal{X}_n$.

A continuous time martingale refers to a random variable $\mathcal{X}(t)$ with expectation value $\mathrm{E}\big(\mathcal{X}(t)\big)$. We first define the conditional expectation value of the random

[9]For convenience we change the numbering of times here and apply the notation of (3.3′).

variable for $\mathcal{X}(t_0) = x_0$ and $\mathrm{E}(|\mathcal{X}(t)|) < \infty$:

$$\mathrm{E}\big(\mathcal{X}(t)|(x_0, t_0)\big) \doteq \int \mathrm{d}x\, p(x, t|x_0, t_0) \; .$$

In a martingale, the conditional mean is simply given by

$$\mathrm{E}\big(\mathcal{X}(t)|(x_0, t_0)\big) = x_0 \; . \tag{3.8}$$

The mean value at time t is identical to the initial value of the process. The martingale property is rather strong but we shall nevertheless use it to characterize specific processes.

As an example of a martingale we consider the unlimited symmetric random walk in one dimension (Fig. 3.3). Equal-sized steps of length l to the right and to the left are taken with equal probability. In the discrete time random walk, the waiting time between two steps is θ [t], we measure time in multiples of the waiting time, $t - t_0 = k\theta$, and the position in multiples of the step length l [l]. The corresponding probability of being at location $x - x_0 = nl$ at time θ is simply expressed in pairs of variables (n, k):

$$\begin{aligned} P\big(n, k+1 \,|\, n_0, k_0\big) &= \frac{1}{2}\Big(P\big(n+1, k \,|\, n_0, k_0\big) + P\big(n-1, k \,|\, n_0, k_0\big)\Big) \; , \\ P_{n,k+1} &= \frac{1}{2}\big(P_{n+1,k} + P_{n-1,k}\big) \; , \quad \text{with } P_{n,k_0} = \delta_{n,n_0} \; , \end{aligned} \tag{3.9}$$

where we expresses the initial conditions by a separate equation in the *short-hand* notation. Our choice of variables allows for simplified initial conditions $n_0 = 0$ and $k_0 = 0$ without loss of generality. Equation (3.9) is readily solved by means of the characteristic function:

$$\phi(s, k) = \mathrm{E}\big(\mathrm{e}^{\mathrm{i}ns}\big) = \sum_{n=-\infty}^{\infty} P(n, k \,|\, 0, 0)\mathrm{e}^{\mathrm{i}ns} = \sum_{n=-\infty}^{\infty} P_{n,k}\mathrm{e}^{\mathrm{i}ns} \; . \tag{2.32$'$}$$

Using (3.9) yields

$$\phi(s, k+1) = \frac{1}{2}\phi(s, k)\big(\mathrm{e}^{\mathrm{i}s} + \mathrm{e}^{-\mathrm{i}s}\big) = \cosh(\mathrm{i}s) \; , \quad \text{with } \phi(s, 0) = 1 \; ,$$

and the solution is calculated to be

$$\phi(s, k) = \cosh^k(\mathrm{i}s) = \frac{1}{2^k}\left(\mathrm{e}^{\mathrm{i}ks} + \binom{k}{1}\mathrm{e}^{\mathrm{i}(k-2)s} + \binom{k}{2}\mathrm{e}^{\mathrm{i}(k-4)s} + \cdots + \mathrm{e}^{-\mathrm{i}ks}\right) . \tag{3.10a}$$

Equating the coefficients of the individual terms e^{ins} in expressions (3.10a) and (2.32′) determines the probabilities

$$P_{n,k} = \begin{cases} \frac{1}{2^k}\binom{k}{\nu}, & \text{if } |n| \le k\,,\ \nu = \frac{k-n}{2} \in \mathbb{N}\,, \\ 0\,, & \text{otherwise}\,. \end{cases} \tag{3.10b}$$

The distribution is binomial with $k+1$ terms and width $2k$, and every other term is equal to zero. It spreads with time according to $t = k\theta$.

Calculation of the first and second moments is straightforward and is best achieved using the derivatives of the characteristic function, as shown in (2.34):

$$\frac{\partial\phi(s,k)}{\partial s} = \mathrm{i}\,n\cosh^{n-1}(\mathrm{i}s)\sinh(\mathrm{i}s)\,,$$

$$\frac{\partial^2\phi(s,k)}{\partial s^2} = -n\Big(\cosh^n(\mathrm{i}s) + (n-1)\cosh^{n-2}(\mathrm{i}s)\sinh^2(\mathrm{i}s)\Big)\,.$$

Inserting $s = 0$ yields $(\partial\phi/\partial s)|_{s=0} = 0$ and $(\partial^2\phi/\partial s^2)|_{s=0} = -n$, and by (2.34), with $n(0) = n_0$ and $k(0) = k_0$, we obtain for the moments:

$$\mathrm{E}\big(\mathcal{X}(t)\big) = x_0 = n_0 l\,, \qquad \mathrm{var}\big(\mathcal{X}(t)\big) = t - t_0 = (k-k_0)\,\theta\,. \tag{3.11}$$

The unlimited, symmetric, and discrete random walk in one dimension is a martingale and the standard deviation $\sigma\big(\mathcal{X}(t)\big)$ increases as $\sqrt{t}$, as predicted in the ground-breaking work of Albert Einstein [133] and Marian von Smoluchowski [559]. This implies that trajectories will in general diverge and approach $\pm\infty$, as is characteristic for a type (i) process.

We remark that the standardized sum of the outcomes of Bernoulli trials, $s(n) = \mathcal{S}(n)/n - \mu$ with $\mathcal{S}_n = \sum_{i=1}^{n}\mathcal{X}_i$ and $\mathcal{X}_i = \pm 1$, which was used to illustrate the law of the iterated logarithm (Fig. 2.13), is itself a martingale, but here the trajectories are confined to the domain $s(n) = \pm 1$ and the long-term limit is zero. A time scale in this case results from the assignment of a time interval between two successive trials.

The somewhat relaxed notion of a *semimartingale* is of importance because it covers most processes that are accessible to modeling by *stochastic differential equations*. A semimartingale is composed of a *local martingale* and an *adapted càdlàg-process*[10] with bounded variation:

$$\mathcal{X}(t) = \mathcal{M}(t) + \mathcal{A}(t)\,.$$

[10]The term *càdlàg* is an acronym from French which stands for *continue à droite, limites à gauche*. The English expression is *right continuous with left limits* (RCLL). It is a common property of step functions in probability theory (Sect. 1.6.2). We shall reconsider the *càdlàg* property in the context of sampling trajectories (Sect. 4.2.1).

A local martingale is a stochastic process that satisfies the martingale property (3.8) locally, while its expectation value $\langle \mathcal{M}(t) \rangle$ may be distorted at long times by large values of low probability. Hence, every martingale is a local martingale and every bounded local martingale is a martingale. In particular, every driftless diffusion process is a local martingale, but need not be a martingale.

An adapted process $\mathcal{A}(t)$ is *nonanticipating* in the sense that it *cannot see into the future*. An informal interpretation [574, Sect. II.25] would say that a stochastic process $\mathcal{X}(t)$ is adapted if and only if, for every realization and for every time t, $\mathcal{X}(t)$ is known at time t and not before. The notion 'nonanticipating' is irrelevant for deterministic processes, but it matters for processes containing fluctuating elements, because only the independence of random or irregular increments makes it impossible to look into the future. The concept of adapted processes is essential for the definition and evaluation of the Itō stochastic integral, which is based on the assumption that the integrand is an adapted process (Sect. 3.4.2).

Two generalizations of martingales are in common use:

(i) A discrete time *submartingale* is a sequence $\mathcal{X}_1, \mathcal{X}_2, \mathcal{X}_3, \ldots,$ of random variables that satisfy

$$\mathrm{E}(\mathcal{X}_{n+1} | \mathcal{X}_1, \ldots, \mathcal{X}_n) \geq \mathcal{X}_n \,, \tag{3.12}$$

while for the continuous time analogue, we have the condition

$$\mathrm{E}\big(\mathcal{X}(t) | \{\mathcal{X}(\theta) : \theta \leq s\}\big) \geq \mathcal{X}(s) \,, \quad \forall\, s \leq t \,. \tag{3.13}$$

(ii) The relations for *supermartingales* are in complete analogy to those for submartingales, except that $\geq$ must be replaced by $\leq$:

$$\mathrm{E}(\mathcal{X}_{n+1} | \mathcal{X}_1, \ldots, \mathcal{X}_n) \leq \mathcal{X}_n \,, \tag{3.14}$$

$$\mathrm{E}\big(\mathcal{X}(t) | \{\mathcal{X}(\theta) : \theta \leq s\}\big) \leq \mathcal{X}(s) \,, \quad \forall\, s \leq t \,. \tag{3.15}$$

A straightforward consequence of the martingale property is this: if a sequence or a function of random variables is a simultaneously a submartingale and a supermartingale, it must be a martingale.

Markov Processes

Markov processes are processes that share the Markov property. In a nutshell, this assumes that knowledge of the present alone is all we need to predict the future, or in other words, information about the past will not improve prediction of the future. Although processes that satisfy the Markov property are only a minority among general stochastic processes [542], they are of particular importance because almost all models in science assume the Markov property, and this assumption facilitates the analysis enormously.

The Markov process is named after the Russian mathematician Andrey Markov[11] and can be formulated in a straightforward manner in terms of conditional probabilities:

$$p(x_1, t_1; x_2, t_2; \ldots \,|\, x_k, t_k; x_{k+1}, t_{k+1}, \ldots) = p(x_1, t_1; x_2, t_2; \ldots \,|\, x_k, t_k) \,. \tag{3.16}$$

As already mentioned, the Markov condition expresses independence from the history of the process prior to time t_k. For example, we have

$$p(x_1, t_1; x_2, t_2; x_3, t_3) = p(x_1, t_1 | x_2, t_2)\, p(x_2, t_2 | x_3, t_3) \,.$$

As we saw in Sect. 1.6.4, any arbitrary joint probability can be simply expressed as products of conditional probabilities:

$$\begin{aligned} &p(x_1, t_1; x_2, t_2; x_3, t_3; \ldots; x_n, t_n) \\ &\qquad = p(x_1, t_1 | x_2, t_2) p(x_2, t_2 | x_3, t_3) \cdots p(x_{n-1}, t_{n-1} | x_n, t_n) p(x_n, t_n) \,, \end{aligned} \tag{3.16$'$}$$

under the assumption of time ordering $t_1 \geq t_2 \geq t_3 \geq \ldots \geq t_{n-1} \geq t_n$. Because of these sequential products of conditional probabilities of two events, one also speaks of a *Markov chain*. The Bernoulli process can now be seen as a special Markov process, in which the next state is not only independent of the past states, but also of the current state.

3.1.4 Stationarity

Stationarity of a deterministic process implies that all observable dependence on time vanishes at a stationary state. In the case of multistep processes, the definition leaves two possibilities open:

(i) At thermodynamic equilibrium, the fluxes of all individual steps vanish, as expressed in the *principle of detailed balance* [531].
(ii) Only the total flux, i.e., the sum of all fluxes, becomes zero, whereas individual fluxes may have nonzero values which balance out in the sum.

---

[11]The Russian mathematician Andrey Markov (1856–1922) was one of the founders of Russian probability theory and pioneered the concept of memory-free processes which is named after him. Among other contributions he expressed the assumptions that were made by Albert Einstein [133] and Marian von Smoluchowski [559] in their derivation of the diffusion process in more precise terms.

Stationarity of stochastic processes in general, and of Markov processes in particular, is more subtle, since random fluctuations do not vanish at equilibrium. Several definitions of stationarity are possible. Three of them are relevant for our purposes here.

Strong Stationarity
A stochastic process is said to be *strictly* or *strongly stationary* if $\mathcal{X}(t)$ and $\mathcal{X}(t+\Delta t)$ obey the same statistics for every Δt. Accordingly, joint probability densities are invariant under time translations:

$$p(x_1, t_1; x_2, t_2; \ldots; x_n, t_n) = p(x_1, t_1 + \Delta t; x_2, t_2 + \Delta t; \ldots; x_n, t_n + \Delta t) . \tag{3.17}$$

In other words, the probabilities are functions of time differences $\Delta t = t_k - t_j$ alone, and this leads to time independent stationary *one-time probabilities*

$$p(x, t) \Longrightarrow \bar{p}(x) , \tag{3.18}$$

and two-time joint or conditional probabilities of the form

$$\begin{aligned} p(x_1, t_1; x_2, t_2) &\Longrightarrow \bar{p}(x_1, t_1 - t_2; x_2, 0) , \\ p(x_1, t_1 | x_2, t_2) &\Longrightarrow \bar{p}(x_1, t_1 - t_2 | x_2, 0) . \end{aligned} \tag{3.19}$$

Since all joint probabilities of a Markov process can be written as products of two-time conditional probabilities and a one-time probability (3.16′), the necessary and sufficient condition for stationarity is cast into the requirement that one should be able to write all one- and two-time probabilities as shown in (3.18) and (3.19). A Markov process that becomes stationary in the limit $t \to \infty$ or $t_0 \to -\infty$ is called a *homogeneous Markov process*.

Weak Stationarity
The notion of weak stationarity or covariance stationarity is used, for example, in signal processing, and relaxes the stationarity condition (3.17) for a process $\mathcal{X}(t)$ to

$$\begin{aligned} \mathrm{E}\big(\mathcal{X}(t)\big) &= \mu_{\mathcal{X}}(t) = \mu_{\mathcal{X}}(t + \Delta t) , \ \ \forall\, \Delta t \in \mathbb{R} , \\ \mathrm{cov}\big(\mathcal{X}(t_1), \mathcal{X}(t_2)\big) &= \mathrm{E}\Big(\big(\mathcal{X}(t_1) - \mu_{\mathcal{X}}(t_1)\big)\big(\mathcal{X}(t_2) - \mu_{\mathcal{X}}(t_2)\big)\Big) \\ &= \mathrm{E}\big(\mathcal{X}(t_1)\mathcal{X}(t_2)\big) - \mu_{\mathcal{X}}(t_1)\mu_{\mathcal{X}}(t_2) \\ &= C_{\mathcal{X}}(t_1, t_2) = C_{\mathcal{X}}(t_1 - t_2, 0) = C_{\mathcal{X}}(\Delta t) . \end{aligned} \tag{3.20}$$

Instead of the entire probability function, only the process mean $\mu_{\mathcal{X}}$ has to be constant, while the *autocovariance function*[12] of the stochastic process $\mathcal{X}(t)$ denoted by $C_{\mathcal{X}}(t_1, t_2)$ does not depend on t_1 and t_2, but only on the difference $\Delta t = t_1 - t_2$.

Second Order Stationarity
The notion of second order stationarity of a process with finite mean and finite autocovariance expresses the fact that the conditions of strict stationarity are applied only to pairs of random variables from the time series. Then the first and second order density functions satisfy:

$$\begin{aligned} f_{\mathcal{X}}(x_1;t_1) &= f_{\mathcal{X}}(x_1;t_1 + \Delta t)\,, \quad \forall\,(t_1, \Delta t)\,, \\ f_{\mathcal{X}}(x_1, x_2; t_1, t_2) &= f_{\mathcal{X}}(x_1, x_2; t_1 + \Delta t, t_2 + \Delta t)\,, \quad \forall\,(t_1, t_2, \Delta t)\,. \end{aligned} \tag{3.21}$$

The definition can be extended to higher orders and then strict stationarity is tantamount to stationarity in all orders. A second order stationary process satisfies the criteria for weak stability, but a process can be stationary in the broad sense without satisfying the criteria of second order stationarity.

3.1.5 Continuity in Stochastic Processes

Continuity in deterministic processes requires the absence of any kind of jump, but it does not require differentiability expressed as continuity in the first derivative. We recall the conventional definition of continuity at $x = x_0$:

$$\forall\,\varepsilon > 0\,,\ \exists\,\delta > 0 \text{ such that } \forall\,x:\ |x - x_0| < \delta \implies |f(x) - f(x_0)| < \varepsilon\,.$$

In other words, we require that $|f(x) - f(x_0)|$ can become arbitrarily small for all $|x - x_0|$, no matter how close x is to x_0, whence no jumps are allowed. The condition of continuity in Markov processes is defined analogously, but requires a more detailed discussion. For this purpose, we consider a process that progresses from location z at time t to location $x = z + \Delta z$ at time $t + \Delta t$, denoted by $(z, t) \to (z + \Delta z, t + \Delta t) = (x, t + \Delta t)$.[13]

---

[12]The notion of *autocovariance* reflects the fact that the process is correlated with itself at another time, while cross-covariance implies the correlation of two different processes (for the relation between autocorrelation and autocovariance, see Sect. 3.1.6).

[13]The notation used for time dependent variables is explained in Fig. 3.4. For convenience and readability, we write x for $z + \Delta z$.

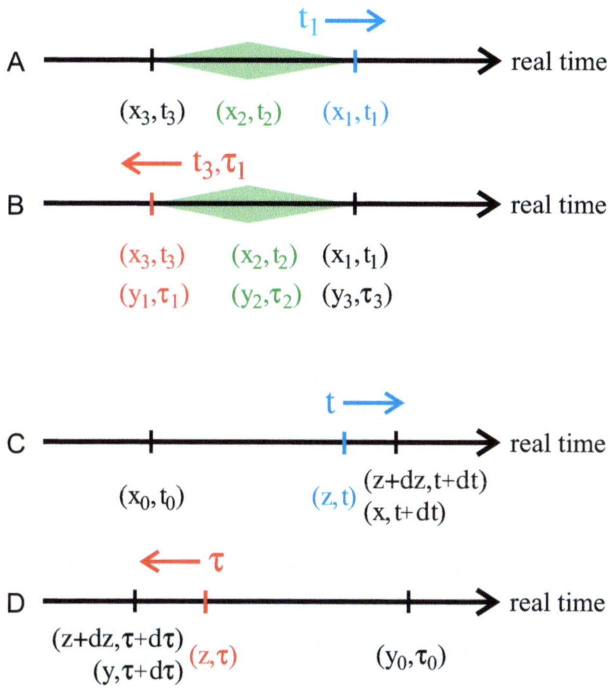

Fig. 3.4 Notation for time dependent variables. In the following sections we shall require several time dependent variables and adopt the following notation. For the Chapman–Kolmogorov equation, we require three variables at different times, denoted by x_1, x_2, and x_3. The variable x_2 is associated with the intermediate time t_2 (*green*) and disappears through integration. In the forward equation, (x_3, t_3) are fixed initial conditions and (x_1, t_1) is moving (A). For backward integration, the opposite relation is assumed: (x_1, t_1) being fixed and (x_3, t_3) moving (B, the lower notation is used for the backward equation in Sect. 3.3). In both cases, *real time* progresses from left to right, while computational time increases in the same direction as real time in the forward evaluation (*blue*), but in the opposite direction for backward evaluation (*red*). The *lower part* of the figure shows the notation used for forward and backward differential Chapman–Kolmogorov equations. In the forward equation (C), $x(t)$ is the variable, the initial conditions are denoted by (x_0, t_0), and (z, t) is an intermediate pair. In the backward equation, the time order is reversed (D): $y(\tau)$ is the variable and (y_0, τ_0) are the final conditions. In both cases, we could use $z + \mathrm{d}z$ instead x or y, respectively, but the equations would then be less clear

The general requirement for consistency and continuity of a Markov process can be cast into the relation

$$\lim_{\Delta t \to 0} p(x, t + \Delta t | \, z, t) = \delta(x - z) \, , \tag{3.22}$$

where $\delta(\cdot)$ is the so-called delta-function (see Sect. 1.6.3). In other words, z becomes x if Δt goes to zero. The process is continuous if and only if, in the limit $\Delta t \to 0$, the probability of z being finitely different from x goes to zero faster than

Δt, as expressed by the equation

$$\lim_{\Delta t \to 0} \frac{1}{\Delta t} \int_{|\Delta z|=|x-z|>\varepsilon} \mathrm{d}x\, p(x, t + \Delta t|z, t) = 0\,,$$

and this convergence is uniform in z, t, and Δt. In other words, the difference in probability as a function of $|z - x|$ approaches zero sufficiently fast to ensure that no jumps occur in the random variable $\mathcal{X}(t)$.

Continuity in Markov processes can be illustrated by means of two examples [194, pp. 65–68] which give rise to trajectories as sketched in Fig. 3.5:

(i) The Wiener process or Brownian motion [69], which is the continuous version of the random walk in one dimension shown in Fig. 3.3.[14] This leads to a normally distributed conditional probability

$$p\big(x, t + \Delta t|z, t\big) = \frac{1}{\sqrt{4\pi D \Delta t}} \exp\left(-\frac{(x-z)^2}{4D\Delta t}\right)\,. \tag{3.23}$$

(ii) The so-called Cauchy process following the Cauchy–Lorentz distribution

$$p\big(x, t + \Delta t|z, t\big) = \frac{1}{\pi} \frac{\Delta t}{(x-z)^2 + \Delta t^2}\,. \tag{3.24}$$

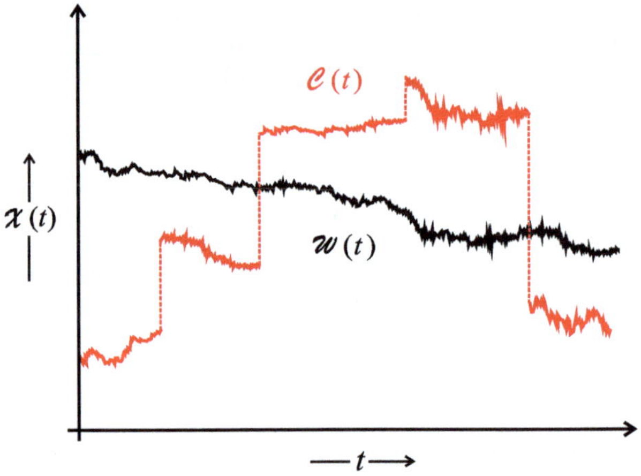

Fig. 3.5 Continuity in Markov processes. Continuity is illustrated by means of two stochastic processes of the random variable $\mathcal{X}(t)$: the Wiener process $\mathcal{W}(t)$ (3.23) (*black*) and the Cauchy process $\mathcal{C}(t)$ (3.24) (*red*). The Wiener process describes Brownian motion and is continuous, but almost nowhere differentiable. The even more irregular Cauchy process also contains steps and is discontinuous

[14]Later on we shall discuss the limit of the random walk for vanishing step size in more detail and call it a *Wiener process* (Sect. 3.2.2.2).

The distribution in the case of the Wiener process follows directly from the binomial distribution of the random walk (3.10b) in the limit of vanishing step size. For the analysis of continuity, we exchange the limit and the integral, introduce $\vartheta = 1/\Delta t$, take the limit $\vartheta \to \infty$, and find

$$\begin{aligned}
&\lim_{\Delta t\to 0} \frac{1}{\Delta t} \int_{|x-z|>\varepsilon} \mathrm{d}x \, \frac{1}{\sqrt{4\pi D}} \frac{1}{\sqrt{\Delta t}} \exp\left(-\frac{(x-z)^2}{4D\Delta t}\right) \\
&\qquad = \int_{|x-z|>\varepsilon} \mathrm{d}x \, \lim_{\Delta t\to 0} \frac{1}{\Delta t} \frac{1}{\sqrt{4\pi D}} \frac{1}{\sqrt{\Delta t}} \exp\left(-\frac{(x-z)^2}{4D\Delta t}\right) \\
&\qquad = \int_{|x-z|>\varepsilon} \mathrm{d}x \, \lim_{\vartheta\to\infty} \frac{1}{\sqrt{4\pi D}} \frac{\vartheta^{3/2}}{\exp\left(\dfrac{(x-z)^2}{4D}\vartheta\right)} ,
\end{aligned}$$

where

$$\lim_{\vartheta\to\infty} \frac{\vartheta^{3/2}}{1 + \dfrac{(x-z)^2}{4D}\vartheta + \dfrac{1}{2!}\left(\dfrac{(x-z)^2}{4D}\right)^2 \vartheta^2 + \dfrac{1}{3!}\left(\dfrac{(x-z)^2}{4D}\right)^3 \vartheta^3 + \cdots} = 0 .$$

Since the power series expansion of the exponential in the denominator increases faster than every finite power of ϑ, the ratio vanishes in the limit $\vartheta \to \infty$, the value of the integral is zero, and the Wiener process is continuous everywhere. Although it is continuous, the trajectory of the Wiener process is extremely irregular, since it is nowhere differentiable (Fig. 3.5).

In the second example, the Cauchy process, we exchange the limit and integral as we did for the Wiener process, and take the limit $\Delta t \to 0$:

$$\begin{aligned}
&\lim_{\Delta t\to 0} \frac{1}{\Delta t} \int_{|x-z|>\varepsilon} \mathrm{d}x \, \frac{\Delta t}{\pi} \frac{1}{(x-z)^2 + \Delta t^2} \\
&= \int_{|x-z|>\varepsilon} \mathrm{d}x \, \lim_{\Delta t\to 0} \frac{1}{\Delta t} \frac{\Delta t}{\pi} \frac{1}{(x-z)^2 + \Delta t^2} \\
&= \int_{|x-z|>\varepsilon} \mathrm{d}x \, \lim_{\Delta t\to 0} \frac{1}{\pi} \frac{1}{(x-z)^2 + \Delta t^2} = \int_{|x-z|>\varepsilon} \frac{1}{\pi(x-z)^2} \mathrm{d}x \neq 0 .
\end{aligned}$$

The value of the last integral $I = \int_{|x-z|>\varepsilon}^{\infty} \mathrm{d}x/(x-z)^2 = 1/\pi(x-z)$ is of the order $I \approx 1/\varepsilon$ and hence finite. Consequently, the curve for the Cauchy process is irregular and only piecewise continuous, since it contains discontinuities in the form of jumps (Fig. 3.5).

The mathematical definition for continuity in Markov processes [194, p. 46] in vector notation for the locations $\boldsymbol{x}$ and $\boldsymbol{z}$ can be encapsulated as follows:

A Markov process has, with probability one, sample paths that are continuous functions of time t, if for any $\varepsilon > 0$ the limit

$$\lim_{\Delta t \to 0} \frac{1}{\Delta t} \int_{|x-z|>\varepsilon} \mathrm{d}\boldsymbol{x}\, p(\boldsymbol{x}, t + \Delta t | \boldsymbol{z}, t) = 0 \tag{3.25}$$

is approached uniformly in $\boldsymbol{z}$, t, and Δt.

In essence, (3.25) expresses the fact that probabilistically the difference between $\boldsymbol{x}$ and $\boldsymbol{z}$ converges to zero faster than Δt.

3.1.6 Autocorrelation Functions and Spectra

Analysis of experimentally recorded or computer created trajectories is often largely facilitated by the usage of additional tools complementing moments and probability distributions, since they can, in principle, be derived from single recordings. These tools are autocorrelation functions and spectra of random variables, which provide direct insight into the dynamics of the process, since they deal with relations between points collected from the same sample path at different times. The autocorrelation is readily accessible experimentally (for the application of the autocorrelation function to fluorescence correlation spectroscopy, see, e.g., Sect. 4.4.2) and represents a basic tool in *time series analysis* (see, for example, [565]).

Convolution, Cross-Correlation and Autocorrelation

These three integral relations between two functions $f(t)$ and $g(t)$ are important in statistics, and in particular in signal processing. The *convolution* is defined as

$$(f * g)(\mathbf{x}) \stackrel{\text{def}}{=} \int_{-\infty}^{\infty} \mathrm{d}\mathbf{y}\, f(\mathbf{y}) g(\mathbf{x} - y) = \int_{-\infty}^{\infty} \mathrm{d}\mathbf{y}\, f(\mathbf{x} - \mathbf{y}) g(\mathbf{y}) \,, \tag{3.26}$$

where $\mathbf{x}$ and $\mathbf{y}$ represent vectors in n-dimensional space, i.e., $(\mathbf{x}, \mathbf{y}) \in \mathbb{R}^n$. Among other properties, the *convolution theorem* is of great practical importance because it allows for straightforward computation of the convolution as the product of two integrals after Fourier transform:

$$\mathcal{F}(f * g) = \mathcal{F}(f)\mathcal{F}(g) \,, \qquad f * g = \mathcal{F}^{-1}\big(\mathcal{F}(f)\mathcal{F}(g)\big) \,, \tag{3.27}$$

where the Fourier transform and its inverse are defined by[15]

$$\tilde{f}(\boldsymbol{\xi}) = \mathcal{F}(f) = \int_{-\infty}^{\infty} f(\mathbf{x}) \exp(-2\pi \mathrm{i}\, \mathbf{x} \cdot \boldsymbol{\xi})\, \mathrm{d}\mathbf{x} \,,$$

$$f(\mathbf{x}) = \mathcal{F}^{-1}(\tilde{f}) = \int_{-\infty}^{\infty} \tilde{f}(\boldsymbol{\xi}) \exp(2\pi \mathrm{i}\, \mathbf{x} \cdot \boldsymbol{\xi})\, \mathrm{d}\boldsymbol{\xi} \,.$$

The convolution theorem can also be inverted to yield

$$\mathcal{F}(fg) = \mathcal{F}(f) * \mathcal{F}(g) \,.$$

Another useful relation is provided by the Laplace transform of a convolution, viz.,

$$\mathcal{L}\left(\int_0^t f(t-\tau)g(\tau)\,\mathrm{d}\tau\right) = \mathcal{L}\big(f(t)\big)\mathcal{L}\big(g(t)\big) = F(s)\,G(s) \,,$$

where $F(s)$ and $G(s)$ are the Laplace transforms of $f(t)$ and $g(t)$, respectively.

The *cross-correlation* is related to the convolution and commonly defined by

$$(f \star g)(\mathbf{x}) \stackrel{\text{def}}{=} \int_{-\infty}^{\infty} \mathrm{d}\mathbf{y}\, f^*(\mathbf{y}) g(\mathbf{x}+\mathbf{y}) \,, \tag{3.28}$$

and in analogy to the convolution theorem, the relation

$$\mathcal{F}(f \star g) = \big(\mathcal{F}(f)\big)^* \mathcal{F}(g)$$

holds for the Fourier transform of the cross-correlation. It is a nice exercise to show that the identity

$$(f \star g) \star (f \star g) = (f \star f) \star (g \star g)$$

is satisfied by the cross-correlation [567]. The autocorrelation,

$$(f \star f)(\mathbf{x}) \stackrel{\text{def}}{=} \int_{-\infty}^{\infty} \mathrm{d}\mathbf{y}\, f^*(\mathbf{y}) f(\mathbf{x}+\mathbf{y}) \,, \tag{3.29}$$

is a special case of the cross-correlation, namely, the cross-correlation of a function f with itself after a shift $\mathbf{x}$.

---

[15] We remark that this definition of the Fourier transform is used in signal processing and differs from the convention used in modern physics (see [568] and Sect. 2.2.3).

Autocorrelation and Spectrum

The autocorrelation function of a stochastic process is defined by (2.9′) as the correlation coefficient $\rho(\mathcal{X},\mathcal{Y})$ of the random variable $\mathcal{X}=\mathcal{X}(t_1)$ at some time t_1 with the same variable $\mathcal{Y}=\mathcal{X}(t_2)$ at another time t_2:

$$\begin{aligned} R(t_1,t_2) &= \rho\big(\mathcal{X}(t_1),\mathcal{X}(t_2)\big) \\ &= \frac{\mathrm{E}\Big(\big(\mathcal{X}(t_1)-\mu_{\mathcal{X}}(t_1)\big)\big(\mathcal{X}(t_2)-\mu_{\mathcal{X}}(t_2)\big)\Big)}{\sigma_{\mathcal{X}}(t_1)\,\sigma_{\mathcal{X}}(t_2)}\,, \quad R\in[-1,1]\,. \end{aligned} \tag{3.30}$$

Thus the autocorrelation function is obtained from the autocovariance function (3.20) through division by the product of the standard deviations:

$$R(t_1,t_2)=\frac{\mathrm{cov}\big(\mathcal{X}(t_1)\mathcal{X}(t_2)\big)}{\sigma_{\mathcal{X}}(t_1)\,\sigma_{\mathcal{X}}(t_2)}\,.$$

Accordingly, the autocorrelation of the random variable $\mathcal{X}(t)$ is a measure of the influence that the value of $\mathcal{X}$ recorded at time t_1 has on the measurement of the same variable at time t_2. Under the assumption that we are dealing with a weak or second order stationary process, the mean and the variance are independent of time, and then the autocorrelation function depends only on the time difference $\Delta t=t_2-t_1$:

$$R(\Delta t)=\frac{\mathrm{E}\Big(\big(\mathcal{X}(t)-\mu_{\mathcal{X}}\big)\big(\mathcal{X}(t+\Delta t)-\mu_{\mathcal{X}}\big)\Big)}{\sigma_{\mathcal{X}}^2}\,, \quad R\in[-1,1]\,. \tag{3.30′}$$

In spectroscopy, the autocorrelation of a spectroscopic signal $\mathcal{F}(t)$ is measured as a function of time. Then we are dealing with

$$\begin{aligned} G(\Delta t) &= \langle\mathcal{F}(t)\mathcal{F}(t+\Delta t)\rangle=\mathrm{E}\big(\mathcal{F}(t)\mathcal{F}(t+\Delta t)\big) \\ &= \lim_{t\to\infty}\frac{1}{t}\int_0^t \mathrm{d}\theta\, x(\theta)x(\theta+\Delta t)\,. \end{aligned} \tag{3.31}$$

Thus, the autocorrelation function is the time average of the product of two values recorded at different times with a given interval Δt.

Another relevant quantity is the *spectrum* or the spectral density of the quantity $x(t)$. In order to derive the spectrum, we construct a new variable $y(\omega)$ by means of the transformation $y(\omega)=\int_0^t \mathrm{d}\theta\,\mathrm{e}^{\mathrm{i}\omega\theta}\,x(\theta)$. The spectrum is then obtained from y by taking the limit $t\to\infty$:

$$S(\omega)=\lim_{t\to\infty}\frac{1}{2\pi t}|y(\omega)|^2=\lim_{t\to\infty}\frac{1}{2\pi t}\left|\int_0^t \mathrm{d}\theta\,\mathrm{e}^{\mathrm{i}\omega\theta}x(\theta)\right|^2\,. \tag{3.32}$$

The autocorrelation function and the spectrum are closely related. After some calculation, one finds

$$S(\omega) = \lim_{t\to\infty} \left(\frac{1}{\pi} \int_0^t \cos(\omega\tau)\, d\tau \frac{1}{t} \int_0^{t-\tau} x(\theta)x(\theta+\tau)\, d\theta \right) .$$

Under certain assumptions which ensure the validity of the interchanges of order, we may take the limit $t \to \infty$ to find

$$S(\omega) = \frac{1}{\pi} \int_0^{\infty} \cos(\omega\tau)\, G(\tau)\, d\tau .$$

This result relates the Fourier transform of the autocorrelation function to the spectrum and can be cast in an even more elegant form by using

$$G(-\tau) = \lim_{t\to\infty} \frac{1}{t} \int_{-\tau}^{t-\tau} d\theta\, x(\theta)x(\theta+\tau) = G(\tau)$$

to yield the *Wiener–Khinchin theorem* named after the American physicist Norbert Wiener and the Russian mathematician Aleksandr Khinchin:

$$S(\omega) = \frac{1}{2\pi} \int_{-\infty}^{+\infty} e^{-i\omega\tau} G(\tau)\, d\tau , \qquad G(\tau) = \int_{-\infty}^{+\infty} e^{i\omega\tau} S(\omega)\, d\omega . \tag{3.33}$$

The spectrum and the autocorrelation function are related to each other by the Fourier transformation and its inverse.

Equation (3.33) allows for a straightforward proof that the Wiener process $\mathcal{W}(t) = W(t)$ gives rise to *white noise* (see Sect. 3.2.2.2). Let $\mathbf{w}$ be a zero-mean random vector with the identity matrix as (auto)covariance or autocorrelation matrix, i.e.,

$$E(\mathbf{w}) = \boldsymbol{\mu} = \mathbf{0} , \qquad \mathrm{cov}(\mathcal{W}, \mathcal{W}) = E(\mathbf{w}\mathbf{w}') = \mathbb{I} .$$

Then the Wiener process $W(t)$ satisfies the relations

$$\mu_W(t) = E\big(\mathcal{W}(t)\big) = 0 ,$$
$$G_W(\tau) = E\big(\mathcal{W}(t)\, \mathcal{W}(t+\tau)\big) = \delta(\tau) ,$$

defining it as a zero-mean process with infinite power at zero time shift. For the spectral density of the Wiener process, we obtain

$$S_W(\omega) = \frac{1}{2\pi} \int_{-\infty}^{+\infty} e^{-i\omega\tau} \delta(\tau)\, d\tau = \frac{1}{2\pi} . \tag{3.34}$$

The spectral density of the Wiener process is a constant and hence all frequencies are represented with equal weight in the noise. Mixing all frequencies of electromagnetic radiation with equal weight yields white light and this property of visible light has given the name *white noise*. In *colored noise*, the noise frequencies do not meet the condition of the uniform distribution. Pink or flicker noise, for example, has a spectrum close to $S(\omega) \propto \omega^{-1}$, while red or Brownian noise satisfies $S(\omega) \propto \omega^{-2}$.

The time average of a signal as expressed by an autocorrelation function is complemented by the *ensemble average* $\langle \mathcal{X} \rangle$, or the expectation value of the corresponding random variable $\mathrm{E}(\mathcal{X})$, which implies an (infinite) number of repeats of the same measurement. Ergodic theory relates the two averages [53, 408, 558]. If the prerequisites of *ergodic behavior* are satisfied, the time average is equal to the ensemble average. Thus we find for a fluctuating quantity $\mathcal{X}(t)$, in the ergodic limit,

$$\mathrm{E}\big(\mathcal{X}(t), \mathcal{X}(t+\tau)\big) = \big\langle x(t)x(t+\tau)\big\rangle = G(\tau) \, .$$

It is straightforward to consider dual quantities which are related by Fourier transformation, leading to

$$x(t) = \frac{1}{2\pi} \int \mathrm{d}\omega \, c(\omega) \, \mathrm{e}^{\mathrm{i}\omega t} \, , \qquad c(\omega) = \int \mathrm{d}t \, x(t) \, \mathrm{e}^{-\mathrm{i}\omega t} \, .$$

We use this relation to derive several important results. Measurements refer to real quantities $x(t)$ and this implies that $c(\omega) = c^*(-\omega)$. From the condition of stationarity, it follows that $\langle x(t)x(t')\rangle = f(t - t')$, so it depends on τ alone and does not depend on t. We then derive

$$\begin{aligned} \big\langle c(\omega)c^*(\omega')\big\rangle &= \frac{1}{(2\pi)^2} \iint \mathrm{d}t \, \mathrm{d}t' \mathrm{e}^{-\mathrm{i}\omega t + \mathrm{i}\omega' t'} \big\langle x(t)x(t')\big\rangle \\ &= \frac{\delta(\omega - \omega')}{2\pi} \int \mathrm{d}\tau \mathrm{e}^{\mathrm{i}\omega\tau} G(\tau) = \delta(\omega - \omega')S(\omega) \, . \end{aligned}$$

The last expression not only relates the mean square $\big\langle |c(\omega)|^2 \big\rangle$ with the spectrum of the random variable; it also shows that stationarity alone implies that $c(\omega)$ and $c^*(\omega')$ are uncorrelated.

3.2 Chapman–Kolmogorov Forward Equations

The basic aim when modeling general stochastic processes is to understand the propagation of probability distributions in time. In particular, the aim is to calculate the probability of going from the random variable $\mathcal{X}_3 = n_3$ at time $t = t_3$ to $\mathcal{X}_1 = n_1$ at time $t = t_1$. It seems natural to assume an intermediate state described by the random variable $\mathcal{X}_2 = n_2$ at $t = t_2$ with the implicit time order $t_1 \geq t_2 \geq t_3$

(Fig. 3.4). The value of the variable $\mathcal{X}_2$, however, need not be unique. In other words, there may be a distribution of values $n_{2i}(i = 1, \ldots, k)$ corresponding to several paths or trajectories leading from (n_3, t_3) to (n_1, t_1). Since we want to model the propagation of a distribution and not a sequence of events leading to a single trajectory, the probability distribution at intermediate times is relevant. Therefore individual values of the random variables are replaced by probabilities, i.e., $\mathcal{X} = n \implies P(\mathcal{X} = n, t) = P(n, t)$, and this yields an equation that encapsulates the full diversity of the various sources of randomness.[16] The only generally assumed restriction in the probability propagating equation is the Markov property of the stochastic process. The equation is called the *Chapman–Kolmogorov equation* after the British geophysicist and mathematician Sydney Chapman and the Russian mathematician Andrey Kolmogorov. In this section we shall be concerned with the various forms of this equation.

The conventional form of the Chapman–Kolmogorov equation considers finite time intervals, for example $\Delta t = t_1 - t_2$, and corresponds therefore to a difference equation at the deterministic level, $\Delta x = G(x, t)\Delta t$. For modeling processes, an equation involving an infinitesimal rather than a finite time interval, viz., $\mathrm{dt} = \lim_{t_2 \to t_1} \Delta t$, is frequently advantageous. In a way, such a differential formulation of basic stochastic processes can be compared to the invention of calculus by Gottfried Wilhelm Leibniz and Isaac Newton, $\lim_{\Delta t \to 0} \Delta x / \Delta t = \mathrm{d}x/\mathrm{dt} = g(x, t)$, which provides the ultimate basis for all modeling by means of differential equations. In analogy we shall derive here a differential form of the Chapman–Kolmogorov equation that represents a prominent node in the tree of models of stochastic processes (Fig. 3.1). Compared to solutions of ODEs, which are commonly continuous and at least once continuously differentiable or $\mathcal{C}^1$ functions, the repertoire of solution curves of stochastic processes is richer and consists of drift, diffusion, and jump processes.

3.2.1 *Differential Chapman–Kolmogorov Forward Equation*

A forward equation predicts the future of a system from given information about the present state, and this is the most common strategy when modeling dynamical phenomena. It allows for direct comparison with experimental data, which in observations are, of course, also recorded in the forward direction. However, there are problems such as the computation of first passage times or the reconstruction of phylogenetic trees that call for an opposite strategy, aiming to reconstruct the past from present day information. In such cases, so-called backward equations facilitate the analysis (see, e.g., Sect. 3.3).

---

[16]Here, we need not yet specify whether the sample space is discrete as in $P_n(t)$, or continuous as in $P(x, t)$, and we indicate this by the notation $P(n, t)$. However, we shall specify the variables in Sect. 3.2.1.

Discrete and Continuous Chapman–Kolmogorov Equations
The relation between the three random variables $\mathcal{A}$, $\mathcal{B}$, and $\mathcal{C}$ can be illustrated by applying set theoretical considerations. Let A, B, and C be the corresponding events and B_k $(k = 1, \ldots, n)$ a partition of B into n mutually exclusive *subevents*. Then, if all events of one kind are included in the summation, the corresponding variable $\mathcal{B}$ is eliminated:

$$\sum_k P(A \cap B_k \cap C) = P(A \cap C) .$$

The relation can be easily verified by means of Venn diagrams. Translating this results into the language of stochastic processes, we assume first that we are dealing with a discrete state space, whence the random variables $\mathcal{X} \in \mathbb{N}$ will be defined on the integers. Then we can simply make use of a state space covering and find for the marginal probability

$$P(n_1, t_1) = \sum_{n_2} P(n_1, t_1; n_2, t_2) = \sum_{n_2} P(n_1, t_1 | \, n_2, t_2) P(n_2, t_2) .$$

Next we introduce a third event (n_3, t_3) (Fig. 3.4) and describe the process by the equations for conditional probabilities, viz.,

$$\begin{aligned} P(n_1, t_1 | \, n_3, t_3) &= \sum_{n_2} P(n_1, t_1; n_2, t_2 | \, n_3, t_3) \\ &= \sum_{n_2} P(n_1, t_1 | \, n_2, t_2; n_3, t_3) P(n_2, t_2 | \, n_3, t_3) . \end{aligned}$$

Both equations are of general validity for all stochastic processes, and the series could be extended further to four, five, or more events. Finally, adopting the Markov assumption and introducing the time order $t_1 \geq t_2 \geq t_3$ provides the basis for dropping the dependence on (n_2, t_2) in the doubly conditioned probability, whence

$$P(n_1, t_1 | n_3, t_3) = \sum_{n_2} P(n_1, t_1 | n_2, t_2) P(n_2, t_2 | n_3, t_3) . \tag{3.35}$$

This is the *Chapman–Kolmogorov equation* in its simplest general form. Equation (3.35) can be interpreted as a matrix multiplication $\mathrm{C} = \mathrm{AB}$ with $c_{ij} = \sum_{k=1}^{m} a_{ik} b_{kj}$, where the eliminated dimension m of the matrices reflects the size of the event space of the eliminated variable n_2, which may even be countably infinite.

The extension from the discrete case to probability densities is straightforward. By the same token we find for the continuous case

$$p(x_1, t_1) = \int dx_2 p(x_1, t_1; x_2, t_2) = \int dx_2 p(x_1, t_1|x_2, t_2) p(x_2, t_2) ,$$

while the extension to three events leads to

$$p(x_1, t_1|x_3, t_3) = \int dx_2\, p(x_1, t_1; x_2, t_2|x_3, t_3)$$
$$= \int dx_2\, p(x_1, t_1|x_2, t_2; x_3, t_3) p(x_2, t_2|x_3, t_3) .$$

For $t_1 \geq t_2 \geq t_3$, and making use of the Markov property once again, we obtain the continuous version of the Chapman–Kolmogorov equation:

$$p(x_1, t_1|x_3, t_3) = \int dx_2 p(x_1, t_1|x_2, t_2) p(x_2, t_2|x_3, t_3) . \tag{3.36}$$

Equation (3.36) is of a very general nature. The only relevant approximation is the assumption of a Markov process, which is empirically well justified for most applications in physics, chemistry, and biology. General validity is commonly accompanied by a variety of different solutions, and the Chapman–Kolmogorov equation is no exception in this respect. The generality of (3.36) in the description of a stochastic process becomes evident when the evolution in time is continued $t_1 \geq t_2 \geq t_3 \geq t_4 \geq t_5 \ldots$, and complete summations over all intermediate states are performed:

$$p(x_1, t_1|x_n, t_n) = \int dx_2 \cdots \int dx_{n-1} p(x_1, t_1|x_2, t_2) \ldots p(x_{n-1}, t_{n-1}|x_n, t_n) .$$

It is sometimes useful to indicate an initial state by the doublet (x_0, t_0) instead of (x_n, t_n) and apply the physical notation of time. We shall adopt this notation here.

Differential Chapman–Kolmogorov Forward Equation

Although the conventional Chapman–Kolmogorov equations in discrete and continuous form as expressed by (3.35) and (3.36), respectively, provide a general definition of Markov processes, they are not always useful for describing temporal evolution. Much better suited and more flexible are equations in differential form for describing stochastic processes, for analyzing the nature and the properties of solutions, and for performing actual calculations. Analytical solution or numerical integration of such a *differential Chapman–Kolmogorov equation* (dCKE) is then expected to provide the desired description of the process. A differential form

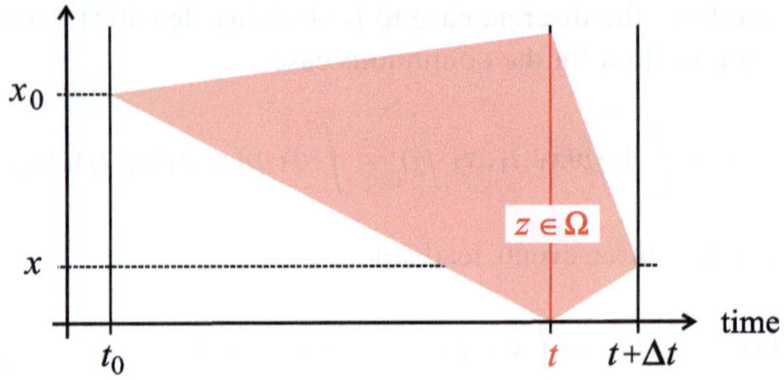

Fig. 3.6 Time order in the differential Chapman–Kolmogorov equation (dCKE). The one-dimensional sketch shows the notation used in the derivation of the forward dCKE. The variable $\mathbf{z}$ is integrated over the entire sample space Ω in order to sum up all trajectories leading from $(\mathbf{x}_0, t_0)$ via $(\mathbf{z}, t)$ to $(\mathbf{x}, t + \Delta t)$

of the Chapman–Kolmogorov equation has been derived by Crispin Gardiner [194, pp. 48–51].[17] We shall follow here, in essence, a somewhat simpler approach given recently by Mukhtar Ullah and Olaf Wolkenhauer [535, 536].

The Chapman–Kolmogorov equation is defined for a sample space Ω and considered on the interval $t_0 \to t + \Delta t$ with $(\mathbf{x}_0, t_0)$ as initial conditions:

$$p\big(\mathbf{x}, t + \Delta t | \, \mathbf{x}_0, t_0\big) = \int_{\Omega} \mathrm{d}\mathbf{z} \, p\big(\mathbf{z}, t + \Delta t | \, \mathbf{x}_0, t_0\big) p\big(\mathbf{z}, t | \, \mathbf{x}_0, t_0\big) \,, \tag{3.36$'$}$$

whereby we assume that the consistency equation (3.22) is satisfied. As illustrated in Fig. 3.6, the probability of the transition from $(\mathbf{x}_0, t_0)$ to $(\mathbf{x}, t + \Delta t)$ is obtained by integrating over all probabilities of occurring via an intermediate, $(\mathbf{x}_0, t_0) \to (\mathbf{z}, t) \to (\mathbf{x}, t + \Delta t)$. In order to simplify the derivation and the notation, we shall assume fixed and sharp initial conditions $(\mathbf{x}_0, t_0)$. In other words, the unconditioned probability of the state $(\mathbf{x}, t)$ is the same as the probability of the transition from $(\mathbf{x}_0, t_0) \to (\mathbf{x}, t)$:

$$p(\mathbf{x}, t) = p(\mathbf{x}, t | \mathbf{x}_0, t_0) \,, \quad \text{with } p(\mathbf{x}, t_0) = \delta(\mathbf{x} - \mathbf{x}_0). \tag{3.37}$$

We write the time derivative by assuming that the probability $p(\mathbf{x}, t)$ is differentiable with respect to time:

$$\frac{\partial}{\partial t} p(\mathbf{x}, t) = \lim_{\Delta t \to 0} \frac{1}{\Delta t} \Big(p(\mathbf{x}, t + \Delta t) - p(\mathbf{x}, t) \Big) \,. \tag{3.38}$$

[17]The derivation is already contained in the first edition of Gardiner's *Handbook of Stochastic Methods* [193], and it was Crispin Gardiner who coined the term *differential Chapman–Kolmogorov equation*.

Introducing the CKE in the form (3.36′) and multiplying $p(\mathbf{x}, t)$ formally by one in the form of the normalization condition of probabilities, i.e.,[18]

$$1 = \int_{\Omega} \mathrm{d}\mathbf{z}\, p\big(\mathbf{z}, t + \Delta t | \, \mathbf{x}, t\big) \, ,$$

we can rewrite (3.38) as

$$\frac{\partial}{\partial t} p(\mathbf{x}, t) = \lim_{\Delta t \to 0} \frac{1}{\Delta t} \int_{\Omega} \mathrm{d}\mathbf{z} \Big(p\big(\mathbf{x}, t + \Delta t | \, \mathbf{z}, t\big) p(\mathbf{z}, t) - p\big(\mathbf{z}, t + \Delta t | \, \mathbf{x}, t\big) p(\mathbf{x}, t) \Big) \, . \tag{3.39}$$

For the purpose of integration, the sample space Ω is divided into parts with respect to an arbitrarily small parameter $\epsilon > 0$: $\Omega = I_1 + I_2$. Using the notion of continuity (Sect. 3.1.5), the region I_1 defined by $\|\mathbf{x} - \mathbf{z}\| < \epsilon$ represents a continuous process.[19] In the second part of the sample space Ω, I_2 with $\|\mathbf{x} - \mathbf{z}\| \geq \epsilon$, the norm cannot become arbitrarily small, indicating a jump process. For the derivative taken on the entire sample space Ω, we get

$$\frac{\partial}{\partial t} p(\mathbf{x}, t) = I_1 + I_2 \, ,$$

with

$$I_1 = \lim_{\Delta t \to 0} \frac{1}{\Delta t} \int_{\|\mathbf{x} - \mathbf{z}\| < \epsilon} \mathrm{d}\mathbf{z} \Big(p\big(\mathbf{x}, t + \Delta t | \, \mathbf{z}, t\big) p(\mathbf{z}, t) - p\big(\mathbf{z}, t + \Delta t | \, \mathbf{x}, t\big) p(\mathbf{x}, t) \Big) \, ,$$

$$I_2 = \lim_{\Delta t \to 0} \frac{1}{\Delta t} \int_{\|\mathbf{x} - \mathbf{z}\| \geq \epsilon} \mathrm{d}\mathbf{z} \Big(p\big(\mathbf{x}, t + \Delta t | \, \mathbf{z}, t\big) p(\mathbf{z}, t) - p\big(\mathbf{z}, t + \Delta t | \, \mathbf{x}, t\big) p(\mathbf{x}, t) \Big) \, . \tag{3.40}$$

In the first region I_1 with $\|\mathbf{x} - \mathbf{z}\| < \epsilon$, we introduce $\mathbf{u} = \mathbf{x} - \mathbf{z}$ with $\mathrm{d}\mathbf{u} = \mathrm{d}\mathbf{x}$ and notice a symmetry in the integral, since $\|\mathbf{x} - \mathbf{z}\| = \|\mathbf{z} - \mathbf{x}\|$, that will be used in the forthcoming derivation:

$$I_1 = \lim_{\Delta t \to 0} \frac{1}{\Delta t} \int_{\|\mathbf{u}\| < \epsilon} \mathrm{d}\mathbf{u} \Big(p\big(\mathbf{x}, t + \Delta t | \, \mathbf{x} - \mathbf{u}, t\big) p(\mathbf{x} - \mathbf{u}, t) - p\big(\mathbf{x} - \mathbf{u}, t + \Delta t | \, \mathbf{x}, t\big) p(\mathbf{x}, t) \Big) \, .$$

---

[18]It is important to note a useful trick in the derivation: by substituting the 1, the time order is reversed in the integral.

[19]The notation $\| \cdot \|$ refers to a suitable vector norm, here the L^1 norm given by $\|\mathbf{y}\| = \sum_k |y_k|$. In the one-dimensional case, we would just use the absolute value $|y|$.

For convenience, we now define

$$f(\mathbf{x};\mathbf{u}) \doteq p\big(\mathbf{x}+\mathbf{u}, t+\Delta t|\mathbf{x}, t\big)p(\mathbf{x}, t)\ .$$

Inserting into the equation for I_1, this yields

$$I_1 = \lim_{\Delta t\to 0}\frac{1}{\Delta t}\int_{\|\mathbf{u}\|<\epsilon} \mathrm{d}\mathbf{u}\Big(f(\mathbf{x}-\mathbf{u},\mathbf{u}) - f(\mathbf{x},-\mathbf{u})\Big) = \lim_{\Delta t\to 0}\frac{1}{\Delta t}\int_{\|\mathbf{u}\|<\epsilon} \mathrm{d}\mathbf{u}\, F(\mathbf{x},\mathbf{u})\ .$$

Next the integrand is expanded in a Taylor series in $-\mathbf{u}$ about $\mathbf{u} = 0$ [20]:

$$\begin{aligned} F(\mathbf{x},\mathbf{u}) =& f(\mathbf{x};\mathbf{u}) - f(\mathbf{x};-\mathbf{u}) - \sum_i u_i \frac{\partial f(\mathbf{x};\mathbf{u})}{\partial x_i} \\ &+ \frac{1}{2!}\sum_{i,j} u_i u_j \frac{\partial^2 f(\mathbf{x};\mathbf{u})}{\partial x_i \partial x_j} - \frac{1}{3!}\sum_{i,j,k} u_i u_j u_k \frac{\partial^3 f(\mathbf{x};\mathbf{u})}{\partial x_i \partial x_j \partial x_k} + \cdots\ . \end{aligned}$$

Insertion into the integral I_1 yields

$$\begin{aligned} I_1 = \lim_{\Delta t\to 0}\frac{1}{\Delta t}\int_{\|\mathbf{u}<\epsilon\|} \mathrm{d}\mathbf{u}\Bigg(&\Big(p\big(\mathbf{x}+\mathbf{u}, t+\Delta t|\mathbf{x}, t\big) - p\big(\mathbf{x}-\mathbf{u}, t+\Delta t|\mathbf{x}, t\big)\Big)p(\mathbf{x}, t) \\ &- \sum_i u_i \frac{\partial}{\partial x_i}\Big(p\big(\mathbf{x}+\mathbf{u}, t+\Delta t|\mathbf{x}, t\big)p(\mathbf{x}, t)\Big) \\ &+ \frac{1}{2!}\sum_{i,j} u_i u_j \frac{\partial^2}{\partial x_i \partial x_j}\Big(p\big(\mathbf{x}+\mathbf{u}, t+\Delta t|\mathbf{x}, t\big)p(\mathbf{x}, t)\Big) \\ &- \frac{1}{3!}\sum_{i,j,k} u_i u_j u_k \frac{\partial^3}{\partial x_i \partial x_j \partial x_k}\Big(p\big(\mathbf{x}+\mathbf{u}, t+\Delta t|\mathbf{x}, t\big)p(\mathbf{x}, t)\Big) + \cdots\Bigg)\ . \end{aligned}$$

Integration over the entire domain $\|\mathbf{u}\| < \epsilon$ simplifies the expression since the term of order zero vanishes by symmetry: $\int f(\mathbf{x};\mathbf{u})\mathrm{d}\mathbf{u} = \int f(\mathbf{x};-\mathbf{u})\mathrm{d}\mathbf{u}$. In addition, all terms of third and higher orders are of $\mathcal{O}(\epsilon)$ and can be neglected [194, pp. 47–48] when we take the limit $\Delta t \to 0$.

[20] Differentiation with respect to $\mathbf{x}$ has to be done with respect to the components x_i. Note that $\mathbf{u}$ vanishes through integration.

In the next step, we compute the expectation values of the increments $\mathcal{X}_i(t+\Delta t)-\mathcal{X}_i(t)$ in the random variables by choosing Δt in the forward direction (different from Fig. 3.6):

$$\big\langle \mathcal{X}_i(t+\Delta t) - \mathcal{X}_i(t) \big| \boldsymbol{\mathcal{X}} = \mathbf{x} \big\rangle = \int_{\|\mathbf{u}<\epsilon\|} \mathrm{d}\mathbf{u}\, u_i p\big(\mathbf{x}+\mathbf{u}, t+\Delta t|\mathbf{x}, t\big) \ ,$$

$$\Big\langle \big(\mathcal{X}_i(t+\Delta t) - \mathcal{X}_i(t)\big)\big(\mathcal{X}_j(t+\Delta t) - \mathcal{X}_j(t)\big) \big| \boldsymbol{\mathcal{X}} = \mathbf{x} \Big\rangle$$

$$= \int_{\|\mathbf{u}<\epsilon\|} \mathrm{d}\mathbf{u}\, u_i u_j p\big(\mathbf{x}+\mathbf{u}, t+\Delta t|\mathbf{x}, t\big) \ .$$

Making use of the differentiability condition (3.25) for continuous processes, $\|\mathbf{x}-\mathbf{z}\| < \epsilon$, we now take the limit $\Delta t \to 0$:

$$\lim_{\Delta t \to 0} \frac{\big\langle \mathcal{X}_i(t+\Delta t) - \mathcal{X}_i(t) | \boldsymbol{\mathcal{X}} = \mathbf{x} \big\rangle}{\Delta t} = A_i(\mathbf{x}, t) + \mathcal{O}(\epsilon) \ . \tag{3.41a}$$

The second order term takes the form

$$\lim_{\Delta t \to 0} \frac{\Big\langle \big(\mathcal{X}_i(t+\Delta t) - \mathcal{X}_i(t)\big)\big(\mathcal{X}_j(t+\Delta t) - \mathcal{X}_j(t)\big) \big| \boldsymbol{\mathcal{X}} = \mathbf{x} \Big\rangle}{\Delta t} = B_{ij}(\mathbf{x}, t) + \mathcal{O}(\epsilon) \ . \tag{3.41b}$$

In the limit $\epsilon \to 0$, the continuous part of the process encapsulated in I_1 becomes equivalent to an equation for the differential increments of the random vector $\boldsymbol{\mathcal{X}}(t)$ describing a single trajectory:

$$\boldsymbol{\mathcal{X}}(t+\mathrm{d}t) = \boldsymbol{\mathcal{X}}(t) + \mathbf{A}\big(\boldsymbol{\mathcal{X}}(t), t\big)\,\mathrm{d}t + \Big(\mathrm{B}\big(\boldsymbol{\mathcal{X}}(t), t\big)\,\mathrm{d}t\Big)^{1/2} \ . \tag{3.42}$$

In the terminology used in physics, $\mathbf{A}$ is the *drift vector* and B is the *diffusion matrix* of the stochastic process. In other words, for $\epsilon \to 0$ and continuity of the process, the expectation value of the increment vector expressed by $\boldsymbol{\mathcal{X}}(t+\mathrm{d}t) - \boldsymbol{\mathcal{X}}(t)$ approaches $\mathbf{A}\big(\boldsymbol{\mathcal{X}}(t), t\big)\,\mathrm{d}t$ and its covariance converges to $\mathrm{B}\big(\boldsymbol{\mathcal{X}}(t), t\big)\,\mathrm{d}t$. Writing $\boldsymbol{\mathcal{X}}(t+\mathrm{d}t) - \boldsymbol{\mathcal{X}}(t) = \mathrm{d}\boldsymbol{\mathcal{X}}(t)$ shows that (3.42) is a stochastic differential equation (SDE) or Langevin equation, named after the French mathematician Paul Langevin. Section 3.4.1 discusses the relationship between the differential Chapman–Kolmogorov equations and stochastic differential equations. Here we point out the fact that the diffusion term of the SDE contains the differential $\sqrt{\mathrm{d}t}$ and the function is the square root of the diffusion matrix, i.e., $\sqrt{\mathrm{B}\big(\boldsymbol{\mathcal{X}}(t), t\big)}$.

As mentioned above, provided the differentiability conditions are satisfied, in the limit $\epsilon \to 0$, the integral I_1 is found to be

$$I_1 = -\sum_i \frac{\partial}{\partial x_i}\big(A_i(\mathbf{x}, t)p(\mathbf{x}, t)\big) + \frac{1}{2}\sum_i\sum_j \frac{\partial^2}{\partial x_i \partial x_j}\big(B_{ij}(\mathbf{x}, t)p(\mathbf{x}, t)\big) . \tag{3.43}$$

These are the expressions that finally show up in the Fokker–Planck equation.

The second part of the integration over sample space Ω involves the probability of jumps:

$$I_2 = \lim_{\Delta t \to 0} \frac{1}{\Delta t} \int_{\|\mathbf{x}-\mathbf{z}\| \geq \epsilon} d\mathbf{z}\Big(p\big(\mathbf{x}, t + \Delta t|\mathbf{z}, t\big)p(\mathbf{z}, t) - p\big(\mathbf{z}, t + \Delta t|\mathbf{x}, t\big)p(\mathbf{x}, t)\Big) .$$

The condition for a jump process is $\|\mathbf{x} - \mathbf{z}\| \geq \epsilon$ (Sect. 3.1.5), and accordingly we have

$$\lim_{\Delta t \to 0} \frac{1}{\Delta t}\Big(p\big(\mathbf{x}, t + \Delta t|\mathbf{z}, t\big)p(\mathbf{z}, t)\Big) = W(\mathbf{x}|\mathbf{z}, t)p(\mathbf{z}, t) , \tag{3.44}$$

where $W(\mathbf{x}|\, \mathbf{z}, t)$ is called the *transition probability* for a jump $\mathbf{z} \to \mathbf{x}$. By the same token, we define a transition probability for the jump in the reverse direction $\mathbf{x} \to \mathbf{z}$. As $\epsilon \to 0$ the integration is extended over the whole of the sample space Ω, and finally we obtain

$$\lim_{\epsilon \to 0} I_2 = ⨍_{\Omega} d\mathbf{z}\Big(W(\mathbf{x}|\mathbf{z}, t)p(\mathbf{z}, t) - W(\mathbf{z}|\mathbf{z}, t)p(\mathbf{x}, t)\Big) . \tag{3.45}$$

Our somewhat simplified derivation of the differential Chapman–Kolmogorov equation is thus complete. It is important to notice that we are using a *principal value integral* here, since the transition probability may approach infinity in the limit $\epsilon \to 0$ or $\mathbf{z} \to \mathbf{x}$, as happens for the Cauchy process, where we have $W(\mathbf{x}|z, t) = 1/\pi(x - z)^2$.

Surface terms at the boundary of the domain of $\mathbf{x}$ have been neglected in the derivation [194, p. 50]. This assumption is not critical for most cases considered here, and it is always correct for infinite domains because the probabilities vanish in the limit $\lim_{\mathbf{x}\to\pm\infty} p(\mathbf{x}, t) = 0$. However, we shall encounter special boundaries in systems with finite sample spaces and discuss the specific boundary effects there.

The evolution of the system is now expressed in terms of functions $\mathbf{A}(\mathbf{x}, t)$ which correspond to the functional relations in conventional differential equations, a diffusion matrix $B(\mathbf{x}, t)$, and a transition matrix for discontinuous jumps $W(\mathbf{x}|\mathbf{z}, t)$:

$$\frac{\partial p(\mathbf{x},t)}{\partial t} = -\sum_i \frac{\partial}{\partial x_i}\Big(A_i(\mathbf{x},t)p(\mathbf{x},t)\Big) \tag{3.46a}$$

$$+\frac{1}{2}\sum_{i,j}\frac{\partial^2}{\partial x_i \partial x_j}\Big(B_{ij}(\mathbf{x},t)p(\mathbf{x},t)\Big) \tag{3.46b}$$

$$+\int_\Omega \mathrm{d}\mathbf{z}\Big(W(\mathbf{x}|\mathbf{z},t)p(\mathbf{z},t) - W(\mathbf{z}|\mathbf{x},t)p(\mathbf{x},t)\Big)\,. \tag{3.46c}$$

Equation (3.46) is called a *forward equation* in the sense of Fig. 3.21.

Properties of the Differential Chapman–Kolmogorov Equation
From a mathematical purist's point of view, it is not clear from the derivation that solutions of the differential Chapman–Kolmogorov equation (3.46) actually exist, and nor is it clear whether they are unique and solutions to the Chapman–Kolmogorov equation (3.36) as well. It is true, however, that the set of conditional probabilities obeying (3.46) does generate a Markov process in the sense that the joint probabilities produced satisfy all the probability axioms. It has been shown that a nonnegative solution to the differential Chapman–Kolmogorov equations exists and satisfies the Chapman–Kolmogorov equation under certain conditions (see [205, Vol. II]):

(i) $\mathbf{A}(\mathbf{x},t) = \big(A_i(\mathbf{x},t);\ i = 1,\ldots\big)$ and $B(\mathbf{x},t) = \big(B_{ij}(\mathbf{x},t);\ i,j = 1,\ldots\big)$ are vectors and positive semidefinite matrices of functions, respectively.[21]
(ii) $W(\mathbf{x}|\mathbf{z},t)$ and $W(\mathbf{z}|\mathbf{x},t)$ are nonnegative quantities.
(iii) The initial condition has to satisfy $p(\mathbf{x},t_0|\mathbf{x}_0,t_0) = \delta(\mathbf{x}_0 - \mathbf{x})$.
(iv) Appropriate boundary conditions have to be satisfied.

General boundary conditions are hard to specify for the full equation, but can be discussed precisely for special cases, for example, in the case of the Fokker–Planck equation [468]. Sharp initial conditions facilitate solution, but a general probability distribution can also be used as initial condition.

[21] A positive definite matrix has exclusively positive eigenvalues $\lambda_k > 0$, whereas a positive semidefinite matrix has nonnegative eigenvalues $\lambda_k \geq 0$.

The nature of the different stochastic processes associated with the three terms in (3.46), viz., $\mathbf{A}(\mathbf{x}, t)$, $\mathrm{B}(\mathbf{x}, t)$, $\mathrm{W}(\mathbf{x}|\mathbf{z}, t)$ and $\mathrm{W}(\mathbf{z}|\mathbf{x}, t)$, is visualized by setting some parameters equal to zero and analyzing the remaining equation. We shall discuss here four cases that are modeled by different equations (for the relations between them, see Fig. 3.1).

1. $\mathrm{B} = 0$, $\mathrm{W} = 0$, deterministic drift process: Liouville equation.
2. $\mathbf{A} = 0$, $\mathrm{W} = 0$, drift-free diffusion or Wiener process: diffusion equation.
3. $\mathrm{W} = 0$, drift and diffusion process: Fokker–Planck equation.
4. $\mathbf{A} = 0$, $\mathrm{B} = 0$, pure jump process: master equation.

The first term (3.46a) in the differential Chapman–Kolmogorov equation is the probabilistic version of a differential equation describing deterministic motion, which is known as the *Liouville equation*, named after the French mathematician Joseph Liouville. It is a fundamental equation of statistical mechanics and will be discussed in some detail Sect. 3.2.2.1. With respect to the theory of stochastic processes (3.46a), it encapsulates the drift of a probability distribution.

The second term in (3.46) deals with the spreading of probability densities by diffusion, and is called a stochastic *diffusion equation*. In pure form, it describes a Wiener process, which can be understood as the continuous time and space limit of the one-dimensional random walk (see Fig. 3.3). The pure diffusion process got its name from the American mathematician Norbert Wiener. The Wiener process is fundamental for understanding stochasticity in continuous space and time, and will be discussed in Sect. 3.2.2.3.

Combining (3.46a) and (3.46b) yields the Fokker–Planck equation, which we repeat here because of its general importance:

$$\frac{\partial p(\mathbf{x}, t)}{\partial t} = -\sum_i \frac{\partial}{\partial x_i}\Big(A_i(\mathbf{x}, t)p(\mathbf{x}, t)\Big) + \frac{1}{2}\sum_{i,j} \frac{\partial^2}{\partial x_i \partial x_j}\Big(B_{ij}(\mathbf{x}, t)p(\mathbf{x}, t)\Big) . \tag{3.47}$$

Fokker–Planck equations are frequently used in physics to model and analyze processes with fluctuations [468] (Sect. 3.2.2.3).

If only the third term (3.46c) of the differential Chapman–Kolmogorov equation has nonzero elements, the variables $\mathbf{x}$ and $\mathbf{z}$ change exclusively in steps and the corresponding differential equation is called a *master equation*. Master equations are the most important tools for describing processes $\mathcal{X}(t) \in \mathbb{N}$ in discrete spaces. We shall devote a whole section to master equations (Sect. 3.2.3) and discuss specific examples in Sects. 3.2.2.4 and 3.2.4. In particular, master equations are indispensable for modeling chemical reactions or biological processes with small particle numbers. Specific applications in chemistry and biology will be presented in two separate chapters (Chaps. 4 and 5).

It is important to stress that the mathematical expressions for the three contributions to the general stochastic process represent a pure formalism that can be applied equally well to problems in physics, chemistry, biology, sociology, economics, or other disciplines. Specific empirical knowledge enters the model in the form of the parameters: the drift vector $\mathbf{A}$, the diffusion matrix B, and the jump transition matrix W. By means of examples, we shall show how physical laws are encapsulated in regularities among the parameters.

3.2.2 *Examples of Stochastic Processes*

In this section we present examples of stochastic processes with characteristic properties that will be useful as references in the forthcoming applications: (1) the Liouville process, (2) the Wiener process, (3) the Ornstein–Uhlenbeck process, and (4) the Poisson process.

3.2.2.1 Liouville Process

The Liouville equation[22] is a straightforward link between deterministic motion and stochastic processes. As indicated in Fig. 3.1 all elements of the jump transition matrix W and the diffusion matrix B are zero and what remains is a differential equation falling into the class of *Liouville equations* from classical mechanics. A Liouville equation is commonly used to describe the deterministic motion of particles in *phase space*.[23] Following [194, p. 54], we show that deterministic trajectories are identical to solutions of the differential Chapman–Kolmogorov equation with $\mathrm{B} = 0$ and $\mathrm{W} = 0$ and relate the result to Liouville's theorem in classical mechanics [352, 353].

First we consider deterministic motion as described by the differential equation

$$\begin{aligned} \frac{\mathrm{d}\boldsymbol{\xi}(t)}{\mathrm{d}t} &= \mathbf{A}\big(\boldsymbol{\xi}(t), t\big) \,, \quad \text{with } \boldsymbol{\xi}(t_0) = \boldsymbol{\xi}_0 \,, \\ \boldsymbol{\xi}(t) &= \boldsymbol{\xi}_0 + \int_{t_0}^{t} \mathrm{d}\boldsymbol{\xi}\; \mathbf{A}\big(\boldsymbol{\xi}(t), t\big) \,. \end{aligned} \tag{3.48}$$

This can be understood as a degenerate Markov process in which the probability distribution degenerates to a Dirac delta function[24] $p(\mathbf{x}, t) = \delta\big(\mathbf{x} - \boldsymbol{\xi}(t)\big)$. We may relax the initial conditions $\boldsymbol{\xi}(t_0) = \boldsymbol{\xi}_0$ or $p(\mathbf{x}, t_0) = \delta(\mathbf{x} - \boldsymbol{\xi}_0)$ to $p(\mathbf{x}, t_0) = p(\mathbf{x}_0)$,

---

[22]The idea of the Liouville equation was first discussed by Josiah Willard Gibbs [202].

[23]Phase space is an abstract space, which is particularly useful for visualizing particle motion. The six independent coordinates of particle S_k are the position coordinates $\mathbf{q}_k = (q_{k1}, q_{k2}, q_{k3})$ and the (linear) momentum coordinates $\mathbf{p}_k = (p_{k1}, p_{k2}, p_{k3})$. In Cartesian coordinates, they are $\mathbf{q}_k = (x_k, y_k, z_k)$ and $\mathbf{p}_k = m_k \mathbf{v}_k$, where $\mathbf{v} = (v_x, v_y, v_z)$ is the velocity vector.

[24]For simplicity, we write $p(\mathbf{x}, t)$ instead of the conditional probability $p(\mathbf{x}, t|\mathbf{x}_0, t_0)$ whenever the initial condition $(\mathbf{x}_0, t_0)$ refers to the sharp density $p(\mathbf{x}, t_0) = \delta(\mathbf{x} - \mathbf{x}_0)$.

and then the result is a distribution migrating through space with unchanged shape (Fig. 3.7) instead of a delta function travelling on a single trajectory (see (3.53′) below).

By setting B = 0 and W = 0 in the dCKE, we obtain for the Liouville process

$$\frac{\partial p(\mathbf{x},t)}{\partial t} = -\sum_i \frac{\partial}{\partial x_i}\Big(A_i(\mathbf{x},t)p(\mathbf{x},t)\Big) . \tag{3.49}$$

The goal is now to show equivalence with the differential equation (3.48) in the form of the common solution

$$p(\mathbf{x},t) = \delta\big(\mathbf{x} - \boldsymbol{\xi}(t)\big) . \tag{3.50}$$

The proof is done by direct substitution:

$$-\sum_i \frac{\partial}{\partial x_i}\Big(A_i(\mathbf{x},t)\delta\big(\mathbf{x}-\boldsymbol{\xi}(t)\big)\Big) = -\sum_i \frac{\partial}{\partial x_i}\Big(A_i(\boldsymbol{\xi}(t),t)\delta\big(\mathbf{x}-\boldsymbol{\xi}(t)\big)\Big)$$
$$= -\sum_i \left(A_i(\boldsymbol{\xi}(t),t)\frac{\partial}{\partial x_i}\delta\big(\mathbf{x}-\boldsymbol{\xi}(t)\big)\right) ,$$

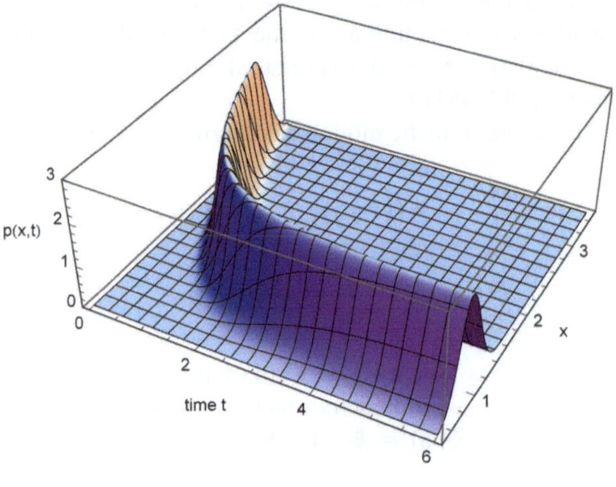

Fig. 3.7 Probability density of a Liouville process. The figure shows the migration of a normal distribution $p(x) = \sqrt{k/\pi s^2}\exp(-k(x-\mu)^2/s^2)$ along a trajectory corresponding to the expectation value of an Ornstein–Uhlenbeck process: $\xi(t) = \mu + (\xi_0 - \mu)\exp(-kt)$ (Sect. 3.2.2.3). The expression for the density is

$$p(x,t) = \sqrt{\frac{k}{\pi s^2}}\mathrm{e}^{-k\left(x-\mu-(\xi_0-\mu)\exp(-kt)\right)/s^2} ,$$

and the long-time limit $p(\overline{x})$ of the distribution is a normal distribution with mean $\mathrm{E}(\overline{x}) = \mu$ and variance $\mathrm{var}(\overline{x}) = \sigma^2 = s^2/2k$. Choice of parameters: $\xi_0 = 3\,[1]$, $k = 1\,[\mathrm{t}]^{-1}$, $\mu = 1\,[1]$, $s = 1/4\,[\mathrm{t}]^{1/2}$

since $\boldsymbol{\xi}$ does not depend on $\mathbf{x}$ and

$$\frac{\partial p(\mathbf{x},t)}{\partial t} = \frac{\partial}{\partial t}\delta\big(\mathbf{x}-\boldsymbol{\xi}(t)\big) = -\sum_i \left(\frac{\mathrm{d}\xi_i(t)}{\mathrm{d}t}\frac{\partial}{\partial x_i}\delta\big(\mathbf{x}-\boldsymbol{\xi}(t)\big)\right) .$$

Making use of (3.48) in component form, viz.,

$$\frac{\mathrm{d}\xi_i(t)}{\mathrm{d}t} = A_i\big(\boldsymbol{\xi}(t),t\big)$$

we see that the sums in the expressions in the last two lines are equal. □

The following part on Liouville's equation illustrates how empirical science, here Newtonian mechanics, enters a formal stochastic equation. In Hamiltonian mechanics [232, 233], dynamical systems may be represented by a *density function* or *classical density matrix* $\varrho(\mathbf{q},\mathbf{p})$ in phase space. The density function allows one to calculate system properties. It is usually normalized so that the expected total number of particles is the integral over phase space:

$$N = \int\cdots\int \varrho(\mathbf{q},\mathbf{p})(\mathrm{d}q)^n(\mathrm{d}p)^n .$$

The evolution of the system is described by a time dependent density that is commonly denoted by $\varrho\big(\mathbf{q}(t),\mathbf{p}(t),t\big)$ with the initial conditions $\varrho\big(\mathbf{q}_0,\mathbf{p}_0,t_0\big)$. For a single particle S_k, the generalized spatial coordinates q_{ki} are related to conjugate momenta p_{ki} by Newton's equations of motion:

$$\frac{\mathrm{d}p_{ki}}{\mathrm{d}t} = f_{ki}(\mathbf{q}) , \quad \frac{\mathrm{d}q_{ki}}{\mathrm{d}t} = \frac{1}{m_k}p_{ki} , \quad i=1,2,3 , \; k=1,\ldots,n , \tag{3.51}$$

where f_{ki} is the component of the force acting on particle S_k in the direction of q_{ki} and m_k the particle mass. Liouville's theorem, which follows from the Hamiltonian mechanics of an n-particle system, makes a statement about the evolution of the density ϱ:

$$\frac{\mathrm{d}\varrho(\mathbf{q},\mathbf{p},t)}{\mathrm{d}t} = \frac{\partial\varrho}{\partial t} + \sum_{k=1}^{n}\sum_{i=1}^{3}\left(\frac{\partial\varrho}{\partial q_{ki}}\frac{\mathrm{d}q_{ki}}{\mathrm{d}t} + \frac{\partial\varrho}{\partial p_{ki}}\frac{\mathrm{d}p_{ki}}{\mathrm{d}t}\right) = 0 . \tag{3.52}$$

The density function does not change with time. It is a constant of the motion and therefore constant along the trajectory in phase space.

We can now show that (3.52) can be transformed into a Liouville equation (3.49). We insert the individual time derivatives and find

$$\frac{\partial\varrho(\mathbf{q},\mathbf{p},t)}{\partial t} = -\sum_{k=1}^{n}\sum_{i=1}^{3}\left(\frac{1}{m_i}p_{ki}\frac{\partial}{\partial q_{ki}}\varrho(\mathbf{q},\mathbf{p},t) + f_{ki}\frac{\partial}{\partial p_{ki}}\varrho(\mathbf{q},\mathbf{p},t)\right) . \tag{3.53}$$

Equation (3.53) already has the form of a differential Chapman–Kolmogorov equation (3.49) with B $= 0$ and W $= 0$, as follows from

$$\varrho(\mathbf{q}, \mathbf{p}, t) \equiv p(\mathbf{x}, t) ,$$
$$\mathbf{x} \equiv (q_{11}, \dots, q_{n3}, p_{11}, \dots, p_{n3}) ,$$
$$\mathbf{A} \equiv \left(\frac{1}{m_1} p_{11}, \dots, \frac{1}{m_n} p_{n3}, f_{11}, \dots, f_{n3} \right) ,$$

where the $6n$ components of $\mathbf{x}$ represent the $3n$ coordinates for the positions and the $3n$ coordinates for the linear momenta of n particles. Finally, we indicate the relationship between the probability density $p(\mathbf{x}, t)$ and (3.48) and (3.49): the density function is the expectation value of the probability distribution, i.e.,

$$\mathbf{x}(t) = \big(\mathbf{q}(t), \mathbf{p}(t)\big) = \mathrm{E}\Big(\varrho\big(\mathbf{q}(t), \mathbf{p}(t), t\big)\Big) , \tag{3.54}$$

and it satisfies the Liouville ODE as well as the Chapman–Kolmogorov equation:

$$\begin{aligned} \frac{\partial p(\mathbf{x}, t)}{\partial t} &\equiv \frac{\partial \varrho(\mathbf{q}, \mathbf{p}, t)}{\partial t} \\ &= -\sum_{i=1}^{3n} \left(\frac{1}{m_i} p_i \frac{\partial}{\partial q_i} \varrho(\mathbf{q}, \mathbf{p}, t) + f_i \frac{\partial}{\partial p_i} \varrho(\mathbf{q}, \mathbf{p}, t) \right) \\ &= -\sum_{i=1}^{6n} \frac{\partial}{\partial x_i} \Big(A_i(\mathbf{x}, t) p(\mathbf{x}, t) \Big) , \end{aligned} \tag{3.53'}$$

$$\frac{\mathrm{d}\mathbf{x}(t)}{\mathrm{d}t} = \mathbf{A}\big(\mathbf{x}(t), t\big) . \tag{3.51'}$$

In other words, the Liouville equation states that the density matrix $\varrho(\mathbf{q}, \mathbf{p}, t)$ in phase space is conserved in classical motion. This result is illustrated for a normal density in Fig. 3.7.

3.2.2.2 Wiener Process

The Wiener process named after the American mathematician and logician Norbert Wiener is fundamental in many respects. The name is often used as a synonym for Brownian motion, and serves in physics at the same time as the basis for diffusion processes due to random fluctuations caused by thermal motion, and also as the model for *white noise*. The fluctuation-driven random variable is denoted by $\mathcal{W}(t)$

and characterized by the cumulative probability distribution

$$P\big(\mathcal{W}(t) \le w\big) = \int_{-\infty}^{w} p(u,t)\,\mathrm{d}u\;,$$

where $p(u,t)$ still has to be determined. From the point of view of stochastic processes, the probability density of the Wiener process is the solution of the differential Chapman–Kolmogorov equation in one variable with a diffusion term $\mathrm{B} = 2D = 1$, zero drift $\mathbf{A} = 0$, and no jumps $\mathrm{W} = 0$:

$$\frac{\partial p(w,t)}{\partial t} = \frac{1}{2}\frac{\partial^2}{\partial w^2}p(w,t)\;, \quad \text{with } p(w,t_0) = \delta(w - w_0)\;. \tag{3.55}$$

Once again, a sharp initial condition (w_0, t_0) is assumed and we write $p(w,t) \equiv p(w,t|w_0,t_0)$ for short.

The closely related deterministic equation

$$\frac{\partial c(x,t)}{\partial t} = D\frac{\partial^2}{\partial x^2}c(x,t)\;, \quad \text{with } c(x,t_0) = c_0(x)\;, \tag{3.56}$$

is called the *diffusion equation*, because $c(x,t)$ describes the spreading of concentrations in homogeneous media driven by thermal molecular motion, also referred to as passive transport through thermal motion (for a detailed mathematical description of diffusion see, for example, [95, 214]). The parameter D is called the *diffusion coefficient*. It is assumed here to be a constant, and this means that it does not vary in space and time. The one-dimensional version of (3.56) is formally identical[25] to (3.55) with $D = 1/2$. The three-dimensional version of (3.55) occurs in physics and chemistry in connection with particle numbers or concentrations $c(\mathbf{r},t)$ which are functions of 3D space and time and satisfy

$$\frac{\partial c(\mathbf{r},t)}{\partial t} = D\nabla^2 c(\mathbf{r},t)\;, \quad \text{with } \mathbf{r} = (x,y,z)\;,\ \nabla^2 = \frac{\partial^2}{\partial x^2} + \frac{\partial^2}{\partial y^2} + \frac{\partial^2}{\partial z^2}\;, \tag{3.57}$$

and the initial condition $c\,(\mathbf{r},t_0) = c_0(\mathbf{r})$. The diffusion equation was first derived by the German physiologist Adolf Fick in 1855 [450]. Replacing the concentration by the temperature distribution in a one-dimensional object $c(x,t) \leftrightarrow u(x,t)$, and the diffusion constant by the thermal diffusivity $D \leftrightarrow \alpha$, the diffusion equation (3.56)

---

[25] We distinguish the two formally identical equations (3.55) and (3.56), because the interpretation is different: the former describes the evolution of a probability distribution with the conservation relation $\int \mathrm{d}w\, p(w,t) = 1$, whereas the latter deals with a concentration profile, which satisfies $\int \mathrm{d}x\, c(x,t) = c_{\mathrm{tot}}$ corresponding to mass conservation. In the case of the heat equation, the conserved quantity is total heat. It is worth considering dimensions here. The coefficient $1/2$ in (3.55) has the dimensions $[\mathrm{t}^{-1}]$ of a reciprocal time, while the diffusion coefficient has dimensions $[\mathrm{l}^2\mathrm{t}^{-1}]$, and the commonly used unit is $[\mathrm{cm}^2/\mathrm{s}]$.

becomes the heat equation, which describes the time dependence of the distribution of heat over a given region.

Solutions of (3.55) are readily calculated by means of the characteristic function:

$$\phi(s,t) = \int_{-\infty}^{+\infty} \mathrm{d}w\, p(w,t)\mathrm{e}^{\mathrm{i}sw} ,$$

$$\frac{\partial \phi(s,t)}{\partial t} = \int_{-\infty}^{+\infty} \mathrm{d}w\, \frac{\partial p(w,t)}{\partial t}\mathrm{e}^{\mathrm{i}sw} = \frac{1}{2}\int_{-\infty}^{+\infty} \mathrm{d}w\, \frac{\partial^2 p(w,t)}{\partial w^2}\mathrm{e}^{\mathrm{i}sw} .$$

First we derive a differential equation for the characteristic function by integrating by parts twice.[26] The first and second partial integration steps yield

$$\int_{-\infty}^{+\infty} \mathrm{d}w \frac{\partial p(w,t)}{\partial w}\mathrm{e}^{\mathrm{i}sw} = p(w,t)\mathrm{e}^{\mathrm{i}sw}\Big|_{-\infty}^{\infty} - \int_{-\infty}^{+\infty} \mathrm{d}w\, p(w,t)\frac{\partial \mathrm{e}^{\mathrm{i}sw}}{\partial w} = \mathrm{i}s\phi(s,t)$$

and

$$\int_{-\infty}^{+\infty} \mathrm{d}w \frac{\partial^2 p(w,t)}{\partial w^2}\mathrm{e}^{\mathrm{i}sw} = -s^2\phi(s,t) .$$

The function $p(w,t)$ is a probability density and accordingly has to vanish in the limits $w \to \pm\infty$. The same is true for the first derivatives $\partial p(w,t)/\partial w$. Differentiating $\phi(s,t)$ in (2.32) with respect to t and applying (3.55), we obtain

$$\frac{\partial \phi(s,t)}{\partial t} = -\frac{1}{2}s^2\phi(s,t) . \tag{3.58}$$

Next we compute the characteristic function by integration and find

$$\phi(s,t) = \phi(s,t_0)\exp\left(-\frac{1}{2}s^2(t-t_0)\right) . \tag{3.59}$$

Inserting the initial condition $\phi(s,t_0) = \exp(\mathrm{i}sw_0)$ completes the characteristic function

$$\phi(s,t) = \exp\left(\mathrm{i}sw_0 - \frac{1}{2}s^2(t-t_0)\right) , \tag{3.60}$$

---

[26] Integration by parts is a standard integration method in calculus. It is encapsulated in the formula

$$\int_a^b u(x)v'(x)\,\mathrm{d}x = u(x)v(x)\Big|_a^b - \int_a^b u'(x)v(x)\,\mathrm{d}x .$$

Characteristic functions are especially well suited to partial integration, because exponential functions $v(x) = \exp(\mathrm{i}sx)$ can be easily integrated, and probability densities $u(x) = p(x,t)$ as well as their first derivatives $u(x) = \partial p(x,t)/\partial x$ vanish in the limits $x \to \pm\infty$.

and finally we find the probability density through inverse Fourier transformation:

$$p(w,t) = \frac{1}{\sqrt{2\pi(t-t_0)}} \exp\left(-\frac{(w-w_0)^2}{2(t-t_0)}\right) , \quad \text{with } p(w,t_0) = \delta(w-w_0) . \tag{3.61}$$

The density function of the Wiener process is a normal distribution with the following expectation value and variance:

$$\mathrm{E}\big(\mathcal{W}(t)\big) = w_0 , \quad \mathrm{var}\big(\mathcal{W}(t)\big) = \mathrm{E}\Big(\big(\mathcal{W}(t) - w_0\big)^2\Big) = t - t_0 , \tag{3.62}$$

or $p(w,t) = \mathcal{N}(w_0, t-t_0)$. The standard deviation $\sigma(t) = \sqrt{t-t_0}$ is proportional to the square root of the time $t - t_0$ elapsed since the start of the process, and perfectly follows the famous $\sqrt{t}$-law. Starting the Wiener process at the origin $w_0 = 0$ at time $t_0 = 0$ yields $\mathrm{E}\big(\mathcal{W}(t)\big) = 0$ and $\sigma\big(\mathcal{W}(t)\big)^2 = t$. An initially sharp distribution spreads in time as illustrated in Fig. 3.8 and this is precisely what is experimentally observed in diffusion. The infinite time limit of (3.61) is a uniform distribution $\mathcal{U}(w) = 0$ on the whole real axis, and hence $p(w,t)$ vanishes in the limit $t \to \infty$.

Although the expectation value $\mathrm{E}\big(\mathcal{W}(t)\big) = w_0$ is well defined and independent of time in the sense of a martingale, the mean square $\mathrm{E}\big(\mathcal{W}(t)^2\big)$ becomes infinite as $t \to \infty$. This implies that the individual trajectories $\mathcal{W}(t)$ are extremely variable and diverge after short times (see, for example, the five trajectories of the forward equation in Fig. 3.3). We shall encounter such a situation, with finite mean but diverging variance, in biology, in the case of pure birth and birth-and-death processes. The expectation value, although well defined loses its meaning in practice, when the standard deviation becomes greater than the mean (Sect. 5.2.2).

The consistency and continuity of sample paths in the Wiener process have already been discussed in Sect. 3.2. Here we present proofs for two more features of the Wiener process:

(i) individual trajectories, although continuous, are nowhere differentiable,
(ii) the increments of the Wiener Process are independent of each other.

The non-differentiability of the trajectories of the Wiener process has a consequence for the physical interpretation as Brownian motion: the moving particle has no well defined velocity. Independence of increments is indispensable for the integration of stochastic differential equations (Sect. 3.4).

In order to show non-differentiability, we consider the convergence behavior of the difference quotient

$$\lim_{h\to 0} \left| \frac{\mathcal{W}(t+h) - \mathcal{W}(t)}{h} \right| ,$$

where the random variable $\mathcal{W}$ has the conditional probability (3.61). Ludwig Arnold [22, p.48] illustrates the non-differentiability in a heuristic way: the difference quotient $\big(\mathcal{W}(t+h) - \mathcal{W}(t)\big)/h$ follows the normal distribution $\mathcal{N}(0, 1/|h|)$, which

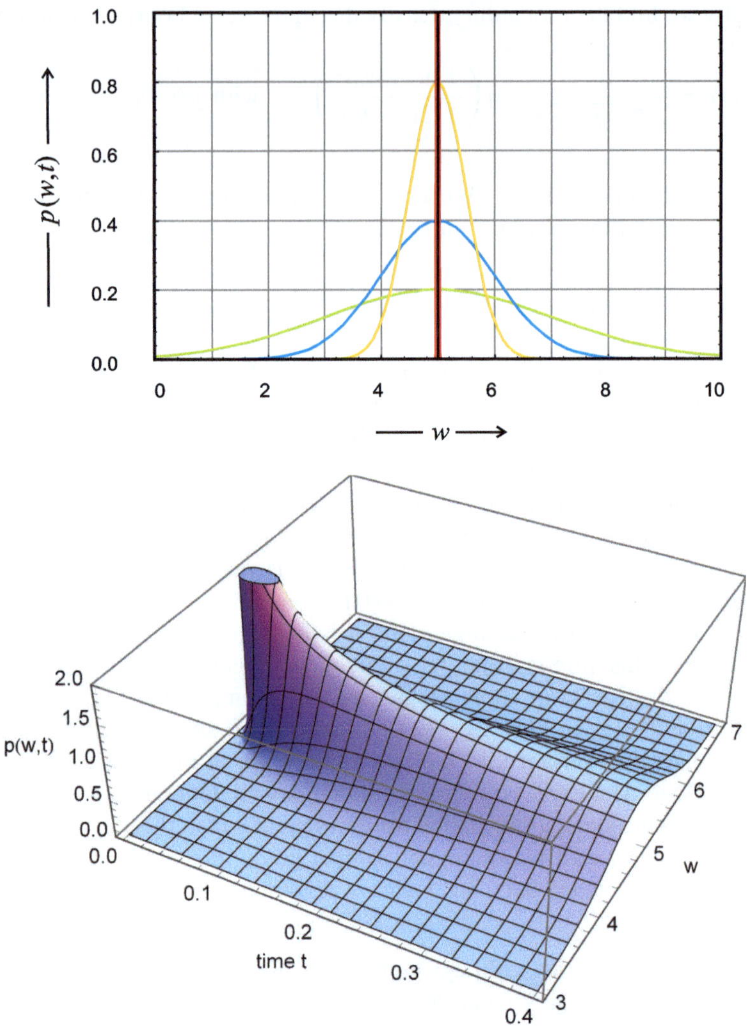

Fig. 3.8 Probability density of the Wiener process. The figure shows the conditional probability density of the Wiener process, which is identical with the normal distribution (Fig. 1.22),

$$p(w,t) \equiv p(w,t|w_0,t_0) = \mathcal{N}(w_0, t-t_0) = \frac{1}{\sqrt{2\pi(t-t_0)}} \mathrm{e}^{-(w-w_0)^2/2(t-t_0)} .$$

The initially sharp distribution $p(w,t_0|w_0,t_0) = \delta(w-w_0)$ spreads with increasing time until it becomes completely flat in the limit $t \to \infty$. Choice of parameters: $w_0 = 5$ [l], $t_0 = 0$, and $t = 0$ (*black*), 0.01 (*red*), 0.5 (*yellow*), 1.0 (*blue*), and 2.0 [t] (*green*). *Lower*: Three-dimensional plot of the density function diverges as $h \downarrow 0$—the limit of a normal distribution with exploding variance is undefined—and hence, for every bounded measurable set S, we have

$$P\Big(\big(\mathcal{W}(t+h)-\mathcal{W}(t)\big)/h \in \mathrm{S}\Big) \longrightarrow 0 \text{ as } h \downarrow 0 .$$

Accordingly, the difference quotient cannot converge with nonzero probability to a finite value.

The convergence behavior can be made more precise by using the law of the iterated logarithm (2.67): for almost every sample function and arbitrary ϵ in the interval $0 < \epsilon < 1$, as $h \downarrow 0$,

$$\frac{\mathcal{W}(t+h)-\mathcal{W}(t)}{h} \geq (1-\epsilon)\sqrt{\frac{2\ln\big(\ln(1/h)\big)}{h}} \text{ infinitely often}$$

and simultaneously

$$\frac{\mathcal{W}(t+h)-\mathcal{W}(t)}{h} \leq (-1+\epsilon)\sqrt{\frac{2\ln\big(\ln(1/h)\big)}{h}} \text{ infinitely often} .$$

Since the expressions on the right-hand side approach $\pm\infty$ as $h \downarrow 0$, the difference quotient $\big(\mathcal{W}(t+h)-\mathcal{W}(t)\big)/h$ has, with probability one and for every fixed t, the extended real line $[-\infty, +\infty]$ as its limit set of cluster points. □

Because of the general importance of the Wiener process, it is essential to present a proof of the statistical independence of nonoverlapping increments of $\mathcal{W}(t)$ [194, pp. 67,68]. We are dealing with a Markov process, and hence can write the joint probability as a product of conditional probabilities (3.16′), where $t_n - t_{n-1}, \ldots, t_1 - t_0$, are subintervals of the time span $t_n \geq t \geq t_0$:

$$p(w_n, t_n; w_{n-1}, t_{n-1}; \ldots; w_0, t_0) = \prod_{i=0}^{n-1} p(w_{i+1}, t_{i+1}|w_i, t_i)p(w_0, t_0) .$$

Next we introduce new variables that are consistent with the partitioning of the process: $\big(\Delta w_i \equiv \mathcal{W}(t_i) - \mathcal{W}(t_{i-1}), \Delta t_i \equiv t_i - t_{i-1}\big)$, $\forall\, i = 1, \ldots, n$. Since $\mathcal{W}(t)$ is also a Gaussian proces, the probability density of any partition is normally distributed, and we express the conditional probabilities in terms of (3.61):

$$p(\Delta w_n, \Delta t_n; \Delta w_{n-1}, \Delta t_{n-1}; \ldots; w_0, t_0) = \prod_{i=i}^{n} \frac{\exp\big(-\Delta w_i^2/2\Delta t_i\big)}{\sqrt{2\pi\Delta t_i}}\, p(w_0, t_0) .$$

The joint probability distribution is factorized into distributions on individual intervals and, provided that the intervals do not overlap, the increments Δw_i are stochastically independent random variables in the sense of Sect. 1.6.4. Accordingly, they are independent of the initial condition $\mathcal{W}(t_0)$. □

The independence relation is readily cast in the precise form

$$\mathcal{W}(t) - \mathcal{W}(s) \text{ is independent of } \big(\mathcal{W}(\tau)\big)_{\tau \le s}\,, \text{ for any } 0 \le s \le t\,, \tag{3.63}$$

which will be used in the forthcoming sections on stochastic differential equations (Sect. 3.4).

Applying (3.62) to the probability distribution within a partition, we find for the interval $\Delta t_k = t_k - t_{k-1}$:

$$\mathrm{E}\big(\mathcal{W}(t_k) - \mathcal{W}(t_{k-1})\big) = \mathrm{E}(\Delta w_k) = w_{k-1}\,, \qquad \mathrm{var}(\Delta w_k) = t_k - t_{k-1}\,.$$

It is now straightforward to calculate the autocorrelation function, which is defined by

$$\begin{aligned}\big\langle \mathcal{W}(t)\mathcal{W}(s)|(w_0,t_0)\big\rangle &= \mathrm{E}\big(\mathcal{W}(t)\mathcal{W}(s)|(w_0,t_0)\big)\\ &= \iint \mathrm{d}w_t \mathrm{d}w_s w_t w_s p(w_t,t;w_s,s|w_0,t_0)\,.\end{aligned} \tag{3.64}$$

Subtracting and adding $\mathcal{W}(s)^2$ inside the expectation value yields

$$\mathrm{E}\big(\mathcal{W}(t)\mathcal{W}(s)|(w_0,t_0)\big) = \mathrm{E}\Big(\big(\mathcal{W}(t) - \mathcal{W}(s)\big)\mathcal{W}(s)\Big) + \mathrm{E}\big(\mathcal{W}(s)^2\big)\,,$$

where the first term vanishes due to the independence of the increments and the second term follows from (3.62):

$$\mathrm{E}\big(\mathcal{W}(t)\mathcal{W}(s)|(w_0,t_0)\big) = \min\big\{t - t_0, s - t_0\big\} + w_0^2\,. \tag{3.65}$$

The latter simplifies to $\mathrm{E}\big(\mathcal{W}(t)\mathcal{W}(s)\big) = \min\{t,s\}$ for $w_0 = 0$ and $t_0 = 0$. This expectation value also reproduces the diagonal element of the covariance matrix, the variance var, since for $s = t$, we find $\mathrm{E}\big(\mathcal{W}(t)^2\big) = t$. In addition, several other useful relations can be derived from the autocorrelation relation. We summarize:

$$\mathrm{E}\big(\mathcal{W}(t) - \mathcal{W}(s)\big) = 0\,, \quad \mathrm{E}\big(\mathcal{W}(t)^2\big) = t\,, \quad \mathrm{E}\big(\mathcal{W}(t)\mathcal{W}(s)\big) = \min\{t,s\}\,,$$

$$\begin{aligned}\mathrm{E}\Big(\big(\mathcal{W}(t) - \mathcal{W}(s)\big)^2\Big) &= \mathrm{E}\big(\mathcal{W}(t)^2\big) - 2\,\mathrm{E}\big(\mathcal{W}(t)\mathcal{W}(s)\big) + \mathrm{E}\big(\mathcal{W}(t)^2\big)\\ &= t - 2\min\{t,s\} + s = |t - s|\,,\end{aligned}$$

and remark that these results are not independent of the *càdlàg* convention for stochastic processes.

The Wiener process has the property of self-similarity. Assume that $\mathcal{W}_1(t)$ is a Wiener process. Then, for every $\gamma > 0$,

$$\mathcal{W}_2(t) = \mathcal{W}_1(\gamma t) = \sqrt{\gamma}\,\mathcal{W}_1(t)$$

is also a Wiener process. Accordingly, we can change the scale at will and the process remains a Wiener process. The power of the scaling factor is called the Hurst factor H (see Sects. 3.2.4 and 3.2.5), and accordingly the Wiener process has $H = 1/2$.

Solution of the Diffusion Equation by Fourier Transform

The Fourier transform is as a convenient tool for deriving solutions of differential equations, because transformation of derivatives results in algebraic equations in Fourier space, which can often be solved easily, and subsequent inverse transformation then yields the desired answer.[27] In addition, the Fourier transform provides otherwise hard-to-obtain insights into problems. Here we shall apply the Fourier transform solution method to the diffusion equation.

Through integration by parts, the Fourier transform of a general derivative yields

$$\begin{aligned}\mathcal{F}\left(\frac{\mathrm{d}p(x)}{\mathrm{d}x}\right) &= \frac{1}{\sqrt{2\pi}}\int_{-\infty}^{\infty}\mathrm{d}x\,p(x)\mathrm{e}^{-\mathrm{i}kx} \\ &= \frac{1}{\sqrt{2\pi}}p(x)\mathrm{e}^{-\mathrm{i}kx}\Big|_{-\infty}^{\infty} + \frac{1}{\sqrt{2\pi}}\int_{-\infty}^{\infty}\mathrm{d}x\,\mathrm{i}kp(x)\mathrm{e}^{-\mathrm{i}kx} \\ &= \mathrm{i}k\tilde{p}(k)\,.\end{aligned}$$

The first term from the integration by parts vanishes as $\lim_{x\to\pm\infty} p(x) = 0$, otherwise the probability could not be normalized. Application of the Fourier transform to higher derivatives requires multiple application of integration by parts and yields

$$\mathcal{F}\left(\frac{\mathrm{d}^n p(x)}{\mathrm{d}x^n}\right) = (\mathrm{i}k)^n\tilde{p}(k)\,. \tag{3.66}$$

[27] Integral transformations, in particular the Fourier and the Laplace transform, are standard techniques for solving ODEs and PDEs. For details, we refer to mathematics handbooks for the scientist such as [149, pp. 89–96] and [467, pp. 449–451, 681–686].

Since t is handled like a constant in the Fourier transformation and in the differentiation by x, and since the two linear operators $\mathcal{F}$ and d/dt can be interchanged without changing the result, we find for the Fourier transformed diffusion equation

$$\frac{\mathrm{d}\tilde{p}(k,t)}{\mathrm{dt}} = -Dk^2\tilde{p}(k,t) . \tag{3.67}$$

The original PDE has become an ODE, which can be readily solved to yield

$$\tilde{p}(k,t) = \tilde{p}(k,0)\sqrt{\frac{Dt}{\pi}}\,\mathrm{e}^{-Dk^2t} . \tag{3.68}$$

This equation corresponds to a relaxation process with a relaxation time $\tau_\mathrm{R} = Dk^2$, where k is the wave number[28] with dimension $[\mathrm{l}^{-1}]$ and commonly measured in units of cm^{-1}. The solution of the diffusion equation is then obtained by inverse Fourier transformation:

$$p(x,t) = \frac{1}{\sqrt{4\pi Dt}}\,\mathrm{e}^{-x^2/4Dt} . \tag{3.69}$$

The solution is, of course, identical with the solution of the Wiener process in (3.61).

Multivariate Wiener Process

The Wiener process is readily extended to higher dimensions. The multivariate Wiener process is defined by

$$\mathcal{W}(t) = \big(\mathcal{W}_1(t), \ldots, \mathcal{W}_n(t)\big) \tag{3.70}$$

and satisfies the Fokker–Planck equation

$$\frac{\partial p(\mathbf{w},t|\mathbf{w}_0,t_0)}{\partial t} = \frac{1}{2}\sum_i \frac{\partial^2}{\partial w_i^2} p(\mathbf{w},t|\mathbf{w}_0,t_0) . \tag{3.71}$$

The solution is a multivariate normal density

$$p(\mathbf{w},t|\mathbf{w}_0,t_0) = \frac{1}{\sqrt{2\pi(t-t_0)}}\exp\left(-\frac{(\mathbf{w}-\mathbf{w}_0)^2}{2(t-t_0)}\right) , \tag{3.72}$$

[28]For a system in 3D space, the wave vector in reciprocal space is denoted by $\mathbf{k}$, and its length $|\mathbf{k}| = k$ is called the wave number.

with mean $\mathrm{E}(\mathcal{W}(t)) = \mathbf{w}_0$ and variance–covariance matrix

$$(\Sigma)_{ij} = \mathrm{E}\Big(\big(\mathcal{W}_i(t) - w_{0i}\big)\big(\mathcal{W}_j(t) - w_{0j}\big)\Big) = (t - t_0)\delta_{ij} ,$$

where all off-diagonal elements, i.e., the proper covariances, are zero. Hence, Wiener processes along different Cartesian coordinates are independent.

Before we consider the Gauss process as a generalization of the Wiener process, it seems useful to summarize the most prominent features.

The Wiener process $W = \big(\mathcal{W}(t), t \geq 0\big)$ is characterized by ten properties and definitions:

1. Initial condition $\mathcal{W}(t_0) = \mathcal{W}(0) \equiv 0$.
2. Trajectories are continuous functions of $t \in [0, \infty[$.
3. Expectation value $\mathrm{E}\big(\mathcal{W}(t)\big) \equiv 0$.
4. Correlation function $\mathrm{E}\big(\mathcal{W}(t)\mathcal{W}(s)\big) = \min\{t, s\}$.
5. The Gaussian property implies that for any $(t_1, \ldots, t_n)$, the random vector $\big(\mathcal{W}(t_1), \ldots, \mathcal{W}(t_n)\big)$ is a Gaussian process.
6. Moments $\mathrm{E}\big(\mathcal{W}(t)^2\big) = t$, $\mathrm{E}\big(\mathcal{W}(t) - \mathcal{W}(s)\big) = 0$, and
$$\mathrm{E}\Big(\big(\mathcal{W}(t) - \mathcal{W}(s)\big)^2\Big) = |t - s| .$$
7. Increments of the Wiener process on non-overlapping intervals are independent, that is, for $(s_1, t_1) \cap (s_2, t_2) = \emptyset$, the random variables $\mathcal{W}(t_2) - \mathcal{W}(s_2)$ and $\mathcal{W}(t_1) - \mathcal{W}(s_1)$ are independent.
8. Non-differentiable trajectories $\mathcal{W}(t)$.
9. Self-similarity of the Wiener process $\mathcal{W}_2(t) = \mathcal{W}_1(\gamma t) = \sqrt{\gamma}\mathcal{W}_1(t)$.
10. The martingale property, i.e., if $\mathcal{W}_0^s = \mathcal{W}(u)$, $\forall\, u$ such that $0 \leq u \leq s$, then $\mathrm{E}\big(\mathcal{W}(t)|\mathcal{W}_0^s\big) = \mathcal{W}(s)$ and $\mathrm{E}\Big(\big(\mathcal{W}(t) - \mathcal{W}(s)\big)^2|\mathcal{W}_0^s\Big) = t - s$.

Out of these ten properties, three will be most important for the goals we shall pursue here: (2) continuity of sample paths, (8) non-differentiability of sample paths, and (7) independence of increments.

Gaussian and AR(n) Processes

A generalization of Wiener processes is the Gaussian process $\mathcal{X}(\mathbf{t})$ with $\mathbf{t} \in \mathcal{T}$, where $\mathcal{T}$ may be a finite index set $\mathcal{T} = \{t_1, \ldots, t_n\}$ or the entire space of real numbers $\mathcal{T} = \mathbb{R}^d$ for continuous time. The integer d is the dimension of the problem, for example, the number of inputs. The condition for a Gaussian process is that any finite linear combination of samples should have a joint normal distribution, i.e., $(\mathcal{X}_{\mathbf{t}}, \mathbf{t} \in \mathcal{T})$ is Gaussian if and only if, for every finite index set $\mathbf{t} = (t_1, \ldots, t_n)$, there exist real numbers μ_k and σ_{kl}^2 with $\sigma_{kk}^2 > 0$ such that

$$\mathrm{E}\Big(\exp\Big(\mathrm{i}\sum_{i=1}^{n} t_i \mathcal{X}_{t_i}\Big)\Big) = \exp\Big(-\frac{1}{2}\sum_{i=1}^{n}\sum_{j=1}^{n}\sigma_{ij}^2 t_i t_j + \mathrm{i}\sum_{i=1}^{n}\mu_i t_i\Big)\,, \tag{3.73}$$

where μ_k $(i = 1, \ldots, n)$ are the mean values of the random variables $\mathcal{X}_i$, and $\sigma_{ij}^2 = \mathrm{cov}(\mathcal{X}_i, \mathcal{X}_j)$ with $i, j = 1, \ldots, n$, are the elements of the covariance matrix Σ. The Wiener process is a nonstationary special case of a Gaussian process, since the variance grows linearly with t. The Ornstein–Uhlenbeck process to be discussed in Sect. 3.2.2.3 is an example of a stationary Gaussian process. After an initial transient period, it settles down to a process with time-independent mean $\overline{\mu}$ and variance $\overline{\mu}_2$. In a nutshell, a Gaussian process can be characterized as a normal distribution migrating in state space and thereby changing shape.

According to Wold's decomposition, named after Herman Wold [578], any stochastic process with stationary covariance can be expressed by a time series that is decomposed into an independent deterministic part and independent stochastic components:

$$\mathcal{Y}_t = \eta_t + \sum_{j=0}^{\infty} b_j \mathcal{Z}_{t-j}\,, \tag{3.74}$$

where η_t is a deterministic process, e.g., the solution of a difference equation, $\mathcal{Z}_{t-j}$ are independent and identically distributed (iid) random variables, and b_j are coefficients satisfying $b_0 = 1$ and $\sum_{j=0}^{\infty} b_j^2 < \infty$. This representation is called the *moving average model*. A stationary Gaussian process $\mathcal{X}_t$ with $t \in \mathcal{T} = \mathbb{N}$ can be written in the form of (3.74), with the condition that the variables $\mathcal{Z}$ are iid normal variables with mean $\overline{\mu} = 0$ and variance $\overline{\mu}_2 = \overline{\sigma}^2$, $\mathcal{Z}_{t-j} = \mathcal{W}_{t-j}$. Since the independent deterministic part can be easily removed, nondeterministic or Gaussian linear processes, i.e.,

$$\mathcal{X}_t = \sum_{j=0}^{\infty} b_j \mathcal{W}_{t-j}\,, \quad \text{with } b_0 = 1\,, \tag{3.75}$$

are frequently used in time series analysis. An alternative representation of times series called autoregression[29] considers the stochastic process by making use of

[29] An autoregressive process of order n is denoted by AR(n). The order n implies that n values of the stochastic variables at previous times are required to calculate the current value. An extension of the autoregressive model is the *autoregressive moving average* (ARMA) model.

past values of the variable itself [231, 565]:

$$\mathcal{X}_t = \varphi_1 \mathcal{X}_{t-1} + \varphi_2 \mathcal{X}_{t-2} + \cdots + \varphi_n \mathcal{X}_{t-n} + \mathcal{W}_t \, . \tag{3.76}$$

The process (3.76) is characterized as autoregressive of order n or as an AR(n) process. Every AR(n) process has a linear representation of the kind shown in (3.75), where the coefficients b_j are obtained as functions of the φ_k values [67]. In other words, for every Gaussian linear process, there exists an AR(n) process such that the two autocovariance functions can be made practically equal for all time differences $t_j - t_{j-1}$. For the first n time lags, the match can be made perfect. An extension to continuous time is possible, and special features of continuous time autoregressive models (CAR) are described, for example, in [68]. Finally, we mention that AR(n) processes provide an excellent possibility for demonstrating the Markov property: an AR(1) process $\mathcal{X}_t = \varphi \mathcal{X}_{t-1} + \mathcal{W}_t$ is Markovian in first order, since knowledge of $\mathcal{X}_{t-1}$ is sufficient to compute $\mathcal{X}_t$ and all future development.

3.2.2.3 Ornstein–Uhlenbeck Process

The Ornstein–Uhlenbeck process is named after the two Dutch physicists Leonard Ornstein and George Uhlenbeck [534] and represents presumably the simplest stochastic process that approaches a stationary state with a definite variance.[30] The Ornstein–Uhlenbeck process has found widespread applications, for example, in economics for modeling the irregular behavior of financial markets [546]. In physics it is among other applications a model for the velocity of a Brownian particle under the influence of friction. In essence, the Ornstein–Uhlenbeck process describes exponential relaxation to a stationary state or to an equilibrium with a Wiener process superimposed on it. Figure 3.9 presents several trajectories of the Ornstein–Uhlenbeck process which nicely show the drift and the diffusion component of the individual runs.

Fokker–Planck Equation and Solution of the Ornstein–Uhlenbeck Process
The one-dimensional Fokker–Planck equation of the Ornstein–Uhlenbeck process for the probability density $p(x,t)$ of the random variable $\mathcal{X}(t)$ with the initial condition $p(x,t_0) = \delta(x - x_0)$ is of the form

$$\frac{\partial p(x,t)}{\partial t} = k \frac{\partial}{\partial x}\big((x - \overline{\mu})p(x,t)\big) + \frac{\sigma^2}{2}\frac{\partial^2 p(x,t)}{\partial x^2} \, , \tag{3.77}$$

with k is the rate parameter of the exponential decay, $\overline{\mu} = \lim_{t\to\infty} \mathrm{E}\big(\mathcal{X}(t)\big)$ is the expectation value of the random variable in the long-time or stationary limit,

[30]The variance of the Wiener process diverges, i.e., $\lim_{t\to\infty} \mathrm{var}\big(\mathcal{W}(t)\big) = \infty$. The same is true for the Poisson process and the random walk, which are discussed in the next two sections.

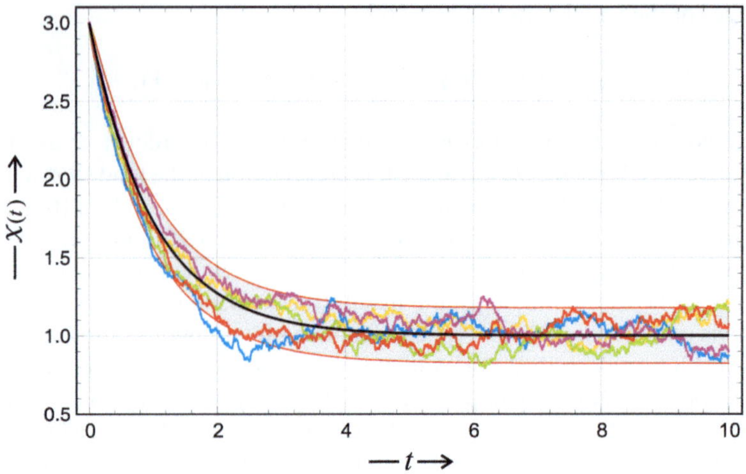

Fig. 3.9 The Ornstein–Uhlenbeck process. Individual trajectories of the process are simulated by

$$\mathcal{X}_{i+1} = \mathcal{X}_i e^{-k\vartheta} + \overline{\mu}(1 - e^{-k\vartheta}) + \sigma \sqrt{\frac{1 - e^{-2k\vartheta}}{2k}} (\mathcal{R}_{0,1} - 0.5) ,$$

where $\mathcal{R}_{0,1}$ is a random number drawn from the uniform distribution on the interval $[0, 1]$ by a pseudorandom number generator [537]. The figure shows several trajectories differing only in the choice of seeds for the Mersenne Twister random number generator. *Lines* represent the expectation value $E(\mathcal{X}(t))$ (*black*) and the functions $E(\mathcal{X}(t)) \pm \sigma(\mathcal{X}(t))$ (*red*). The *gray shaded area* is the confidence interval $E \pm \sigma$. Choice of parameters: $\mathcal{X}(0) = 3, \overline{\mu} = 1, k = 1, \sigma = 0.25, \vartheta = 0.002$ or a total time for the computation of $t_f = 10$. Seeds: 491 (*yellow*), 919 (*blue*), 023 (*green*), 877 (*red*), and 733 (*violet*). For the simulation of the Ornstein–Uhlenbeck model, see [210, 537]

and $\overline{\mu}_2 = \lim_{t\to\infty} \text{var}(\mathcal{X}(t)) = \sigma^2/(2k)$ is the stationary variance. For the initial condition $p(x, 0) = \delta(x - x_0)$, the probability density can be obtained by standard techniques:

$$p(x, t) = \sqrt{\frac{k}{\pi\sigma^2(1 - e^{-2kt})}} \exp\left(-\frac{k}{\sigma^2} \frac{\left(x - \overline{\mu} - (x_0 - \mu)e^{-kt}\right)^2}{1 - e^{-2kt}}\right) . \qquad (3.78)$$

This expression can be easily checked by performing the two limits $t \to 0$ and $t \to \infty$. The first limit has to yield the initial conditions and it does indeed if we recall a common definition of the Dirac delta function:

$$\delta_\alpha(x) = \lim_{\alpha\to 0} \frac{1}{\alpha\sqrt{\pi}} e^{-x^2/\alpha^2} . \qquad (3.79)$$

Inserting $\alpha^2 = \sigma^2(1 - \mathrm{e}^{-2kt})/k$ leads to

$$\lim_{t \to 0} p(x, t) = \delta(x - x_0) .$$

The long-time limit of the probability density is calculated straightforwardly:

$$\lim_{t \to \infty} p(x, t) = \overline{p}(x) = \sqrt{\frac{k}{\pi \sigma^2}}\, \mathrm{e}^{-k(x-\overline{\mu})^2/\sigma^2} , \quad (3.80)$$

which is a normal density with expectation value $\overline{\mu}$ and variance $\sigma^2/2k$. □

The evolution of the probability density $p(x, t)$ from the δ-function at $t = 0$ to the stationary density $\lim_{t \to \infty} p(x, t)$ is shown in Fig. 3.10. The Ornstein–Uhlenbeck process is a stationary Gaussian process and has a representation as a first-order autoregressive $\big(AR(1)\big)$ process, which implies that it fulfils the Markov condition. It is instructive to compare the three 3D plots in Figs. 3.7, 3.8, and 3.10:

(i) The probability density of the Liouville process migrates according to the drift term $\mu(t)$, but does not change shape, i.e., the variance remains constant.
(ii) The Wiener density stays in state space, $\mu = 0$, but changes shape as the variance increases $\mu_2^2 = \sigma(t)^2 = t - t_0$.
(iii) Finally, the density of the Ornstein–Uhlenbeck process drifts and changes shape.

The Ornstein–Uhlenbeck process can also be efficiently modeled by a stochastic differential equation (SDE) (see Sect. 3.4.3):

$$\mathrm{d}x(t) = k\big(\overline{\mu} - x(t)\big)\, \mathrm{d}t + \sigma\, \mathrm{d}W(t) . \quad (3.81)$$

The individual trajectories shown in Fig. 3.9 [210, 537] were simulated by means of the following equation:

$$\mathcal{X}_{i+1} = \mathcal{X}_i \mathrm{e}^{-k\vartheta} + \overline{\mu}(1 - \mathrm{e}^{-k\vartheta}) + \sigma \sqrt{\frac{1 - \mathrm{e}^{-2k\vartheta}}{2k}} (\mathcal{R}_{0,1} - 0.5) ,$$

where $\vartheta = \Delta t / n_{st}$, and n_{st} is the number of steps per unit time interval. The probability density can be computed, for example, from a sufficiently large ensemble of numerically simulated trajectories. The expectation value and variance of the random variable $\mathcal{X}(t)$ can be calculated directly from the solution of the SDE (3.81), as shown in Sect. 3.4.3.

Stationary Solutions of Fokker–Planck Equations

Often one is mainly interested in the long-time solution of a stochastic process and then the stationary solution of a Fokker–Planck equation, provided it exists, may be calculated directly. At stationarity, the time independence of the two functions

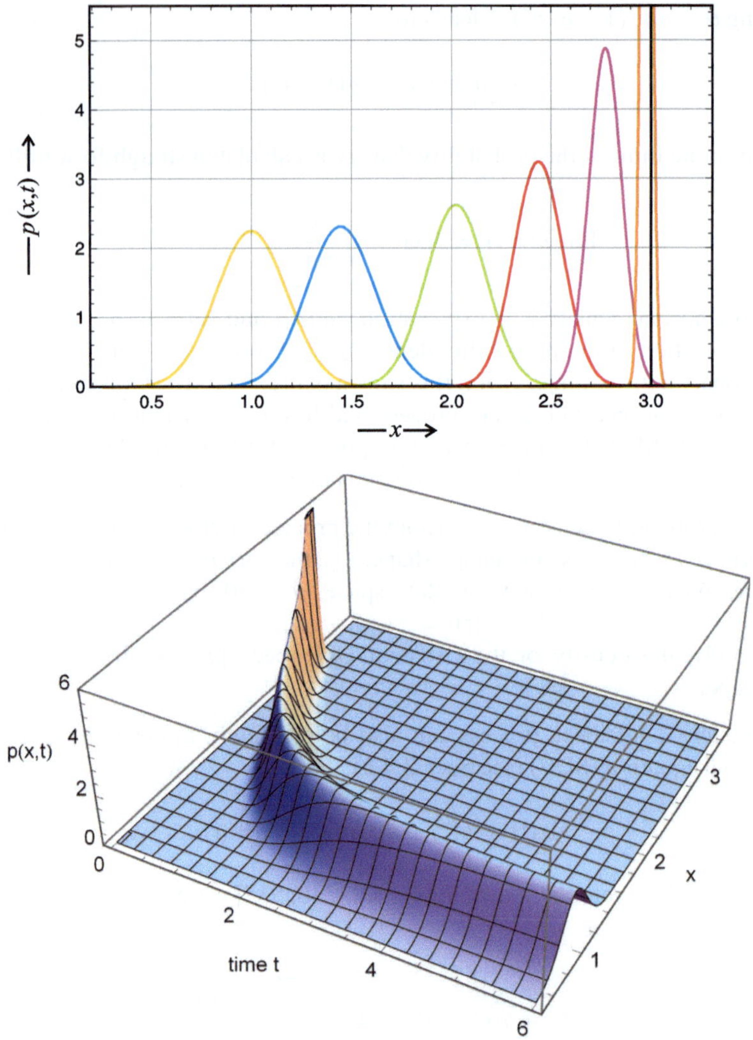

Fig. 3.10 The probability density of the Ornstein–Uhlenbeck process. Starting from the initial condition $p(x, t_0) = \delta(x - x_0)$ (*black*), the probability density (3.78) broadens and migrates until it reaches the stationary distribution (*yellow*). Choice of parameters: $x_0 = 3$, $\overline{\mu} = 1$, $k = 1$, and $\sigma = 0.25$. Times: $t = 0$ (*black*), 0.12 (*orange*), 0.33 (*violet*), 0.67 (*green*), 1.5 (*blue*), and 8 (*yellow*). The *lower plot* presents an illustration in 3D

$A(x, t) = A(x)$ and $B(x, t) = B(x)$ is assumed. We shall be dealing here with the one-dimensional case and consider the Ornstein–Uhlenbeck process as an example. We start by setting the time derivative of the probability density equal to zero:

$$\frac{\partial p(x,t)}{\partial t} = 0 = -\frac{\partial}{\partial x}\Big(A(x)p(x,t)\Big) + \frac{1}{2}\frac{\partial^2}{\partial x^2}\Big(B(x)p(x,t)\Big) ,$$

yielding

$$A(x)\overline{p}(x) = \frac{1}{2}\frac{\mathrm{d}}{\mathrm{d}x}\Big(B(x)\overline{p}(x)\Big) .$$

By means of a little trick we get an easy to integrate expression [468, p. 98]:

$$A(x)\overline{p}(x) = \frac{A(x)}{B(x)}\big(B(x)\overline{p}(x)\big) = \frac{1}{2}\frac{\mathrm{d}}{\mathrm{d}x}\big(B(x)\overline{p}(x)\big) ,$$

$$\frac{\mathrm{d}\ln\big(B(x)\overline{p}(x)\big)}{\mathrm{d}x} = \frac{2A(x)}{B(x)} , \qquad B(x)\overline{p}(x) = \gamma \exp\left(2\int_0^x \frac{A(\xi)}{B(\xi)}\,\mathrm{d}\xi\right) ,$$

where the factor γ arises from the integration constants. Finally, we obtain

$$\overline{p}(x) = \frac{N}{B(x)}\exp\left(2\int_0^x \frac{A(\xi)}{B(\xi)}\,\mathrm{d}\xi\right) , \tag{3.82}$$

with the integration constant absorbed into the normalization factor N which ensures that the probability conservation relation $\int_{-\infty}^{\infty}\overline{p}(x)\mathrm{d}x = 1$ holds. As a rule the calculation of N is straightforward in specific examples.

As an illustrative example, we calculate the stationary probability density of the Ornstein–Uhlenbeck process. For $A(x) = -k(x-\overline{\mu})$ and $B(x) = \sigma^2$, we find

$$\overline{p}(x) = \frac{N}{\sigma^2}\mathrm{e}^{-k(x-\overline{\mu})^2/\sigma^2} \quad \text{and} \quad N = 1\Big/\left(\int_{-\infty}^{\infty}\mathrm{d}x\mathrm{e}^{-k(x-\overline{\mu})^2/\sigma^2}\Big/\sigma^2\right) .$$

Making use of $\int_{-\infty}^{\infty}\mathrm{d}x\,\mathrm{e}^{-k(x-\overline{\mu})^2/\sigma^2} = \sigma\sqrt{(\pi/k)}$, we obtain the final result, which naturally reproduces the previous calculation from the time dependent density by taking the limit $t \to \infty$:

$$\overline{p}(x) = \sqrt{\frac{k}{\pi\sigma^2}}\,\mathrm{e}^{-k(x-\overline{\mu})^2/\sigma^2} . \tag{3.80$'$}$$

We emphasize once again that we got this result without making use of the time dependent probability density $p(x,t)$, and the approach also allows for the calculation of stationary solutions in cases where $p(x,t)$ is not available.

3.2.2.4 Poisson Process

The three processes discussed so far in this section all dealt with continuous random variables and their probability densities. We continue by presenting one example of a process involving discrete variables and pure jump processes according to (3.46c), which are modeled by master equations: the Poisson process. We stress once again that master equations and related techniques are tailored to analyze and model stochasticity at low particle numbers, and are therefore of particular importance in biology and chemistry.

The master equation (3.46c) is rewritten for the discrete case by replacing the integral by a summation[31]:

$$\frac{\partial p(x,t)}{\partial t} = \fint \mathrm{d}z \Big(\mathrm{W}(x|z,t)p(z,t) - \mathrm{W}(z|x,t)p(x,t) \Big) \tag{3.83}$$

$$\Longrightarrow \quad \frac{\mathrm{d}P_n(t)}{\mathrm{d}t} = \sum_{m=0}^{\infty} \Big(\mathrm{W}(n|m,t)P_m(t) - \mathrm{W}(m|n,t)P_n(t) \Big) ,$$

where we are assuming $n, m \in \mathbb{N}$, continuous time, $t \in \mathbb{R}_{\geq 0}$, and sharp initial conditions (n_0, t_0) or $P_n(t_0) = \delta_{n,n_0}$.[32] The matrix $W(m|n,t)$ is called the transition matrix. It contains the probabilities attributed to jumps in the variables. From the two equations, it follows that the diagonal elements $W(n|n,t)$ cancel. The domain of the random variable is implicitly included in the range of integration or summation, respectively.

The Poisson process is commonly applied to model cumulative independent random events. These may be, for example, electrons arriving at an anode, customers entering a shop, telephone calls arriving at a switchboard, or e-mails being registered on an account (see also Sect. 2.6.4). Aside from independence, the requirement is an unstructured time profile of events or, in other words, the probability of occurrence of events is constant and does not depend on time, i.e., $W(m|n,t) = W(m|n)$. The cumulative number of these events is denoted by the random variable $\mathcal{X}(t) \in \mathbb{N}$. In other words $\mathcal{X}(t)$ is counting the number of arrivals and hence can only increase. The probability of arrival is assumed to be γ per unit time, so $\gamma \Delta t$ is the expected number of events recorded in a time interval of length Δt.

Solutions of the Poisson Process

The Poisson process can also be seen as a *one-sided* random walk in the sense that the walker takes steps only to the right, for example, with a probability γ per unit

---

[31]From here on, unless otherwise stated, we shall consider cases in which the limits $\lim_{|x-z|\to 0} W(x|z,t)$ and $\lim_{|x-z|\to 0} W(z|x,t)$ of the transition probabilities are finite and the principal value integral can be replaced by a conventional integral. Riemann–Stieltjes integration converts the integral into a sum, and since we are dealing exclusively with discrete events, we use an index on the probability $P_n(t)$.

[32]The notation δ_{ij} denotes the Kronecker delta, named after the German mathematician Leopold Kronecker, which means

$$\delta_{ij} = \begin{cases} 1, & \text{if } i = j, \\ 0, & \text{if } i \neq j. \end{cases}$$

It is the discrete analogue of the Dirac delta function.

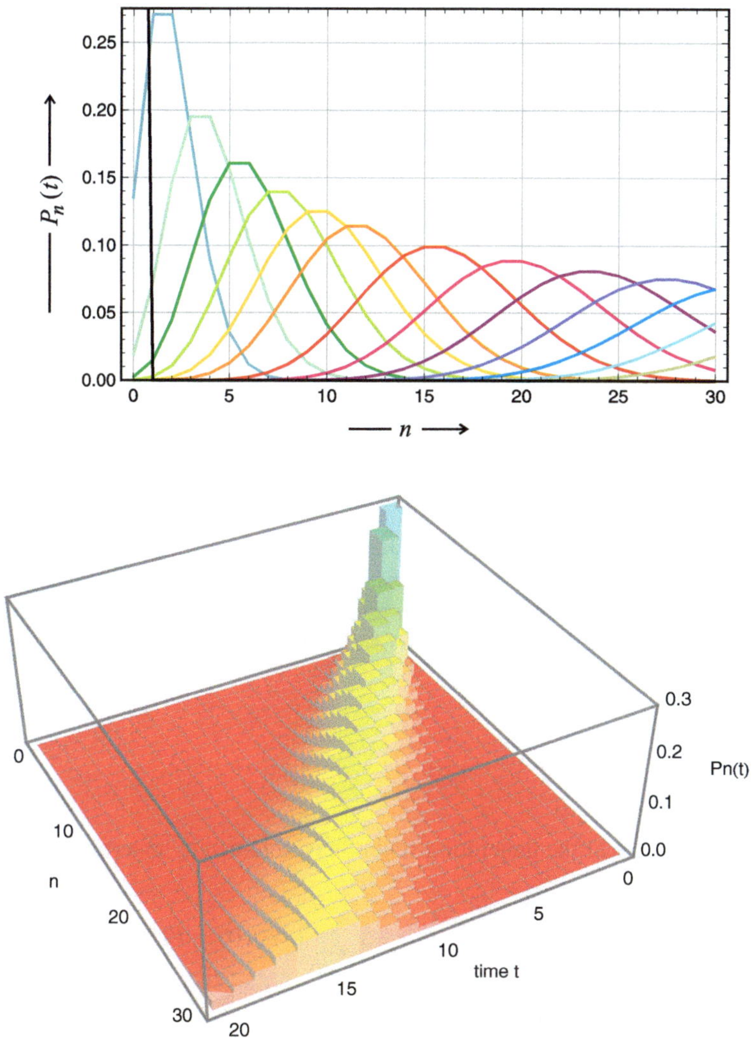

Fig. 3.11 Probability density of the Poisson process. The figures show the spreading of an initially sharp Poisson density $P_n(t) = (\gamma t)^n e^{-\gamma t}/n!$ with time: $P_n(t) = p(n, t|n_0, t_0)$, with the initial condition $p(n, t_0|n_0, t_0) = \delta(n - n_0)$. In the limit $t \to \infty$, the density becomes completely flat. The values used are $\gamma = 2\,[\mathrm{t}]^{-1}$, $n_0 = 0$, $t_0 = 0$, and $t = 0$ (*black*), 1 (*sea green*), 2 (*mint green*), 3 (*green*), 4 (*chartreuse*), 5 (*yellow*), 6 (*orange*), 8 (*red*), 10 (*magenta*), 12 (*blue purple*), 14 (*electric blue*), 16 (*sky blue*), 18 (*turquoise*), and 20 [t] (*martian green*). The *lower picture* shows a discrete 3D plot of the density function

time interval, yielding for the elements of the transition matrix:

$$W(m|n) = \begin{cases} \gamma \,, & \text{if } m = n+1 \,, \\ 0 \,, & \text{otherwise} \,, \end{cases} \tag{3.84}$$

where the probability that two or more arrivals occur within the differential time interval dt is of measure zero. In other words, simultaneous arrivals of two or more events have zero probability. According to (3.46c′), the master equation has the form

$$\frac{\mathrm{d}P_n(t)}{\mathrm{d}t} = \gamma\big(P_{n-1}(t) - P_n(t)\big) \,, \tag{3.85}$$

with the initial condition $P_n(t_0) = \delta_{n,n_0}$. In other words, the number of arrivals recorded before $t = t_0$ is n_0. The interpretation of (3.85) is straightforward: the increase in the probability of recording n events between times t and $t + \mathrm{d}t$ is proportional to the difference in probabilities between $n-1$ and n recorded events, because the elementary single arrival processes $(n-1 \to n)$ and $(n \to n+1)$ increase or decrease, respectively, the probability of having recorded n events at time t.

The method of probability generating functions (Sect. 2.2.1) is now applied to derive solutions of the master equation (3.85). The probability generating function for the Poisson process is

$$g(s,t) = \sum_{n=0}^{\infty} P_n(t) s^n \,, \quad |s| \le 1 \,, \text{ with } g(s,t_0) = s^{n_0} \,. \tag{2.27′}$$

The time derivative of the generating function is obtained by inserting (3.85):

$$\frac{\partial g(s,t)}{\partial t} = \sum_{n=0}^{\infty} \frac{\partial P_n(t)}{\partial t} s^n = \gamma \sum_{n=0}^{\infty} \big(P_{n-1}(t) - P_n(t)\big) s^n \,.$$

The first sum is readily evaluated as

$$\sum_{n=0}^{\infty} \frac{\partial P_{n-1}(t)}{\partial t} s^n = s \sum_{n=0}^{\infty} \frac{\partial P_{n-1}(t)}{\partial t} s^{n-1} = s\, g(s,t) \,,$$

and the second sum is identical to the definition of the generating function. This yields the following equation for the generating function:

$$\frac{\partial g(s,t)}{\partial t} = \gamma(s-1)\, g(s,t) \,. \tag{3.86}$$

Since the equation does not contain a derivative with respect to the dummy variable s, we are dealing with a simple ODE, and the solution by conventional calculus is straightforward:

$$\int_{\ln g(s,t_0)}^{\ln g(s,t)} \mathrm{d}\ln g(s,t) = \int_{t_0}^{t} \gamma(s-1)\,\mathrm{d}t\,,$$

which yields

$$g(s,t) = s^{n_0} \mathrm{e}^{\gamma(s-1)(t-t_0)}\,, \quad \text{or } g(s,t) = \mathrm{e}^{\gamma(s-1)t} \text{ for } (n_0 = 0, t_0 = 0)\,, \tag{3.87}$$

with $g(s,0) = s^{n_0}$. The assumption $(n_0 = 0, t_0 = 0)$ is meaningful, because it implies that the counting of arrivals starts at time $t = 0$, and the expressions become especially simple: $g(0,t) = \exp(-\gamma t)$ and $g(s,0) = 1$. The individual probabilities $P_n(t)$ are obtained by expanding the exponential function and equating the coefficients of the powers of s:

$$\exp\bigl(\gamma(s-1)t\bigr) = \exp\bigl(s\gamma t\bigr)\mathrm{e}^{-\gamma t}\,,$$

$$\exp\bigl(s\gamma t\bigr) = 1 + s\frac{\gamma t}{1!} + s^2\frac{(\gamma t)^2}{2!} + s^3\frac{(\gamma t)^3}{3!} + \cdots\,.$$

Finally, we obtain the solution

$$P_n(t) = \mathrm{e}^{-\gamma t}\frac{(\gamma t)^n}{n!} = \mathrm{e}^{-\alpha}\frac{\alpha^n}{n!}\,, \tag{3.88}$$

which is the well-known Poisson distribution (2.35) with the expectation value $\mathrm{E}\bigl(\mathcal{X}(t)\bigr) = \gamma t = \alpha$ and variance $\mathrm{var}\bigl(\mathcal{X}(t)\bigr) = \gamma t = \alpha$. Since the standard deviation is $\sigma\bigl(\mathcal{X}(t)\bigr) = \sqrt{\gamma t} = \sqrt{\alpha}$, the Poisson process perfectly satisfies the $\sqrt{N}$ law for fluctuations (For an illustrative example see Fig. 3.11).

It is easy to check that the expectation value and variance can be obtained directly from the generating function by differentiating (2.28):

$$\begin{aligned} \mathrm{E}\bigl(\mathcal{X}(t)\bigr) &= \left.\frac{\partial g(s,t)}{\partial s}\right|_{s=1} = \gamma t\,, \\ \mathrm{var}\bigl(\mathcal{X}(t)\bigr) &= \left.\frac{\partial g(s,t)}{\partial s}\right|_{s=1} + \left.\frac{\partial^2 g(s,t)}{\partial s^2}\right|_{s=1} - \left(\left.\frac{\partial g(s,t)}{\partial s}\right|_{s=1}\right)^2 = \gamma t\,. \end{aligned} \tag{3.89}$$

We note that (3.85) can also be solved using the characteristic function (Sect. 2.2.3), which will be applied for the purpose of illustration in deriving the solution of the master equation of the one-dimensional random walk (Sect. 3.2.4).

Arrival and Waiting Times

The Poisson process can be viewed from a slightly different perspective by considering the *arrival times*[33] of the individual independent events as random variables $\mathcal{T}_1, \mathcal{T}_2, \ldots$, where the random counting variable takes on the values $\mathcal{X}(t) \geq 1$ for $t \geq \mathcal{T}_1$ and, in general, $\mathcal{X}(t) \geq k$ for $t \geq \mathcal{T}_k$. All arrival times $\mathcal{T}_k$ with $k \in \mathbb{N}_{>0}$ are positive if we assume that the process started at time $t = 0$. The number of arrivals before some fixed time ϑ is less than k if and only if the waiting time until the kth arrival is greater than ϑ. Accordingly, the two events $\mathcal{T}_k > \vartheta$ and $n(\vartheta) < k$ are equivalent and their probabilities are the same

$$P(\mathcal{T}_k > \vartheta) = P\big(n(\vartheta) < k\big) .$$

Now we consider the time before the first arrival, which is trivially the time until the first event happens:

$$P(\mathcal{T}_1 > \vartheta) = P\big(n(\vartheta) < 1\big) = P\big(n(\vartheta) = 0\big) = \mathrm{e}^{-\vartheta/\tau_\mathrm{w}} \frac{(\vartheta/\tau_\mathrm{w})^0}{0!} = \mathrm{e}^{-t/\tau_\mathrm{w}} ,$$

where we used (3.88) to calculate the distribution of first-arrival times. It is straightforward to show that the same relation holds for all inter-arrival times $\Delta\mathcal{T}_k = \mathcal{T}_k - \mathcal{T}_{k-1}$. After normalization, these follow an exponential density $\varrho(t, \tau_\mathrm{w}) = \mathrm{e}^{-t/\tau_\mathrm{w}}/\tau_\mathrm{w}$ with $\tau_\mathrm{w} > 0$ and $\int_0^\infty \varrho(t, \tau_\mathrm{w})\,\mathrm{d}t = 1$, and thus for each index k, we have

$$P(\Delta\mathcal{T}_k \leq t) = 1 - \mathrm{e}^{-t/\tau_\mathrm{w}} , \quad \text{and thus } P(\Delta\mathcal{T}_k > t) = \mathrm{e}^{-t/\tau_\mathrm{w}} , \quad t \geq 0 .$$

Now we can identify the parameter of the Poisson distribution as the reciprocal *mean waiting time* for an event τ_w^{-1}, with

$$\tau_\mathrm{w} = \int_0^\infty \mathrm{d}t\, t\, \varrho(t, \tau_\mathrm{w}) = \int_0^\infty \mathrm{d}t\, \frac{t}{\tau_\mathrm{w}} \mathrm{e}^{-\vartheta/\tau_\mathrm{w}} .$$

We shall use the exponential density in the calculation of expected times for the occurrence of chemical reactions modeled as first arrival times $\mathcal{T}_1$. Independence of the individual events implies the validity of

$$\begin{aligned} P(\Delta\mathcal{T}_1 > t_1, \ldots, \Delta\mathcal{T}_n > t_n) &= P(\Delta\mathcal{T}_1 > t_1) \cdots P(\Delta\mathcal{T}_n > t_n) \\ &= \mathrm{e}^{-(t_1 + \cdots + t_n)/\tau_\mathrm{w}} , \end{aligned}$$

[33] In the literature both expressions, *waiting time* and *arrival time*, are common. An inter-arrival time is a waiting time.

which determines the joint probability distribution of the *inter-arrival* times $\Delta\mathcal{T}_k$. The expectation value of the incremental arrival times, or times between consecutive arrivals, is simply given by $\mathrm{E}(\Delta\mathcal{T}_k) = \tau_{\mathrm{w}}$. Clearly, the greater the value of τ_{w}, the longer will be the mean inter-arrival time, and thus $1/\tau_{\mathrm{w}}$ can be taken as the intensity of flow. Compared to the previous derivation, we have $1/\tau_{\mathrm{w}} \equiv \gamma$.

For $\mathcal{T}_0 = 0$ and $n \geq 1$, we can readily calculate the cumulative random variable, the arrival time of the the n th arrival:

$$\mathcal{T}_n = \Delta\mathcal{T}_1 + \cdots + \Delta\mathcal{T}_n = \sum_{k=1}^{n} \Delta\mathcal{T}_k \,.$$

The event $\mathcal{I} = (\mathcal{T}_n \leq t)$ implies that the n th arrival has occurred before time t. The connection between the arrival times and the cumulative number of arrivals $\mathcal{X}(t)$ is easily made and illustrates the usefulness of the dual point of view:

$$P(\mathcal{I}) = P(\mathcal{T}_n \leq t) = P\big(\mathcal{X}(t) \geq n\big) \,.$$

More precisely, $\mathcal{X}(t)$ is determined by the whole sequence $\Delta\mathcal{T}_k (k \geq 1)$, and depends on the elements ω of the sample space through the individual inter-arrival times $\Delta\mathcal{T}_k$. In fact, we can compute the number of arrivals exactly as the joint probability of having recorded $n-1$ arrivals until time t and recording one arrival in the interval $[t, t+\Delta t]$ [536, pp. 70–72]:

$$P(t \leq \mathcal{T}_n \leq t + \Delta t) = P\big(\mathcal{X}(t) = n-1\big) P\big(\mathcal{X}(t+\Delta t) - \mathcal{X}(t) = 1\big) \,.$$

Since the two time intervals $[0, t[$ and $[t, t+\Delta t]$ do not overlap, the two events are independent and the joint probability can be factorized. For the first factor, we use the probability of a Poissonian distribution, while the second factor follows simply from the definition of the parameter γ:

$$P\big(t \leq \mathcal{T}_n \leq t + \Delta t\big) = \frac{\mathrm{e}^{-\gamma t}(\gamma t)^{n-1}}{(n-1)!} \gamma \Delta t \,.$$

In the limit $\Delta t \to \mathrm{d}t$, we obtain the probability density of the n th arrival time as

$$f_{\mathcal{T}_n}(t) = \frac{\gamma^n t^{n-1}}{(n-1)!} \mathrm{e}^{-\gamma t} \,, \tag{3.90}$$

which is known as the *Erlang distribution*, named after the Danish mathematician Agner Karup Erlang. It is straightforward now to compute the expectation value of the n th waiting time:

$$\mathrm{E}(\mathcal{T}_n) = \int_0^{\infty} t \frac{\gamma^n t^{n-1}}{(n-1)!} \mathrm{e}^{-\gamma t}\, \mathrm{d}t = \frac{n}{\gamma} \,, \tag{3.91}$$

which is another linear relation. The n th waiting time is proportional to n, with the proportionality factor being the reciprocal rate parameter $1/\gamma$.

The Poisson process is characterized by three properties:

(i) The observations occur one at a time.
(ii) The numbers of observations in disjoint time intervals are independent random variables.
(iii) The distribution of $\mathcal{X}(t + \Delta t) - \mathcal{X}(t)$ is independent of t.

Then there exists a constant $\alpha > 0$ such that, for $\Delta t = t - \tau > 0$, the difference $\mathcal{X}(t) - \mathcal{X}(\tau)$ is Poisson distributed with parameter $\alpha\Delta t$, i.e.,

$$P\big(\mathcal{X}(t + \Delta t) = k\big) = \frac{\big(\alpha\Delta t\big)^k}{k!}\mathrm{e}^{-\alpha\Delta t} .$$

For $\alpha = 1$, the process $\mathcal{X}(t)$ is a *unit* or rate one Poisson process, and the expectation value is $\mathrm{E}\big(\mathcal{Y}(t)\big) = \Delta t$. In other words the mean number of events per unit time is one, $\Delta t = 1$. If $\mathcal{Y}(t)$ is a unit Poisson process and $\mathcal{Y}_\alpha(t) \equiv \mathcal{Y}(\alpha t)$, then $\mathcal{Y}_\alpha$ is a Poisson process with parameter α. A Poisson process is an example of a *counting process* $\mathcal{X}(t)$ with $t \geq 0$ that satisfies three properties:

1. $\mathcal{X}(t) \geq 0$,
2. $\mathcal{X}(t) \in \mathbb{N}$, and
3. if $\tau \leq t$, then $\mathcal{X}(\tau) \leq \mathcal{X}(t)$.

The number of events occurring during the time interval $[\tau, t]$ with $\tau < t$ is $\mathcal{X}(t) - \mathcal{X}(\tau)$.

3.2.3 *Master Equations*

Master equations are used to model stochastic processes on discrete sample spaces, $\mathcal{X}(t) \in \mathbb{N}$, and we have already dealt with one particular example, the occurrence of independent events in the form of the Poisson process (Sect. 3.2.2.4). Because of their general importance, in particular in chemical kinetics and population dynamics in biology, we shall present here a more detailed discussion of the properties and the different versions of master equations.

General Master Equations

The master equations we are considering here describe continuous time processes, i.e., $t \in \mathbb{R}$. Then, the starting point is the dCKE (3.46c) for pure jump processes, with the integral converted into a sum by Riemann–Stieltjes integration (Sect. 1.8.2):

$$\frac{\mathrm{d}P_n(t)}{\mathrm{d}t} = \sum_{m=0}^{\infty}\Big(W(n|m,t)P_m(t) - W(m|n,t)P_n(t)\Big) , \quad n, m \in \mathbb{N} , \tag{3.83}$$

where we have implicitly assumed sharp initial conditions $P_n(t_0) = \delta_{n,n_0}$. The individual terms $W(k|j,t)P_j(t)$ of (3.83) have a straightforward interpretation as transition rates from state Σ_j to state Σ_k in the form of the product of the transition probability and the probability of being in state Σ_j at time t (Fig. 3.22). The *transition probabilities* $W(n|m,t)$ form a possibly infinite transition matrix. In all realistic cases, however, we shall be dealing with a finite *state space*: $m,n \in \{0,1,\dots,N\}$. This is tantamount to saying that we are always dealing with a finite number of molecules in chemistry or to stating that population sizes in biology are finite. Since the off-diagonal elements of the transition matrix represent probabilities, they are nonnegative by definition: $\mathrm{W} \doteq (W_{nm}; n,m \in \mathbb{N}_{\geq 0})$ (Fig. 3.12). The diagonal elements $W(n|n,t)$ cancel in the master equation and hence can be defined at will, without changing the dynamics of the process. Two definitions are in common use:

(i) Normalization of matrix elements:

$$\sum_m W_{mn} = 1\,, \quad W_{nn} = 1 - \sum_{m \neq n} W_{mn}\,, \tag{3.92a}$$

and accordingly W is a stochastic matrix. This definition is applied, for example, in the mutation selection problem [130].

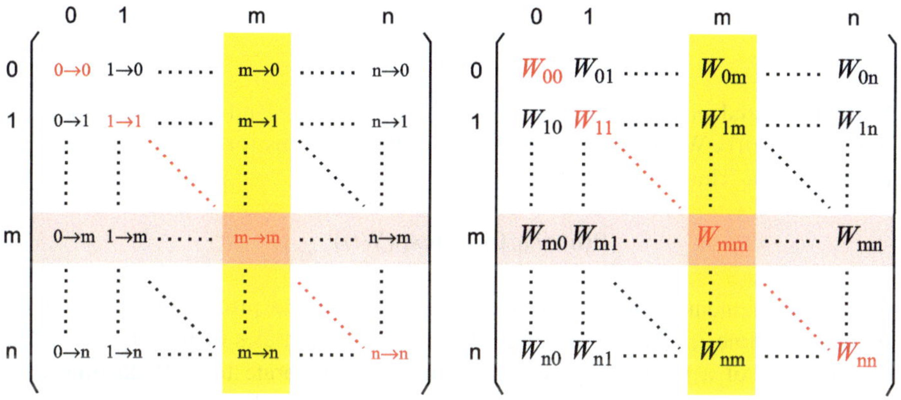

Fig. 3.12 The transition matrix of the master equation. The figure is intended to clarify the meaning and handling of the elements of transition matrices in master equations. The matrix on the *left-hand side* shows the individual transitions that are described by the corresponding elements of the transition matrix $\mathrm{W} = (W_{ij}; i,j = 0,1,\dots,n)$. The elements in a given row (*shaded light red*) contain all transitions going into one particular state m, and they are responsible for the differential change in probabilities: $\mathrm{d}P_m(t)/\,\mathrm{d}t = \sum_k W_{mk}P_k(t)$. The elements in a column (*shaded yellow*) quantify all probability flows going out from state m, and their sums are involved in conservation of probabilities. The diagonal elements (*red*) cancel in master equations (3.83), so they do not change probabilities and need not be specified explicitly. To write master equations in compact form (3.83′), the diagonal elements are defined by the annihilation convention $\sum_k W_{km} = 0$. The summation of the elements in a column is also used in the definition of *jump moments*

(ii) Annihilation of diagonal elements:

$$\sum_m W_{mn} = 0 \,, \qquad W_{nn} = -\sum_{m,m\neq n} W_{mn} \,, \tag{3.92b}$$

which is used, for example, in the compact from of the master equation (3.83′) and in several applications, for example, in phylogeny.

Transition probabilities in the general master equation (3.83) are assumed to be time dependent. Most frequently we shall, however, assume that they do not depend on time and use $W_{nm} = W(n|m)$. Then a Markov process in general and a master equation in particular are said to be *time homogeneous* if the transition matrix W does not depend on time.

Formal Solution of the Master Equation
Inserting the annihilation condition (3.92b) into (3.83) leads to a compact form of the master equation:

$$\frac{\mathrm{d}P_n(t)}{\mathrm{d}t} = \sum_m W_{nm} P_m(t) \,. \tag{3.83′}$$

Introducing vector notation $\mathbf{P}(t)^{\mathrm{t}} = \big(P_1(t), \ldots, P_n(t), \ldots\big)$, we obtain

$$\frac{\mathrm{d}\mathbf{P}(t)}{\mathrm{d}t} = \mathrm{W} \times \mathbf{P}(t) \,. \tag{3.83″}$$

With the initial condition $P_n(0) = \delta_{n,n_0}$ stated above and a time independent transition matrix W, we can solve (3.83″) in formal terms for each n_0 by applying linear algebra. This yields

$$P(n,t|n_0,0) = \big(\exp(\mathrm{W}t)\big)_{n,n_0} \,,$$

where the element (n, n_0) of the matrix $\exp(\mathrm{W}t)$ is the probability of having n particles at time t, $\mathcal{X}(t) = n$, when there were n_0 particles at time $t_0 = 0$. The computation of a matrix exponential is quite an elaborate task. If the matrix is diagonalizable, i.e., there is a matrix T such that $\Lambda = \mathrm{T}^{-1}\mathrm{W}\mathrm{T}$ with

$$\Lambda = \begin{pmatrix} \lambda_1 & 0 & \ldots & 0 \\ 0 & \lambda_2 & \ldots & 0 \\ \vdots & \vdots & \ddots & \vdots \\ 0 & 0 & \ldots & \lambda_n \end{pmatrix} ,$$

and the exponential can be obtained by $e^{W} = Te^{\Lambda}T^{-1}$. Apart from special cases, a matrix can be diagonalized analytically only in rather few low-dimensional cases, and in general, one has to rely on numerical methods.

Jump Moments

It is often convenient to express changes in particle numbers in terms of the so-called *jump moments* [415, 503, 541]:

$$\alpha_p(n) = \sum_{m=0}^{\infty} (m-n)^p W(m|n) \,, \quad p = 1, 2, \ldots \,. \tag{3.93}$$

The usefulness of the first two jump moments with $p = 1, 2$ is readily demonstrated. We multiply (3.83) by n and obtain by summation

$$\begin{aligned}
\frac{\mathrm{d}\langle n\rangle}{\mathrm{d}t} &= \sum_{n=0}^{\infty} n \sum_{m=0}^{\infty} \Big(W(n|m)P_m(t) - W(m|n)P_n(t) \Big) \\
&= \sum_{m=0}^{\infty}\sum_{n=0}^{\infty} mW(m|n)P_n(t) - \sum_{n=0}^{\infty}\sum_{m=0}^{\infty} nW(m|n)P_n(t) \\
&= \sum_{m=0}^{\infty}\sum_{n=0}^{\infty} (m-n)W(m|n)P_n(t) = \langle \alpha_1(n) \rangle \,.
\end{aligned}$$

Since the variance $\mathrm{var}(n) = \big\langle (n - \langle n\rangle)^2 \big\rangle$ involves $\langle n^2 \rangle$, we need the time derivative of the second raw moment $\hat{\mu}_2 = \langle n^2 \rangle$, and obtain it by (i) multiplying (3.93) for $p = 2$ by n^2 and (ii) summing:

$$\begin{aligned}
\frac{\mathrm{d}\langle n^2\rangle}{\mathrm{d}t} &= \sum_{m=0}^{\infty}\sum_{n=0}^{\infty} (m^2 - n^2)W(m|n)P_n(t) \\
&= \langle \alpha_2(n) \rangle + 2\, \langle n\alpha_1(n) \rangle \,.
\end{aligned}$$

Adding the term $\mathrm{d}\,\langle n\rangle^2 / \mathrm{d}t = 2\,\langle n\rangle\, \mathrm{d}\,\langle n\rangle / \mathrm{d}t$ yields the expression for the evolution of the variance, and finally we obtain for the first two moments:

$$\frac{\mathrm{d}\langle n\rangle}{\mathrm{d}t} = \langle \alpha_1(n) \rangle \,, \tag{3.94a}$$

$$\frac{\mathrm{d}\,\mathrm{var}(n)}{\mathrm{d}t} = \langle \alpha_2(n) \rangle + 2\Big(\langle n\alpha_1(n) \rangle - \langle n \rangle \langle \alpha_1(n) \rangle \Big) \,. \tag{3.94b}$$

The expression (3.94a) is not a closed equation for $\langle n \rangle$, since its solution involves higher moments of n. Only if $\alpha_1(n)$ is a linear function can the two summations, $\sum_{m=0}^{\infty}$ for the jump moment and $\sum_{n=0}^{\infty}$ for the expectation value, be interchanged. Then, after the swap, we obtain a single standalone ODE

$$\frac{\mathrm{d}\langle n \rangle}{\mathrm{d}t} = \alpha_1\big(\langle n \rangle\big) , \tag{3.94a$'$}$$

which can be integrated directly to yield the expectation value $\langle n(t) \rangle$. The latter coincides with the deterministic solution in this case (see birth-and-death master equations). Otherwise, in nonlinear systems, the expectation value does not coincide with the deterministic solution (see, for example, Sect. 4.3), or in other words initial values of moments higher than the first are required to compute the time course of the expectation value.

Nico van Kampen [541] also provides a straightforward approximation derived from a series expansion of $\alpha_1(n)$ in $n - \langle n \rangle$, with truncation after the second derivative:

$$\frac{\mathrm{d}\langle n \rangle}{\mathrm{d}t} = \alpha_1\big(\langle n \rangle\big) + \frac{1}{2}\mathrm{var}(n)\frac{\mathrm{d}^2}{\mathrm{d}n^2}\alpha_1\big(\langle n \rangle\big) . \tag{3.94a$''$}$$

A similar and consistent approximation for the time dependence of the variance reads

$$\frac{\mathrm{d}\,\mathrm{var}(n)}{\mathrm{d}t} = \alpha_2\big(\langle n \rangle\big) + 2\,\mathrm{var}(n)\frac{\mathrm{d}}{\mathrm{d}n}\alpha_1\big(\langle n \rangle\big) . \tag{3.94b$''$}$$

The two expressions together provide a closed equation for calculating the expectation value and variance. They show directly the need to know initial fluctuations when computing the time course of expectation values.

Birth-and-Death Master Equations

In the derivation of the dCKE and the master equation, we made the realistic assumption that the limit of infinitesimal time steps $\lim \Delta t \rightarrow 0$ excludes the simultaneous occurrence of two or more jumps. The general master equation (3.83), however, allows for simultaneous jumps of all sizes, viz., $\Delta n = n - m$, $m = 0, \ldots, \infty$, and this introduces a dispensable complication. In this paragraph we shall make use of a straightforward simplification in the form of *death-and-birth* processes, which restricts the size of jumps, reduces the number of terms in the master equation, and makes the expressions for the jump moments much easier to handle.

The idea of birth-and-death processes was invented in biology (Sect. 5.2.2) and is based on the assumption that constant and finite numbers of individuals are produced (born), or disappear (die), in single events. Accordingly the jump size is a

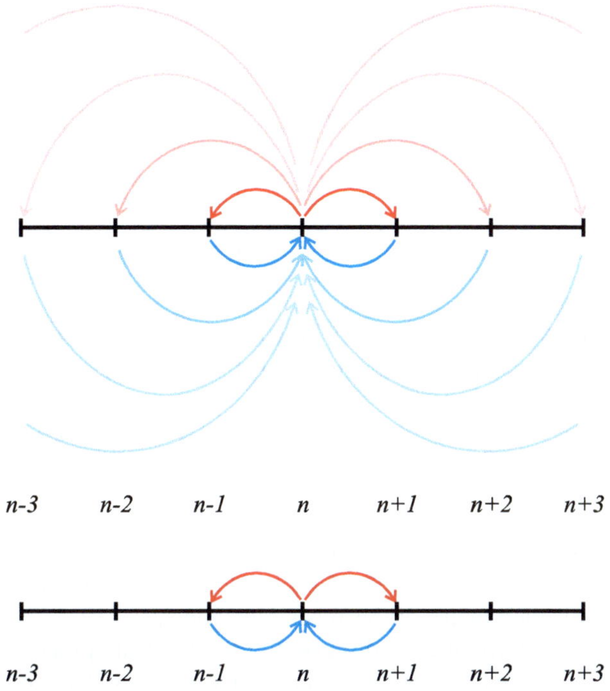

Fig. 3.13 Sketch of the transition probabilities in master equations. In the general master equation, steps of any size are admitted (*upper diagram*), whereas in birth-and-death processes, all jumps have the same size. The simplest and most common case concerns the condition that the particles are born and die one at a time (*lower diagram*), which is consistent with the derivation of the differential Chapman–Kolmogorov equation (Sect. 3.2.1)

matter of the application, be it in physics, chemistry, or biology, and the information about it has to come from empirical observations. To give examples, in chemical kinetics the jump size is determined by the stoichiometry of the process, and in population biology the jump size for birth is the litter size,[34] and it is commonly one for natural death.

Here we shall consider jump size as a feature of the mathematical characterization of a stochastic process. The jump size determines the handling of single events, and we adopt the same procedure that we used in the derivation of the dCKE, i.e., we choose a sufficiently small time interval Δt for recording events such that the simultaneous occurrence of two events has probability measure zero. The resulting models are commonly called single step birth-and-death processes and the time step Δt is referred to as the *blind interval*, because the time resolution does not go beyond Δt. The difference in choosing steps between general and birth-and-death master equations is illustrated in Fig. 3.13 (see also Sect. 4.6). In this chapter we shall restrict analysis and discussion to processes with a single variable and postpone the discussion of multivariate cases to chemical reaction networks, dealt with in Chap. 4.

[34]The litter size is defined as the mean number of offspring produced by an animal in a single birth.

Within the single step birth-and-death model, the transition probabilities are reduced to neighboring states and we assume time independence

$$W(n|m) = W_{nm} = w_m^+ \delta_{n,m+1} + w_m^- \delta_{n,m-1} ,$$
$$\text{or} \quad W_{nm} = \begin{cases} w_m^+ , & \text{if } m = n-1 , \\ w_m^- , & \text{if } m = n+1 , \\ 0 , & \text{otherwise} , \end{cases} \tag{3.95}$$

since we are dealing with only two allowed processes out of and into each state n in the unit step size transition probability model, viz.,[35]

$$w_n^+ \equiv \lambda_n \text{ for } n \rightarrow n+1 , \tag{3.96a}$$

$$w_n^- \equiv \mu_n \text{ for } n \rightarrow n-1 , \tag{3.96b}$$

respectively. The notations for *step-up* and *step-down* transitions for these two classes of events are self-explanatory. As a consequence of this simplification, the transition matrix W becomes tridiagonal.

We have already discussed birth-and-death processes in Sect. 3.2.2.4, where we considered the Poisson process. This can be understood as a birth-and-death process with zero death rate, or simply a birth process, on $n \in \mathbb{N}$. The one-dimensional random walk (Sect. 3.2.4) is a birth-and-death process with equal birth and death rates when the population variable is interpreted as the spatial coordinate and negative values are admitted, i.e., $n \in \mathbb{Z}$. Modeling of chemical reactions by birth-and-death processes will turn out to be a very useful approach.

Within the single step model the stochastic process can be described by a *birth-and-death master equation*

$$\frac{\mathrm{d}P_n(t)}{\mathrm{d}t} = w_{n-1}^+ P_{n-1}(t) + w_{n+1}^- P_{n+1}(t) - (w_n^+ + w_n^-) P_n(t) . \tag{3.97}$$

There is no general technique that allows one to find the time-dependent solutions of (3.97). However, special cases are important in chemistry and biology, and we shall therefore present several examples later on. In Sect. 5.2.2, we shall also give a detailed overview of the exactly solvable single step birth-and-death processes [216]. Nevertheless, it is possible to analyze the stationary case in full generality.

[35]Exceptions with only one transition are the lowest and the highest state, $n = n_{\min}$ and $n = n_{\max}$, which are the boundaries of the system. In biology, the notation $w_n^+ \equiv \lambda_n$ and $w_n^- \equiv \mu_n$ for death and birth rates is common.

Stationary Solutions

Provided there exists a stationary solution of the birth-and-death master equation (3.97), $\lim_{t\to\infty} P_n(t) = \overline{P}_n$, we can compute it in a straightforward manner. We define a probability current $\varphi(n)$ for the n th step in the series involving $n-1$ and n:

$$\begin{array}{lcccccccc} \text{particle number } 0 & \rightleftharpoons 1 & \rightleftharpoons \ldots & \rightleftharpoons & n-1 & \rightleftharpoons n & \rightleftharpoons & n+1 \ldots \\ \text{reaction step} & 1 & 2 \ldots n-1 & & & n & n+1 & \ldots \end{array}$$

which attributes a positive sign to the direction of increasing n:

$$\varphi_n = w_{n-1}^+ P_{n-1} - w_n^- P_n \ , \qquad \frac{\mathrm{d}P_n(t)}{\mathrm{d}t} = \varphi_n - \varphi_{n+1} \ . \tag{3.98}$$

Restriction to nonnegative particle numbers $n \in \mathbb{N}$ implies $w_0^- = 0$ and $P_n(t) = 0$ for $n < 0$, which in turn leads to $\varphi_0 = 0$. The conditions for the stationary solution are given by

$$\frac{\mathrm{d}P_n(t)}{\mathrm{d}t} = 0 = \overline{\varphi}_n - \overline{\varphi}_{n+1} \ , \qquad \overline{\varphi}_{n+1} = \overline{\varphi}_n \ . \tag{3.99}$$

We now sum the vanishing flow terms according to (3.99). From the telescopic sum with $n_{\min} = l = 0$ and $n_{\max} = u = N$, we obtain

$$0 = \sum_{n=0}^{N-1} (\overline{\varphi}_n - \overline{\varphi}_{n+1}) = \overline{\varphi}_0 - \overline{\varphi}_N \ .$$

Thus we find $\overline{\varphi}_n = 0$ for arbitrary n, which leads to

$$\overline{P}_n = \frac{w_{n-1}^+}{w_n^-} \overline{P}_{n-1} \ , \quad \text{and finally,} \quad \overline{P}_n = \overline{P}_0 \prod_{m=1}^{n} \frac{w_{m-1}^+}{w_m^-} \ . \tag{3.100}$$

The probability $\overline{P}_0$ is obtained from normalization $\sum_{n=0}^{N} \overline{P}_n = 1$ (for example, see Sects. 4.6.4 and 5.2.2).

The vanishing flow condition $\overline{\varphi}_n = 0$ for every reaction step at equilibrium is known in chemical kinetics as the principle of detailed balance. It is commonly attributed to the American mathematical physicist Richard Tolman [531], although it was already known and applied earlier [340, 564] (see also, for example, [194, pp. 142–158]).

So far we have not yet asked how a process might be confined to the domain $n \in [l, u]$. This issue is closely related to the problem of boundaries for birth-and-death processes that will be analyzed in a separate section (Sect. 3.3.4). In essence, we distinguish two classes of boundaries: (i) *absorbing boundaries* and (ii) *reflecting boundaries*. If a stochastic process hits an absorbing boundary, it ends there. A reflecting boundary sends arriving processes back into the allowed domain of the variable, $n \in [l, u]$. The existence of an absorbing boundary at $n = 0$ implies $\lim_{t\to\infty} \mathcal{X}(t) = 0$ and only reflecting boundaries are compatible with nontrivial stationary solutions. The conditions

$$w_l^- = 0\,, \quad w_u^+ = 0\,, \tag{3.101}$$

are sufficient for the existence of reflecting boundaries on both sides of the domain $n \in [l, u]$, and thus represent a prerequisite for a stationary birth-and-death process (for details see Sect. 3.3.4).

Calculating Moments Directly from Master Equations

The simplification of the general master equation (3.83) introduced through the restriction to single step jumps (3.97) provides the basis for the derivation of fairly simple expressions for the time derivatives of first and second moments.[36] All calculations are facilitated by the trivial but important equalities[37]

$$\sum_{n=-\infty}^{+\infty}(n-1)w_{n-1}^{\pm}P_{n-1}(t) = \sum_{n=-\infty}^{+\infty} n w_n^{\pm} P_n(t) = \sum_{n=-\infty}^{+\infty}(n+1)w_{n+1}^{\pm}P_{n+1}(t)\,,$$

and we shall make use of these shifts in summation indices later on when solving master equations by means of probability generating functions. Multiplying $\mathrm{d}P_n/\,\mathrm{d}t$ by n, summing over n, and making use of

$$\sum_{n=-\infty}^{\infty}(n+1)w_n^{\pm}P_n(t) = \sum_{n=-\infty}^{\infty} n w_n^{\pm}P_n(t) + \sum_{n=-\infty}^{\infty} w_n^{\pm}P_n(t)\,,$$

we obtain for the expectation value:

$$\frac{\mathrm{d}\langle n\rangle}{\mathrm{d}t} = \sum_{n=-\infty}^{\infty} n\frac{\mathrm{d}P_n(t)}{\mathrm{d}t} = \langle w_n^+\rangle - \langle w_n^-\rangle = \langle w_n^+ - w_n^-\rangle\,. \tag{3.102a}$$

[36] An excellent tutorial on this subject by Bahram Houchmandzadeh can be found at http://www.houchmandzadeh.net/cours/Master_Eq/master.pdf. Retrieved 2 May 2014.

[37] In general these equations hold also for summations from 0 to $+\infty$ if the corresponding physically meaningless probabilities are set equal to zero by definition: $P_n(t) = 0\,,\ \forall\, n \in \mathbb{Z}_{<0}$.

The second raw moment $\hat{\mu}_2 = \langle n^2 \rangle$ and the variance are derived by an analogous procedure, namely, multiplication by n^2, summation, and substitution:

$$\frac{\mathrm{d}\langle n^2 \rangle}{\mathrm{d}t} = \sum_{n=-\infty}^{\infty} n^2 \frac{\mathrm{d}P_n(t)}{\mathrm{d}t} = 2\left\langle n(w_n^+ - w_n^-)\right\rangle + \left\langle w_n^+ + w_n^- \right\rangle ,$$

$$\begin{aligned}\frac{\mathrm{d}\,\mathrm{var}(n)}{\mathrm{d}t} &= \frac{\mathrm{d}\left(\langle n^2 \rangle - \langle n \rangle^2\right)}{\mathrm{d}t} = \frac{\mathrm{d}\langle n^2 \rangle}{\mathrm{d}t} - \frac{\mathrm{d}\langle n \rangle^2}{\mathrm{d}t} \\ &= \frac{\mathrm{d}\langle n^2 \rangle}{\mathrm{d}t} - 2\,\langle n \rangle\,\frac{\mathrm{d}\langle n \rangle}{\mathrm{d}t} \\ &= 2\left\langle \left(n - \langle n \rangle\right)(w_n^+ - w_n^-)\right\rangle + \left\langle w_n^+ + w_n^- \right\rangle . \end{aligned} \tag{3.102b}$$

Jump Moments
Jump moments are substantially simplified by the assumption of single birth-and-death events as well:

$$\alpha_p(n) = \sum_{n=0}^{\infty} (m-n)^p W_{mn} = (-1)^p w_n^- + w_n^+ .$$

Neglect of the fluctuation part in the first jump moment $\alpha_1(n)$ results in a rate equation for the deterministic variable $\hat{n}(t)$ corresponding to $\langle n \rangle$:

$$\frac{\mathrm{d}\hat{n}}{\mathrm{d}t} = w_{\hat{n}}^+ - w_{\hat{n}}^- , \quad \text{with } w_{\hat{n}}^{\pm} = w_{\langle n \rangle}^{\pm} = \sum_{n=0}^{\infty} w_n^{\pm} P_n(t) . \tag{3.103a}$$

The first two jump moments, $\alpha_1(n)$ and $\alpha_2(n)$, together with the two simplified coupled equations (3.94a″) and (3.94b″) yield

$$\frac{\mathrm{d}\langle n \rangle}{\mathrm{d}t} = w_{\langle n \rangle}^+ - w_{\langle n \rangle}^- + \frac{1}{2}\mathrm{var}(n)\frac{\mathrm{d}^2}{\mathrm{d}n^2}\left(w_{\langle n \rangle}^+ - w_{\langle n \rangle}^-\right) , \tag{3.103b}$$

$$\frac{\mathrm{d}\,\mathrm{var}(n)}{\mathrm{d}t} = w_{\langle n \rangle}^+ + w_{\langle n \rangle}^- + 2\mathrm{var}(n)\frac{\mathrm{d}}{\mathrm{d}n}\left(w_{\langle n \rangle}^+ - w_{\langle n \rangle}^-\right) . \tag{3.103c}$$

It is now straightforward to show by example how linear jump moments simplify the expressions. In the case of a linear birth-and-death process, for step-up and step-down transitions, and for jump moments, respectively, we have

$$w_n^+ = \lambda n , \quad w_n^- = \mu n , \quad \alpha_p(n) = \left(\lambda + (-1)^p \mu\right) n .$$

Differentiating $w_n^{\pm}$ or $\alpha_p(n)$ twice with respect to n yields zero, the differential equations (3.103a) and (3.103b) are identical, and the solution is of the form

$$\langle n(t)\rangle = \hat{n}(t) = \hat{n}(0)\mathrm{e}^{(\lambda-\mu)t} .$$

The expectation value of the stochastic variable $\langle n\rangle$ coincides with the deterministic variable $\hat{n}$. We stress again that this coincidence requires linear step-up and step-down transition probabilities (see also Sect. 4.2.2). More details on the linear birth-and-death process can be found in Sect. 5.2.2.

Extinction Probabilities and Extinction Times

The state Σ_0 with $n = 0$ is an absorbing state in most master equations describing autocatalytic reactions or birth-and-death processes in biology. Then two quantities, the probability of absorption or extinction and the time to extinction, from state Σ_m, $\mathcal{Q}_m$ and $\mathcal{T}_m$, are of particular interest in biology, and their calculation represents a standard problem in stochastic processes. Straightforward derivations are given in [290, pp. 145–150] and we repeat them briefly here.

We consider a process $\big(\mathcal{X}(t)\big)$ with probability $P_n(t) = P\big(\mathcal{X}(t) = n\big)$, which is defined on the natural numbers $n \in \mathbb{N}$, and which satisfies the sharp initial condition $\mathcal{X}(0) = m$ or $P_n(0) = \delta_{n,m}$. The birth-and-death rates are $w_n^+ = \lambda_n$ and $w_n^- = \mu_n$, both for $n = 1, 2, \ldots$, and the value $w_0^+ = \lambda_0 = 0$ guarantees that, once it has reached the state of extinction Σ_0, the process gets absorbed and will stay there forever. First we calculate the probabilities of absorption from Σ_m into Σ_0 that we denote by $\mathcal{Q}_m$. Two transitions starting from Σ_i are allowed, and we get for the first step

$$i \to i-1 \text{ with probability } \frac{\mu_i}{\lambda_i + \mu_i} , \qquad i \to i+1 \text{ with probability } \frac{\lambda_i}{\lambda_i + \mu_i} .$$

Consecutive transitions can be turned into a recursion formula[38]:

$$\mathcal{Q}_i = \frac{\mu_i}{\lambda_i + \mu_i}\mathcal{Q}_{i-1} + \frac{\lambda_i}{\lambda_i + \mu_i}\mathcal{Q}_{i+1} , \quad i \geq 1 , \tag{3.104a}$$

where $\mathcal{Q}_0 = 1$. Rewriting the equation and introducing differences between consecutive extinction probabilities, viz., $\Delta\mathcal{Q}_i = \mathcal{Q}_{i+1} - \mathcal{Q}_i$, yields

$$(\mathcal{Q}_{i+1} - \mathcal{Q}_i) = \frac{\mu_i}{\lambda_i}(\mathcal{Q}_i - \mathcal{Q}_{i-1}) , \quad \text{or } \Delta\mathcal{Q}_i = \frac{\mu_i}{\lambda_i}\Delta\mathcal{Q}_{i-1} .$$

[38]The probability of extinction from state Σ_i is the probability of proceeding one step down multiplied by the probability of extinction from state Σ_{i-1} plus the probability of going one step up times the probability of becoming extinct from Σ_{i+1}.

The last expression can be iterated and yields

$$(Q_{i+1} - Q_i) = \Delta Q_i = \Delta Q_0 \prod_{j=1}^{i} \frac{\mu_j}{\lambda_j} = (Q_1 - 1) \prod_{j=1}^{i} \frac{\mu_j}{\lambda_j} . \tag{3.104b}$$

Summing all terms from $i = 1$ to $i = m$ yields a telescopic sum of the form

$$Q_{m+1} - Q_1 = (Q_1 - 1) \sum_{i=1}^{m} \left(\prod_{j=1}^{i} \frac{\mu_j}{\lambda_j} \right) , \quad m \geq 1 . \tag{3.104c}$$

By definition probabilities are bounded by one and so is the left-hand side of the equation, viz., $|Q_{m+1} - Q_1| \leq 1$. Hence, $Q_1 = 1$ has to hold, whenever the sum diverges, $\sum_{i=1}^{\infty} (\prod_{j=1}^{i} (\mu_j/\lambda_j)) = \infty$. From $Q_1 - 1 = \Delta Q_0 = 0$, it follows directly that $Q_m = 1$ for all $m \geq 2$, so extinction is certain from all initial states.

Alternatively, from $0 < Q_1 < 1$, it follows directly that

$$\sum_{i=1}^{\infty} \left(\prod_{j=1}^{i} \frac{\mu_j}{\lambda_j} \right) < \infty .$$

In addition, it is straightforward to see from (3.104b) that Q_m decreases when m increases from zero to m, i.e., $Q_0 = 1 > Q_1 > Q_2 > \ldots > Q_m$. Furthermore, we claim that $Q_m \to 0$ as $m \to \infty$, as can be shown by rebuttal of the opposite assumption that Q_m is bounded away from zero: $Q_m \geq \alpha > 0$, which can be satisfied only by $\alpha = 1$. The solution is obtained by considering the limit $m \to \infty$:

$$Q_m = \frac{\sum_{i=m}^{\infty} \left(\prod_{j=1}^{i} \mu_j/\lambda_j \right)}{1 + \sum_{i=1}^{\infty} \left(\prod_{j=1}^{i} \mu_j/\lambda_j \right)} , \quad m \geq 1 . \tag{3.104d}$$

As a particularly simple example, we consider the linear birth-and-death process with $\lambda_n = \lambda n$ and $\mu_n = \mu n$. The summations lead to geometric series and the final result is

$$Q_m = \begin{cases} (\mu/\lambda)^m , & \text{if } \lambda > \mu , \\ 1 , & \text{if } \lambda \leq \mu , \end{cases} \quad m \geq 1 . \tag{3.105}$$

Extinction is certain if the parameter of the birth rate is less than or equal to the parameter of the death rate, i.e., $\lambda \leq \mu$. We shall encounter this result and its consequences several times in Sect. 5.2.2.

The mean time until absorption from state Σ_m, denoted by $E(\mathcal{T}_m) = \vartheta_m$, is derived in a similar way. We start from state Σ_i and consider the first transition $\Sigma_i \to \Sigma_{i\pm1}$. As outlined in the case of the Poisson process (Sect. 3.2.2.4), the time until the first event happens is exponentially distributed, and this leads to a mean waiting time of $\tau_\mathrm{w} = (\lambda_i + \mu_i)^{-1}$. Inserting the mean extinction times from the two neighboring states yields

$$\vartheta_i = \frac{1}{\lambda_i + \mu_i} + \frac{\lambda_1}{\lambda_i + \mu_i}\vartheta_{i+1} + \frac{\mu_1}{\lambda_i + \mu_i}\vartheta_{i-1}\,, \quad i \geq 1\,, \tag{3.106a}$$

with $\vartheta_0 = 0$. As in the derivation of the absorption probabilities, we introduce differences between consecutive extinction times, viz., $\Delta\vartheta_i = \vartheta_i - \vartheta_{i+1}$, then rearrange to obtain, for the recursion and the first iteration,

$$\Delta\vartheta_i = \frac{1}{\lambda_i} + \frac{\mu_i}{\lambda_i}\Delta\vartheta_{i-1}\,, \quad \Delta\vartheta_i = \frac{1}{\lambda_i} + \frac{\mu_i}{\lambda_i\lambda_{i-1}} + \frac{\mu_i\mu_{i-1}}{\lambda_i\lambda_{i-1}}\Delta\vartheta_{i-2}\,, \quad i \geq 1\,.$$

Finally, with the convention $\prod_{j=m+1}^{m} \mu_j/\lambda_j = 1$, we find:

$$\begin{aligned}\vartheta_m - \vartheta_{m+1} = \Delta\vartheta_m &= \sum_{i=1}^{m} \frac{1}{\lambda_i} \prod_{j=i+1}^{m} \frac{\mu_j}{\lambda_j} + \Delta\vartheta_0 \prod_{j=1}^{i} \frac{\mu_j}{\lambda_j} \\ &= \prod_{j=1}^{m} \frac{\mu_i}{\lambda_i} \sum_{i=1}^{m} \rho_i - \vartheta_1 \prod_{j=1}^{m} \frac{\mu_i}{\lambda_i}\,,\end{aligned} \tag{3.106b}$$

where

$$\sum_{i=1}^{m} \frac{1}{\lambda_i} \prod_{j=i+1}^{m} \frac{\mu_j}{\lambda_j} = \prod_{j=1}^{m} \frac{\mu_i}{\lambda_i} \sum_{i=1}^{m} \rho_i\,, \quad \text{with } \rho_i = \frac{\lambda_1\lambda_2\cdots\lambda_{i-1}}{\mu_1\mu_2\cdots\mu_{i-1}\mu_i}\,.$$

Multiplying both sides by the product $\prod_{i=1}^{m}(\lambda_i/\mu_i)$ yields an equation that is suitable for analysis:

$$(\vartheta_m - \vartheta_{m+1}) \left(\prod_{i=1}^{m} \frac{\lambda_i}{\mu_i} \right) = \sum_{i=1}^{m} \rho_i - \vartheta_1\,. \tag{3.106c}$$

Similarly, as when deriving the extinction probabilities, the assumption of divergence $\sum_{i=1}^{\infty} \rho_i = \infty$ can only be satisfied with $\vartheta_1 = \infty$, and since $\vartheta_m < \vartheta_{m+1}$, all mean extinction times are infinite. If, however, $\sum_{i=1}^{\infty} \rho_i$ remains finite, (3.106c) can be used to calculate ϑ_1. To do this, one has to show that the term $(\vartheta_m - \vartheta_{m+1}) \prod_{i=1}^{m}(\lambda_i/\mu_i)$ vanishes as $m \to \infty$. The proof follows essentially the same lines as in the previous case of the extinction probabilities, but it is more elaborate

[290, p. 149]. One then obtains

$$\vartheta_1 = \sum_{i=1}^{\infty} \rho_i \,. \tag{3.106d}$$

Equations (3.106b) and (3.106c) imply the final result:

$$\vartheta_m = \begin{cases} \infty \,, & \text{if } \sum_{i=1}^{\infty} \rho_i = \infty \,, \\ \sum_{i=1}^{\infty} \rho_i + \sum_{i=1}^{m-1} \left(\prod_{k=1}^{i} \frac{\mu_k}{\lambda_k} \right) \sum_{j=i+1}^{\infty} \rho_j \,, & \text{if } \sum_{i=1}^{\infty} \rho_i < \infty \,. \end{cases} \tag{3.106e}$$

Again we use the linear birth-and-death process, $\lambda_n = \lambda n$ and $\mu_n = \mu n$, for illustration and calculate the mean time to absorption from the state Σ_1:

$$\begin{aligned} \vartheta_1 &= \sum_{i=1}^{\infty} \rho_i = \frac{1}{\mu} \sum_{i=1}^{\infty} \left(\frac{\lambda}{\mu} \right)^{i-1} = \frac{1}{\lambda} \sum_{i=1}^{\infty} \left(\frac{\lambda}{\mu} \right)^{i} = \frac{1}{\lambda} \sum_{i=0}^{\infty} \int_0^{\lambda/\mu} \xi^i \mathrm{d}\xi \\ &= \frac{1}{\lambda} \int_0^{\lambda/\mu} \frac{1}{1-\xi} \mathrm{d}\xi = -\frac{1}{\lambda} \log\left(1 - \frac{\lambda}{\mu} \right) . \end{aligned} \tag{3.107}$$

3.2.4 Continuous Time Random Walks

The term *random walk* goes back to Karl Pearson [444] and is generally used for stochastic processes describing a walk in physical space with random increments. We have already used the concept of a random walk in one dimension several times to illustrate specific properties of stochastic processes (see, for example, Sects. 3.1.1 and 3.1.3). Here we focus on the random walk itself and its infinitesimal step size limit, the Wiener process. For the sake of simplicity and accessibility by analytical methods, we shall be dealing here predominantly with the 1D random walk, although 2D and 3D walks are of similar or even greater importance in physics and chemistry.

In one and two dimensions, the random walk is *recurrent*. This implies that each sufficiently long trajectory will visit every point in phase space, and it does this infinitely often if the trajectory is of infinite length. In particular, every trajectory will return to its origin. In three and more dimensions, this is not the case and the process is thus said to be *transient*. A 3D trajectory revisits the origin only in 34 % of the cases, and this value decreases further in higher dimensions. Somewhat humoristically, one may say a drunken sailor will find his way back home for sure, but a drunken pilot only in roughly one out of three trials.

Discrete Random Walk in One Dimension

The 1D random walk in its simplest form is a classic problem of probability theory and science. A walker moves along a line by taking steps to the left or to the right with equal probability and length l, and regularly, after a constant waiting time τ. The location of the walker is thus nl, where n is an integer $n \in \mathbb{Z}$. We used the 1D random walk in discrete space n with discrete time intervals τ to illustrate the properties of a martingale in Sect. 3.1.3. Here we relax the condition of synchronized discrete time intervals and study a *continuous time random walk* (CTRW) by keeping the step size discrete, but assuming time to be continuous. In particular, the probability that the walker takes a step is well defined and the random walk is modeled by a master equation.

For the master equation we require transition probabilities per unit time, which are simply defined to have a constant value ϑ for single steps and to be zero otherwise:

$$W(m|n,t) = \begin{cases} \vartheta\,, & \text{if } m = n+1\,, \\ \vartheta\,, & \text{if } m = n-1\,, \\ 0\,, & \text{otherwise}\,. \end{cases} \tag{3.108}$$

The master equation falls into the birth-and-death class and describes the evolution of the probability that the walker is at location nl at time t:

$$\frac{\mathrm{d}P_n(t)}{\mathrm{d}t} = \vartheta\Big(P_{n+1}(t) + P_{n-1}(t) - 2P_n(t)\Big)\,, \tag{3.109}$$

provided he started at location $n_0 l$ at time t_0, i.e., $P_n(t_0) = \delta_{n,n_0}$.

The master equation (3.109) can be solved by means of the time dependent characteristic function (see equations (2.32) and (2.32′)):

$$\phi(s,t) = \mathrm{E}(\mathrm{e}^{\mathrm{i}sn(t)}) = \sum_{n=-\infty}^{\infty} P_n(t)\,\exp(\mathrm{i}sn)\,. \tag{3.110}$$

Combining (3.109) and (3.110) yields

$$\frac{\partial\phi(s,t)}{\partial t} = \vartheta\left(\mathrm{e}^{\mathrm{i}s} + \mathrm{e}^{-\mathrm{i}s} - 2\right)\phi(s,t) = 2\vartheta\left(\cosh(\mathrm{i}s) - 1\right)\phi(s,t)\,.$$

Accordingly, the solution for the initial condition $n_0 = 0$ at $t_0 = 0$ is

$$\begin{aligned} \phi(s,t) &= \phi(s,0)\exp\Big(2\vartheta t\big(\cosh(\mathrm{i}s) - 1\big)\Big) \\ &= \exp\Big(2\vartheta t\big(\cosh(\mathrm{i}s) - 1\big)\Big) = \mathrm{e}^{-2\vartheta t}\exp\Big(2\vartheta t\big(\cosh(\mathrm{i}s) - 1\big)\Big)\,. \end{aligned} \tag{3.111}$$

Inserting the expansion

$$\cosh(\mathrm{i}s) - 1 = \frac{(\mathrm{i}s)^2}{2!} + \frac{(\mathrm{i}s)^4}{4!} + \frac{(\mathrm{i}s)^6}{6!} + \cdots = -\frac{s^2}{2!} + \frac{s^4}{4!} - \frac{s^6}{6!} + \cdots$$

and comparing coefficients of equal powers of s, we obtain the individual probabilities

$$P_n(t) = I_n(2\vartheta t)\mathrm{e}^{-2\vartheta t}, \quad n \in \mathbb{Z}, \tag{3.112}$$

where the pre-exponential term is written in terms of modified Bessel functions $I_k(\theta)$ with $\theta = 2\vartheta t$ (for details, see [21, p. 208 ff.]), which are defined by

$$\begin{aligned} I_k(\theta) &= \sum_{j=0}^{\infty} \frac{(\theta/2)^{2j+k}}{j!(j+k)!} = \sum_{j=0}^{\infty} \frac{(\theta/2)^{2j+k}}{j!\Gamma(j+k+1)} \\ &= \sum_{j=0}^{\infty} \frac{(\vartheta t)^{2j+k}}{j!(j+k)!} = \sum_{j=0}^{\infty} \frac{(\vartheta t)^{2j+k}}{j!\Gamma(j+k+1)} . \end{aligned} \tag{3.113}$$

The probability that the walker is found at his initial location $n_0 l$, for example, is given by

$$P_0(t) = I_0(2\vartheta t)\,\mathrm{e}^{-2\vartheta t} = \left(1 + (\vartheta t)^2 + \frac{(\vartheta t)^4}{4} + \frac{(\vartheta t)^6}{36} + \cdots\right)\mathrm{e}^{-2\vartheta t} .$$

Illustrative numerical examples are shown in Fig. 3.14. It is straightforward to calculate the first and second moments from the characteristic function $\phi(s,t)$, using (2.34) and the result is

$$\mathrm{E}\big(\mathcal{X}(t)\big) = n_0 , \quad \mathrm{var}\big(\mathcal{X}(t)\big) = 2\vartheta(t - t_0) . \tag{3.114}$$

The expectation value is constant, coinciding with the starting point of the random walk, and the variance increases linearly with time. The continuous time 1D random walk is a martingale.

The density function $P_n(t)$ allows for straightforward calculation of practically all interesting quantities. For example, we might like to know the probability that the walker reaches a given point at distance nl from the origin within a predefined time span, which is simply obtained from $P_n(t)$ with $P_n(t_0) = \delta_{n,0}$ (Fig. 3.14). This probability distribution is symmetric because of the symmetric initial condition $P_n(t_0) = \delta_{n,0}$, and hence $P_n(t) = P_{-n}(t)$. For long times the probability density $P(n,t)$ becomes flatter and flatter and eventually converges to the uniform distribution over the spatial domain. For $n \in \mathbb{Z}$, all probabilities vanish, i.e., $\lim_{t\to\infty} P_n(t) = 0$ for all n.

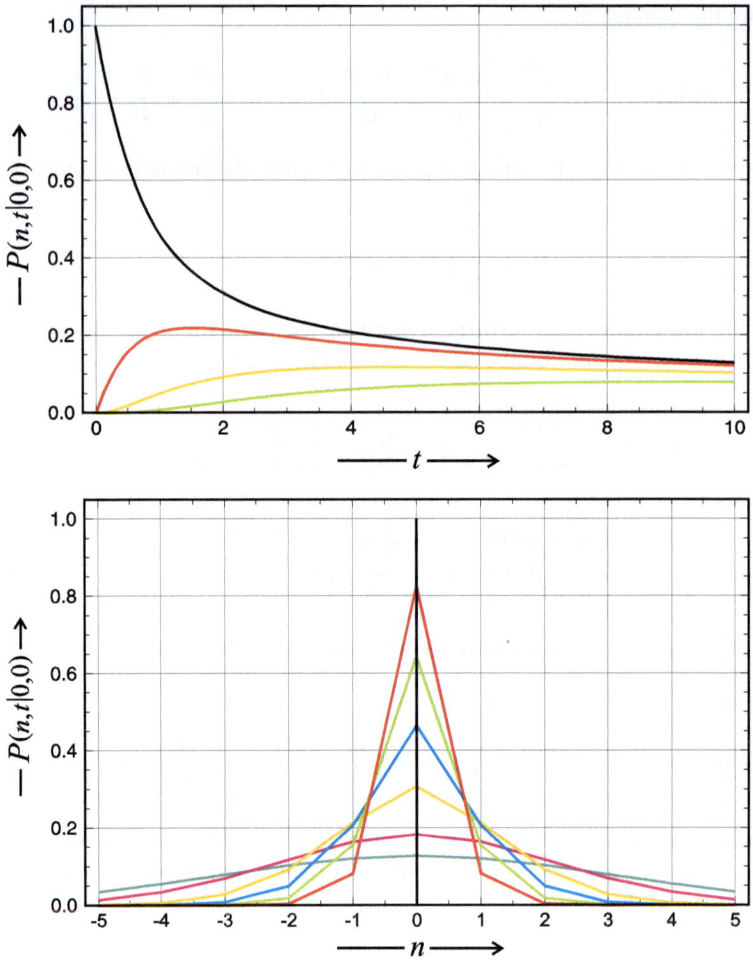

Fig. 3.14 Probability distribution of the random walk. The figure presents the conditional probabilities $P_n(t)$ of a random walker to be in location $n \in \mathbb{Z}$ at time t, for the initial condition of being at $n = 0$ at time $t = t_0 = 0$. *Upper*: Dependence on t for given values of n: $n = 0$ (*black*), $n = 1$ (*red*), $n = 2$ (*yellow*), and $n = 3$ (*green*). *Lower*: Probability distribution as a function of n at a given time t_k. Parameter choice: $\vartheta = 0.5$; $t_k = 0$ (*black*), 0.2 (*red*), 0.5 (*green*), 1 (*blue*), 2 (*yellow*), 5 (*magenta*), and 10 (*cyan*)

From Random Walks to Diffusion

In order to derive the stochastic diffusion equation (3.55), we start from a discrete time random walk of a single particle on an infinite one-dimensional lattice, where the lattice sites are denoted by $n \in \mathbb{Z}$. Since the transition to diffusion is of general importance, we present two derivations:

(i) from the discrete time and space random walk model presented and solved in Sect. 3.1.3, and
(ii) from the continuous time discrete space random walk (CTRW) discussed in the previous paragraph.

The particle is assumed to be at position n at time t and within a discrete time interval Δt it is obliged to jump to one of the neighboring sites, $n+1$ or $n-1$. The time elapsed between two jumps is called the *waiting time*. Spatial isotropy demands that the probabilities of jumping to the right or to the left are the same and equal to one half. The probability of being at site n at time $t+\Delta t$ is therefore given by[39]

$$P_n(t+\Delta t) = \frac{1}{2}P_{n-1}(t) + \frac{1}{2}P_{n+1}(t)\,. \tag{3.9$'$}$$

Next we make a Taylor series expansion in time and truncate after the linear term in Δt, assuming that t is a continuous variable:

$$P_n(t+\Delta t) = P_n(t) + \Delta t\,\frac{\mathrm{d}P_n(t)}{\mathrm{d}t} + \mathcal{O}\big((\Delta t)^2\big)\,.$$

Now we convert the discrete site number into a continuous spatial variable, i.e., $n \to x$ and $P_n(t) \to p(x,t)$, and find

$$P_{n\pm 1}(t) = p(x,t) \pm \Delta x\frac{\partial p(x,t)}{\partial x} + \frac{(\Delta x)^2}{2}\frac{\partial^2 p(x,t)}{\partial x^2} + \mathcal{O}\big((\Delta x)^3\big)\,.$$

Here we truncate only after the quadratic term in Δx, because the term with the first derivatives will cancel. Inserting in (3.9$'$) and omitting the residuals, we obtain

$$p(x,t) + \Delta t\,\frac{\partial p(x,t)}{\partial t} = p(x,t) + \frac{(\Delta x)^2}{2}\frac{\partial^2 p(x,t)}{\partial x^2}\,.$$

The next and final task is to carry out the simultaneous limits to infinitesimal differences in time and space[40]:

$$\lim_{\Delta t\to 0,\,\Delta x\to 0} \frac{(\Delta x)^2}{2\Delta t} = D\,, \tag{3.115}$$

[39] It is worth pointing out a subtle difference between (3.109) and (3.9): the term containing $-2P_n(t)$ is missing in the latter, because motion is obligatory in the discrete time model. The walker is not allowed to take a rest.

[40] The most straightforward way to take the limit is to introduce a *scaling assumption*, using a variable γ such that $\Delta x = \gamma\Delta x_0$ and $\Delta t = \gamma^2\Delta t_0$. Then we have $\Delta x^2/2\Delta t = \Delta x_0^2/2\Delta t_0 = D$ and the limit $\gamma \to 0$ is trivial.

where D is called the diffusion coefficient, which, as already mentioned in Sect. 3.2.2.2, has the dimension $[D] = [\mathrm{l}^2\mathrm{t}^{-1}]$.

Eventually, we obtain the stochastic version of the diffusion equation

$$\frac{\partial p(x,t)}{\partial t} = D\frac{\partial^2 p(x,t)}{\partial x^2}\,, \tag{3.55$'$}$$

which is fundamental in physics and chemistry for the description of diffusion (see also (3.56) in Sect. 3.2.2.2).

It is also straightforward to consider the continuous time random walk in the limit of continuous space. This is achieved by setting the distance traveled to $x = nl$ and performing the limit $l \to 0$. For that purpose we start from the characteristic function of the distribution in x, viz.,

$$\phi(s,t) = \mathrm{E}\big(\mathrm{e}^{\mathrm{i}sx(t)}\big) = \Phi(ls,t) = \exp\Big(2\vartheta t\big(\cosh(\mathrm{i}ls) - 1\big)\Big)\,,$$

where ϑ is again the transition probability to neighboring positions per unit time, and make use of the series expansion of the cosh function, viz.,

$$\cosh y = \sum\nolimits_{k=0}^{\infty} \frac{y^{2k}}{(2k)!} = 1 + \frac{y^2}{2!} + \frac{y^4}{4!} + \frac{y^6}{6!} + \cdots\,.$$

We then take the limit of infinitesimally small steps $\lim l \to 0$:

$$\begin{aligned}\lim\nolimits_{l\to 0} \exp\Big(2\vartheta t\big(\cosh(\mathrm{i}ls) - 1\big)t\Big) &= \lim\nolimits_{l\to 0} \exp\big(\vartheta t(-l^2s^2 + \cdots)\big)\\ &= \lim\nolimits_{l\to 0} \exp(-s^2l^2\vartheta t)\\ &= \exp(-s^2Dt)\,,\end{aligned}$$

where we have used the definition $D = \lim_{l\to 0}(l^2\vartheta)$ for the diffusion coefficient D (Fig. 3.15). Since this is the characteristic function of the normal distribution, we obtain for the probability density the well-known equation

$$p(x,t) = \frac{1}{\sqrt{4\pi Dt}}\exp\big(-x^2/4Dt\big) \tag{2.45}$$

for the sharp initial condition $\lim_{t\to 0} p(x,t) = p(x,0) = \delta(x)$. We could also have proceeded directly from (3.109) and expanded the right-hand side as a function of x up to second order in l, which yields once again the stochastic diffusion equation

$$\frac{\partial p(x,t)}{\partial t} = D\frac{\partial^2 p(x,t)}{\partial x^2}\,, \tag{3.56}$$

with $D = \lim_{l\to 0}(l^2\vartheta)$ as before.

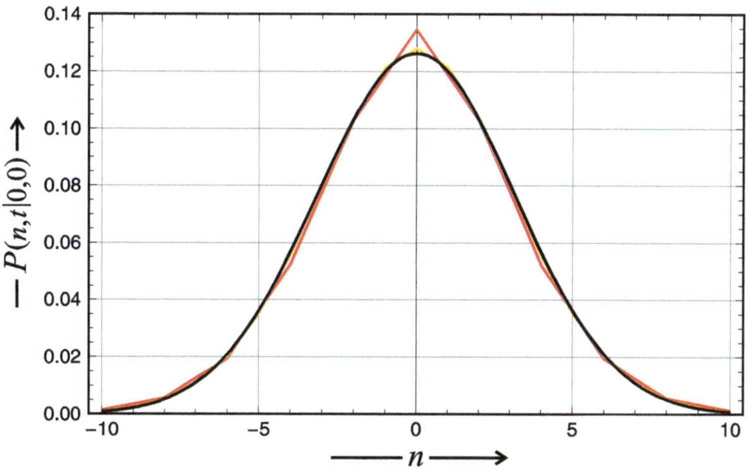

Fig. 3.15 Transition from random walk to diffusion. The figure presents the conditional probabilities $P(n, t|0, 0)$ during convergence from a discrete space random walk to diffusion. The *black curve* is the normal distribution (2.45) resulting from the solution of the stochastic diffusion equation (3.55′) with $D = 2\lim_{l\to 0}(l^2\vartheta) = 2$. The *yellow curve* is the random walk approximation with $l = 1$ and $\vartheta = 1$, and the *red curve* was calculated with $l = 2$ and $\vartheta = 0.25$. A smaller step width of the random walk, viz., $l \leq 0.5$, leads to curves that are indistinguishable within the thickness of the line from the normal distribution. In order to obtain comparable curves, the probability distributions were scaled by a factor $\gamma = l^{-1}$. Choice of other parameters: $t = 5$

Random Walks with Variable Increments

In order to prepare for the discussion of anomalous diffusion in Sect. 3.2.5, we generalize the 1D continuous time random walk (CTRW) and analyze it from a different perspective [61, 396]. The random variable $\mathcal{X}(t)$ is defined as the sum of previous step increments ξ_k, i.e.,

$$\mathcal{X}_n(t) = \sum_{k=1}^{n} \xi_k \,, \quad \text{with } t_n = \sum_{k=1}^{n} \tau_k \,,$$

and the time t_n is the sum of all earlier waiting times τ_k. This discrete random walk differs from the case we analyzed previously (Sect. 3.1.3) by the assumption that both the *jump increments* or jump lengths, $\xi_k \in \mathbb{R}$, and the time intervals between two jumps referred to as *waiting times*, $\tau_k \in \mathbb{R}_{\geq 0}$, are variable (Fig. 3.16). Since jump lengths and waiting times are real quantities, the random variable is real as well, i.e., $\mathcal{X}(t) \in \mathbb{R}$. At time t_k, the probability that the next jump occurs at time $t_k + \Delta t = t_k + \tau_{k+1}$ and that the jump length will be $\Delta x = \xi_{k+1}$ is given by the joint density function

$$P\big(\Delta x = \xi_{k+1} \wedge \Delta t = \tau_{k+1} \,|\, \mathcal{X}(t_k) = x_k\big) = \varphi(\xi, \tau) \,, \tag{3.116}$$

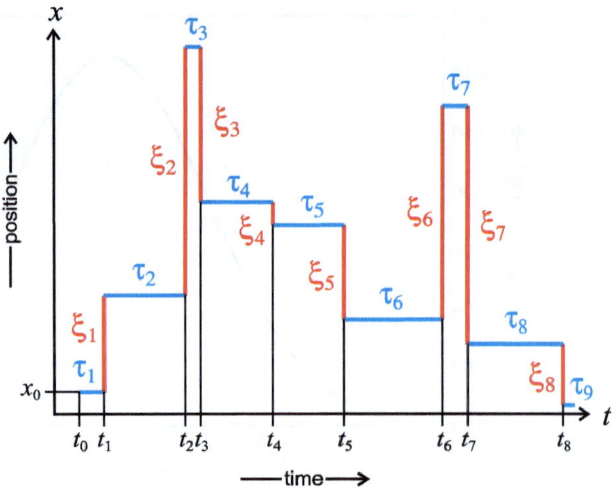

Fig. 3.16 A random walk with variable step sizes. Both, the jump lengths, ξ_k, and the waiting times, τ_k, are assumed to be variable. The jumps occur at times t_1, t_2, ..., and both jump length and waiting times are drawn form the distributions $f(\xi)$ and $w(\tau)$, respectively

where

$$\psi(\tau) = \int_{-\infty}^{+\infty} \mathrm{d}\xi\, \varphi(\xi, t) \;\text{ and }\; f(\xi) = \int_0^{\infty} \mathrm{d}\tau\, \varphi(\xi, \tau)$$

are the two marginal distributions. Since $\varphi(\xi, \tau)$ does not depend on the time t, the process is homogeneous. We assume that waiting times and jump lengths are independent random variables and that the joint density is factorizable[41]:

$$\varphi(\xi, \tau) = f(\xi)\psi(\tau) \,. \tag{3.117}$$

In the case of Brownian motion or normal diffusion, the marginal densities in space and time are Gaussian and exponential distributions modeling normal distributed jump lengths and Poissonian waiting times:

$$f(\xi) = \frac{1}{\sqrt{4\pi\sigma^2}} \exp\left(-\frac{\xi^2}{4\sigma^2}\right) \;\text{ and }\; \psi(\tau) = \frac{1}{\tau_w} \exp\left(-\frac{\tau}{\tau_w}\right) . \tag{3.118}$$

It is worth recalling that (3.118) is sufficient to predict the nature of the probability distributions of $\mathcal{X}_n$ and t_n. Since the spatial increments are independent and identically distributed (iid) Gaussian random variables, the sum is normally distributed by the central limit theorem (CLT), and since the temporal increments follow an exponential distribution, the probability distribution of the sum is Poissonian.

[41] If the jump lengths and waiting times were coupled, we would have to deal with $\varphi(\xi, \tau) = \varphi(\xi|\tau)\psi(\tau) = \varphi(\tau|\xi)f(\xi)$. Coupling between space and time could arise, for example, from the fact that it is impossible to jump a certain distance within a time span shorter than some minimum time.

The task is now to express the probability $p(x,t) = P\big(\mathcal{X}(t) = x|\mathcal{X}(0) = x_0\big)$ that the random walk is in position x at time t, using the functions $f(\xi)$ and $\psi(\tau)$. For this goal, we first calculate the probability of the walk arriving at position x at time t under the condition that it was at position z at time ϑ:

$$\eta(x,t) = p(x,t|z,\vartheta) = \int_{-\infty}^{x} \mathrm{d}z \int_0^{\infty} \mathrm{d}\vartheta f(x-z)\psi(t-\vartheta)\eta(z,\vartheta) + \delta(x)\delta(t) \ ,$$

with $\psi(t) = 0\,,\ \forall\, t \leq \vartheta$. The last term takes into account the fact that the random walk started at the origin $x = 0$ at time $t = 0$, as expressed by $p(x,0) = \delta(x)$, and defines the initial condition $\eta(0,0) = 1$.

Next we consider the condition that the step $(z,\vartheta) \to (x,t)$ was the last step in the walk until t, and introduce the probability that no step occurred in the time interval $[0,t]$:

$$\Theta(t) = 1 - \int_0^t \mathrm{d}\vartheta\ \psi(\vartheta) \ .$$

Now we can write down the probability density we are searching for:

$$p(x,t) = \int_0^t \mathrm{d}\vartheta\ \Theta(t-\vartheta)\eta(x,\vartheta) \ .$$

It is important to realize that the expression for $\eta(x,t)$ is a convolution of $f(x)$ and η with respect to space x and of $\psi(t)$ and η with respect to time t, while $p(x,t)$ is finally a convolution of Θ and η with respect to t alone.

Making use of the convolution theorem (3.27), which turns convolutions in (x,k) space into products in (k,u) or Fourier–Laplace space, we can readily write down the expressions for the transformed probability distributions:

$$\hat{\tilde{p}}(k,u) = \widehat{\Theta}(u)\hat{\tilde{\eta}}(k,u) \ ,$$

with

$$\hat{\tilde{\eta}}(k,u) = \hat{\psi}(u)\tilde{f}(k) + 1 \implies \hat{\tilde{\eta}}(k,u) = \frac{1}{1-\tilde{f}(k)\hat{\psi}(u)} \ ,$$

and

$$\mathcal{L}\left(\frac{\mathrm{d}\Theta(t)}{\mathrm{d}t}\right) = \mathcal{L}\big(\delta(t) - \psi(t)\big) \implies u\widehat{\Theta}(u) = 1 - \tilde{f}(k) \ ,$$

$$\widehat{\Theta}(u) = \frac{1-\tilde{f}(k)}{u} \ ,$$

where we use the following notation for the Fourier–Laplace transform:

$$\mathcal{L}\Big(\mathcal{F}\big(f(\xi,\tau)\big)\Big)(k,u) = \hat{\tilde{f}}(k,u) = \frac{1}{\sqrt{2\pi}} \int_0^\infty \int_{-\infty}^\infty \mathrm{e}^{-u\tau} \mathrm{e}^{\mathrm{i}k\xi} f(\xi,\tau)\, \mathrm{d}\xi\, \mathrm{d}\tau \, .$$

As can be shown straightforwardly, the transformed probability satisfies the well-known Montroll–Weiss equation [407], named after the American mathematicians Elliot Montroll and George Weiss:

$$\hat{\tilde{p}}(k,u) = \frac{1-\hat{\psi}(u)}{u} \frac{1}{1-\tilde{f}(k)\hat{\psi}(u)} \, . \tag{3.119}$$

This provides the desired relation between the increment densities and the probability distribution of the position of the walk as a function of time. What remains to do is to calculate the Fourier and Laplace transformed increment functions, which are expanded for small values of k and u corresponding to long distances and long times, respectively. The Laplace transform of $\psi(\tau)$ and the Fourier transform of $f(\xi)$ have the asymptotic forms

$$\hat{\psi}(u) = \int_0^\infty \mathrm{d}\tau\, \psi(\tau)\mathrm{e}^{-u\tau} = \frac{1}{1+\tau_\mathrm{w} u} = 1 - \tau_\mathrm{w} u + \mathcal{O}(u^2) \, , \tag{3.120a}$$

$$\sqrt{2\pi}\tilde{f}(k) = \int_{-\infty}^{+\infty} \mathrm{d}\xi\, f(\xi)\mathrm{e}^{\mathrm{i}k\xi} = \mathrm{e}^{-k^2\lambda} = 1 - \lambda k^2 + \mathcal{O}(k^4) \, , \tag{3.120b}$$

where $\lambda = \sigma^2/2$ and the diffusion coefficient $D = \lambda/\tau_\mathrm{w} = \sigma^2/2\tau_\mathrm{w}$. The exponents $\alpha = 2$ and $\theta = 1$ of the leading terms in the expansions (3.120b) and (3.120a), respectively, determine the nature of the diffusion process and are called *universality exponents*. Inserting in the Montroll–Weiss equation yields

$$\hat{\tilde{p}}(k,u) = \frac{1}{\sqrt{2\pi}} \frac{\tau_\mathrm{w}}{\tau_\mathrm{w} u + \lambda k^2} = \frac{1}{\sqrt{2\pi}} \frac{1}{u + Dk^2} \, . \tag{3.120c}$$

As expected, consecutive inverse transformations yield the density distribution $p(x,t)$ of the Wiener process (3.61):

$$\mathcal{F}^{-1}\left(\frac{1}{\sqrt{2\pi}} \frac{1}{u+Dk^2}\right) = \frac{1}{\sqrt{4Du}} \mathrm{e}^{-\sqrt{u/D}|x|}$$

$$\mathcal{L}^{-1}\left(\frac{1}{\sqrt{4Du}} \mathrm{e}^{-\sqrt{u/D}|x|}\right) = \frac{1}{\sqrt{4\pi Dt}} \mathrm{e}^{-x^2/4Dt} \, ,$$

where D is the diffusion coefficient and $x_0 = 0$, $t_0 = 0$ are the initial conditions. It is a good exercise to show that reversing the order of the transformations yields the same result:

$$\mathcal{L}^{-1}\left(\frac{1}{\sqrt{2\pi}}\frac{1}{u+Dk^2}\right) = \frac{1}{\sqrt{2\pi}}\mathrm{e}^{-Dk^2t} ,$$

$$\mathcal{F}^{-1}\left(\frac{1}{\sqrt{2\pi}}\mathrm{e}^{-Dk^2t}\right) = \frac{1}{\sqrt{4\pi Dt}}\mathrm{e}^{-x^2/4Dt} .$$

If one were only interested in the solution for the normal distribution, the derivation of the solution presented here would be a true case of *overkill*. We shall, however, extend the analysis to anomalous diffusion with generalized exponents $0 < \alpha \leq 2$ and $0 < \theta \leq 1$, which are non-integer quantities and thus lead us into the realm of fractals (Sect. 3.2.5).

Before we give an interpretation of the two expansions, we visualize the meaning of the two transformed variables u and k. The exponents $u\tau$ and ξk are dimensionless quantities, so the dimensions of u and k are reciprocal time $[\mathrm{t}^{-1}]$ and reciprocal length $[\mathrm{l}^{-1}]$ respectively. The values $u = 0$ and $k = 0$ of the transformed variables refer to infinite time and infinite space, and accordingly, expansions around these points are valid for long times and long distances. Commonly, specific properties of the problem are dominant at short times and small distances and universal behaviour is expected to be found at the opposite ends of the scales of time and space, as expressed by vanishing u and k. Both transformed probability distributions in (3.120a) and (3.120b) are given in expressions that allow for direct readout of the so-called *universality exponents*, which are $\alpha = 2$ for the spatial density $\tilde{f}(k)$ and $\theta = 1$ for the temporal density $\hat{\psi}(u)$. Random walks can be classified by the variance $\mathrm{var}(\xi)$ of jump lengths and by the expectation value $\mathrm{E}(\tau)$ of waiting times:

(i) The *variance of the jump length*[42] $\langle \xi^2 \rangle = 2\sigma^2 = \int_{-\infty}^{+\infty} \mathrm{d}\xi\, \xi^2 f(\xi)$.
(ii) The *characteristic* or *mean waiting time* $\langle \tau \rangle = \tau_\mathrm{w} = \int_0^\infty \mathrm{d}\tau\, \tau\psi(\tau)$.

These are both finite quantities that do not diverge in the integration limits $\xi \to \infty$ and $\tau \to \infty$, in contrast to Lévy processes with $0 < \alpha < 2$ and $0 < \theta < 1$, which will be discussed in Sect. 3.2.5. As a matter of fact, any pair of probability density functions with finite τ_w and σ^2 leads to the same asymptotic result and this is a beautiful manifestation of the central limit theorem (Sect. 2.4.2): in the inner part of the transformed densities, all representatives of the universality class of CTRWs with finite mean waiting times and positional variances satisfy (3.120a) and (3.120b), and the individuality of the densities comes into play only in the higher order terms $\mathcal{O}(\tau^2)$ and $\mathcal{O}(k^4)$.

[42]As in the previous examples, we assume that the random walk is symmetric and started at the origin. Then the expectation value of the location of the particle stays at the origin and we have $\langle \xi \rangle = 0$, $\langle \xi \rangle^2 = 0$, and hence $\mathrm{var}(\xi) = \langle \xi^2 \rangle$.

Finally, we mention a feature that will be brought up again and generalized in Sect. 3.2.5: the Wiener process is self-similar. In general, a stochastic process is self-similar with Hurst exponent H, named after the British hydrologist Harold Edwin Hurst, if the two processes

$$(\mathcal{X}(a\,t), t \geq 0) \text{ and } (a^H \mathcal{X}(t), t \geq 0)$$

with the same initial condition $\mathcal{X}(0) = 0$ have the same finite-dimensional distribution for all $a \geq 0$. Expressed in popular language, if you look on a self-similar process with a magnifying glass, it looks the same as without magnification, no matter how large the magnification factor is. The expectation value of the generalized or fractal Brownian process $\mathcal{B}_H(t)$ at two different times, t_1 and t_2, and with Hurst exponent $0 \leq H \leq 1$, is

$$\mathrm{E}\big(\mathcal{B}_H(t_1)\mathcal{B}_H(t_2)\big) = \frac{1}{2}\big(|t_1|^{2H} + |t_2|^{2H} - |t_2 - t_1|^{2H}\big) .$$

For $H = 1/2$ we are dealing with conventional Brownian motion or Wiener processes, where the expectation value $E\big(\mathcal{B}_H(t_1)\mathcal{B}_H(t_2)\big) = \min\{t_1, t_2\}$ is exactly the value for the Wiener process, and increments at different times are uncorrelated, as we showed in Sect. 3.2.2.2.

3.2.5 Lévy Processes and Anomalous Diffusion

Lévy processes are the simplest conceivable stochastic processes that comprise all three components contained in the differential Chapman–Kolmogorov equation: drift, diffusion, and jumps. They were defined in precise mathematical terms and analyzed in detail by the famous French mathematician Paul Lévy. Drift and diffusion correspond to a linear Liouville process and a Wiener process, respectively, and the discontinuous part can be seen as a generalized continuous-time random walk with jumps of random size occurring at random times. The probability distribution of the jumps allows for a classification of the process, and it can be quite general except that it has to be infinitely divisible. Lévy processes within the realm of general stochastic processes are often interpreted as the analogues to linear functions in function space.

Lévy processes constitute a core theme of financial mathematics [18, 136, 475, 480] and they are indispensable constituents of every course in theoretical economy (see, e.g., [436]). Many stochastic processes from other fields and also from science fall into this class. From the examples of stochastic processes we have already encountered here, Brownian motion (Sect. 3.2.2.2), the Poisson process (Sect. 3.2.2.4), the random walk (Sect. 3.2.4), and the Cauchy process (Sect. 3.1.5) are special cases of Lévy processes. Among other applications, Lévy processes are used in the mathematical theory of anomalous diffusion [61, 396] and other forms of fractional kinetics, but also in Lévy flights, based on probability densities with heavy tails, which have been and still are applied, for example, in behavioral biology to model the foraging strategies of animals.

We are also interested here in Lévy processes, because they allow for a general analytic treatment combining all three classes of process appearing in the differential Chapman–Kolmogorov equation (dCKE): drift, diffusion, and jumps. This is possible because of the simplifying assumption that all random variables $\mathcal{X}(t)$ are independent and identically distributed (iid) and all increments $\mathcal{Z} = \Delta\mathcal{X}(t)$ depend only on Δt and not explicitly on t. The time dependence is restricted to the probability densities $p(x,t)$, and the functions $A(x,t) = A(x) = a$ and $B(x,t) = B(x) = \sigma^2/2$ as well as the transition probabilities $W(x|z,t) = W(x|z) = w(z-x)$ are strictly time independent.

A Lévy process $\mathcal{X} = (\mathcal{X}(t), t \geq 0)$ is a stochastic process that satisfies the following four properties:

1. The random variable $\mathcal{X}(t)$ has independent increments, as expressed by the property that the variables $\mathcal{Z}_k = \mathcal{X}(t_k) - \mathcal{X}(t_{k-1})$ with $k = 1, 2, \ldots$, are statistically independent.
2. The increments $\mathcal{Z}_k$ of the random variable $\mathcal{X}(t)$ are stationary in the sense that the probability distributions of the increments $\mathcal{Z}_k$ depend only on the length of the time interval $\Delta t = t_k - t_{k-1}$, but do not depend explicitly on time t, while increments on equal time intervals are identically distributed.
3. The process starts at the origin, i.e., $\mathcal{X}_0 = 0$, with probability one.
4. The trajectory of the random variable $\mathcal{X}(t)$ is at least piecewise *stochastically continuous* in the sense that it satisfies the relation

$$\lim_{t_1 \to t_2} P\big(|\mathcal{X}(t_2) - \mathcal{X}(t_1)| > \epsilon\big) = 0 ,$$

for all $\epsilon > 0$ and for all $t_2 - t_1 \geq 0$.

The conditions (1), (2), and (3) simplify the general dCKE substantially. Condition (2) in particular allows for the replacement of functions by parameters:

$$A(x,t) \text{ is replaced by } a , \quad B(x,t) \text{ is replaced by } \frac{\sigma^2}{2} , \qquad (3.121)$$
$$\text{and } W(x|z,t) \text{ is replaced by } w(x-z) ,$$

where $w(x-z)$ is a *transition function* replacing the transition matrix. For the initial condition $p(x,t_0) = \delta(x_0)$, the dCKE has the form[43]

$$\frac{\partial p(x,t)}{\partial t} = -a\frac{\partial p(x,t)}{\partial x} + \frac{1}{2}\sigma^2\frac{\partial^2 p(x,t)}{\partial x^2} + \int \mathrm{d}z\big(w(x-z)\,p(z,t) - w(z-x)\,p(x,t)\big)\,. \tag{3.122}$$

Lévy processes are thus fully characterized by the Lévy–Khinchin triplet (a, σ^2, w) which is named after Paul Lévy and the Russian mathematician Aleksandr Khinchin. It follows from condition (2) that a Lévy process is a homogeneous Markov process.

The replacement of the functions $A(x,t)$ and $B(x,t)$ by the constants a and $\sigma^2/2$, and the elimination of time from the jump probability $W(z|x,t)$, leads to a remarkable analogy to linearity in deterministic dynamical systems. Indeed, the Liouville equation corresponding to the dCKE (3.122) gives rise to a linear time dependence, viz.,

$$\frac{\partial p(x,t)}{\partial t} = -a\frac{\partial p(x,t)}{\partial x} \implies \frac{\mathrm{d}x}{\mathrm{d}t} = a \text{ and } x(t) = x(0) + at = at\,,$$

the diffusion part is a Wiener process with a linearly growing variance, i.e.,

$$\frac{\partial p(x,t)}{\partial t} = \frac{1}{2}\sigma^2\frac{\partial^2 p(x,t)}{\partial x^2} \implies \mathrm{var}\big(\mathcal{X}(t)\big) = \sigma^2\mathrm{var}\big(\mathcal{W}(t)\big) = \sigma^2 t\,,$$

and the jumps have time independent transition probabilities, but here the analogy with a linear process appears a little bit far-fetched.

Characteristic Functions of Lévy Processes

Equation (3.122) describing a Lévy process starting with $p(x,0) = \delta(x_0)$ at $t_0 = 0$ is solved using the characteristic function as defined in Sect. 2.2.3:

$$\frac{\partial}{\partial t}\phi(s,t) = \frac{\partial}{\partial t}\int_{-\infty}^{+\infty} \mathrm{d}x\,\mathrm{e}^{\mathrm{i}sx}p(x,t) = \int_{-\infty}^{+\infty} \mathrm{d}x\,\mathrm{e}^{\mathrm{i}sx}\frac{\partial p(x,t)}{\partial t}\,.$$

Inserting in (3.122) and integrating the first two terms by parts yields the differential equation (see Sect. 3.2.2.2)

$$\frac{\partial \phi(s,t)}{\partial t} = \Big(\mathrm{i}as - \frac{1}{2}\sigma^2 s^2 + J(s)\Big)\phi(s,t)\,.$$

[43]For Lévy processes in general it will be necessary to replace the integral by a principal value integral because they may lead to a singularity at the origin, i.e., $\lim_{z\to x} w(x-z) = \infty$, and this prohibits conventional integration.

The third term in the square brackets, the jump term $J(s)$, is calculated using a little trick. We substitute $z - x \Rightarrow u$, apply $\mathrm{d}z = \mathrm{d}u$, and find for the second summand in the integral of (3.122)

$$\int \mathrm{d}z\, w(z-x) \int \mathrm{d}x\, \mathrm{e}^{\mathrm{i}sx} p(x,t) = \int \mathrm{d}u\, w(u) \int \mathrm{d}x\, \mathrm{e}^{\mathrm{i}sx} p(x,t) = \int \mathrm{d}u\, w(u) \phi(s,t) ,$$

while the first summand is calculated by means of a shift in the variable:

$$\begin{aligned}\int \mathrm{d}x \int \mathrm{d}z\, w(x-z) \mathrm{e}^{\mathrm{i}sx} p(z,t) &= \int \mathrm{d}(-u)\, w(-u) \int \mathrm{d}z\, \mathrm{e}^{\mathrm{i}s(z-u)} p(z,t) \\ &= \int \mathrm{d}(-u)\, w(-u) \mathrm{e}^{\mathrm{i}s(-u)} \int \mathrm{d}z\, \mathrm{e}^{\mathrm{i}sz}(z,t) \\ &= \int \mathrm{d}u\, w(u) \mathrm{c}^{isu} \phi(s,t) .\end{aligned}$$

Collecting all terms and reintroducing the principal value integral yields the differential equation for the characteristic function:

$$\frac{\partial \phi(s,t)}{\partial t} = \left(\mathrm{i}as - \frac{1}{2}\sigma^2 s^2 + ⨍ \mathrm{d}u\, w(u)(\mathrm{e}^{\mathrm{i}su} - 1) \right) \phi(s,t) , \tag{3.123}$$

which may be solved in general terms. We recall that the principal value integral takes care of a singularity of $w(u)$ at $u = 0$ [194, pp. 248–252]:

$$\begin{aligned}\phi(s,t) &= \int_{-\infty}^{+\infty} \mathrm{d}x\, \mathrm{e}^{\mathrm{i}sx}\, p(x,t) \\ &= \exp\left(\left(\mathrm{i}as - \frac{1}{2}\sigma^2 s^2 + ⨍_{-\infty}^{+\infty} \mathrm{d}u\, (\mathrm{e}^{\mathrm{i}su} - 1) w(u) \right) t \right) .\end{aligned} \tag{3.124}$$

The density of the Lévy process can be obtained, in principle, by inverse Fourier transform, although no analytical expressions are available.

The first factor in the exponent of the exponential function is called the characteristic exponent:

$$\psi(s) = \frac{\ln \phi(s,t)}{t} = \mathrm{i}as - \frac{1}{2}\sigma^2 s^2 + ⨍_{-\infty}^{+\infty} \mathrm{d}u (\mathrm{e}^{\mathrm{i}su} - 1) w(u) . \tag{3.124'}$$

In practice, it is appropriate to circumvent the sophistication caused by the potential singularity in the integral of (3.124) at $u = 0$ by separating the region around the singularity from the rest of the integral. The zone to be treated separately can be the (doubled) unit interval, $u \in [-1, 1]$, as chosen, for example, in the famous

Lévy–Khinchin formula:

$$\int_{-\infty}^{+\infty} \mathrm{d}u\, w(u)\left(\mathrm{e}^{\mathrm{i}su}-1\right)$$

$$\equiv \lim_{\epsilon\to 0}\left(\int_{-\infty}^{-\delta(\epsilon)} \mathrm{d}u\, w(u)\big(\mathrm{e}^{\mathrm{i}su}-1\big) + \int_{\epsilon}^{+\infty} \mathrm{d}u\, w(u)\big(\mathrm{e}^{\mathrm{i}su}-1\big)\right)$$

$$= \mathrm{i}sA_{\mathcal{L}} + I_{\mathcal{L}}(s)\ , \qquad (3.125)$$

with

$$I_{\mathcal{L}}(s) = \int_{-\infty}^{+\infty} \mathrm{d}u\big(\mathrm{e}^{\mathrm{i}su} - 1 - \mathrm{i}su\,\mathbf{1}_{|u|\le 1}\big)\, w(u)\ ,$$

and

$$\mathrm{i}sA_{\mathcal{L}} = \lim_{\epsilon\to 0}\left(\int_{-1}^{+\delta(\epsilon)} \mathrm{d}u\ \mathrm{i}su\, w(u) + \int_{\epsilon}^{+1} \mathrm{d}u\ \mathrm{i}su\, w(u)\right) + \kappa\ .$$

Here, κ is the integration constant that is obtained, for example, by normalization, while the function $\delta(\epsilon)$ allows for asymmetry in the interval used to calculate the limit in the evaluation of the principal value integral. The latter satisfies $\delta(\epsilon) \to 0$ if $\epsilon \to 0$ and is to be derived from the transition function $w(u)$. Finiteness of both parts of the integral imposes restrictions on the acceptable probability densities $w(u)$: (i) $I_{\mathcal{L}}(s)$ diverges when there are too many *long* jumps, and (ii) $A_{\mathcal{L}}$ becomes infinite when there are too many *short* jumps, although a finite value of the integral for countably infinitely many short jumps is possible (see the discussion of *Pareto processes*).

With these modifications, the characteristic exponent of a general Lévy process becomes

$$\psi(s) = \mathrm{i}(a + A_{\mathcal{L}})s - \frac{1}{2}\sigma^2 s^2 + \int_{-\infty}^{+\infty} \mathrm{d}u\ \big(\mathrm{e}^{\mathrm{i}su} - 1 - \mathrm{i}su\,\mathbf{1}_{|u|\le 1}\big)\, w(u)\ , \qquad (3.124'')$$

where the evaluation of the principal value integral by residue calculus is shifted into the calculation of the parameter $A_{\mathcal{L}}$.

In the following paragraphs we present a few examples of special Lévy processes.

Poisson Processes

The conventional Poisson process and two modifications of it are discussed as examples of simple Lévy processes. As mentioned before, the Poisson process is a Lévy process with the parameters $a = \sigma = 0$ and the transition function $w(u) = \gamma\,\delta(u-1)$. First we calculate the characteristic function from the Lévy–Khinchin formula (3.125) and find[44]

$$\phi(s,t) = \exp\left(\left(\int_{-\infty}^{\infty} du\,(e^{isu}-1)\gamma\delta(u-1)\right)t\right) = \exp\bigl(\gamma t(e^{is}-1)\bigr)\,.$$

This is the characteristic function of the Poisson process (Sect. 3.2.2.4) and the corresponding probability mass function represents a Poisson distribution (Sect. 2.3.1) with parameter $\alpha = \gamma t$:

$$P_n(t) = e^{-\alpha}\,\frac{\alpha^n}{n!} = e^{-\gamma t}\,\frac{(\gamma t)^n}{n!}\,. \tag{3.88}$$

The parameter γ is often referred to as the intensity of the process. It represents the reciprocal of the mean time between two jumps.

The *compensated Poisson process* is another example of a Lévy process. The stochastic growth of the Poisson process is compensated by a linear deterministic term. The two parameters and the transition function are $a = -\gamma$, $\sigma = 0$, and $w(u) = \gamma\,\delta(u-1)$. The process is described by the random variable $\mathcal{X}(t) = \mathcal{Z}(t)-\gamma t$ with expectation value $E\bigl(\mathcal{X}(t)\bigr) = 0$, where $\mathcal{Z}(t)$ is a conventional Poisson process. Accordingly, the compensated Poisson process is a martingale (Sect. 3.1.3).

An important generalization of the conventional Poisson process is the *compound Poisson process*, which is a Poisson process with variable step sizes expressed as random variables $\mathcal{X}_k$ drawn from a probability density $w(u)/\gamma$:

$$f(u)\,du = P\bigl(u < \mathcal{X}_k < u+du\bigr) = \frac{w(u)}{\gamma}\,du\,, \tag{3.126}$$

where the transition function is assumed to be normalizable, i.e.,

$$\gamma = \int_{-\infty}^{\infty} w(u)\,du < \infty\,,$$

and γ is again the intensity of the process. The number of events occurring in a compound Poisson process up until time t is described by the random variable $\mathcal{Z}(t) = \sum_{k=1}^{n(t)} \mathcal{X}_k$. The Poisson process and the compound Poisson process are the one-sided analogues of the constant and variable step size random walks (Fig. 3.16).

[44]The discreteness of u would require here a Stieltjes integral with $dW(u)$ but the trick with a Dirac delta function allows one to use the differential expression $w(u)du$.

Wiener Process

The Wiener process follows trivially from (3.122) by choosing $a = 0$ and $\sigma = 1$, and setting the jump probability to zero. This leads to the characteristic function

$$\phi(s,t) = \mathrm{e}^{-s^2 t/2} ,$$

and this is is the characteristic function of a Wiener process (3.60) with $x_0 = w_0 = 0$ and $t_0 = 0$. The probability density is obtained by inverse Fourier transform:

$$\mathcal{N}(0,1) = f(x,t) = \frac{1}{\sqrt{2\pi t}} \mathrm{e}^{-x^2/2t} .$$

Interestingly, we shall see that the Wiener process can also be obtained from (3.122) with $a = \sigma = 0$ and a special transition function $w(u) \propto |u|^{-(\alpha+1)}$ with $\alpha = 2$ (see next paragraph).

Pareto Processes

Pareto or Paretian processes are special pure jump Lévy processes with $a = \sigma = 0$ and a transition function of the type

$$w(u) = \begin{cases} \theta_- |u|^{-(\alpha+1)} , & \text{for } -\infty < u < 0 , \\ \theta_+ u^{-(\alpha+1)} , & \text{for } 0 < u < \infty , \end{cases} \tag{3.127}$$

with a *stability parameter* $0 \leq \alpha < 2$. The process is named after the Italian civil engineer and economist Vilfredo Pareto. It makes use of a transition function that is closely related to the Pareto distribution and which satisfies the same functional relationship for positive values of the variable on the support $u \in [u_{\mathrm{m}}, +\infty[= [\tilde{\mu}, +\infty[$ (Sect. 2.5.5).

Figure 3.17 shows the singularity of the transition functions (3.127) for Pareto processes at $u = 0$. The larger the value of the stability parameter α, the broader the peak embracing the singularity, and this has a strong influence on the frequency of occurrence of infinitesimally small jumps.

We are now in a position to choose an appropriate function for evaluating the principal value integral by means of the Lévy–Khinchin formula as expressed in (3.125) [194, p. 251]: $\delta(\epsilon) = \big((\theta_+ \epsilon^{1-\alpha} + \kappa)/\theta_-\big)^{1/(\alpha-1)}$. Clearly, the case $\alpha = 1$ cannot be handled in this way, but there is the possibility of a direct integration without using the Lévy–Khinchin formula.

For Pareto processes, the principal value integral can be calculated directly using Cauchy's integration with analytic continuation in the complex plane, $z = u +$

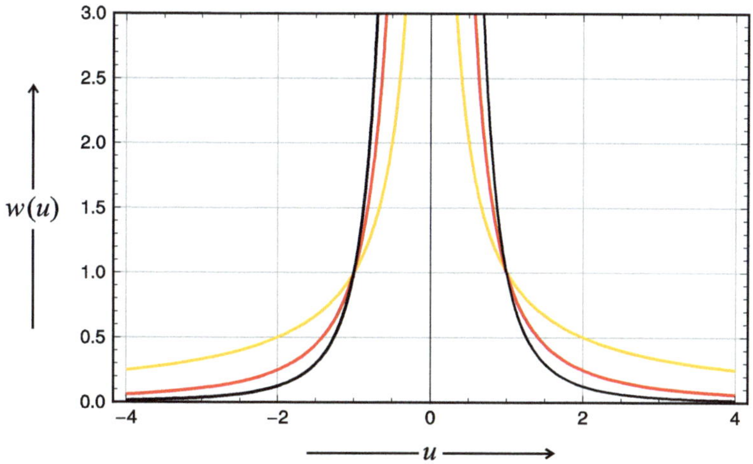

Fig. 3.17 Transition functions of Pareto processes. The transition functions $w(u) = |u|^{-(\alpha+1)}$ of the Pareto processes with $\alpha = 0$ (*yellow*), $\alpha = 1$ (*red*), and $\alpha = 2$ (*black*) are plotted against the variable u. The curve for $\alpha = 2$ is the reference corresponding to normal diffusion. All curves with $2 > \alpha \geq 0$ have heavier tails and this implies a larger probability for longer jumps

$\mathrm{i}v = |z|\,\mathrm{e}^{\mathrm{i}\vartheta}$, and residue calculus [21, Chaps. 6 and 7]:

$$\oint_{\varrho} f(z)\,\mathrm{d}z = 2\pi\mathrm{i}\,\mathrm{Res}\big(f(z), z_0\big) = 2\pi\mathrm{i}a_{-1}$$
$$= 2\pi\mathrm{i}\lim_{z\to z_0}\big((z-z_0)f(z)\big)\,, \tag{3.128a}$$

where ϱ is a closed contour encircling the (isolated) pole at $z = z_0$ in a region where $f(z)$ is analytical, $\mathrm{Res}\big(f(z), z_0\big)$ is the residue of $f(z)$ at this pole, and a_{-1} is the coefficient of $(z - z_0)^{-1}$ in the Laurent series[45]

$$f(z) = \sum_{n=-\infty}^{\infty} a_n(z-z_0)^n\,, \quad \text{with } a_n = \frac{1}{2\pi\mathrm{i}}\oint_{\gamma}\frac{f(z)}{(z-z_0)^n}\,. \tag{3.128b}$$

If $f(z)$ has a pole of order m at $z = z_0$, all coefficients a_n for $n < -m < 0$ with $n \in \mathbb{Z}$ are zero and the expression for a_{-m}, viz.,

$$2\pi\mathrm{i}\,a_{-m} = 2\pi\mathrm{i}\frac{1}{(m-1)!}\lim_{z\to z_0}\frac{\mathrm{d}^{m-1}}{\mathrm{d}z^{m-1}}\big((z-z_0)^m f(z)\big)\,, \tag{3.128c}$$

defines the first non-vanishing term of the series.

[45]The Laurent series is an extension of the Taylor series to negative powers of $(z - z_0)$, named in honor of the French mathematician Pierre Alphonse Laurent.

As can be seen from (3.127), the transition function $w(u)$ has a pole of order $\alpha+1$ at $u = u_0 = 0$. Since α need not be an integer, the analysis of Pareto processes opens the door to the world of fractals. It is worth noting that the Γ function and also the factorials contain an infinite number of factors for non-integer arguments, i.e., $\alpha! = \alpha(\alpha-1)(\alpha-2)(\alpha-3)\cdots$.

Evaluating the characteristic exponent yields for $\kappa = 0$:

$$\begin{aligned}\psi(s) &= |s|^\alpha \Gamma(-\alpha)\left((\theta_+ + \theta_-)\cos\frac{\alpha\pi}{2} - \mathrm{i}\frac{s}{|s|}(\theta_+ - \theta_-)\sin\frac{\alpha\pi}{2}\right)\\ &= -|s|^\alpha\gamma^\alpha\left(1 - \mathrm{i}\beta\frac{s}{|s|}\,\omega(s,\alpha)\right),\end{aligned}$$

with

$$\omega(s,\alpha) = \begin{cases}\tan\dfrac{\alpha\pi}{2}, & \text{if } \alpha \neq 1,\ 0<\alpha<2,\\ -\dfrac{2}{\pi}\ln|s|, & \text{if } \alpha = 1,\end{cases}$$

$$\beta = \frac{\theta_+ - \theta_-}{\theta_+ + \theta_-}, \qquad \gamma^\alpha = -(\theta_+ + \theta_-)\Gamma(-\alpha)\cos\frac{\alpha\pi}{2}.$$

For the characteristic function of the Pareto process, we finally obtain

$$\phi(s,t) = \exp\left(-|s|^\alpha\gamma^\alpha\left(1 - \mathrm{i}\beta\frac{s}{|s|}\,\omega(s,\alpha)\right)t\right). \tag{3.129}$$

The parameter α is called the stability parameter since transition functions $w(u)$ with α values outside $]0,2[$ lead to divergence. The skewness parameter β determines the symmetry of the transition function and the process:

(i) $\beta = 0$ implies invariance with respect to inversion at the the origin, i.e., $u \to -u$ or $s \to -s$, respectively, and $\theta_+ = \theta_- = \theta$.
(ii) $\beta \neq 0$ expresses the skewness of the characteristic function and the density.

Distributions with $\beta = \pm 1$ are said to be extremal, and for $\alpha < 1$ and $\beta \pm 1$, they are one-sided (see, for example, the Lévy distribution in Sect. 2.5.8 and Fig. 2.23).

Previously, we encountered the two symmetric processes $\beta = 0$ and $\theta_+ = \theta_- = \theta$, with $\alpha = 1$ and $\alpha = 2$, for which analytical probability densities are available:

1. The Wiener Process ($\alpha = 2$ and $\gamma = \theta = 1/2$)[46]:

$$\phi(s,t) = \exp(-s^2\gamma t) \ \text{ and } \ p(x,t) = \frac{1}{\sqrt{4\pi\gamma t}}\exp\frac{x^2}{4\gamma t}.$$

[46]Although the value $\alpha = 2$ leads to divergence in the regular derivation, applying $\alpha = 2$, $\beta = 0$, and $\gamma = 1/2$ yields the probability density of the normal diffusion process.

2. The Cauchy process ($\alpha = 1$ and $\gamma = \theta\pi$)[47]:

$$\phi(s,t) = \exp(-\pi|s|\theta t) \text{ and } p(x,t) = \frac{1}{\pi}\frac{\theta t}{(\theta t)^2 + x^2} .$$

The probability densities were normalized after the Fourier transform.

Levy Processes with Analytic Probability Densities
Although the characteristic function can be written down for any Lévy process, the probability density need not be expressible in terms of analytic functions. The three examples of Lévy processes where full analytical access is possible are shown in Fig. 2.23 and repeated here:

1. The normal distribution (Sect. 2.3.3) with $\alpha = 2$ and $\beta = 0$.
2. The Cauchy distribution (Sect. 2.5.7) with $\alpha = 1$ and $\beta = 0$.
3. The Lévy distribution (Sect. 2.5.8) with $\alpha = 1/2$ and $\beta = 1$.

In practice, this is hardly a restriction since numerical computation allows one to handle all cases, and most mathematics packages contain fast routines for Lévy and Pareto distributions.

Interpretation of Singularities in the Transition Functions
Lévy processes and Pareto processes in particular are also defined for cases where the transition functions have singularities at the origin $u = u_0 = 0$. In other words, we are considering examples with $\lim_{u\to 0} w(u) = \infty$. How can we visualize such a situation? Clearly, the condition modeled by the singularity implies the occurrence of shorter and shorter jumps at higher and higher rates until we are dealing in the limit $u \to 0$ with an infinite number of steps of infinitesimal size. Actually, such a limit is not unfamiliar to us, because in the transition from random walk to diffusion precisely the same problem arose and was solved straightforwardly. We mention again that diffusion appears twice in Lévy processes: (i) in the diffusion term of (3.122), and (ii) at the singularity of the transition function $w(u)$.

As shown in Fig. 3.17, the higher order $m = \alpha + 1$ of the pole at $u = 0$ is accompanied by a *broader* singularity and a less heavy tail. For the unbiased mind, it may seem strange to describe a process whose continuity was demonstrated earlier (see Sect. 3.1.5) exclusively by infinitesimally small jumps. The probabilistic interpretation, however, is straightforward: longer jumps of finite length also occur, but their frequency is of measure zero. The parameter α is confined to the range $0 < \alpha < 2$, and we can expect more diffusion-like behavior the closer α approaches

[47] We use the relations $\lim_{\alpha\to 1}\Gamma(-\alpha) = \pm\infty$, but $\lim_{\alpha\to 1}\Gamma(-\alpha)\cos(\pi\alpha/2) = -\pi/2$, which are easy to check.

the value $\alpha = 2$, which is the limit of normal diffusion. Smaller values of α result in higher probabilities for longer jumps (see Lévy flights).

Stable Distributions in Pareto Processes

Stable distributions $\mathcal{S}(\alpha, \beta, \gamma, \mu)$ are characterized by the following four parameters (Sect. 2.5.9):

(i) a stability parameter $\alpha \in]0, 2]$,
(ii) a skewness parameter $\beta \in [-1, 1]$,
(iii) a scale parameter $\gamma \geq 0$,
(iv) a location parameter $\mu \in \mathbb{R}$.

These parameters are identical to α, β, and γ as appearing in the transition function (3.129) of the Pareto process. The location parameter was chosen so that $\mu = 0$ by definition. Important properties of stable distributions are stability and infinite divisibility

Infinite Divisibility and Stability

The property of infinite divisibility is defined for classes of probability densities $p(x)$ and requires that a random variable $\mathcal{S}$ with this density $p(x)$ can be partitioned into any arbitrary number $n \in \mathbb{N}_{>0}$ independent and identically distributed (iid) random variables such that all individual variables $\mathcal{X}_k$, the sum $\mathcal{S}_n = \mathcal{X}_1 + \mathcal{X}_2 + \cdots + \mathcal{X}_n$, and all possible partial sums have the same probability density $p(x)$. Lévy processes are homogeneous Markov processes and they are infinitely divisible. In general, however, the probability distributions of the individual parts $\mathcal{X}_k$ will be different and they will differ from the density $p(x)$. Stable distributions are infinitely divisible, but the opposite is not true: there are infinitely divisible distributions outside the class of stable distributions (Sect. 2.5.9).

As for the normal distribution, we define standard stable distributions with only two parameters by setting $\gamma = 1$ and $\mu = 0$:

$$p_{\alpha,\beta}(x) = p_{\alpha,\beta,1,0}(x) = p_{\alpha,\beta,1,0}\left(\frac{\xi - \mu}{\gamma}\right) = \gamma p_{\alpha,\beta,\gamma,\mu}(\xi) \,.$$

As mentioned earlier, the characteristic exponent α is also called the *index of stability*, since it determines the order of the singularity at $u = 0$ and at the same time, the long-distance scaling of the probability density [43, 81, 454]:

$$p_{\alpha,\beta}(x) \approx \frac{C(\alpha)}{|x|^{\alpha+1}} \,, \quad \text{for } x \to \pm\infty \,.$$

This scaling law is determined by the scaling parameter α, which turns out to be the spatial universality exponent ($\alpha = 2$ for conventional diffusion as described in Sect. 3.2.4).

Universality and Self-Similarity

Self-similarity and shapes of objects fitting fractal or non-integer dimensions are the major topics of Benoît Mandelbrot's seminal book *Fractal Geometry of Nature* [366]. The self-similarity of stochastic processes was already mentioned in Sect. 3.2.4 in the context of continuous-time random walks. In a nutshell, looking at a self-similar object with a magnifying glass, we see the same pattern no matter how large the magnification factor may be. Needless to say, real objects can be self-similar only over a few orders of magnitude, because resolutions cannot be increased without limit.

The notion of *universality* [14, 150] was developed in statistical physics, in particular in the theory of phase transitions, where large collectives of atoms or molecules exhibit characteristic properties near critical points that are independent of the specific material parameters. Commonly *power laws* with critical exponents $\alpha, f(s) = f(s_0)|s - s_{\text{crit}}|^{\alpha}$, are observed, and when they are valid over several orders of magnitude, the patterns become independent of the sizes of objects. Diffusion of molecules and condensation through aggregation are indeed examples of universal phenomena with the critical exponents $\alpha = 2$ in length and $\theta = 1$ in time. As already mentioned, universality concerns the fact that all random walks with finite variance and finite waiting times fall into this universality class. With the experience gained from stable distributions and Lévy processes, we can generalize the phenomena and compare the properties of processes with other universality exponents, $0 < \alpha \leq 2$ in space and $0 < \theta \leq 1$ in time. Higher exponents $\alpha > 2$ are incompatible with normalizable probability densities. In particular, convergence of the principal value integral cannot be achieved by a proper choice of $\delta(\epsilon)$ and ϵ [194, pp. 251–252].

The continuous time random walk (Fig. 3.16) is revisited here under the assumption of Lévy distributed jump lengths and waiting times. The calculation of the probability density $p(x, t)$ is full analogous to the one in Sect. 3.2.4, and starts from the joint probability distribution $\varphi(\xi, \tau) = f(\xi)\psi(\tau)$, where independence according to (3.116) is assumed. The spatial increments are now derived from a stable Lévy distribution $f_{\alpha,0,\gamma,0}(\xi)$, which is symmetric ($\beta = 0$) and specified by the characteristic exponent α, the scale parameter γ, and the location at the origin ($\mu = 0$). Since $f_{\alpha,0,\gamma,0}(\xi)$ need not be expressible in analytic form, we define it in terms of its characteristic function, which we write here in the form of the limit of long distances or small k values[48]:

$$\begin{aligned}\tilde{f}_{\alpha,\gamma}(|k|) = \mathrm{E}\big(\exp(\mathrm{i}|k|\mathcal{X}_{\alpha,\gamma})\big) &= \int_{-\infty}^{\infty} \mathrm{d}\xi \mathrm{e}^{\mathrm{i}|k|\xi} f_{\mathcal{X}_{\alpha,\gamma}}(\xi) \\ &= \exp(-\gamma^{\alpha}|k|^{\alpha}) = 1 - \gamma^{\alpha}|k|^{\alpha} + \mathcal{O}(|k|^{2\alpha}) \,.\end{aligned} \tag{3.130}$$

The condition for obtaining an acceptable probability density, i.e., nonnegative everywhere and normalizable, by inverse Fourier transform defines the range of possible values for the universality exponent to be $0 < \alpha \leq 2$. We illustrate by means of two examples. The obvious first example is the continuous-time random walk with $\alpha = 2$ and normally distributed jump length:

$$f(\xi) = \frac{\mathrm{e}^{-\xi^2/4D}}{\sqrt{4\pi D}} \quad \text{and} \quad \tilde{f}(k) = \exp(-\gamma^2 k^2) = 1 - \gamma^2 k^2 + \mathcal{O}(k^2) \,.$$

As the second illustrative example, we mention the Cauchy distribution with universality exponent $\alpha = 1$:

$$f(\xi) = \frac{\gamma}{\pi(\gamma^2 + \xi^2)} \quad \text{and} \quad \tilde{f}(k) = \exp(-\gamma|k|) = 1 - \gamma|k| + \mathcal{O}(|k|^2) \,.$$

Since $-k^2$ decays faster than $-|k|$, the Cauchy distribution has heavier tails and sustains longer jumps, something we already showed earlier.

The property of infinite divisibility can be used for a straightforward calculation of the mean length $l = \overline{\mathcal{X}_0 \mathcal{X}_n} = x_n = \sum_i \xi_i$ of a CTRW, as expressed in terms of the *width* of the density $f_{\alpha,\gamma}(\xi/n)$. Stability of the Lévy distribution requires a linear combination of independent copies of the variable to have the same distribution as the copy itself, and this yields for the sum

$$f_{n,\alpha,\gamma}\Big(\sum_i \xi_i\Big) = f_{\alpha,\gamma}(\xi_1) * f_{\alpha,\gamma}(\xi_2) * \cdots * f_{\alpha,\gamma}(\xi_n) \,.$$

Application of the convolution theorem yields

$$\tilde{f}_n(|k|) = \prod_{i=1}^{n} \tilde{f}_{\alpha,\gamma}(|k_i|) = \exp\big(-\gamma n^{1/\alpha}|k|^{\alpha}\big) \,.$$

[48]The absolute value of the wave number $|k|$ is sometimes used in all expressions, which is necessary when complex k values are admitted, or in the multidimensional case where $\mathbf{k}$ is the wave vector. Here we use real k values in one dimension and we need $|k|$ only to express a cusp at $k = 0$.

Transforming back into physical space yields a generalization of the central limit theorem:

$$f_{n,\alpha,\gamma}\Big(\sum_i x_i\Big) = f_{n,\alpha,\gamma}\big(x/n^{1/\alpha}\big) \,. \tag{3.131}$$

The length of the random walk is related to the width of the distribution, and (3.131) yields the scaling $l = \langle x(n)\rangle \propto n^{1/\alpha}$ of mean walk lengths. In normal diffusion, $\alpha = 2$ and the length grows as $\sqrt{n}$. For Lévy stable distributions with $\alpha < 2$, the walks become longer because of heavier tails compared to the normal distribution. The corresponding trajectories are called Lévy flights and will be discussed in the last part of this section. In polymer theory, the length of the walk corresponds to the end-to-end distance of the polymer chain, for which analytic probability densities are available [506]. For polymers with Gaussian distributions, which follow from CLT for sufficiently long chains, the mean of the end-to-end distance satisfies a square root law, i.e., $l \propto \sqrt{n}$.

The density of the waiting times is modified according to the empirical evidence. There are well-documented processes with waiting times that deviate from the expected exponential distribution, in the sense that longer waiting times have higher probabilities or, in other words, the tails of the probability densities decay more slowly than exponentially. These deviations may have different origins. They are referred to as *subdiffusion*, and novel mathematical methods have been developed in order to deal with them properly [218, 396, 477]. In particular, adequate modeling of subdiffusion requires fractional calculus, and since we shall not need this elegant but rather involved technique in this monograph, we dispense here with further details of this discipline.

In order to take into account long rests corresponding to the long tails of the distribution of waiting times τ_{w}, we assume an asymptotic behavior of the form [396]

$$\psi(\tau) \approx A_\theta \left(\frac{\tau_{\mathrm{w}}}{t}\right)^{1+\theta} \,, \quad \text{with } 0 < \theta \le 1 \,. \tag{3.132}$$

After Laplace transformation, this yields

$$\hat{\psi}(u) \approx \frac{1}{1 + (\tau_{\mathrm{w}}u)^\theta} = 1 - (\tau_{\mathrm{w}}u)^\theta + \mathcal{O}(u^{1+\theta}) \,. \tag{3.133}$$

The transformed joint distribution function is obtained from the Montroll–Weiss equation (3.119) and is of the form:

$$\hat{\tilde{p}}(|k|, u) = \frac{1 - \hat{\psi}(u)}{u}\,\frac{1}{1 - \hat{\psi}(u)\tilde{f}(|k|)} \approx \frac{\tau_{\mathrm{w}}u^{\theta-1}}{\tau_{\mathrm{w}}u^\theta + \lambda|k|^\alpha} \,. \tag{3.134}$$

As we saw in Sect. 3.2.4, the expression on the right-hand side of (3.134) with $\alpha = 2$ and $\theta = 1$ can be subjected straightforwardly to inverse Laplace and Fourier transform, and this yields the density function of normal diffusion $p(x,t) = \mathcal{L}^{-1}\mathcal{F}^{-1}\big(\hat{\tilde{p}}(|k|,u)\big) = \mathrm{e}^{-x^2/4Dt}/\sqrt{4\pi Dt}$ as given in (3.61).

For general Pareto processes, the inverse Laplace and inverse Fourier transform of the expression of the right-hand side of (3.134) is much more involved and cannot be completed in closed form. We can only indicate how one might proceed in the fractal case. For the inverse Laplace transform, we get

$$\begin{aligned} p(x,t) &\approx \int_0^{\infty} \mathrm{d}u \int_{-\infty}^{+\infty} \mathrm{d}k\, \mathrm{e}^{-\mathrm{i}|k|x+ut} \frac{\tau_{\mathrm{w}} u^{\theta-1}}{\tau_{\mathrm{w}} u^{\theta} + \lambda |k|^{\alpha}} \\ &= \int_{-\infty}^{+\infty} \mathrm{d}k\, \mathrm{e}^{-\mathrm{i}|k|x} E_{\theta}(-|k|^{\alpha} t^{\theta}) \,, \end{aligned} \tag{3.135}$$

where we have used the Mittag-Leffler function $E_\theta(-|k|^\alpha t^\theta)$, which is named after Magnus Gösta Mittag-Leffler. It occurs in inverse Laplace transforms of functions with a parameter $p^\alpha(a + bp^\beta)$ for the Laplace transform [374], and is represented by the infinite series [399]

$$E_\alpha(z) = \sum_{k=0}^{\infty} \frac{z^k}{\Gamma(1+\alpha k)} \,, \quad \alpha \in \mathbb{C} \,,\ \Re(\alpha) > 0 \,,\ z \in \mathbb{C} \,.$$

This leads to quite involved expressions, except in some simple cases, e.g., $E_1(z) = \exp(z)$ or $E_0(z) = 1/(1-z)$ [244]. The evaluation of the inverse Fourier transform (3.135) is even more complicated, but we shall only need to consider the form of the leading terms. The function of the form $\tilde{p}\,(t^\theta |k|^\alpha)$ in the integrand becomes a function $p(x^\alpha/t^\theta)$ after the inverse Fourier transform. If we express distance as a function of time, we eventually obtain $x^\alpha/t^\theta = c \rightarrow x(t) \propto t^{\theta/\alpha}$. The expression covers normal diffusion with $\alpha = 2$ and $\theta = 1$, leading to the relation $x(t) \propto \sqrt{t}$, and fractional diffusion with $\alpha = 2$ and $\theta < 1$, resulting in $x(t) \propto t^{\theta/2}$.

Figure 3.18 summarizes the results on Lévy processes in space and time. All continuous-time random walks (CTRW) may be characterized by two universality exponents, $0 < \alpha \leq 2$ and $0 < \theta \leq 1$, for scaling behavior in space and time. Normal diffusion is the limiting case with $\alpha = 2$ and $\theta = 1$. The probability densities of waiting times and jump length, the Poisson distribution and the normal distribution, respectively, both have finite expectation values and variances. In *anomalous diffusion*, one or both of the two variances diverge, or do not exist. Lévy stable distributions with $\alpha < 2$ have heavy tails and the variance of the jump length diverges. Heavy tails make larger jump increments more probable, and the processes are characterized by longer walk lengths, since $x(n) \propto n^{1/\alpha}$. Alternatively the variance of the step size is kept finite in *fractal Brownian motion*, but the jumps are delayed and the waiting times diverge. The inner part of the square is filled by

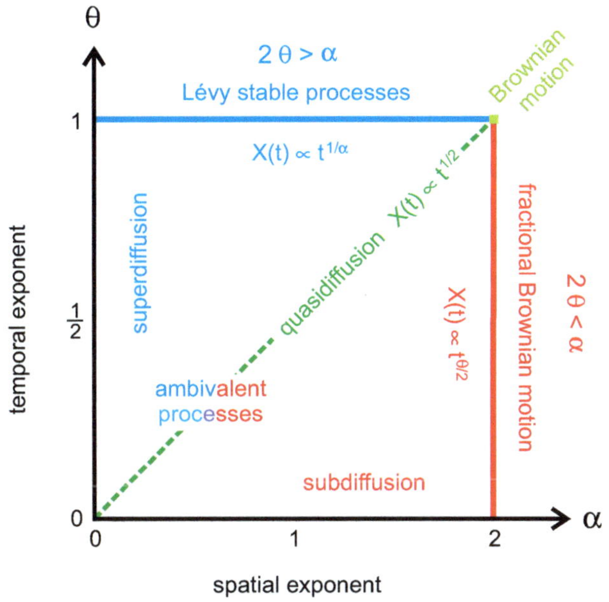

Fig. 3.18 Normal and anomalous diffusion. The figure sketches continuous-time random walks (CTRW) with space and time universality exponents in the ranges $0 < \alpha \le 2$ and $0 < \theta \le 1$, respectively. The limiting cases of characteristic asymptotic behavior are (1) Lévy flights with $(0 < \alpha < 2, \theta = 1)$ (*blue*), (2) normal diffusion with $(\alpha = 2, \theta = 1)$ (*chartreuse*), and (3) fractional Brownian motion with $(\alpha = 2, 0 < \theta < 1)$ (*red*). In the interior of the square, we find the general class of ambivalent processes. Processes situated along the diagonal satisfying $\theta = \alpha/2$ (*green*) are referred to as quasidiffusion. Adapted from [66, Suppl. 2]

so-called *ambivalent processes* which are characterized by divergence of variances in space and time (for details see [61, 396]).

Lévy processes derived from transition functions (3.127) with $0 < \alpha < 2$ correspond to densities with heavy tails and diverging variances. They were called *Lévy flights* by Benoît Mandelbrot [366]. Lévy flights with $\alpha = 2$, which Mandelbrot called *Rayleigh flights*, turned out to be almost indistinguishable from conventional random walks with constant step size, and accordingly both processes are suitable models for Brownian motion (Fig. 3.19). Since the Pareto transition function coincides with the normal, the Cauchy, and the Lévy distribution only in the asymptotic tails ($x \to \infty$), this similarity is a nice demonstration of the relevance of asymptotic behavior. In the limit $t \to \infty$, as already mentioned, 1D and 2D random walks lead to complete coverage of the line and the plane, respectively. Compared to the tails of the normal distribution, the tails of all other Pareto transition functions, $\alpha < 2$, are heavier, and this implies higher probabilities for longer steps. In the special classes of Lévy flights with $\alpha = 1$ and $\alpha = 0.5$, for example, the step lengths may be drawn randomly from Cauchy or Lévy distributions, or derived from the power laws (3.127). The higher probability of long steps completely changes the appearance of the trajectories. In the 2D plots, densely visited zones are interrupted by occasional wide jumps that initiate a new local diffusion-like process in another part of the plane. In Fig. 3.19, we compare trajectories of 100,000 individual steps calculated by a random walk routine with those computed for Lévy flights with $\alpha = 2$ and $\alpha = 0.5$. The 2D pattern calculated for the Lévy flight with $\alpha = 2$ is very similar to the random walk pattern,[49] whereas the Lévy flight with $\alpha = 0.5$ shows

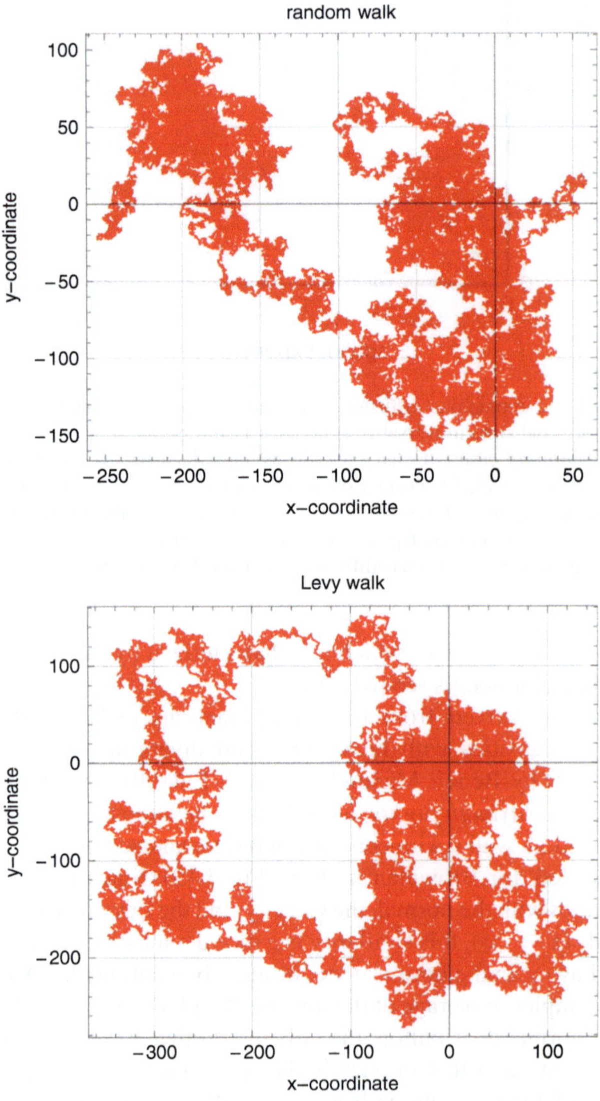

Fig. 3.19 Brownian motion and Lévy flights in two dimensions. Continued on next page

[49]Because of this similarity we called the $\alpha = 2$ Pareto process a *Lévy walk*.

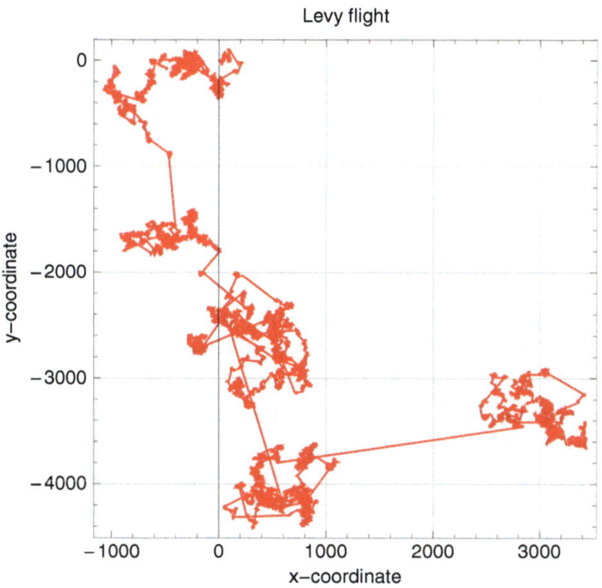

Fig. 3.19 Brownian motion and Lévy flights in two dimensions. The figure compares three trajectories of processes in the (x, y)-plane. Each trajectory consists of 100,000 incremental steps, and each step combines a direction that is randomly chosen from a uniform distribution $\vartheta \in \mathcal{U}_{\Omega}$, $\Omega = [0, 2\pi]$ with a step length l. For the simulation of the random walk, the step length was chosen to be $l = 1$ [l] and for the Lévy flights the length was taken as a second set of random variables $l = \ell$, drawn from a density function $f_{\ell}(u) = u^{-(\alpha+1)}$ from (3.127). The components of the trajectory in the x and y directions were $x_{k+1} = x_k + l\cos\vartheta$ and $y_{k+1} = y_k + l\sin\vartheta$, respectively. The random variable ℓ is calculated from a uniformly distributed random variable v on [0, 1] via the inverse cumulative distribution [106]:

$$\ell = F^{-1}(v) = u_{\mathrm{m}}(1 - v)^{-1/(\alpha+1)}.$$

For a uniform density on [0, 1], there is no difference in distribution between the random variables $1 - v$ and v, and hence we used the simpler expression $\ell \propto v^{-1/(\alpha+1)}$. (The computation of pseudorandom numbers following a predefined distribution will be mentioned again in Sect. 4.6.3.) The factor u_{m} is introduced as a lower bound for u, in order to allow for normalization of the probability density. It can also be understood as a scaling factor. Here we used $u_{\mathrm{m}} = 1$ [l]. The examples shown were calculated with $\alpha = 2$ and $\alpha = 0.5$ and characterized as a *Lévy walk* and a *Lévy flight*, respectively. Apparently, there is no appreciable observable difference between the random walk and the Lévy walk. Random number generator: Mersenne Twister with seed 013 for the random walk, 016 for the Lévy walk, and 327 for the Lévy flight

the expected small, more or less densely covered patches that are separated by long jumps. It is instructive to consider the physical dimensions of the area visited by the 2D processes. The random walk and the Lev́y walk cover areas of approximately 300×300 [l^2] and 400×400 [l^2], but the Lévy flight ($\alpha = 0.5$) takes place in a much larger domain, 4000×4000 [l^2].

The trajectories shown in Fig. 3.19 suggest using the mean walk length for the classification of processes. Equation (3.131) implies a mean walk length $\langle x(n) \rangle \propto n^{1/\alpha}$, where n is the number of steps of the walk. Using the *mean square displacement* to characterize walk lengths, for normal diffusion starting at the origin $\mathbf{x}(0) = \mathbf{x}_0 = 0$, we find

$$\left\langle \mathbf{r}(t)^2 \right\rangle_{\text{normal diffusion}} = \left\langle \left(\mathbf{x}(t) - \mathbf{x}_0\right)^2 \right\rangle = 2Dt \propto t^{\theta} \;, \quad \text{with } \theta = 1 \,.$$

Anomalous diffusion is classified with respect to the asymptotic time dependence (Fig. 3.20). Subdiffusion is characterized by $\theta < 1$, and as mentioned above, it deals with diffusion processes that are slowed down or delayed by structured environments. Superdiffusion with $\theta > 1$ is faster than normal diffusion and this is caused, for example, by a higher probability of longer jumps in a random walk.

The trajectory of the Lévy flight in Fig. 3.19 suggests an optimized search strategy: a certain area is searched thoroughly and after some time, for example, when the territory has been exhaustively harvested, the search is continued in a rather distant zone. Prey foraging strategies of marine predators, for example, those of sharks, were found to come close to Lévy flights. An optimal strategy consists in the combination of local searches by a Brownian motion type of movements and long jumps into distant regions where the next local search can start. The whole trajectory of such a combined search resembles the path of a Lévy flight [269, 551].

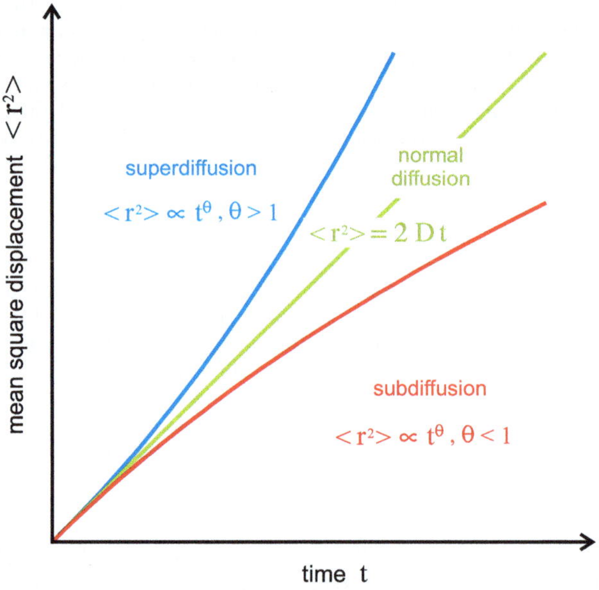

Fig. 3.20 Mean square displacement in normal and anomalous diffusion. The mean square displacement in normal diffusion is $\langle \mathbf{r}^2 \rangle = \langle (\mathbf{x} - \mathbf{x}_0)^2 \rangle = 2Dt$, and the generalization to anomalous diffusion allows for a classification of the processes according to the time exponent of the mean square displacement, i.e., θ given by $\langle \mathbf{r}^2 \rangle \propto t^{\theta}$: $\theta < 1$ characterizes *subdiffusion* and $\theta > 1$ *superdiffusion*

3.3 Chapman–Kolmogorov Backward Equations

Time inversion in a conventional differential equation changes the direction in which trajectories are traversed, and this has only minor consequences for the phase portrait of the dynamical system: ω limits become α limits and vice versa, stable equilibrium points and limit cycles become unstable, and so on, but the trajectories, without the *arrow of time*, remain unchanged. Apart from the direction, integrating forward yields precisely the same results as integrating backward from the endpoint of the forward trajectory. The same is true, of course, for a Liouville equation, but it does not hold for a Wiener process or a Langevin equation. As sketched in Fig. 3.21 (lower), time reversal results in trajectories that diverge in the backward direction. In other words, the commonly chosen reference conditions are such that a forward process has sharp initial conditions at the beginning of the ordinary time scale, i.e., t_0 for t progressing into the future, whereas a backward process has sharp final conditions at the end, i.e., τ_0 for a virtual *computational time* τ progressing backwards into the past. Accordingly, the Chapman–Kolmogorov equation can be interpreted in two different ways giving rise to forward and backward equations that are equivalent to each other, and the basic difference between them concerns the set of variables that is held fixed. For the forward equation, we hold $(\mathbf{x}_0, t_0)$ fixed, and consequently solutions exist for $t \geq t_0$, so that $p(\mathbf{x}, t_0|\mathbf{x}_0, t_0) = \delta(\mathbf{x} - \mathbf{x}_0)$ is an *initial condition* for the forward equation. The backward equation on the other hand has solutions for $t \leq t_0$ corresponding to $\tau \geq \tau_0$, and hence describes the evolution in τ. Accordingly, $p(\mathbf{y}, \tau_0|\mathbf{y}_0, \tau_0) = \delta(\mathbf{y} - \mathbf{y}_0)$ is an appropriate *final* condition (rather than an initial condition).[50]

Naïvely we could expect to find full symmetry between forward and backward computation. However, there is one fundamental difference between calculations progressing in opposite directions, which will become evident when we consider backward equations in detail: in addition to the two different computational time scales for forward and backward equations in Fig. 3.21, i.e., t and τ, respectively, we have the real or physical time of the process, which has the same direction as t. Unless we use some scaling factor, it is measured in the same units as t, and we shall not distinguish the two time scales unless it is necessary. The computational time τ, however, runs opposite to physical time, and the basic symmetry-breaking difference between forward and backward equations concerns the arrow of computational time. The difference can also be expressed by saying that forward equations make predictions of the future, while backward equations reconstruct the past. As we shall see, the backward equation is (somewhat) better defined in the eyes of mathematicians than its forward analogue (see [158] and [161, pp. 321 ff.]).

[50] In order to avoid confusion we shall reserve the variable $y(\tau)$ and $y(0) = y_0$ for backward computation.

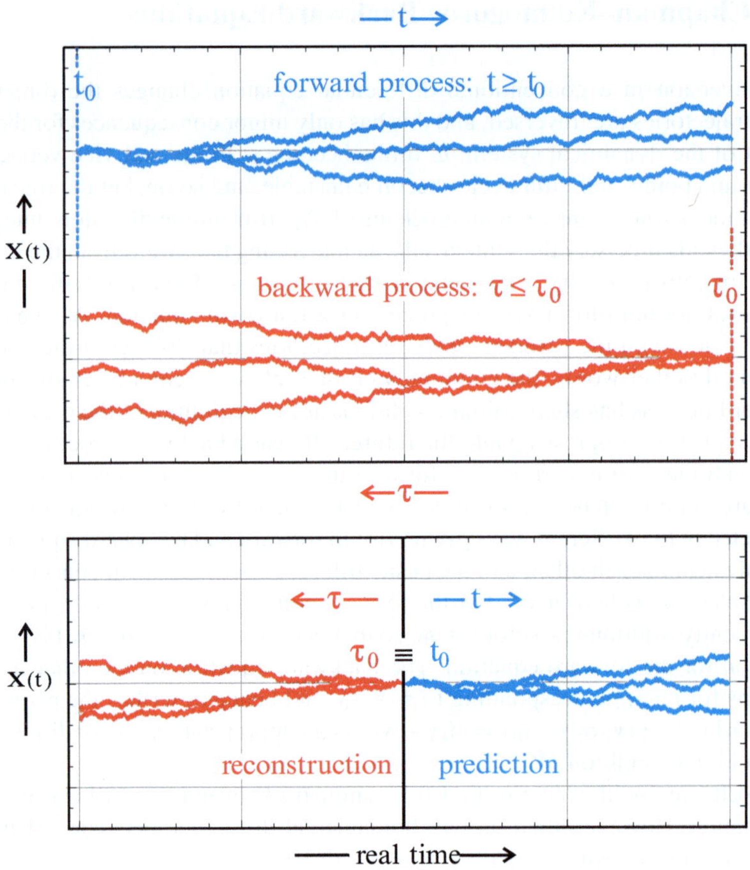

Fig. 3.21 Illustration of forward and backward equations. The forward differential Chapman–Kolmogorov equation is used to calculate the future development of ensembles or populations. The trajectories (*blue*) start from an initial condition $(\mathbf{x}_0, t_0)$, commonly corresponding to the sharp distribution $p(\mathbf{x}, t_0) = \delta(\mathbf{x} - \mathbf{x}_0)$, and the probability density unfolds with time $t \geq t_0$. The backward equation is commonly applied to calculate first passage times or to solve exit problems. In order to minimize the risk of confusion, in backward equations, we choose the notation $\mathbf{y}$ and τ for the variable and the time, respectively, and we have the correspondence $(\mathbf{y}(\tau), \tau) \Leftrightarrow (\mathbf{x}(t), t)$. In backward equations, at the latest time τ_0, the corresponding value of the variables at this time, viz., $(\mathbf{y}_0, \tau_0)$, are held constant, and sharp initial conditions—better called *final conditions* in this case—are applied, i.e., $p(\mathbf{y}, t_0 | \mathbf{y}, t) = \delta(\mathbf{y} - \mathbf{y}_0)$, and the time dependence of the probability density corresponds to samples unfolding into the past, i.e., $\tau \geq \tau_0$ (trajectories in *red*). In the *lower part* of the figure an alternative interpretation is given. The forward and the backward process start at the same time $t_0 \equiv \tau_0$, but progress in different time directions: computation of the forward process predicts the future, whereas computation of the backward process reconstructs the past

3.3.1 Differential Chapman–Kolmogorov Backward Equation

The Chapman–Kolmogorov equations (3.35 and 3.36) are interpreted in two different ways, giving rise to the two formulations known as forward and backward equations. In the forward equation, the pair (x_3, t_3) is considered to be fixed and (x_1, t_1) expresses the variable in the sense of $x_1(t)$, where the time t_1 proceeds in the direction of positive real time (see Fig. 3.4). The backward equation, on the other hand, is exploring the past of a given situation. Here, the pair (x_1, t_1) is fixed and (x_3, t_3) is propagating backwards in time. This difference is taken into account by the alternative notation: (y_3, τ_3) is fixed and (y_1, τ_1) is moving. The fact that real time always proceeds in the forward direction manifests itself through the somewhat different forms of forward and backward equations. Both Chapman–Kolmogorov differential expressions, the forward and the backward versions, are useful in their own rights. The forward equation gives the values of measurable quantities directly as functions of the *measured* (real) time. Accordingly, it is preferentially used to describe the dynamics of processes and to model experimental systems, and it is well suited to predicting future probabilities. The backward equation finds applications in the computation of the evolution towards given events, for example, *first passage times* or *exit problems*, which seek the probability that a particle leaves a region at a certain time, and in reconstructions of the past.

Since the difference in the derivation of the forward and backward equations is essential for the interpretation of the results, we make a brief digression into the derivation of the backward equation, which is similar, but not identical to the procedure for the forward equation. The starting point again is the conditional probability of a Markov process from a recording $(\mathbf{y}, \tau)$ in the past given the final condition $(\mathbf{y}_0, \tau_0)$ at present: $p(\mathbf{y}_0, \tau_0|\mathbf{y}, \tau_0) = \delta(\mathbf{y}_0 - \mathbf{y})$. However, as the term *backward* indicates, we shall assume that the computational time τ proceeds from τ_0 into the past (Fig. 3.21) and the difference with the forward equation comes from the fact that computational and real time progress in opposite directions.

The derivation of the backward equation proceeds essentially in the same way as in Sect. 3.2.1, except τ runs opposite to t. We begin by writing down the infinitesimal limit of the difference equation:

$$
\begin{aligned}
&\frac{\partial p(\mathbf{y}_0, \tau_0|\mathbf{y}, \tau)}{\partial \tau} \\
&= \lim_{\Delta\tau \to 0} \frac{1}{\Delta\tau}\Big(p(\mathbf{y}_0, \tau_0|\mathbf{y}, \tau+\Delta\tau) - p(\mathbf{y}_0, \tau_0|\mathbf{y}, \tau)\Big) \\
&= \lim_{\Delta\tau \to 0} \frac{1}{\Delta\tau} \int_\Omega \mathrm{d}\mathbf{z}\, p(\mathbf{z}, \tau+\Delta\tau|\mathbf{y}, \tau)\Big(p(\mathbf{y}_0, \tau_0|\mathbf{y}, \tau+\Delta\tau) - p(\mathbf{y}_0, \tau_0|\mathbf{z}, \tau+\Delta\tau)\Big) ,
\end{aligned}
$$

where we have applied the same two operations as used for the derivation of (3.39). The first is resolution of unity, i.e.,

$$
1 = \int_\Omega \mathrm{d}\mathbf{z}\, p(\mathbf{z}, \tau+\Delta\tau|\mathbf{y}, \tau) ,
$$

and the second, insertion of the Chapman–Kolmogorov equation in the second term with $\mathbf{z}$ as the intermediate variable:

$$p(\mathbf{y}_0, \tau_0|\mathbf{y}, \tau) = \int_\Omega \mathrm{d}\mathbf{z}\, p(\mathbf{y}_0, \tau_0|\mathbf{z}, \tau+\Delta\tau)p(\mathbf{z}, \tau+\Delta\tau|\mathbf{y}, \tau) .$$

Further steps parallel those in the derivation of the forward case:

(i) Separation of the domain of integration into two parts, with integrals I_1 and I_2 with $\|\mathbf{z}-\mathbf{y}\| < \epsilon$ and $\|\mathbf{z}-\mathbf{y}\| \geq \epsilon$, respectively.
(ii) Expansion of I_1 into a Taylor series.
(iii) Neglect of higher order residual terms.
(iv) Introduction of transition probabilities for jumps in the limit of vanishing $\Delta\tau$:

$$\lim_{\Delta\tau\to 0} \frac{1}{\Delta\tau} p(\mathbf{z}, \tau+\Delta\tau|\mathbf{y}, t) = W(\mathbf{z}|\mathbf{y}, \tau) . \tag{3.136}$$

(v) Consideration of boundary effects if there are any.

Eventually, we obtain [194, pp. 55, 56]:

$$\frac{\partial p(\mathbf{y}_0, \tau_0|\mathbf{y}, \tau)}{\partial \tau} = \sum_i A_i(\mathbf{y}, \tau)\frac{\partial p(\mathbf{y}_0, \tau_0|\mathbf{y}, \tau)}{\partial y_i} + \frac{1}{2}\sum_{i,j} B_{ij}(\mathbf{y}, \tau)\frac{\partial^2 p(\mathbf{y}_0, \tau_0|\mathbf{y}, \tau)}{\partial y_i \partial y_j}$$
$$+ \int_\Omega \mathrm{d}\mathbf{z}\, W(\mathbf{z}|\mathbf{y}, \tau)\Big(p(\mathbf{y}_0, \tau_0|\mathbf{z}, \tau) - p(\mathbf{y}_0, \tau_0|\mathbf{y}, \tau)\Big) .$$

Next we reintroduce real time $\Delta\tau = -\Delta t$, and obtain the *backward differential Chapman–Kolmogorov equation*, which complements the previously derived forward equation (3.46):

$$\frac{\partial p(\mathbf{y}_0, t_0|\mathbf{y}, t)}{\partial t} = -\sum_i A_i(\mathbf{y}, t)\frac{\partial p(\mathbf{y}_0, t_0|\mathbf{y}, t)}{\partial y_i} \tag{3.137a}$$

$$-\frac{1}{2}\sum_{i,j} B_{ij}(\mathbf{y}, t)\frac{\partial^2 p(\mathbf{y}_0, t_0|\mathbf{y}, t)}{\partial y_i \partial y_j} \tag{3.137b}$$

$$+ \int_\Omega \mathrm{d}\mathbf{z}\, W(\mathbf{z}|\mathbf{y}, t)\Big(p(\mathbf{y}_0, t_0|\mathbf{y}, t) - p(\mathbf{y}_0, t_0|\mathbf{z}, t)\Big) . \tag{3.137c}$$

The appropriate final condition replacing (3.37) is

$$p(\mathbf{y}_0, t_0|\mathbf{y}, t) = \delta(\mathbf{y}_0 - \mathbf{y}) ,$$

which expresses a sharp condition for $t = t_0$, namely, $p(\mathbf{y}, t_0) = \delta(\mathbf{y}_0 - \mathbf{y})$ (Figs. 3.4 and 3.21). Apart from the change in sign due to the fact that $\Delta t = -\Delta\tau$, we realize changes in the structure of the PDE that make the equation in essence easier to handle than the forward equation. In particular, we find for the three terms:

(i) The Liouville equation (Sect. 3.2.2.1) is a partial differential equation whose physically relevant solutions coincide with the solution of an ordinary differential equation, and therefore the trajectories are invariant under time reversal. Only the direction of the process is reversed: going backwards in time changes the signs of all components of $\mathbf{A}$ and the particle travels in the opposite direction along the same trajectory that is determined by the initial or final conditions, $(\mathbf{x}_0, t_0)$ or $(\mathbf{y}_0, \tau_0)$, respectively.
(ii) The diffusion process described by (3.137b) spreads in the opposite direction as a consequence of the reversed arrow of time. The mathematics of time reversal in diffusion was studied extensively in the 1980s [10, 135, 245, 500], and rigorous mathematical proofs were derived which confirmed that time reversal does indeed lead to a diffusion process in the time-reversed direction in the sense of the backward processes sketched in Fig. 3.21: starting from a sharp final condition, the trajectories diverge in the direction of $\tau = -t$.
(iii) The third term (3.137c) describes jump processes and will be handled in Sect. 3.3.2, which deals with backward master equations.

3.3.2 Backward Master Equations

The backward master equation follows directly from the backward dCKE by setting $\mathbf{A} = 0$ and $\mathrm{B} = 0$, and this is tantamount to considering only the third term (3.137c). Since the difference in forward and backward equations is essential for the interpretation of the results, we consider the backward master equation in some detail. The starting point is the conditional probability $p(\mathbf{y}_0, \tau_0|\mathbf{y}, \tau_0) = \delta(\mathbf{y}_0 - \mathbf{y})$ of a Markov step process recorded from $(\mathbf{y}, \tau)$ in the past to the final condition $(\mathbf{y}_0, \tau_0)$ at present time τ_0. However, as the term *backward* indicates, we shall assume that the computational time τ progresses from τ_0 into the past. Some care is needed in applications to problem solving, because the direction of the time axis influences the appearance and interpretation of transition probabilities. In computational time τ, the jumps go in the opposite direction (Fig. 3.22).

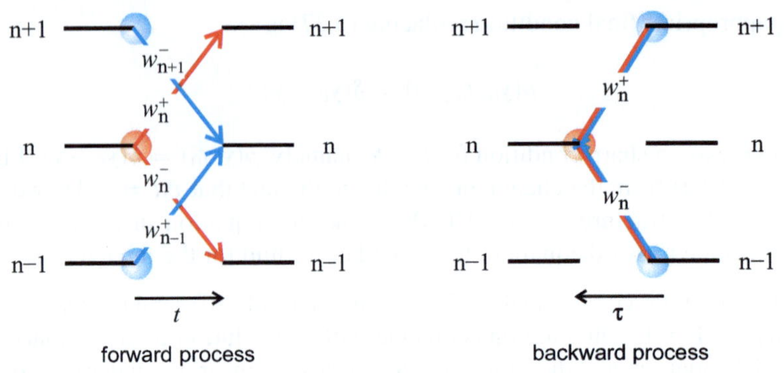

Fig. 3.22 Jumps in the single event master equations. The sketch on the *left-hand side* shows the four single steps in the forward birth-and-death master equations, which are determined by the four transition probabilities w_n^+, w_{n-1}^+, w_{n+1}^-, and w_n^-. Transitions leading to a gain in probability P_n are indicated in *blue*, while those reducing P_n are shown in *red*. On the right-hand side, we sketch the situation in the backward master equation, which is less transparent than in the forward equation [208, p. 355]. Only two transition probabilities, w_n^+ and w_n^-, enter the equations, and as a result of computational time progressing in the opposite direction to real time, the probabilities determining the amount of gain or loss in P_n are given at the final jump destinations rather than the beginnings

Using Riemann–Stieltjes integration, (3.137c) in real time t yields

$$\frac{\partial p(y_0,t_0|y,t)}{\partial t} = \int_\Omega \mathrm{d}z\, W(z|y,t)\Big(p(y_0,t_0|y,t) - p(y_0,t_0|z,t)\Big)$$
$$= \sum_{z=0}^{\infty} W(z|y,t)\Big(p(y_0,t_0|y,t) - p(y_0,t_0|z,t)\Big)\,.$$

We now introduce the notation for discrete particle numbers, $y \Leftrightarrow n \in \mathbb{N}$, $z \Leftrightarrow m \in \mathbb{N}$, and $y_0 \Leftrightarrow n_0 \in \mathbb{N}$:

$$\frac{\mathrm{d}P_n(n_0,t_0|n,t)}{\mathrm{d}t} = \sum_{m=0}^{\infty} W(m|n,t)\Big(P(n_0,t_0|n,t) - P(n_0,t_0|m,t)\Big)\,. \tag{3.138}$$

As previously, we now assume time independent transition rates and restrict transitions to single births and deaths:

$$W(m|n,t) = W_{mn} = w_n^+\delta_{n+1,n} + w_n^-\delta_{n-1,n}\,,$$

$$\text{or } W_{mn} = \begin{cases} w_n^+\,, & \text{if } m = n+1\,,\\ w_n^-\,, & \text{if } m = n-1\,,\\ 0\,, & \text{otherwise}\,. \end{cases} \tag{3.95'}$$

Then, the backward single-step master equation is of the form

$$\begin{aligned}\frac{\partial P(n_0,t_0|n,t)}{\partial t} &= w_n^+\Big(P(n_0,t_0|n,t) - P(n_0,t_0|\,n+1,t)\Big)\\ &\quad + w_n^-\Big(P(n_0,t_0|n,t) - P(n_0,t_0|n-1,t)\Big)\\ &= -w_n^+P(n_0,t_0|n+1,t) - w_n^-P(n_0,t_0|n-1,t)\\ &\quad + (w_n^+ + w_n^-)P(n_0,t_0|n,t)\;.\end{aligned} \tag{3.139}$$

As in the case of the forward equation (3.97), the notation is simplified by eliminating the final state (n_0, t_0) and making use of the discreteness of n. Then we find for the backward equation:

$$\begin{aligned}\frac{\mathrm{d}P_n(t)}{\mathrm{d}t} &= w_n^+\Big(P_{n+1}(t) - P_n(t)\Big) + w_n^-\Big(P_{n-1}(t) - P_n(t)\Big)\\ &= w_n^+P_{n+1}(t) + w_n^-P_{n-1}(t) - (w_n^+ + w_n^-)P_n(t)\;.\end{aligned} \tag{3.139$'$}$$

For the purpose of comparison, we repeat the forward equation:

$$\frac{\mathrm{d}P_n(t)}{\mathrm{d}t} = w_{n+1}^-P_{n+1}(t) + w_{n-1}^+P_{n-1}(t) - (w_n^+ + w_n^-)P_n(t)\;. \tag{3.97}$$

The differences are easy to visualize (Fig. 3.22). In the forward equation we need four transition probabilities, w_{n-1}^+, w_n^-, w_n^+, and w_{n+1}^- to describe the time derivative of the probability $P_n(t)$, and an interpretation of the terms as jump rates is straightforward (Sect. 3.2.3). The transition rates $w_k^\pm$ $(k = n-1, n, n+1)$ are multiplied by the probabilities of being in the state before the jump at the instant of hopping. Calculations with the backward equation are simpler, but the interpretation of the individual terms is more involved, since the different directions of real time and computational time in the backward process change the situation: only two transition probabilities appear in the backward equation, and the probability terms are differences between the densities of two neighboring states, n and $n+1$ or n and $n-1$, respectively. Daniel Gillespie writes [208, p. 355]: the backward master equation "does not admit a simple 'rate' interpretation of the kind described above for the forward master equation".

The backward master equation will now be applied to two different problems: (i) the backward Poisson process, where we compute the first passage time of reaching the absorbing barrier of zero events, and (ii) a more general calculation of first passage times by means of the backward master equation.

3.3.3 Backward Poisson Process

The relation between the solutions of the backward and forward master equations is illustrated for the Poisson process, which is sufficiently simple to be handled completely in closed form. The backward master equation of the Poisson process is, as expected, closely related to the forward equation, since the transition probabilities are constant, viz., $w_k^+ = \gamma$ and $w_k^- = 0$, $\forall\, k$:

$$\frac{\mathrm{d}P_n(\tau)}{\mathrm{d}t} = \gamma\Big(P_n(\tau) - P_{n+1}(\tau)\Big) , \quad \text{with } P_n(\tau_0) = \delta(n_0 - n) . \tag{3.140}$$

Indeed, the forward equation is obtained in this simple case just by replacing $P_n(\tau)$ by $P_{n-1}(\tau)$ and $P_{n+1}(\tau)$ by $P_n(\tau)$. The backward Poisson process describes how n_0 events recorded at time t_0 could result from independent arrivals when the probability distribution of events follows an exponential density.

The solution of the master equation (3.140) is straightforward. Nevertheless, we repeat the technique based on the probability generating function $g(s,\tau)$, because a few instructive tricks are involved. The expansion of $g(s,\tau)$ is not limited to positive n values [216, pp. 8–12]

$$g(s,\tau) = \sum_{n=-\infty}^{\infty} P_n(\tau)\, s^n ,$$

where the range of acceptable n values, $n_\mathrm{l} \le n \le n_\mathrm{h}$, $n \in \mathbb{Z}$, is introduced by setting $P_n(\tau) = 0$, $\forall\, n \notin [n_\mathrm{l}, n_\mathrm{h}]$. Inserting (3.140) and making use of the relation

$$\sum_{n=-\infty}^{\infty} P_{n+1}(\tau)s^n = \frac{1}{s}\sum_{n=-\infty}^{\infty} P_{n+1}(\tau)s^{n+1} = \frac{1}{s}\sum_{n=-\infty}^{\infty} P_n(\tau)s^n$$

yields the differential equation for the generating function, which turns out to be a simple ODE because the expression does not contain derivatives with respect to the dummy variable s:

$$\frac{\mathrm{d}g(s,\tau)}{\mathrm{d}\tau} = \gamma\left(\frac{1}{s} - 1\right) g(s,\tau) \quad \text{and} \quad g(s,\tau) = s^{n_0}\mathrm{e}^{\gamma\tau/s}\mathrm{e}^{-\gamma\tau} .$$

Taylor expanding the second factor in $g(s,\tau)$ in powers of $n_0 - n$ and equating coefficients yields the solution

$$P_n(\tau) = \frac{(\gamma\tau)^{n_0-n}}{(n_0-n)!}\,\mathrm{e}^{-\gamma\tau} . \tag{3.141}$$

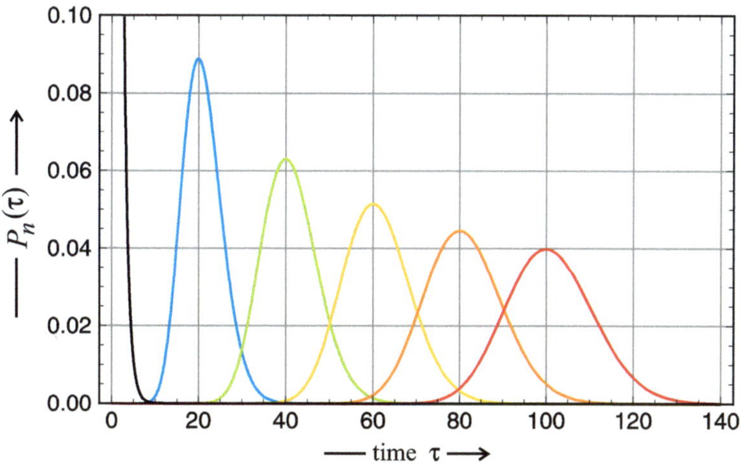

Fig. 3.23 Probability density of the backward Poisson process. The plot shows the probability density of the backward Poisson process $P_n(\tau)$ (3.141) for different numbers of events $n = 0$ (*black*), 20 (*blue*), 40 (*chartreuse*), 60 (*yellow*), 80 (*orange*), and 100 (*red*). Further parameters: $n_0 = 100$ and $\gamma = 1\,[\mathrm{t}^{-1}]$

Figure 3.23 shows the evolution of the probability density in the domain $0 \le \tau \le t_0$ corresponding to $t_0 \ge t \ge 0$. The computation of the moments of the random variable $\mathcal{X}(t)$ using 2.28 is a straightforward exercise in calculus, giving

$$\mathrm{E}\big(\mathcal{X}(\tau)\big) = n_0 - \gamma\tau \quad \text{and} \quad \mathrm{var}\big(\mathcal{X}(\tau)\big) = \gamma\tau \,. \tag{3.142}$$

The linear time dependence is beautifully reflected by the trajectories shown in Fig. 3.24.

Next we make an attempt to calculate the time to reach $n = 0$ in a backward Poisson process. We shall call this kind of first passage time the *extinction time* $\mathcal{T}_0$ of the process, in analogy to the notion of extinction times in biology. The state Σ_0 with $n = 0$ represents an absorbing barrier. Once the process has reached this state, it ends here, because the jumps in the backward Poisson are defined to satisfy $\Delta n = -1$. For the first passage times, we adopt the notation of Sect. 3.2.2.4 and find for the random variable

$$\mathcal{T}_0 = \sum_{k=1}^{n_0} \Delta t_k \,. \tag{3.143}$$

Here $\Delta\tau_k$ is the time span during which exactly $n = k$ events appeared on the record. The probability of $\mathcal{T}_0$ lying between τ and $\tau + \Delta\tau$ is given by the simultaneous occurrence of two events [536, pp. 71, 72]: (i) one event is on the record, $P\big(\mathcal{X}(\tau) = 1\big)$, which implies that $n_0 - 1$ events have already taken place, and (ii) one further

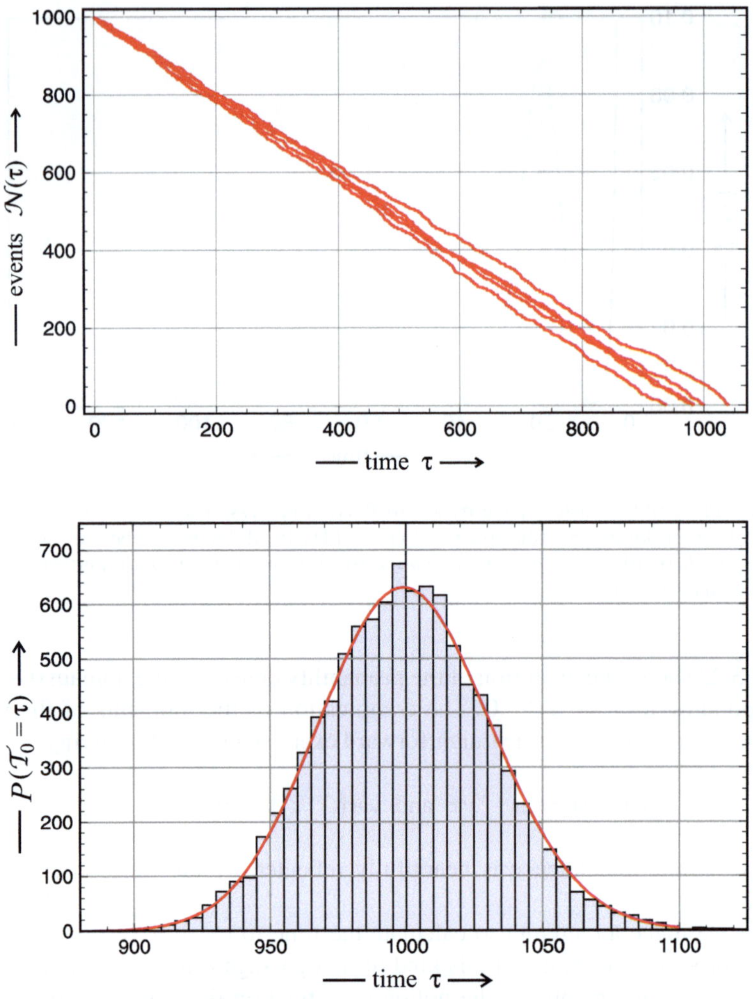

Fig. 3.24 The extinction time of the backward Poisson process. *Upper*: Five trajectories of the backward Poisson process starting from $n_0 = 1000$ with $\gamma = 1$ and seeds 013, 091, 491, 512, and 877, respectively. *Lower*: Histogram of the extinction times $\mathcal{T}_0$ obtained from 10,000 individual trajectories. The histogram is compared with the probability density of $\mathcal{T}_0$, which follows an Erlang distribution. Parameter values: $n_0 = 1000$ and $\gamma = 1$

jump occurs, $P\big(\Delta\mathcal{X}(\tau) = -1\big)$:

$$\begin{aligned} P(t \le \mathcal{T}_0 \le \tau + \Delta\tau) &= P\big(\mathcal{X}(\tau) = 1\big) P\big(\Delta\mathcal{X}(\tau) = -1\big) \\ &= \frac{e^{-\gamma\tau}(\gamma\,\tau)^{n_0-1}}{(n_0-1)!}\gamma\,\Delta t \\ &= \frac{\gamma^{n_0}\tau^{n_0-1}}{(n_0-1)!}e^{-\gamma\,\tau}\Delta\tau\ . \end{aligned}$$

Now we take the limit $\Delta\tau \to \mathrm{d}\tau$ and find that the extinction time is distributed according to

$$f_{\mathcal{T}_0} = \frac{\gamma^{n_0}\tau^{n_0-1}}{(n_0-1)!}\mathrm{e}^{-\gamma\tau} , \tag{3.144}$$

which is known as the Erlang distribution. It is straightforward to compute the expectation value and the variance of the extinction time:

$$\mathrm{E}(\mathcal{T}_0) = \int_0^\infty \tau f_{\mathcal{T}_0}\,\mathrm{d}t = \frac{n_0}{\gamma} , \qquad \mathrm{var}(\mathcal{T}_0) = \frac{n_0}{\gamma^2} . \tag{3.145}$$

Numerically simulated extinction times and the analytical Erlang distribution are compared in Fig. 3.24. The numerical data agree with the analytical expression as well as they could. At the same time we see that the backwards process provides natural access to the distribution of initial values which lead to the final state (n_0, τ_0).

3.3.4 Boundaries and Mean First Passage Times

A *first passage time* is a random variable $\mathcal{T}$ that indicates the instant when a particle passes a predefined location or state for the first time, and its expectation value $\mathrm{E}(\mathcal{T})$ is called the *mean first passage time*. We need to stress *first*, because in the processes we are discussing here, the variables may assume certain values a finite number of times or even infinitely often. Most processes, and all processes in reality, are confined by boundaries, which in the case of master equations fall into two classes: (i) *absorbing boundaries*[51] and (ii) *reflecting boundaries*. When a particle hits an absorbing boundary it disappears, when a process reaches the boundary, it ends there, whereas a reflecting boundary automatically sends the particle or the process back into the domain of allowed values. Accordingly, an absorbing boundary can be reached only once, whereas reflecting boundaries can be hit many times. First passage times are not restricted to boundaries. Consider, for example, a random walk. As we pointed out every point on a straight line or a 2D plane is visited an infinite number of times by any trajectory of infinite length.

Boundaries in Birth-and-Death Master Equations

It is a straightforward matter to implement boundary conditions in single-step birth-and-death master equations. For example, if the process is assumed to be restricted to the interval $l \le n \le u$, $n \in \mathbb{Z}$, we only need to choose the appropriate transition

[51] Boundaries are also called *barriers* in the literature and the notions are taken to be synonymous. We shall use here exclusively the word *boundary*. The term *barrier* will be reserved for obstacles to motion inside the domain of the random variable.

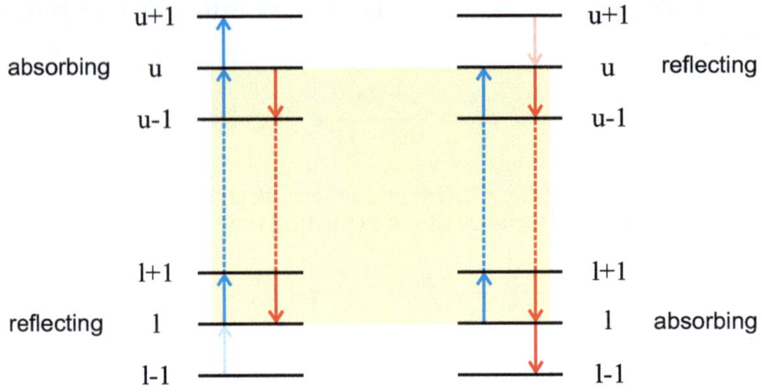

Fig. 3.25 Boundaries in single-step birth-and-death master equations. The figure on the *left-hand side* shows an interval $l \leq n \leq u$ (indicated by the *yellow* background), with a reflecting boundary at $n = l$ and an absorbing boundary at $n = u$, whereas the interval on the *right-hand side* has the absorbing boundary at $n = l$ and the reflecting boundary at $n = u$. The step-up transition probabilities w_n^+ are shown in *blue*, the step-down transition probabilities w_n^- in *red*, a reflecting boundary has a zero outgoing probability, w_l^- or w_u^+, and the incoming probabilities, w_{l-1}^+ or w_{u+1}^-, are zero at an absorbing boundary. The incoming transition probabilities at the reflecting boundaries are shown in *light colors* and play no role in the stochastic process, because the probabilities of the corresponding virtual states Σ_{l-1} and Σ_{u+1} are zero by definition, i.e., $P_{l-1}(t) = 0$ and $P_{u+1}(t) = 0$

probabilities that prevent exit from the interval in the case of a reflecting boundary or return to the interval for an absorbing boundary (see Fig. 3.25 and for reflecting boundaries see also Sect. 3.2.3). For the confinement of a process to the domain $n \in [l, u]$, we need a lower boundary at $n = l$ and an upper one at $n = u$. The boundary at $n = l$ is absorbing when the particle, once it has left the domain $[l, u]$, cannot return to it in future jumps. This can be achieved by setting $w_{l-1}^+ = 0$. The reflecting boundary results from the assumption $w_l^- = 0$, since the particle cannot then leave the domain. By symmetry, we have $w_{u+1}^- = 0$ and $w_u^+ = 0$ for the absorbing and the reflecting upper boundary.

It is instructive to calculate the probability current (3.98) across boundaries. For the lower boundary in the forward master equation, we find

$$\frac{dP_l(t)}{dt} = w_{l+1}^- P_{l+1}(t) - w_l^+ P_l(t) + w_{l-1}^+ P_{l-1}(t) - w_l^- P_l(t) = \varphi_l - \varphi_{l+1} \, .$$

Thus the current across the lower boundary is φ_l and we find

$$\varphi_l = \begin{cases} w_{l-1}^+ P_{l-1}(t) = 0 \, , & \text{reflecting boundary} \, , \\ -w_l^- P_l(t) \, , & \text{absorbing boundary} \, , \end{cases}$$

and by analogy we obtain for the upper boundary

$$\varphi_u = \begin{cases} -w_{u+1}^- P_{u+1}(t) = 0 \,, & \text{reflecting boundary} \,, \\ w_u^+ P_u(t) \,, & \text{absorbing boundary} \,. \end{cases}$$

For absorbing boundaries we obtain an outflux with negative sign to states with $n < l$ at the lower boundary and analogously a positive probability current to states with $n > u$ (Fig. 3.25). For reflection at the boundary, the condition is that nothing flows out of the domain, and this is satisfied by $w_l^- = 0$. If the reflecting boundary is combined with no influx, either w_{l-1}^+ or $P_{l-1}(t)$ (or both) must be zero. At the upper bound, the conditions have to be replaced by $w_u^+ = 0$, $w_{u+1}^- = 0$, and $P_{u+1}(t) = 0$, respectively. The reflecting boundary conditions are reminiscent of the no-flux or Neumann boundary conditions used in the theory of partial differential equations: the flux at the boundary has to vanish and this requires $P_{l-1}(t) = 0$ or $P_{u+1}(t) = 0$, which trivially implies that the probability of finding the particle outside the domain $[l, u]$ is zero.

Crispin Gardiner [194, pp. 283–284] shows that the boundary conditions can also be satisfied by the assumption of virtual states Σ_{l-1} and Σ_{u+1}, provided that the following relations hold:

$$w_{l-1}^+ P_{l-1}(t) = w_l^- P_l(t) \,, \quad w_u^+ P_u(t) = w_{u+1}^- P_{u+1}(t) \tag{3.146}$$

for reflecting boundaries and

$$P_{l-1}(t) = 0 \,, \quad P_{u+1}(t) = 0 \tag{3.147}$$

for absorbing boundaries.

In the case of the backward master equation (3.139)

$$\begin{aligned} \frac{\mathrm{d}P(n_0, t_0 | l, t)}{\mathrm{d}t} &= w_l^+ P(n_0, t_0 | l+1, t) - w_l^+ P(n_0, t_0 | l, t) \\ &\quad + w_l^- P(n_0, t_0 | l-1, t) - w_l^- P(n_0, t_0 | l, t) \,, \end{aligned}$$

for $n_0 \in [l, u]$, the general conditions $w_{l-1}^+ = 0$ and $w_{u+1}^- = 0$ for absorbing boundaries and $w_l^- = 0$ and $w_u^+ = 0$ for reflecting boundaries can be replaced by alternative expressions. As an equivalent to setting $w_l^-(t) = 0$ at the lower boundary $n = l$, we introduce a reflecting lower boundary by setting

$$P(n_0, t_0 | l-1, t) = P(n_0, t_0 | l, t) \,, \tag{3.148}$$

whence the second line in the master equation is equal to zero. It is a little more tricky to introduce an absorbing lower boundary since the transition rate w_{l-1}^{+} does not appear in the backward master equation. Clearly, the condition $P(n_0, t_0|n, t) = 0$ with $n_0 \in [l, u]$ and $n < l$ will have the same effect as $w_{l-1}^{+} = 0$. In single-step birth-and-death processes, only the term with the highest value of $n < l$ will be relevant for the process confined to the domain $[l, u]$, so $P(n_0, t_0|l-1, t) = 0$ is sufficient. At the upper boundary, the corresponding two equations having the same effect as $w_u^{+} = 0$ and $w_{u+1}^{-} = 0$ are $P(n_0, t_0|u+1, t) = P(n_0, t_0|u, t)$ and $P(n_0, t_0|u+1, t) = 0$ for the reflecting and the absorbing boundary, respectively.

Here it should be mentioned that, in the case of the chemical master equation, we shall encounter *natural boundaries* where reaction kinetics itself takes care of reflecting or absorbing boundaries. If we are dealing with a reversible chemical reaction approaching a thermodynamic equilibrium in a system with a total number of N molecules, the states $\Sigma_{n_\mathsf{K}=0}$ and $\Sigma_{n_\mathsf{K}=N}$ are reflecting for each molecular species K, whereas in an irreversible reaction, the state $n_\mathsf{K} = 0$ is absorbing when the reactant K is in short supply. Similarly, in the absence of migration, the state of extinction $n_\mathsf{S} = 0$ is an absorbing boundary for species S.

First Passage Times in Birth-and-Death Master Equations

The calculation of a mean first passage time is illustrated by a simple example, namely, the escape of a particle from a domain $[l, u]$ with a reflecting boundary at $n = l$ and an absorbing boundary at $n = u$ [364, pp. 90–92]. We make use of the backward master equation (3.139) and, according to last paragraph, we adopt the following conditions for the boundaries:

$$P(n_0, t_0|l-1, t) = P(n_0, t_0|l, t)\,, \qquad P(n_0, t_0|u+1, t) = 0\,.$$

The probability that the particle is still in the interval $[l, u]$ is calculated by summing over all states in the accessible domain:

$$I_n(t) = \sum_{m=l}^{u} P(m, t|n, 0)\,, \quad m \in \mathbb{Z}\,. \tag{3.149}$$

Inserting the individual terms from the backward master equation (3.139) yields for the time derivative

$$\begin{aligned} -\frac{\mathrm{d}I_n(t)}{\mathrm{d}t} &= \sum_{m=l}^{u} \frac{\mathrm{d}P(m, t|n, 0)}{\mathrm{d}t} \\ &= w_n^{+}\Big(I_n(t) - I_{n+1}(t)\Big) + w_n^{-}\Big(I_n(t) - I_{n-1}(t)\Big) \end{aligned} \tag{3.150}$$

with the condition $I_{l-1}(t) = I_l(t)$ for the reflecting boundary at $n = l$ and $I_{u+1}(t) = 0$ for the absorbing boundary at $n = u$. The minus sign expresses the decrease in the probability of remaining within the interval $[l, u]$ in real time, and is a consequence of the two time scales in backward processes, i.e., $\mathrm{d}t = -\mathrm{d}\tau$.

The probability of leaving the interval $[l, u]$, i.e., the probability of absorption, within an infinitesimal interval of time $[t, t + \mathrm{d}t]$ is calculated to be

$$I_n(t) - I_n(t + \Delta t) = -\frac{\partial I_n}{\partial t}\,\mathrm{d}t\,,$$

and by integration, we can now obtain the mean first passage time $\langle \mathcal{T}_n \rangle$ for escape from state n:

$$\langle \mathcal{T}_n \rangle = -\int_0^\infty t \frac{\partial I_n}{\partial t}\,\mathrm{d}t = \int_0^\infty I_n\,\mathrm{d}t\,, \tag{3.151}$$

where the last expression results from integration by parts. Integrating (3.150) yields

$$\int_0^\infty -\frac{\partial I_n(t)}{\partial t}\,\mathrm{d}t = 1$$

for the left-hand side since absorption of the particle or escape from the domain is certain. Integration of the right-hand side yields the mean passage times, and finally we obtain

$$1 = w_n^+ \Big(\langle \mathcal{T}_n \rangle - \langle \mathcal{T}_{n+1} \rangle \Big) + w_n^- \Big(\langle \mathcal{T}_n \rangle - \langle \mathcal{T}_{n-1} \rangle \Big)\,, \tag{3.152}$$

the equation for calculating $\langle \mathcal{T}_n \rangle$. The boundary conditions are $\langle \mathcal{T}_{l-1} \rangle = \langle \mathcal{T}_l \rangle$ and $\langle \mathcal{T}_{u+1} \rangle = 0$.

The solution of (3.152) for $\langle \mathcal{T}_n \rangle$ is facilitated by the introduction of new variables S_n and auxiliary functions φ_n:

$$S_n = \frac{\langle \mathcal{T}_{n+1} \rangle - \langle \mathcal{T}_n \rangle}{\varphi_n}\,, \quad n \in [l, u]\,,$$

$$\varphi_n = \prod_{m=l+1}^{n} \frac{w_m^-}{w_m^+}\,, \quad n \in [l+1, u]\,, \quad \text{with } \varphi_l = 1\,.$$

Using the new variables, (3.152) takes the form

$$-1 = w_n^+ \varphi_n \big(S_n - S_{n-1} \big)\,,$$

which allows one to derive a solution for the new variables:

$$S_k = -\sum_{m=l}^{k} \frac{1}{w_m^+ \varphi_m} \, .$$

From $\varphi_k S_k = \langle \mathcal{T}_{k+1} \rangle - \langle \mathcal{T}_k \rangle$, considering the telescopic sum from $k = n$ to $k = u$, we obtain

$$\sum_{k=n}^{u} \Big(\langle \mathcal{T}_{k+1} \rangle - \langle \mathcal{T}_k \rangle \Big) = \langle \mathcal{T}_{n+1} \rangle - \langle \mathcal{T}_n \rangle + \langle \mathcal{T}_{n+2} \rangle - \langle \mathcal{T}_{n+1} \rangle + \cdots + \langle \mathcal{T}_{u+1} \rangle - \langle \mathcal{T}_u \rangle$$

$$= - \langle \mathcal{T}_n \rangle \, ,$$

because of the boundary condition $\langle \mathcal{T}_{u+1} \rangle = 0$, and we obtain the desired result

$$\langle \mathcal{T}_n \rangle = \sum_{k=n}^{u} \varphi_k \sum_{m=l}^{k} \frac{1}{w_m^+ \varphi_m} = \sum_{k=n}^{u} \frac{1}{w_k^+ \overline{P}_k} \sum_{m=l}^{k} \overline{P}_m \, , \tag{3.153}$$

where we have used the stationary probabilities $\overline{P}$ from (3.100), instead of the functions φ, to calculate the mean passage times.

For the purpose of illustration we choose an example that yields simple analytical expressions for the mean first passage times. The simplification is made with the transition probabilities:

$$w_k^+ = w_k^- = \kappa \, , \;\; \forall \, k = 1, \ldots, N \, , \quad w_0^+ = \kappa \, , \;\; w_0^- = 0 \, , \;\; w_{u+1}^- = 0 \, .$$

The number of states n in the interval $[l, u]$ with $l, u, n \in \mathbb{Z}$ is $N = u - l + 1$. Since $\overline{P}_n = \overline{P}_0$ follows from (3.100), the normalization condition $\sum_n \overline{P}_n = 1$ implies the same probability $\overline{P}_n = 1/N \, , \; \forall \, n$. Inserting in (3.153) yields the expression

$$\langle \mathcal{T}_n \rangle = \frac{1}{2\kappa} (u + n - 2l + 2)(u - n + 1) \, , \tag{3.154}$$

which has the leading term $-n^2$ in n. Numerical results are given in Fig. 3.26, and indeed the curves approach a negative quadratic function for large N.

Mean first passage times find widespread applications in chemistry and biology. Important study cases are escape from potential traps, for example, classical motion in the double well potential, fixation of alleles in a population, and extinction times. We shall discuss examples in Chaps. 4 and 5.

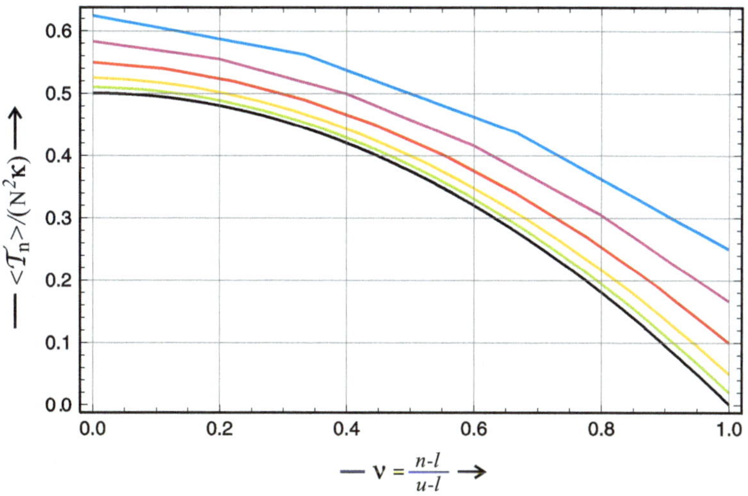

Fig. 3.26 Mean first passage times of a single-step birth-and-death process. Mean first passage times are computed from (3.154). In order to be able to compare the results for different sizes of the interval $[l, u]$, the interval is normalized: $l = 0$ and $u = 1$, or $\nu = (n - l)/(u - l)$. Computed mean first passage times are scaled by a factor $(N^2\kappa)^{-1}$ with $N = u - l + 1$. The values for N chosen in the computations and the color code are: 4 (*blue*), 6 (*violet*), 10 (*red*), 20 (*yellow*), 50 (*green*), and 1000 (*black*).

3.4 Stochastic Differential Equations

The Chapman–Kolmogorov equation was conceived in order to model the propagation of probabilities in sample space. An alternative modeling approach (Fig. 3.1) starts out from a deterministic description by means of a difference or differential equation and superimposes a fluctuating random element. The idea of introducing stochasticity into deterministic modeling of processes goes back to the beginning of the twentieth century. In 1900 the French mathematician Louis Bachelier conceived and analyzed the stochastic difference equation

$$\mathcal{X}(t_{n+1}) = \mathcal{X}(t_n) + \mu\,\Delta t + \sigma\sqrt{\Delta t}\,\mathcal{W}_{n+1}$$

in order to model the fluctuating prices in the Paris stock exchange. Herein $\mu(\mathcal{X}_t, t)$ is a function related to the foreseeable development, $\sigma(\mathcal{X}_t, t)$ describes the amplitude of the random fluctuations, and the $\mathcal{W}_n$ are independent normally distributed random variables with mean zero and variance one in the sense of Brownian increments [31]. Remarkably, Bachelier's thesis preceded Einstein's and von Smoluchowski's famous contributions by 5 and 6 years, respectively.

The concept of stochastic differential equations is commonly attributed to the French mathematician Paul Langevin who proposed an equation named after him that allows for the introduction of random fluctuations into conventional differential equations [325]. The idea was to find a sufficiently simple approach to model Brownian motion. In its original form the Langevin equation was written as

$$m\frac{\mathrm{d}^2\mathbf{r}}{\mathrm{d}t^2} = -\gamma\,\frac{\mathrm{d}\mathbf{r}}{\mathrm{d}t} + \boldsymbol{\xi}(t)\,, \quad \text{or} \quad \frac{\mathrm{d}\mathbf{p}(t)}{\mathrm{d}t} = -\frac{\gamma}{m}\mathbf{p}(t) + \boldsymbol{\xi}(t)\,. \tag{3.155}$$

It describes the motion of a Brownian particle of mass m, where $\mathbf{r}(t)$ and $\mathrm{d}\mathbf{r}/\,\mathrm{d}t = \mathbf{v}(t)$ are the location and velocity of the particle, respectively. The term $\mathrm{d}\mathbf{p}/\,\mathrm{d}t$ on the left-hand side describes the Newtonian gain in linear momentum $\mathbf{p}$ due to the force, while the first term on the right-hand side is the loss of momentum due to friction and the second term $\boldsymbol{\xi}(t)$ represents the irregularly fluctuating *random Brownian force*. The Langevin equation can be written in terms of the momentum $\mathbf{p}$ and then it takes on the more familiar form. The parameter $\gamma = 6\pi\eta r$ is the friction coefficient according to Stokes law, with η the viscosity coefficient of the medium and r the radius of the particle, which is assumed to have spherical geometry. The analogy between (3.155) and Newton's equation of motion is clear: the deterministic force $f(x) = -\partial V/\partial x$ derived from the potential energy $V(x)$ is replaced by $\xi(t)$.

In Fig. 3.1, stochastic differential equations were shown as an alternative to the Chapman–Kolmogorov equation when modeling Markov processes. As mentioned, the basic difference between the Chapman–Kolmogorov and the Langevin approach is the object whose time dependence is investigated. The Langevin equation (3.155) considers a single particle moving in a physical 3D space where it is exposed to thermal motion, and the integration yields a single stochastic trajectory. The Chapman–Kolmogorov equation of continuous motion leads to a Fokker–Planck equation (3.47), which describes the migration of a probability density in the same 3D space as the one where the trajectory is defined. The *equivalence* between the two approaches expresses the fact that a sample of trajectories of a Langevin equation converges in distribution to the time dependent probability density of the corresponding Fokker–Planck equation in the limit of infinitely large samples. The equivalence of the Langevin and the Chapman–Kolmogorov approaches is discussed in more detail in Sect. 3.4.1. When an analytical solution to a stochastic differential equation is available, the solution can be used to calculate moments of the probability distribution of $\mathcal{X}(t)$ and their time-dependence (Sect. 3.4.3), especially the mean and variance, which in practice often suffice to describe a process.

In the literature one can find an enormous variety of detailed treatises on stochastic differential equations. We mention here the monograph [22] and two books that are available on the internet: [388, 431]. The forthcoming short sketch of stochastic differential equations follows in essence the line of thought chosen by Crispin Gardiner [194, pp. 77–96].

3.4.1 Mathematics of Stochastic Differential Equations

The generalization of (3.155) from Brownian motion to an arbitrary stochastic process yields

$$\frac{dx}{dt} = a(x,t) + b(x,t)\xi(t) , \tag{3.156}$$

where x is the variable under consideration and $\xi(t)$ is an irregularly fluctuating term, often called *noise*. If the fluctuating term is independent of x, one speaks of *additive noise*. The two functions $a(x,t)$ and $b(x,t)$ are defined by the process to be investigated and the letters are chosen in order to remind us of the equivalence between stochastic differential equations and Fokker–Planck equations (3.47).[52]

The theory of stochastic processes is based upon the assumption of statistical independence for $\xi(t_1)$ and $\xi(t_2)$ if and only if $t_1 \neq t_2$, and furthermore we require the martingale condition $\langle \xi(t) \rangle = 0$, which is introduced without loss of generality, since any drift term can be absorbed into $a(x,t)$. The two conditions are reminiscent of the Wiener process discussed in Sect. 3.2.2.2, and will become fundamental in the mathematical formulation and analysis of stochastic differential equations. All requirements are encapsulated in the irregularity condition

$$\big\langle \xi(t_1)\,\xi(t_2) \big\rangle = \delta(t_1 - t_2) . \tag{3.157}$$

The Dirac δ-function diverges as $|t_1 - t_2| \to 0$ and this has the consequence that $\big\langle \xi(t)\,\xi(t) \big\rangle$ and the variance $\mathrm{var}\big(\xi(t)\big) = \big\langle \xi(t)\,\xi(t) \big\rangle - \langle \xi(t) \rangle^2$ are infinite for $t_1 = t_2 = t$.

In order to be able to search for solutions of the differential equation (3.156), we require the existence of the integral

$$\omega(t) = \int_0^t \xi(t)\, dt .$$

If $\omega(t)$ is a continuous function of time, it has the Markov property, as can be proven by partitioning the integral

$$\begin{aligned}\omega(t_2) = \omega(t_1) + \int_{t_1}^{t_2} \xi(\tau_2)\, d\tau_2 &= \int_0^{t_1} \xi(\tau_1)\, d\tau_1 + \int_{t_1}^{t_2} \xi(\tau_2)\, d\tau_2 \\ &= \lim_{\varepsilon \to 0} \int_0^{t_1 - \varepsilon} \xi(\tau_1)\, d\tau_1 + \int_{t_1}^{t_2} \xi(\tau_2)\, d\tau_2 .\end{aligned}$$

[52]We remark that the relation between the diffusion functions in the Fokker–Planck equation and the SDE, $B(x,t)$ and $b(x,t)$, is more subtle, and involves a square root according to (3.42), as will be discussed later on here in this section.

By (3.157), for every $\varepsilon > 0$, the variable $\xi(\tau_1)$ in the first integral is independent of $\xi(\tau_2)$ in the second integral. Continuity requires that $\omega(t_1)$ and $\omega(t_2) - \omega(t_1)$ be statistically independent in the limit $\varepsilon \to 0$ as well, and furthermore, $\omega(t_2) - \omega(t_1)$ is independent of all $\omega(\theta)$ with $\theta < t_1$. In other words, $\omega(t_2)$ is completely determined in probabilistic terms by the value $\omega(t_1)$ and no information on past values is required, whence $\omega(t)$ is Markovian. □

Using the experience gained from the derivation of the differential Chapman–Kolmogorov equation (3.46) and the postulated continuity of $\omega(t)$, we conjecture the existence a Fokker–Planck equation that describes $\omega(t)$. Computation of the drift and diffusion term is indeed straightforward [194, pp. 78, 79] and yields $A(t) = 0$ and $B(t) = 1$:

$$\frac{\partial p(\omega,t)}{\partial t} = \frac{1}{2}\frac{\partial^2}{\partial \omega^2} p(\omega,t) \,, \quad \text{with } p(\omega,t_0) = \delta(\omega - \omega_0) \,. \tag{3.55}$$

Accordingly, the Fokker–Planck equation describing the noise term in the Langevin equation is identical to that of the Wiener process (3.55):

$$\int_0^t \xi(\tau)\mathrm{d}\tau = \omega(t) = \mathcal{W}(t) \,.$$

This innocent looking equation confronts us with a dilemma: as we know from the discussion of the Wiener process, the solution of (3.55) $\mathcal{W}(t)$ is continuous, but *nowhere differentiable*, and this has the consequence that neither $\mathrm{d}\mathcal{W}(t)/\,\mathrm{d}t$, nor $\xi(t)$, nor the Langevin equation (3.155), nor the stochastic differential equation (3.156) exist in strict mathematical terms. Only the integral equation

$$x(t) - x(0) = \int_0^t a\big(x(\tau),\tau\big)\mathrm{d}\tau + \int_0^t b\big(x(\tau),\tau\big)\xi(\tau)\mathrm{d}\tau \tag{3.158}$$

is accessible to consistent interpretation in this rigorous sense. The relationship with the Wiener process becomes more visible by writing

$$\mathrm{d}\mathcal{W}(t) \equiv \mathcal{W}(t + \mathrm{d}t) - \mathcal{W}(t) = \xi(t)\,\mathrm{d}t \,,$$

whereupon insertion into (3.158) yields the correct formulation:

$$x(t) - x(0) = \int_0^t a\big(x(\tau),\tau\big)\mathrm{d}\tau + \int_0^t b\big(x(\tau),\tau\big)\mathrm{d}\mathcal{W}(\tau) \,. \tag{3.159}$$

The first integral is a conventional Riemann integral or a Riemann–Stieltjes integral if the function $a\big(x(\tau),\tau\big)$ contains steps, and the second integral is a stochastic Stieltjes integral, the evaluation of which will be discussed in Sect. 3.4.2. In differential form, we obtain a formulation of the stochastic differential equation which is compatible with standard mathematics:

$$dx = a\big(x(t), t\big)\,dt + b\big(x(t), t\big)d\mathcal{W}(t)\,. \qquad (3.160)$$

The nature of the fluctuations is given implicitly by the differential Wiener process $d\mathcal{W}(t)$: the probability density is Gaussian corresponding to *white noise*. White noise is an idealization, but if more information on the physical background of the fluctuations is available $d\mathcal{W}(t)$, it can be readily replaced by a more realistic noise term.

3.4.2 Stochastic Integrals

A *stochastic integral* requires additional definitions compared to ordinary Riemann integration. We shall explain this rather unexpected fact and sketch some practical recipes for integration (for more details see [460]).

Definition of the Stochastic Integral

Let $G(t)$ be an arbitrary function of time and $\mathcal{W}(t)$ the Wiener process. Then the stochastic integral $I(t, t_0)$ is defined as a Riemann–Stieltjes integral (1.56) of the form

$$I(t, t_0) = \int_{t_0}^{t} G(\tau)d\mathcal{W}(\tau)\,. \qquad (3.161)$$

The integral is partitioned into n subintervals, which are separated by the points t_i with $t_0 \leq t_1 \leq t_2 \leq \ldots \leq t_{n-1} \leq t$ (Fig. 3.27). Intermediate points τ_i, which will be used for the evaluation of the function $G(\tau_i)$, are defined within each of the subintervals $t_{i-1} \leq \tau_i \leq t_i$ and, as will be shown below, the value of the integral depends on the position chosen for τ_i within the subintervals.

As in case of the Riemann integral and the Darboux sum, the stochastic integral $\int_0^t G(\tau)d\mathcal{W}(\tau)$ is approximated by summation

$$S_n = \sum_{i=1}^{n} G(\tau_i)\Big(\mathcal{W}(t_i) - \mathcal{W}(t_{i-1})\Big)\,,$$

and it is not difficult to recognize that the integral is different for different choices of the intermediate point τ_i. As an illustrative example, we consider the case of the

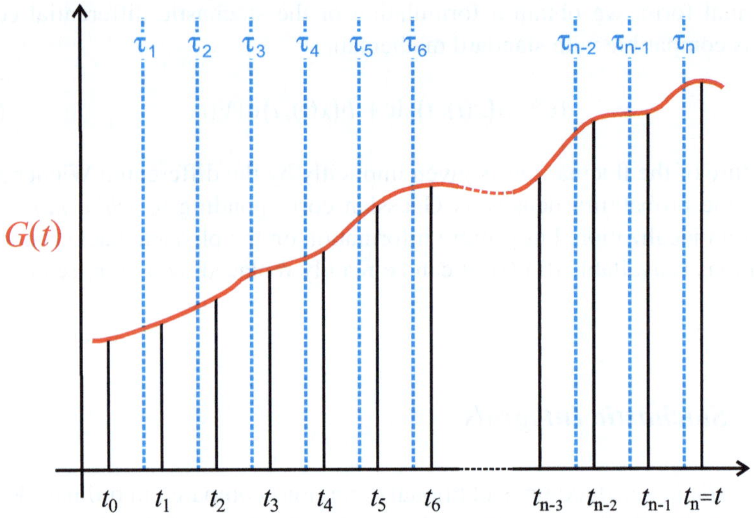

Fig. 3.27 Stochastic integral. The time interval $[t_0, t]$ is partitioned into n segments and an intermediate point τ_i is defined within each segment: $t_{i-1} \le \tau_i \le t_i$

Wiener process itself, where $G(t) = \mathcal{W}(t)$, and make use of (3.62):

$$\begin{aligned}\langle S_n \rangle &= \left\langle \sum_{i=1}^{n} \mathcal{W}(\tau_i)\Big(\mathcal{W}(t_i) - \mathcal{W}(t_{i-1})\Big) \right\rangle \\ &= \sum_{i=1}^{n} \langle \mathcal{W}(\tau_i)\mathcal{W}(t_i) \rangle - \sum_{i=1}^{n} \langle \mathcal{W}(\tau_i)\mathcal{W}(t_{i-1}) \rangle \\ &= \sum_{i=1}^{n} \Big(\min(\tau_i, t_i) - \min(\tau_i, t_{i-1})\Big) = \sum_{i=1}^{n} (\tau_i - t_{i-1}) \, .\end{aligned}$$

As indicated in Fig. 3.27, we choose equivalent intermediate positions τ_i for all equally sized subintervals i:

$$\tau_i = \alpha t_i + (1 - \alpha) t_{i-1} \, , \quad \text{with } 0 \le \alpha \le 1 \, , \tag{3.162}$$

and obtain for the telescopic sum[53]

$$\langle S_n \rangle = \sum_{i=1}^{n} (t_i - t_{i-1})\alpha = (t - t_0)\alpha \, .$$

[53]In a telescopic sum, all terms except the first and the last summand cancel.

Accordingly, the mean value of the integral may take any value between zero and $t - t_0$, depending on the choice of the position of the intermediate points as expressed by the parameter α. Out of all possible choices, two of these, the Itō and Stratonovich stochastic integrals, are applied in practice.

Itō Stochastic Integral

The most frequently used and most convenient definition of the stochastic integral is due to the Japanese mathematician Kiyoshi Itō [272, 273]. Semimartingales (Sect. 3.1.3), in particular local martingales, are the most common stochastic processes that allow for straightforward application of Itō's formulation to stochastic calculus. Itō chooses $\alpha = 0$ or $\tau_i = t_{i-1}$, and this leads to

$$\int_{t_0}^{t} G(\tau)\,\mathrm{d}\mathcal{W}(\tau) = \lim_{n\to\infty} \sum_{i=1}^{n} G(t_{i-1})\Big(\mathcal{W}(t_i) - \mathcal{W}(t_{i-1})\Big)\,, \tag{3.163}$$

for the Itō stochastic integral of a function $G(t)$, where the limit is a mean square limit (1.47). We sketch the procedure and leave the details to the reader. The calculation of the integral $\int_{t_0}^{t} \mathcal{W}(\tau)\mathrm{d}\mathcal{W}(\tau)$ begins with the sum S_n:

$$\begin{aligned} S_n &= \sum_{i=1}^{n} \mathcal{W}_{i-1}(\mathcal{W}_i - \mathcal{W}_{i-1}) \equiv \sum_{i=1}^{n} \mathcal{W}_{i-1}\Delta\mathcal{W}_i \\ &= \frac{1}{2}\sum_{i=1}^{n}\Big((\mathcal{W}_{i-1} + \Delta\mathcal{W}_i)^2 - \mathcal{W}_{i-1}^2 - \Delta\mathcal{W}_i^2\Big) \\ &= \frac{1}{2}\Big(\mathcal{W}(t)^2 - \mathcal{W}(t_0)^2\Big) - \frac{1}{2}\sum_{i=1}^{n}\Delta\mathcal{W}_i^2\,, \end{aligned}$$

where the expression $2\sum ab = \sum(a+b)^2 - \sum a^2 - \sum b^2$ is used in the second line.[54] The expectation value of $\sum_{i=1}^{n} \Delta\mathcal{W}_i^2$ is

$$\left\langle \sum_{i=1}^{n} \Delta\mathcal{W}_i^2 \right\rangle = \sum_i \Big\langle (\mathcal{W}_i - \mathcal{W}_{i-1})^2 \Big\rangle = \sum_i (t_i - t_{i-1}) = t - t_0\,, \tag{3.164}$$

[54] To derive this relation, we use the fact that the stochastic variables of the Wiener process at different times are uncorrelated, i.e., $\big\langle \mathcal{W}(t_i)\mathcal{W}(t_j)\big\rangle = 0$, and the variance is $\mathrm{var}\big(\mathcal{W}(t_i)\big) = \big\langle \mathcal{W}(t_i)^2\big\rangle - \big\langle \mathcal{W}(t_i)\big\rangle^2 = \big\langle \mathcal{W}(t_i)^2\big\rangle - (t_i - t_0)^2$. We simplify the expressions in the derivation and write from here on $\mathcal{W}_i$ for $\mathcal{W}(t_i)$ and $\Delta\mathcal{W}_i$ for $\Delta\mathcal{W}(t_i)$.

where the second equality results from the Gaussian nature of the probability density (3.61):

$$\langle (\mathcal{W}_i - \mathcal{W}_j)^2 \rangle = \langle \mathcal{W}_i^2 \rangle - \langle \mathcal{W}_j^2 \rangle = \mathrm{var}(\mathcal{W}_i) - \mathrm{var}(\mathcal{W}_j) = t_i - t_j \, .$$

What remains to be done is to show that the mean square deviation in (3.164) does indeed vanish in the mean square limit $n \to \infty$:

$$\lim_{n\to\infty} \left\langle \left(\sum_{i=1}^{n} (\mathcal{W}_i - \mathcal{W}_{i-1})^2 - (t - t_0) \right)^2 \right\rangle = 0 \, .$$

Three hints may be useful:

(i) The fourth moment of a Gaussian variable can be expressed in terms of the variance:

$$\left\langle (\mathcal{W}_i - \mathcal{W}_{i-1})^4 \right\rangle = 3 \left\langle (\mathcal{W}_i - \mathcal{W}_{i-1})^2 \right\rangle^2 = 3(t_i - t_{i-1})^2 \, .$$

(ii) The independence of Gaussian variables implies

$$\left\langle (\mathcal{W}_i - \mathcal{W}_{i-1})^2 (\mathcal{W}_j - \mathcal{W}_{j-1})^2 \right\rangle = (t_i - t_{i-1})(t_j - t_{j-1}) \, .$$

(iii) It is often useful to partition the square of a sum into diagonal and off-diagonal terms:

$$\left(\sum_{i=1}^{n} (a_i - b_i) \right)^2 = \sum_{i=1}^{n} (a_i - b_i)^2 + 2 \sum_{i=1}^{n-1} \sum_{j=2, j<i}^{n} (a_i - b_i)(a_j - b_j) \, .$$

Evaluation of the expression confirms the mean square limit of the expectation value.

The Itō stochastic integral of the Wiener process finally yields

$$\int_{t_0}^{t} \mathcal{W}(\tau) \mathrm{d}\mathcal{W}(\tau) = \frac{1}{2} \Big(\mathcal{W}(t)^2 - \mathcal{W}(t_0)^2 - (t - t_0) \Big) \, . \tag{3.165}$$

Clearly, the Itō integral differs from the conventional Riemann–Stieltjes integral, where the term $t - t_0$ is absent. This unusual behavior of the limit of the sum S_n can be explained by the fact that the quantity $|\mathcal{W}(t + \Delta t) - \mathcal{W}(t)|$ is almost always of order $\sqrt{t}$, and hence, in contrast to what happens in ordinary integration, the terms of second order in $\Delta \mathcal{W}(t)$ do not vanish on taking the limit.

We remark that the expectation value of the integral (3.165) vanishes:

$$\left\langle \int_{t_0}^{t} \mathcal{W}(\tau)\mathrm{d}\mathcal{W}(\tau) \right\rangle = \frac{1}{2}\Big(\langle \mathcal{W}(t)^2 \rangle - \langle \mathcal{W}(t_0)^2 \rangle - (t - t_0) \Big) = 0 \,, \qquad (3.166)$$

since the intermediate terms $\langle \mathcal{W}(t_{i-1})\Delta\mathcal{W}(t_i) \rangle$ vanish due to the fact that $\Delta\mathcal{W}(t_i)$ and $\mathcal{W}(t_{i-1})$ are statistically independent.

Nonanticipation

As already mentioned in Sect. 3.1.3, a stochastic process $\mathcal{X}(t)$ is adapted or nonanticipating if and only if, for every trajectory and for every time t, the value of the random variable $\mathcal{X}(t)$ is known at time t and *not before*.[55] In other words, a nonanticipating or adapted process *cannot look into the future*, so a function $G(t)$ is nonanticipating or adapted to the process $\mathrm{d}\mathcal{W}(t)$ if the value of $G(t)$ at time t depends only on the random increments $\mathrm{d}\mathcal{W}(\tau)$ for $t \le \tau$. Here we shall use this property for the solution of certain classes of Itō stochastic integrals, which can be expressed as functions or functionals[56] of the Wiener process $\mathcal{W}(t)$ by means of a stochastic differential or integral equation of the form

$$x(t) - x(t_0) = \int_{t_0}^{t} a\big(x(\tau), \tau\big)\mathrm{d}\tau + \int_{t_0}^{t} b\big(x(\tau), \tau\big)\mathrm{d}\mathcal{W}(\tau) \,. \qquad (3.159)$$

A function $G(t)$ is *nonanticipating* with respect to t if $G(t)$ is probabilistically independent of $\mathcal{W}(s) - \mathcal{W}(t)$ for all s and t with $s > t$. In other words, $G(t)$ is independent of the behavior of the Wiener process in the future $s > t$. This is a natural and physically reasonable requirement for a solution of (3.159), because it boils down to the condition that $x(t)$ involves $\mathcal{W}(\tau)$ only for $\tau \le t$. Important nonanticipating functions are [194, p. 83]:

(1) $\mathcal{W}(t)$,

(2) $\int_{t_0}^{t} F\big(\mathcal{W}(\tau)\big)\mathrm{d}\tau$,

(3) $\int_{t_0}^{t} F\big(\mathcal{W}(\tau)\big)\mathrm{d}\mathcal{W}(\tau)$,

(4) $\int_{t_0}^{t} G(\tau)\mathrm{d}\tau$, when $G(t)$ itself is nonanticipating,

(5) $\int_{t_0}^{t} G(\tau)\mathrm{d}\mathcal{W}(\tau)$, when $G(t)$ itself is nonanticipating.

[55] Every deterministic process is nonanticipating: in order to calculate the value $G(t + \mathrm{d}t)$ of a function $t \to G(t)$, no value $G(\tau)$ with $\tau > t$ is required.

[56] A function assigns a value to the argument of the function, $x_0 \to f(x_0)$, whereas a functional relates a function to the value of a function, $f \to f(x_0)$.

Items (3) and (5) depend on the fact that in Itō's calculus the stochastic integral is defined as the limit of a sequence in which $G(\tau)$ and $\mathcal{W}(\tau)$ are involved exclusively for $\tau < t$.

There are three important reasons for the specific discussion of nonanticipating functions:

(i) Results can be derived that are only valid for nonanticipating functions.
(ii) Nonanticipating functions occur naturally in situations in which *causality* can be expected, in the sense that the future cannot affect the present.
(iii) The definition of stochastic differential equations requires nonanticipating functions.

In conventional calculus we never encounter situations in which the future acts back on the present or even on the past.

Several differential relations collecting terms up to order t^1 are useful and required in Itō calculus:

$$d\mathcal{W}(t)^2 = dt\,, \tag{3.167a}$$

$$d\mathcal{W}(t)^{2+n} = 0\,, \quad \text{for } n > 0\,, \tag{3.167b}$$

$$d\mathcal{W}(t)dt = 0\,, \tag{3.167c}$$

$$\int_{t_0}^{t} \mathcal{W}(\tau)^n d\mathcal{W}(\tau) = \frac{1}{n+1}\Big(\mathcal{W}(t)^{n+1} - \mathcal{W}(t_0)^{n+1}\Big) - \frac{n}{2}\int_{t_0}^{t} \mathcal{W}(\tau)^{n-1} d\tau\,, \tag{3.167d}$$

$$df\big(\mathcal{W}(t), t\big) = \left(\frac{\partial f}{\partial t} + \frac{1}{2}\frac{\partial^2 f}{\partial \mathcal{W}^2}\right) dt + \frac{\partial f}{\partial \mathcal{W}} d\mathcal{W}(t)\,, \tag{3.167e}$$

$$\left\langle \int_{t_0}^{t} G(\tau) d\mathcal{W}(\tau) \right\rangle = 0\,, \tag{3.167f}$$

$$\left\langle \int_{t_0}^{t} G(\tau) d\mathcal{W}(\tau) \int_{t_0}^{t} H(\tau) d\mathcal{W}(\tau) \right\rangle = \int_{t_0}^{t} \big\langle G(\tau) H(\tau) \big\rangle d\tau\,. \tag{3.167g}$$

The expressions are easier to memorize when we assign a dimension $[\mathrm{t}^{1/2}]$ to $\mathcal{W}(t)$ and discard all terms of order t^{1+n} with $n > 0$.

Stratonovich Stochastic Integral

As we saw before, the value of a stochastic integral depends on the particular choice of the intermediate points τ_i. The Russian physicist and engineer Ruslan Leontevich Stratonovich [512] and the American mathematician Donald LeRoy Fisk [178] simultaneously developed an alternative approach to Itō's stochastic integration, which is called Stratonovich integration[57]:

$$\mathop{\,\,\$\!\!\!\!\int}_{t_0}^{t} G(\tau)\,\mathrm{d}\mathcal{W}(\tau)\;.$$

The intermediate points τ_i are chosen such that the unconventional term $t - t_0$ in the integral $\int_{t_0}^{t} \mathcal{W}(\tau)\mathrm{d}\mathcal{W}(\tau)$ vanishes. For the purpose of illustration, the integrand is chosen once again to be $G(t) = \mathcal{W}(t)$, but now it is evaluated precisely in the middle of the interval, namely at time $\tau_i = (t_i - t_{i-1})/2$. Then, the mean square limit converges to the expression for the Stratonovich integral over $\mathcal{W}(t)$:

$$\begin{aligned}\mathop{\,\,\$\!\!\!\!\int}_{t_0}^{t} \mathcal{W}(\tau)\mathrm{d}\mathcal{W}(\tau) &= \lim_{n\to\infty}\sum_{i=1}^{n}\frac{\mathcal{W}(t_i)+\mathcal{W}(t_{i-1})}{2}\Big(\mathcal{W}(t_i)-\mathcal{W}(t_{i-1})\Big)\\ &= \frac{1}{2}\Big(\mathcal{W}(t)^2-\mathcal{W}(t_0)^2\Big)\;.\end{aligned} \tag{3.168}$$

In contrast to the Itō integrals, Stratonovich integration obeys the rules of conventional calculus, but it is *not* nonanticipating, because the value of the function $\mathcal{W}\big((t_{i-1} + t_i)/2\big)$ is already required at time t_{i-1}.

We compare here the derivation of the Stratonovich and Itō integrals [277, pp. 85–89], because additional insights can be gained into the nature of stochastic processes. The starting point is the general Itō difference equation

$$\Delta x = F(x,t)\Delta t + G(x,t)\Delta\mathcal{W}\;, \tag{3.169}$$

where $F(x,t)$ and $G(x,t)$ are functions defining drift and diffusion of the process under consideration, and Δt and $\Delta\mathcal{W}$ are the time interval and the random increment, respectively. Next we choose equal time intervals as in Fig. 3.27 and have $t_k = k\,\Delta t + t_0$ with $x_k = x(t_k)$, $\Delta x_{k-1} = x_k - x_{k-1}$, and $\Delta\mathcal{W}_{k-1} = \mathcal{W}_k - \mathcal{W}_{k-1}$, where the starting point of the integration is $t_0 = 0$, $x(0) = x_0$, $\Delta x_0 = x_1 - x_0$, and $\Delta\mathcal{W}_0 = \mathcal{W}_1 - \mathcal{W}_0$ is the first random increment. Equation (3.169) takes on the

[57] In order to distinguish the two versions of stochastic integrals we use the symbol $\int_{t_0}^{t}$ for the Itō integral and $\mathop{\,\,\$\!\!\!\!\int}_{t_0}^{t}$ for the Stratonovich integral [277, p. 86]. The distinction from ordinary integrals is automatically provided by the differential $\mathrm{d}\mathcal{W}(t)$.

precise form

$$\Delta x_{k-1} = F(x_{k-1}, t_{k-1})\Delta t + G(x_{k-1}, t_{k-1})\Delta \mathcal{W}(t_{k-1}) \,, \quad k = 1, \ldots, n \,.$$

Now we choose the starting point $t_0 = 0$ and $x(0) = x_0$, and find the general solution of the difference equation at $t = t_n$:

$$x(t_n) = x_n = x_0 + \sum_{k=0}^{n-1} F(x_k, t_k)\Delta t + \sum_{k=0}^{n-1} G(x_k, t_k)\Delta \mathcal{W}(t_k) \,. \tag{3.170}$$

Equation (3.170) represents the explicit formula for the Cauchy–Euler integration (Fig. 3.28) and is commonly used in numerical SDE evaluation.

We use it here as a basis for the explicit comparison of the Itō integral

$$\int_0^t G(x,t)\mathrm{d}\mathcal{W} = \lim_{n\to\infty} \sum_{k=0}^{n-1} G(x_k, t_k)\, \Delta \mathcal{W}(t_k) \tag{3.163$'$}$$

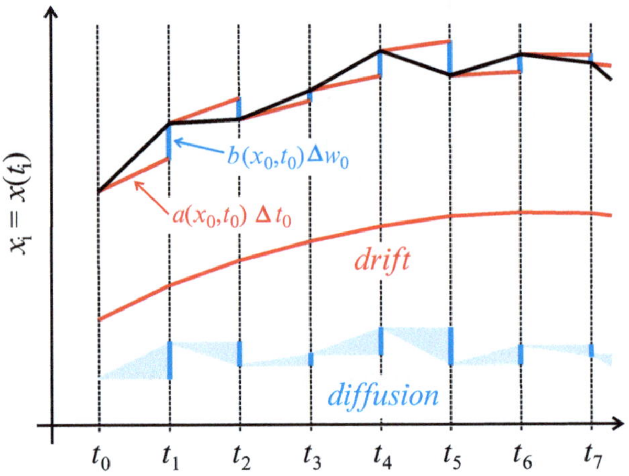

Fig. 3.28 Stochastic integration. The figure sketches the Cauchy–Euler integration procedure for the construction of an approximate solution of the stochastic differential equation (3.156′). The stochastic process consists of two different components: (1) the drift term, which is the solution of the ODE in the absence of diffusion (*red*, $b(x_i, t_i) = 0\,,\ \forall\, i$) and (2) the diffusion term representing a Wiener process $\mathcal{W}(t)$ (*blue*, $a(x_i, t_i) = 0\,,\ \forall\, i$). The superposition of the two terms yields the stochastic process (*black*). The *two lower plots* show the two components separately. The increments $\Delta\mathcal{W}(t_i)$ of the Wiener process are uncorrelated and independent. An approximation to a particular solution of the stochastic process is constructed by choosing the mesh size sufficiently small, whereas the exact trajectory requires $\lim \Delta t \to 0$

and the Stratonovich integral

$$\mathcal{S}\!\!\int_0^t G(x,t)\mathrm{d}\mathcal{W} = \lim_{n\to\infty}\sum_{k=0}^{n-1} G\Big(\frac{x_{k+1}+x_k}{2}, t_k\Big)\Delta\mathcal{W}(t_k)\,, \tag{3.171}$$

and calculate the relationship between them. First we expand the function $G(x,t)$ in the Stratonovich analogue of the noise term in (3.170), viz.,

$$G\Big(\frac{x_{k+1}+x_k}{2}, t_k\Big)\Delta\mathcal{W}(t_k) = G\Big(x_k + \frac{\Delta x_k}{2}, t_k\Big)\Delta\mathcal{W}(t_k)\,,$$

in a power series around the point (x_n, t_n), and simplify the notation by defining $F_n \equiv F(x_n,t_n)$ and $G_n \equiv G(x_n,t_n)$ for the expansion:

$$G\Big(x_k + \frac{\Delta x_k}{2}, t_k\Big) = G_n + \frac{\partial G_n}{\partial x}\frac{\Delta x_n}{2} + \frac{1}{2}\frac{\partial^2 G_n}{\partial x^2}\Big(\frac{\Delta x_n}{2}\Big)^2 + \cdots\,.$$

Next we insert $\Delta x_n = F_n\Delta t + G_n\Delta\mathcal{W}(t_n)$ and recall that $\Delta\mathcal{W}(t)^2 = \Delta t$. We omit the higher order terms, which will not contribute since all differences with higher powers, $(\Delta t)^\gamma$ with $\gamma > 1$ and $\big(\Delta\mathcal{W}(t)\big)^\alpha$ with $\alpha > 2$ (3.167), vanish in the continuum limits $\Delta t \to \mathrm{d}t$ and $\Delta\mathcal{W} \to \mathrm{d}\mathcal{W}(t)$. Then we find

$$G\Big(x_k + \frac{\Delta x_k}{2}, t_k\Big) = G_n + \Big(\frac{F_n}{2}\frac{\partial G_n}{\partial x} + \frac{G_n^2}{4}\frac{\partial^2 G_n}{\partial x^2}\Big)\Delta t + \frac{G_n}{2}\frac{\partial G_n}{\partial x}\Delta\mathcal{W}(t_n)\,.$$

Next we insert this result into the discrete sum for the Stratonovich integral (3.171), omit the term with $\Delta t\Delta\mathcal{W}$ since $\Delta t\Delta\mathcal{W} \to \mathrm{d}t\,\mathrm{d}\mathcal{W}(t) = 0$, and find

$$\sum_{k=0}^{n-1} G\Big(x_k + \frac{\Delta x_k}{2}, t_k\Big)\Delta\mathcal{W}(t_k) = \sum_{k=0}^{n-1} G_k\Delta\mathcal{W}(t_k) + \sum_{k=0}^{n-1}\frac{G_k}{2}\frac{\partial G_k}{\partial x}\Delta t\,.$$

Taking the continuum limit, we obtain the desired relation between the Itō and Stratonovich integrals, viz.,

$$\mathcal{S}\!\!\int_0^t G(x,t)\mathrm{d}\mathcal{W}(t) = \int_0^t G(x,t)\mathrm{d}\mathcal{W}(t) + \frac{1}{2}\int_0^t \frac{\partial G(x,t)}{\partial x}G(x,t)\,\mathrm{d}t\,. \tag{3.172}$$

The Stratonovich integral is equal to the Itō integral plus an additional contribution that can be assimilated as the drift term.

In summary, we have derived two integration methods for the stochastic differential equation

$$\mathrm{d}x = F(x,t)\,\mathrm{d}t + G(x,t)\mathrm{d}\mathcal{W}(t)\;. \tag{3.173}$$

(i) The Itō method yielding

$$x(t) = x(0) + \int_0^t F(x,t)\,\mathrm{d}t + \int_0^t G(x,t)\mathrm{d}\mathcal{W}(t)\;.$$

(ii) The Stratonovich method resulting in a different solution, which we denote by $z(t)$ for the purpose of distinction:

$$\begin{aligned} z(t) &= z(0) + \int_0^t F(z,t)\,\mathrm{d}t + \oint_0^t G(z,t)\mathrm{d}\mathcal{W}(t) \\ &= z(0) + \int_0^t \left(F(z,t) + \frac{G(z,t)}{2}\frac{\partial G(z,t)}{\partial z}\right)\mathrm{d}t + \int_0^t G(z,t)\mathrm{d}\mathcal{W}(t)\;. \end{aligned}$$

On the other hand we would obtain the same solution $z(t)$ if we applied the Itō calculus to the stochastic differential equation

$$\mathrm{d}z = \left(F(z,t) + \frac{G(z,t)}{2}\frac{\partial G(z,t)}{\partial z}\right)\mathrm{d}t + G(z,t)\mathrm{d}\mathcal{W}(t)\;. \tag{3.174}$$

Since the Stratonovich calculus is much more involved than the Itō calculus, we can readily see a strategy for obtaining Stratonovich solutions: use (3.174) and derive the solution by means of Itō calculus. It is worth mentioning that a stand-alone Stratonovich integral bears no relationship to a stand-alone Itō integral. In other words, there is no connection between the two classes of integrals for an arbitrary function $G(t)$. Only when we know the stochastic differential equation to which the two integrals refer can a formula be derived, as we did here, that relates the Itō integral to the Stratonovich integral.

Itō's Lemma
Itō's formula was derived to extend stochastic calculus to arbitrary functions $f\big(x(t)\big)$. The basis of the expansion is the stochastic differential equation

$$\mathrm{d}x = a\big(x(t), t\big)\,\mathrm{d}t + b\big(x(t), t\big)\mathrm{d}\mathcal{W}(t)\;, \tag{3.160}$$

which is truncated after second order in dt and $\mathrm{d}\mathcal{W}(t)$. Since we need only the terms linear in dt, we neglect the two higher order contributions in $\mathrm{dt}\,\mathcal{W}(t)$ and dt^2:

$$\begin{aligned}
\mathrm{d}f\big(x(t)\big) &= \frac{\partial f\big(x(t)\big)}{\partial x}\,\mathrm{d}x(t) + \frac{\partial^2 f\big(x(t)\big)}{\partial x^2}\,\mathrm{d}x(t)^2 + \cdots \\
&= \frac{\partial f\big(x(t)\big)}{\partial x}\Big(a\big(x(t), t\big)\,\mathrm{d}t + b\big(x(t), t\big)\mathrm{d}\mathcal{W}(t)\Big) \\
&\qquad + \frac{1}{2}\frac{\partial^2 f\big(x(t)\big)}{\partial x^2} b\big(x(t), t\big)^2 \mathrm{d}\mathcal{W}(t)^2 + \cdots\;.
\end{aligned}$$

Next we substitute dt for $\mathrm{d}\mathcal{W}(t)^2$ according to (3.167a) and introduce $\mathrm{d}t$ into the last line of this equation to obtain Itō's formula:

$$\begin{aligned}
\mathrm{d}f\big(x(t)\big) &= \left(a\big(x(t), t\big)\frac{\partial f\big(x(t)\big)}{\partial x} + \frac{1}{2} b\big(x(t), t\big)^2 \frac{\partial^2 f\big(x(t)\big)}{\partial x^2}\right)\mathrm{d}t \\
&\quad + b\big(x(t), t\big)\frac{\partial f\big(x(t)\big)}{\partial x}\,\mathrm{d}\mathcal{W}(t)\;.
\end{aligned} \tag{3.175}$$

It is worth noting that Itō's formula and ordinary calculus lead to different results unless $f(x)$ is linear in $x(t)$ and the second derivative $\partial^2 f(x)/\partial x^2$ vanishes accordingly.

As an exercise we suggest calculating the expression for the simple function $f(x) = x^2$. The result is

$$\mathrm{d}(x^2) = \big(2xa(x, t) + b(x, t)^2\big)\,\mathrm{d}t + 2b(x, t)\mathrm{d}\mathcal{W}(t)\;,$$

which is useful, for example, to calculate the time derivative of the variance:

$$\frac{\mathrm{d}\,\mathrm{var}\big(x(t)\big)}{\mathrm{d}t} = \frac{\mathrm{d}\,\big\langle x^2\big\rangle}{\mathrm{d}t} + 2\,\langle x\rangle\,\frac{\mathrm{d}\,\langle x\rangle}{\mathrm{d}t}\;.$$

Itō Calculus in Two or More Dimensions

In general, the extension of the Itō formalism to many dimensions becomes very complicated. The most straightforward simplification is the derivation of Itō calculus for the multivariate case by introducing an n-dimensional Wiener process as described in (3.71), viz., $\mathcal{W}(t) = \big(\mathcal{W}_1(t), \mathcal{W}_2(t), \ldots, \mathcal{W}_n(t)\big)$, and making use of the rule that $\mathrm{d}\mathcal{W}(t)$ is an infinitesimal of order $t^{1/2}$ for which the following relations hold in addition to (3.167):

$$\mathrm{d}\mathcal{W}_i(t), \mathrm{d}\mathcal{W}_j(t) = \delta_{ij}\,\mathrm{d}t\,, \tag{3.176a}$$

$$\mathrm{d}\mathcal{W}_i(t)^{2+N} = 0\,, \qquad (N > 0)\,, \tag{3.176b}$$

$$\mathrm{d}\mathcal{W}_i(t)\,\mathrm{d}t = 0\,, \tag{3.176c}$$

$$\mathrm{d}t^{1+N} = 0\,, \qquad (N > 0)\,. \tag{3.176d}$$

The first relation is a consequence of the independence of increments of Wiener processes, which also applies for different coordinate axes $\mathrm{d}\mathcal{W}_i(t)$ and $\mathrm{d}\mathcal{W}_j(t)$, $i \neq j$.

Making use of the drift vector $\mathbf{A}(\mathbf{x}, t)$ and the diffusion matrix $\mathsf{B}(\mathbf{x}, t)$, the multidimensional stochastic differential equation is[58]

$$\mathrm{d}\mathbf{x} = \mathbf{A}(\mathbf{x}, t)\,\mathrm{d}t + \mathsf{B}(\mathbf{x}, t)\mathrm{d}\mathcal{W}(t)\,. \tag{3.177}$$

By extending Itō's lemma to two or more dimensions, we obtain for an arbitrary well-behaved function $f\big(\mathbf{x}(t)\big)$ the result

$$\mathrm{d}f(\mathbf{x}) = \Big(\sum_i A_i(x,t)\,\frac{\partial}{\partial x_i} f(\mathbf{x}) + \frac{1}{2}\sum_{i,j}\big(\mathsf{B}(x,t)\;\mathsf{B}^{\mathrm{t}}(x,t)\big)_{ij}\frac{\partial^2}{\partial x_i \partial x_j} f(\mathbf{x})\Big)\mathrm{d}t$$
$$+ \sum_{i,j}(\mathsf{B})_{ij}\frac{\partial}{\partial x_i} f(\mathbf{x})\mathrm{d}\mathcal{W}_j(t)\,. \tag{3.178}$$

Again we observe the additional term in line two, which is introduced through the definition of the Itō integral.

[58] We use here the sanserif font for the diffusion matrix in order to distinguish it from the conventional diffusion matrix $B = \mathsf{B}^{\mathrm{t}}\mathsf{B}$ in the Fokker–Planck equation.

SDEs and Fokker–Planck Equations

The expectation value of an arbitrary function $f\big(x(t)\big)$ can be calculated by means of Itō's formula. We begin with a single variable:

$$\left\langle \frac{\mathrm{d}f\big(x(t)\big)}{\mathrm{d}t} \right\rangle = \frac{\mathrm{d}}{\mathrm{d}t} \big\langle f\big(x(t)\big)\big\rangle$$

$$= \left\langle a\big(x(t),t\big)\frac{\partial f\big(x(t)\big)}{\partial x} + \frac{1}{2} b\big(x(t),t\big)^2 \frac{\partial^2 f\big(x(t)\big)}{\partial x^2} \right\rangle .$$

The term containing the Wiener differential $\mathrm{d}\mathcal{W}(t)$ vanishes when we take the expectation value $\mathrm{E}\big(\mathrm{d}\mathcal{W}(t)\big)$. The stochastic variable $\mathcal{X}(t)$ is described by the conditional probability density $p(x,t|x_0,t_0)$ and we can compute its expectation value by integration, again simplifying the notation by writing $f(x) \equiv f\big(x(t)\big)$ and $p(x,t) \equiv p(x,t|x_0,t_0)$:

$$\frac{\mathrm{d}}{\mathrm{d}t}\langle f(x)\rangle = \int \mathrm{d}x f(x) \frac{\partial}{\partial t} p(x,t)$$

$$= \int \mathrm{d}x \left(a(x,t)\frac{\partial f(x)}{\partial x} + \frac{1}{2} b(x,t)^2 \frac{\partial^2 f(x)}{\partial x^2} \right) p(x,t) .$$

The rest of the derivation follows the procedure that is used in the derivation of the differential Chapman–Kolmogorov equation [194, 48–51], in particular integration by parts and neglect of surface terms. We thus obtain

$$\int \mathrm{d}x f(x) \frac{\partial}{\partial t} p(x,t)$$

$$= \int \mathrm{d}x f(x) \left(-\frac{\partial}{\partial x}\Big(a(x,t)p(x,t)\Big) + \frac{1}{2}\frac{\partial^2}{\partial x^2}\Big(b(x,t)^2 p(x,t)\Big) \right) .$$

Since the choice of a function $f(x)$ was arbitrary, we can drop it now and finally obtain a Fokker–Planck equation:

$$\frac{\partial p(x,t|x_0,t_0)}{\partial t} = -\frac{\partial}{\partial x}\Big(a(x,t)\,p(x,t|x_0,t_0)\Big) + \frac{1}{2}\frac{\partial^2}{\partial x^2}\Big(b(x,t)^2 p(x,t|x_0,t_0)\Big) . \tag{3.179}$$

The probability density $p(x,t)$ thus obeys an equation that is completely equivalent to the equation for a diffusion process, characterized by a drift function $a(x,t) \equiv A(x,t)$ and a diffusion function $b(x,t)^2 \equiv B(x,t)$, as derived from the Chapman–Kolmogorov equation. Hence, Itō's stochastic differential equation does indeed provide a local approximation to a drift and diffusion process in probability space. It is useful to compare the dimensions with respect to time of the individual terms in the stochastic differential and the Fokker–Planck equation. In the SDE (3.160), the term dx on the left-hand side does not contain time and has the dimension $[\mathrm{t}^0]$. Both terms on the right-hand side have dimension $[\mathrm{t}^0]$ as well, and this implies the time dimensions $[\mathrm{t}^{-1}]$ for $a(x,t)$ and $[\mathrm{t}^{-1/2}]$ for $b(x,t)$, since the dimensions of dt and $d\mathcal{W}(t)$ are $[\mathrm{t}^1]$ and $[\mathrm{t}^{1/2}]$, respectively. The terms of the Fokker–Planck equation (3.179) have the dimension $[\mathrm{t}^{-1}]$, and this confirms the above-mentioned dimensions for the functions $a(x,t)$ and $b(x,t)$. The functions $b(x,t)^2$ and $B(x,t)$ also have dimension $[\mathrm{t}^{-1}]$.

The extension to the multidimensional case based on Itō's formula is straightforward, and we obtain the following Fokker–Planck equation for the conditional probability density $p \equiv p(\mathbf{x},t|\mathbf{x}_0,t_0)$:

$$\frac{\partial p}{\partial t} = -\sum_i \frac{\partial}{\partial x_i}\big(A_i(\mathbf{x},t)p\big) + \frac{1}{2}\sum_{i,j}\frac{\partial}{\partial x_i}\frac{\partial}{\partial x_j}\Big(\big(\mathsf{B}(\mathbf{x},t)\,\mathsf{B}^{\mathrm{t}}(\mathbf{x},t)\big)_{i,j}p\Big) \,. \qquad (3.180)$$

Here, we derive one additional property, which is relevant in practice. The stochastic differential equation

$$d\mathbf{x} = \mathbf{A}(\mathbf{x},t)\,dt + \mathsf{B}(\mathbf{x},t)d\mathcal{W}(t) \qquad (3.177')$$

is mapped into a Fokker–Planck equation that depends only on the matrix product BB^{t}, and accordingly, the same Fokker–Planck equation arises from all matrices B that give rise to the same product BB^{t}. Thus, the Fokker–Planck equation is invariant under the replacement $\mathsf{B} \Rightarrow \mathsf{BS}$ for any orthogonal matrix S, i.e., such that $\mathrm{SS}^{\mathrm{t}} = \mathbb{I}$. If S satisfies the orthogonality relation, it may depend on $\mathbf{x}(t)$, but for the stochastic purposes it has to be *nonanticipating*.

It is straightforward to prove the redundancy of BB^{t} directly from the SDE by means of a transformed Wiener process

$$d\mathcal{V}(t) = \mathrm{S}(t)d\mathcal{W}(t) \,.$$

The random vector $\mathcal{V}(t)$ is a normalized linear combination of Gaussian variables $d\mathcal{W}_i(t)$ and $\mathrm{S}(t)$ is nonanticipating. Hence, $d\mathcal{V}(t)$ is itself Gaussian with the same correlation matrix. Averages $d\mathcal{W}_i(t)$ to various powers and taken at different times factorize, and the same is true for the $d\mathcal{V}_i(t)$. Accordingly, the infinitesimal elements $d\mathcal{V}(t)$ are increments of a Wiener process: the orthogonal transformation mixes trajectories, but without changing the stochastic nature of the process, and (3.177) can be rewritten to yield

$$\begin{aligned} d\mathbf{x} &= \mathbf{A}(\mathbf{x},t)\,dt + \mathsf{B}(\mathbf{x},t)\mathrm{S}^{\mathrm{t}}(t)\mathrm{S}(t)d\mathcal{W}(t) \\ &= \mathbf{A}(\mathbf{x},t)\,dt + \mathsf{B}(\mathbf{x},t)\mathrm{S}^{\mathrm{t}}(t)d\mathcal{V}(t) \\ &= \mathbf{A}(\mathbf{x},t)\,dt + \mathsf{B}(\mathbf{x},t)d\mathcal{W}(t) \,, \end{aligned}$$

since $\mathcal{V}(t)$ is as good a Wiener process as $\mathcal{W}(t)$, and both SDEs give rise to the same Fokker–Planck equation. □

At the end of this section we are left with a dilemma: the Itō integral is mathematically and technically the most satisfactory, but the more natural choice would be the Stratonovich integral which enables one to use conventional calculus. In addition, the noise term in the Stratonovich interpretation can be real noise with finite correlation time, whereas the idealized white noise required as reference in Itō's formalism gives rise to divergence of variances and correlations. For example, only the Stratonovich calculus, and not the Itō calculus, can deal with multiplicative noise in physical systems.

3.4.3 Integration of Stochastic Differential Equations

A stochastic variable $x(t)$ is consistent with an Itō stochastic differential equation (SDE)

$$\mathrm{d}x(t) = a\big(x(t), t\big)\,\mathrm{dt} + b\big(x(t), t\big)\mathrm{d}\mathcal{W}(t) , \tag{3.160}$$

if for all t and t_0 the integral equation (3.159) is satisfied. Time is ordered, i.e.,

$$t_0 < t_1 < t_2 < \ldots < t_n = t ,$$

and the time axis is split into equal or unequal increments $\Delta t_i = t_{i+1} - t_i$ (Fig. 3.27). We visualize a particular solution curve of the SDE for the initial condition $x(t_0) = x_0$ in the spirit of a discretized equation

$$x_{i+1} = x_i + a(x_i, t_i)\Delta t_i + b(x_i, t_i)\Delta\mathcal{W}(t_i) , \tag{3.159$'$}$$

where $x_i = x(t_i)$, $\Delta t_i = t_{i+1} - t_i$, and $\Delta\mathcal{W}(t_i) = \mathcal{W}(t_{i+1}) - \mathcal{W}(t_i)$. Figure 3.28 illustrates the partitioning of the stochastic process into a deterministic drift component, which is the discretized solution curve of the ODE obtained by setting $b\big(x(t), t\big) = 0$ in (3.159′), and a stochastic diffusion component, which is a Wiener process $\mathcal{W}(t)$ that is derived by setting $a\big(x(t), t\big) = 0$ in the SDE. The increment $\Delta\mathcal{W}(t_i)$ of the Wiener process is independent of x_i, provided (i) x_0 is independent of all $\mathcal{W}(t) - \mathcal{W}(t_0)$ for $t > t_0$ and (ii) $a(x, t)$ is a nonanticipating function of t for any fixed x. Condition (i) is tantamount to the requirement that any random initial condition must be nonanticipating.

In the construction of approximate solutions $x(t)$, the value $x_i = x(t_i)$ is always independent of $\Delta\mathcal{W}(t_j)$ for $j \geq i$, as is easily checked by inspecting (3.159′) or considering Fig. 3.28:

$$x_i = x_{i-1} + a(x_{i-1}, t_{i-1})\Delta t_{i-1} + b(x_{i-1}, t_{i-1})\Delta\mathcal{W}(t_{i-1}) .$$

A particular solution to (3.160) is obtained by taking the limit $\lim \Delta t \to 0$ of vanishing mesh size, which implies $\lim n \to \infty$. Uniqueness of solutions refers to individual trajectories in the sense that a particular solution is uniquely obtained for a given sample function $\mathcal{W}_k(t)$ of the Wiener process $\mathcal{W}(t)$. The existence of a solution, however, is defined for the whole ensemble of sample functions: a solution of (3.160) exists if a particular solution exists with probability one for any choice of a sample function of the Wiener process $\mathcal{W}_k(t)\,,\ \ \forall\, k \in \mathbb{N}_{>0}$. Existence and uniqueness of a nonanticipating solution $x(t)$ of an Itō stochastic differential equation within the time interval $[t_0, t]$ can be proven whenever the following conditions are satisfied [22, pp. 100–115]:

(i) *Lipschitz condition.* There exists a Lipschitz constant $K > 0$ such that

$$\big|a(x,\tau) - a(y,t)\big| + \big|b(x,t) - b(y,t)\big| \le K|x - y|\ ,$$

for all x and y and $t \in [t_0, t]$.

(ii) *Linear growth condition.* There exists κ such that, for all $t \in [t_0, t]$,

$$|a(x,t)|^2 + |b(x,t)|^2 \le \kappa^2(1 + |x|^2)\ .$$

The *Lipschitz condition*, named after the German mathematician Rudolf Lipschitz, is violated, for example, when we are dealing with a function that has a vertical tangent. It ensures existence and uniqueness of the solution and is almost always satisfied for stochastic differential equations in practice, because it is in essence a smoothness condition.

The *linear growth condition* guarantees boundedness of the solution (for details see, for example, [431, pp. 68–71]). The growth condition may be violated in abstract model equations, for example, when a solution *explodes* and progresses to infinity at finite time. A simple example is given by the ODE

$$\frac{\mathrm{d}x}{\mathrm{d}t} = x^2\ ,\ \ x(0) = 1 \ \Longrightarrow\ x(t) = \frac{1}{1-t}\ ,\ \ \text{for } 0 \le t < 1\ ,$$

which is unbounded at $t = 1$ and has no global solution defined for all values of t (see Sect. 5.1.3), i.e., the value of x becomes infinite at some finite time. We shall encounter such situations in Chap. 5. As a matter of fact this is a typical model behavior since no population or spatial variable can approach infinity at finite times in a finite world.

Several other properties known to apply to solutions of ordinary differential equations can be shown without major modifications to apply also to SDEs: continuity in the dependence on parameters and boundary conditions, as well as the Markov property (for proofs, we refer to [22]).

In order to show how stochastic differential equations can be handled in practice, we first calculate the expectation value and the variance of stochastic differential equations, and then consider three specific cases:

(i) general linear stochastic differential equations,
(ii) chemical Langevin equations,
(iii) the Ornstein–Uhlenbeck process, already discussed in Sect. 3.2.2.3.

The First Two Moments of Stochastic Differential Equations
In many cases, it is sufficient to know the expectation value and the variance of the stochastic variable of a process as a function of time. These two moments can be calculated without solving the stochastic equations explicitly. We consider the general SDE

$$\mathrm{d}x = a(x,t)\,\mathrm{d}t + b(x,t)\mathrm{d}\mathcal{W}(t)\ ,$$

and compute the mean value by taking the average and recalling that the second term on the right-hand side vanishes because $\langle \mathrm{d}\mathcal{W}(t)\rangle = 0$:

$$\mathrm{d}\,\langle x\rangle = \langle \mathrm{d}x\rangle = \langle a(x,t)\rangle\ \mathrm{d}t\ ,\quad \text{or}\quad \frac{\mathrm{d}\langle x\rangle}{\mathrm{d}t} = \langle a(x,t)\rangle\ . \tag{3.181}$$

Thus, the calculation of the expectation value boils down to solving an ODE. To derive an expression for the second moment and the variance, we have to calculate the differential of the square of the variable. Using

$$\mathrm{d}(x^2) = \big(2xa(x,t) + b(x,t)^2\big)\,\mathrm{d}t + 2b(x,t)\mathrm{d}\mathcal{W}(t)\ ,$$

and forming the average again using $\langle \mathrm{d}\mathcal{W}(t)\rangle = 0$, we obtain

$$\big\langle \mathrm{d}(x^2)\big\rangle = \mathrm{d}\,\big\langle x^2\big\rangle = \big\langle 2xa(x,t) + b(x,t)^2\big\rangle\ \mathrm{d}t\ .$$

If we knew the expectation values, a differential equation for the variance would be given by

$$\frac{\mathrm{d}\,\mathrm{var}(x)}{\mathrm{d}t} = \frac{\mathrm{d}\big\langle x^2\big\rangle}{\mathrm{d}t} - \frac{\mathrm{d}\langle x\rangle^2}{\mathrm{d}t} = \frac{\mathrm{d}\big\langle x^2\big\rangle}{\mathrm{d}t} - 2\,\langle x\rangle\,\frac{\mathrm{d}\langle x\rangle}{\mathrm{d}t}\ .$$

Further calculation requires knowledge of the functions $a(x,t)$ and $b(x,t)$.

As an example, we consider the simple linear SDE with $a(x,t) = \alpha x$ and $b(x,t) = \beta x$:

$$\mathrm{d}x = \alpha x\,\mathrm{d}t + \beta x\,\mathrm{d}\mathcal{W}(t) = x\big(\alpha\,\mathrm{d}t + \beta\,\mathrm{d}\mathcal{W}(t)\big)\ ,$$

and find for the expectation value

$$\langle x(t)\rangle = \langle x(0)\rangle\,\mathrm{e}^{\alpha t} = x_0\mathrm{e}^{\alpha t}\ ,\qquad \text{for } p(x,0) = \delta(x - x_0)\ , \tag{3.182a}$$

and for the variance

$$\begin{aligned}
\mathrm{var}\big(x(t)\big) &= \big\langle x(t)^2\big\rangle - \langle x(t)\rangle^2 \\
&= \big\langle x(0)^2\big\rangle\, \mathrm{e}^{(2\alpha+\beta^2)t} - \langle x(0)\rangle^2\, \mathrm{e}^{2\alpha t} \\
&= x_0^2 \left(\mathrm{e}^{(2\alpha+\beta^2)t} - \mathrm{e}^{2\alpha t}\right), \quad \text{for } p(x,0) = \delta(x - x_0)\,.
\end{aligned} \tag{3.182b}$$

These expressions are easily generalized to time dependent coefficients $\alpha(t)$ and $\beta(t)$, as we shall see in the paragraph on linear SDEs.

In this last part we have shown that analytical expressions for the most important quantities of stochastic processes can be derived from stochastic differential equations, which in this sense are as useful in practice as Fokker–Planck equations.

The Linear Stochastic Differential Equation

For the next example, we consider once again the linear SDE, but allow for time dependent parameters

$$\mathrm{d}x = \alpha(t)x\,\mathrm{d}t + \beta(t)x\,\mathrm{d}\mathcal{W}(t) = x\big(\alpha(t)\,\mathrm{d}t + \beta(t)\mathrm{d}\mathcal{W}(t)\big)\,.$$

Now we make the substitution $y = \ln x$, expand up to second order, viz.,

$$\mathrm{d}y = \frac{\mathrm{d}x}{x} - \frac{\mathrm{d}x^2}{x^2} = \alpha(t)\,\mathrm{d}t + \beta(t)\mathrm{d}\mathcal{W}(t) - \frac{1}{2}\beta(t)^2\,\mathrm{d}t\,,$$

and find the solution by integration and resubstitution:

$$x(t) = x(0)\exp\left(\int_0^t \Big(\alpha(\tau) - \frac{1}{2}\beta(\tau)^2\Big)\mathrm{d}\tau + \int_0^t \beta(\tau)\mathrm{d}\mathcal{W}(\tau)\right)\,. \tag{3.183}$$

All Gaussian variables satisfy the relation[59] $\langle \mathrm{e}^z\rangle = \exp\big(\langle z^2\rangle/2\big)$, and applying it to (3.183), we find for the nth raw moment [194, p. 109]:

$$\begin{aligned}
\langle x(t)^n\rangle &= \langle x(0)^n\rangle \left\langle \exp\left(n\int_0^t \Big(\alpha(\tau) - \frac{1}{2}\beta(\tau)^2\Big)\mathrm{d}\tau + n\int_0^t \beta(\tau)\mathrm{d}\mathcal{W}(\tau)\right)\right\rangle \\
&= \langle x(0)^n\rangle \exp\left(n\int_0^t \alpha(\tau)\mathrm{d}\tau + \frac{1}{2}n(n-1)\int_0^t \beta(\tau)^2\mathrm{d}\tau\right)\,.
\end{aligned} \tag{3.184}$$

[59]In order to prove the conjecture, one makes use of the fact that all cumulants κ_n with $n > 2$ vanish (see Sect. 2.3.3). The reader is encouraged to complete the proof.

All moments can be calculated from this expression, and for the two lowest moments we obtain

$$\langle x(t)\rangle = \langle x(0)\rangle \exp\left(\int_0^t \alpha(\tau)\mathrm{d}\tau\right) , \tag{3.185a}$$

$$\mathrm{var}\big(x(t)\big) = \mathrm{var}\big(x(0)\big) \exp\left(2\int_0^t \alpha(\tau)\mathrm{d}\tau\right) + \big\langle x(0)^2\big\rangle \exp\left(\int_0^t \beta(\tau)^2\mathrm{d}\tau\right) . \tag{3.185b}$$

Analytical solutions have also been derived for the inhomogeneous case $a(x,t) = \alpha_0 + \alpha_1 x$ and $b(x,t) = \beta_0 + \beta_1 x$, and the raw moments are readily calculated [194, p. 109].

Langevin Equation in Chemical Kinetics

Although the chemical Langevin equation [211] will be discussed extensively in Sect. 4.2.4, we already mention a few fundamental properties here. In order to keep the analysis simple, we consider only the case of a single reaction channel, and postpone reaction networks to Chap. 4. The conventional Langevin equation models a process in the presence of some random external force, which is expressed by the noise term $b\big(x(t),t\big)\mathrm{d}\mathcal{W}(t)$.

In chemical kinetics, such external forces may exist, but chemists are primarily interested in the internal fluctuations of particle numbers that ultimately result from thermal motion, and find their expression in the Poissonian nature of reaction events. Single reaction events are assumed to occur independently. The time interval between two reaction events is taken to follow an exponential distribution, and this implies that the total number of events denoted by m obeys a Poissonian distribution. In particular, $\mathcal{M}\big(\alpha(\tau),\tau\big)$ is the integer random variable counting the number of reaction events that took place in the interval $t \in [0,\tau[$, while $P_m\big(\alpha(t),t\big) = P\big(\mathcal{M}(\alpha(\tau),\tau) = m\big)$ is its probability density, and $\alpha(\tau)$ a function returning the probability or *propensity* of the chemical reaction to take place. Then we have[60]:

$$P_m(\alpha\tau) = \frac{(\alpha\tau)^m}{m!}\mathrm{e}^{-\alpha\tau} , \quad \text{with } \mathrm{E}\big(\mathcal{P}(\alpha\tau)\big) = \alpha\tau , \quad \mathrm{var}\big(\mathcal{P}(\alpha\tau)\big) = \alpha\tau .$$

In (2.52), we have already shown that the Poisson distribution can be approximated by a normal distribution for large values of $\alpha\tau$. This result follows from the central

[60] A Poissonian distribution depends on a single parameter $\big(\alpha(\tau),\tau\big) \equiv \alpha(\tau)\tau$. For simplicity, we shall write α instead of $\alpha(\tau)$ from now on.

limit theorem and can be derived from moment generating functions (Sect. 2.3.3), or by direct computation making use of Stirling's formula,

$$\pi(\alpha\tau) \doteq \frac{(\alpha\tau)^k}{k!}\mathrm{e}^{-\alpha\tau} \approx \frac{1}{\sqrt{2\pi\alpha\tau}}\mathrm{e}^{-(k-\alpha t)^2/2\alpha\tau} \doteq \mathcal{N}(\alpha\tau, \alpha\tau) \;, \quad \text{for } \alpha\tau \gg 1 \;. \tag{2.52'}$$

The function α depends on the numbers of molecules that can undergo the reaction, with large particle numbers giving rise to large values of α. Hence the condition can be met either by large particle numbers or by long time intervals τ, or both (Fig. 3.29).

The number of molecules of species A at time t in the reaction volume V is modeled by the random variable $\mathcal{X}_\mathsf{A}(t) = n(t)$. Two relations from chemical kinetics

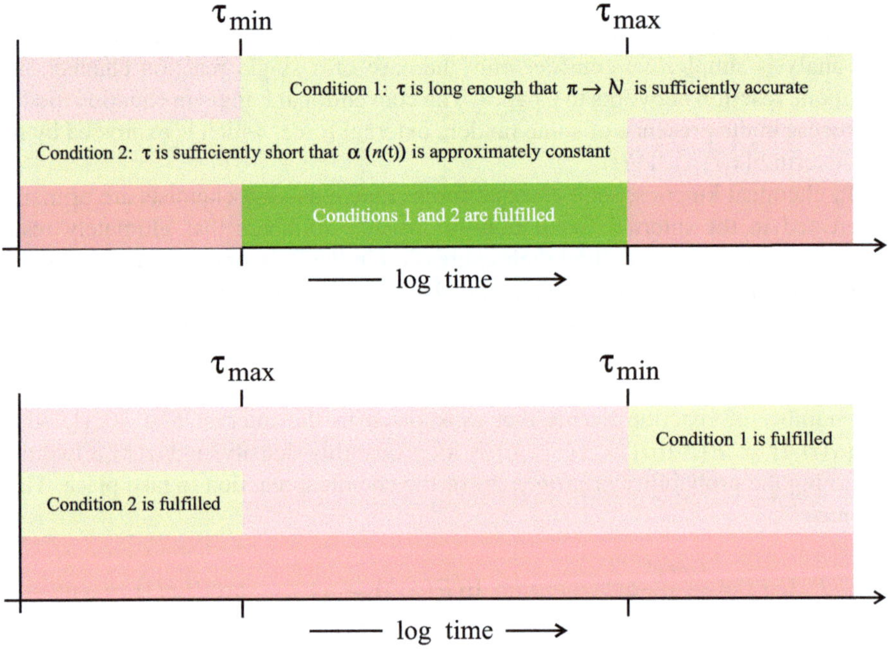

Fig. 3.29 Chemical Langevin equation. The chemical Langevin equation [211] is understood as an approximation to the master equation for modeling chemical reactions (Sect. 4.2.2). The approximation is built upon two contradictory conditions concerning the time leap interval: (1) τ has to be sufficiently long to satisfy the relation $\alpha\tau \gg 1$, and (2) τ has to be short enough to ensure that the function $\alpha\big(n(t)\big)$ does not change appreciably within the interval $[t, t+\tau]$. The *upper diagram* shows a situation where the use of the chemical Langevin equation is justified in the range $\tau_{\min} < \tau < \tau_{\max}$, whereas the Langevin equation is nowhere suitable under the conditions shown in the *lower diagram*

are basic to stochastic modeling of the process [211]:

(i) The chemical rate function or propensity function $\alpha\big(n(t)\big) = \chi\big(n(t)\big)$ for the reaction channel[61]:

$$\chi\big(n(t)\big)\mathrm{d}t = \begin{array}{l}\text{probability, given } n(t), \text{ that one reaction}\\ \text{will occur within } [t, t+\mathrm{d}t[\ .\end{array} \tag{3.186a}$$

(ii) The stoichiometric coefficient s, which determines the change in $\mathcal{X}_\mathsf{A}(t)$ by a single reaction event:

$$\mathcal{X}_\mathsf{A}(t) = n \implies \mathcal{X}_\mathsf{A}(t+\mathrm{d}t) = n + s\,, \quad \text{with } s \in \mathbb{Z}\,. \tag{3.186b}$$

It is worth noting that, in contrast to an ordinary Poisson process, where we had $\Delta m = +1$ in a chemical reaction, s might also be a negative integer, since molecules may be formed and may disappear, and it need not be $s = \pm 1$ because, for example, molecules can be formed two or more at a time, and they may also disappear two or more at a time (see Sect. 4.1). The function $\chi(n)$ is the product of two factors:

$$\chi(n) = \gamma h(n)\,. \tag{3.186c}$$

The reaction rate parameter γ depends on external conditions like temperature, pressure, etc., does not depend on the number of particles or collision events, and is in general independent of time.[62] The factor $h(n)$ which depends on particle number counts the number of events that can give rise to reaction events. It may be simply $h(n) = n$ for spontaneous reactions involving one molecule, or, for example, $n(n-1)/2$ if a collision between two molecules is required. Now we record the particle number in regular time intervals $\Delta t = \tau$. The number of reactions occurring in the time interval $[t, t+\tau]$ is another random variable denoted by $\mathcal{K}(n, \tau)$. For the time dependence of the number of molecules A, we formulate the Markov process

$$\mathcal{X}_\mathsf{A}(t+\tau) = n(t+\tau) = n(t) + s\mathcal{K}(n, \tau)\,, \tag{3.186d}$$

which is as hard to solve as the corresponding master equation. However, a straightforward approximation leading to a chemical Langevin equation can be achieved, provided that the following two conditions are satisfied:

Condition (1) requires that the time leap interval τ is small enough to ensure that the chemical rate function does not change appreciably, viz.,

[61] We use here $\alpha\big(n(t)\big)$ for the propensity function in order to show the relationship with a Poisson process. In chemical kinetics (Chap. 4), it will be denoted by $\chi\big(n(t)\big)$.

[62] Under certain rather rare circumstances, modeling reactions with time dependent reaction rate parameters may be advantageous.

$$\alpha\big(\mathcal{X}_\mathsf{A}(\theta)\big) \approx \alpha\big(n(t)\big) = \chi\big(n(t)\big)\,, \quad \forall\,\theta \in [t, t+\tau]\,. \tag{3.187a}$$

Usually, the change in particle numbers hardly exceeds $|s| \leq 2$ and the condition is readily met by sufficiently large molecule populations of the kind involved in conventional chemical kinetics. At the same time, for practical purposes, constancy of the rate functions guarantees independence of all reaction events within the interval $[t, t+\tau]$.

Condition (2) is required for the approximation of the Poissonian probability by a Gaussian distribution in the sense of (2.52′):

$$\mathrm{E}\Big(\mathcal{M}\Big(\alpha\big(n(t), \tau\big)\Big)\Big) = \alpha\big(n(t)\big)\tau \gg 1\,. \tag{3.187b}$$

The two conditions pull in opposite directions, and the existence of a range of validity for both at the same time is not automatic. Figure 3.29 illustrates two different situations: one where there is a range of suitable values of τ satisfying the two conditions so that an approximation of the master equation by a Langevin-type equation is possible, and one where such a range does not exist.

If the two conditions are satisfied, we can rewrite (3.186d) and find

$$\mathcal{X}_\mathsf{A}(t+\tau) = n(t) + s\mathcal{N}\Big(\alpha\big(n(t)\big)\tau, \alpha\big(n(t)\big)\tau\Big)\,.$$

Next we make use of the linear combination theorem for normal random variables, viz.,

$$\mathcal{N}(\mu, \sigma^2) = \mu + \sigma\mathcal{N}(0,1) = \mu + \sigma\varphi\,, \quad \varphi(x) = \frac{1}{\sqrt{2\pi}}\exp\left(-\frac{x^2}{2}\right)\,,$$

to obtain

$$\mathcal{X}_\mathsf{A}(t+\tau) = n(t) + s\alpha\big(n(t)\big)\tau + s\sqrt{\alpha\big(n(t)\big)\tau}\,\mathcal{N}(0,1)\,.$$

The next step is an approximation in the same spirit as conditions (1) and (2): a time interval τ that satisfies the two conditions may be considered as *macroscopic infinitesimal*, and we therefore treat τ as if it were a true infinitesimal dt, replacing the discrete variable n by the continuous x, and inserting $\mathrm{d}\mathcal{W} = \varphi(t)\sqrt{\mathrm{d}t}$. We thus find

$$\mathrm{d}x(t) = s\alpha\big(x(t)\big)\,\mathrm{d}t + s\Big(\alpha\big(x(t)\big)\Big)^{1/2}\varphi(t)\sqrt{\mathrm{d}t}\,. \tag{3.188}$$

Comparing the two equations (3.160) and (3.188), we may identify the coefficient functions:

$$\begin{aligned} a\big(x(t),t\big) \quad &\Longleftrightarrow \quad s\alpha\big(x(t),t\big) \;, \\ b\big(x(t),t\big) \quad &\Longleftrightarrow \quad s\sqrt{\alpha\big(x(t),t\big)} \;. \end{aligned} \tag{3.189}$$

The chemical Langevin equation does not contain an external noise term, but there are internal fluctuations resulting from the Poissonian nature of the chemical reaction events: $b(x(t),t) = s\sqrt{\alpha\big(x(t)\big)}$. The corresponding Fokker–Planck equation with $A\big(x(t),t\big) \equiv s\alpha\big(x(t),t\big)$ and $B\big(x(t),t\big) \equiv s^2\alpha\big(x(t),t\big)$ takes the form

$$\frac{\partial P(x,t)}{\partial t} = -\frac{\partial}{\partial x}\Big(s\alpha\big(x(t),t\big)P(x,t)\Big) + \frac{\partial^2}{\partial x^2}\Big(s^2\alpha\big(x(t),t\big)P(x,t)\Big) \;, \tag{3.190}$$

with the initial conditions (x_0, t_0).

We repeat what was said earlier: the occurrence of chemical reaction events is modeled by a Poisson process, but the consequences of the event itself follow from the specific laws of chemical reactions. The conversion of a master equation into a Langevin equation depends on being able to satisfy two potentially contradictory conditions, and it may happen that under certain circumstances, no acceptable Langevin approximation exists. In this paragraph, we have outlined only the basic features of modeling chemical reactions by means of Langevin equations. The formalism will be extended to reactions involving several molecules and reaction networks in Sect. 4.2.4, where we shall also present applications.

The Ornstein–Uhlenbeck Process

The general SDE for the Ornstein–Uhlenbeck process was given in (3.81). Without loss of generality, but simplifying the solution, we shift the long-time expectation value to the origin, i.e., setting $\mu = 0$, to obtain

$$\mathrm{d}x = -k\,x\,\mathrm{d}t + \sigma\,\mathrm{d}\mathcal{W}(t) \;. \tag{3.81$'$}$$

The solution of the deterministic equation is an exponential decay or relaxation to the long-time value $\lim_{t\to\infty} x(t) = 0$:

$$\mathrm{d}x = -kx\,\mathrm{d}t \;, \quad x(t) = x(0)\mathrm{e}^{-kt} \;,$$

and we make a substitution that compensates for the exponential decay, viz.,

$$x(t) = y(t)\mathrm{e}^{-kt} \;, \quad y(t) = x(t)\mathrm{e}^{kt} \;, \quad y(0) = x(0) \;.$$

Now we expand $\mathrm{d}y$ up to second order:

$$\mathrm{d}y = \mathrm{d}x\,\mathrm{e}^{kt} + x\mathrm{d}(\mathrm{e}^{kt}) + (\mathrm{d}x)^2 + \mathrm{d}x\,\mathrm{d}(\mathrm{e}^{kt}) + \big(\mathrm{d}(\mathrm{e}^{kt})\big)^2\,, \quad \mathrm{d}\big(\mathrm{e}^{kt}\big) = k\mathrm{e}^{kt}\,\mathrm{dt}\,.$$

All higher order terms vanish, because the expansion contains no term of order $\mathrm{d}\mathcal{W}(t)^2$ and all other terms vanish when we take the limit $\Delta t \to 0$. By integration, we find

$$\mathrm{d}y = \sigma \mathrm{e}^{kt}\,\mathrm{d}\mathcal{W}(t)\,, \quad y(t) = y(0) + \sigma \int_0^t \mathrm{e}^{k\tau}\,\mathrm{d}\mathcal{W}(\tau)\,,$$

and resubstitution yields the solution

$$x(t) = x(0)\mathrm{e}^{-kt} + \sigma \int_0^t \mathrm{e}^{-k(t-\tau)}\,\mathrm{d}\mathcal{W}(\tau)\,. \tag{3.191}$$

Calculation of the expectation value and variance is straightforward:

$$\big\langle x(t)\big\rangle = \left\langle x(0)\mathrm{e}^{-kt} + \sigma \int_0^t \mathrm{e}^{-k(t-\tau)}\,\mathrm{d}\mathcal{W}(\tau)\right\rangle = \big\langle x(0)\big\rangle \mathrm{e}^{-kt}\,, \tag{3.192a}$$

and with

$$\begin{aligned}\big\langle x(t)^2\big\rangle &= \left\langle\left(x(0)\mathrm{e}^{-kt} + \sigma \int_0^t \mathrm{e}^{-k(t-\tau)}\,\mathrm{d}\mathcal{W}(\tau)\right)^2\right\rangle \\ &= \big\langle x(0)^2\big\rangle\,\mathrm{e}^{-2kt} + \frac{\sigma^2}{2k}\big(1 - \mathrm{e}^{-2kt}\big)\,,\end{aligned}$$

we obtain

$$\mathrm{var}\big(x(t)\big) = \left(\mathrm{var}\big(x(0)\big) - \frac{\sigma^2}{2k}\right)\mathrm{e}^{-2kt} + \frac{\sigma^2}{2k}\,. \tag{3.192b}$$

With sharp initial conditions $p(x,0) = \delta(x - x_0)$, we find

$$\mathrm{var}\big(x(t)\big) = \frac{1}{2k}\big(1 - \mathrm{e}^{-2kt}\big)\,. \tag{3.192c}$$

Finally, we mention that the analysis of the Ornstein–Uhlenbeck process can be readily extended to many dimensions and time dependent parameters $k(t)$ and $\sigma(t)$ [194].

Chapter 4
Applications in Chemistry

There is nothing so practical as a good theory.
Kurt Lewin 1952

Abstract In chemistry the master equation is the best suited and most commonly used tool to model stochasticity in reaction kinetics. We review the common elementary reactions in mass action kinetics and discuss Michaelis–Menten kinetics as an example of combining several elementary steps into an overall reaction. Multistep reactions or reaction networks are considered and a formal mathematical theory that provides tools for the derivation of general properties of networks is presented. Then we digress into theory and empirical determination of rate parameters. The chemical master equation is introduced as the most popular tool for modeling stochasticity in chemical reactions, and the single reaction step approach is extended to reaction networks. The chemical Langevin equation is discussed as an alternative to the master equation: it has a number of convenient features but is not always applicable. Then, a selection of one-step reactions is presented for which the master equation can be solved exactly. The exact solutions are also used to illustrate the relation between the mathematical approach and the recorded data. A separate chapter deals with correlation functions, fluctuation spectroscopy, single molecule data, and stochastic modeling. Deterministic and stochastic parts of solutions can be separated by means of size expansions. Most reaction mechanisms are not accessible to the analytical approach and therefore we present a simulation method that is exact within the concept of the chemical master equation, and apply it to some selected examples of chemical reactions.

Conventional chemical reaction kinetics commonly does not require a stochastic approach because the numbers of particles are very large. There are exceptions when the particle numbers of certain species become very small during reactions—oscillations of species may serve as examples. Such cases will be mentioned and discussed in this and in the next chapter, but even more important is the requirement of a stochastic approach for direct measurements of fluctuations, which have become possible due to progress in spectroscopy leading to a spectacular increase in sensitivity. Single molecule techniques comprise another not completely unrelated and also rapidly developing field in which a stochastic approach is indispensable. On the other hand, if one wants to resolve reaction dynamics at the molecular level,

P. Schuster, *Stochasticity in Processes*, Springer Series in Synergetics,
DOI 10.1007/978-3-319-39502-9_4

the situation is different, because conventional statistical mechanics blurs the details of interest. Molecules are involved in huge numbers of collisions and crude but proper averaging is essential. Considered individually, however, collisions in the vapor phase can be calculated by the methods of classical mechanics or quantum mechanics, and experimental approaches for the study of single collisions in vacuum are well developed, although we have to admit that a detailed computational approach to reactive collisions in solution where molecules are densely packed is hopeless.

Stochastic chemical kinetics is based on the assumption that knowledge about the transformation of molecules in chemical reactions is not accessible in full atomistic detail, and that, even if it were, the information would be overwhelming and would obscure the essential features. Thus, it is assumed that chemical reactions have a probabilistic element and can be modeled properly by means of stochastic processes. The random processes are caused by thermal noise, which expresses itself through random encounters of molecules in collisions. Fluctuations, therefore, play an important role and they are responsible for the limitations in the reproduction of experiments. This concept is not substantially different from the ideas underlying equilibrium statistical mechanics, although statistics applied to thermodynamic equilibrium is on safer grounds than statistical mechanics applied to chemical reaction kinetics. On the other hand, the current theory of chemical reaction rates has been around for more than fifty years and it has not yet been replaced by any better founded or more applicable theory [324].

Particle numbers necessarily change in jumps of integer numbers and thus require discrete stochastic modeling. As we mentioned in Sect. 3.2.3 the most popular and general approach is the description by means of master equations. Alternative models are branching processes, birth-and-death processes and other special stochastic processes, some of which will be discussed in Chap. 5, because they are more frequently applied in biology than in chemistry. In general, different approaches do not exclude each other since, for example, birth-and-death processes can be solved by applying precisely the same techniques as master equations. Chemical master equations (Sect. 4.2.2) and birth-and-death master equations, which were analyzed in Sect. 3.2.3, are the same except for the step size, which in chemical kinetics is given by molecular stoichiometry. Approximation of chemical reactions by continuous variables, as discussed, for example, in the section on chemical Langevin equations (3.188), serves in essence two purposes:

(i) It provides a natural transition from the stochastic to the deterministic treatment through increasing particle numbers $n(t)$, thereby reducing the relative contribution of the Wiener process $s\sqrt{\alpha\big(n(t)\big)}\mathrm{d}\mathcal{W}$, until it becomes zero in the limit $n(t) \to \infty$.
(ii) It is indispensable for population size expansions, which provide the basis for a separation of the deterministic part of the solution from a diffusion term.

Conventional chemical reaction kinetics can be formally understood as the Liouville approach (Sect. 3.2.2.1) to stochastic chemical kinetics. It deals, in essence, with two classes of problems[1]:

1. *Forward problems*,which deal with the determination of time dependent concentrations as solutions of kinetic model equations, where the kinetic parameters are assumed to be known (for an introduction to traditional chemical kinetics, see [323] or the modern textbook [265]).
2. *Inverse problems*, which aim to determine parameters from measured data, where the kinetic model is commonly assumed to be known [139, 524].

The first problem boils down to deriving the solution curves or performing qualitative analysis of a kinetic ODE, or a PDE when the spatial distribution is nonhomogeneous. The inverse task is often referred to as the *parameter identification problem*. Qualitative analysis of differential equations aims to reconstruct bifurcation patterns in dynamical systems, and here too we encounter an inverse problem: the determination of the regions in parameter space where parameter combinations give rise to a certain dynamic behavior [138].

The chapter starts with an introduction to conventional chemical kinetics (Sect. 4.1), consisting of short reviews of elementary step kinetics (Sect. 4.1.1), Michaelis–Menten kinetics, which is discussed as an example of a reaction mechanism merging two or more single elementary steps into overall reactions (Sect. 4.1.2), and a formal theory of reaction networks developed for the qualitative analysis of multidimensional kinetic differential equations (Sect. 4.1.3). Then, we shall answer questions concerning the origin and empirical determination of rate parameters (Sects. 4.1.4 and 4.1.5). Stochasticity in chemical reactions is introduced in terms of the chemical master equation. We discuss how reaction networks can be analyzed stochastically, and come back to chemical Langevin equations when handling multidimensional cases (Sect. 4.2). Examples of exactly solvable chemical master equations are then presented:

(i) The equilibration of particle numbers or concentrations in the flow reactor.
(ii) Irreversible and reversible monomolecular reactions.
(iii) Bimolecular reactions that can still be solved exactly, but where the solutions become so complicated that practical work with them has to rely on numerical computation (Sect. 4.3).

A separate section treats correlation functions, fluctuation spectroscopy, and single molecule techniques and their implications as challenges for stochastic modeling (Sect. 4.4). The next section deals with the transition from microscopic to macroscopic systems by means of power series expansions in an extensive physical parameter Ω, e.g., the size of the system or the total number of particles. Size

[1]We remark that notions of *forward* and *inverse* in the context of handling differential equations have nothing to do with the direction of computational time in forward and backward Chapman–Kolmogorov equations.

expansions are particularly useful when the particle numbers are sufficiently large (Sect. 4.5). Most reaction mechanisms involve many reactions steps, and in many cases exact analytical solutions are available neither for the conventional deterministic approach nor for stochastic methods. The last sections deal with a numerical approach to stochastic chemical kinetics in which probability distributions or low moments are obtained by sampling a sufficiently large number of numerically calculated individual trajectories (Sect. 4.6).

4.1 A Glance at Chemical Reaction Kinetics

Chemical reactions will be modeled as Markov processes and analyzed in the form of the corresponding master equations later in this chapter. In a few cases Langevin or Fokker–Planck equations will be applied too. Chemical reaction mechanisms are typically networks of several reaction steps. A reaction step will be called *elementary* if no further resolution is possible at the level of atoms or molecules.[2] Appropriate criteria for the classification of elementary steps are the molecularity of reactions[3] and the complexity of the reaction dynamics. Molecularity is discussed in Sect. 4.1.1. With regard to reaction dynamics we shall distinguish reactions and reaction networks with (i) linear behavior, (ii) nonlinear behavior with simple dynamics in the sense of a monotonic approach to thermodynamic equilibrium or towards a unique stationary state, and (iii) complex behavior as exhibited by dynamical systems showing multiple stable stationary states, oscillations, or deterministic chaos.

The stochastic approach to chemical reaction kinetics is not so recent. It was initiated in the late 1950s in two different initiatives:

(i) Approximation of the complex vibrational relaxation in small molecules and its application to chemical reactions [40, 405, 501].
(ii) Direct modeling of chemical reactions as stochastic processes [35–37].

The latter approach can be viewed in the sense of limited information about reaction details, in the spirit of the initially mentioned characteristic of stochasticity. It was taken up and developed further by several groups [100, 101, 271, 301, 329, 381, 384]. The major part of the early work was summarized in an early review [382], which is recommended here for further reading. Anthony Bartholomay's

[2]In modern spectroscopy, further resolution into different molecular or atomic states can be achieved, and then the different states have to be treated as individual entities in reaction kinetics. A simple example of such a higher resolution dealing with different states of a molecule is applied when modeling monomolecular reactions (Sect. 4.1.4).

[3]The *molecularity* of a reaction is the number of molecules that are involved in the reaction, for example two in a reactive collision between molecules or one in a conformational change. An elementary step is a reaction at the molecular level that cannot be further resolved in mass action kinetics (Sect. 4.1.1). We shall distinguish elementary steps and elementary processes: the latter are more general and need not refer to the level of molecules.

studies are also highly relevant for biological models of evolution, because he modeled reproduction as a linear birth-and-death process. Exact solutions to master, Langevin, and Fokker–Planck equations can only be derived for particularly simple one-step reactions or for a few special cases. Approximations are often used, or the calculations are restricted to low moments, namely, the expectation values and variances of the stochastic variables. Later on, computer-assisted approximation techniques and numerical simulation methods were developed, which allow one to handle stochastic phenomena in chemical kinetics in a more general manner [194, 213, 541].

4.1.1 Elementary Steps of Chemical Reactions

Chemical reactions at the level of mass action kinetics are defined by mechanisms which can be decomposed into *elementary steps*. Elementary steps describe the transformation of reactant molecules into products and are written as stoichiometric equations.[4] Common elementary steps involving zero, one, or two reactant molecules are:

$$\star \longrightarrow \mathsf{A} \tag{4.1a}$$

$$\mathsf{A} \longrightarrow \oslash \tag{4.1b}$$

$$\mathsf{A} \longrightarrow \mathsf{B} \tag{4.1c}$$

$$\mathsf{A} \longrightarrow 2\mathsf{B} \tag{4.1d}$$

$$\mathsf{A} \longrightarrow \mathsf{B}+\mathsf{C} \tag{4.1e}$$

$$\mathsf{A}+\mathsf{B} \longrightarrow \mathsf{C} \tag{4.1f}$$

$$\mathsf{A}+\mathsf{X} \longrightarrow 2\mathsf{X} \tag{4.1g}$$

$$\mathsf{A}+\mathsf{E} \longrightarrow \mathsf{C}+\mathsf{E} \tag{4.1h}$$

$$\mathsf{A}+\mathsf{B} \longrightarrow \mathsf{C}+\mathsf{D} \tag{4.1i}$$

$$2\mathsf{A} \longrightarrow \mathsf{B} \tag{4.1j}$$

$$2\mathsf{A} \longrightarrow 2\mathsf{B} \tag{4.1k}$$

$$2\mathsf{A} \longrightarrow \mathsf{B}+\mathsf{C} \tag{4.1l}$$

[4] Stoichiometry deals with the relative quantities of reactants and products in chemical reactions. Reaction stoichiometry, in particular, determines the molar ratios of the reactants, which are converted into products, and the products that are formed. For example, in the reaction $2\mathbf{H}_2+\mathbf{O}_2 \rightarrow 2\mathbf{H}_2\mathbf{O}$ the stoichiometric ratios of $\mathbf{H}_2 : \mathbf{O}_2 : \mathbf{H}_2\mathbf{O}$ are $2 : 1 : 2$ (see also the stoichiometric matrix defined in Sect. 4.1.3).

For abstract molecules we use the first letters of the alphabet, i.e., A, B, C, and D, while catalysts are denoted by E and autocatalysts by X. The molecularity of a reaction is defined by the number of—different or identical—molecules on the reactant side of the stoichiometric equation, and we distinguish *zero-molecular*, *monomolecular*, *bimolecular*, or *termolecular*, reactions and so on.

The above list contains one zero-molecular reaction (4.1a), four monomolecular reactions (4.1b)–(4.1e), and seven bimolecular reactions (4.1f)–(4.1l). Nonreactive events, which occur in open systems, for example, in flow reactors, are the creation of molecules through inflow (4.1a)[5] or the annihilation of molecules through outflow (4.1b). They were included in the list because we shall need them to produce examples of open systems. Elementary steps with molecularities of three and higher are not included in the list, because simultaneous encounters of three or more molecules are extremely improbable, and apart from exceptions, elementary steps involving three or more molecules are not considered in conventional chemical kinetics.[6]

With two exceptions, namely, (4.1g) and (4.1h), all elementary steps in the list are characterized by the fact that molecular species represent either reactants or products. In other words, no molecules show up on both sides of the reaction arrow. The two exceptions are catalysis (4.1h) and autocatalysis (4.1g). These two processes have specific features that make them unique among other chemical reactions. Catalysis implies the presence of a catalyst, E in reaction (4.1h), which is consumed and produced in the same amounts during the reaction.[7] The presence of more catalyst increases the reaction rate. Catalysis by protein enzymes is the basis of biochemical kinetics. It is described by a multistep reaction, the Michaelis–Menten mechanism being the most popular example (Sect. 4.1.2). The elementary step shown in (4.1g) is an example of an *autocatalytic* elementary process. The unique feature of autocatalysis is self-enhancement of chemical species, $X \rightarrow 2X$, leading to amplification of fluctuations and exponential growth of particle numbers in the case of (4.1g). In practice, autocatalytic reactions almost always involve many elementary steps and obey complex reaction mechanisms (see, e.g., the review [472]). The formal kinetics of reproduction meets the conditions of autocatalysis, and in this case a complex multistep process is once again subsumed under a one-step overall reaction, here of type (4.1g). Because of its fundamental importance in biology, we shall discuss autocatalysis also in Sect. 5.1, in the chapter dealing with applications in biology.

In order to model and analyze basic features of autocatalysis and chemical self-enhancement, single-step autocatalytic reactions are used in case studies, rather than autocatalytic multistep reaction networks. Despite its termolecular nature, one

---

[5]The simplest kind of inflow, as indicated here by the *star*, is an arrival process with independent events, and hence follows a Poissonian probability law.

[6]Such exceptions are reactions involving surfaces as third partner, which are important in gas phase kinetics and, for example, biochemical reactions involving macromolecules.

[7]The notation E is inspired by *enzymes* in biochemistry, which are protein catalysts.

particular termolecular autocatalytic process, viz.,

$$\mathsf{A} + 2\mathsf{X} \longrightarrow 3\mathsf{X}\,, \tag{4.1m}$$

has become very popular [421], although it is unlikely to occur in pure form in real systems. The elementary step (4.1m) is the essential step in the so-called *Brusselator model* developed by Ilya Prigogine and Gregoire Nicolis at the Free University of Brussels. It can be studied straightforwardly by rigorous mathematical analysis. A more realistic model was conceived by the American chemists Richard Noyes and Richard Field from the University of Oregon in Eugene and has been called *Oregonator model* [166, 167, 429]. In Sect. 4.6.4 we shall analyze the results of numerical computations for both mechanisms. The Brusselator and the Oregonator models give rise to complex dynamical phenomena in space and time, which are otherwise rarely observed in chemical reaction systems. Among other features such special phenomena are: (i) multiple stationary states, (ii) chemical hysteresis, (iii) oscillations in concentrations, (iv) deterministic chaos, and (v) spontaneous formation of spatial structures. The last example is known as Turing instability [533] and is frequently used as a model for pattern formation or morphogenesis in biology [389]. An excellent review on *nonlinear phenomena* in chemistry can be found in the literature [472].

Stoichiometry and Chemical Equilibria

Although chemists were intuitively familiar with mass action throughout the nineteenth century, the precise formulation of a *law of mass action* (ma) is due to two Norwegians, the mathematician and chemist Cato Maximilian Guldberg and the chemist Peter Waage [560]. For the reaction (4.1f), for example, the mass action rate law yields

$$\frac{\mathrm{d}[\mathsf{A}]}{\mathrm{d}t} = \frac{\mathrm{d}[\mathsf{B}]}{\mathrm{d}t} = -\frac{\mathrm{d}[\mathsf{C}]}{\mathrm{d}t} = v^{(\mathrm{ma})}([\mathsf{A}],[\mathsf{B}]) = k[\mathsf{A}][\mathsf{B}]\,.$$

The rate of the reaction $v^{(\mathrm{ma})}$ is proportional to a rate parameter k and to the amounts $[\mathsf{A}]$ and $[\mathsf{B}]$ at which the two reactants A and B are present in the reaction system.[8] It is worth noticing that products—here the species C—do not show up in the reaction rate of *irreversible* chemical reactions,[9] which are reactions progressing exclusively in one direction, e.g., $\mathsf{A} + \mathsf{B} \rightarrow \mathsf{C}$. In other words the reaction in the

[8]We shall use the notation $[\mathsf{A}]$ and $[\mathsf{B}]$ with square brackets when we want to leave it open which units are to be used.

[9]The notions *reversible* and *irreversible* for chemical reactions are used differently from the notions in thermodynamics. In chemical kinetics a reaction is *irreversible* if the occurrence of the reaction in the opposite direction is not observable on realistic time scales and hence can be neglected. Strict chemical irreversibility causes an instability in thermodynamics. All chemical reactions that proceed with nonzero velocity are *irreversible* in the sense of thermodynamics, where reversibility requires infinitely slow progress of all processes including chemical reactions.

opposite direction $\mathsf{C} \rightarrow \mathsf{A} + \mathsf{B}$ has zero rate by definition. In the single reaction step (4.1f), $v^{(\mathrm{ma})}$ is described by three different but linearly dependent kinetic differential equations, and this redundancy will be eliminated in an appropriate manner by introducing the extent of reaction ξ in the next section.

Depending on the nature of the reaction system the chemists prefer to use different units:

(i) Particle numbers N_{A} and N_{B} which count the numbers of molecules and which are dimensionless.
(ii) Molar concentrations $c_{\mathsf{A}} = [c^0]N_{\mathsf{A}}/N_{\mathrm{L}}V$ and $c_{\mathsf{B}} = [c^0]N_{\mathsf{B}}/N_{\mathrm{L}}V$ (Sect. 1.1), where $c^0 = 1\ [\mathrm{mol\,l^{-1}}]$=1 [M] is used for the unit concentration.
(iii) Dimensionless molar ratios $x_{\mathsf{A}} = N_{\mathsf{A}}/(N_{\mathsf{A}} + N_{\mathsf{B}})$ and $x_{\mathsf{B}} = N_{\mathsf{B}}/(N_{\mathsf{A}} + N_{\mathsf{B}})$ with $x_{\mathsf{A}} + x_{\mathsf{B}} = 1$, when only A and B are present in the reaction system.

In precise terms, the law of mass action states that the rate of any given chemical reaction is proportional to the product of the concentrations or activities of the reactants.[10] In particular, the numbers of identical molecules that are consumed in a reaction step—called the stoichiometric coefficients ν_{A} and ν_{B}—appear as exponents of concentrations. Further, $v^{(\mathrm{ma})}$ is the reaction rate and k is a rate parameter:

$$\begin{aligned} &\nu_{\mathsf{A}}\mathsf{A} + \nu_{\mathsf{B}}\mathsf{B} \xrightarrow{\ k\ } \text{products} \\ \Longrightarrow\quad &\text{reaction rate} = v^{(\mathrm{ma})} = -\frac{1}{\nu_{\mathsf{A}}}\frac{\mathrm{d}a}{\mathrm{d}t} = -\frac{1}{\nu_{\mathsf{B}}}\frac{\mathrm{d}b}{\mathrm{d}t} = k a^{\nu_{\mathsf{A}}} b^{\nu_{\mathsf{B}}}\ . \end{aligned} \tag{4.2}$$

In a *reversible* reaction, which consists of two chemical reaction steps and which can be understood as a special combination of two elementary steps compensating each other, the reverse reaction is accounted for by a negative sign and we obtain for the case of two disjunct reactants and two products:

$$\begin{aligned} &\nu_{\mathsf{A}}\mathsf{A} + \nu_{\mathsf{B}}\mathsf{B} \underset{l}{\overset{k}{\rightleftharpoons}} \nu_{\mathsf{C}}\mathsf{C} + \nu_{\mathsf{D}}\mathsf{D} \\ \Longrightarrow\quad &v^{(\mathrm{ma})} = -\frac{1}{\nu_{\mathsf{A}}}\frac{\mathrm{d}a}{\mathrm{d}t} = -\frac{1}{\nu_{\mathsf{B}}}\frac{\mathrm{d}b}{\mathrm{d}t} = \frac{1}{\nu_{\mathsf{C}}}\frac{\mathrm{d}c}{\mathrm{d}t} = \frac{1}{\nu_{\mathsf{D}}}\frac{\mathrm{d}d}{\mathrm{d}t} \\ &\qquad = v_{\rightarrow}^{(\mathrm{ma})} - v_{\leftarrow}^{(\mathrm{ma})} = k a^{\nu_{\mathsf{A}}} b^{\nu_{\mathsf{B}}} - l c^{\nu_{\mathsf{C}}} d^{\nu_{\mathsf{D}}}\ . \end{aligned} \tag{4.3}$$

[10] Several idealized regularities hold only in the limit $\lim c \rightarrow 0$ of vanishing concentrations. The idealized laws can be retained by replacing concentrations by activities, i.e., $a_{\mathsf{A}} = f_{\mathsf{A}} c_{\mathsf{A}}$. Here we shall approximate activities by concentrations unless stated otherwise, and for the sake of simplicity, use lower case letters to indicate the species: $f_{\mathsf{A}} \approx 1$ and $c_{\mathsf{A}} = a\,[\mathrm{mol\,l^{-1}}]$ and $f_{\mathsf{B}} \approx 1$ and $c_{\mathsf{B}} = b\,[\mathrm{mol\,l^{-1}}]$.

The condition of zero net reaction rate, i.e., $v_{\rightarrow}^{(\mathrm{ma})} - v_{\leftarrow}^{(\mathrm{ma})} = 0$, yields an expression for the *equilibrium constant*, which was already described by Guldberg and Waage for the notion of mass action at equilibrium:

$$K = \frac{k}{l} = \frac{\bar{c}^{\nu_\mathrm{C}}\,\bar{d}^{\nu_\mathrm{D}}}{\bar{a}^{\nu_\mathrm{A}}\bar{b}^{\nu_\mathrm{B}}}\,, \tag{4.4}$$

where the bar denotes equilibrium concentrations. Later derivations of mass action use the chemical potentials of reactants and products, and were first introduced by Josiah Willard Gibbs around 1900 [202, 203].

Generalized One-Step Kinetics

In order to generalize the expressions, we introduce a set of M chemical species, viz., $\mathcal{S} = \{\mathsf{X}_1, \ldots, \mathsf{X}_\mathrm{M}\}$, in a single reaction step[11] and allow species to appear on both sides of the reaction arrows:

$$\nu_1\mathsf{X}_1 + \nu_2\mathsf{X}_2 + \cdots + \nu_\mathrm{M}\mathsf{X}_\mathrm{M} \underset{l}{\overset{k}{\rightleftharpoons}} \nu_1'\mathsf{X}_1 + \nu_2'\mathsf{X}_2 + \cdots + \nu_\mathrm{M}'\mathsf{X}_\mathrm{M}\,. \tag{4.5}$$

In the following we shall use column vector notation $\mathbf{x} = (x_1, \ldots, x_\mathrm{M})^\mathrm{t}$ for the concentrations, and $\boldsymbol{\nu} = (\nu_1, \ldots, \nu_\mathrm{M})^\mathrm{t}$ and $\boldsymbol{\nu}' = (\nu_1', \ldots, \nu_\mathrm{M}')^\mathrm{t}$ for the stoichiometric coefficients. Then the rate functions boil down to

$$\begin{aligned} v_{\rightarrow}^{(\mathrm{ma})}(\mathbf{x}) &= k x_1^{\nu_1} x_2^{\nu_2} \cdots x_\mathrm{M}^{\nu_\mathrm{M}} = k\mathbf{x}^{\boldsymbol{\nu}}\,, \\ v_{\leftarrow}^{(\mathrm{ma})}(\mathbf{x}) &= l x_1^{\nu_1'} x_2^{\nu_2'} \cdots x_\mathrm{M}^{\nu_\mathrm{M}'} = l\mathbf{x}^{\boldsymbol{\nu}'}\,, \end{aligned} \tag{4.6}$$

where we apply the *multi-index notation* $\mathbf{x}^{\boldsymbol{\nu}} = x_1^{\nu_1} x_2^{\nu_2} \cdots x_\mathrm{M}^{\nu_\mathrm{M}}$, and where unprimed and primed coefficients $\boldsymbol{\nu}$ and $\boldsymbol{\nu}'$ refer to the reactant and product sides, respectively. At equilibrium we have $\bar{v}^{(\mathrm{ma})} = \bar{v}_{\rightarrow}^{(\mathrm{ma})} = \bar{v}_{\leftarrow}^{(\mathrm{ma})}$, or $k\bar{\mathbf{x}}^{\boldsymbol{\nu}} = l\bar{\mathbf{x}}^{\boldsymbol{\nu}'}$. The stoichiometric coefficients are reformulated by accounting for the net production of a compound in a reversible reaction: $s_j = \nu_j' - \nu_j$, which yields the differential equation

$$\frac{1}{s_j}\frac{\mathrm{d}x_j}{\mathrm{d}t} = v_{\rightarrow}^{(\mathrm{ma})}(\mathbf{x}) - v_{\leftarrow}^{(\mathrm{ma})}(\mathbf{x})\,, \quad \forall j = 1, \ldots, \mathrm{M}\,.$$

[11] Later on we shall be dealing with multistep reaction networks of irreversible and reversible reactions and apply a notation that allows for straightforward identification of reaction steps by choosing k_j and l_j as reaction parameters for the reaction $\mathbf{R}_j$. The stoichiometric coefficients of the reactants in the reaction $\mathbf{R}_j$ will be denoted by $\nu_{\mathsf{X}_1 j}, \nu_{\mathsf{X}_2 j}, \ldots$, we shall use $\nu_{\mathsf{X}_1 j}', \nu_{\mathsf{X}_2}', \ldots$ for the reaction products, and the elements of the stoichiometric matrix are $\mathrm{S} = \{s_{ij} = \nu_{ij}' - \nu_{ij}\}$, $i = 1, \ldots, \mathrm{M}$ (see Sect. 4.1.3).

This equation is equally valid for a reversible reaction and both irreversible reactions related to it, which are obtained by setting either $l = 0$ or $k = 0$, respectively.

For the analysis of near-equilibrium kinetics, it is useful to define new variables that vanish at equilibrium:

$$\boldsymbol{\eta} = (\eta_1, \ldots, \eta_{\mathrm{M}})^{\mathrm{t}} = \mathbf{x} - \bar{\mathbf{x}} = (x_1 - \bar{x}_1, \ldots, x_{\mathrm{M}} - \bar{x}_{\mathrm{M}})^{\mathrm{t}} ,$$

and one common variable $\xi = \eta_i/(\nu_i' - \nu_i) = \eta_i/s_i, \forall\, i = 1, \ldots, \mathrm{M}$.[12] The thermodynamics of irreversible processes requires that every process sufficiently close to the equilibrium state approaches stationarity by a simple exponential relaxation process.[13] After linearization around equilibrium, we obtain

$$\frac{\mathrm{d}\xi}{\mathrm{d}t} = -\frac{1}{\tau_{\mathrm{R}}}\xi \,, \quad \xi(t) = \xi(0)\exp\bigl(-t/\tau_{\mathrm{R}}\bigr) \,, \tag{4.7}$$

where τ_{R} is the so-called relaxation time of the chemical reaction [435, 582]:

$$\tau_{\mathrm{R}}^{-1} = \sum_{i=1}^{\mathrm{M}} (\nu_i' - \nu_i)\frac{\nu_i \bar{v}_{\rightarrow} - \nu_i' \bar{v}_{\leftarrow}}{\bar{x}_i} = \sum_{i=1}^{\mathrm{M}} \frac{(\nu_i' - \nu_i)^2 \bar{v}}{\bar{x}_i} \,. \tag{4.8}$$

For the elementary steps in (4.1), the relaxation times are simple expressions, for example,

$$\mathsf{A} \rightleftharpoons \mathsf{B} \rightarrow \tau_{\mathrm{R}}^{-1} = k + l \,, \quad 2\mathsf{A} \rightleftharpoons \mathsf{B} \rightarrow \tau_{\mathrm{R}}^{-1} = 4k\bar{a} + l \,,$$

$$\mathsf{A} + \mathsf{B} \rightleftharpoons \mathsf{C} \rightarrow \tau_{\mathrm{R}}^{-1} = k(\bar{a} + \bar{b}) + l \,, \quad \mathsf{A} + 2\mathsf{X} \rightleftharpoons 3\mathsf{X} \rightarrow \tau_{\mathrm{R}}^{-1} = \bigl(k(\bar{a} + \bar{x}) + l\bar{x}\bigr)\bar{x}.$$

Relaxation in the multi-dimensional case will be discussed in Sect. 4.1.3.

Equation (4.7) is the linearized solution of a general nonlinear ODE. The solution of the nonlinear kinetic equation of a one-step reaction can be cast into a straightforward integral equation

$$\mathbf{x}(t) = \mathbf{x}(0) + \left(\int_0^t v\bigl(\mathbf{x}(\tau)\bigr)\mathrm{d}\tau\right)\bigl(\boldsymbol{\nu}' - \boldsymbol{\nu}\bigr) = \mathbf{x}(0) + \left(\int_0^t v\bigl(\mathbf{x}(\tau)\bigr)\mathrm{d}\tau\right)\mathbf{s} \,. \tag{4.9}$$

[12]The International Union of Pure and Applied Chemistry (IUPAC) has recommended using the term *rate of reaction* exclusively for the differential quotient $\mathrm{d}\xi/\mathrm{d}t = |1/(v_i' - v_i)|(\mathrm{d}[\mathsf{X}_i]/\mathrm{d}t)$, where ξ is the *degree of advancement* or the *extent of reaction* with the initial conditions as reference state: $\xi = ([\mathsf{X}_i] - [\mathsf{X}_i]_0)/s_i$. Here we define the variable ξ differently, with the thermodynamic equilibrium as reference state. The variable ξ is independent of stoichiometric coefficients in both definitions.

[13]Linear laws near an equilibrium point are generally valid and not restricted to chemistry. Hooke's law, named after the English natural philosopher Robert Hooke, may serve as an example from mechanics.

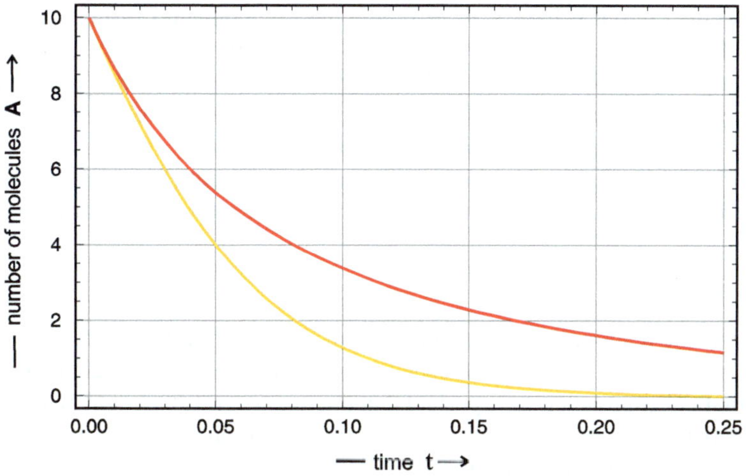

Fig. 4.1 The role of stoichiometry in kinetic equations. Reaction dynamics is determined by stoichiometry in an indirect way, too. The two reactions shown in the figure correspond to the elementary step reactions (4.1i) and (4.1g), and formally follow the same rate law: $v = k[\mathsf{A}][\mathsf{B}]$ and $v = k[\mathsf{A}][\mathsf{X}]$. The two reactions differ in the stoichiometry on the product side and this leads to different conservation relations and also to different ODEs: $\mathrm{d}a/\,\mathrm{d}t = -ka(n_0 - a)$ and $\mathrm{d}a/\,\mathrm{d}t = -ka(\vartheta_0 + a)$, respectively, where n_0 and ϑ_0 are two different constants (for details, see Sects. 4.1.3 and 4.3.5). Parameter choice: $k = 1.0\,[\mathrm{M}^{-1}\mathrm{t}^{-1}]$, $[\mathsf{A}]_{t=0} = 10$, $[\mathsf{B}]_{t=0} = [\mathsf{X}]_{t=0} = 15$. Color code: *red* bimolecular conversion, and *yellow* autocatalytic reaction

The extension of the deterministic model to networks of arbitrary numbers of reactions is presented in Sect. 4.1.3.

It is important to point out that the product side may also exert influence on the reaction dynamics in an irreversible reaction if one or more reactants appear among the products. The best and simplest examples are autocatalytic reactions. For the purpose of illustration we compare the elementary autocatalytic reaction (4.1g) with the bimolecular conversion (4.1i). The kinetic differential equations and their solution are

$$\frac{\mathrm{d}a}{\mathrm{d}t} = kax = ka(n_0 - a)\,, \qquad a(t) = \frac{n_0 a_0}{a_0 + (n_0 - a_0)\exp(n_0 kt)}\,,$$

$$\frac{\mathrm{d}a}{\mathrm{d}t} = kab = ka(\vartheta_0 + a)\,, \qquad a(t) = \frac{\vartheta_0 a_0}{(\vartheta_0 + a_0)\exp(n_0 kt) - a_0}\,,$$

where a_0, x_0, and b_0 are the initial values of the variables a, x, and b at time $t = 0$, while $n_0 = a_0 + x_0$ and $\vartheta_0 = b_0 - a_0$.[14] In Fig. 4.1, we compare the two reactions

[14]Since neither A nor B appear on the product side, it would make no difference to compare with (4.1f) or with $\mathsf{A} + \mathsf{B} \rightarrow 2\mathsf{C}$, which is the inversion of (4.1l).

with identical initial values of A, $a(0)$, and the same tangents $\mathrm{d}a/\,\mathrm{d}t|_{t=0} = -ka_0x_0$ or $\mathrm{d}a/\,\mathrm{d}t|_{t=0} = -ka_0b_0$, respectively. We observe the buildup of a difference in rate that grows in time, which is due to self-enhancement of the autocatalyst: an increase in the concentration [X] leads to an increase in the reaction rate, a steady acceleration of the reaction, and faster consumption of A.

Another generalization in the notation of the differential reaction rate will turn out useful later on when handling chemical reactions as stochastic processes:

$$\mathrm{d}\mathbf{x} = (v\,\mathrm{d}t)\mathbf{s} = \big(kh(\mathbf{x})\,\mathrm{d}t\big)\mathbf{s}\,, \tag{4.10}$$

where k is the reaction parameter and the function $h(\mathbf{x})$ expresses the concentration dependence of the differential change in the concentration vector $\mathbf{x}$. For the mathematical approach it is important that the reaction rate v should be independent of dt and that it should be a scalar quantity, expressing the fact that in a single reaction step there is one common reaction variable for all M molecular species. The function $h(\mathbf{x})$ contains the contribution of the concentrations of reactants, and in mass action kinetics simply takes the form $h(\mathbf{x}) = \mathbf{x}^{\boldsymbol{\nu}'-\boldsymbol{\nu}}$.

Strictly speaking, the resolution to the level of elementary steps implies the application of mass action kinetics, and this means that no further resolution is assumed to be achievable for molecules. As already mentioned, advances in spectroscopy have made it possible to distinguish between different states of molecules, in particular between ground states and various excited states in quantum molecular physics or between minimum free energy structures and suboptimal conformations in biopolymers, and then the ultimate resolution has to be pushed further down to individual states in order to be able to describe chemical reactions adequately.

Elementary step resolution and mass action kinetics often lead to complex reaction networks with a great number of variables. These are hard to analyze and yield results that are difficult to interpret. It is sometimes useful to reduce the number of variables and to introduce some simpler *higher level kinetics*. The difference between mass action and higher level kinetics is illustrated by means of an old and well studied example, the Michaelis–Menten reaction kinetics of enzyme catalyzed reactions in biochemistry.

4.1.2 Michaelis–Menten Kinetics

Chemical kinetics was already relevant in biology at the end of the nineteenth century when biochemical processes gained a quantitative perspective. Biochemical kinetics became a discipline in its own right, and has been revived recently in the form of systems biology, which is chasing the ambitious goal of modeling all processes in cells and whole organisms at the molecular level. In particular, enzyme catalyzed reactions have been in the focus of biochemists since the very beginning, and indeed biochemical kinetics as we understand it today was initiated by the ground-breaking work of Leonor Michaelis and Maud Menten [397]. General enzyme catalysis is modeled by three elementary steps, which at first are assumed to be reversible:

$$\mathbf{S} + \mathbf{E} \rightleftharpoons \mathbf{S} \cdot \mathbf{E} \rightleftharpoons \mathbf{E} \cdot \mathbf{P} \rightleftharpoons \mathbf{E} + \mathbf{P} . \tag{4.11}$$

These are (i) binding of the substrate **S** to the enzyme **E**, (ii) conversion of the substrate into product, both being bound to the enzyme, and (iii) the release of the product **P** through dissociation of the enzyme–product complex. Then, the full mechanism of simple enzyme catalyzed reaction consists of six elementary steps (Fig. 4.6):

$$\mathbf{S} + \mathbf{E} \xrightarrow{k_1} \mathbf{S} \cdot \mathbf{E} , \tag{4.12a}$$

$$\mathbf{S} \cdot \mathbf{E} \xrightarrow{l_1} \mathbf{S} + \mathbf{E} , \tag{4.12b}$$

$$\mathbf{S} \cdot \mathbf{E} \xrightarrow{k_2} \mathbf{E} \cdot \mathbf{P} , \tag{4.12c}$$

$$\mathbf{E} \cdot \mathbf{P} \xrightarrow{l_2} \mathbf{S} \cdot \mathbf{E} , \tag{4.12d}$$

$$\mathbf{E} \cdot \mathbf{P} \xrightarrow{k_3} \mathbf{P} + \mathbf{E} , \tag{4.12e}$$

$$\mathbf{P} + \mathbf{E} \xrightarrow{l_3} \mathbf{E} \cdot \mathbf{P} . \tag{4.12f}$$

For an efficient enzyme reaction, it is essential that the steps (4.12d) and (4.12f) should be negligibly slow. In particular, the latter reaction (4.12f) can lead to a substantial reduction in the production of the product **P** at high concentrations, a phenomenon known as *product inhibition* in biochemistry. It is necessary for high catalytic efficiency that the reaction (4.12b) be slow too.

Here we present a brief analysis of the Michaelis–Menten mechanism by conventional kinetics. In Sects. 4.2.3 and 4.4 we shall come back to stochastic Michaelis–Menten kinetics with single enzyme molecules [355, 462].

Simple Michaelis–Menten Kinetics

The full Michaelis–Menten reaction scheme (4.12) cannot be solved analytically, and accordingly it was already simplified in the original publication [397]: only the binding reaction (i) consisting of the two steps (4.12a) and (4.12b) is assumed to be reversible, whereas the catalytic reaction (ii) is modeled by an irreversible step, which is combined with step (iii) to yield directly the product **P**:

$$\mathbf{S} + \mathbf{E} \underset{l_1}{\overset{k_1}{\rightleftharpoons}} \mathbf{S} \cdot \mathbf{E} \xrightarrow{k_2} \mathbf{E} + \mathbf{P} . \tag{4.13a}$$

Simple Michaelis–Menten enzyme kinetics deals with four molecular species, namely, the substrate, enzyme, substrate–enzyme complex, and product, denoted **S**, **E**, **S · E**, and **P**, respectively. The enzyme–product complex **E·P** is not considered explicitly, and the concentration of the product can be interpreted as the total product concentration: $p = [\mathbf{P}] + [\mathbf{E} \cdot \mathbf{P}] \approx [\mathbf{P}]$. Again we denote concentrations by small letters, $[\mathbf{S}] = s$, $[\mathbf{E}] = e$, $[\mathbf{P}] \approx p$, and for the complex we use $[\mathbf{S} \cdot \mathbf{E}] = c$. Total concentrations give rise to conservation relations: $e_0 = e + c$, $s_0 = s + c + p$, provided $p(0) = 0$. In Michaelis–Menten kinetics, the stoichiometric equations and the rate $v^{(\mathrm{mm})}$ are of the form

$$\mathbf{S} + \mathbf{E} \rightleftharpoons \mathbf{S} \cdot \mathbf{E} \rightarrow \mathbf{E} + \mathbf{P} \Longrightarrow \text{reaction rate} = v^{(\mathrm{mm})} = \frac{\mathrm{d}[\mathbf{P}]}{\mathrm{d}t} = \frac{v_{\max} s}{K_{\mathrm{M}} + s} \,.$$

The parameters $v_{\max}$ and K_{M} denote the maximal reaction rate and the Michaelis constant, respectively. The Michaelis constant is the free substrate concentration s at half the maximal reaction rate $v_{\max}/2$.

In order to derive the Michaelis–Menten equation we start from the mechanism (4.13) given above. For two reaction steps we expect two independent kinetic differential equations and the derivation of an analytical solution in closed form is expected to be difficult if not impossible. Two independent variables also follow from four molecular species and the two conservation relations for e_0 and s_0. We choose the two variables e and c to be substituted:

$$e = e_0 - s_0 + s + p \,, \quad c = s_0 - s - p \,,$$

and we are left with the problem of solving two ODEs:

$$\begin{aligned}\frac{\mathrm{d}s}{\mathrm{d}t} &= k_1 s(e_0 - s_0 + s + p) - (l_1 + k_2)(s_0 - s - p) \\ &= k_1(e_0 - c)(s_0 - c - p) - (l_1 + k_2)c \,, \end{aligned} \tag{4.13b}$$

$$\frac{\mathrm{d}p}{\mathrm{d}t} = k_2(s_0 - s - p) \,. \tag{4.13c}$$

The choice of the concentrations s and p as variables corresponds to the interests of biochemists, since the conversion of substrate into product is the primary goal of biotechnology. Because of the irreversibility of the second step, all substrate is converted into product in the Michaelis–Menten model. Indeed, calculation of the stationary state defined by $\mathrm{d}s/\mathrm{d}t = 0$ and $\mathrm{d}p/\mathrm{d}t = 0$ yields $\bar{s} = 0$ and $\bar{p} = s_0$: all substrate has been converted into product. Results from computer integration of (4.13b) and (4.13c) are shown in Fig. 4.2. A fast binding reaction leading to a quasi-equilibrium is followed by relatively slow conversion of substrate into product that is characterized by an approximately constant concentration of the enzyme–substrate complex **S·E** that is tantamount to a constant rate of product synthesis. The

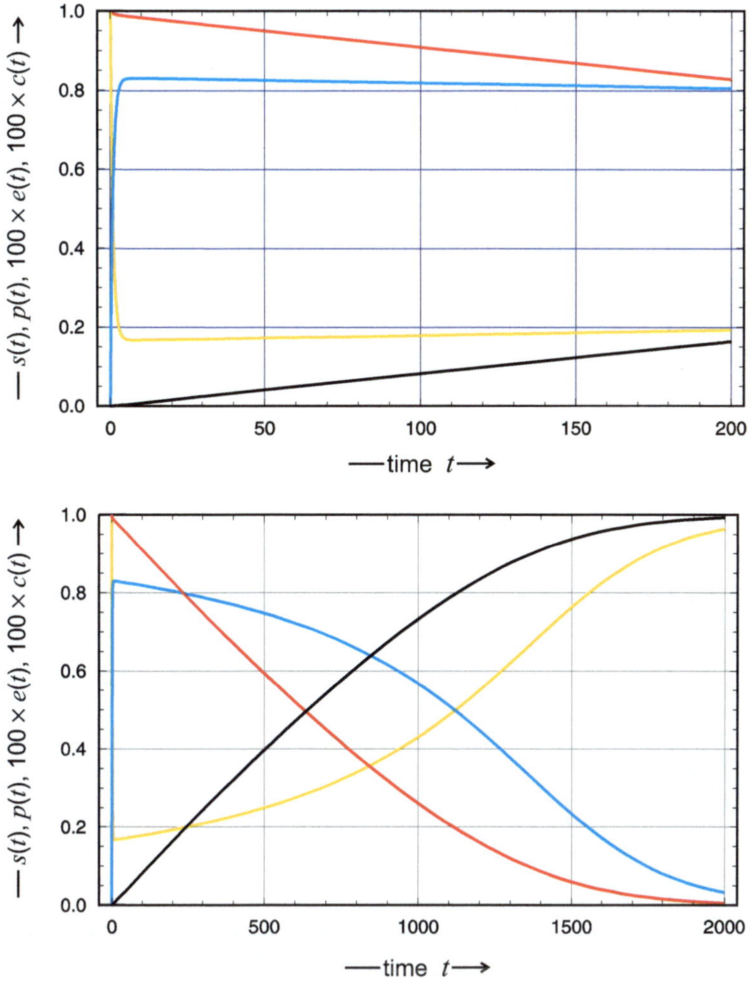

Fig. 4.2 The Michaelis–Menten mechanism for an enzyme catalyzed reaction. The plot shows a numerical integration of the reaction scheme (4.13) leading to the ODE (4.13b,c). Choice of parameters: $k_1 = 1.0\,\mathrm{s}^{-1}\mu\mathrm{M}^{-1}$, $l_1 = k_2 = 0.1\,\mathrm{s}^{-1}$ and the initial concentrations $e_0 = 0.01\,\mu\mathrm{M}$ and $s_0 = 1\,[\mu\mathrm{M}]$. Color code: $s(t)$ *red*, $p(t)$ *black*, $e(t)$ *yellow*, and $c(t)$ *blue*. Concentrations $e(t)$ and $c(t)$ are multiplied by a factor 100

Michaelis constant is obtained straightforwardly from the substrate concentration s at half-maximal reaction rate $v_{\max}/2$ and the adjusted value agrees perfectly with the value from the mathematical analysis, i.e., $K_\mathrm{M} = (l_1 + k_2)/k_1$ (Fig. 4.3).

Three historic approaches introducing simplifications into the mechanism (4.13) are of general interest because they can be applied to all consecutive reactions

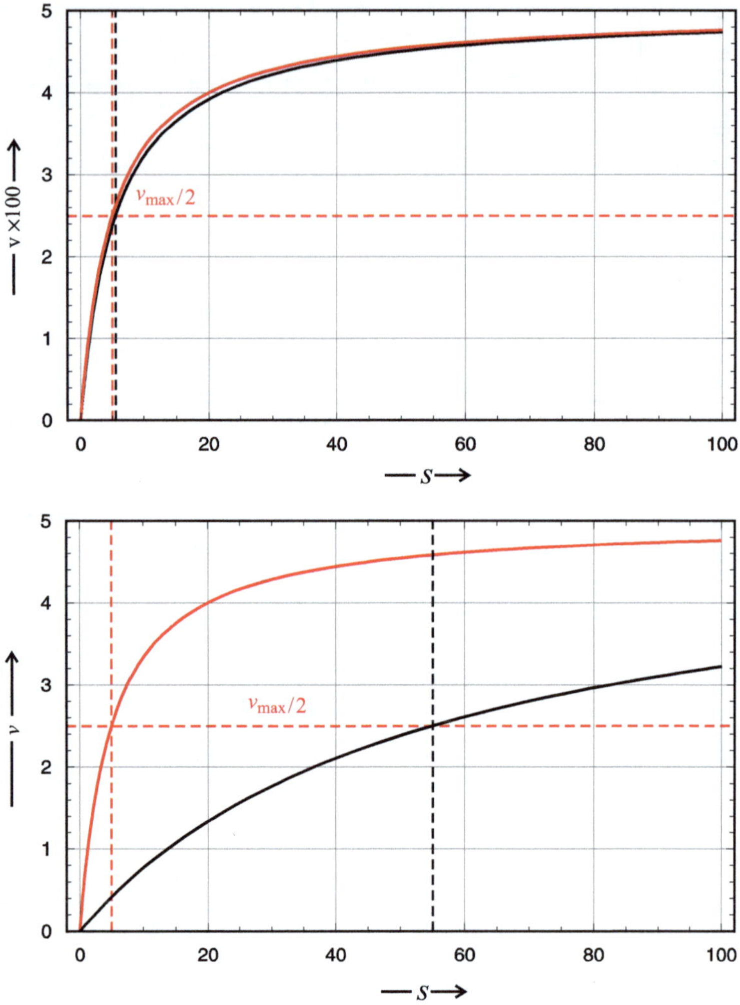

Fig. 4.3 Michaelis–Menten kinetics for enzyme catalyzed reaction. In the plots we compare the quasi-steady state approximation [65] and the pre-equilibrium approximation [397] for two different enzymatic reactions, which are characterized by slow and fast catalysis. The *upper plot* shows an example of the case $k_2 \ll l_1$ where the pre-equilibrium approximation (red) gives almost the same result as the quasi-steady state approximation (black), $K_M = 5.5\,\mu\text{M} \approx L_1$, whereas the *lower plot* presents a case with $k_2 \gg l_1$ where the pre-equilibrium approximation (red) fails to make a correct prediction, $K_M = 55\,\mu\text{M} \gg L_1$. Parameter choice: $k_1 = 10\,\mu\text{M}^{-1}\text{s}^{-1}$, $l_1 = 50\,\text{s}^{-1}$, and $L_1 = 5\,\mu\text{M}$; $k_2 = 5\,\text{s}^{-1}$ (*upper plot*) and $k_2 = 500\,\text{s}^{-1}$ (*lower plot*) were chosen as rate parameters for the catalytic reaction step

consisting of a reversible bimolecular addition reaction followed by an irreversible monomolecular conversion:

(1) The *pre-equilibrium* approximation originally made by Michaelis and Menten [397].
(2) The *irreversible substrate binding* approximation introduced one year later by the American biochemists Donald van Slyke and Glenn Cullen [545].
(3) The *quasi-steady state approximation* was proposed about ten years later by the English biologists George Briggs and J.B.S. Haldane [65]. It represents the currently used assumption for the enzyme substrate complex in the Michaelis–Menten mechanism sketched in the last section.

We present case (3) first and introduce (1) and (2) afterwards. The formation of product in the quasi-steady state approximation is given by

$$\frac{\mathrm{d}p}{\mathrm{d}t} = k_2 \frac{e_0 s}{K_M + s} = \frac{v_{\max} s}{K_M + s} = v^{(\mathrm{mm})} , \tag{4.13d}$$

where $v_{\max}$ is the maximal product formation rate at excess substrate and $K_M = (l_1 + k_2)/k_1$ the Michaelis constant.

In order to prove the Michaelis–Menten equation, we assume a small concentration $c(t)$ and its even smaller changes with time are neglected by setting $\mathrm{d}c/\mathrm{d}t = 0$ (for details, see the next section on the steady state approximation)[15]:

$$\frac{\mathrm{d}c}{\mathrm{d}t} = k_1 e s - (k_2 + l_1) c = 0 \implies (l_1 + k_2)\hat{c} = k_1 \hat{e}\hat{s} .$$

Next we introduce the Michaelis constant and substitute $e_0 = e + c$ for the total enzyme concentration in order to eliminate the free enzyme concentration variable e:

$$\frac{l_1 + k_2}{k_1} = K_M = \frac{(e_0 - \hat{c})\hat{s}}{\hat{c}} \implies \hat{c} = \frac{e_0 \hat{s}}{K_M + \hat{s}} .$$

The rate of product formation is obtained by multiplying by the rate constant of the catalytic reaction:

$$v^{(\mathrm{mm})} = \frac{\mathrm{d}p}{\mathrm{d}t} = k_2 \frac{e_0 s}{K_M + s} = \frac{v_{\max} s}{K_M + s} , \quad \text{with } v_{\max} = k_2 e_0 ,$$

and the result is the equation reported above. □

[15] In order to distinguish quasi-stationary states from true stationary or equilibrium states, we use a hat rather than an overbar, e.g., $\hat{c}$ instead of $\bar{c}$.

The pre-equilibrium case (1) assumes that the catalytic reaction is so slow that the complex **S · E** is at equilibrium with the substrate **S** and the enzyme **E**:

$$K_1^{-1} = L_1 = \frac{l_1}{k_1} = \frac{\bar{e}\bar{s}}{\bar{c}} \implies \bar{c} = \frac{e_0 \bar{s}}{L_1 + \bar{s}}, \quad v^{(\mathrm{pe})} = \frac{v_{\max} s}{L_1 + s}.$$

The condition of validity of the pre-equilibrium assumption is clearly $k_2 \ll l_1$ and $L_1 \approx K_M$. Figure 4.3 compares the pre-equilibrium approximation with the conventional Michaelis–Menten quasi-steady state model.

The irreversible binding case (2) assumes $l_1 = 0$ and gives rise to the simple consecutive reaction **S** + **E** → **S · E** → **P**. After introducing a quasi-steady state assumption, we obtain

$$v^{(\mathrm{ib})} = \frac{k_2 e_0 s}{k_2/k_1 + s} = \frac{v_{\max} s}{k_2/k_1 + s}.$$

Figure 4.4 shows that the irreversible binding mechanism can also perfectly satisfy the condition of an approximately constant concentration of the complex **S · E** over long time spans, giving rise to a constant rate of product formation.

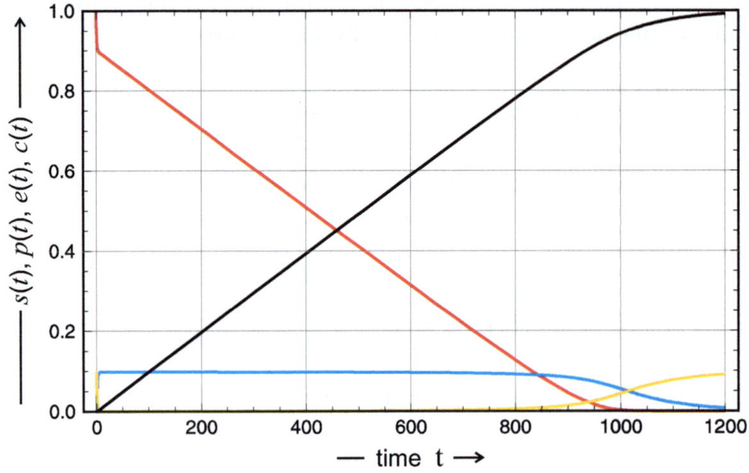

Fig. 4.4 Irreversible binding kinetics for the enzyme catalyzed reaction. The plot shows the concentrations of product p = [**P**] (*black*), substrate s = [**S**] (*red*), enzyme e = [**E**] (*yellow*), and enzyme-substrate complex c = [**S · E**] (*blue*) as a function of time. The mechanism applied is the irreversible consecutive reaction **S** + **E** → **S · E** → **P** [545]. The plot shows three phases of the reaction: a fast initial *binding phase* during which the enzyme–substrate complex is formed, a long *production phase* of practically linear substrate consumption and product formation, and a fast final *enzyme release phase* where the complex is consumed through net release of enzyme. The figure provides an impressive example for the existence of a long linear production phase: the concentration $c(t)$ of the enzyme–substrate complex is roughly constant during the production phase and therefore we observe a constant rate of product formation. Choice of parameters: $k_1 = 1.0\,[\mu\mathrm{M}^{-1}\mathrm{t}^{-1}]$, $k_2 = 0.01\,[\mathrm{t}^{-1}]$, $s_0 = 1.0\,[\mu\mathrm{M}]$, $e_0 = 0.1\,[\mu\mathrm{M}]$, $p_0 = 0$

It is remarkable that all three simplifications give rise to the same formal expression for the reaction rate v, the only difference being the relative size of the rate constants:

(i) $l_1 \gg k_2$ for the pre-equilibrium.
(ii) $l_1 \ll k_2$ for the irreversible binding.
(iii) Insensitivity to the relative size of l_1 and k_2 for the quasi-steady state approximation.

It is important to mention that the validity of the Michaelis–Menten approach requires the condition $e_0/(K_M + s_0) \ll 1$ to be satisfied. If this condition is not satisfied, evaluation of the v/s plot yields an s value at $v_{max}/2$, K_s, which is different from the Michaelis constant, i.e., $K_s \not\approx K_M = (k_2 + l_1)/k_1$.

Finally, we mention some practical aspects of measurement in conventional enzyme kinetics. The total concentrations s_0 and e_0 are determined in the stock solutions, the free substrate concentration s is either measured or calculated in cases where the dissociation constant of the enzyme–substrate complex is known, the rate of product formation $\mathrm{d}p/\mathrm{d}t = v^{(mm)}$ is measured for different substrate concentrations $v(s)$, v_{max} follows from the limiting rate at large excess of substrate, and the Michaelis constant is obtained from the $v^{(mm)}/s$ plot. For more than half a century since the pioneering work of Michaelis and Menten, the Michaelis constant K_M has been the most important quantitative parameter of enzymes, and it has been used, for example, to determine the purity of enzyme preparations.

The most important results of the Michaelis–Menten analysis of enzyme catalyzed reactions are:

(i) A small value of the Michaelis constant K_M means that the enzyme already reaches its maximal turnover at small substrate concentrations.
(ii) A large value of K_M implies the opposite: the maximal reaction rate is achieved only at high substrate concentrations.
(iii) The Michaelis constant K_M is proportional to the sum $l_1 + k_2$, so large K_M does not necessarily imply a high catalytic rate parameter $k_2 = k_{cat}$. It can also indicate weak binding of the substrate.

Michaelis–Menten kinetics has seen a recent revival with the possibility of single-molecule studies of enzyme catalysis (see Sect. 4.4).

Quasi-Steady State Approximation

There are many forms of simplified kinetics in which the number of independent variables is reduced at the expense of more complicated expressions. The Michaelis–Menten approach has been mentioned as an example in the last section. Here we shall consider the quasi-steady state approximation in more detail, but we will call it the steady state approximation for short. The simplest example is the two step reaction

$$\mathsf{A} \xrightarrow{\;k_1\;} \mathsf{B} \xrightarrow{\;k_2\;} \mathsf{C}\,, \tag{4.14a}$$

which is described by three kinetic differential equations. Only two are independent since we have the conservation relation $a + b + c = \text{const.}$:

$$\frac{\mathrm{d}a}{\mathrm{d}t} = -k_1 a\,, \quad \frac{\mathrm{d}b}{\mathrm{d}t} = k_1 a - k_2 b\,, \quad \frac{\mathrm{d}c}{\mathrm{d}t} = -k_2 b\,. \tag{4.14b}$$

Solution curves for the initial conditions $a(0) = a_0$, $b(0) = b_0$, and $c(0) = 0$, and $k_1 \neq k_2$ are readily obtained. Since they involve a nice little trick, we report the calculation here. We first solve the equation for $a(t)$ and find $a(t) = a_0 \mathrm{e}^{-k_1 t}$, then substitute in the equation for $b(t)$ to obtain

$$\frac{\mathrm{d}b}{\mathrm{d}t} = k_1 a_0 \mathrm{e}^{-k_1 t} - k_2 b \implies \frac{\mathrm{d}b}{\mathrm{d}t} + k_2 b = k_1 a_0 \mathrm{e}^{-k_1 t}\,.$$

The left-hand side of this equation can be transformed into a single differential

$$\frac{\mathrm{d}}{\mathrm{d}t}\left(b(t)\mathrm{e}^{k_2 t}\right) = \frac{\mathrm{d}b(t)}{\mathrm{d}t}\mathrm{e}^{k_2 t} + b(t) k_2 \mathrm{e}^{k_2 t}\,,$$

and we obtain the differential equation

$$\frac{\mathrm{d}}{\mathrm{d}t}\left(b(t)\mathrm{e}^{k_2 t}\right) = k_1 a_0 \mathrm{e}^{-(k_1 - k_2)t}\,.$$

Upon integration from $\tau = 0$ to $\tau = t$, this yields

$$b(t)\mathrm{e}^{k_2 t} - b(0) = \frac{k_1 a_0}{k_2 - k_1}\left(\mathrm{e}^{(k_2 - k_1)t} - 1\right),$$

whence

$$b(t) = \frac{k_1 a_0}{k_2 - k_1}\left(\mathrm{e}^{-k_1 t} - \mathrm{e}^{-k_2 t}\right) + b(0)\mathrm{e}^{-k_2 t}\,.$$

The concentration $c(t)$ is straightforwardly obtained from the conservation relation $c(t) = a_0 - a(t) - b(t)$. For $b(0) = 0$, we have

$$a(t) = a_0 \mathrm{e}^{-k_1 t}\,, \quad b(t) = a_0 \frac{k_1}{k_2 - k_1}\left(\mathrm{e}^{-k_1 t} - \mathrm{e}^{-k_2 t}\right),$$

$$c(t) = a_0 \left(1 + \frac{k_1 \mathrm{e}^{-k_2 t} - k_2 \mathrm{e}^{-k_1 t}}{k_2 - k_1}\right),$$

which are well known from radioactive decay series.

The steady state approximation is based on the assumption that the concentration of the intermediate does not change, i.e., $\mathrm{d}b/\mathrm{d}t = 0$, which allows elimination of the variable $b(t)$ and the result

$$c(t) \approx a_0 \left(1 - \frac{k_1 + k_2}{k_2} \mathrm{e}^{-k_1 t}\right) . \tag{4.14c}$$

Figure 4.5 illustrates the validity of the steady state approximation as a function of the ratio k_2/k_1: the larger this ratio, the better the agreement between the approximation and the exact solution.

Coming back to the general solutions for the initial conditions $a(0) = a_0$, $b(0) = b_0$, and $c(0) = c_0$ we recall that different treatment is required for the parameter relations $k_1 \neq k_2$ and $k_1 = k_2 = k$. The results are:

$$a(t) = a_0 \mathrm{e}^{-k_1 t} ,$$

$$b(t) = \begin{cases} a_0 \dfrac{k_1}{k_1 - k_2} \left(\mathrm{e}^{-k_2 t} - \mathrm{e}^{-k_1 t}\right) + b_0 \mathrm{e}^{-k_2 t} , & \text{if } k_1 \neq k_2 , \\ a_0 k t \mathrm{e}^{kt} + b_0 \mathrm{e}^{-kt} , & \text{if } k_1 = k_2 , \end{cases}$$

$$c(t) = \begin{cases} a_0 \left(1 - \dfrac{k_1 \mathrm{e}^{-k_2 t} - k_2 \mathrm{e}^{-k_1 t}}{k_1 - k_2}\right) + b_0 (1 - \mathrm{e}^{-k_2 t}) + c_0 , & \text{if } k_1 \neq k_2 , \\ a_0 (1 - \mathrm{e}^{-kt} - k t \mathrm{e}^{-kt}) + b_0 (1 - \mathrm{e}^{-kt}) + c_0 , & \text{if } k_1 = k_2 . \end{cases} \tag{4.14d}$$

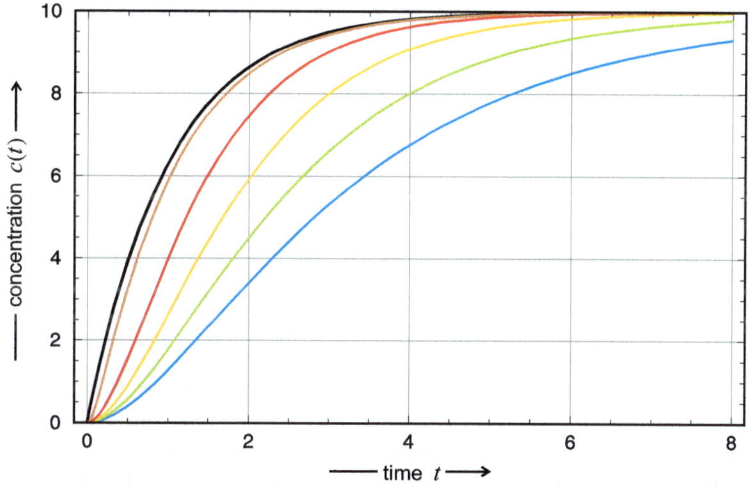

Fig. 4.5 The steady state approximation for multistep reactions. A test of the validity of the steady state approximation for the chain of irreversible first order reactions A → B → C (4.14). The concentration of the reaction product C is plotted as a function of time. The larger the value of k_2, the better the steady state solution (*black*) approximates the exact curves. Parameter choice: $a_0 = 10\,[\mathrm{M}]$, $b_0 = c_0 = 0$, $k_1 = 1\,[\mathrm{t}^{-1}]$, $k_2 = 0.4\,[\mathrm{t}^{-1}]$ (*blue*), $0.6\,[\mathrm{t}^{-1}]$ (*green*), $1.0001\,[\mathrm{t}^{-1}]$ (*yellow*), $2\,[\mathrm{t}^{-1}]$ (*red*), and $10\,[\mathrm{t}^{-1}]$ (*brown*)

It is straightforward to show that both conditions $k_1 \ll k_2$ and $k_1 \gg k_2$ simplify the equations for product formation:

$$b(t) \approx \begin{cases} a_0 \dfrac{k_1}{k_2} \, \mathrm{e}^{-k_1 t} \,, & \text{if } k_1 \ll k_2 \,, \\ a_0 \mathrm{e}^{-k_2 t} \,, & \text{if } k_1 \gg k_2 \,, \end{cases}$$

$$c(t) \approx \begin{cases} a_0 \left(1 - \mathrm{e}^{-k_1 t}\right) \,, & \text{if } k_1 \ll k_2 \,, \\ a_0 \left(1 - \mathrm{e}^{-k_2 t}\right) \,, & \text{if } k_1 \gg k_2 \,. \end{cases} \tag{4.14e}$$

What we observe in this simple example is a manifestation of the rule for the rate determining step. The overall kinetics of a chain of reactions is determined by the slowest step called the rate determining step: This is step 1 and the parameter k_1 for $k_1 \ll k_1$ and step 2 with k_2 for $k_2 \ll k_1$.

A complete mathematical analysis can be extended to a mechanism in which the first reaction step is reversible:

$$\mathsf{A} \underset{l_1}{\overset{k_1}{\rightleftharpoons}} \mathsf{B} \xrightarrow{k_2} \mathsf{C} \,. \tag{4.15}$$

However, the solutions are very complicated and are derived by means of symbolic computation. We dispense here from listing the rather clumsy equations and refer to the reference [314, pp. 35–72], which is an excellent introduction to analytic chemical kinetics on the computer.

Extended Michaelis–Menten Kinetics

Beginning in the 1960s, new spectroscopic and kinetic techniques were developed that allowed for the resolution of reaction kinetics into individual reaction steps. The simple Michaelis–Menten mechanism (4.13) deals with only two states of the enzyme, free **E** and substrate bound **S** · **E**, and gives rise to a single chemical relaxation mode (see Sect. 4.4). Such simple kinetics is only observed in a few enzyme catalyzed reactions, whereas most enzymes exhibit more complex kinetics with two or more relaxation modes [355, 462], or even oscillations [126]. Based on this empirical evidence, two extended versions of the original Michaelis–Menten mechanism are in common use (Fig. 4.6a and b). We shall apply here the natural extension by explicit consideration of the enzyme–product complex shown in (4.12), and find for the five kinetic equations using $[\mathbf{S} \cdot \mathbf{E}] = c$ and $[\mathbf{E} \cdot \mathbf{P}] = d$ for the

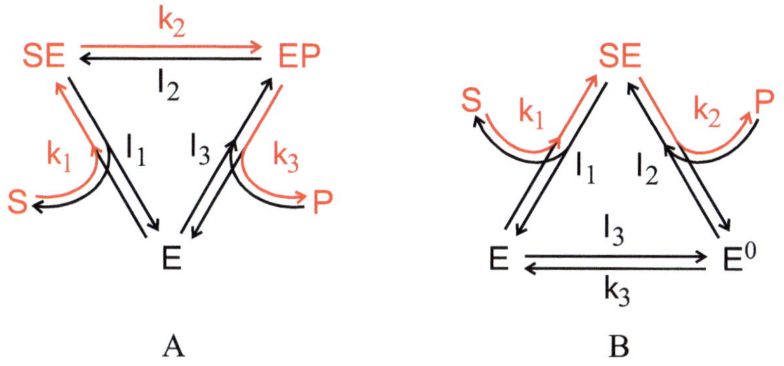

Fig. 4.6 The extended Michaelis–Menten mechanism. Two extended versions of the simplest Michaelis–Menten mechanism are consistent with empirical data for the majority of enzyme catalyzed reactions: (i) The mechanism on the left-hand side (**A**) explicitly includes the enzyme–product complex **EP** and its dissociation into free enzyme **E** and product **P** (see (4.12) and, for example, [462]). (ii) Another extension of the simple Michaelis–Menten mechanism (4.13) includes an additional conformational state $\mathbf{E}^0$ of the enzyme after release of the product from the complex (**B**). This mechanism is used, for example, in single molecule enzyme kinetics (see [316] and Sect. 4.4.1). The highlighted path (*red*) illustrates the conversion of substrate **S** into product **P**

protein–substrate complexes:

$$\frac{\mathrm{d}e}{\mathrm{d}t} = -(k'_1 s + l'_3 p)e + l_1 c + k_3 d \,, \tag{4.16a}$$

$$\frac{\mathrm{d}c}{\mathrm{d}t} = -(k_2 + l_1)c + k'_1 se + l_2 d \,, \tag{4.16b}$$

$$\frac{\mathrm{d}d}{\mathrm{d}t} = -(k_3 + l_2)d + k_2 c + l'_3 p\, e \,, \tag{4.16c}$$

$$\frac{\mathrm{d}s}{\mathrm{d}t} = -k'_1 se + l_1 c \,, \tag{4.16d}$$

$$\frac{\mathrm{d}p}{\mathrm{d}t} = -l'_3 pe + k_3 d \,, \tag{4.16e}$$

where we choose primed symbols for the second order rate constants in order to facilitate the forthcoming change in notation: $k_1 = k'_1 s$ and $l_3 = l'_3 p$. The concentrations in the mechanism (4.16) converge to a thermodynamic equilibrium (see Sect. 4.2.3):

$$\frac{\bar{p}}{\bar{s}} = \frac{[\bar{\mathbf{P}}]}{[\bar{\mathbf{S}}]} = \frac{k'_1 k_2 k_3}{l_1 l_2 l'_3} = K_1 K_2 K_3 \,. \tag{4.17}$$

The individual equilibrium concentrations of the extended Michaelis–Menten mechanism are readily computed, and for the initial condition of zero product, viz., $p(0) = 0$, the conservation relations are

$$s(0) = s_0 = s + c + d + p\,, \qquad e(0) = e_0 = e + c + d\,. \tag{4.18}$$

With $\alpha = 1 + K$, $\beta = K_1(1 + K_2)$, $K_1 = k_1'/l_1$, $K_2 = k_2/l_2$, $K_3 = k_3/l_3'$, and $K = K_1K_2K_3$, we obtain:

$$\begin{aligned}\bar{s} &= \frac{\beta(s_0 - e_0) - \alpha + \sqrt{4\alpha\beta s_0 + \big(\beta(s_0 - e_0) - \alpha\big)^2}}{2\alpha\beta}\,,\\ \bar{e} &= \frac{e_0}{1 + \beta\bar{s}}\,,\\ \bar{p} &= K\bar{s}\,,\\ \bar{c} &= K_1\bar{s}\bar{e}\,,\\ \bar{d} &= K_3^{-1}\bar{p}\bar{e}\,.\end{aligned} \tag{4.19}$$

The expression for the equilibrium concentration $\bar{s}$ of the substrate makes (4.19) prohibitive for analytical work, but these equations can be readily computed numerically. The results, however, are mainly of academic interest since in the two cases of general importance the equilibrium conditions are never fulfilled in experiments:

(i) If product formation is the goal, efficient synthesis under conditions far from equilibrium are required.
(ii) In single molecule studies (see Sect. 4.4), the turnover of enzyme conformations occurs under conditions where equilibrium thermodynamics cannot be applied.

Numerical integration of (4.16) will be discussed in Sect. 4.6.4.

For many experimental investigations, in particular for single molecule experiments, the assumption of constant concentrations of substrate and product, i.e., $[\mathbf{S}] = s_0 =$ const. and $[\mathbf{P}] = p_0 =$ const., is realistic. Then the nonlinear ODE (4.16) becomes a three-dimensional linear ODE with $k_1 = k_1' s_0$ and $l_3 = l_3' p_0$:

$$\frac{\mathrm{d}}{\mathrm{d}t}\begin{pmatrix} e \\ c \\ d \end{pmatrix} = \begin{pmatrix} -(k_1 + l_3) & l_1 & k_3 \\ k_1 & -(k_2 + l_1) & l_2 \\ l_3 & k_2 & -(k_3 + l_2) \end{pmatrix} \begin{pmatrix} e \\ c \\ d \end{pmatrix}\,. \tag{4.20}$$

Now the analysis is straightforward [462], and the computation of the eigenvalues yields[16]:

$$\lambda_{1,2} = -\frac{1}{2}\left((k_1 + k_2 + k_3 + l_1 + l_2 + l_3) \pm \sqrt{\Delta}\right) ,$$
$$\Delta = (k_1 - k_2 - k_3 + l_1 - l_2 + l_3)^2 - 4(k_2 - l_3)(k_3 - l_1) , \tag{4.21}$$
$$\lambda_3 = 0 .$$

The zero eigenvalue indicates a conservation relation that is given by the total enzyme concentration $e_0 = e + c + d$. The commonly chosen experimental conditions are no product $[\mathbf{P}] = 0 \Rightarrow l_3 = 0$, or at least the initial condition $p(0) = 0$, and excess substrate $[\mathbf{S}] = s \approx s_0 = [\mathbf{S}]_0$, where the total concentration $[\mathbf{S}]_0$ is the sum of the concentrations of all complexes containing substrate or product: $s(0) = [\mathbf{S}]_0 + [\mathbf{P}]_0 = [\mathbf{S}] + [\mathbf{S} \cdot \mathbf{E}] + [\mathbf{E} \cdot \mathbf{P}] + [\mathbf{P}]$. The two nonzero eigenvalues are complex in the range [462]

$$l_2 + \left(\sqrt{k_2} - \sqrt{k_3 - l_1}\right)^2 < k_1 = k_1'[\mathbf{S}] < l_2 + \left(\sqrt{k_2} + \sqrt{k_3 - l_1}\right)^2 ,$$

and damped oscillations have indeed been observed in single enzyme molecule experiments [126]. The oscillations are heavily damped because the ratio $\Im(\lambda)/\Re(\lambda)$ is small for this three-state system $(\mathbf{E}, \mathbf{S} \cdot \mathbf{E}, \mathbf{E} \cdot \mathbf{P})$.

Generalized Rate Functions

It is often useful to define a common rate function for different kinetics, which allows for insertion of specific *association functions* $\theta_i(x_i)$:

$$v_j(\mathbf{x}) = k_j \prod_{i=1}^{M} \theta_i(x_i) , \tag{4.22}$$

$$\theta_i^{(\mathrm{ma})}(x_i) = x_i^{\nu_{i(j)}} \text{ for mass action kinetics} ,$$

$$\theta_i^{(\mathrm{mm})}(x_i) = \frac{v_{\mathrm{max}}^{(ij)} x_i}{K_{\mathrm{M}}^{(ij)} + x_i} \text{ for Michaelis–Menten kinetics} ,$$

where a combination of different frequency functions is possible.

[16] We remark that, for a linear n-dimensional ODE $\mathrm{d}\mathbf{x}^{\mathrm{t}}/\,\mathrm{d}t = \mathrm{A}\mathbf{x}^{\mathrm{t}}$, the matrix A is identical to the Jacobian matrix $\mathrm{J} = \{J_{ij} = \partial f_i/\partial x_j,\ i,j = 1,\dots,n\}$ for a general ODE $\mathrm{d}\mathbf{x}^{\mathrm{t}}/\,\mathrm{d}t = \mathbf{f}(\mathbf{x})^{\mathrm{t}}$, and its eigenvalues λ_k, $k = 1,\dots,n$, determine the (here global) stability of the system.

Mass action and Michaelis–Menten kinetics are used here as widely known and commonly used examples, but many other cases of simplified mechanisms represent conceivable kinetics.

4.1.3 *Reaction Network Theory*

So far either we have considered chemical reactions as single step processes or we have discussed techniques that approximated multistep mechanisms by a single *overall* step.[17] Almost all interesting chemical systems, however, consist of networks of reactions that are characterized by a variety of interacting molecular species, and this leads to dynamical systems with more than one, often many variables, for which analytical solutions are only very rarely available.[18] In the second half of the twentieth century, when chemists and physicists began to consider kinetic differential equations as dynamical systems and to apply qualitative analysis, new questions became relevant in addition to forward and inverse problems. These new questions are concerned with general properties of reaction networks, e.g., to ascertain:

(i) whether or not a network can sustain multiple steady states in the positive orthant of concentration space,
(ii) whether or not undamped oscillations resulting from a stable limit cycle are possible,
(iii) whether of not a specific reaction network can display deterministic chaos.

Some of these questions can be answered by the deterministic theory of chemical reaction networks described here, which is also the basis for the stochastic approach. A complementary but general technique that can be applied to answer these questions consists in the inversion of qualitative analysis [138, 356, 357]: inverse bifurcation analysis aims to explore the domains in parameter space that give rise to certain forms of complex dynamics. However, there may be substantial differences between deterministic and stochastic bifurcations [23].

A formal deterministic theory of chemical reaction networks has been developed already in the 1970s by Fritz Horn, Roy Jackson, and Martin Feinberg [152, 263] in order to complement conventional chemical kinetics by providing tools that allow for the derivation of general results for entire classes of reaction networks. The theoretical approach became really popular only recently when chemical reaction kinetics was applied to systems biology and when it was realized that stochastic

[17]Two trivial exceptions were the inflow and outflow of a compound A in the flow reactor and the reversible reaction $A \rightleftharpoons B$. In both cases, however, we were dealing with a single stochastic variable counting the numbers of molecules A.

[18]An exception was the two-step irreversible reaction $A \rightarrow B \rightarrow C$ (4.14a).

modeling of extended chemical reaction networks is required for any deeper understanding of regulation and control of cellular dynamics and cellular metabolism [87, 226]. Before we consider modeling stochastic chemical reaction networks (SCRNs) in Sect. 4.2.3, we present a brief introduction to the Feinberg–Horn–Jackson theory, which allows for straightforward answers to otherwise difficult-to-predict properties of chemical reaction networks, e.g., the nonexistence of multiple steady states or the absence of oscillating concentrations. The theory does not aim to deduce the properties of networks for given sets of rate parameters, but it derives tools for studying features of families of networks irrespective of the particular choice of parameters.

Formal Stoichiometry

For the forthcoming discussions, it will be necessary to formalize the concept of stoichiometry in order to make it accessible to operations based on linear algebra. To this end we assume a set of M chemical species $\mathcal{S} = \{\mathsf{X}_1, \mathsf{X}_2, \ldots, \mathsf{X}_\mathrm{M}\}$ which are interconverted by K chemical reactions $\mathbf{R}_1, \mathbf{R}_2, \ldots, \mathbf{R}_\mathrm{K}$. It is useful to define a row vector of species, namely, $\mathbf{X} = (\mathsf{X}_1, \mathsf{X}_2, \ldots, \mathsf{X}_\mathrm{M})$. Each individual chemical reaction $\mathbf{R}_j$

$$\sum_{i=1}^{\mathrm{M}} \nu_{ij}\mathsf{X}_i \rightarrow \sum_{i=1}^{\mathrm{M}} \nu'_{ij}\mathsf{X}_i \tag{4.23}$$

is characterized by two column vectors containing the stoichiometric coefficients $\boldsymbol{\nu}_j = \left(\nu_{1j}, \nu_{2j}, \ldots, \nu_{\mathrm{M}j}\right)^{\mathrm{t}}$ and $\boldsymbol{\nu}'_j = \left(\nu'_{1j}, \nu'_{2j}, \ldots, \nu'_{\mathrm{M}j}\right)^{\mathrm{t}}$ of reactants and products, respectively. Now we can write the stoichiometric equation of reaction $\mathbf{R}_j$ (4.23) in the compact form

$$\mathbf{R}_j : \ \mathbf{X} \cdot \boldsymbol{\nu}_j \rightarrow \mathbf{X} \cdot \boldsymbol{\nu}'_j \quad \text{and} \quad \mathbf{X} \cdot (\boldsymbol{\nu}'_j - \boldsymbol{\nu}_j) = \mathbf{X} \cdot \mathbf{s}_j \ . \tag{4.23'}$$

A linear combination of species as defined by the stoichiometry of a chemical reaction is characterized as a *reaction complex* (Sect. 4.1.3)[19]: $\mathsf{C}_j = \mathbf{X} \cdot \boldsymbol{\nu}_j$ and $\mathsf{C}_{j'} = \mathbf{X} \cdot \boldsymbol{\nu}'_j$ being the reactant complex and the product complex of the reaction $\mathbf{R}_j$, respectively. The stoichiometric coefficients of all N complexes appearing in a chemical reaction network together form the M×N *matrix of complex compositions*:

$$\mathsf{C} = \left(\mathbf{X} \cdot \boldsymbol{\nu}_1 \ \ \mathbf{X} \cdot \boldsymbol{\nu}_2 \ \ \ldots \ \ \mathbf{X} \cdot \boldsymbol{\nu}_\mathrm{N}\right) \ .$$

[19]The notion of *reaction complex* needs clarification, since it is different from an association complex like the enzyme–substrate complex in the Michaelis–Menten reaction. A reaction complex is a combination of molecules in the correct stoichiometric ratio as it appears at the reactant side or at the product side of a stoichiometric equation.

As indicated already in (4.23′), we combine the stoichiometric vectors belonging to the reactants and the products of the same reaction, counting reactant coefficients as negative in order to provide a measure of the change introduced by the reaction. The stoichiometry of the entire reaction network is properly encapsulated in the $\mathrm{M} \times \mathrm{K}$ *stoichiometric matrix*:

$$\mathrm{S} = (\mathbf{s}_1, \mathbf{s}_2, \ldots, \mathbf{s}_\mathrm{K}) = \{s_{ij},\ i = 1, \ldots, \mathrm{M},\ j = 1, \ldots, \mathrm{K}\}\ . \tag{4.24}$$

The stoichiometric matrix allows for a compact form of the kinetic differential equations and their solutions:

$$\frac{\mathrm{d}\mathbf{x}(t)}{\mathrm{d}t} = \mathrm{S} \cdot \mathbf{v}\ , \qquad \mathbf{x}(t) - \mathbf{x}_0 = \sum_{j=1}^{\mathrm{K}} \left(\int_0^t v_j\big(\mathbf{x}(\tau)\big)\, \mathrm{d}\tau \right) \mathbf{s}_j\ , \tag{4.25}$$

where $\mathbf{v} = \Big(v_1\big(\mathbf{x}(t)\big), v_2\big(\mathbf{x}(t)\big), \ldots, v_\mathrm{K}\big(\mathbf{x}(t)\big)\Big)^\mathrm{t}$ is the vector of reaction rates, here mass action rates $v^{(\mathrm{ma})}$ according to (4.3), the variables are concentrations described by a vector $\mathbf{x}(t) = \big(x_1(t), x_2(t), \ldots, x_\mathrm{M}(t)\big) \in \mathbb{R}^\mathrm{M}$ with $x_i(t) = [\mathsf{X}_i(t)]$ the concentration of compound X_i at time t, and $\mathbf{x}_0 = \big(x_1(0), x_2(0), \ldots, x_\mathrm{M}(0)\big)$ is the vector of initial concentrations. Equation (4.25) is the straightforward extension of (4.9) to an arbitrary number of reactions.

A number of restrictions apply to chemical kinetics:

(i) Concentrations are positive real numbers[20] $x_j(t) \in \mathbb{R}_{>0}\ ,\ \forall j = 1, \ldots, \mathrm{M}$.
(ii) The solutions have to satisfy the stoichiometric relations for all reactions $\mathbf{R}_j$ ($j = 1, \ldots, \mathrm{K}$), and this is encapsulated in the restriction to *stoichiometric compatibility classes*.

We define the *stoichiometric subspace* of a reaction system by

$$\mathbf{S} = \mathrm{span}\big\{\mathbf{s}_j \,|\, j = 1, \ldots, \mathrm{K}\big\} \subset \mathbb{R}^\mathrm{M}\ , \qquad R \doteq \dim(\mathbf{S})\ . \tag{4.26}$$

The stoichiometric compatibility class contains the stoichiometric subspace shifted by some constant vector, i.e., $\mathbf{c} + \mathbf{S}$, and we restrict the variables to positive values of the concentrations of the form

$$\mathcal{D} = \Big(\mathbf{c} + \mathrm{span}\big\{\mathbf{s}_j \,|\, j = 1, \ldots, \mathrm{K}\big\}\Big) \cap \mathbb{R}^\mathrm{M}_{>0} = (\mathbf{c} + \mathbf{S}) \cap \mathbb{R}^\mathrm{M}_{>0}\ . \tag{4.27}$$

Figure 4.7 shows a simple example of a 1D compatibility class embedded in a 3D concentration space. Since the linear span is built from all reaction vectors $\mathbf{s}_j$, linear dependencies will occur in most cases. The number of independent vectors

[20] In chemistry concentrations of molecular species are commonly required to be positive quantities, whereas extinction corresponding to concentration zero is often an important issue in biology. Then, *positive* has to be replaced only by *nonnegative*, i.e., $\mathbb{R}_{>0} \rightarrow \mathbb{R}_{\geq 0}$.

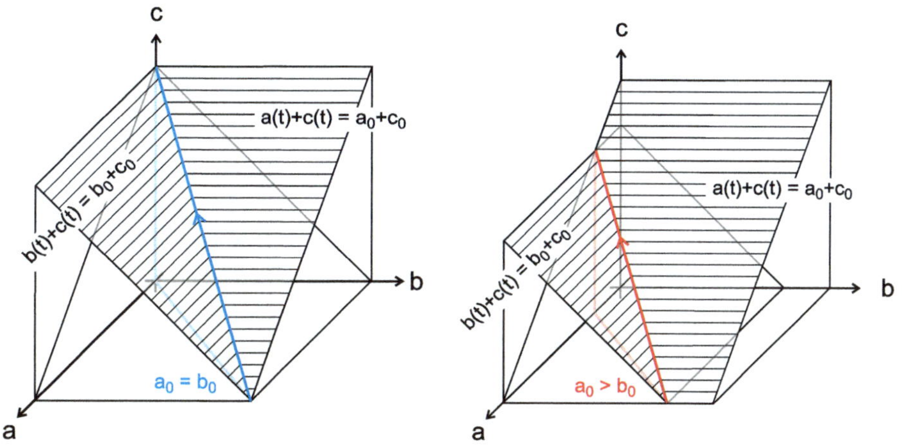

Fig. 4.7 Stoichiometric subspace and compatibility class. The *right-hand side* shows the stoichiometric subspace $\mathbf{S} = \text{span}_j\{\mathbf{s}_j\}$ of the irreversible reaction $\mathsf{A} + \mathsf{B} \rightarrow \mathsf{C}$. The concentration space $X = \{a, b, c\} \in \mathbb{R}^3$ is 3D, and two independent conservation relations, viz., $a(t) = a_0 + c_0 - c(t)$ and $b(t) = b_0 + c_0 - c(t)$, introduce linear dependencies, so the stoichiometric subspace is in fact 1D. The stoichiometric compatibility class is formed by adding a constant vector $\mathbf{c} \in \mathbb{R}^{\mathrm{M}}$, such as the initial conditions $\mathbf{x}_0 = (a_0, b_0, c_0)$, to the stoichiometric subspace: $\mathbf{x}_0 + \mathbf{S}$. The two initial conditions applied here are: (i) $\mathbf{x}_0 = (a_0, b_0 = a_0, 0)$ shown on the left-hand side, and (ii) $\mathbf{x}_0 = (a_0, b_0 < a_0, 0)$ on the right-hand side

in $\text{span}_j(\mathbf{s}_j)$, the dimension or the rank R of the stoichiometric subspace, is the number of independent concentration variables or the number of degrees of freedom in the kinetic reaction system. The rank R of the stoichiometric matrix represents the number of degrees of freedom of the kinetic system and is either determined analytically or computed by routine software. For small systems, like the examples presented here, it is useful and instructive to reduce the degrees of freedom by means of easy to find conservation relations, but for larger systems with several hundred variables and more, a stable numerical procedure is commonly to be preferred.

Chemical Reaction Networks

The notion of a chemical reaction network stands at the center of the reaction network theory. Each network consists of three usually finite sets of objects:

(i) A set of M molecular species, $\mathcal{S} = \{\mathsf{X}_1, \mathsf{X}_2, \ldots, \mathsf{X}_{\mathrm{M}}\}$, which interact through a finite number of chemical reactions.
(ii) A set of N complexes, $\mathcal{C} = \{\mathsf{C}_1, \mathsf{C}_2, \ldots, \mathsf{C}_{\mathrm{N}}\}$, which are linear combinations of species, i.e., $\mathsf{C}_j = \sum_{i=1}^{\mathrm{M}} \nu_{ij}\mathsf{X}_i$, with $\nu_{ij} \in \mathbb{N}$.
(iii) A set of K molecular reactions $\mathcal{R} = \{\mathbf{R}_1, \mathbf{R}_2, \ldots, \mathbf{R}_{\mathrm{K}}\}$, with $\mathcal{R} \subset \mathcal{C} \times \mathcal{C}$ in the sense of individual elements being directed combinations of two complexes: $(\mathsf{C}_{\mathrm{R}}, \mathsf{C}_{\mathrm{P}}) \in \mathcal{R}$ is written as $\mathsf{C}_{\mathrm{R}} \rightarrow \mathsf{C}_{\mathrm{P}}$, where R and P stand for *reactants* and *products*, respectively.

Restrictions are imposed on the sets $\mathcal{S}$ and $\mathcal{C}$: each element of $\mathcal{S}$ has to be found in at least one reaction complex or, in other words, there are no superfluous species. Condition (iii) is supplemented by two exclusions: no complex may react with itself, i.e., $\mathsf{C}_\mathrm{R} \neq \mathsf{C}_\mathrm{P}$, and isolated complexes are not allowed, in the sense that every element of $\mathcal{C}$ must be the reactant or the product complex of some reaction. It is worth remembering that a reversible reaction (see, e.g., Sect. 4.3.2) is represented by two reactions: $\mathsf{C}_\mathrm{R} \rightarrow \mathsf{C}_\mathrm{P}$ and $\mathsf{C}_\mathrm{P} \rightarrow \mathsf{C}_\mathrm{R}$.

The above-mentioned restriction can be cast in a somewhat different form, presented here in order to clarify the definitions. Complexes and species are related through:

(i) $\mathcal{C} \subset \mathbb{R}^{\mathcal{S}}$, where $\mathbb{R}^{\mathcal{S}}$ stands for a vector space spanned by unit vectors representing individual species. Commonly, the coefficients in the linear combinations of species called complexes are natural numbers, $\nu_{ij} \in \mathbb{N}$. These can be zero, but then the species is not considered to be part of the complex,

(ii) The union $\bigcup_{\mathrm{supp},\, \mathsf{C}_j \in \mathcal{C}} \mathsf{C}_j = \mathcal{S}$ of the species in all complexes is the species set, and no species can exist in $\mathcal{S}$ which does not appear in at least one complex.[21]

Species X_i and reactions $\mathbf{R}_j$ are directly related by the stoichiometric matrix $\mathrm{S} = \{s_{ij} = \nu'_{ij} - \nu_{ij}\}$. The columns of S refer to reactions and the rows to species. We shall also make use of S in Sect. 4.6 to implement a simulation tool for chemical master equations.

The fourth component of a reaction system is the kinetics $\mathcal{K}$ of the reactions. Mass action kinetics ($v^{(\mathrm{ma})}$) has been discussed in Sect. 4.1.1 and Michaelis–Menten kinetics ($v^{(\mathrm{mm})}$) as an example of higher-level kinetics in Sect. 4.1.2. In the majority of the examples discussed here, mass action will be applied. We repeat the basic equation (4.2) for reaction $\mathbf{R}_j$:

$$s_{1j}\mathsf{X}_1 + s_{2j}\mathsf{X}_2 + \cdots \Longrightarrow v_j = k_j[\mathsf{X}_1]^{s_{1j}}[\mathsf{X}_2]^{s_{2j}} \ldots = k_j x_1^{s_{1j}} x_2^{s_{2j}} \ldots . \tag{4.28}$$

In mass action kinetics $v^{(\mathrm{ma})}$, we need one reaction parameter k_j for every elementary step, so the number of rate parameters is equal to K, the number of reactions.[22] So finally, a reaction system consists of the four components $\{\mathcal{S}, \mathcal{C}, \mathcal{R}, \mathcal{K}\}$, and the evolution in time of the reaction system can be encapsulated in an ODE or in a master equation in the case of a stochastic description.

[21]The notion supp stands for the support of a vector, which is the subset of unit vectors for which the vector has nonzero coefficients.

[22]In order to make the notation clearer for reversible reactions, we use two symbols and the same index for both reactions, i.e., k_j and l_j for the forward and the reverse reaction, respectively.

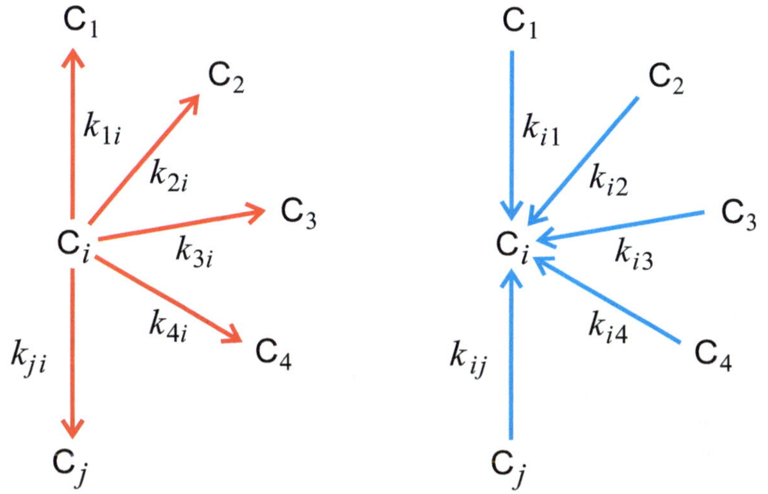

Fig. 4.8 Complex balancing. Balancing of complexes is achieved when the inflow into every complex C_i (*blue*) is precisely compensated by the outflow from it (*red*). Complex balancing is a relaxation of the constraint of detailed balance that requires all individual reaction steps to be at equilibrium

Stationary States

It is important to distinguish two kinds of stationarity: (i) equilibria with detailed balance and (ii) complex-balanced equilibria. Detailed balance follows from statistical thermodynamics and implies that the flow for every individual reaction step $\mathbf{R}_j$ vanishes at equilibrium [531]: $v_j^{(\mathrm{ma})} = 0 = k_j \bar{\mathbf{x}}^{\boldsymbol{\nu}_j} - l_j \bar{\mathbf{x}}^{\boldsymbol{\nu}_j{}'}$, $\forall j = 1, \ldots, \mathrm{K}$. The limit of chemical kinetics is realized at thermodynamic equilibrium (3.100). The weaker condition of complex balancing [151, 262, 416] requires that, for all complexes, the net inflow into a complex C_i is compensated by the net outflow (Fig. 4.8):

$$C_i : \frac{d[C_i]}{dt} = 0 \,, \quad \text{or} \quad \sum_{j,j \neq i}^{N} k_{ij} \widehat{\mathbf{x}}^{\boldsymbol{\nu}_j} = \widehat{\mathbf{x}}^{\boldsymbol{\nu}_i} \sum_{j,j \neq i}^{N} k_{ji} \,, \; \forall\, C_i \in \mathcal{C} \,. \tag{4.29}$$

For the definition and illustration of complex balancing, we apply here a different notation of the rate parameter. For the reaction $C_i \to C_j$, we use k_{ji}. In order to facilitate the distinction, equilibrium concentrations are indicated by an overbar and stationary concentrations obtained from complex balancing by a hat.

Reaction Graphs

Some general properties of reaction networks can be predicted directly from the reaction graph (Fig. 4.9), which is a directed graph containing the *complexes* $C_k \in$

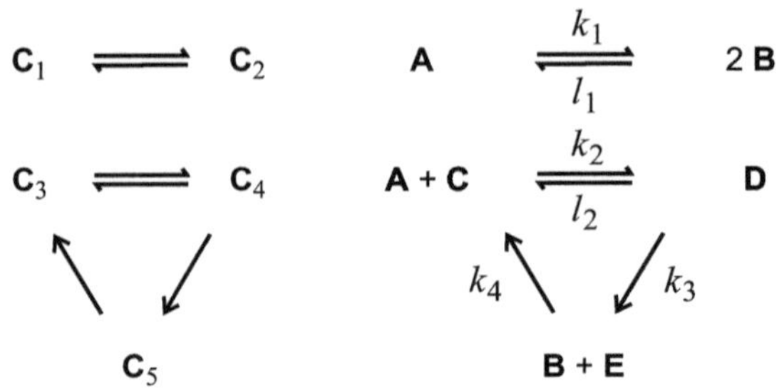

Fig. 4.9 The graph corresponding to the chemical reaction network (4.32a). Each node of the graph (*left*) corresponds to a reaction complex. Three different symbols characterize the directed edges: $\rightarrow$, $\leftarrow$, and $\rightleftharpoons$ for forward, backward, and reversible reaction, respectively. This graph consists of $L = 2$ linkage classes. On the *right-hand side* we show the Feinberg mechanism, which is an implementation of the reaction graph on the *left-hand side*. The mechanism differs from the graph by additional information: (i) the molecular realization of the reaction complexes and (ii) the rate parameters

$\mathcal{C}$ $(k = 1, \ldots, \mathrm{N})$ as nodes and three symbols indicating forward ($\rightarrow$), backward ($\leftarrow$), and reversible ($\rightleftharpoons$) reactions as edges. A reaction graph may have several components called *linkage classes*. Different linkage classes have no common node and no edge connecting them. Two properties are important for reaction graphs: (i) every complex appears only once as a node of the graph and (ii) different linkage classes do not share complexes.

The network in Fig. 4.9 has two linkage classes since the two clusters don't share a single complex. The information on the number of complexes and the number of linkage classes is contained in the reaction graph. The same is true for the classification of a network as reversible, weakly reversible, or not reversible. A (strongly) reversible network contains exclusively reversible reactions in the strict thermodynamic sense. Weak reversibility relaxes the condition of (*strong*) *reversibility*: a network is weakly reversible when, for every pair of complexes, there exists a directed arc leading from one complex to the other. The network in Fig. 4.9 satisfies the condition of weak reversibility, while for strong reversibility it has to be supplemented by the arrows $C_3 \rightarrow C_5$ and $C_5 \rightarrow C_4$. For the determination of linkage classes, only the existence or absence of arrows between complexes matters. Clearly, the direction of arrows is required for the classification of reversibility.

A reaction graph differs from a reaction mechanism in three respects:

(i) The reaction complexes are not defined in terms of chemical compounds and therefore the reaction graph does not consider stoichiometry.
(ii) The reaction graph does not specify the algebraic relations of reaction rates in the form of mass action, Michaelis–Menten, or other frequency functions.

(iii) The reaction graph does not contain weighting factors of edges in the sense of rate parameters.

The reaction graph represents nothing more than the topology of a reaction network and general properties derived from the graph are valid for a large number of specific cases, irrespective of stoichiometries, frequency functions, and rate constants.

Examples of Reactions and Networks
We illustrate chemical reaction network theory by means of a few examples.

The Irreversible Association Reaction $\mathbf{A}+\mathbf{B}\rightarrow\mathbf{C}$
The first example is the irreversible association reaction[23] of (4.1f):

$$\mathsf{A}+\mathsf{B}\xrightarrow{\quad k\quad}\mathsf{C}\,. \tag{4.30a}$$

For the three sets of the chemical reaction network we have

$$\mathcal{S}=\{\mathsf{A},\mathsf{B},\mathsf{C}\}\,, \tag{4.30c}$$

$$\mathcal{C}=\{\mathsf{C}_1=\mathsf{A}+\mathsf{B},\mathsf{C}_2=\mathsf{C}\}\,, \tag{4.30d}$$

$$\mathcal{R}=\{\mathbf{R}_1=\mathsf{C}_1\rightarrow\mathsf{C}_2\}\,. \tag{4.30e}$$

The stoichiometric matrix S is of dimension 3×1:

$$\mathrm{S}=\begin{pmatrix}-1\\-1\\+1\end{pmatrix}. \tag{4.30f}$$

In deterministic mass action kinetics $v^{(\mathrm{ma})}$, the variables are the concentrations of the molecular species, i.e., $[\mathsf{A}]=a(t)$, $[\mathsf{B}]=b(t)$, and $[\mathsf{C}]=c(t)$. In order to solve the kinetic differential equation we require a rate parameter k and three initial conditions $a(0)=a_0$, $b(0)=b_0$, and $c(0)=c_0$. The three variables are stoichiometrically related by two conservation relations derived from (4.30a), which can be used to eliminate two variables, $b(t)$ and $c(t)$, for example, yielding the remaining single

[23]In narrative chemical kinetics, distinctions are made for notions concerning the association–dissociation reaction $\mathsf{A}+\mathsf{B}\rightleftharpoons\mathsf{C}$ that are synonyms in formal kinetics: the word *addition* is used when A and B are of similar size and *binding* is preferred for molecules of very different size, for example, a substrate is *bound* to an enzyme.

degree of freedom as $\frac{\mathrm{d}a}{\mathrm{d}t} = \frac{\mathrm{d}b}{\mathrm{d}t} = -\frac{\mathrm{d}c}{\mathrm{d}t}$ corresponding to $R = 1$ (see Fig. 4.7):

$$a(t) + c(t) = a_0 + c_0 = \vartheta_0^{(ac)} \,,$$

$$b(t) + c(t) = b_0 + c_0 = \vartheta_0^{(bc)} \,,$$

$$b(t) - a(t) = b_0 - a_0 = \vartheta_0^{(b)} \,.$$

One of these three conditions is dependent, since the second line minus the first line yields the third. Eventually, one finds

$$\frac{\mathrm{d}a}{\mathrm{d}t} = -kab = -ka\big(\vartheta_0^{(b)} + a\big) \,. \tag{4.30g}$$

The ODE is solved by standard techniques and we obtain the following solutions by direct integration:

$$a(t) = \frac{a_0\vartheta_0^{(b)}\exp(-\vartheta_0^{(b)}kt)}{\vartheta_0 + a_0\Big(1 - \exp(-\vartheta_0^{(b)}kt)\Big)} \,, \quad \text{for } \vartheta_0^{(b)} > 0 \,, \ (b_0 > a_0) \,,$$

$$a(t) = \frac{a_0|\vartheta_0^{(b)}|}{a_0 - \big(a_0 - |\vartheta_0^{(b)}|\big)\exp(-|\vartheta_0^{(b)}|kt)} \,, \quad \text{for } \vartheta_0^{(b)} < 0 \,, \ (b_0 < a_0) \,, \tag{4.30h}$$

$$a(t) = \frac{a_0}{1 + a_0kt} \,, \quad \text{for } \vartheta_0^{(b)} = 0 \,, \ (b_0 = a_0) \,.$$

The three cases differ in their long-time behavior:

(i) $\lim_{t\to\infty} a(t) = 0$ for $\vartheta_0^{(b)} \geq 0$, $b_0 > a_0$.
(ii) $\lim_{t\to\infty} a(t) = |b_0 - a_0|$ for $\vartheta_0^{(b)} < 0$, $b_0 > a_0$.

The Reversible Bimolecular Conversion Reaction $\mathsf{A} + \mathsf{B} \rightleftharpoons \mathsf{C} + \mathsf{D}$
The second case simply consists of a reversible bimolecular conversion reaction that is decomposed into two elementary reactions of type (4.1i):

$$\mathsf{A} + \mathsf{B} \xrightarrow{\ k\ } \mathsf{C} + \mathsf{D} \,, \tag{4.31a}$$

$$\mathsf{C} + \mathsf{D} \xrightarrow{\ l\ } \mathsf{A} + \mathsf{B} \,. \tag{4.31b}$$

For the three sets of the chemical reaction network, we have

$$\mathcal{S} = \{\mathsf{A}, \mathsf{B}, \mathsf{C}, \mathsf{D}\} , \tag{4.31c}$$

$$\mathcal{C} = \{\mathsf{C}_1 = \mathsf{A} + \mathsf{B},\ \mathsf{C}_2 = \mathsf{C} + \mathsf{D}\} , \tag{4.31d}$$

$$\mathcal{R} = \{\mathbf{R}_1 = \mathsf{C}_1 \to \mathsf{C}_2,\ \mathbf{R}_2 = \mathsf{C}_2 \to \mathsf{C}_1\} . \tag{4.31e}$$

The stoichiometric matrix S is of dimension 4×2:

$$\mathrm{S} = \begin{pmatrix} -1 & +1 \\ -1 & +1 \\ +1 & -1 \\ +1 & -1 \end{pmatrix} . \tag{4.31f}$$

In deterministic mass action kinetics $v^{(\mathrm{ma})}$, the variables are the concentrations of the molecular species, i.e., $[\mathsf{A}] = a(t)$, $[\mathsf{B}] = b(t)$, $[\mathsf{C}] = c(t)$, and $[\mathsf{D}] = d(t)$. In order to solve the kinetic differential equation, we require two rate parameters, k and l, and four initial conditions: $a(0) = a_0$, $b(0) = b_0$, $c(0) = c_0$, and $d(0) = d_0$. The four variables are stoichiometrically related by three conservation relations in (4.31a) and (4.31b):

$$a(t) + b(t) + c(t) + d(t) = a_0 + b_0 + c_0 + d_0 ,$$

$$a(t) - b(t) = a_0 - b_0 ,$$

$$c(t) - d(t) = c_0 - d_0 ,$$

and only one degree of freedom remains, corresponding to the rank $R = 1$ of the stoichiometric matrix: $\mathrm{d}a/\mathrm{d}t = \mathrm{d}b/\mathrm{d}t = -\mathrm{d}c/\mathrm{d}t = -\mathrm{d}d/\mathrm{d}t$. Hence, we can substitute $b(t) = b_0 - a_0 + a(t)$, $c(t) = c_0 + a_0 - a(t)$, and $d(t) = d_0 + a_0 - a(t)$ and the ODE for the last remaining variable $a(t)$ takes the form

$$\begin{aligned} \frac{\mathrm{d}a}{\mathrm{d}t} &= -kab + lcd = -ka\big(\vartheta_0^{(b)} + a\big) + l\big(\vartheta_0^{(c)} - a\big)\big(\vartheta_0^{(d)} - a\big) \\ &= (l-k)a^2 - \big(k\vartheta_0^{(b)} + l\vartheta_0^{(c)} + l\vartheta_0^{(d)}\big)a + l\vartheta_0^{(c)}\vartheta_0^{(d)} , \end{aligned} \tag{4.31g}$$

where the initial conditions are contained in the quantities $\vartheta_0^{(b)} = b_0 - a_0$, $\vartheta_0^{(c)} = c_0 + a_0$, and $\vartheta_0^{(d)} = d_0 + a_0$.

Equation (4.31g) can be integrated by standard methods to yield an implicit solution of the form $t = f(a)$, but the expression is so clumsy that we dispense from listing it here. The analytical solutions for the irreversible forward reaction are identical with the solutions of the association reaction (4.30h) treated in the previous example, since the kinetic ODEs of an irreversible reaction do not depend on the concentrations on the product side. Clearly, the expressions are also valid for the irreversible backward reaction by replacing $a \leftrightarrow c$, $b \leftrightarrow d$, and $k \leftrightarrow l$.

The Feinberg Mechanism

Our third example is taken directly from Feinberg [152, 154] and deals with six elementary reactions involving five chemical species related by the following mechanism (Fig. 4.9):

$$\begin{aligned}
\mathsf{A} &\xrightarrow{k_1} \mathsf{2B}\,,\\
\mathsf{2B} &\xrightarrow{l_1} \mathsf{A}\,,\\
\mathsf{A+C} &\xrightarrow{k_2} \mathsf{D}\,,\\
\mathsf{D} &\xrightarrow{l_2} \mathsf{A+C}\,,\\
\mathsf{D} &\xrightarrow{k_3} \mathsf{B+E}\,,\\
\mathsf{B+E} &\xrightarrow{k_4} \mathsf{A+C}\,.
\end{aligned} \tag{4.32a}$$

The three sets defining the chemical reaction network are:

$$\mathcal{S} = \{\mathsf{A, B, C, D, E}\}\,, \tag{4.32c}$$

$$\mathcal{C} = \{\mathsf{C}_1 = \mathsf{A},\ \mathsf{C}_2 = \mathsf{2B},\ \mathsf{C}_3 = \mathsf{A+C},\ \mathsf{C}_4 = \mathsf{D},\ \mathsf{C}_5 = \mathsf{B+E}\}\,, \tag{4.32d}$$

$$\begin{aligned}
\mathcal{R} = \{&\mathbf{R}_1 = \mathsf{C}_1 \to \mathsf{C}_2,\ \mathbf{R}_2 = \mathsf{C}_2 \to \mathsf{C}_1,\ \mathbf{R}_3 = \mathsf{C}_3 \to \mathsf{C}_4,\\
&\mathbf{R}_4 = \mathsf{C}_4 \to \mathsf{C}_3,\ \mathbf{R}_5 = \mathsf{C}_4 \to \mathsf{C}_5,\ \mathbf{R}_6 = \mathsf{C}_5 \to \mathsf{C}_3\}\,.
\end{aligned} \tag{4.32e}$$

The stoichiometric matrix S for the mechanism (4.32a) is readily obtained:

$$\mathrm{S} = \begin{pmatrix} -1 & +1 & -1 & +1 & +1 & 0 \\ +2 & -2 & 0 & 0 & -1 & +1 \\ 0 & 0 & -1 & +1 & +1 & 0 \\ 0 & 0 & +1 & -1 & 0 & -1 \\ 0 & 0 & 0 & 0 & -1 & +1 \end{pmatrix}\,. \tag{4.32f}$$

It has the dimension 5×6 and its rank is $R = 3$. The reaction graph corresponding to this mechanism is also shown in Fig. 4.9. Comparison between the two graphs provides a nice illustration of a property of reaction graphs mentioned earlier: the graph visualizes only the interconversions between reaction complexes and contains no information about the molecular realization of the kinetic reaction network, whereas the graphical representation of the reaction mechanism contains all the information except for the specific initial conditions. Analytical solutions for the reaction network (4.32a) are not available, but numerical integration for given initial conditions is easily achieved. Some qualitative properties will be derived in the following sections.

Multidimensional Relaxation

Chemical relaxation theory was applied to a single-step reaction in Sect. 4.1.1. It can be readily extended to an arbitrary number of chemical reactions [488]. For K reactions, the elements of the relaxation matrix are of the form

$$\mathrm{A} = \left(a_{ij} = -\sum_{k=1}^{\mathrm{K}} (\nu_{ki} - \nu'_{ki}) \frac{\nu_{kj}(\bar{v}_k)_{\rightarrow} - \nu'_{kj}(\bar{v}_k)_{\leftarrow}}{\bar{x}_j} \right) , \tag{4.33}$$

and we expect to find more than one relaxation mode in the approach towards equilibrium, corresponding to more than one relaxation time. In vector notation with $\boldsymbol{\eta} = \mathbf{x} - \bar{\mathbf{x}}$ and the thermodynamic equilibrium as reference state as before, the relaxation equation is of the form

$$\frac{\mathrm{d}\boldsymbol{\eta}}{\mathrm{d}t} = \mathrm{A}\boldsymbol{\eta} , \qquad \boldsymbol{\eta}(t) = \exp(\mathrm{A}t)\boldsymbol{\eta}(0) . \tag{4.7$'$}$$

The formal exponential function of matrix A in (4.7′) is readily solved by means of an eigenvalue problem. It is instructive to symmetrize the matrix A by means of a similarity transformation, viz.,

$$\mathrm{D} = \mathrm{G}\mathrm{A}\mathrm{G}^{-1} , \quad \text{with } \mathrm{G} = \left(g_{ij} = \frac{\delta_{ij}}{\sqrt{\bar{x}_i}} \right) ,$$

$$\mathrm{D} = \left(d_{ij} = -\sum_{k=1}^{\mathrm{K}} \frac{(\nu_{ki} - \nu'_{ki})(\nu_{kj} - \nu'_{kj})\bar{v}_k}{\sqrt{\bar{x}_i \bar{x}_j}} \right) ,$$

since $(\bar{v}_k)_{\rightarrow} = (\bar{v}_k)_{\leftarrow} = \bar{v}_k$ at equilibrium. The matrices D and A have the same eigenvalues and the fact that A can be transformed to a symmetric matrix has the consequence that all its eigenvalues are real. Simple numerical diagonalization routines can thus be applied. For the original matrix A, the diagonalization yields

$$\Lambda = \mathrm{B}^{-1}\mathrm{A}\mathrm{B} , \quad \mathrm{A}\mathrm{B} = \mathrm{B}\Lambda , \quad \text{with } \Lambda = \begin{pmatrix} \tau_1^{-1} & 0 & \dots & 0 \\ 0 & \tau_2^{-1} & \dots & 0 \\ \vdots & \vdots & \ddots & \vdots \\ 0 & 0 & \dots & \tau_n^{-1} \end{pmatrix} .$$

The diagonal matrix Λ contains the eigenvalues and the matrix $\mathrm{B} = \left(b_{ij}\right)$ collects together the eigenvectors:

$$\mathbf{b}_j = (b_{1j}, \ldots, b_{nj})^{\mathrm{t}} \,, \quad \text{with } \mathbf{A}\mathbf{b}_j = \lambda_j \mathbf{b}_j = \frac{1}{\tau_j}\mathbf{b}_j \,,$$

which have the simple exponential time dependence

$$\mathbf{b}_j(t) = \mathbf{b}_j(0)\mathrm{e}^{-t/\tau_j} \,.$$

Expressed in the original variables η_j and using $\mathrm{B}^{-1} = \mathrm{H} = \left(h_{ij}\right)$ for simplicity, the result is

$$\eta_j(t) = \frac{\sum_{k=1}^{n} b_{jk}\beta_k(0)\exp(-t/\tau_k)}{\sum_{i=1}^{n}\sum_{k=1}^{n} b_{ik}\beta_k(0)\exp(-t/\tau_k)} \,, \quad \text{with } \beta_k(0) = \sum_{l=1}^{n} h_{kl}\eta_l(0) \,. \tag{4.34}$$

An example for the superposition of three relaxation curves is shown in Fig. 4.10. Since the rank R of the reaction network is commonly smaller than the number of

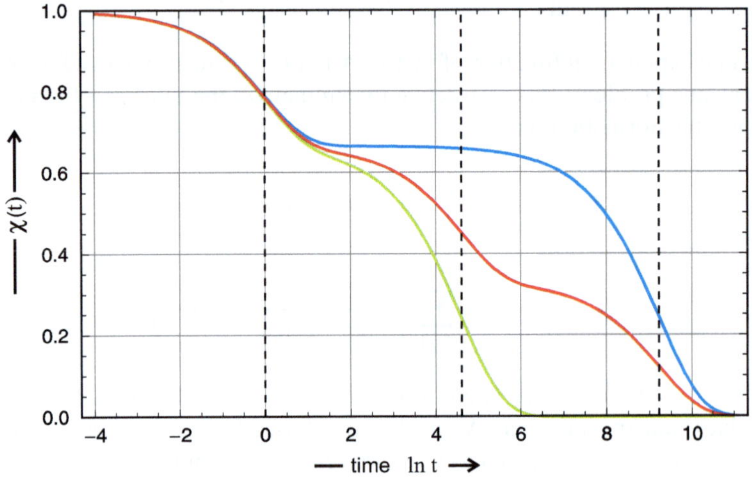

Fig. 4.10 Relaxation times from multimode relaxation. Relaxation times are readily detected as points of inflection in the plot of $\eta(t)$ against $\ln t$. Multiple relaxation times are easily found when the relaxation processes appear well separated on the time axis. The three curves show two relaxation processes separated by factors of 100 (*green curve*) or 10,000 (*blue curve*) on the time axis, and three relaxations with relaxation times 1/100/10,000 (*red curve*). All amplitudes are chosen 1/3 or 2/3 (second process in the *green* and the *blue curve*). The time scale is $\ln t$. In cases where the individual processes are not so well separated on the time axis, the problem of calculating the relaxation times may be ill-posed [4, p. 252] (see also Sect. 4.1.5)

chemical species M, some of the eigenvalues will vanish: $\lambda = 0$. The corresponding eigenvectors then represent conservation relations. Alternatively, the constraints can be used to reduce the number of ODEs.

Definition of Deficiency
First, we repeat the basic definitions of chemical reaction network theory and point out how the relevant properties can be obtained:

(i) A *linkage class* is a subset of complexes that are linked by reactions, and the number of linkage classes is denoted by L.
(ii) A reaction network is *weakly reversible* if and only if a directed arc leads from every complex to every complex of the network.
(iii) The *reaction vectors* combine reactants and products in the stoichiometric way, i.c., $\mathbf{R} = -\mathcal{C}_{\mathrm{R}} + \mathcal{C}_{\mathrm{P}}$.
(iv) The *rank* R of a reaction network is the largest linearly independent set that can be found among its reaction vectors.

The linkage classes of a reaction network are obtained straightforwardly: each complex is displayed exactly once in the sketch of the network, the complexes are joined by introducing the reaction arrows into the sketch, and linkage classes comprise all complexes joined together. The network in Fig. 4.9, for example, has $L = 2$ linkage classes.

Strong and weak reversibility can be seen directly in the reaction graph. In a strongly reversible network, all reactions $\mathbf{R} \in \mathcal{R}$ are reversible:

$$(\mathsf{C}_j \to \mathsf{C}_k \in \mathcal{R}) \implies (\mathsf{C}_k \to \mathsf{C}_j \in \mathcal{R}) \,, \quad \forall\, (\mathsf{C}_j, \mathsf{C}_k) \in \mathcal{C} \,. \tag{4.35}$$

Weak reversibility relaxes the condition for strong reversibility in the sense that it is sufficient to be able to reach every species from every species by a sequence of reactions. The network in Fig. 4.9 is weakly reversible.

The rank of a chemical reaction network is defined as

$$R \doteq \mathrm{rank}\Big\{\mathsf{C}_{\mathrm{P}} - \mathsf{C}_{\mathrm{R}} \in \mathbb{R}^{\mathcal{S}} : \mathsf{C}_{\mathrm{R}} \to \mathsf{C}_{\mathrm{P}} \in \mathcal{R}\Big\} \,. \tag{4.36}$$

We illustrate by means of a simple example. The six reaction vectors of the network (4.32a), viz.,

$$\Big\{2\mathsf{B} - \mathsf{A},\ \mathsf{A} - 2\mathsf{B},\ \mathsf{D} - (\mathsf{A} + \mathsf{C}),\ (\mathsf{A} + \mathsf{C}) - \mathsf{D},\ (\mathsf{B} + \mathsf{E}) - \mathsf{D},\ (\mathsf{A} + \mathsf{C}) - (\mathsf{B} + \mathsf{E})\Big\} \,,$$

can be contracted to a linearly independent subset of dimension three:

$$\left\{2\mathrm{B} - \mathrm{A},\ (\mathrm{A} + \mathrm{C}) - \mathrm{D},\ (\mathrm{B} + \mathrm{E}) - \mathrm{D}\right\} .$$

Although the network (4.32a) consists of six reactions, only three of them are linearly independent, and accordingly it has rank $R = 3$. It is straightforward to see that every reversible reaction consists of two reactions, but only one of them can be linearly independent. The determination of the rank R in small systems is properly done by means of the conservation relations, but for larger systems a numerical computation of the rank of the stoichiometric matrix S is usually much faster.

The most important quantity in reaction network theory is the *deficiency* of a reaction system, which is defined by

$$\text{Deficiency } \delta \doteq \mathrm{N} - L - R , \tag{4.37}$$

where N is the number of complexes, L the number of linkage classes, and R the number of degrees of freedom or the rank of the reaction kinetics.

The deficiency of a chemical reaction network is a nonnegative quantity [153] and it determines essential features of the reaction system like the existence of unique equilibria and stationary states.

The Deficiency Zero Theorem
The deficiency zero theorem holds for all chemical reaction networks $\{\mathcal{S}, \mathcal{C}, \mathcal{R}\}$ that have deficiency zero, and it makes three statements [153]:

(i) If the network is not weakly reversible, then the ODEs for the reaction system $\{\mathcal{S}, \mathcal{C}, \mathcal{R}, \mathcal{K}\}$ with arbitrary kinetics $\mathcal{K}$ cannot admit a positive equilibrium, i.e., a stationary point in $\mathbb{R}^{\mathrm{M}}_{>0}$.
(ii) If the network is not weakly reversible, then the ODEs for the reaction system $\{\mathcal{S}, \mathcal{C}, \mathcal{R}, \mathcal{K}\}$ with arbitrary kinetics $\mathcal{K}$ cannot admit a cyclic trajectory containing a positive composition, i.e., a point in $\mathbb{R}^{\mathrm{M}}_{>0}$.
(iii) If the network is weakly reversible (or reversible) then, for any mass action kinetics $v^{(\mathrm{ma})} = \kappa \in \mathbb{R}^{R}_{>0}$, the ODEs for the mass action system $\{\mathcal{S}, \mathcal{C}, \mathcal{R}, \kappa\}$ have the following properties: within each positive stoichiometric compatibility class, there exists exactly one equilibrium, this equilibrium is asymptotically stable, and there cannot exist a nontrivial cyclic trajectory in $\mathbb{R}^{\mathrm{M}}_{>0}$.

The third property is a highly important extension of equilibrium thermodynamics because existence and uniqueness of a stable equilibrium in the interior of the positive orthant of concentration space is extended from strictly reversible to weakly reversible systems, and from closed systems to closed and open systems of deficiency zero. It is worth stressing once again that the statements hold for arbitrary finite dimensions of the reaction system, irrespective of the particular choice of rate parameters, provided they are nonnegative.

The Deficiency One Theorem

The results of the deficiency one theorem hold for a much wider class of networks than those with deficiency zero. The extension of the range of validity is encapsulated in the *deficiency one theorem*. For the formulation of the theorem, it is important to extend the notion of deficiency to individual linkage classes, which are denoted by $\mathcal{L} = \{\mathbf{L}_1, \mathbf{L}_2, \ldots, \mathbf{L}_L\}$. The number of complexes in linkage class $\mathbf{L}_j$ is denoted by N_j, and since a complex can appear in only one linkage class, we have $\sum_{j=1}^{L} \mathrm{N}_j = \mathrm{N}$. The number of independent degrees of freedom of the ODE or the rank of a linkage class $\mathbf{L}_j$ is denoted by R_j:

$$R_j \doteq \mathrm{rank}\Big\{\mathsf{C}_\mathrm{P} - \mathsf{C}_\mathrm{R} \in \mathbb{R}^{\mathcal{S}} : \mathsf{C}_\mathrm{R} \to \mathsf{C}_\mathrm{P} \in \mathcal{R} \wedge \mathcal{C}_\mathrm{R} \in \mathbf{L}_j\Big\} .$$

We make the following definition:

$$\text{Deficiency of class } \mathbf{L}_j : \quad \delta_j = \mathrm{N}_j - 1 - R_j . \tag{4.38}$$

The class deficiency δ_j is a nonnegative integer like δ. The ranks of the subsystems need not be additive but they satisfy $\sum_{j=1}^{L} R_j \geq R$, and this yields for the deficiency of the total network

$$\delta \geq \sum_{j=1}^{L} \delta_j = \mathrm{N} - L - \sum_{j=1}^{L} R_j . \tag{4.37'}$$

It is instructive to consider zero deficiency networks because they are precisely those networks that satisfy both conditions:

$$\delta_j = 0 , \ \forall j = 1, 2, \ldots, L , \qquad \delta = \sum_{j=1}^{L} \delta_j = 0 .$$

We are now in a position to introduce the deficiency one theorem [153].

Let $\{\mathcal{S}, \mathcal{C}, \mathcal{R}\}$ be a reaction network with L linkage classes, let the deficiency of the network be denoted by $\delta = \mathrm{N} - L - R$, and let the deficiencies of the individual linkage classes be denoted by $\delta_j = \mathrm{N}_j - 1 - R_j, j = 1, \ldots, L$, and then assume that the following two conditions are satisfied:

$$\delta_j \leq 1\,, \;\; \forall j = 1, 2, \ldots, L\,, \;\; \text{and } \delta = \sum_{j=1}^{L} \delta_j\,.$$

If the network is weakly reversible and, in particular, if it is strongly reversible, then the ODEs for the mass action system $\{\mathcal{S}, \mathcal{C}, \mathcal{R}, \kappa\}$ sustain precisely one equilibrium inside each positive stoichiometric compatibility class for any mass action kinetics $\kappa \in \mathbb{R}^R_{>0}$.

For networks with just one linkage class the theorem states that multiple steady states within a stoichiometric compatibility class are sustained only if the deficiency of the network exceeds one $\delta > 1$.

Concerning multiple steady states of mass action systems, the deficiency one theorem is much more general than the deficiency zero theorem, and in this sense it is a true extension of the latter.

Thus, the deficiency one theorem is a powerful tool for recognizing reaction systems lacking multiple stationary states. In later work, the existence of multiple stationary states came into focus [93, 155], and these studies make a bridge between applications in chemistry and applications in biology. We shall come back to reaction systems with multiple steady states and complex dynamics in Chap. 5.

4.1.4 Theory of Reaction Rate Parameters

Although this monograph aims to analyze the problems of stochasticity without touching upon the question where the input parameters for the models come from, we make one exception in the case of chemical kinetics. The reason is twofold:

(i) In this case, there exists a well developed theory that allows one to trace the chemical kinetics down to first principles from theoretical physics.
(ii) At the same time, chemical reaction kinetics is built upon a true wealth of empirical data that provide a unique testing ground for stochastic models.

For a comprehensive understanding of chemical reactions, the solutions of kinetic differential equations or chemical master equations have to be complemented by detailed knowledge of the processes at the molecular level. In particular, we need the frequencies or probabilities $\pi(t, \mathrm{d}t)$ in order to quantify the event that a reactant molecule or a reaction complex C, which has been randomly selected at time t, will react and yield the products of some reaction **R** within the next infinitesimal time interval $[t, t + \mathrm{d}t]$. Under two general assumptions, viz.,

1. spatial homogeneity assumed to be achieved by fast mixing,
2. thermal equilibrium,

virtually all chemical reactions satisfy the condition

$$\pi(t, \mathrm{d}t) = \gamma \, \mathrm{d}t \,, \tag{4.39}$$

where γ is the reaction-specific *deterministic* or *probabilistic rate parameter*.[24] If γ is independent of t, π is simply proportional to dt. The two basic conditions (1) and (2) are likewise satisfied for chemical reactions in the vapor phase and in dilute aqueous solutions. The rate parameter is a function of external conditions like temperature and pressure, i.e., $\gamma = \gamma(T, p, \ldots)$. In solution, in particular in aqueous solution, other external parameters like pH, ionic strength, viscosity, and others are important.

The task here is to design a theory that allows us to derive rate parameters and their dependence on external parameters like temperature and pressure from first principles of physics and firm empirical data. Such an ambitious undertaking has been successful, or is at least promising, for some disciplines of physics as well as for chemical kinetics, but such a background theory does not exist for most other probabilistic concepts, and in particular not for most of biology, sociology, or economics. The rate parameters of chemical kinetics and their dependence on external parameters can be deduced, in principle, from quantum mechanics. Here we present a brief digression into the molecular theory of reaction rates in order to illustrate how rate parameters originate from a physical background. Although the rigorous approaches were conceived and tested for reactions in dilute gases, they are with some modifications regularly and successfully applied to reactions in solutions.

We start by considering two model equations for the calculation of rate parameters:

(i) The empirical Arrhenius equation, derived in the nineteenth century.
(ii) The Eyring equation, based on quantum mechanics, and in particular *transition state theory*.

[24] The probabilistic rate parameter γ is almost always identical to the conventional deterministic parameter k, and we shall assume that γ can be interchanged with k whenever necessary.

Then we give an overview of rate parameter calculations from the collision theory of chemical reactions, and finally, glance at reactive scattering in the *semiclassical* and quantum mechanical approach. In classical mechanics, the motions of particles satisfy Newton's laws, whereas in quantum mechanics, particles are described by the quantum wavefunction, which is a solution of Schrödinger's equation. The full quantum mechanical calculations are consistent, but require expensive computer time, so are tractable only in the simplest cases. An approach is said to be semiclassical if one part of a system is described by quantum mechanics while the rest is modeled classically. In the semiclassical theory of chemical reactions, quantum mechanics is used to describe molecules and molecular interactions, whereas classical trajectories are used to describe the reaction. In an advanced form, each trajectory is given a quantum phase so that quantum effects such as interference and tunneling can be described using only classical information.

Model Equations

Two equations modeling the temperature dependence of rate parameters have found widespread application: the empirical Arrhenius equation, obtained in the nineteenth century, and the Eyring equation which results from the *transition state theory*.

The Arrhenius Equation

For reaction **R** the probabilistic rate parameter and its temperature dependence are given by

$$\gamma(T) = k(T) = A \exp\left(-\frac{\mathrm{e_a}}{k_\mathrm{B} T}\right) , \tag{4.40}$$

where $\mathrm{e_a}$ is the *activation energy*[25] and A is the so-called pre-exponential factor. Equation (4.40) was first proposed for the temperature dependence of the deterministic rate parameter as early as 1884 by the Swedish physicist and chemist Svante Arrhenius. It is still used for the evaluation of rate parameters from the known temperature dependence of reaction rates, in particular, in the form $\ln k = \ln A - \mathrm{e_a}/k_\mathrm{B}T$. The assumption of a temperature independent pre-exponential factor A can be challenged, and more flexible equations are known as the modified Arrhenius equation:

$$\gamma(T) = k(T) = A \left(\frac{T}{T_0}\right)^n \exp\left(-\frac{\mathrm{e_a}}{k_\mathrm{B} T}\right) , \tag{4.40'}$$

$$\gamma(T) = k(T) = A \exp\left(-\frac{\mathrm{e_a}}{k_\mathrm{B} (T - T_0)}\right) , \tag{4.40''}$$

[25]The activation energy $\mathrm{e_a}$ is given in joule per molecule and this implies use of the Boltzmann constant $k_\mathrm{B} = 1.380,648,8 \times 10^{-23}\,\mathrm{J\,K^{-1}}$. In chemistry it is common to use kilojoule (kJ) per mole instead of molecule, which we indicate by using E_a for the activation energy, whence k_B has to be replaced by the gas constant $R = N_\mathrm{L} k_\mathrm{B}$.

where T_0 is a reference temperature. The dimensionless exponent n commonly lies in the range $-1 \leq n \leq 1$.

Transition State Theory

The theory of the *transition state* dates back to the early applications of quantum mechanics to chemistry. Roughly a decade after the formulation of quantum mechanics by Erwin Schrödinger and Werner Heisenberg the American physicist Henry Eyring [148] proposed a theory of chemical reactions which can be used to calculate rate parameters and which is still in use more than 65 years after its invention. It provides an alternative to the fully empirical reaction parameters of the Arrhenius equation [324].

The theory deals with reactant molecules which come together and form an unstable complex called the *transition state*, whereupon the reaction proceeds to yield products. The approach can be understood as a kind of semi-empirical theory: the reactant molecules and the transition state are described by quantum mechanics, but the motion along the reaction coordinate ϱ (Fig. 4.11) is treated by classical mechanics. Pure quantum effects like tunneling are not included, although they can be added (see, e.g., [369]). In order to be activated for the reaction, the reaction complex has to be driven up the reaction coordinate[26] ϱ by energy transfer from other molecular degrees of freedom or from the environment, until the local maximum—called the transition state—is reached. The reaction complex then travels down the product valley and loses energy that is transferred to other degrees of freedom.

The transition state is symbolized by ‡ and is treated like a molecular entity, except for one *unstable vibrational mode*, understood as translational motion along the reaction coordinate ϱ. Thermodynamics is applied to calculate the reaction rate parameter for the reaction

$$\mathsf{A} + \mathsf{B} \overset{K^{\ddagger}}{\rightleftharpoons} [\mathsf{AB}]^{\ddagger} \xrightarrow{\;\;k^{\ddagger}\;\;} \text{products}$$

by making a quasi-equilibrium assumption for the transition state, viz.,

$$K^{\ddagger} = \frac{[\mathsf{AB}^{\ddagger}]}{[\mathsf{A}][\mathsf{B}]} \,. \tag{4.41}$$

[26]The reaction coordinate is a combination of atomic movements that leads from reactants to products over the lowest conceivable pass on the energy landscape.

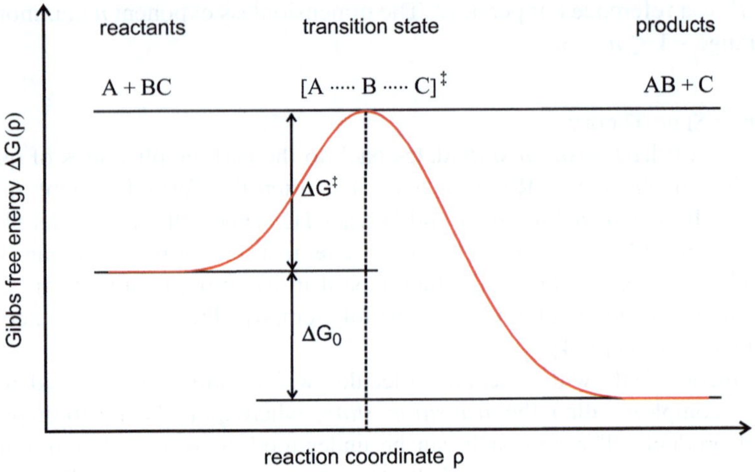

Fig. 4.11 The transition state for the reaction A + BC → AB + C. Reaction dynamics is visualized as a process along a single coordinate ϱ called the *reaction coordinate*. The Gibbs free energy $\Delta G(\varrho)$ of the reaction complex is plotted against the reaction coordinate and increases during the approach of the reactants until it reaches a (local) maximum on the energy landscape (see Fig. 4.14) denoted as transition state. Then through dissipation of free energy to the environment, the reaction complex progresses downward in the product valley until it reaches the stable product state. The example presented is an exergonic reaction, since $\Delta G_0 = \Delta G_{products} - \Delta G_{\text{reactants}} < 0$

The conventional rate parameter is then obtained from $k = k^{\ddagger} K^{\ddagger}$ and it remains to find an expression for the rate $k^{\ddagger}$ with which the transition state is converted into products.

The transition state is considered as a molecular complex with one special degree of freedom consisting of the unstable motion along the reaction coordinate ϱ, which leads to products. All other $3n-7$ degrees of freedom, or $3n-6$ in the case of linear geometries, are handled by conventional statistical mechanics, and the equilibrium constant for complex formation is of the form

$$K^{\ddagger} = \frac{q_{\mathrm{AB}^{\ddagger}}}{q_{\mathrm{A}}\, q_{\mathrm{B}}} \mathrm{e}^{-\Delta H_0^{\ddagger}/RT} ,$$

where the individual partition functions are denoted by q and the enthalpy difference[27] between the transition state and the reactants is $\Delta H_0^{\ddagger}$. The remaining degree of freedom is responsible for product formation and has the partition function $q^{(\varrho)}_{\mathrm{AB}^{\ddagger}}$.

[27]At constant pressure, for example in solution, where the volume change ΔV_0 of a reaction is small, the reaction enthalpy ΔH_0 takes on practically the same values as the reaction energy ΔE_0.

No matter whether this mode is interpreted as a degenerate vibration with a negative harmonic potential or as a translational degree of freedom, we find $k^{\ddagger} q^{(\varrho)}_{\mathrm{AB}^{\ddagger}} = k_{\mathrm{B}}T/h$, where h is Planck's constant and the final result is the same:

$$k = k^{\ddagger} K^{\ddagger} = \kappa \frac{k_{\mathrm{B}}T}{h} \mathrm{e}^{\Delta S_0^{\ddagger}/R} \mathrm{e}^{-\Delta H_0^{\ddagger}/RT} . \tag{4.42}$$

By κ we denote an empirical transmission factor measuring the probability that the vibrating activated complex decomposes into the product valley, and the activation entropy and activation enthalpy are related to the equilibrium constant through

$$RT \ln K^{\ddagger} = -\Delta G_0^{\ddagger} = -\Delta H_0^{\ddagger} + T \Delta S_0^{\ddagger} .$$

Equation (4.42) is Eyring's formula for the value of the reaction rate parameter that corresponds to the rate probability $\gamma_{(\mathrm{A+B \to AB^{\ddagger}})}$. The value of the formula is twofold: (i) it shows how reaction rate parameters can be derived from first principles, and (ii) it provides a thermodynamic interpretation of the steric factor ρ in equation (4.40′) by means of an activation entropy $\Delta S_0^{\ddagger}$. Direct calculations of rate constants, however, are highly inaccurate, since energy surfaces cannot be obtained with sufficient accuracy, apart from a few special cases like the $H+H_2$ reaction (Fig. 4.14). It should be mentioned that the simpler Arrhenius approach is often preferred over the application of transition state theory for interpretations of temperature dependencies in mechanisms involving biopolymer molecules [576]. There are many possible pitfalls in cases where the reaction mechanisms of the experimental systems are not known in sufficient detail.

Molecular Collisions

Molecules or atoms have to come together before they can react, so molecular collisions play a key role in chemical reactions [82], and we present here a short account of molecular collisions in order to illustrate the relation between the kinematics of molecules and chemical kinetics. (For an excellent introduction to the statistical physics of molecular reactions see, e.g., [46, pp. 803–1018].) A vapor phase reaction mixture in which the molecules behave according to Maxwell–Boltzmann theory is assumed. This theory is based on *classical collisions*, which implies that the molecules obey the laws of Newtonian mechanics, and further it is assumed that the gas is at thermal equilibrium. It has to be remarked, however, that the application of classical collision theory to molecular details of chemical reactions can only be an illustrative and useful heuristic, because the molecular domain falls into the realm of quantum phenomena and any theory that aims at a derivation of reaction probabilities from first principles has to be built upon a firm quantum mechanical basis (see quantum mechanical reaction dynamics).

Molecules change their motions, their internal states, and their natures in collisions, which are classified as elastic, inelastic, or reactive, respectively. In an elastic collision the collision partners exchange linear momentum and kinetic energy, and only the directions and the absolute values of the velocities of the collision partners before and after the collision are different (Fig. 4.13). In an inelastic collision internal energy, rotational and/or vibrational, and in exceptional cases also electronic energy, is transferred between the reaction partners. Finally, reactive collisions describe chemical reactions between the reaction partners, and the molecular species before and after the collision are different. In collision theory it is often assumed that the colliding objects have spherical geometry, which is clearly a very crude approximation. Corrections can be made by introducing a *geometric factor* or by much more elaborate calculations.

In order to be able to handle the specific properties of individual molecules, it is necessary to distinguish molecular species, e.g., A, and individual molecules A.[28] In the latter case knowledge of the detailed molecular state Λ_A may be required, for example,

$$A_{\Lambda_A}\,, \quad \text{with } \Lambda_A = (N_A, \Sigma_A, n_A, J_A; m_A, \mathbf{r}_A, \mathbf{v}_A)\,, \tag{4.43}$$

where $(N_A, \Sigma_A, n_A, J_A)$ stands for a complete set of molecular quantum numbers characterizing the electronic and spin state (N_A, Σ_A), vibrational state (n_A), and rotational state (J_A) of molecule A. The mass of the molecule is m_A, and its position $(\mathbf{r}_A)$ and velocity coordinates $(\mathbf{v}_A)$ are commonly measured in a Cartesian labor coordinate system: $\mathbf{r}_A(t) = (x_A, y_A, z_A)$ and $\mathbf{v}_A(t) = \left(v_x^{(A)}, v_y^{(A)}, v_z^{(A)}\right)$. In the spirit of classical mechanics, the position vector is—apart from spontaneous changes in collisions—a linear function of time, i.e., $\mathbf{r}(t) = \mathbf{r}_0 + \mathbf{v}t$, and the velocity is constant, i.e., $\mathbf{v} = \mathbf{v}_0$. In other words, the molecules travel on straight lines with constant speed between collisions. On this basis we can easily identify the different classes of bimolecular collisions $A + B \rightarrow$ by means of examples where primes are used to indicate the state after the collision:

(i) Elastic collisions: $A_{\Lambda_A} + B_{\Lambda_B} \rightarrow A_{\Lambda_A} + B_{\Lambda_B}$ with conservation of linear momentum and kinetic energy, viz., $m_A\mathbf{v}_A + m_B\mathbf{v}_B = m_A\mathbf{v}'_A + m_B\mathbf{v}'_B$ and $m_A|\mathbf{v}_A|^2 + m_B|\mathbf{v}_B|^2 = m_A|\mathbf{v}'_A|^2 + m_B|\mathbf{v}'_B|^2$, respectively. The set of internal quantum numbers remains unchanged in both molecules.
(ii) Inelastic collisions: $A_{\Lambda_A} + B_{\Lambda_B} \rightarrow A_{\Lambda'_A} + B_{\Lambda'_B}$, where the set of quantum numbers for internal motions has been changed by the collision.
(iii) Reactive collisions: $A+B \rightarrow \ldots$, where the two molecules undergo a chemical reaction in which the nature of at least one molecule is changed.

[28]For molecular species we shall also use the notation X_i when we refer to reaction networks, for example $\mathcal{S} = (X_1, X_2, \ldots)$.

The correct description of translational motion in a macroscopic reaction vessel does not require quantum mechanical treatment,[29] and hence elastic collisions are just an exercise in Newtonian mechanics. Internal energy of molecules is converted into translational energy in inelastic collisions, and a quantum mechanical approach is needed for detailed modeling. The same is true for reactive collisions when one is interested in reactions of molecules in specific states, otherwise the reaction can described by a mean reaction probability that averages over a Boltzmann ensemble (for the theory of molecular collisions see, e.g., [82]).

Maxwell–Boltzmann Distribution

The two conditions (i) *perfect mixture* and (ii) *thermal equilibrium* can now be cast into precise physical meanings. Premise (i), *spatial homogeneity*, requires that the probability of finding the center of an arbitrarily chosen molecule inside a container subregion with a volume ΔV is equal to $\Delta V/V$, where V is the total volume. The system is spatially homogeneous on macroscopic scales but it allows for random fluctuations from homogeneity. Formally, requirement (i) asserts that the position of a randomly selected molecule is described by a random variable, which is uniformly distributed over the interior of the container. Premise (ii), thermal equilibrium, implies that the Cartesian coordinates of the velocity $\mathbf{v} = \big(v_x, v_y, v_z\big)$, with $v = \sqrt{v^2} = \sqrt{v_x^2 + v_y^2 + v_z^2}$, of a randomly chosen particle with mass m are normally distributed with mean $\mu = 0$ and variance $\sigma^2 = k_\mathrm{B}T/m$ (k_B being Boltzmann's constant):

$$f_\mathrm{MB}(v_i)\,\mathrm{d}v_i = \left(\frac{m}{2\pi k_\mathrm{B}T}\right)^{1/2} \mathrm{e}^{-mv_i^2/2k_\mathrm{B}T}\,\mathrm{d}v_i\ , \quad i = x, y, z\ . \tag{4.44}$$

At zero absolute temperature, the velocity is a delta function at $\mathbf{v} = 0$ and it becomes steadily broader with increasing temperature. The extension to the 3D case is straightforward: (i) the velocity densities along the three Cartesian coordinate axes are independent and the expressions are identical by the equipartition theorem [47], and (ii) the 3D volume element

$$\mathrm{d}\mathbf{v}^3 = \mathrm{d}v_x\,\mathrm{d}v_y\,\mathrm{d}v_z = v^2\mathrm{d}v\ \sin\theta\ \mathrm{d}\theta\ \mathrm{d}\phi\ , \quad \text{with} \quad \int_0^{2\pi}\int_0^{\pi} \sin\theta\ \mathrm{d}\theta\ \mathrm{d}\phi = 4\pi$$

[29]The individual energy levels of the translational partition function are so close together that the quantum mechanical summation can be replaced by an integral.

is evaluated in polar coordinates, taking into account spherical symmetry. Then, we obtain the *Maxwell–Boltzmann velocity distribution*

$$f_{\mathrm{MB}}(v)\,\mathrm{d}\mathbf{v}^3 = \sqrt{\frac{2}{\pi}\frac{v^2}{\alpha^3}}\mathrm{e}^{-v^2/2a^2}\,\mathrm{d}\mathbf{v}^3\;,$$
$$F_{\mathrm{MB}}(v) = \mathrm{erf}\left(\frac{v}{\sqrt{2}\alpha}\right) - \sqrt{\frac{2}{\pi}\frac{v}{\alpha}}\mathrm{e}^{-v^2/2a^2}\;, \tag{4.45}$$

where $\alpha = \sqrt{k_{\mathrm{B}}T/m}$. The velocity of molecules is commonly characterized by several averaged values: (i) the mode or the most probable value of the distribution $\widetilde{v}$, (ii) the expectation value $\mathrm{E}(v) = \langle v \rangle$, and (iii) the root mean square velocity derived from the second raw moment, i.e., $(\widehat{\mu}_2)^{1/2} = \sqrt{\langle v^2 \rangle}$:

$$\widetilde{v} = \sqrt{2k_{\mathrm{B}}}\left(\frac{T}{m}\right)^{1/2}\;,\quad \langle v \rangle = \sqrt{\frac{8k_{\mathrm{B}}}{\pi}}\left(\frac{T}{m}\right)^{1/2}\;,\quad \sqrt{\langle v^2 \rangle} = \sqrt{3k_{\mathrm{B}}}\left(\frac{T}{m}\right)^{1/2}\;,$$

with $\widetilde{v} < \langle v \rangle < \sqrt{\langle v^2 \rangle}$.

It is worth considering the density of the energy as well because it provides a rational explanation for the empirical Arrhenius factor. The total energy is a sum of three equal independent contributions for the three coordinate axes: $\mathrm{e} = \varepsilon_x + \varepsilon_y + \varepsilon_z = 3\varepsilon$. In one dimension, we find

$$f_{\mathrm{el}}(\varepsilon)\,\mathrm{d}\varepsilon = \frac{1}{\sqrt{\pi k_{\mathrm{B}}T}}\frac{1}{\sqrt{\varepsilon}}\mathrm{e}^{-\varepsilon/k_{\mathrm{B}}T}\,\mathrm{d}\varepsilon\;, \tag{4.46a}$$

which, by inserting $\varepsilon = xk_{\mathrm{B}}T/2$, is easily shown to be a χ^2-density of dimension one:

$$f_{\chi_1^2}(x)\,\mathrm{d}x = \frac{1}{\sqrt{2\pi}}\frac{1}{\sqrt{x}}\mathrm{e}^{-x/2}\,\mathrm{d}x\;.$$

Extension to three dimensions yields

$$f_{\mathrm{e}}(\mathrm{e})\,\mathrm{de} = 2\sqrt{\frac{\mathrm{e}}{\pi}}\left(\frac{1}{k_{\mathrm{B}}T}\right)^{3/2}\mathrm{e}^{-\mathrm{e}/k_{\mathrm{B}}T}\,\mathrm{de}\;, \tag{4.46b}$$

and this expression is equivalent to a χ^2-density of dimension three:

$$f_{\chi_3^2}(x)\,\mathrm{d}x = \frac{1}{\sqrt{2\pi}}\sqrt{x}\mathrm{e}^{-x/2}\,\mathrm{d}x\;.$$

The equivalence with the χ^2-distribution is not surprising, since the total energy results from $\mathrm{e} = m\mathbf{v}^2/2 = m(v_x^2 + v_y^2 + v_z^2)/2$, a sum of three squares.

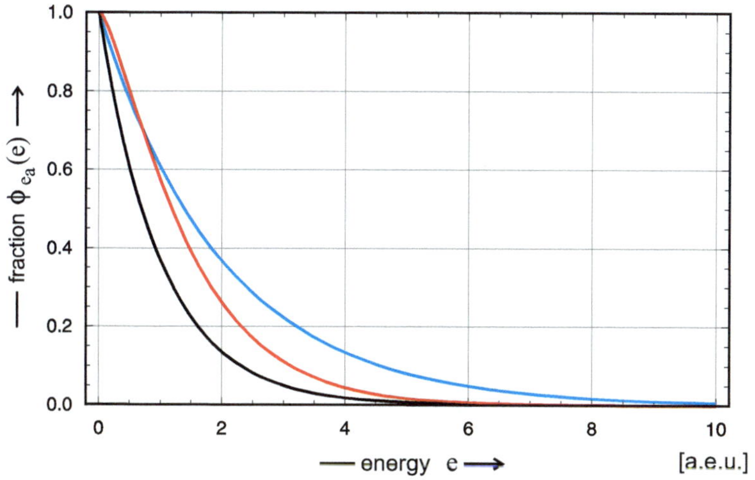

Fig. 4.12 Interpretation of the Arrhenius factor. The fraction of molecules $\phi_{e_a}(e)$, which have a kinetic energy greater than the activation energy e_a (*red*), calculated using (4.46c), is shown together with two simple exponential functions $f_1(e) = \exp(-e_a/k_B T)$ (*black*) and $f_2(e) = \exp(-e_a/2k_B T)$ (*blue*)

Equations (4.46) can be used to calculate the percentage of molecules which have a kinetic energy that is greater than a given reference level e_a :

$$\phi_{e_a}(e) = \int_{e_a}^{\infty} f_e(e)\, de = 1 - F_e(e_a) , \tag{4.46c}$$

where $F_e(e)$ is the cumulative distribution function associated with the density $f_e(e)$. Figure 4.12 shows that $\phi_{e_a}(e)$ is not substantially different from the Arrhenius factor $\exp(-e_a/k_B T)$.

To sum up, premises (i) and (ii) assert that the distribution of molecular velocities is isotropic and a function of mass m and temperature T alone. Implicitly, the two conditions also guarantee that the molecular position and velocity components are all statistically independent of each other. For practical purposes, we expect the two premises to be valid for any dilute gas at constant temperature in which *non-reactive* molecular collisions occur much more frequently than *reactive* molecular collisions. The extension to dilute solutions is straightforward, although difficult in practice [47].

Bimolecular Reactive Collisions

The theory of molecular collisions in dilute gases is the best developed microscopic model for chemical reactions, apart from the quantum mechanical approach. It is well suited for providing a rigorous link between molecular motion and chemical

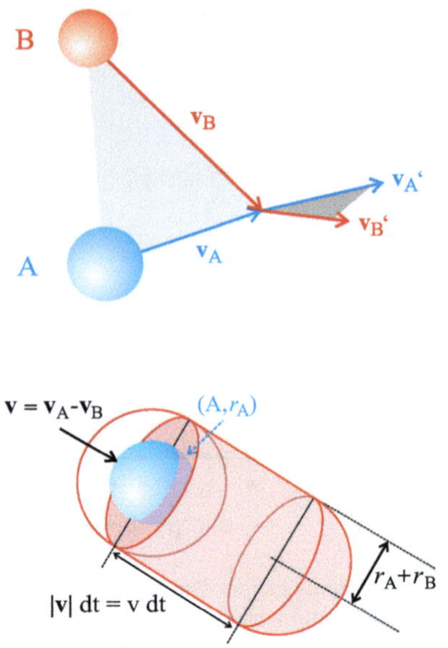

Fig. 4.13 Sketch of molecular collisions in the vapor phase. A spherical molecule A with radius r_A moves with a velocity $\mathbf{v} = \mathbf{v}_A - \mathbf{v}_B$ relative to a spherical molecule B with radius r_B. *Upper*: Geometry of a typical elastic collision, for which linear angular momentum $\mathbf{p} = m\mathbf{v}$ and kinetic energy $E_{kin} = mv^2/2$ are conserved: $\mathbf{p}_A + \mathbf{p}_B = \mathbf{p}'_A + \mathbf{p}'_B$ and $m_A|\mathbf{v}_A|^2 + m_B|\mathbf{v}_B|^2 = m_A|\mathbf{v}'_A|^2 + m_B|\mathbf{v}'_B|^2$, where the primed quantities refer to the situation after the collision. *Lower*: Geometry of the collision in the coordinate system of B. If the two spherical molecules are to collide within the next infinitesimal time interval dt, the center of B has to lie inside a cylinder of radius $r = r_A + r_B$ and height $|\mathbf{v}|$ dt $= v$ dt. The upper and lower surfaces of the cylinder are deformed into identically oriented hemispheres of radius r, and therefore the infinitesimal collision reaction volume, i.e., the volume $\mathrm{d}V_{col}$ dt of the deformed cylinder, is identical with the volume $\mathrm{d}V = (r_A + r_B)^2 \pi\, v\, \mathrm{d}t$ of a non-deformed infinitesimal cylinder

kinetics. The rate parameters of general bimolecular reactions are calculated using classical mechanics (Fig. 4.13) and the Maxwell–Boltzmann distribution (for a comprehensive review see [163]).

The occurrence of a bimolecular reaction

$$A + B \longrightarrow C + \cdots \tag{4.47}$$

has to be preceded by an encounter between a molecule A and a molecule B. First we calculate the probability of a collision within the reaction volume V_R. For simplicity, molecular species are regarded as spheres with specific masses and radii, m_A and r_A for A and m_B and r_B for B, respectively. A collision occurs whenever r_{AB}, the center-to-center distance between the two molecules, becomes as small as the sum $(r_{AB})_{min} = r_A + r_B$ of the two radii. The probability that, at time t, a randomly selected pair $\boldsymbol{\mu} = (\mathrm{A}, \mathrm{B})$ of $\mathbf{R}_\mu$ reactant molecules will collide within the next infinitesimal time interval $[t, t + \mathrm{d}t]$ is defined by $\Pi_\mu(t, \mathrm{d}t)$, and the geometry of such a collision is shown diagramatically in Fig. 4.13. The probability density of the relative velocity $\Delta\mathbf{v} = \widehat{\mathbf{v}} = \mathbf{v}_A - \mathbf{v}_B$ for the randomly selected pair of reactant molecules, precisely the probability of $\widehat{\mathbf{v}}$ lying within an infinitesimal volume element $d\widehat{\mathbf{v}}^3$ around $\widehat{\mathbf{v}}$ at time t, is denoted by $f(\widehat{\mathbf{v}}(t), \boldsymbol{\mu})$ and obtained from Maxwell–Boltzmann theory (4.45):

$$f(\widehat{\mathbf{v}}(t), \boldsymbol{\mu})\, d\widehat{\mathbf{v}}^3 = \left(\frac{\widehat{m}}{2\pi k_B T}\right)^{3/2} e^{-\widehat{m}\,\widehat{v}^2/2k_B T}\, d\widehat{\mathbf{v}}^3 .$$

Here, $\widehat{v} = |\widehat{\mathbf{v}}| = \sqrt{\widehat{v}_x^2 + \widehat{v}_y^2 + \widehat{v}_z^2}$ is the absolute value of the relative velocity and $\widehat{m}$ is the reduced mass of the two molecules A and B.[30]

Next we define a set of all combinations of velocities for the reaction partners in the reaction $\mathbf{R}_\mu$ at time t: $\mathcal{R}_\mu(t) = \{\omega_\mu(\widehat{\mathbf{v}}(t))\}$. Two properties of the probability densities $f(\widehat{\mathbf{v}}(t), \boldsymbol{\mu})$ for different velocities $\widehat{\mathbf{v}}$ are important:

(i) The elements of the set $\mathcal{R}_\mu(t)$ of all combinations of velocities of the reactant molecules are mutually exclusive.
(ii) They are collectively exhaustive since $\widehat{\mathbf{v}}$ is varied over the entire 3D velocity space: $-\infty < (\widehat{v}_x, \widehat{v}_y, \widehat{v}_z) < +\infty$.

The probability density $f(\widehat{\mathbf{v}}(t), \boldsymbol{\mu})$ is related to the probability of a collision event between the reaction partners of the reaction $\mathbf{R}_\mu$ denoted by $\omega_{col,\mu}$ through the conditional probability $P(\omega_{col,\mu}(t + dt)|\omega_\mu(\widehat{\mathbf{v}}(t)))$. In other words this is the conditional probability that two molecules defined by the reaction complex $\boldsymbol{\mu}$ and moving with the relative velocity $\widehat{\mathbf{v}}(t)$ at time t will collide within the next moment dt. Figure 4.13 shows the geometry of the collision event of two randomly selected spherical molecules A and B, which are assumed to collide within the infinitesimal time interval $[t, t + dt]$.[31] A randomly selected molecule A moves along the vector $\widehat{\mathbf{v}}$ between A and B and a collision between the two molecules will take place in the interval $[t, t + dt]$ if and only if the center of molecule B at time t is situated inside the spherically distorted cylinder shown in Fig. 4.13, so the probability of a collision is tantamount to the probability $P(\omega_{col,\mu}(t + dt)|\omega_\mu(\widehat{\mathbf{v}}(t)))$ that the center of a randomly selected molecule B is situated within the subregion of V defined by moving A at time t. This subregion has volume $V_{col} = |\widehat{\mathbf{v}}|\sigma_{AB}\, dt$, where $\sigma_{AB} = (r_A + r_B)^2\pi$ is the *reaction cross-section*, and by scaling with the total

[30] In order to handle the relative motion of two particles, the original system consisting of particle A with mass m_A and velocity $\mathbf{v}_A$, and particle B with mass m_B and velocity $\mathbf{v}_B$, respectively, is transformed into a system with center of mass (cm) motion and relative or *internal* motion, where the center of mass has mass $M = m_A + m_B$ moving with the velocity $\mathbf{v}_{cm} = (m_A\mathbf{v}_A + m_B\mathbf{v}_B)/(m_A + m_B)$ and the internal motion with reduced mass $\widehat{m} = m_A m_B/(m_A + m_B)$ moving with the velocity $\widehat{\mathbf{v}}$.

[31] The absolute time t comes into play because the positions $\mathbf{r}_A$ and $\mathbf{r}_B$ of the molecules and their velocities $\mathbf{v}_A$ and $\mathbf{v}_B$ depend on t.

volume V we obtain[32]:

$$P\Big(\omega_{\mathrm{col},\mu}(t+\mathrm{d}t)|\omega_\mu\big(\widehat{\mathbf{v}}(t)\big)\Big)\,\mathrm{d}t = \frac{|\widehat{\mathbf{v}}(t)|\sigma_{\mathrm{AB}}}{V}\,\mathrm{d}t\,. \tag{4.48}$$

The desired probability is calculated by substitution and integration over the entire velocity space, i.e.,

$$\Pi_\mu(t,\,\mathrm{d}t) = \iiint_{\mathbf{v}=-\infty}^{\infty}\left(\frac{\widehat{m}}{2\pi k_{\mathrm{B}}T}\right)^{3/2} \mathrm{e}^{-\widehat{m}\,\widehat{v}^2/2k_{\mathrm{B}}T}\,\frac{\widehat{v}(t)\,\mathrm{d}t\,\sigma_{\mathrm{AB}}}{V}\,\mathrm{d}\widehat{\mathbf{v}}^3\,.$$

The evaluation of the integral is straightforward and yields

$$\Pi_\mu(t,\,\mathrm{d}t) = \left(\frac{8k_{\mathrm{B}}T}{\pi V^2}\right)^{1/2}\frac{\sigma_{\mathrm{AB}}}{\sqrt{\widehat{m}}}\,\mathrm{d}t\,. \tag{4.49}$$

The first factor contains only constants and two macroscopic quantities, the volume V and the temperature T, whereas the molecular parameters, the radii r_{A} and r_{B} and the reduced mass $\widehat{m}$, appear in the second factor.

A collision is a necessary but not a sufficient condition for a reaction to take place, and therefore we introduce a *collision-conditioned reaction probability* p_μ, which is the probability that a randomly selected pair of colliding $\mathbf{R}_\mu$ reactant molecules will indeed react according to $\mathbf{R}_\mu$. By multiplication of independent probabilities and taking into account (4.39), we find

$$\pi_\mu(t,\,\mathrm{d}t) = \gamma_\mu\,\mathrm{d}t = p_\mu\Pi_\mu(t,\,\mathrm{d}t) = p_\mu\left(\frac{8k_{\mathrm{B}}T}{V}\right)^{1/2}\frac{\sigma_{\mathrm{AB}}}{\sqrt{\widehat{m}}}\,\mathrm{d}t\,. \tag{4.50}$$

The probability γ_μ is independent of dt, as required by (4.39), and this will be the case if and only if the reaction probability p_μ does not depend on dt. This is highly plausible for the derivation given above, and it is supported by an empirical test through the detailed examination of bimolecular reactions, which can be found, for example, in [209, pp. 413–417].

The Arrhenius factor can be illustrated within the framework of collision theory if we make the assumption that the collision energy has to exceed the activation energy $\mathrm{e_a}$. The fraction $\phi(\mathrm{e_a})$ of molecules whose kinetic energies exceed this energy threshold is readily calculated from the energy distribution function (4.46b) and obtained in the form of (4.46c). Figure 4.12 compares $\phi(\mathrm{e_a})$ with the conventional Arrhenius factor $\exp(-\mathrm{e_a}/k_{\mathrm{B}}T)$ and the factor $\exp(-\mathrm{e_a}/2k_{\mathrm{B}}T)$. The second case is

[32]In the derivation we implicitly made use of the infinitesimally small size of dt. Only if the distance $|\widehat{\mathbf{v}}|\,\mathrm{d}t$ is vanishingly small can the possibility of collisional interference by a third molecule be neglected.

rationalized by the idea that both reaction partners contribute an equal share $e_a/2$ to the reaction energy. Although there are recognizable differences between the three curves in the figure, the entirely empirical Arrhenius equation nicely parallels the factor $\phi(e_a)$ derived from collision theory.

The results of collision theory for reactive bimolecular encounters can be summarized in a commonly used form for the rate parameter and its temperature dependence:

$$\gamma_\mu(T) = A\left(\frac{T}{T_0}\right)^n \exp\left(-\frac{e_a}{k_B T}\right) = \zeta(T)\rho \exp\left(-\frac{e_a}{k_B T}\right) . \tag{4.40'}$$

Here $\zeta(T)$ is the collision frequency as calculated above, i.e.,

$$\zeta(T) = \sigma_{AB}\sqrt{\frac{8k_B T}{\pi \hat{m}}} , \quad \text{with } \sigma_{AB} = (r_A + r_B)^2\pi . \tag{4.51}$$

The factor ρ is sometimes referred to as a steric factor and e_a is the activation energy of the reaction, measured here as energy per molecule. The actual number of collisions in the volume V per time unit is $Z = N_L V\zeta$. Comparison with the Arrhenius equation (4.40′) yields $n = 1/2$. The exponential temperature dependence of the rate parameter on temperature is often satisfied with astonishing accuracy, but the interpretation of the steric factor ρ is often unsatisfactory and therefore some chemists prefer to stay away from any rationalization of the steric factor and define it simply as the ratio between the pre-exponential factor and the collision frequency, i.e., $\rho = A/\zeta(T)$.

Monomolecular Reactions

In the strict sense, a *monomolecular reaction* refers to the spontaneous conversion

$$\mathsf{A} \longrightarrow \mathsf{C} . \tag{4.52}$$

One molecule A is spontaneously converted into one molecule C. The monomolecular reaction was initially considered to be particularly simple because only one type of reactant molecule is involved, but this expectation turned out to be wrong. Most apparently monomolecular reactions follow a bimolecular rate law, at least at sufficiently low concentrations, and have to be distinguished from true monomolecular conversions. It is worth mentioning that even a class of spontaneous dissociation reactions of small cluster ions, such as $(H_3O^+)(H_2O)_n$ or $Cl^-(H_2O)_n$ with $n = 2$–4, which were considered as the prototypes of truly monomolecular processes, are not strictly spontaneous, because the loss of ligands seems to be initiated by collisions with the wall of the reaction vessel [453].

In the absence of interaction with an environment, the true monomolecular conversion (4.52) is necessarily driven by some quantum mechanical mechanism, similar to the radioactive decay of a nucleus. Time-dependent perturbation theory in quantum mechanics [395, pp. 724–739] shows that almost all weakly perturbed energy-conserving transitions have linear probabilities of occurrence in time intervals δt, when δt is *microscopically large* but *macroscopically small*. Therefore, to a good approximation, the probability for a radioactive nucleus to decay within the next infinitesimal time interval dt is of the form $\alpha\,dt$, were α is some time-independent constant. On the basis of analogy, we may expect the probability $\pi_\mu(t,\,dt)$ for a genuine monomolecular conversion to be approximately of the form $\gamma_\mu\,dt$, with γ_μ independent of dt.

The vast majority of apparently monomolecular reactions, however, follow a different mechanism and involve a reaction partner in the sense of a catalyzed bimolecular conversion:

$$\mathsf{A} + \mathsf{B} \longrightarrow \mathsf{C} + \mathsf{B}\,, \tag{4.47$'$}$$

$$\mathsf{A} + \mathsf{A} \longrightarrow \mathsf{C} + \mathsf{A}\,. \tag{4.47$''$}$$

In (4.47′), the conversion $\mathsf{A} \longrightarrow \mathsf{C}$ is initiated by a collision between a molecule A and a molecule B which acts as a catalyst since it is not consumed by the process.[33] When the collision partner is another molecule A (4.47″), we are dealing with a monomolecular reaction in the conventional sense, which is described straightforwardly as a special class of bimolecular processes.

The first proposal of a mechanism for the conversion (4.47″) was made as early 1922 by Frederick Lindemann [351]. He suggested that the seemingly monomolecular conversion is a two-step mechanism of the form

$$\mathsf{A} + \mathsf{A} \underset{l_1}{\overset{k_1}{\rightleftharpoons}} \mathsf{A} + \mathsf{A}^*\,, \qquad \mathsf{A}^* \xrightarrow{\;k_2\;} \mathsf{C}\,, \tag{4.53}$$

with $k_2 \ll l_1$. The Lindemann mechanism with a conventional rate parameter k_1 did not fit the experimental data and was improved by Cyril Hinshelwood [252], using a different interpretation of the activation of molecule A that was extended to a range of energy values $k_{1(E_0 \to E_1)} \to k_{1(E_0 \to E_1 + \delta E)}$. Later on, the molecular mechanistic details were improved and the Lindemann–Hinshelwood mechanism was substantially extended by Oscar Rice, Herman Ramsperger [465],

33 Formally, we are dealing with a reaction that is catalyzed by a molecule of the same or another molecular species, and the reaction is related to the spontaneous conversion by rigorous thermodynamics: whenever a catalyzed reaction appears in a mechanism, the uncatalyzed process has to be considered as well, no matter how slow it is.

and Louis Kassel [291] by explicit introduction of a transition state $\mathbf{A}^{\ddagger}$:

$$\begin{aligned} \mathsf{A} + \mathsf{A} \; &\underset{h_1}{\overset{k_1}{\rightleftharpoons}} \; \mathsf{A} + \mathsf{A}^* \,, \\ \mathsf{A}^* \; &\xrightarrow{k_{2a}} \; \mathsf{A}^{\ddagger} \xrightarrow{k^{\ddagger}} \mathsf{C} \,. \end{aligned} \tag{4.54}$$

As in transition state theory, the rate parameter $k^{\ddagger}$ corresponds to the fast process associated with the reactive mode of the transition state. Since $k^{\ddagger}$ is thought to be greater than any other rate parameter, the rate-limiting step in the formation of the product C is the conversion $\mathsf{A}^* \rightarrow \mathsf{A}^{\ddagger}$, and comparing the Lindemann and Rice–Ramsperger–Kassel (RRK) mechanisms, we have $k_2 \approx k_{2a}$ and $k_{2a} = k^{\ddagger}[\mathsf{A}^{\ddagger}]/[\mathsf{A}^*]$ from the steady state assumption. Eventually, the theory of monomolecular reactions got its present form through a reformulation of the transition state by Rudolph Marcus and Oscar Rice [368, 370, 371]. The current version of the so-called Rice–Ramsperger–Kassel–Marcus (RRKM) theory of monomolecular reactions theory allows for a highly accurate and very detailed description of reactions, and it can be readily converted into a stochastic model [348].

Termolecular and Other Reactions

Termolecular or trimolecular reactions of the form

$$\mathsf{A} + \mathsf{B} + \mathsf{C} \longrightarrow \mathsf{D} + \cdots \tag{4.55}$$

are rare and need not be considered, because collisions of three particles occur only with a probability of measure zero. Exceptions are two classes of reactions: (i) vapor phase association reactions where a third body is required as collision partner removing energy, and (ii) the reaction of nitrogen monoxide with oxygen or halogens. A characteristic example of a class (i) reaction is the formation of ozone

$$\mathbf{O} + \mathbf{O}_2 + \mathbf{N}_2 \longrightarrow \mathbf{O}_3 + \mathbf{N}_2 \,,$$

where the nitrogen molecule removes energy so that a bound state of ozone can be reached [440]. The typical class (ii) reaction is the oxidation of nitrogen oxide with molecular oxygen [432]:

$$2\mathbf{NO} + \mathbf{O}_2 \longrightarrow 2\mathbf{NO}_2 \,.$$

Although the oxidation of nitric oxide by oxygen is considered as the prototype of a termolecular reaction, two competing two-step mechanisms involving only bimolecular collisions are also discussed:

$$\mathsf{NO} + \mathsf{NO} \rightleftharpoons (\mathsf{NO})_2\,, \quad (\mathsf{NO})_2 + \mathsf{O}_2 \rightarrow 2\mathsf{NO}_2\,,$$

$$\mathsf{NO} + \mathsf{O}_2 \rightleftharpoons \mathsf{NO}_3\,, \quad \mathsf{NO}_3 + \mathsf{NO} \rightarrow 2\mathsf{NO}_2\,.$$

A comparison of the data for all three mechanistic variants of the reaction can be found in the review [532]. However, there may also be special situations where approximations of complicated processes by termolecular events are justified. One example is a set of three coupled reactions with four reactant molecules [208, pp. 359–361], where $\pi_\mu(t,\ \mathrm{d}t)$ is essentially linear in $\mathrm{d}t$.

Zero-Molecular Reactions

The last class of reaction to be considered here is not a proper chemical reaction but, for example, an inflow of material into the reactor. It is often referred to as a zeroth order reaction (4.1a):

$$* \longrightarrow \mathsf{A}\,. \tag{4.56}$$

Here, the assumption that the inflow is accompanied by efficient mixing so as to satisfy the homogeneity condition is essential, because it guarantees that the number of molecules entering the homogeneous system per time unit is a constant, and does not depend on $\mathrm{d}t$.

Quantum Mechanical Reaction Dynamics

For any detailed understanding of chemical reactions from first physical principles, knowledge of quantum mechanics is indispensable. Here we direct readers to the wide range of existing textbooks (recently published monographs are [30, 487] for basic quantum mechanics and applications in chemistry, while a quite elaborate text on dynamics can be found in [373]) and sketch only the basic idea because of its general importance. In conventional quantum chemistry, the fast motions of electrons are separated from the slow motions of atomic nuclei, and the stationary Schrödinger equation of a molecule or a reaction complex is partitioned into two equations:

$$\mathcal{H}_{\mathrm{el}}\Psi_{\mathrm{el}}^{(n)}(\mathbf{r}) = E_n(\mathbf{R})\Psi_{\mathrm{el}}^{(n)}(\mathbf{r})\,, \quad \text{with } \mathcal{H}_{\mathrm{el}} = \mathcal{T}_{\mathrm{el}} + V(\mathbf{r},\mathbf{R})\,, \tag{4.57a}$$

$$\big(\mathcal{T}_{\mathrm{nuc}} + E_n(\mathbf{R})\big)\,\Xi_{\mathrm{nuc}}^{(k;n)}(\mathbf{R}) = W_{k,n}\,\Xi_{\mathrm{nuc}}^{(k;n)}(\mathbf{R})\,. \tag{4.57b}$$

The positions of all electrons are subsumed in the vector $\mathbf{r}$, and likewise the nuclei occupy positions denoted by $\mathbf{R}$. Both equations are partial differential equations and they are coupled through the energy hypersurface $E_n(\mathbf{R})$ (see Fig. 4.14). The

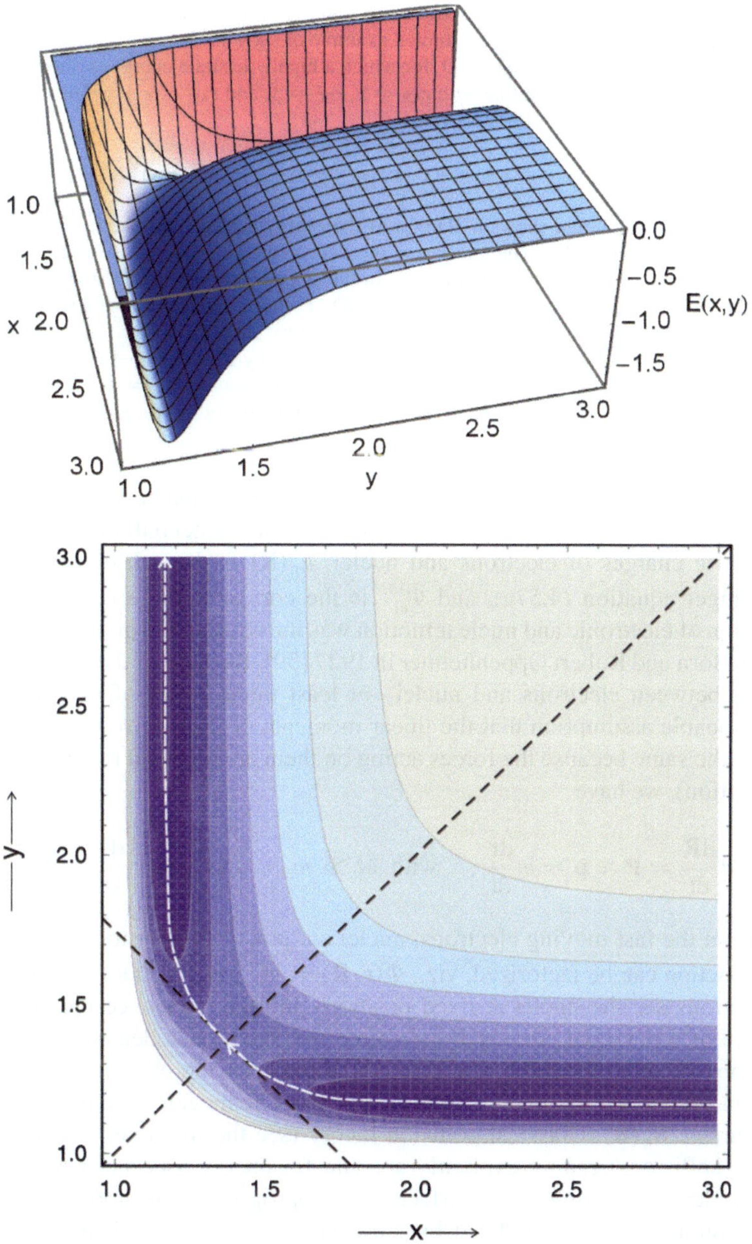

Fig. 4.14 Energy surface of the symmetric bimolecular triatomic exchange reaction A + BC → AB + C. Caption on next page

Fig. 4.14 Energy surface of the symmetric bimolecular triatomic exchange reaction A + BC → AB + C (see previous page). The best studied example of such a reaction is the hydrogen isotope exchange reaction D + HD → DH + D, for which a highly accurate energy surface is available. The three atoms lie on a straight line with $r_{AB} = x$, $r_{BC} = y$, and $r_{AC} = x + y$. The model surface plotted here is

$$E(x, y) = a/x^{12} - b/x^6 + a/y^{12} - b/y^6 + c/(x + y)^{12} .$$

The *upper part* of the figure shows a 3D plot of the energy surface with the reaction path being recognizable as a steep valley. The *lower part* presents a contour plot of this surface. The *broken white line* indicates the reaction coordinate ϱ : in the steep horizontal valley at the *bottom* of the figure, the atom D is approaching the molecule HD. Then the bond becomes longer and, at the saddle point, the two bonds are of equal length. Parameters: $a = 10$, $b = 8$, and $c = 1.5 \times 10^5$, leading to a bond length of $r_e = 1.165$ [l] and a bond energy of $\Delta E = -1.6$ [e]. At the saddle point the distance is $x = y = 1.3856$ [l] and the energy amounts to $\Delta E = -1.1303$ [e]. Length and energy are given in arbitrary units: [l] stands for length unit and [e] for energy unit

Hamilton operator $\mathcal{H}_{el}$ describes the motion of electrons and consists of the kinetic energy operator of electrons $\mathcal{T}_{el}$ and the electrostatic potential $V(\mathbf{r}, \mathbf{R})$ caused by the electric charges of electrons and nuclei, $E_n(\mathbf{R})$ is the nth eigenvalue of the Schrödinger equation (4.57a), and $\Psi_{el}^{(n)}$ is the corresponding eigenfunction. The separation of electronic and nuclear motion was introduced into quantum mechanics by Max Born and Robert Oppenheimer in 1927 [59]. Because of the large difference in mass between electrons and nuclei—at least three orders of magnitude—and the reasonable assumption that the linear momenta of the electrons and nuclei are roughly the same because the forces acting on them are identical (equality of action and reaction), we have

$$M\frac{\mathrm{d}\mathbf{R}}{\mathrm{d}t} = \mathbf{P} \approx \mathbf{p} = m\frac{\mathrm{d}\mathbf{r}}{\mathrm{d}t} , \quad \text{with } M \gg m , \quad \text{and hence } \frac{\mathrm{d}\mathbf{R}}{\mathrm{d}t} \ll \frac{\mathrm{d}\mathbf{r}}{\mathrm{d}t} .$$

Seen from the fast moving electrons, nuclei are practically immobile and the total wave function can be factorized, viz., $\Phi(\mathbf{r}, \mathbf{R}) = \Psi_{el}^{(n)}(\mathbf{r})\Xi_{nuc}^{(k;n)}(\mathbf{R})$. In other words, the electrons see the nuclei at fixed positions and the nuclei see the electrons in the form of a potential coming from a time-averaged mean density. In the Born–Oppenheimer approximation, the connection between the electron density in the quantum state n and the nuclear motion, but also chemical reactions, is the energy hypersurface $E_n(\mathbf{R})$. Classical collision theory (see the discussion of bimolecular reactive collisions) cannot explicitly account for energy aspects of reactions and the consideration of an energy surface is an appropriate and important extension. Nuclear motion can be modeled by Newtonian mechanics and the combination of an energy surface of quantum mechanical origin and classical dynamics is often referred to as semiclassical collision theory, in contrast to the full quantum mechanical approach based on scattering theory [82].

Despite the spectacular progress in numerical quantum chemistry, many chemical reaction systems and most biologically relevant structures are too large for systematic computational studies, which frequently have to handle the motions of up to 100,000 atoms on time scales of tens of nanoseconds. Hybrid methods combining quantum mechanical calculations with molecular mechanics simulations based on Newtonian mechanics (QM/MM) seem to be the most promising at present [347, 494, 495].

4.1.5 Empirical Rate Parameters

The rate parameters—often called rate constants despite the fact that they are not actually constants and the fact that their dependence on external quantities like temperature, pressure, pH, ionic strengths, and other quantities provides insights into reaction mechanisms—are the first quantities derived from measured data, and as such they make sense only for a given mechanism. Very often the mechanism of a reaction is not precisely known and then we are confronted with the difficult task of determining the reaction mechanism and the rate parameters simultaneously. Three different approaches are in common use:

(i) Traditional parameter fitting by means of linearized functions of time dependencies of signals.
(ii) Parameter fitting by means of computer-assisted minimization of a cost function commonly adapted for a given mechanism.
(iii) The mathematically and computationally more expensive but professional method of treating parameter fitting as an inverse problem, which, because of its ill-posedness, requires regularization in the search for a solution.

Until the second half of last century traditional parameter fitting was done by hand, and we mention here the analysis of first order reactions and binding equilibria as characteristic examples. At present more elaborate methods of parameter evaluation replace the human eye by conventional statistics, employing generalized mean least square fits or maximum likelihood methods (Sect. 2.6.4).

Regression Analysis and the Method of Least Squares
Regression analysis searches for relations among variables in a statistical manner. Commonly a dependent variable is modeled as a parameterized function of one or more independent variables, where the nature of the function is known and the parameters are to be determined from empirical data. In mathematical fitting, a large number of noisy experimental data points is used to determine a few parameters in the sense of a massively overdetermined problem. Several standard techniques like, for example, the method of least squares [54, 577], are available for fitting data to linear relations. Historically the first documented use of the least square method is due to the French mathematician Adrien-Marie Legendre, who published it in a monograph in 1805 [336]. Carl Friedrich Gauss published his version of regression analysis in 1809 [60], but claimed priority by contending that he had already used the method in 1795.

There is an enormous literature on regression analysis and we can only give a very brief sketch of the methods here. For details, a monograph covering the theory and computational approaches is recommended [489].

The mathematical problem underlying regression analysis and most other optimization methods is overdetermination: many individual measurements create n data points, which have to be reproduced as well as possible by a model with m parameters. In other words we have data $\omega_i = (x_i, y_i)$ with $i = 1, \ldots, n$, where $y_i = f(x_i; \beta_1, \ldots, \beta_m) + \varrho_i = f(x_i; \boldsymbol{\beta}) + \varrho_i$ with $n \gg m$. Here, y_i is the measured value, x_i is the predetermined condition of the experiment ω_i, ϱ_i is the random error, and $\boldsymbol{\beta} = (\beta_1, \ldots, \beta_m)^{\mathrm{t}}$ is a column vector of parameters.

Linear Regression Analysis

Linear regression implies that $f(x; \boldsymbol{\beta})$ is a linear function of $\boldsymbol{\beta}$. Linearity in x is not required and the polynomial

$$f(x; \boldsymbol{\beta}) = \beta_0 + \beta_1 x + \beta_2 x^2 + \beta_3 x^3 + \cdots$$

is perfect for linear regression.[34] In general, we write the linear regression function as

$$f(x; \boldsymbol{\beta}) = \sum_{i=1}^{m} \beta_j \phi_j(x) \ , \quad \text{with} \ \frac{\partial f(x; \boldsymbol{\beta})}{\partial \beta_j} = \phi_j(x) \ ,$$

and for the individual data point we use $\phi_j(x_i) = X_{ij}$. The dependent variables y_i that correspond to the measured data are expressed by

$$y_j = \sum_{i=1}^{m} \beta_i X_{ji} + \varrho_j \ , \ j = 1, \ldots, n \ , \quad \text{or } \mathbf{y} = \mathrm{X} \cdot \boldsymbol{\beta} + \boldsymbol{\varrho} \ , \tag{4.58a}$$

where $\mathbf{y} = (y_1, \ldots, y_n)^{\mathrm{t}}$ and $\boldsymbol{\varrho} = (\rho_1, \ldots, \rho_n)^{\mathrm{t}}$ is the so-called vector of residuals, containing the differences between the calculated values $f(x_i; \boldsymbol{\beta})$ and the measured or actual values y_i. As mentioned above, $\boldsymbol{\varrho}$ accounts for all types of errors. The rectangular $n \times m$ matrix X is called the *design matrix*:

$$\mathrm{X} = \{X_{ij}\} = \begin{pmatrix} X_{11} & X_{12} & \ldots & X_{1m} \\ X_{21} & X_{22} & \ldots & X_{2m} \\ X_{31} & X_{32} & \ldots & X_{3m} \\ X_{41} & X_{42} & \ldots & X_{4m} \\ \vdots & \vdots & \ddots & \vdots \\ X_{n1} & X_{n2} & \ldots & X_{nm} \end{pmatrix} .$$

[34] The parameters are renamed in order to fit $f(x; \boldsymbol{\beta}) = \sum_{j=0}^{m-1} \beta_j x^j$.

It contains the values of the *explanatory variables* $\mathbf{X}_j$ with $j = 1\ldots, m$ in individual columns. For the purpose of illustration, we show (4.58a) for the polynomial function $f(x;\boldsymbol{\beta}) = \sum_{j=0}^{m-1} \beta_j x^j$ with $m = 4$:

$$\mathbf{y} = \begin{pmatrix} y_1 \\ y_2 \\ \vdots \\ y_n \end{pmatrix} = \begin{pmatrix} 1 & x_1 & x_1^2 & x_1^3 \\ 1 & x_2 & x_2^2 & x_2^3 \\ \vdots & \vdots & \vdots & \vdots \\ 1 & x_n & x_n^2 & x_n^3 \end{pmatrix} \begin{pmatrix} \beta_0 \\ \beta_1 \\ \beta_2 \\ \beta_3 \end{pmatrix} + \begin{pmatrix} \varrho_1 \\ \varrho_2 \\ \vdots \\ \varrho_n \end{pmatrix} .$$

The goal of the optimization method is to find the parameter values that fit the data best or which minimize a cost function S of the residuals. In the most popular and most frequently used method called *least squares regression* or *least squares fitting*, the overdetermined equations are solved by minimizing a cost function consisting of the sum of the squares of the residuals ϱ_j:

$$S(\boldsymbol{\beta}) = \sum_{j=1}^{n} \varrho_j^2 \,, \quad \text{with } \varrho_j = y_j - f(\mathbf{x}_j, \boldsymbol{\beta}) \,. \tag{4.58b}$$

In formal terms the minimization can be written as $\widehat{\boldsymbol{\beta}} = \arg\min_{\boldsymbol{\beta}\in\mathcal{B}} S(\boldsymbol{\beta})$, where $\mathcal{B}$ is the entire parameter space, and $\widehat{\boldsymbol{\beta}}$ represents the *best* choice of parameters within the framework of the least sum of squares. The minimum is found by calculating the parameter values where the gradient vanishes:

$$\frac{\partial S}{\partial \beta_j} = 2\sum_{i=1}^{n} \varrho_i \frac{\partial \varrho_i}{\partial \beta_j} = 0 \,, \quad \frac{\partial \varrho_i}{\partial \beta_j} = -X_{ij} \,, \; j = 1, \ldots, m \,,$$

$$\frac{\partial S}{\partial \beta_j} = -2\sum_{i=1}^{n} \left(y_i - \sum_{k=1} X_{ik}\widehat{\beta}_k \right) X_{ij} = 0 \,, \; j = 1, \ldots, m \,,$$

$$\left(y_i - \sum_{k=1}^{m} X_{ik}\widehat{\beta}_k \right) X_{ij} = 0 \implies \sum_{i=1}^{n}\sum_{k=1}^{m} X_{ij}X_{ik}\widehat{\beta}_k = \sum_{i=1}^{m} X_{ij}y_i \,, \; j = 1, \ldots, m \,.$$

The expressions can be readily rewritten in the matrix form

$$(\mathrm{X}^{\mathrm{t}}\mathrm{X})\widehat{\boldsymbol{\beta}} = \mathrm{X}^{\mathrm{t}}\mathbf{y} \,, \tag{4.58c}$$

which contains the derivatives of $f(x;\boldsymbol{\beta})$ in the form of the design matrix X. These equations are called *normal equations*, and they are the common starting point for the development of numerical techniques for solving linear regression problems. The matrix $\mathrm{G} = \mathrm{X}^{\mathrm{t}}\mathrm{X}$ is called the Gramian matrix, after the Danish mathematician Jørgen Pedersen Gram. It can be understood as the $m \times m$ matrix of all scalar products, viz., $G = \{G_{ij} = \mathbf{X}_i \cdot \mathbf{X}_j\}$, of m vectors of arbitrary size n. A Gramian matrix is positive semidefinite, and every Gramian matrix is derivable from a set of vectors. The Gram determinant is nonzero if and only if the m vectors are linearly independent, and this property is used for testing linear independence.

The solution of (4.58c) is straightforward and leads to

$$\widehat{\boldsymbol{\beta}} = (X^tX)^{-1}X^t\mathbf{y} = X^+\mathbf{y} \,, \tag{4.58d}$$

where $X^+ = (X^tX)^{-1}X^t$ is the so-called Moore–Penrose *pseudoinverse matrix*. It got its name from the American mathematician Eliakin Hastings Moore and the English mathematical physicist Roger Penrose, and has been used independently by the Swedish geodesist Arne Bjerhammar. The pseudoinverse matrix plays a similar role for rectangular matrices as the conventional inverse matrix does for square matrices:

$$\mathbf{y} = A\mathbf{x} \implies \mathbf{x} = A^{-1}\mathbf{y} \,,$$

$$\mathbf{y} = B\mathbf{x} \implies \mathbf{x} = B^+\mathbf{y} = (B^tB)^{-1}B^t\mathbf{y} \,,$$

where A is a non-singular square matrix and B a rectangular matrix.

In the case of multiple regression, the function to be fitted to the data $(\mathbf{x}_i, y_i)$ with $\mathbf{x}_i = \big((x_1)_i, \ldots, (x_k)_i\big)$ and $i = 1, \ldots, n$, is not a curve but a surface in k dimensions. The design matrix is not substantially different from the one in one-dimensional regression, for example:

$$\mathbf{y} = \begin{pmatrix} y_1 \\ y_2 \\ \vdots \\ y_n \end{pmatrix} = \begin{pmatrix} 1 & (x_1)_1 & (x_1)_1^2 & \ldots & (x_2)_1 & (x_2)_1^2 \\ 1 & (x_1)_2 & (x_1)_2^2 & \ldots & (x_2)_2 & (x_2)_2^2 \\ \vdots & \vdots & \vdots & \ddots & \vdots & \vdots \\ 1 & (x_1)_n & (x_1)_n^2 & \ldots & (x_2)_n & (x_2)_n^2 \end{pmatrix} \begin{pmatrix} \beta_0 \\ \beta_1^{(1)} \\ \beta_2^{(1)} \\ \vdots \\ \beta_1^{(2)} \\ \beta_2^{(2)} \end{pmatrix} + \begin{pmatrix} \varrho_1 \\ \varrho_2 \\ \vdots \\ \varrho_n \end{pmatrix} .$$

Noise Terms and Error Distributions

As already explained, the noise term $\boldsymbol{\varrho}$ covers all errors in the measured $\mathbf{y}$ values. Commonly, the errors are assumed to be independent and normally distributed, i.e., $f(\varrho_j) = \mathcal{N}(\varrho_j; 0, \sigma_j^2)$. We distinguish homoscedastic random variables with identical variances, $\mathrm{var}(\varrho_j) = \sigma^2$, $\forall\, j$, and the heteroscedastic case with different variances. In the simplest method called *ordinary least squares* and sketched above, the error terms ϱ_j are considered uncorrelated with the independent variables X_{ji} and among each other, and they are assumed to have finite and identical variances. If the assumptions of uncorrelatedness and homoscedasticity of noise are relaxed, the regression problem called *generalized least squares* can be readily handled by linear algebra, provided that the covariance matrix of the noise terms, viz., $\Sigma = (\Sigma_{ij} = \text{cov}(\varrho_i, \varrho_j))$ (Sect. 2.3.4), is known:

$$\widehat{\boldsymbol{\beta}} = (\mathrm{X}^{\mathrm{t}} \Sigma^{-1} \mathrm{X})^{-1} \mathrm{X}^{\mathrm{t}} \Sigma^{-1} \mathbf{y} . \tag{4.58e}$$

The matrix Σ covers both deviations from the idealized *ordinary* case: the diagonal elements $\Sigma_{jj} = \sigma_j^2$ take care of heteroscedasticity in the case of uncorrelatedness, and the off-diagonal elements Σ_{ji} cover correlations between the noise terms in different data. Often the assumption that the errors in all measured points have the same normal distribution is not justified, and then (4.58e) provides a useful tool for heteroscedastic data sets.

Single and Multiple Exponentials

First order reactions and in particular all relaxation processes (4.7) are described by an exponential function:

$$\frac{\mathrm{d}\eta}{\mathrm{d}t} = \frac{1}{\tau_\mathrm{R}} \eta , \quad \eta(t) = \eta_0 \exp\left(-\frac{t}{\tau_\mathrm{R}}\right) , \quad \ln \eta(t) = \log \eta_0 - \frac{t}{\tau_\mathrm{R}} , \tag{4.59}$$

Plotted on a semi-logarithmic or log-linear coordinate system, this yields a straight line.[35] We consider the determination of the relaxation time τ_R of a single exponential in the light of linear regression. The calculation is straightforward if we substitute $\log \eta(t_j) \leftrightarrow y_j$, $t_j \leftrightarrow x_{j1}$, $\log \eta_0 \leftrightarrow \beta_0$, and $1/\tau_\mathrm{R} \leftrightarrow \beta_1$ in (4.59). Then the residuals reduce to

$$\varrho_j = y_j - \beta_0 - \beta_1 t_j ,$$

allowing for direct evaluation by (4.58). The superposition of two or more exponentials, as in the case of multiple relaxations (Fig. 4.10) and in many other cases, gives rise to substantial fitting problems. The American computer scientist Forman Sinnickson Acton [4, 5] characterizes the task of fitting two exponentials as a notoriously ill-posed problem when the two relaxation times differ by less than a factor of five. We refer to several original papers dealing with this subject [80, 111, 268, 322]. Finally, we mention a method that allows one to fit parameters to a continuous spectrum of infinitely many relaxation processes [461].

[35] We remark that the plot shown in Fig. 4.10, which was used there to detect several relaxation processes on different time scales, was a $\eta(t)/\log t$ plot, whereas the semi-logarithmic plot used here is a $\log \eta(t)/t$ plot.

Binding Equilibria

The study of binding equilibria is an important tool in physical biochemistry. Enzyme–substrate binding, for example,

$$\mathbf{S} + \mathbf{E} \underset{l_1}{\overset{k_1}{\rightleftharpoons}} \mathbf{S} \cdot \mathbf{E} \,, \tag{4.60a}$$

is the first step in Michaelis–Menten kinetics, and the assumption of a pre-equilibrium was one of several simplification strategies (Sect. 4.1.2).

The condition for the binding equilibrium with $[\mathbf{S} \cdot \mathbf{E}] = c$ is

$$k_1 \bar{s}\bar{e} = l_1 \bar{c} \,, \quad K_1 = \frac{k_1}{l_1} = \frac{\bar{c}}{\bar{s}\bar{e}} \,, \quad \text{with } s_0 = \bar{s} + \bar{c} \,, \quad e_0 = \bar{e} + \bar{c} \,, \tag{4.60b}$$

where s_0 and e_0 are the total concentrations of substrate and enzyme, respectively. Chemical and biochemical kinetics often require knowledge of the free substrate concentration s, which can be calculated from known binding constants K:

$$\bar{s} = \frac{1}{2}\left((s_0 - e_0 - L_1) + (s_0 - e_0 + L_1)\sqrt{1 + \frac{4\, e_0 L_1}{(s_0 - e_0 + L_1)^2}} \right) , \tag{4.60c}$$

where $L_1 = K_1^{-1} = l_1/k_1$ is the dissociation constant of the enzyme–substrate complex $\mathbf{S} \cdot \mathbf{E}$.

The most common strategy for determining the binding constants K consists in measuring the free equilibrium substrate concentrations $\bar{s}$ for given input concentrations s_0 and e_0. The idea is to calculate K by some parameter fitting method and, for example, ordinary least squares may be applied. For the purposes of illustration, we show here the determination of the binding constant by means of a linearized plot, called the Scatchard plot [476].

Linearization of Binding Equilibria

The Scatchard equation is presented as an example of a nonlinear relation transformed in such a way as to obtain a linear plot (Fig. 4.15). It was invented to analyze the binding equilibria of small ligands to macromolecules. The plot is named after George Scatchard, who was a chemist at the Massachusetts Institute of Technology (MIT). Today it is mainly of historical interest, but it still has the advantage that the quality of data is easily checked by visual inspection. The binding equlibrium $\mathbf{S}{+}\mathbf{E} \rightleftharpoons \mathbf{S}{\cdot}\mathbf{E}$ involves a ligand $\mathbf{S}$, a macromolecule $\mathbf{E}$, usually a protein, and the association complex $\mathbf{C} \equiv \mathbf{S} \cdot \mathbf{E}$, with the free concentrations $[\mathbf{S}] = s$, $[\mathbf{E}] = e$, and $[\mathbf{S} \cdot \mathbf{E}] = c$, and the total concentrations of ligand and protein, $s_0 = s + c$

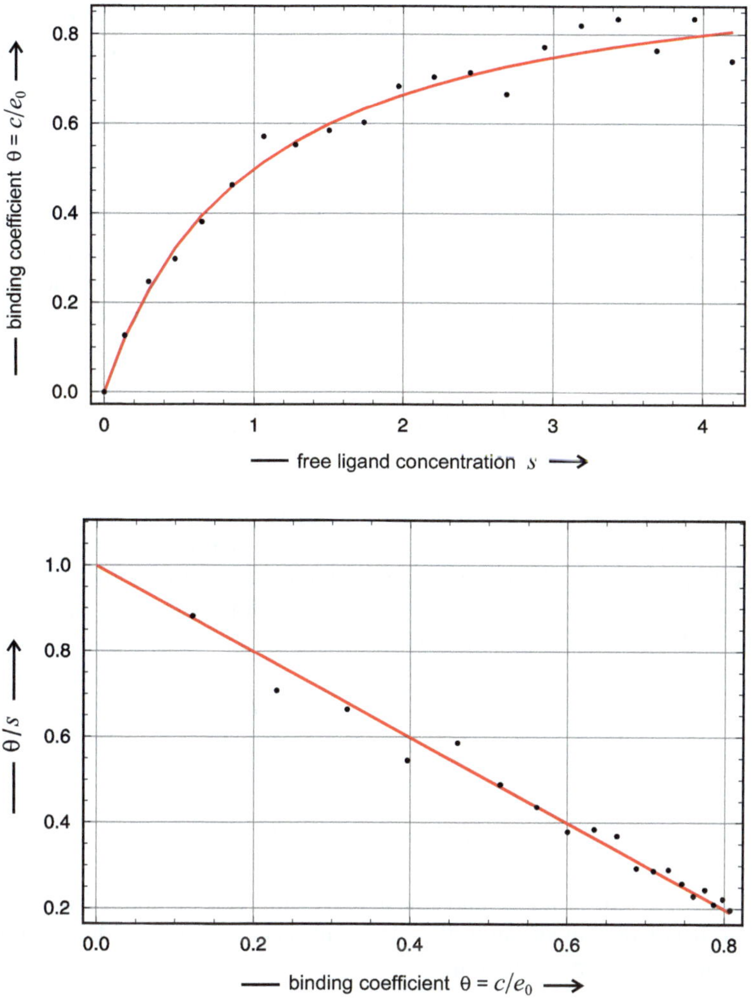

Fig. 4.15 The Scatchard plot and fitting of binding constants. *Upper*: Hyperbolic binding isotherm of the binding equilibrium $\mathbf{S}+\mathbf{E} \rightleftharpoons \mathbf{S}\cdot\mathbf{E}$ according to (4.60d). The degree of saturation or the binding coefficient $\theta = c/e_0$ is plotted against the free ligand concentration s. Random scatter is introduced, for example, through errors in the determination of the free concentrations. *Lower*: Scatchard plot of the same data according to (4.60e): θ/s is plotted against θ. Parameter choice: $K = 1$, $s_0 = e_0 = 1$

and $e_0 = e + c$, respectively. The equilibrium constant is formulated as a binding constant K_1 or as a dissociation constant L_1. Biochemists prefer the latter choice, and we shall also adopt it here and drop the subscript: $L_1 \equiv L$.

The experimental investigation of binding is conventionally carried out in terms of the *degree of saturation* or the *binding coefficient* $\theta = c/s_0$ with the property

$0 \leq \theta \leq 1$, where $\theta = 0$ refers to no binding and $\theta = 1$ to complete saturation of the protein. The function $\theta(s)$ is generally characterized as the *binding isotherm*[36]:

$$\theta(s) = \frac{c}{e_0} = \frac{s}{L+s}, \quad s = \frac{1}{2}\left((s_0 - e_0 - L) + \sqrt{4Ls_0 + (s_0 - e_0 - L)^2}\right), \tag{4.60d}$$

where the free concentration s of the ligand is nonlinearly related to the initial concentrations s_0 and e_0. The nonlinear relation (4.60d) is not suitable to determine the asymptotic maximum of the binding curve by visual inspection and to adjust a value to the dissociation constant L (Fig. 4.15), but it can be transformed and one of the resulting linear relations is known as a Scatchard plot:

$$\frac{\theta}{s} = \frac{1}{L}(1-\theta) = \frac{1}{L} - \frac{1}{L}\theta. \tag{4.60e}$$

The binding constant is obtained from the slope of the straight line, i.e., $\alpha = K$ in the plot. Figure 4.15 shows a typical example. The scatter of points was obtained by superimposing a random component on the concentration c of the complex. A derivation of the Scatchard equation starts by dividing both sides of (4.60d) by s, i.e., $\theta/s = 1/(L+s)$ and proceeds as follows:

$$\frac{1}{L+s} = \frac{1}{L} - \frac{1}{L} + \frac{1}{L+s} = \frac{1}{L} - \frac{L+s-s}{L(L+s)}$$

$$= \frac{1}{L} - \frac{s}{L(L+s)} = \frac{1}{L} - \frac{\theta}{L}. \qquad \square$$

The $(\theta/s, \theta/L)$ plot is linear and linear regression can be applied to the binding problem. It is worth mentioning that the result is exact, and we did not perform a linearization as an approximation.

For current methods in parameter estimation, we refer to two monographs as samples of the enormous literature [413, 538]. Present day numerical analysis of measured data is mostly based on the application of inverse methods. We give here a few references to reviews and monographs [28, 138, 139, 524] and mention one recent paper on parameter analysis of the multistep reaction of chlorite with iodide that aims to determine the data-sensitive parameters by means of sparsity regularization [320].

[36] The notion of *isotherm* refers to the fact that the curve is recorded at constant temperature, thereby indicating the existence of a pronounced temperature dependence of the equilibrium parameter.

4.2 Stochasticity in Chemical Reactions

There are two frequently applied techniques for analyzing stochasticity in chemical reaction systems:

(i) Modeling and simulation of sample trajectories that correspond to single experimental recordings.
(ii) Analysis of chemical master equations that provide a probability distribution over the admissible states as functions of time.

Simulations of kinetic trajectories (Sect. 4.6) constitute the computer-based counterpart of recording experiments. The simulations involve relatively easy but time-consuming numerical computations which provide the desired results but are not generally very insightful. Modeling of sample trajectories through the search for solutions of chemical Langevin equations (Sect. 3.188) is not particularly popular, but it provides another powerful technique for the analysis of chemical reactions. Solving chemical master equations (Sect. 4.2.2) would be the method of choice, were there not a lack of general methods for solving nonlinear partial differential equations. The probability densities once derived are the probabilistic counterparts to analytical solutions for kinetic differential equations and provide direct access to full information on the underlying processes.

Chemical Langevin and Fokker–Planck equations are based on continuous stochastic variables which correspond to the concentrations in the deterministic equations. Chemical master equations and numerical simulations use discrete stochastic variables $\mathcal{X}(t)$ which represent particle numbers. Hence, they can adopt only nonnegative integer values, i.e., $n \in \mathbb{N}$, and the probabilities are given by $P_n(t) = P\big(\mathcal{X}(t) = n\big)$.[37] Some conventions and notational simplifications are introduced. We shall use forward equations unless stated otherwise and assume infinitely sharp initial densities $P(n, 0|n_0, 0) = \delta_{n,n_0}$ with $n_0 = n(0)$. The full notation will be simplified by putting $P(n, t|n_0, 0) \rightarrow P_n(t)$. In addition, the notation $P_n(t)$ already indicates that t is a continuous variable, whereas n is discrete. The expectation value of the stochastic variable $\mathcal{X}(t)$ will be denoted by

$$E\big(\mathcal{X}(t)\big) = \langle n(t) \rangle = \sum_{n=0}^{\infty} n P_n(t) \,, \tag{4.61}$$

and its stationary value, provided it exists, will be expressed as

$$\bar{n} = \lim_{t \to \infty} \langle n(t) \rangle \,. \tag{4.62}$$

[37] If a second running index for integers is needed, it will be denoted by m. In cases where more than two running indices are required, we shall use n', m', etc.

Often, but not always, the stationary expectation value $\bar{n}$ will be identical with the long-time value of the corresponding deterministic variable.

4.2.1 Sampling of Trajectories

Sampling of trajectories is the most popular computer simulation technique for stochastic processes. The computation of individual trajectories is described extensively in Sect. 4.6.3. Here we present only a brief survey of the basic facts of the concept of sampling, but many specific examples will follow in Sect. 4.6.4 and in Chap. 5.

Individual Trajectories
The concept of trajectories in stochastic processes was introduced in equation (3.3). Here we adapt it to chemical reactions. The composition of the reaction mixture is described by the random vector $\boldsymbol{\mathcal{X}}(t)$, which changes when a chemical reaction takes place. Accordingly, the stochastic process is discrete in particle numbers and in time, and the spontaneous changes or jumps are determined by stoichiometry. An individual trajectory is a step function and can be written as

$$\mathcal{T}_j = \Big(\big(\boldsymbol{\mathcal{X}}_j(t_0^{(j)}), t_0^{(j)}\big), \big(\boldsymbol{\mathcal{X}}_j(t_1^{(j)}), t_1^{(j)}\big), \ldots, \big(\boldsymbol{\mathcal{X}}_j(t_n^{(j)}), t_n^{(j)}\big) \Big) ,$$

where $\boldsymbol{\mathcal{X}}_j(t) = \big(\mathcal{X}_{1j}(t), \ldots, \mathcal{X}_{Mj}(t)\big)$ is the random vector of particle numbers of M different chemical species at time t. In a specific experiment, we record $\boldsymbol{\mathcal{X}}_j(t) = \mathbf{n}_j(t) = \big(\mathrm{n}_{1j}(t), \ldots, \mathrm{n}_{Mj}(t)\big)$, with $\mathrm{n}_{ij} \in \mathbb{N}$. Reaction events occur at times $t_k^{(j)}$ with $k \in \mathbb{N}$, and accordingly the number of X_i molecules at the end of the $k^{(j)}$th interval $\big[t_{k-1}^{(j)}, t_k^{(j)} = t_{k-1}^{(j)} + \Delta t_k^{(j)}\big]$ will be given by

$$\mathbf{n}(t_k^{(j)}) = \mathbf{n}(t_{k-1}^{(j)}) + \mathbf{s} , \quad \text{or } \mathrm{n}_i(t_k^{(j)}) = \mathrm{n}_i(t_{k-1}^{(j)}) + s_i , \ i = 1, \ldots, M , \tag{4.63}$$

where s_i is the ith component of the single reaction stoichiometric vector $\mathbf{s} = (s_1, \ldots, s_M)^{\mathrm{t}}$ (Fig. 4.16 upper). It is important to recall that the jumps occur at different times in individual trajectories, as indicated by the notation $t_k^{(j)}$ (Fig. 4.16 lower). This is a result of the independence of random intervals in different trajectories, which, as outlined in Sect. 2.3.1, implies that the number of events as a function of time follows a Poisson distribution, and lengths of intervals $[t_{k-1}^{(j)}, t_k^{(j)}]$ are exponentially distributed. As an illustrative example, we show a single trajectory and a bundle of five trajectories for the reaction $\mathsf{A} + \mathsf{B} \rightleftharpoons \mathsf{C} + \mathsf{D}$ in Fig. 4.16.

Sampling and Evaluation
An ensemble of L trajectories is computed, yielding a bundle of step functions $\mathcal{T} = \{\mathcal{T}_j , \ j = 1, \ldots, L\}$ that can be evaluated at predefined times t_k according to (3.4). The simulation may be called 'exact' because the probability distribution

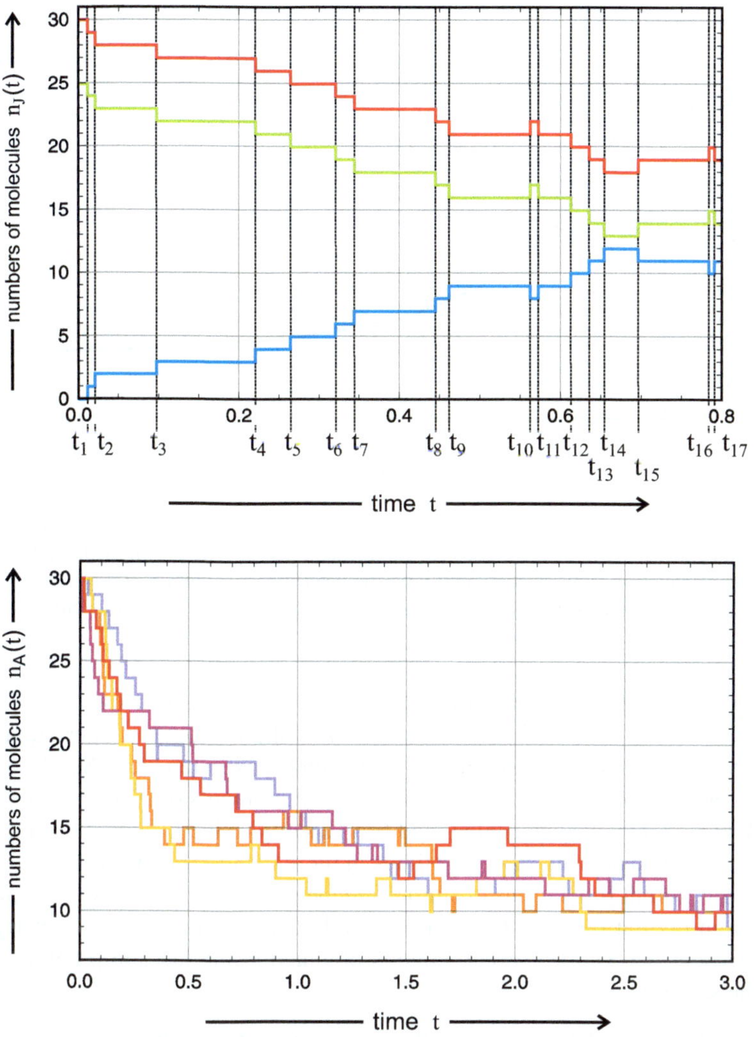

Fig. 4.16 Stochastic trajectories of the reaction $\mathsf{A} + \mathsf{B} \to \mathsf{C} + \mathsf{D}$. The plot in the *upper part* of the figure shows a trajectory of the irreversible bimolecular conversion reaction. The four variables are $\mathcal{X}_{\mathsf{A}}(\mathrm{t}) = n_{\mathsf{A}}(\mathrm{t})$ (*red*), $\mathcal{X}_{\mathsf{B}}(\mathrm{t}) = n_{\mathsf{B}}(\mathrm{t})$ (*green*), $\mathcal{X}_{\mathsf{C}}(\mathrm{t}) = n_{\mathsf{C}}(\mathrm{t}) = \mathcal{X}_{\mathsf{D}}(\mathrm{t}) = n_{\mathsf{D}}(\mathrm{t})$ (*blue*). Reaction events occur at times t_k ($k = 1, \ldots$) and give rise to jumps determined by stoichiometry: $\Delta n_{\mathsf{A}}(\mathrm{t}_k) = \Delta n_{\mathsf{B}}(\mathrm{t}_k) = \mp 1$, and $\Delta n_{\mathsf{C}}(\mathrm{t}_k) = \Delta n_{\mathsf{D}}(\mathrm{t}_k) = \pm 1$. The intervals $\Delta \mathrm{t}_k = \mathrm{t}_k - \mathrm{t}_{k-1}$ have an exponential distribution. The *lower plot* presents the superposition of five trajectories $(\mathcal{X}_{\mathsf{A}})_j(\mathrm{t})$ with $j = 1, \ldots, 5$. The trajectories in different colors illustrate the uncorrelatedness of the reaction events. Nevertheless an approximate general shape of the trajectories can be easily recognized. Choice of parameters: $k = 0.06$, $l = 0.01$, $n_{\mathsf{A}}(0) = 30$, $n_{\mathsf{B}}(0) = 25$, and $n_{\mathsf{C}}(0) = n_{\mathsf{D}}(0) = 0$

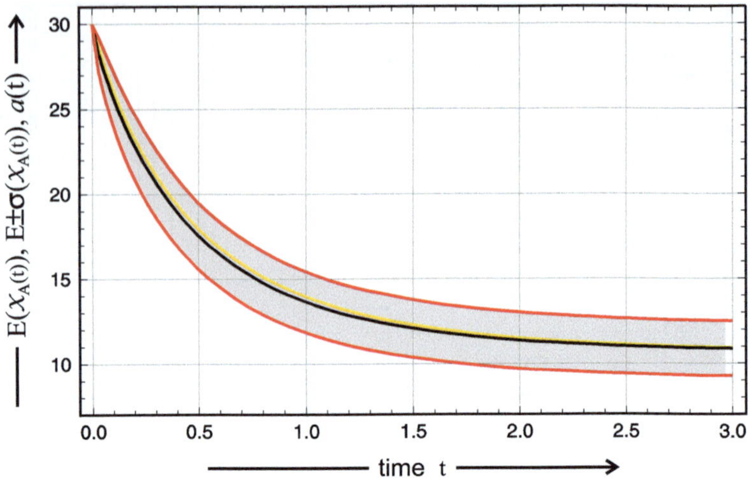

Fig. 4.17 Expectation value and variance of the reaction $\mathsf{A} + \mathsf{B} \rightarrow \mathsf{C} + \mathsf{D}$. The figure shows the expectation value $\mathrm{E}\big(\mathcal{X}_\mathsf{A}(t)\big)$ (*black*) and the confidence interval (*gray*) with the boundaries $\mathrm{E} \pm \sigma$ (*red*). The expectation value is compared with the deterministic solution $a(t)$ (*yellow*). Choice of parameters: $k = 0.06$, $l = 0.01$, $a_0 = 30$, and $b_0 = 25$. Sample size 10000 trajectories

obtained by sampling the trajectories $\{\boldsymbol{\mathcal{X}}_j(t); j = 1, \ldots, L\}$ at any time t_k converges in distribution to the corresponding solution of the analogue master equation for $L \to \infty$. Provided L is sufficiently large, finite size statistics (Sect. 2.6) yields perfect results. For example, the first and second moments of $\mathcal{X}_\mathsf{A}(t)$ for the reaction $\mathsf{A} + \mathsf{B} \rightleftharpoons \mathsf{C} + \mathsf{D}$ have been calculated from a sample size $L = 10{,}000$ (Fig. 4.17).

It is worth considering sampling times in more detail. The instances at which the jumps in different trajectories occur are independent, and in general the values are taken from the flat plateaus between jumps. Because of the càdlàg property of the process (Sect. 3.1.3), the correct value is obtained at the last jump before and not after the sampling time.

In addition, Fig. 4.17 compares the computed mean value $m\big(\mathcal{X}_\mathsf{A}(t)\big)$, which is in excellent agreement with the expectation value $\mathrm{E}\big(\mathcal{X}_\mathsf{A}(t)\big)$, and the deterministic concentration $a(t)$. Although the curves are very close, they do not coincide, which is not unexpected since the reaction $\mathsf{A} + \mathsf{B} \rightleftharpoons \mathsf{C} + \mathsf{D}$ has a nonlinear rate $v^{\mathrm{ma}} = -k_1 a(b_0 - a_0 + a) + l_1(a_0 - a)^2$ (see Sect. 3.2.3).

4.2.2 The Chemical Master Equation

The *chemical master equation* has been shown to be based on a rigorous microscopic concept of chemical reactions in the vapor phase within the framework of classical collision theory [209], provided that two general requirements are fulfilled:

(i) A homogeneous mixture, assumed to exist after thorough stirring.
(ii) Thermal equilibrium implying that the velocities of molecules have a Maxwell–Boltzmann distribution.

Daniel Gillespie's approach focuses on chemical reactions rather than molecular species and is well suited to handling reaction networks. In addition, the algorithm can be easily implemented for computer simulation. In Sect. 4.6, we shall discuss the Gillespie algorithm together with computer implementations. Although the numerical approach is straightforward and yields excellent results for specific examples and small population sizes there is, at the same time, a need for an analytical approach in order to find answers to general questions that cannot be given by the numerical simulations.

Chemical Master Equations
The birth-and-death forward master equation is written in the form

$$\frac{\mathrm{d}P_n(t)}{\mathrm{d}t} = w_{n-1}^{+} P_{n-1}(t) + w_{n+1}^{-} P_{n+1}(t) - (w_n^{+} + w_n^{-}) P_n(t) \,, \tag{3.97}$$

where transitions are restricted to neighboring states, i.e., $\varDelta n = \pm 1$, and transition probabilities are assumed to be independent of time:

$$w_n^{+} \equiv \lambda_n \,, \quad \text{for } n \to n+1 \,, \tag{3.96a}$$

$$w_n^{-} \equiv \mu_n \,, \quad \text{for } n \to n-1 \,. \tag{3.96b}$$

The adaptation to chemical reactions is straightforward (see, e.g., [209]). The restriction to neighboring states is translated into the occurrence of single reaction events, and the single step is replaced by stoichiometry in the sense $\varDelta n = \pm 1 \iff \Delta \mathbf{n} = \mathbf{s}$. The differential reaction probability $\chi(\mathbf{n})$ is the product of a rate parameter γ and a frequency function $h(\mathbf{n})$ which expresses the number of distinct combinations of $\mathbf{R}$ reactant molecules in the system:

$$\chi(\mathbf{n})\,\mathrm{d}t = \gamma h(\mathbf{n})\,\mathrm{d}t \,. \tag{4.64}$$

The restriction to single events is commonly formulated in terms of three conditions which are readily adapted for applications to chemical kinetics:

Condition 1. If $\mathcal{X}(t) = \mathbf{n}$, then the probability that *exactly one reaction* $\mathbf{R}$ will occur in the system within the time interval $[t, t + \mathrm{d}t[$ is equal to

$$\chi(\mathbf{n})\,\mathrm{d}t + o(\mathrm{d}t) = \gamma h(\mathbf{n})\,\mathrm{d}t + o\,(\mathrm{d}t) \,,$$

where $o(\mathrm{d}t)$ denotes terms that approach zero with $\mathrm{d}t$ faster than $\mathrm{d}t$.

Condition 2. If $\mathcal{X}(t) = \mathbf{n}$, then the probability that *no* reaction will occur within the time interval $[t, t + \mathrm{d}t[$ is equal to

$$1 - \chi(\mathbf{n})\,\mathrm{d}t + o(\mathrm{d}t) = 1 - \gamma h(\mathbf{n})\,\mathrm{d}t + o(\mathrm{d}t)\;.$$

Condition 3. The probability of more than one reaction occurring in the system within the time interval $[t, t + \mathrm{d}t[$ is of order $o(\mathrm{d}t)$.

Based on these three conditions, the description of the process can be derived in terms of the population vector $\mathcal{X}(t)$. The initial state of the system at some initial time t_0 is fixed: $\mathcal{X}(t_0) = \mathbf{n}_0$. The probability $P(\mathbf{n}, t + \mathrm{d}t|\mathbf{n}_0, t_0)$ is expressed as the sum of the probabilities of several mutually exclusive and collectively exhaustive *routes* from $\mathcal{X}(t_0) = \mathbf{n}_0$ to $\mathcal{X}(t + \mathrm{d}t) = \mathbf{n}$. These routes are distinguished from one another with respect to the three conditions assigned to events that happened in the last time interval $[t, t + \mathrm{d}t[$:

$$\begin{aligned} P(\mathbf{n}, t + \mathrm{d}t|\mathbf{n}_0, t_0) &= P(\mathbf{n}, t|\mathbf{n}_0, t_0) \times \big(1 - \chi(\mathbf{n})\,\mathrm{d}t + o(\mathrm{d}t)\big) \\ &\quad + P(\mathbf{n} - \mathbf{s}, t|\mathbf{n}_0, t_0) \times \big(\chi(\mathbf{n} - \mathbf{s})\,\mathrm{d}t + o(\mathrm{d}t)\big) \\ &\quad + o(\mathrm{d}t)\;. \end{aligned} \tag{4.65}$$

Only in a few cases is it possible to derive an exact solution for the time evolution of the probability function $P(\mathbf{n}, t|\mathbf{n}_0, t_0)$ by solving the chemical master equation, but a deterministic function for the differential change in the probability for $t \geq t_0$ is readily obtained. The three different routes from $\mathcal{X}(t_0) = \mathbf{n}_0$ to $\mathcal{X}(t + \mathrm{d}t) = \mathbf{n}$ are obvious from the balance equation (4.65):

(i) One route from $\mathcal{X}(t_0) = \mathbf{n}_0$ to $\mathcal{X}(t + \mathrm{d}t) = \mathbf{n}$ is given by the first term on the right-hand side of the equation: *no reaction* occurs in the time interval $[t, t+ \mathrm{d}t[$, and hence $\mathcal{X}(t) = \mathbf{n}$ was also satisfied at time t. The joint probability for this route is therefore *the probability of being in* $\mathcal{X}(t) = \mathbf{n}$, *conditioned by* $\mathcal{X}(t_0) = \mathbf{n}_0$ times *the probability that no reaction has occurred in* $[t, t + \mathrm{d}t[$. In other words, the probability for this route is the probability of going from $\mathbf{n}_0$ at time t_0 to $\mathbf{n}$ at time t, and to stay in this state during the next interval $\mathrm{d}t$.

(ii) An alternative route from $\mathcal{X}(t_0) = \mathbf{n}_0$ to $\mathcal{X}(t + \mathrm{d}t) = \mathbf{n}$ is accounted for by the second term on the right-hand side of the equation: *exactly one reaction* **R** occurs in the time interval $[t, t + \mathrm{d}t[$, and hence $\mathcal{X}(t) = \mathbf{n} - \mathbf{s}$ is satisfied at time t. The joint probability for this route is therefore *the probability of being in* $\mathcal{X}(t) = \mathbf{n} - \mathbf{s}$, *conditioned by* $\mathcal{X}(t_0) = \mathbf{n}_0$ times *the probability that exactly one reaction* **R** *has occurred in* $[t, t + \mathrm{d}t[$. In other words, the probability for this route is the probability of going from $\mathbf{n}_0$ at time t_0 to $\mathbf{n} - \mathbf{s}$ at time t by undergoing a reaction yielding $\mathbf{n}$ during the next interval $\mathrm{d}t$.

(iii) The third possibility, neither *no* reaction nor *exactly one* reaction, must inevitably invoke *more than one reaction within the time interval* $[t, t + \mathrm{d}t[$. The probability for such events, however, is $o(\mathrm{d}t)$ or of measure zero.

The three routes are mutually exclusive since different events take place within the last interval $[t, t + \mathrm{d}t[$.

The last step to derive the *chemical master equation* is straightforward: $P(\mathbf{n}, t|\mathbf{n}_0, t_0)$ is subtracted from both sides in (4.65), then both sides are divided by dt, the limit $\mathrm{d}t \downarrow 0$ is taken, whence all $o(\mathrm{d}t)$ terms vanish, and finally we obtain

$$\frac{\mathrm{d}}{\mathrm{d}t}P(\mathbf{n}, t|\mathbf{n}_0, t_0) = \gamma h(\mathbf{n} - \mathbf{s})P(\mathbf{n} - \mathbf{s}|\mathbf{n}_0, t_0) - \gamma h(\mathbf{n})P(\mathbf{n}, t|\mathbf{n}_0, t_0) . \tag{4.66}$$

Initial conditions are required to calculate the time evolution of the probability $P(\mathbf{n}, t|\mathbf{n}_0, t_0)$ and, for sharp initial conditions, we can easily express them in the form

$$P(\mathbf{n}, t_0|\mathbf{n}_0, t_0) = \delta_{\mathbf{n},\mathbf{n}_0} = \begin{cases} 1 , & \text{if } \mathbf{n} = \mathbf{n}_0 , \\ 0 , & \text{if } \mathbf{n} \neq \mathbf{n}_0 , \end{cases} \tag{4.66$'$}$$

which is precisely the initial condition used in the derivation of (4.65): $P\big(n_k, t_0|n_k^{(0)}, t_0\big) = \delta(n_k - n_k^{(0)}) , \quad \forall\, k$. The assumption of extended initial distributions is, of course, also possible, but the corresponding master equations become more sophisticated.

Two Examples

First, we write down the master equation for the simple monomolecular chemical reaction (4.1c), viz.,

$$\mathsf{A} \xrightarrow{\;k\;} \mathsf{B} ,$$

with the constraint of a constant total number of molecules. The two random variables $\mathcal{X}_{\mathsf{A}}$ and $\mathcal{X}_{\mathsf{B}}$ satisfy the condition

$$\mathcal{X}_{\mathsf{A}}(t) + \mathcal{X}_{\mathsf{B}}(t) = n_0 .$$

For $\mathcal{X}_{\mathsf{A}} = n$ and $\mathcal{X}_{\mathsf{B}} = n_0 - n$, it is straightforward to write down the single step master equation for the initial conditions $P(n, 0) = \delta(n - n_0)$:

$$\frac{\mathrm{d}P_n(t)}{\mathrm{d}t} = k(n + 1)P_{n+1}(t) - knP_n(t) , \quad n = 0, 1, \ldots, n_0 ,$$

with $P_{n+1}(t) = 0, \forall\, t \in \mathbb{R}_{\geq 0}$, which is identical with the simple death master equation. Further discussion of the equation and its solution can be found in Sect. 4.3.2. For direct comparison with birth-and-death processes, it is interesting to characterize the elementary chemical steps in this simple case as step-up and step-down transitions: $w_n^- = kn$ and $w_n^+ = 0$. We notice that the process has a built-in or *natural* absorbing boundary at $n = 0$, since $w_0^- = 0$ and $w_0^+ = 0$ without any further assumption.

The second example deals with the stochastic description of the reversible bimolecular conversion reaction (4.31):

$$\mathsf{A} + \mathsf{B} \xrightarrow{\ k\ } \mathsf{C} + \mathsf{D} ,$$

$$\mathsf{C} + \mathsf{D} \xrightarrow{\ l\ } \mathsf{A} + \mathsf{B} .$$

The four random variables $\mathcal{X}_\mathsf{A}(t)$, $\mathcal{X}_\mathsf{B}(t)$, $\mathcal{X}_\mathsf{C}(t)$, and $\mathcal{X}_\mathsf{D}(t)$, are combined with three conservation relations, viz.,

$$\mathcal{X}_\mathsf{A}(t) + \mathcal{X}_\mathsf{B}(t) + \mathcal{X}_\mathsf{C}(t) + \mathcal{X}_\mathsf{D}(t) = \mathcal{X}_\mathsf{A}(0) + \mathcal{X}_\mathsf{B}(0) + \mathcal{X}_\mathsf{C}(0) + \mathcal{X}_\mathsf{D}(0) ,$$

$$\mathcal{X}_\mathsf{A}(t) - \mathcal{X}_\mathsf{B}(t) = \mathcal{X}_\mathsf{A}(0) - \mathcal{X}_\mathsf{B}(0) ,$$

$$\mathcal{X}_\mathsf{C}(t) - \mathcal{X}_\mathsf{D}(t) = \mathcal{X}_\mathsf{C}(0) - \mathcal{X}_\mathsf{D}(0) ,$$

which leave only one degree of freedom. Once again, we choose $\mathcal{X}_\mathsf{A}(t)$ as the independent variable: $P_n(t) = P(\mathcal{X}_\mathsf{A} = n)$. In order to simplify, we assume as initial conditions that only A and B are present at time $t = 0$, and that they have sharp values: n_0 molecules A, $P_n(0) = \delta_{n,n_0}$, and b_0 molecules B, $P\big(\mathcal{X}_\mathsf{B}(0) = b\big) = \delta_{b,b_0}$, and we have $\mathcal{X}_\mathsf{B}(t) = \vartheta_0 + \mathcal{X}_\mathsf{A}(t)$ with $\vartheta_0 = b_0 - n_0$, and $\mathcal{X}_\mathsf{C}(t) = \mathcal{X}_\mathsf{D}(t) = n_0 - \mathcal{X}_\mathsf{A}(t)$. Under these conditions the master equation becomes

$$\begin{aligned}\frac{\mathrm{d}P_n(t)}{\mathrm{d}t} &= k(n+1)(\vartheta_0 + n + 1)P_{n+1}(t) + l(n_0 - n + 1)^2 P_{n-1} \\ &\quad -\big(kn(\vartheta_0 + n) + l(n_0 - n)^2\big)P_n(t) .\end{aligned}$$

This master equation is based on the step-up and step-down transitions

$$w_n^- = kn(\vartheta_0 + n) , \qquad w_n^+ = l(n_0 - n)^2 ,$$

which satisfy

$$\big(w_{n_0}^- = kn_0(\vartheta_0 + n_0) > 0 ,\ w_{n_0}^+ = 0\big) , \qquad \big(w_0^- = 0 ,\ w_0^+ = ln_0^2 > 0\big) ,$$

and sustain two in-built reflecting boundaries at $n = n_0$ and $n = 0$, respectively, thus leaving $0 \le n \le n_0$, $n \in \mathbb{N}$, as the physically meaningful accessible domain. The master equation for the irreversible reaction (4.31a) has been solved and will be discussed in Sect. 4.3.3. The full reversible reaction is rather hard to solve and we dispense from further analysis because size expansion, numerical simulation, and the chemical Langevin equation are to be preferred for practical purposes.

Chemical Master Equation and Stoichiometry
As before, reactants and reaction products are described by a random vector $\boldsymbol{\mathcal{X}}$ with M components for M molecular species, viz.,

$$\boldsymbol{\mathcal{X}} = (\mathcal{X}_1, \ldots, \mathcal{X}_M)\,, \quad \text{with } \mathcal{X}_i \in \mathbb{N}\,.$$

A birth-and-death master equation (3.97) with only two classes of allowed changes, $n \to n \pm 1$, is assumed, and these steps imply changes $\mathbf{n} \to \mathbf{n} + \mathbf{s}$ for the chemical reaction $\mathbf{R}$. In other words, if a reaction $\mathbf{R}$ occurs at time t, then the random vector $\boldsymbol{\mathcal{X}} \in \mathbb{N}^M$ changes in accordance with the stoichiometry: $\boldsymbol{\mathcal{X}}(t) = \boldsymbol{\mathcal{X}}(t - \mathrm{d}t) + \mathbf{s}$. Like the rate function $v^{(\mathrm{ma})}(\mathbf{n}) = k\,h(\mathbf{n})$ in conventional kinetics, its counterpart $\chi^{(\mathrm{ma})}(\mathbf{n})$ is fundamental in stochastic reaction kinetics.[38] In mass action kinetics, it has the general form

$$\chi^{(\mathrm{ma})}(\mathbf{n}) = \gamma h(\mathbf{n}) = \gamma \prod_{i=1}^{M} \frac{n_i!}{(n_i - \nu_i)!}\,. \tag{4.67}$$

The expression differs in two respects from the deterministic rate functions $v^{(\mathrm{ma})}$ in (4.6):

(i) γ is meant here as a probabilistic rate coefficient, which replaces the deterministic rate parameter k (Sect. 4.1.4), but as mentioned, the two rate parameters are almost always assumed to be the same, i.e., $\gamma \equiv k$.[39]
(ii) The functions $h(\mathbf{n})$ are different, because the stoichiometric coefficients $|\nu_i| > 1$ require explicit consideration of the number of distinct subsets of molecules.[40]

Alternatively, we may consider the reaction as a sequence of events modeled by a continuous time Markov chain, where $p(\tau|\mathbf{n}, t)\,\mathrm{d}\tau$ is the so-called *next-reaction density function*, which expresses the probability that the next reaction $\mathbf{R}$ will occur in the infinitesimal time interval $[t + \tau, t + \tau + \mathrm{d}\tau[$. Every reaction event changes the random variables by $\mathbf{s}$, and individual trajectories follow an equation with changes

[38] In stochastic kinetics, the differential reaction rate $\chi^{(\mathrm{ma})}(\mathbf{n}) = \gamma h(\mathbf{n})$ is also known as the *intensity* or *propensity function*. The product term in (4.67) results from the combinatorics of molecular encounters leading to $\binom{n_i}{\nu_i} = n_i!/\nu_i!(n_i - \nu_i)!$ combinations of molecules, whereby the factor $1/\nu_i!$ is absorbed into the stochastic rate coefficient γ in order to obtain an expression that comes as close as possible to deterministic kinetics. We remark that the alternative notation with intact binomial factors and $\gamma \longleftrightarrow \gamma\nu_i!$ is also common.

[39] Unless otherwise stated, we shall also use (k_j, l_j) as rate parameters in the stochastic models. Care has to be taken over concentration units. Concentrations commonly given in $[\mathrm{mol\,l^{-1}}]$ have to be converted into dimensionless units counting particle numbers (see also Sect. 4.6).

[40] For example, the rate for dimerization $2\mathsf{A} \to \mathsf{C}$ is $v^{(\mathrm{ma})} = -ka^2$ in the deterministic case and $\chi_n^{(\mathrm{ma})} = -kn(n-1)$ in the stochastic case. The two expressions become identical in the limit of large numbers where we have $n \gg 1$.

at random times:

$$\mathcal{X}(t) = \mathcal{X}(0) + \mathbf{s}\mathcal{Y}\left(\int_0^t \gamma_{\mathbf{n}} \mathrm{d}\tau\right) , \tag{4.68}$$

where $\mathcal{Y}(t)$ is an unit-rate Poisson process with the probability density $\pi_k(\tau) = \mathrm{e}^{-\tau}\tau^k/k!$. The unit Poisson process $\mathcal{Y}(t)$ is a counting process and provides the times when the jumps in the variable $\mathcal{X}(t)$ occur, whereas the stoichiometric parameters $\mathbf{s} = \boldsymbol{\nu}' - \boldsymbol{\nu}$ give the size including the sign.

Master Equation and Deterministic Kinetics

The chemical master equation (4.66) allows for computation of expectation values. We illustrate by means of our simple example $\mathsf{A} \to \mathsf{B}$:

$$\begin{aligned}
\sum_{n=0}^{n_0} n \frac{\mathrm{d}P_n(t)}{\mathrm{d}t} &= \frac{\mathrm{d}}{\mathrm{d}t} \sum_{n=0}^{n_0} nP_n(t) = \frac{\mathrm{d}\langle n \rangle}{\mathrm{d}t} \\
&= \sum_{n=0}^{n_0} n\big(k(n+1)P_{n+1}(t) - knP_n(t)\big) \\
&= k\sum_{n=0}^{n_0} n(n+1)P_{n+1}(t) - k\sum_{n=0}^{n_0} n^2 P_n(t) \\
&= k \sum_{n'=n+1=1}^{n_0+1} (n'-1)n'P_{n'}(t) - k\sum_{n=0}^{n_0} n^2 P_n(t) \\
&= k\left\langle n^2 \right\rangle - k\left\langle n \right\rangle - k\left\langle n^2 \right\rangle = -k\left\langle n \right\rangle .
\end{aligned}$$

In this case the macroscopic rate equation is readily derived from the master equation by interpreting the expectation value: $\langle n \rangle$ is a real number and up to the factor $1/VN_{\mathrm{L}}$ represents the concentration of a molecular species: $x = \langle n \rangle / VN_{\mathrm{L}} \in \mathbb{R}_{\geq 0}$, and $\mathrm{d}x/\mathrm{d}t = -kx$.

Coincidence of the expectation value $\langle n \rangle$ and the deterministic particle number $\widehat{n} = x(VN_{\mathrm{L}})$ is restricted to cases where the reaction rate function is linear. A comparison of linear and nonlinear cases has already been presented and discussed in the closely related situation which occurs with jump moments (Sect. 3.2.3 and Fig. 4.17).

In nonlinear examples, the same procedure yields the deterministic equation from the master equation through multiplication of both sides by n_i and summation over all values of $\mathbf{n}$ in the limit of large particle numbers[41]:

$$\frac{\mathrm{d}\langle n_i(t)\rangle}{\mathrm{d}t} = s_i \langle \gamma_{\mathbf{n}} \rangle \quad \Longrightarrow \quad \frac{\mathrm{d}x_i(t)}{\mathrm{d}t} = s_i v\big(\mathbf{x}(t)\big) . \tag{4.69}$$

Accordingly, the master equation takes on the form of the conventional kinetic equations in this limit. It is important, however, to stress that it is not sufficient that the total number of particles is large. Every molecular species X_i has to be present in sufficient amounts at all times t in order to make the influence of fluctuations sufficiently small.

4.2.3 Stochastic Chemical Reaction Networks

Although most studies on stochastic chemical reaction networks (SCRNs) are carried out by computer simulation, the combined analytical and numerical approach is more promising, since it can give answers to general questions that cannot be addressed by pure computer analysis. As an example, we mention the generalization of the deficiency zero theorem to master equations [12]. General texts on stochastic modeling can be found in [217, 572, 573]. We begin here with the multivariate chemical master equation, presenting a general method to calculate the equilibrium probability densities. For a comparison of analytical results with numerical simulations, we refer to Sect. 4.6 .

Multivariate Chemical Master Equation

The master equation for many variables is readily derived by an extension of (4.66) to M molecular species involved in K reactions:

$$\frac{\mathrm{d}P_{\mathbf{n}}(t)}{\mathrm{d}t} = \sum_{\mu=1}^{K} \chi_\mu(\mathbf{n}-\mathbf{s}_\mu) P_{\mathbf{n}-\mathbf{s}_\mu}(t) - P_{\mathbf{n}}(t) \sum_{\mu=1}^{K} \chi_\mu(\mathbf{n}) , \tag{4.70}$$

where we have introduced the vector notation $\mathbf{n} = (n_1, \ldots, n_M)^{\mathrm{t}}$ for the particle numbers and $\mathbf{s}_\mu = \boldsymbol{\nu}'_\mu - \boldsymbol{\nu}_\mu$ with $\boldsymbol{\nu}'_\mu = (\nu'_{1\mu}, \ldots, \nu'_{M\mu})^{\mathrm{t}}$, and $\boldsymbol{\nu}_\mu = (\nu_{1\mu}, \ldots, \nu_{M\mu})^{\mathrm{t}}$ for the stoichiometric coefficients. The index $\boldsymbol{\mu}$ refers to individual reactions, $\chi_\mu(\mathbf{n}) = \gamma_\mu h_\mu(\mathbf{n})$ is the differential reaction probability from (4.64), and $\nu_{i\mu}$ and $\nu'_{i\mu}$ are the stoichiometric coefficients for species X_i in the $\boldsymbol{\mu}$ th reaction. For general considerations, it is common to treat reversible reactions as two reaction steps. The

[41]Because of the in-built or natural boundaries, it makes no difference whether the summation runs over a finite or infinite state space.

system of equations (4.71) can be solved as shown in Sect. 3.2.3. Now we can also generalize the expression for the trajectory to the reaction network

$$\mathcal{X}(t) = \mathcal{X}(0) + \sum_{\mu=1}^{K} s_\mu \mathcal{Y}_\mu \left(\int_0^t \mathrm{d}\tau k_\mu(\mathcal{X}(\tau)) \right) , \qquad (4.68')$$

where the processes $\mathcal{Y}_\mu(t)$ are independent unit-rate Poisson processes. Examples of reaction networks will be discussed in Sect. 4.6.

Stochastic mass action kinetics for the reaction $\mathbf{R}_\mu$ is modeled by the rate function

$$\chi_\mu^{(\mathrm{ma})}(\mathbf{n}) = k_\mu \prod_{i=1}^{M} \frac{n_i!}{(n_i - \nu_{i\mu})!} = k_\mu \frac{\mathbf{n}!}{(\mathbf{n} - \boldsymbol{\nu}_\mu)!} , \quad \mu = 1, \ldots, K , \qquad (4.67')$$

where we use the multi-index notation and the notation $h_\mu(\mathbf{n}) = \mathbf{n}!/(\mathbf{n} - \boldsymbol{\nu}_\mu)!$ and $k_\mu \equiv \gamma_\mu$. If it exists, a stationary distribution satisfies

$$\overline{P}_{\mathbf{n}} \sum_{\mu=1}^{K} \chi_\mu(\mathbf{n}) = \sum_{\mu=1}^{K} \chi_\mu(\mathbf{n} - \mathbf{s}_\mu) \overline{P}_{\mathbf{n}-\mathbf{s}_\mu} , \quad \forall\, \mathbf{n} \in \Omega . \qquad (4.71)$$

Equilibrium Densities and the Stochastic Deficiency Zero Theorem
Finally, we mention that the deficiency zero theorem has been extended to stochastic chemical reaction networks [12]. Assume a stochastic reaction network $\{\mathcal{S}, \mathcal{C}, \mathcal{R}\}$ with rate functions (4.67′), viz.,

$$\chi_\mu^{(\mathrm{ma})}(\mathbf{n}) = k_\mu \prod_{i=1}^{M} n_i(n_i - 1) \cdots (n_i - \nu_{i\mu} + 1) ,$$

for which the associated deterministic mass action system with the same rate functions k_μ ($\mu = 1, \ldots, K$) has a complex-balanced equilibrium $\bar{\mathbf{n}} \in \mathbb{R}^M_{\geq 0}$. Then the stochastically modeled network sustains a stationary probability distribution which is a product of Poisson distributions, provided that the variables n_i or $\bar{n}_i$ are independent[42]:

$$\overline{P}_{\mathbf{n}} = \prod_{i=1}^{M} \frac{\bar{n}_i^{n_i}}{n_i!} \mathrm{e}^{-\bar{n}_i} , \quad \mathbf{n} \in \mathbb{N}^M . \qquad (4.72a)$$

[42]For simplicity, we denote the equilibrium values of the particle numbers here by $\bar{n}_i$, although they are non-negative real numbers and not integers.

If the domain of the variable $\mathbf{n}$ in $\mathbb{N}^M$ is *irreducible* (Fig. 4.18), then (4.72a) represents the unique stationary distribution.

For dependent variables the starting point is once again the expression

$$\overline{P}_{\mathbf{n}} = N \prod_{i=1}^{M} \frac{\bar{n}_i^{n_i}}{n_i!} , \quad \mathbf{n} \in \mathbb{N}^M , \tag{4.72b}$$

where N is a normalization factor that has to be determined from the condition $\sum_{\mathbf{n}} \overline{P}_{\mathbf{n}} = 1$. Conservation relations and stoichiometry have to be introduced into the variables $\bar{\mathbf{n}}$ and $\mathbf{n}$. We illustrate by means of a simple standard example, the reversible monomolecular reaction $\mathsf{A} \rightleftharpoons \mathsf{B}$ with k and l as reaction rate parameters. The conservation relation is $n_{\mathsf{A}}+n_{\mathsf{B}} = a_0+b_0 = n_0$, and with $a = n$ and $b = n-n_0$, the linear dependence is eliminated and we obtain

$$\overline{P}_n = N \frac{\bar{n}^n}{n!} \frac{(n_0-\bar{n})^{n_0-n}}{(n_0-n)!} = N \frac{n_0^{n_0}}{n_0!} \binom{n_0}{n} \frac{k^{n_0-n} l^n}{(k+l)^{n_0}} , \quad N = \frac{n_0!}{n_0^{n_0}} ,$$

for the probability of the single random variable n, leading to a binomial equilibrium distribution:

$$\overline{P}_{\mathbf{n}} = \binom{n_0}{n} \frac{1}{(k+l)^{n_0}} k^{n_0-n} l^n . \tag{4.72c}$$

Other examples will be discussed in Sects. 4.3.3 and 4.6.4. For more than two variables and first order reaction kinetics, the equilibrium densities are described by the multinomial distribution (Sect. 2.3.2).

If the domain is not irreducible, the situation is more involved (Fig. 4.18). Then there exist two or more closed, irreducible communicating equivalence classes $\mathcal{C}$ which have their own probability densities:

$$\overline{P}_n^{(\mathcal{C})} = N_{\mathcal{C}} \prod_{i=1}^{M} \frac{\bar{n}_i^{n_i}}{n_i!} , \quad n_i \in \mathcal{C} , \tag{4.72d}$$

where $N_{\mathcal{C}}$ is a normalization factor. Such a situation occurs in reactions involving two or more molecules of the same species, e.g., in the bimolecular conversion reaction $2\mathsf{A} \rightleftharpoons 2\mathsf{B}$ (Fig. 4.18). For odd total numbers of molecules, i.e., $n_0 = 2\mu+1$ with $\mu \in \mathbb{N}$, there are two systems of states $(n_{\mathsf{A}}, n_{\mathsf{B}})$ or equivalence classes that do not communicate:

$$\mathcal{C}_1 : \quad (0, n_0) \rightleftharpoons (2, n_0-2) \rightleftharpoons (4, n_0-4) \rightleftharpoons \ldots \rightleftharpoons (n_0-1, 1)$$

$$\mathcal{C}_2 : (1, n_0-1) \rightleftharpoons (3, n_0-3) \rightleftharpoons (5, n_0-5) \rightleftharpoons \ldots \rightleftharpoons (n_0, 0)$$

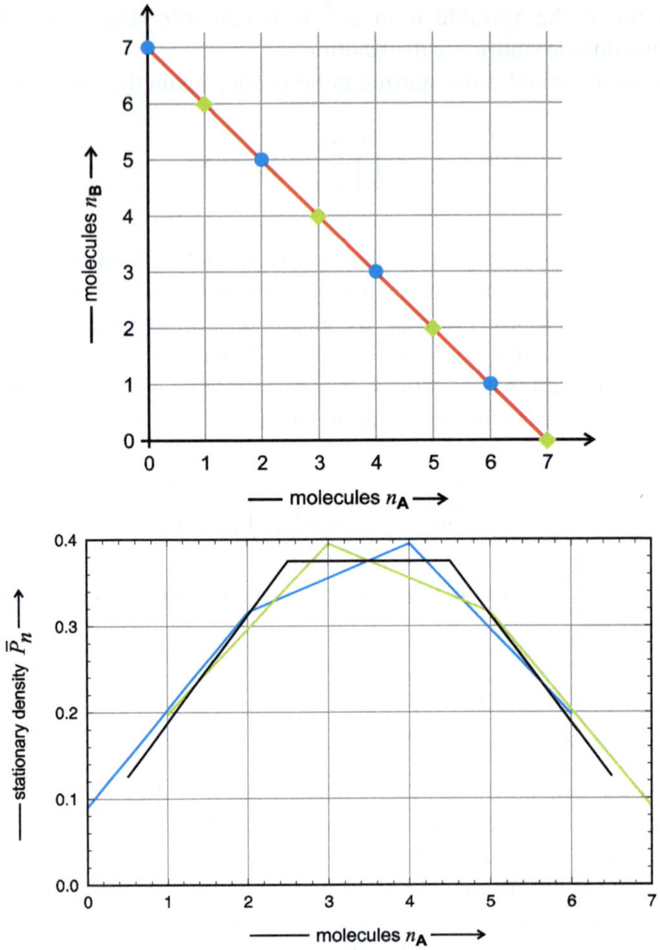

Fig. 4.18 Irreducible communicating equivalence classes. *Upper*: State space for the reaction

$$2A \underset{l}{\overset{k}{\rightleftharpoons}} 2B \quad \text{with } n_A(0) + n_B(0) = 7.$$

Because of the simultaneous conversion of two molecules, the domain is partitioned into two closed, irreducible communicating classes: $\Gamma_1 = \{(0,7),(2,5),(4,3),(6,1)\}$ (*blue circles*) and $\Gamma_2 = \{(1,6),(3,4),(5,2),(7,0)\}$ (*green diamonds*). *Lower*: Comparing the probability densities of the two irreducible classes (Γ_1 *blue* and Γ_2 *green*, $k = 2$, $h = 2$) with the density of the corresponding case with an even number of molecules ($n_A + n_B = 6$ *black*) which has only one irreducible class. (For direct comparison, the last curve has been shifted along the abscissa axis by $\Delta n = 1/2$) Continued on next page

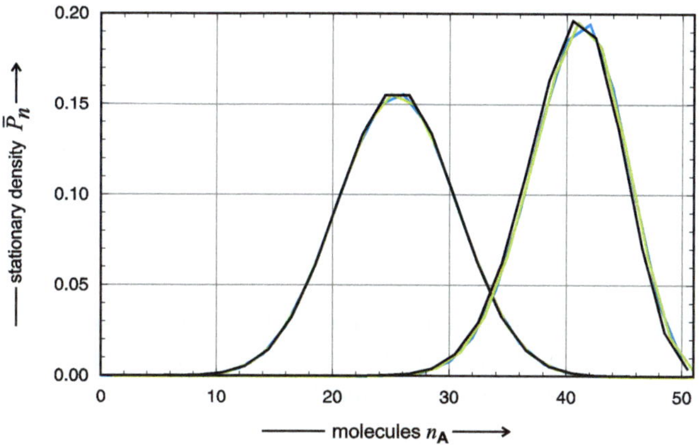

Fig. 4.18 (Cont.) Irreducible communicating equivalence classes. Comparing the probability densities for $n_A(0)+n_B(0) = 51$ and 50, for two parameter choices: $k = 2, l = 2$ and $k = 1, l = 4$

Although the distinction of irreducible classes is highly important for methodological reasons, it is hardly of empirical significance for chemical reactions, because the probability densities for different classes are already almost identical at fairly small particle numbers (see Fig. 4.18 and also the dimerization reaction in Sects. 4.3.3 and 4.3.4). This does not preclude its importance in biology, where the numbers of regulatory molecules can be extremely small.

It is important to remember another convenient way to compute stationary densities that makes use of the expression for the probability density of stationary Markov processes (3.100). We shall come back to it in Sect. 4.6.4.

The Monomolecular Triangle Reaction

The monomolecular triangle reaction (Fig. 4.19) is one of the simplest mechanisms with two independent degrees of freedom. It is straightforward to check that the deficiency of the mechanism is zero, i.e., $\delta = 0$, whence (4.72) applies and the stationary probability distributions can be calculated. The kinetic equations are

$$\begin{aligned}\frac{dx_1}{dt} &= -(k_1 + l_3)x_1 + l_1x_2 + k_3x_3 \;, \\ \frac{dx_2}{dt} &= -(k_2 + l_1)x_2 + l_2x_3 + k_1x_1 \;, \\ \frac{dx_3}{dt} &= -(k_3 + l_2)x_3 + l_3x_1 + k_2x_2 \;.\end{aligned} \tag{4.73}$$

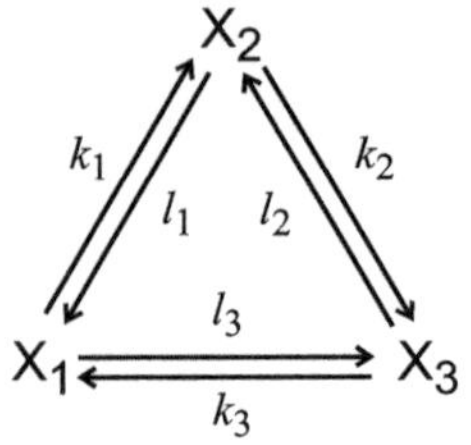

Fig. 4.19 The monomolecular triangle reaction. The sketch shows the fully reversible mechanism. For weak reversibility, one cycle, either $(k_1, k_2, k_3) > 0$ or $(l_1, l_2, l_3) > 0$, is sufficient

The sum of the concentrations $c(t) = x_1(t) + x_2(t) + x_3(t)$ satisfies the conservation relation $c(t) = c_0 = \text{const.}$, and the stationary concentrations defined by $\mathrm{d}x_1/\mathrm{dt} = \mathrm{d}x_2/\mathrm{dt} = \mathrm{d}x_3/\mathrm{dt} = 0$ are readily calculated:

$$\begin{aligned}
\bar{x}_1 &= (k_2k_3 + k_3l_1 + l_1l_2)\frac{c_0}{\Sigma} , \\
\bar{x}_2 &= (k_3k_1 + k_1l_2 + l_2l_3)\frac{c_0}{\Sigma} , \\
\bar{x}_3 &= (k_1k_2 + k_2l_3 + l_3l_1)\frac{c_0}{\Sigma} , \\
\Sigma &= k_1k_2 + k_2k_3 + k_3l_1 + k_1l_2 + k_2l_3 + k_3l_1 + l_1l_2 + l_2l_3 + l_3l_1 ,
\end{aligned} \tag{4.74}$$

which yields $\bar{x}_1 = \bar{x}_2 = \bar{x}_3 = c_0/3$ for the symmetric case in which $k_1 = k_2 = k_3 = k$ and $l_1 = l_2 = l_3 = l$. For thermodynamic equilibrium with concentrations $[\mathrm{X}_i] = \bar{x}_i$, we have the condition

$$K_1 = \frac{k_1}{l_1} , \quad K_2 = \frac{k_2}{l_2} , \quad K_3 = \frac{k_3}{l_3} , \quad K_1K_2K_3 = \frac{k_1k_2k_3}{l_1l_2l_3} = 1 . \tag{4.75}$$

The stationary distribution is calculated from (4.72):

$$\overline{P}_{n_1,n_2,n_3} = N\frac{\bar{n}_1^{n_1}}{n_1!}\frac{\bar{n}_2^{n_2}}{n_2!}\frac{\bar{n}_3^{n_3}}{n_3!} , \quad \text{with } \bar{n}_k = \bar{x}_k N_\mathrm{L} V , \ k = 1, 2, 3 ,$$

where the stationary concentrations $\bar{x}_k$ have been converted into stationary particle numbers $\bar{n}_k$. Figure 4.20 shows plots of the 2D probability density, which is centered as expected around the stationary point.

The cyclic closure of the mechanism introduces one constraint into the equilibrium or rate parameters. In addition, we see immediately that the existence of a thermodynamic equilibrium requires that none of the six rate parameters should vanish, i.e., $(k_1, k_2, k_3, l_1, l_2, l_3) > 0$, and this is a consequence of the principle of detailed balance, which demands that the flow of each individual reaction step should vanish, i.e., $k_1\bar{x}_1 = l_2\bar{x}_2 = 0$, $k_2\bar{x}_2 = l_3\bar{x}_3 = 0$, and $k_3\bar{x}_3 = l_1\bar{x}_1 = 0$.

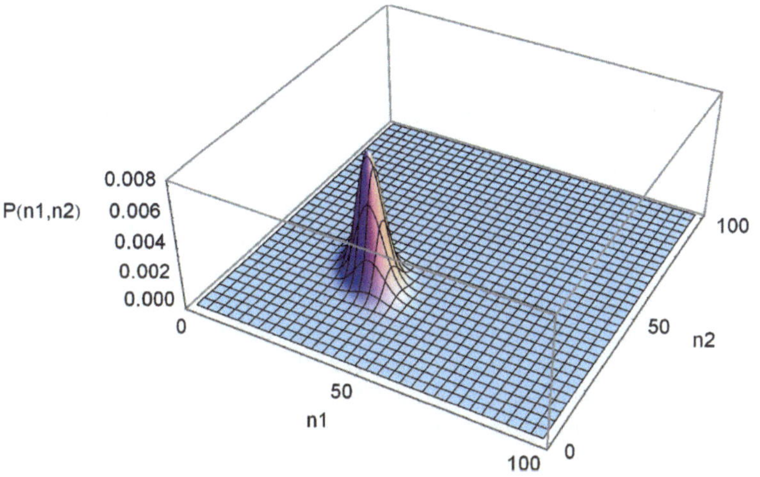

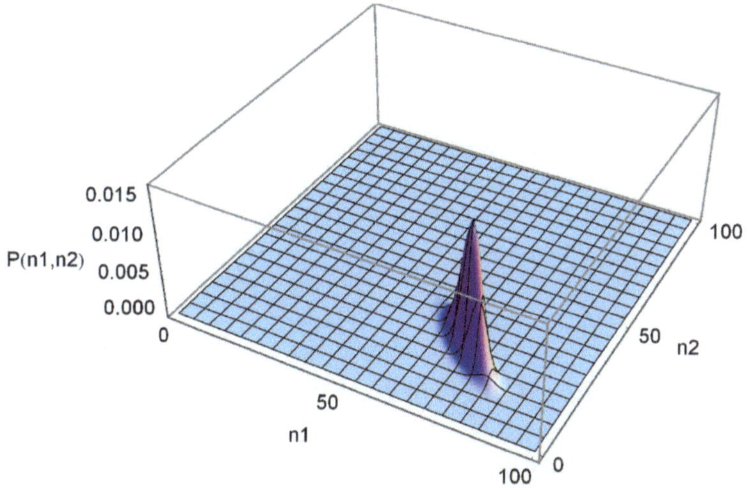

Fig. 4.20 Stationary density of the monomolecular triangle reaction. The plots show the stationary joint density $\overline{P}_{n_1,n_2}$ of the triangle reaction

$$X_1 \rightleftharpoons X_2 \rightleftharpoons X_3 \rightleftharpoons X_1 .$$

Upper: Density for the symmetric case $k_1 = k_2 = k_3 = l_1 = l_2 = l_3 = 1$. *Lower*: An asymmetric example: $k_1 = 1.0$, $k_2 = 2.0$, $k_3 = 10.0$, $l_1 = 1.0$, $l_2 = 0.2$, and $l_3 = 0.1$

Stochastic Model of the Michaelis–Menten Mechanism

The simple Michaelis–Menten mechanism, viz., $\mathbf{S} + \mathbf{E} \rightleftharpoons \mathbf{S} \cdot \mathbf{E} \rightarrow \mathbf{E} + \mathbf{P}$, has been studied by Péter Arányi and János Tóth [20], who solved the corresponding master equation. They obtained an exact solution under the assumption that the reaction occurs in a sufficiently small compartment that contains only a single enzyme molecule E. We remark that single molecule enzyme kinetics has become experimentally accessible thanks to modern spectroscopic techniques (Sect. 4.4) and we shall discuss the model again in Sect. 4.3.6. Earlier attempts to analyze the Michaelis–Menten mechanism by stochastic methods are also acknowledged [38, 274].

The extended Michaelis–Menten mechanism $\mathbf{S}+\mathbf{E} \rightleftharpoons \mathbf{S}\cdot\mathbf{E} \rightleftharpoons \mathbf{E}\cdot\mathbf{P} \rightleftharpoons \mathbf{E}+\mathbf{P}$ is readily analyzed. The system has one linkage class and consists of the five species **S**, **E**, $\mathbf{S}\cdot\mathbf{E}$, $\mathbf{E}\cdot\mathbf{P}$, and **P**, in four complexes $\mathbf{S}+\mathbf{E}$, $\mathbf{S}\cdot\mathbf{E}$, $\mathbf{E}\cdot\mathbf{P}$, and $\mathbf{E}+\mathbf{P}$. The rank of the stoichiometric matrix is three, since there are five concentration variables $[\mathbf{S}] = s$, $[\mathbf{E}] = e$, $[\mathbf{S}\cdot\mathbf{E}] = c$, $[\mathbf{E}\cdot\mathbf{P}] = d$, and $[\mathbf{E}\cdot\mathbf{P}] = p$, constrained by two conservation relations for the total enzyme and total substrate plus product, whence $\delta = 4-1-3 = 0$. The deficiency zero theorem applies, and we are dealing with one unique stable stationary state. The equilibrium concentrations are given in (4.19) and the probability densities can be obtained from the stochastic deficiency zero theorem (4.72). The expressions, however, are too sophisticated to be used in analytical work, and numerical calculations are also of limited usefulness, because as already mentioned equilibrium conditions are rarely applied in experimental studies or in biotechnology. Stochastic Michaelis–Menten kinetics will be discussed again in Sect. 4.3.6 and later in Sect. 4.4, where we deal with single molecule enzyme kinetics.

Finally, we mention that we have presented here only the simplest examples of reactions for the purpose of illustration. However, the *stochastic chemical reaction network approach* (SCRN) has turned out to be very useful for modeling of *real* networks in systems biology and we shall encounter examples in the forthcoming sections.

4.2.4 The Chemical Langevin Equation

Stochastic differential equations (SDEs) were introduced in Sect. 3.4 as an alternative to the Chapman–Kolmogorov approach to handle stochastic processes (Fig. 3.1). We presented the chemical Langevin equation and mentioned some of its most salient features already there. The Chapman–Kolmogorov approach in its various forms, i.e., master equation, Fokker–Planck equation, and others, aims to model the evolution of probability densities, whereas stochastic differential or Langevin equations (SDEs) concern single trajectories. Here we present a short derivation of the Langevin equation applied to chemical kinetics [211] in order to illustrate the approximations implicit in this approach.

Equation (4.63) is first generalized to a network of K different reactions $\mathbf{R}_j$ ($j = 1,\ldots,K$) and we define a time interval $\Delta t = [t, t+\tau]$[43] for recording reaction

[43]The time interval is the same for all reactions. It is predetermined, so differs in nature from the time interval in (4.63).

events. The random vector of particle numbers at the beginning of the time interval is $\boldsymbol{\mathcal{X}}(t) = \mathbf{n}(t)$, and at the end of the interval we have

$$\mathbf{n}(t+\tau) = \mathbf{n}(t) + \mathrm{S}\cdot\boldsymbol{\eta}\big(\mathbf{n}(t),\tau\big)\,,$$
$$\mathrm{n}_i(t+\tau) = \mathrm{n}_i(t) + \sum_{j=1}^{K} s_{ij}\eta_j\big(\mathbf{n}(t),\tau\big)\,,\quad i=1,\ldots,M\,, \tag{4.76}$$

where S is the stoichiometric matrix and $\boldsymbol{\eta} = (\eta_1,\ldots,\eta_K)^{\mathrm{t}}$ is a random vector counting the number of $\mathbf{R}_j$ reaction events which occurred during the interval $[t, t+\tau]$. These random numbers obviously depend on the time-dependent particle numbers and the length of the interval, so we write $\boldsymbol{\eta} = \boldsymbol{\eta}\big(\mathbf{n}(t),\tau\big)$.

The full-fledged computation of $\mathcal{X}_i(t+\tau)$ for arbitrary $\tau > 0$ is definitely as difficult as solving the corresponding master equation (4.66), but (4.76) allows for the introduction of transparent approximations provided that the two conditions (3.187) are satisfied. We repeat here them here for convenience:

1. The interval τ has to be sufficiently short to ensure that the rate functions do not change appreciably during $[t, t+\tau]$:
$$\chi_j\big(\boldsymbol{\mathcal{X}}(\theta)\big) \approx \chi_j\big(\mathbf{n}(t)\big)\,,\quad \forall\,\theta\in[t,t+\tau]\,,\quad \forall j=1,\ldots,K.$$
2. The interval τ has to be long enough to ensure that the expected number of occurrences of reactions in each reaction channel during the time interval $[t, t+\tau]$ is greater than one:
$$\big\langle \mathcal{P}_j\big(\chi_j(\mathbf{n}(t)),\tau\big)\big\rangle = \chi_j\big(\mathbf{n}(t)\big)\tau \gg 1\,,\quad \forall j=1,\ldots,K,$$
where the quantities $\mathcal{P}_j(\alpha)$ are random variables with Poissonian densities (Sect. 2.3.1).

Condition (1) substantially simplifies the equation for the stochastic trajectory. Because of the assumed constancy of $\chi_j\big(\mathbf{n}(t)\big)$ during the interval $[t, t+\tau]$, the random variables $\eta_j\big(\mathbf{n}(t),\tau\big)$ become statistically independent Poisson variables $\mathcal{P}_j\big(\chi_j\big((\mathbf{n}(t)\big),\tau\big)$ and (4.76) can be approximated by

$$\mathrm{n}_i(t+\tau) = \mathrm{n}_i(t) + \sum_{j=1}^{K} s_{ij}\mathcal{P}_j\big(\chi_j\big(\mathbf{n}(t)\big),\tau\big)\,,\quad i=1,\ldots,M\,. \tag{4.76$'$}$$

Condition (2) allows one to approximate the discrete Poisson variable $\mathcal{P}$ by a continuous random variable $\boldsymbol{\mathcal{X}}(t) = \mathbf{x}(t)$ with normal distribution $\mathcal{N}(\mu,\sigma)$ (Sect. 2.3.3), whereby the mean and variance remain the same:

$$x_i(t+\tau) = x_i(t) + \sum_{j=1}^{K} s_{ij}\mathcal{N}_j\Big(\chi_j\big(\mathbf{x}(t)\big)\tau\,,\ \chi_j\big(\mathbf{x}(t)\big)\tau\Big)\,,\quad i=1,\ldots,M\,. \tag{4.76$''$}$$

At the same time this approximation changed the originally discrete random variables $\mathcal{X}_j(t) = n_j(t)$ into continuous variables $\mathcal{X}_j(t) = x_j(t)$ on the domain of the nonnegative real numbers: $x_j \in \mathbb{R}_{\geq 0}$, $\forall j = 1, \ldots, M$. Using the linear combination theorem for normal variables, i.e., $\mathcal{X}(m, \sigma^2) = m + \sigma\mathcal{X}(0, 1)$, we can rewrite this equation as

$$x_i(t+\tau) = x_i(t) + \sum_{j=1}^{K} s_{ij}\chi_j\big(\mathbf{x}(t)\big)\tau + \sum_{j=1}^{K} s_{ij}\sqrt{\chi_j\big(\mathbf{x}(t)\big)\tau}\mathcal{N}(0,1)\,, \quad i = 1, \ldots, M\,.$$

Recalling that the probability density of the Wiener process W is the standard normal distribution, we obtain, with $\sqrt{\mathrm{dt}} = \mathrm{d}\mathcal{W}$,

$$\begin{aligned}\mathcal{X}_i(t+\mathrm{dt}) = \mathcal{X}_i(t) + \sum_{j=1}^{K} s_{ij}\chi_j\big(\boldsymbol{\mathcal{X}}(t)\big)\,\mathrm{dt} \\ + \sum_{j=1}^{K} s_{ij}\sqrt{\chi_j\big(\boldsymbol{\mathcal{X}}(t)\big)}\mathcal{N}(0,1)\mathrm{d}\mathcal{W}_j\,, \quad i = 1, \ldots, M\,.\end{aligned}$$

After rewriting, this yields the chemical Langevin equation

$$\mathrm{d}x_i = \sum_{j=1}^{K} s_{ij}\chi_j\big(\boldsymbol{\mathcal{X}}(t)\big)\,\mathrm{dt} + \sum_{j=1}^{K} s_{ij}\sqrt{\chi_j\big(\boldsymbol{\mathcal{X}}(t)\big)}\mathrm{d}\mathcal{W}_j(t)\,, \quad i = 1, \ldots, M\,. \tag{4.77}$$

Each reaction $\mathbf{R}_j (j = 1, \ldots, K)$ contributes to fluctuations of particle numbers $\mathcal{X}_i$ $(i = 1, \ldots, M)$ through the Wiener process $\mathcal{W}_j$. The K contributions are temporarily uncorrelated, statistically independent white noise terms.

At first glance, the two approximations appear to be contradictory (see Fig. 3.29), since τ has to be *sufficiently small* for approximation (1), but at the same it has to be *large enough* to satisfy condition (2). However, when we are dealing with particle numbers of 10^{20} and more that are typical in conventional chemistry, we realize that there is indeed enough room to have variables whose size allows for a sufficiently accurate approximation of a Poissonian by a normal distribution and for which changes $\Delta\mathrm{n} = \pm 1, \pm 2, \ldots,$ lead to negligibly small relative variations. This approximation method, also known as τ-leaping, will be discussed in more detail in Sect. 4.6.2.

Equation (4.77) corresponds to a forward Fokker–Planck equation (see Sect. 3.4.1). It describes the evolution of the multivariate probability density of the random vector $\mathcal{X}(t)$ [211]:

$$\begin{aligned}\frac{\mathrm{d}P(\mathbf{x},t)}{\mathrm{d}t} &= -\sum_{i=1}^{M}\frac{\partial}{\partial x_i}\left(\left(\sum_{k=1}^{K} s_{ij}\chi_k(\mathbf{x})\right)P(\mathbf{x},t)\right)\\ &+\frac{1}{2}\sum_{i=1}^{M}\frac{\partial^2}{\partial x_i^2}\left(\left(\sum_{k=1}^{K} s_{ik}^2\chi_k(\mathbf{x})\right)P(\mathbf{x},t)\right)\\ &+\sum_{i,j=1,i<j}^{M}\frac{\partial^2}{\partial x_i\partial x_j}\left(\left(\sum_{k=1}^{K} s_{ik}s_{jk}\chi_k(\mathbf{x})\right)P(\mathbf{x},t)\right).\end{aligned} \tag{4.78}$$

The initial conditions are $P(\mathbf{x}_0, t_0) = \delta(\mathbf{x} - \mathbf{x}_0)$.

As shown in Fig. 3.1 the equivalence of the Langevin and Fokker–Planck equations is a bridge built by rigorous mathematics between the single trajectory and the probability density approach in chemical kinetics. This equivalence is based on the use of continuous variables, which in the case of reaction kinetics is almost always well justified by the large particle numbers in chemical applications. We mention that a second bridge exists between the numerical simulation of stochastic trajectories and chemical master equations that is mediated by numerical mathematics on the level of discrete variables [209, 213].

4.3 Examples of Chemical Reactions

In this section, we shall present exact solutions of chemical master equations for study cases from three classes of chemical reactions:

(i) Zero-molecular reactions in the form of the flow in a reactor.
(ii) Monomolecular reactions with one reactant.
(iii) Bimolecular reactions involving two reactants.

The molecularity of a reaction is commonly reflected by the chemical rate law of reaction kinetics in the form of the reaction order. In particular, we distinguish first order and second order kinetics, which are typically observed with monomolecular and bimolecular reactions, respectively. Exceptions are conditions like an excess of one reactant, which leads to an observed reaction order that is smaller than the molecularity. The examples most frequently encountered are *pseudo-first order reactions* (see Sect. 4.3.3). Because of its fundamental importance in chemistry and biology, the autocatalytic elementary step (4.1g) will be discussed separately in Sect. 4.3.5.

4.3.1 The Flow Reactor

The flow reactor is introduced as an experimental device that allows for investigations of systems under controlled conditions away from thermodynamic equilibrium. The establishment of a stationary state or *flow equilibrium* in a *continuous flow stirred tank reactor* (CFSTR or CSTR) (see Fig. 4.21) is a suitable case study to illustrate the search for the solution of a birth-and-death master equation. At the same time the nonreactive flow of a single compound represents the simplest conceivable process in such a reactor. The stock solution contains A at the concentration $[\mathsf{A}]_{\mathrm{in}} = \widehat{a} = a_0 = \bar{a}\,[\mathrm{mol\,l^{-1}}]$. The inflow concentration $\widehat{a}$ is equal to the concentration of A in the stock solution a_0 and the stationary concentration $\bar{a}$, because no reaction is assumed to take place in the reactor.

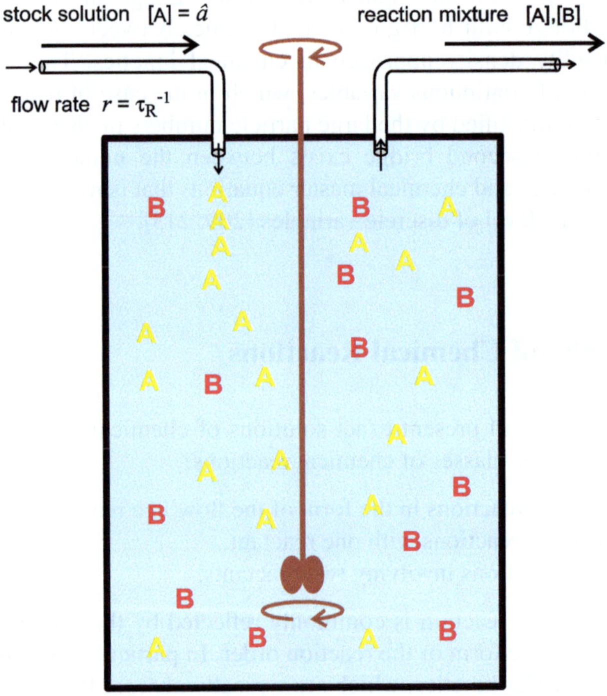

Fig. 4.21 The flow reactor. The reactor shown in the sketch is a device for experimental and theoretical chemical reaction kinetics, used to carry out chemical reactions in an open system. The stock solution contains materials, for example A at the concentration $[\mathsf{A}]_{\mathrm{in}} = \widehat{a}$, which are usually consumed during the reaction to be studied. The reaction mixture is stirred in order to guarantee a spatially homogeneous reaction medium. Constant volume implies an outflow from the reactor that compensates the inflow in volume. The flow rate r is equivalent to the reciprocal mean residence time of solution in the reactor multiplied by the reactor volume $\tau_v^{-1} V = r$. The reactor shown here is commonly called a continuously stirred tank reactor (CSTR)

For the flow reactor in Fig. 4.21, the *flow* is understood as volume flow and expressed in terms of the volume flow rate r, measured in the units $[\mathrm{l\,sec^{-1}}]$.[44] Accordingly the inflow of compound A into the reactor is $a_0 r$ [mol sec^{-1}] expressed as a concentration after instantaneous mixing with the content of the reactor. The outflow of the mixture in the reactor occurs with the same flow rate r.[45] The reactor has volume V [l], so we have a mean residence time of $\tau_{\mathrm{v}} = Vr^{-1}$ [sec] for a volume element $\mathrm{d}V$ in the reactor. The inflow and outflow of compound A into and out of the reactor are modeled by two formal elementary steps or pseudoreactions:

$$\begin{aligned} \star &\longrightarrow \mathsf{A}\,, \\ \mathsf{A} &\longrightarrow \oslash\,. \end{aligned} \tag{4.79a}$$

In chemical kinetics the differential equations are almost always formulated in terms of molecular concentrations. For the stochastic treatment, the concentrations are replaced by particle numbers, $n = aVN_{\mathrm{L}}$, where $n \in \mathbb{N}$ and N_{L} is Avogadro's constant.

The particle number of A in the reactor is a stochastic variable $\mathcal{X}(t)$ with probability $P_n(t) = P\big(\mathcal{X}(t) = n\big)$. The time derivative of the probability density is described by means of the master equation

$$\frac{\partial P_n(t)}{\partial t} = r\big(a_0 P_{n-1}(t) + (n+1)P_{n+1}(t) - (a_0 + n)P_n(t)\big)\,, \quad n \in \mathbb{N}\,, \tag{4.79b}$$

where the flow rate can be absorbed in the time scale by a redefinition of the time axis. Equation (4.79b) is equivalent to a birth-and-death process with the step-up and step-down transition probabilities $w_n^+ = ra_0$ and $w_n^- = rn(t)$, respectively. Thus we have a constant birth rate and a death rate which is proportional to n. Solutions of the master equation (4.79b) can be found in textbooks listing stochastic processes with known solutions, such as [216].

Here we derive the solution illustrating the use of probability generating functions as introduced in (2.27) (Sect. 2.2.1):

$$g(s,t) = \sum_{n=0}^{\infty} P_n(t)s^n\,. \tag{2.27$'$}$$

[44] Volume flow is to be distinguished from *mass flow*, the measure of which is a mass flow rate $\tilde{r}$ [kg sec^{-1}]. Mass flow is a scalar quantity, and when it is measured with respect to a unit area through which the transport takes place, it is called a *flux* $\boldsymbol{\phi}$, measured in [kg/m^2 sec^1]. In contrast to flow, the flux is a vector perpendicular to the reference unit area [504].

[45] The assumption of equal inflow and outflow rate is required because we are dealing with a flow reactor of constant volume V (CSTR) (see Fig. 4.21).

The initial state $n(0) = n_0$ is encapsulated in the expression $g(s,0) = s^{n_0}$, which implies $P_n(0) = \delta_{n,n_0}$. Partial derivatives with respect to time t and the dummy variable s are readily computed:

$$\frac{\partial g(s,t)}{\partial t} = \sum_{n=0}^{\infty} \frac{\partial P_n(t)}{\partial t} s^n$$

$$= r \sum_{n=0}^{\infty} \Big(a_0 P_{n-1}(t) + (n+1) P_{n+1}(t) - (a_0 + n) P_n(t) \Big) s^n ,$$

$$\frac{\partial g(s,t)}{\partial s} = \sum_{n=0}^{\infty} n P_n(t)\, s^{n-1} .$$

Proper collection of terms and rearrangement of summations, taking into account the fact that $w_0^- = 0$, yields

$$\frac{\partial g(s,t)}{\partial t} = r a_0 \sum_{n=0}^{\infty} \Big(P_{n-1}(t) - P_n(t) \Big) s^n + r \sum_{n=0}^{\infty} \Big((n+1)\, P_{n+1}(t) - n P_n(t) \Big) s^n .$$

Evaluating the four infinite sums

$$\sum\nolimits_{n=0}^{\infty} P_{n-1}(t) s^n = s \sum\nolimits_{n=0}^{\infty} P_{n-1}(t) s^{n-1} = s g(s,t) ,$$

$$\sum\nolimits_{n=0}^{\infty} P_n(t) s^n = g(s,t) ,$$

$$\sum\nolimits_{n=0}^{\infty} (n+1) P_{n+1}(t) s^n = \frac{\partial g(s,t)}{\partial s} ,$$

$$\sum\nolimits_{n=0}^{\infty} n P_n(t) s^n = s \sum\nolimits_{n=0}^{\infty} n P_n(t) s^{n-1} = s \frac{\partial g(s,t)}{\partial s} ,$$

and regrouping terms yields a linear partial differential equation of first order:

$$\frac{\partial g(s,t)}{\partial t} = r \left(a_0 (s-1) g(s,t) - (s-1) \frac{\partial g(s,t)}{\partial s} \right) . \tag{4.79c}$$

There is a general method for deriving solutions called the *method of characteristics* for linear first-order PDEs [585, pp. 390–396]. The trick applied in this approach is to reduce the problem of solving a PDE to the task of solving an ODE. Although the method of characteristics is a standard technique in mathematics, we illustrate the method briefly because many scientists are not very familiar with it.

The PDE (4.79c) is brought into standard form

$$\alpha(s,t)\frac{\partial g}{\partial t} + \beta(s,t)\frac{\partial g}{\partial s} + \gamma(s,t)g = 0\,, \tag{4.79d}$$

with $\alpha(s,t) = 1$, $\beta(s,t) = -r(1-s)$, and $\gamma(s,t) = ra_0(1-s)$, and the initial condition $g(s,0) = f(s_0) = s_0^{n_0}$. A change of coordinates $(s,t) \to (s_0, v)$ is performed such that the PDE becomes an ODE along the characteristic curves or *characteristics*:

$$\big\{\big(s(v), t(v)\big) : 0 \leq v < \infty\big\} \text{ in the } (s,t)\text{-plane}\,.$$

Here v is the variable and s_0 is constant for one particular characteristic, but changes along the initial curve $v = 0$ in the (s,t)-plane. In order to find the characteristic manifold, we choose $\mathrm{d}t/\mathrm{d}v = \alpha(s,t)$ and $\mathrm{d}s/\mathrm{d}v = \beta(s,t)$ and obtain the ODE

$$\frac{\mathrm{d}g}{\mathrm{d}v} = \frac{\mathrm{d}t}{\mathrm{d}v}\frac{\partial g}{\partial t} + \frac{\mathrm{d}s}{\mathrm{d}v}\frac{\partial g}{\partial s} = \alpha(s,t)\frac{\partial g}{\partial t} + \beta(s,t)\frac{\partial g}{\partial s}$$

along the manifold where the initial conditions determine the individual curves. Inserting in (4.79d) yields the ODE for calculating the solution:

$$\frac{\mathrm{d}g(s_0, v)}{\mathrm{d}v} + \gamma(s,t)g(s_0, v) = 0\,. \tag{4.79e}$$

Changing variables back, i.e., $(s_0, v) \to (s,t)$, we obtain the desired solution $g(s,t)$.

The procedure is commonly carried out in three steps, which are illustrated by means of our specific example:

(i) Solution of the characteristic equations for (4.79c) with the initial conditions $v_0 = 0$ and $t(v_0) = 0$ and $s(v_0) = s_0$ yields

$$\frac{\mathrm{d}t}{\mathrm{d}v} = 1\,, \quad t(v, s_0) = v\,, \quad \frac{\mathrm{d}s}{\mathrm{d}v} = r(s-1)\,, \quad s(v, s_0) = 1 + (s_0 - 1)\mathrm{e}^{rv}\,.$$

(ii) The ODE along the characteristic manifold is solved next with the initial condition $g(0) = s_0^{n_0}$, yielding

$$\frac{\mathrm{d}g}{\mathrm{d}v} = -a_0 r(1 - s_0)g\,, \quad g(v, s_0) = s_0^{n_0}\exp\big(a_0(1-s_0)(1-\mathrm{e}^{-rv})\big)\,.$$

(iii) Resubstitution completes the procedure and gives the final solution as

$$g(s,t) = \big(1 + (s-1)\mathrm{e}^{-rt}\big)^{n_0}\exp\big(a_0(s-1)(1-\mathrm{e}^{-rt})\big)\,. \tag{4.79f}$$

From the generating function, we compute with somewhat tedious but straightforward algebra the probability distribution

$$P_n(t) = \sum_{k=0}^{\min\{n_0,n\}} \binom{n_0}{k} a_0^{n-k} \frac{e^{-krt}\,(1-e^{-rt})^{n_0+n-2k}}{(n-k)!} e^{-a_0(1-e^{-rt})} , \qquad (4.79g)$$

with $n, n_0, a_0 \in \mathbb{N}^0$. In the limit $t \to \infty$, we find a non-vanishing contribution to the stationary probability only from the first term $k = 0$, and obtain

$$\lim_{t\to\infty} P_n(t) = \bar{P}_n(t) = \frac{a_0^n}{n!} \exp(-a_0) . \qquad (4.79h)$$

This is a Poisson distribution with parameter and expectation value $\alpha = a_0$. The Poisson distribution has a variance which is numerically identical to the expectation value, i.e., $\mathrm{var}(\mathcal{X}) = \mathrm{E}(\mathcal{X}) = a_0$, and in the stationary state the distribution of particle numbers perfectly satisfies a $\sqrt{n}$-law.

The time dependent probability distribution allows us to compute the expectation value and the variance of the particle number as a function of time:

$$\begin{aligned} \mathrm{E}\big(\mathcal{X}(t)\big) &= a_0 + (n_0 - a_0)e^{-rt} , \\ \mathrm{var}\big(\mathcal{X}(t)\big) &= \big(a_0 + n_0 e^{-rt}\big)\big(1 - e^{-rt}\big) . \end{aligned} \qquad (4.79i)$$

As expected for linear transition probabilities, the expectation value coincides with the solution curve of the deterministic differential equation

$$\frac{dn}{dt} = w_n^+ - w_n^- = r\,(a_0 - n) , \quad n(t) = a_0 + (n_0 - a_0)e^{-rt} . \qquad (4.79j)$$

Since we start from sharp initial densities, the variance and standard deviation are zero at time $t = 0$. The qualitative time dependence of $\mathrm{var}\big(\mathcal{X}(t)\big)$, however, depends on the sign of $n_0 - a_0$:

(i) For $n_0 \le a_0$, the standard deviation increases monotonically until it reaches the stationary value $\sqrt{a_0}$ in the limit $t \to \infty$.
(ii) For $n_0 > a_0$, the standard deviation increases until it passes through a maximum at

$$t(\sigma_{\max}) = \frac{1}{r}\big(\ln 2 + \ln n_0 - \ln(n_0 - a_0)\big) ,$$

and approaches the long-time value $\sqrt{a_0}$ from above.

Figure 4.22 shows an example for the evolution of the probability density (4.79g). In addition, the figure contains a plot of the expectation value $\mathrm{E}\big(\mathcal{X}(t)\big)$ inside the band $\mathrm{E} - \sigma < \mathrm{E} < \mathrm{E} + \sigma$. For a normally distributed stochastic variable, we find 68.3 % of all values within this *confidence interval*. In the interval $\mathrm{E} - 2\sigma < \mathrm{E} < \mathrm{E} + 2\sigma$, we would even find 95.4 % of all stochastic trajectories (Sect. 2.3.3).

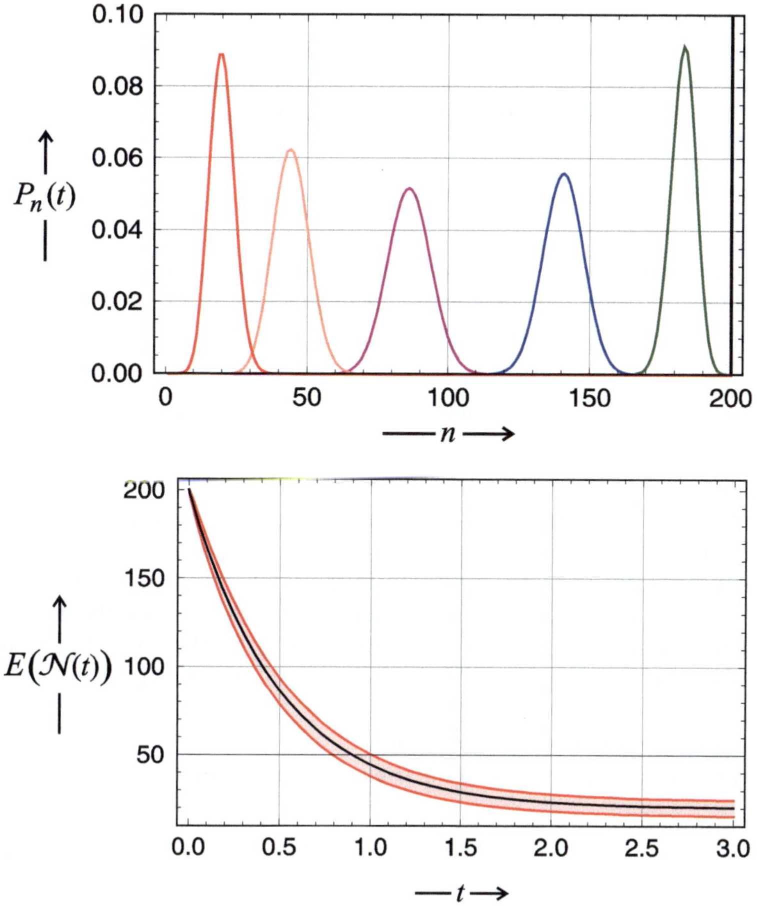

Fig. 4.22 Establishment of the flow equilibrium in the CSTR. *Upper*: Evolution of the probability density $P_n(t)$ of the number of molecules of a compound A which flows through a reactor of the type illustrated in Fig. 4.21. The initially infinitely sharp density becomes broader with time, until the variance reaches its maximum and then sharpens again until it reaches stationarity. The stationary density is Poissonian with expectation value and variance $\mathrm{E}(\mathcal{X}) = \mathrm{var}(\mathcal{X}) = \bar{n} = a_0$. *Lower*: Expectation value $\mathrm{E}\big(\mathcal{X}(t)\big)$ in the confidence interval $\mathrm{E} \pm \sigma$. Parameters used: $a_0 = 20$, $n_0 = 200$, and $V = 1$. Sampling times (*upper*): $\tau = rt = 0$ (*black*), 0.05 (*green*), 0.2 (*blue*), 0.5 (*violet*), 1 (*pink*), and ∞ (*red*)

4.3.2 Monomolecular Chemical Reactions

The reversible mono- or unimolecular chemical conversion or isomerization reaction can be split into two irreversible elementary reactions

$$\mathsf{A} \xrightarrow{\;k\;} \mathsf{B}\,, \tag{4.80a}$$

$$\mathsf{A} \xleftarrow{\;l\;} \mathsf{B}\,, \tag{4.80b}$$

where k and l are the *reaction rate parameters*, which depend on temperature, pressure, and other environmental factors. At equilibrium, the rate of the forward reaction (4.80a) is precisely compensated by the rate of the reverse reaction (4.80b), i.e., $k[\mathsf{A}] = l[\mathsf{B}]$, leading to the following condition for thermodynamic equilibrium:

$$K = \frac{k}{l} = \frac{[\mathsf{B}]}{[\mathsf{A}]}, \tag{4.80c}$$

where the parameter K is the *equilibrium constant*, which again depends on temperature, pressure, and other environmental factors. In an isolated or in a closed system, we have a conservation law:

$$\frac{\mathcal{X}_\mathsf{A}(t) + \mathcal{X}_\mathsf{B}(t)}{VN_\mathrm{L}} = [\mathsf{A}] + [\mathsf{B}] = c(t) = c_0 = \bar{c} = \text{constant}, \tag{4.80d}$$

where c is the constant total concentration.

The two irreversible reactions are characterized by vanishing rate parameters, $\lim l \rightarrow 0$ or $\lim k \rightarrow 0$, respectively. It is worth mentioning that zero rate parameters correspond to an instability in the Gibbs free energy at equilibrium, viz., $\Delta G_0 = -RT \ln K$, and are incompatible with rigorous thermodynamics. Nevertheless, the assumption of irreversibility is a good approximation in cases where equilibria are lying almost completely on the side of reactants or products.

Irreversible Monomolecular Reaction

We start by discussing the case of the irreversible reaction,

$$\mathsf{A} \xrightarrow{\;k\;} \mathsf{B}, \tag{4.80a$'$}$$

which can be modeled and analyzed in full analogy to the previous case of the flow equilibrium. We are dealing with two molecular species A and B, but the process is fully described by a single stochastic variable $\mathcal{X}_\mathsf{A}(t) = n$, since we have the conservation relation (4.80d) $\mathcal{X}_\mathsf{A}(t) + \mathcal{X}_\mathsf{B}(t) = n_0$, with $n_0 = n(0)$ the initial number of A molecules and $[\mathsf{B}(0)] = 0$. The reaction is tantamount to a single-step pure death process with $w_n^+ = 0$ and $w_n^- = kn$.[46] The probability density is defined by $P_n(t) = P(\mathcal{X}_A = n)$, and the time dependence obeys the master equation

$$\frac{\partial P_n(t)}{\partial t} = k(n+1)P_{n+1}(t) - knP_n(t). \tag{4.81a}$$

[46]We remark that $w_n^- = 0$ and $w_n^+ = 0$, the conditions for a natural absorbing boundary (Sect. 5.2.2), are satisfied at $n = 0$.

Equation (4.81a) can be solved once again using the probability generating function

$$g(s,t) = \sum_{n=0}^{\infty} P_n(t) s^n \,, \quad |s| \le 1 \,,$$

which satisfies the first order linear PDE

$$\frac{\partial g(s,t)}{\partial t} - k(1-s)\frac{\partial g(s,t)}{\partial s} = 0 \,.$$

This is solved by the same technique as shown in Sect. 4.3.1 and yields

$$g(s,t) = \left(1 + (s-1)\mathrm{e}^{-kt}\right)^{n_0} \,. \tag{4.81b}$$

This expression is expanded in binomial form, which introduces an ordering with respect to increasing powers of s:

$$g(s,t) = (1-\mathrm{e}^{-kt})^{n_0} + \binom{n_0}{1} s\mathrm{e}^{-kt}(1-\mathrm{e}^{-kt})^{n_0-1} + \binom{n_0}{2} s\mathrm{e}^{-2kt}(1-\mathrm{e}^{-kt})^{n_0-2}$$

$$+ \cdots + \binom{n_0}{n_0-1} s^{n_0-1}\mathrm{e}^{-(n_0-1)kt}(1-\mathrm{e}^{-kt}) + s^{n_0}\mathrm{e}^{-n_0 kt} \,.$$

Comparison of coefficients yields the time dependent probability density

$$P_n(t) = \binom{n_0}{n} \left(\exp(-kt)\right)^n \left(1-\exp(-kt)\right)^{n_0-n} \,. \tag{4.81c}$$

It is straightforward to compute the expectation value of the stochastic variable $\mathcal{X}_A$, which again coincides with the deterministic solution, and its variance

$$\begin{aligned} \mathrm{E}\big(\mathcal{X}_\mathrm{A}(t)\big) &= n_0 \mathrm{e}^{-kt} \,, \\ \mathrm{var}\big(\mathcal{X}_\mathrm{A}(t)\big) &= n_0 \mathrm{e}^{-kt}\left(1-\mathrm{e}^{-kt}\right) \,. \end{aligned} \tag{4.81d}$$

The half-life of a population of n_0 particles, i.e.,

$$t_{1/2} : \mathrm{E}\{\mathcal{X}_\mathrm{A}(t)\} = \frac{n_0}{2} = n_0 \mathrm{e}^{-kt_{\max}} \quad \Longrightarrow \quad t_{1/2} = t_{\max} = \frac{1}{k}\ln 2 \,,$$

is the time of maximum variance or standard deviation, $\mathrm{dvar}\big(\mathcal{X}_\mathrm{A}(t)\big)/\,\mathrm{dt} = 0$ or $\mathrm{d}\sigma\big(\mathcal{X}_\mathrm{A}(t)\big)/\,\mathrm{dt} = 0$, respectively. An example of the time course of the probability density of an irreversible monomolecular reaction is shown in Fig. 4.23.

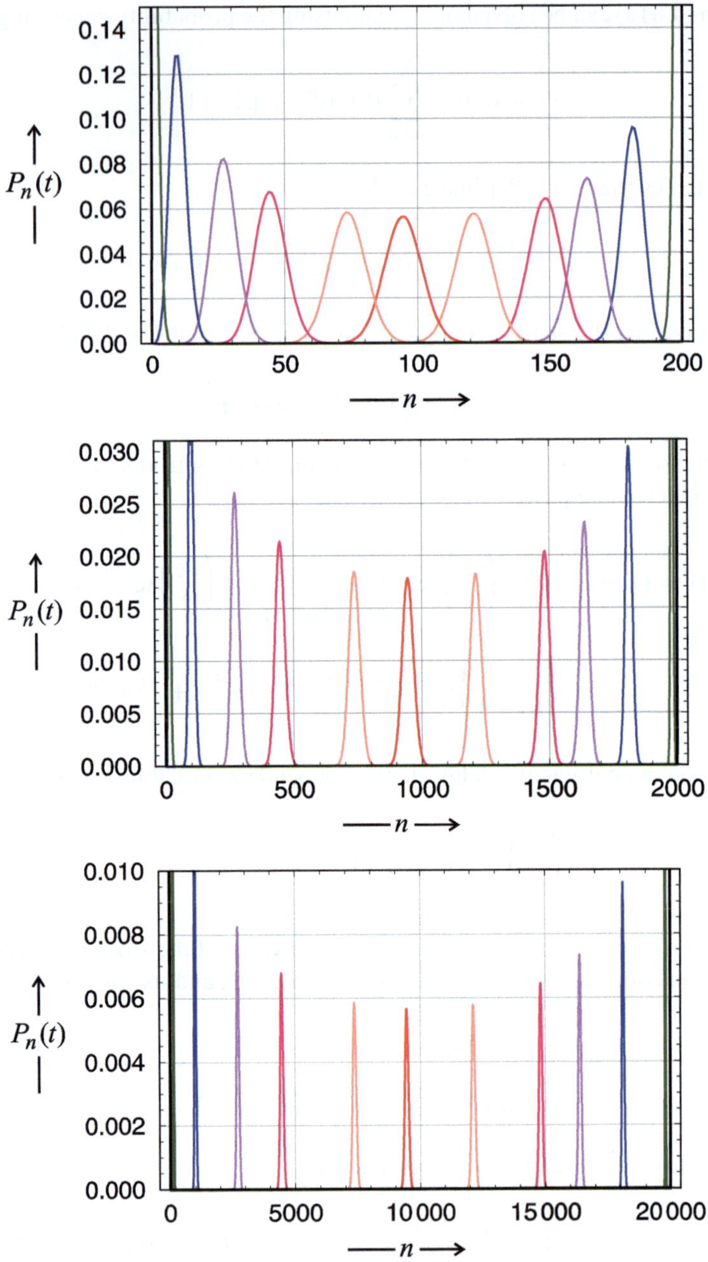

Fig. 4.23 Probability density of an irreversible monomolecular reaction. Three plots showing the evolution of the probability density $P_n(t)$ of the number of molecules of a compound A which undergo a reaction A→B. The initially infinitely sharp density $P_n(0) = \delta_{n,n_0}$ becomes broader with time, until the variance reaches its maximum at time $t = t_{1/2} = t_{\max} = \ln 2/k$, and then sharpens again until it approaches full transformation, i.e., $\lim_{t\to\infty} P_n(0) = \delta_{n,0}$. Continued on next page

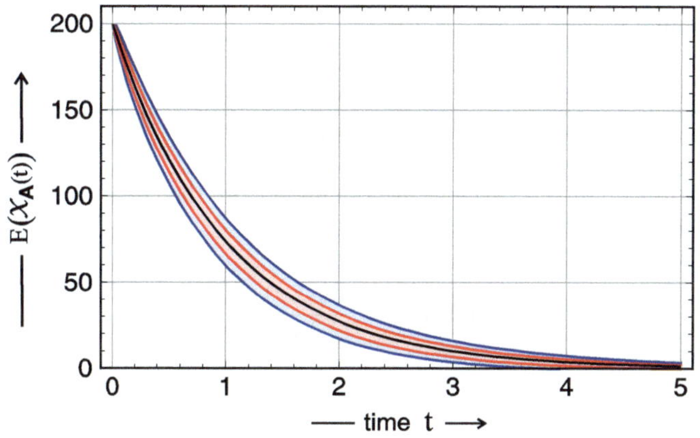

Fig. 4.23 (Cont.) Probability density of an irreversible monomolecular reaction. Expectation value $\mathrm{E}(\mathcal{X}_A(t))$ and confidence intervals $\mathrm{E} \pm \sigma$ (68.3 % *red*) and $\mathrm{E} \pm 2\sigma$ (95.4 % *blue*), where $\sigma^2 = \mathrm{var}(\mathcal{X}_A(t))$ is the variance. Parameters used: $n_0 = 200$ (*topmost plot*), 2,000 (*middle plot*), and 20,000 (*bottom plot*), $k = 1\,[\mathrm{t}^{-1}]$. Sampling times: 0 (*black*), 0.01 (*green*), 0.1 (*blue*), 0.2 (*violet*), (0.3) (*magenta*), 0.5 (*pink*), 0.75 (*red*), 1 (*pink*), 1.5 (*magenta*), 2 (*violet*), 3 (*blue*), and 5 (*green*). The initial condition for the time dependence of the expectation value was $n_0 = 200$

Reversible Monomolecular Reaction

The analysis of the irreversible reaction is readily extended to the reversible case (4.80), which corresponds to a single step birth-and-death process that can be solved exactly [381]. Again we are dealing with a closed system. It satisfies the conservation relation $\mathcal{X}_\mathsf{A}(t) + \mathcal{X}_\mathsf{B}(t) = n_0$ and the step-up and step-down transition probabilities are $w_n^+ = l(n_0 - n)$ and $w_n^- = kn$. We note the existence of boundaries at $n = 0$ and $n = n_0$, which are characterized by $w_0^- = 0$, $w_0^+ = ln_0 > 0$, and by $w_{n_0}^+ = 0$, $w_{n_0}^- = k\,n_0 > 0$, respectively. These equations satisfy the conditions for reflecting natural boundaries (Sect. 5.2.2). The master equation is of the form

$$\frac{\partial P_n(t)}{\partial t} = l(n_0 - n + 1)P_{n-1}(t) + k(n+1)P_{n+1}(t) - \big(kn + l(n_0 - n)\big)P_n(t)\,, \qquad (4.82\text{a})$$

and we shall solve it for the initial condition $P_n(0) = \delta_{n,a_0}$, or exactly a_0 molecules A and $n_0 - a_0$ molecules B initially present. Making use of the probability generating function $g(s,t)$, we derive the PDE

$$\frac{\partial g(s,t)}{\partial t} = \big(k + (l-k)s - ls^2\big)\frac{\partial g(s,t)}{\partial s} + n_0 l(s-1)g(s,t)\,. \qquad (4.82\text{b})$$

The solution of the PDE is facilitated by using the parameter combination $\kappa = k + l$ and the equilibrium constant $K = k/l$. For the initial condition $P_n(0) = \delta_{n,a_0}$, we find, in the same way as in Sect. 4.3.1,[47]

$$g(s,t) = \frac{\left(K + s - K(1-s)e^{-\kappa t}\right)^{a_0}\left(K + s + (1-s)e^{-\kappa t}\right)^{n_0 - a_0}}{(1+K)^{n_0}} \,. \tag{4.82c}$$

The expectation value and variance are obtained from the generating function in the conventional way, i.e.,

$$E\big(\mathcal{X}_A(t)\big) = \left.\frac{dg(s,t)}{ds}\right|_{s=1} ,$$

and

$$\mathrm{var}\big(\mathcal{X}_A(t)\big) = \left.\frac{d^2 g(s,t)}{ds^2}\right|_{s=1} + \left.\frac{dg(s,t)}{ds}\right|_{s=1} - \left(\left.\frac{dg(s,t)}{ds}\right|_{s=1}\right)^2 ,$$

with the result

$$\begin{aligned} E\big(\mathcal{X}_A(t)\big) &= \frac{a_0(1 + Ke^{-\kappa t}) + (n_0 - a_0)(1 - e^{-\kappa t})}{1+K} , \\ \mathrm{var}\big(\mathcal{X}_A(t)\big) &= \frac{\big(a_0(K^2-1)e^{-\kappa t} + n_0(K + e^{-\kappa t})\big)(1 - e^{-\kappa t})}{(1+K)^2} . \end{aligned} \tag{4.82d}$$

In the literature, we commonly find the somewhat simpler results obtained for the special case $a_0 = n_0$ or $b_0 = 0$, i.e., no B present at the beginning of the reaction. These are easily derived form the equations given above:

$$\begin{aligned} g(s,t) &= \left(\frac{K(1-e^{-\kappa t}) + s(Ke^{-\kappa t}+1)}{1+K}\right)^{n_0} \\ &= \sum_{n=0}^{n_0}\left(\binom{n_0}{n}\left(Ke^{-\kappa t}+1\right)^n K(1-e^{-\kappa t})\right)^{n_0-n} \frac{s^n}{(1+K)^{n_0}} . \end{aligned} \tag{4.82e}$$

The probability density for the reversible reaction is then obtained as

$$P_n(t) = \binom{n_0}{n}\frac{1}{(1+K)^{n_0}}\left(Ke^{-\kappa t}+1\right)^n \left(K(1-e^{-\kappa t})\right)^{n_0-n} . \tag{4.82f}$$

[47]The derivation of the solutions involves quite complicated substitutions, which are facilitated enormously by computer assistance using symbolic computation.

Taking the limit to infinite time yields the equilibrium density

$$\lim_{t\to\infty} P_n(t) = \overline{P}_n = \binom{n_0}{n} \frac{K^{n_0-n}}{(1+K)^{n_0}} = \binom{n_0}{n} \frac{1}{(k+l)^{n_0}} k^{n_0-n} l^n ,$$

which is, of course, identical with the expression (4.72c) derived earlier. The expectation value and variance for $a_0 = n_0$ expressed in terms of the function $\omega(t) = K \exp(-\kappa t) + 1$ are:

$$\begin{aligned} \mathrm{E}\big(\mathcal{X}_A(t)\big) &= \frac{n_0}{1+K}\omega(t) , \\ \mathrm{var}\big(\mathcal{X}_A(t)\big) &= \frac{n_0\omega(t)}{1+K}\left(1 - \frac{\omega(t)}{1+K}\right) . \end{aligned} \tag{4.82g}$$

Since the equilibrium does not depend on the initial values, the stationary values obtained in the limit of (4.82c) or (4.82e) are:

$$\begin{aligned} \lim_{t\to\infty} \mathrm{E}\big(\mathcal{X}_A(t)\big) &= n_0 \frac{1}{1+K} = n_0 \frac{l}{k+l} , \\ \lim_{t\to\infty} \mathrm{var}\big(\mathcal{X}_A(t)\big) &= n_0 \frac{K}{(1+K)^2} = n_0 \frac{kl}{(k+l)^2} , \\ \lim_{t\to\infty} \sigma\big(\mathcal{X}_A(t)\big) &= \sqrt{n_0}\sqrt{\frac{K}{(1+K)^2}} = \sqrt{n_0}\frac{\sqrt{kl}}{k+l} . \end{aligned} \tag{4.82h}$$

This result shows that the $\sqrt{N}$-law is satisfied up to a factor that is independent of N: $\mathrm{E}/\sigma = \sqrt{n_0}l/\sqrt{kl}$.

Starting from a sharp distribution $P_n(0) = \delta_{n,a0}$, the variance increases, passes through a maximum for $k > l$, and eventually reaches the equilibrium value $\bar{\sigma}^2 = kln_0/(k+l)^2$. The time of maximal fluctuations is easily calculated from the condition $\mathrm{d}\sigma^2/\,\mathrm{dt} = 0$, and one obtains

$$t_{\max(\mathrm{var})} = \frac{1}{k+l} \ln\left(\frac{2k}{k-l}\right) . \tag{4.82i}$$

The sign of $k - l$ decides whether the approach towards equilibrium passes a maximum value or not. The maximum is readily detected from the height of the mode of $P_n(t)$, as seen in Fig. 4.24, where a case with $k > l$ is presented.

In order to illustrate fluctuations and their sizes under equilibrium conditions, the Austrian physicist Paul Ehrenfest designed a game called Ehrenfest's urn model [129], which was indeed played in order to verify the $\sqrt{N}$-law. A total of $2N$ balls are numbered consecutively $1, 2, \ldots, 2N$, and distributed arbitrarily between two containers, say A and B. A lottery machine draws lots, which carry the

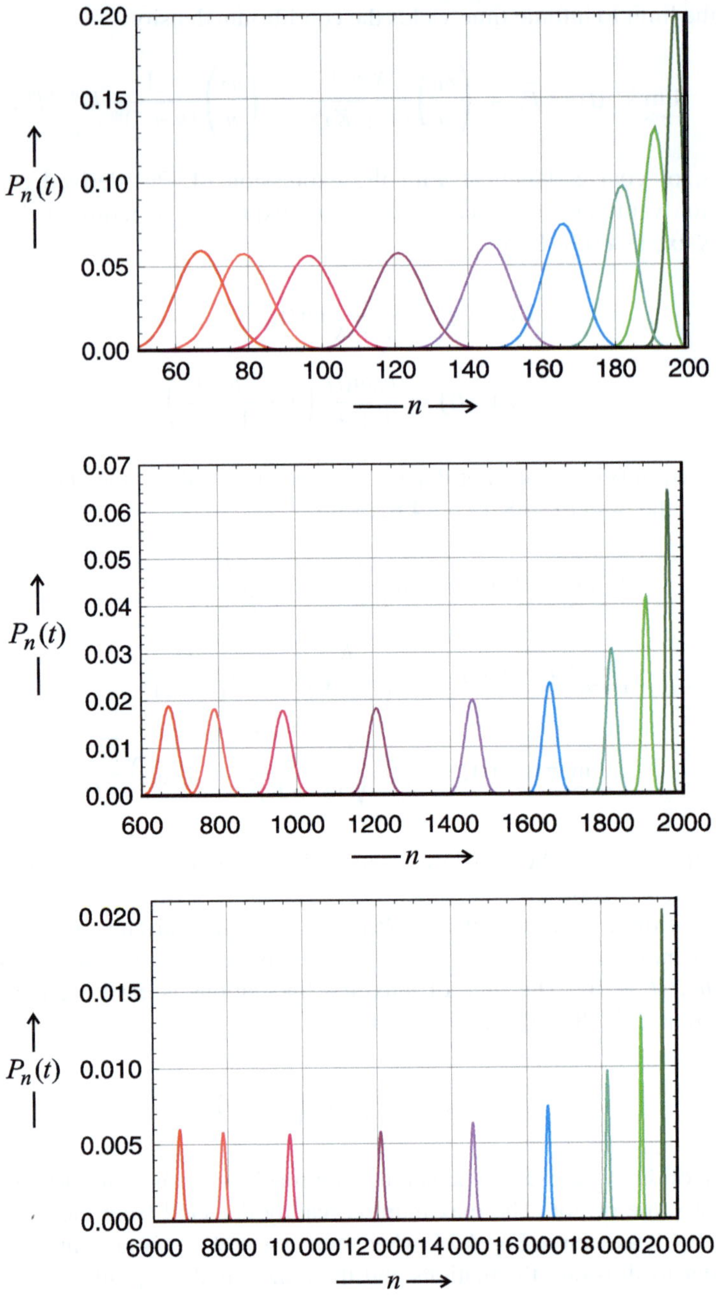

Fig. 4.24 Probability density of a reversible monomolecular reaction. Evolution of the probability density $P_n(t)$ of the number of molecules of a compound A which undergo a reaction $A \rightleftharpoons B$. The initially infinitely sharp density $P_n(0) = \delta_{n,n_0}$ becomes broader with time, until the variance settles down at the equilibrium value, eventually passing a point of maximum variance. Continued on next page

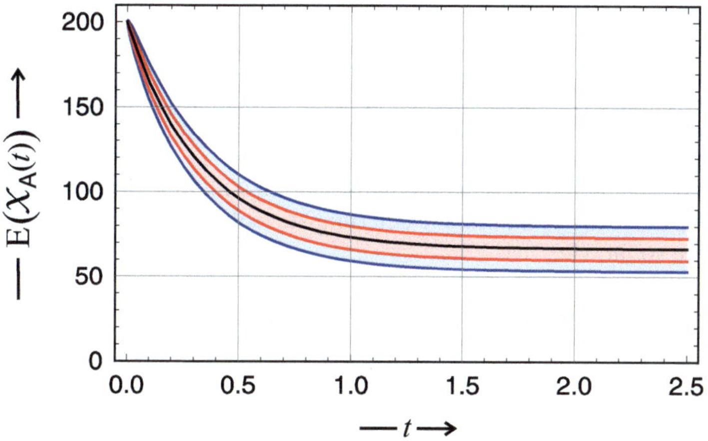

Fig. 4.24 (Cont.) Probability density of a reversible monomolecular reaction. Expectation value $E(\mathcal{X}_A(t))$ and confidence intervals $E \pm \sigma$ (68.3 % *red*) and $\pm 2\sigma$ (95.4 % *blue*), where $\mathrm{var}(\mathcal{X}_A(t))$ is the variance. Parameters used: $n_0 = 200$, 2,000, and 20,000, $k = 2l = 1\,[t^{-1}]$. Sampling times: 0 (*black*), 0.01 (*dark green*), 0.025 (*green*), 0.05 (*turquoise*), 0.1 (*blue*), 0.175 (*blue violet*), 0.3 (*purple*), 0.5 (*magenta*), 0.8 (*deep pink*), 2 (*red*). The initial condition for the time dependence of the expectation value was $n_0 = 200$

numbers of the balls. When the number of a ball is drawn, the ball is put from one container into the other. This setup is already sufficient for a simulation of the equilibrium condition. The more balls there are in a container, the more likely it is that the number of one of its balls will be drawn and a transfer will occur into the other container. Just as happens with chemical reactions, we have self-controlling fluctuations: whenever a fluctuation becomes large, it creates a force for compensation which is proportional to the size of the fluctuation.

First-Order Chemical Reaction Networks

Networks of first order chemical reactions (4.1c) including inflow (4.1a) and outflow (4.1b) can be analyzed in great generality [190, 278]. The chemical master equation of coupled linear reactions has been solved for several initial distributions including sharp values and probability distributions. Evolution equations for first and second moments are found in [190]. Tobias Jahnke and Wilhelm Huisinga present explicit equations for the time dependent probability densities [278] and show that the solutions can be expressed as convolutions of multinomial and product Poissonian distributions with time dependent parameters. The results can be encapsulated in the catchphrase: *multinomial stays multinomial for closed systems and Poissonian distributions stay Poissonian.* As we have seen already in Sect. 4.2.3, the equilibrium distributions of finite closed systems are multinomial, whereas those of open systems they are product Poisson distributions [190]. This is easily checked using (4.82f), which can be written as

$$P_n(t) = \binom{n_0}{n} \left(\frac{K\mathrm{e}^{-\kappa t} + 1}{1+K} \right)^n \left(\frac{K - K\mathrm{e}^{-\kappa t}}{1+K} \right)^{n_0-n} . \tag{4.82f'}$$

This has the standard form of a binomial equation with time dependent parameters.

Finally, we stress again that the deterministic solution curves for all three cases shown here so far, viz.,

$$\text{Flow reactor: } a(t) = a_0 - (a_0 - n_0)\mathrm{e}^{-rt} ,$$

$$\mathsf{A} \rightarrow \mathsf{B}: \ a(t) = n_0 \mathrm{e}^{-kt} ,$$

$$\mathsf{A} \rightleftharpoons \mathsf{B}: \ a(t) = a_0 \mathrm{e}^{-kt} + n_0 \frac{1}{1+K} \left(1 - \mathrm{e}^{-\kappa t}\right) ,$$

coincide exactly with the expectation values $\mathrm{E}\big(\mathcal{X}_\mathsf{A}(t)\big)$, a consequence of the linearity in n of the first jump moments α_1.

4.3.3 Bimolecular Chemical Reactions

Three classes of bimolecular reactions are accessible to full stochastic analysis in the irreversible limit $l \rightarrow 0$, corresponding to the elementary steps (4.1j), (4.1f), and (4.1g) [26, 99, 100, 271, 329, 384]:

$$2\mathsf{A} \underset{l}{\overset{k}{\rightleftharpoons}} \mathsf{C} , \tag{4.83a}$$

$$\mathsf{A} + \mathsf{B} \underset{l}{\overset{k}{\rightleftharpoons}} \mathsf{C} , \tag{4.83b}$$

$$\mathsf{A} + \mathsf{X} \underset{l}{\overset{k}{\rightleftharpoons}} 2\mathsf{X} . \tag{4.83c}$$

Bimolecularity gives rise to nonlinearities in the kinetic differential equations and in the master equations, and this substantially complicates the analysis. Bimolecular reactions from the first two classes do not show substantial differences in their qualitative behavior compared to the corresponding monomolecular case $\mathsf{A} \rightarrow \mathsf{B}$.

The exact coincidence of the expectation value and the deterministic solution, however, is no longer valid.

For the solution of the master equations, we present here the direct PDE approach as before and compare it to another technique based on Laplace transforms. Autocatalysis in the form of the reaction (4.83c) gives rise to intrinsic rate enhancement (Sect. 4.1.1) and different behavior of fluctuations, but the reaction dynamics remains simple in the sense that it approaches unique stationary states monotonically. Autocatalytic processes of higher molecularity like, for example, the termolecular step in the Brusselator reaction $\mathsf{A} + 2\mathsf{X} \rightarrow 3\mathsf{X}$ (4.1m), may give rise to multiple steady states, oscillations in concentrations, and deterministic chaos (Sect. 5.1).

A few analytic solutions for PDEs derived for the calculation of generating functions are shown here in order to illustrate the enormous complications resulting from nonlinear reaction rates. An alternative method to find solutions of the master equation is provided by the Laplace transform (Sect. 4.3.4).

Dimerization Reaction $2\mathsf{A} \rightarrow \mathsf{C}$

Like the irreversible monomolecular reaction $\mathsf{A} \rightarrow \mathsf{B}$, the dimerization reaction (4.83a) is a pure death process, and when modeled by means of a master equation [384], we have to take into account stoichiometry, which determines that two molecules A vanish at a time in order to form one C molecule. Thus the stochastic variable $\mathcal{X}_{\mathsf{A}}(t)$ with the probability density $P_n(t) = P\big(\mathcal{X}_{\mathsf{A}}(t) = n\big)$ makes exclusively jumps of size $|\Delta\mathcal{X}_{\mathsf{A}}| = 2$. This creates two irreducible communicating equivalence classes (see Fig. 4.18) comprising odd or even numbers of A molecules, respectively (Fig. 4.25). In other words, when the initial number of A molecules is odd, $a_0 = n_0 = 2\lambda + 1$ with $\lambda \in \mathbb{N}$, $\mathcal{X}_{\mathsf{A}}$ will always stay odd and the last A molecule will be unable to react. For an initially even number of A molecules, i.e.,

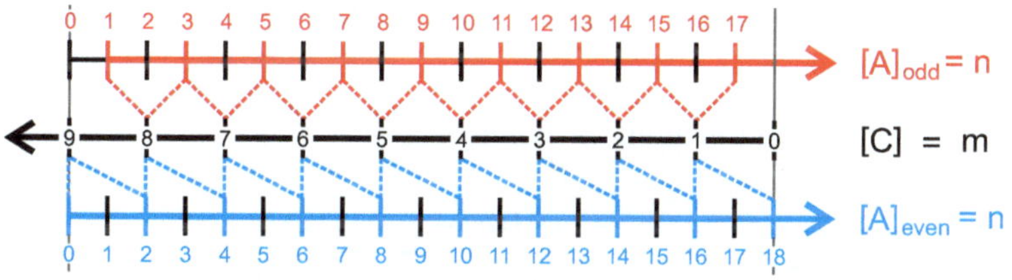

Fig. 4.25 Particle numbers $[\mathsf{A}] = n$ and $[\mathsf{C}] = m$ in the dimerization reaction $2\mathsf{A} \rightarrow \mathsf{C}$. Two molecules A react to yield one molecule C, so not all molecules A can react when a_0 is odd. We call this irreversible equivalence class the *odd class* (*red*), while the other irreducible equivalence class, the *even class* (*blue*), has an even number of A molecules and is disjunct. Forbidden states are marked in *black*. The conservation law is $\mathcal{X}_{\mathsf{A}} + 2\mathcal{X}_{\mathsf{C}} = n + 2m = a_0$

$a_0 = n_0 = 2\lambda$, $\mathcal{X}_A$ will always be even and the state $\mathcal{X}_A = 0$ is allowed:

$$\mathcal{X}_A(0) = a_0 = n_0 \implies \lim_{t\to\infty} \mathcal{X}_A(t) = \begin{cases} 1\,, & \text{if } n_0 = 2\lambda + 1\,,\ \lambda \in \mathbb{N}\,, \\ 0\,, & \text{if } n_0 = 2\lambda\,,\ \lambda \in \mathbb{N}\,. \end{cases}$$

The master equation taking into account the jump size $\Delta n = -2$ is of the form

$$\frac{\mathrm{d}P_n(t)}{\mathrm{d}t} = k(n+2)(n+1)P_{n+2}(t) - kn(n-1)P_n(t)\,, \tag{4.84a}$$

with the initial conditions $P_n(0) = \delta_{n,n_0}$ and $P(\mathcal{X}_C(0) = m) = \delta_{m,0}$, and the final condition $\lim_{t\to\infty} \mathcal{X}_C = m_0$ with $m_0 = \lfloor \frac{a_0}{2} \rfloor$. Jumps $|\Delta n| = 2$ give rise to two disjoint stochastic processes for even and odd initial numbers of molecules n_0. In the former case, we have $P_{2\lambda+1}(t) = 0$, $\forall\, t \in \mathbb{R}_{\geq 0}$, and in the latter case $P_{2\lambda}(t) = 0$, $\forall\, t \in \mathbb{R}_{\geq 0}$ for all $\lambda \in \mathbb{N}$.

The master equation gives rise to the following PDE for the probability generating function: [384]:

$$\frac{\partial g(s,t)}{\partial t} = k(1-s^2)\frac{\partial^2 g(s,t)}{\partial s^2}\,. \tag{4.84b}$$

The analysis of this PDE is more involved than it might appear at first glance, but it allows for separation of variables.

For the initial condition $P_n(0) = \delta_{n,a_0}$ and proper boundary conditions, exact solutions of (4.84b) in terms of auxiliary functions with separation of variables are of the form

$$g(s,t) = \sum_{j=0}^{\infty} A_j C_j^{(-1/2)}(s) T_j(t)\,, \tag{4.84c}$$

where the constants and functions are defined by

$$\begin{aligned} A_j &= \frac{2j-2}{2^j}\,\frac{\Gamma(a_0+1)\Gamma\big((a_0-j+1)/2\big)}{\Gamma(a_0-j+1)\Gamma\big((a_0+j+1)/2\big)}\,, \\ C_j^{(-1/2)}(s)&:\ (1-s^2)\,\frac{\mathrm{d}^2 C_j^{(-1/2)}(s)}{\mathrm{d}s^2} + j(j-1)C_j^{(-1/2)}(s) = 0\,, \\ T_j(t) &= \exp\big(-kj(j-1)t\big)\,. \end{aligned} \tag{4.84d}$$

Individual terms of the generating function are factorized into a factor, a function of the variable s, and a function of time t. The factors A_j are a result of combinatorics.

The functions $C_j^{\alpha}(s)$ are Gegenbauer polynomials, named after the German mathematician Leopold Gegenbauer [2, Chap. 22, pp. 773–802]. The generating function

$$\frac{1}{(1-2sw+w^2)^{\alpha}} = \sum_{j=0}^{\infty} C_j^{(\alpha)}(s) w^j$$

applies, and the normalization integral for $\alpha > -1/2$ is

$$\int_{-1}^{1} (1-x^2)^{\alpha-1/2} \left(C_j^{(\alpha)}\right)^2 \mathrm{d}x = \frac{2^{1-2\alpha}\pi\Gamma(j+2\alpha)}{(j+\alpha)\big(\Gamma(\alpha)\big)^2\Gamma(j+1)} .$$

The Gegenbauer polynomials are particular solutions of the Gegenbauer differential equation

$$(1-s^2)\frac{\mathrm{d}^2 C_j^{(\alpha)}(s)}{\mathrm{d}s^2} - (2\alpha+1)s\frac{\mathrm{d}C_j^{(\alpha)}(s)}{\mathrm{d}s} + j(j+2\alpha)C_j^{(\alpha)}(s) = 0 ,$$

which, for $\alpha = -1/2$, becomes (4.84d). There are several useful relations between Gegenbauer polynomials and other special functions, for example, Jacobi polynomials or hypergeometric functions [2], which like the normalization are strictly valid only for $\alpha > -1/2$. Nevertheless, the members of the series $C_j^{(-1/2)}(s)$ $(j = 0, 1, \ldots)$

$$C_0^{(-1/2)}(s) = 1 , \quad C_1^{(-1/2)}(s) = -s , \quad C_2^{(-1/2)}(s) = \frac{1}{2}(1-s^2) ,$$

$$C_3^{(-1/2)}(s) = \frac{s}{2}(1-s^2) , \quad C_4^{(-1/2)}(s) = -\frac{1}{8}\left(1-6x^2+5x^4\right) ,$$

and so on, represent the solutions of our PDF. The time dependence of $g(s,t)$ is given by a superposition of exponential functions containing the rate parameter k. Although the summation in (4.84c) is extended to infinity, finite numbers of initial molecules cause the sums to terminate by conservation laws, since $P_n(t) = 0, \forall\, n > 2m_0 + 1$.

The calculation of closed expressions for the probability densities $P_n(t)$ from the probability generating function is quite involved and we postpone it to the next Sect. 4.3.4 where we shall obtain it much easier by inverse Laplace transform. It is straightforward, nevertheless, to calculate expressions for the expectation values and the variance of the stochastic variable $\mathcal{X}_A(t)$ [271, 384] where λ stands for an integer running index, $\lambda \in \mathbb{N}$:

$$\mathrm{E}\big(\mathcal{X}_A(t)\big) = \sum_{j=2\lambda=2}^{2\lfloor a_0/2 \rfloor} A_j T_j(t) = \sum_{j=2\lambda=2}^{2\lfloor a_0/2 \rfloor} \frac{(2j-1)a_0!(a_0-j-1)!!}{(a_0-j)!(a_0+j-1)!!} \mathrm{e}^{-kj(j-1)t} ,$$

$$\mathrm{E}\big(\mathcal{X}_A(t)^2\big) = \sum_{j=2\lambda=2}^{2\lfloor a_0/2 \rfloor} \frac{(2j-1)(j^2-j+2)a_0!(a_0-j-1)!!}{2(a_0-j)!(a_0+j-1)!!} \mathrm{e}^{-kj(j-1)t} ,$$

$$\mathrm{var}\big(\mathcal{X}_A(t)\big) = \mathrm{E}\big(\mathcal{X}_A(t)^2\big) - \mathrm{E}\big(\mathcal{X}_A(t)\big)^2 . \tag{4.84e}$$

In order to obtain specific results these expressions can be evaluated numerically, but care is needed since the factorials lead to very large numbers in the numerator and denominator when calculated naïvely. The time course of the probability density function is shown in Fig. 4.26. For a comparison of the relative widths of the densities of the three bimolecular reactions, we refer to Fig. 4.34. In addition the figure presents the expectation value within the one-sigma confidence interval. As expected, for nonlinear step-down transition probabilities w_n^-, the expectation value does not coincide with the solution of the kinetic ODEs. The difference between the stochastic expectation values and the deterministic solutions of the dimerization reaction will be brought up in detail in Sect. 4.3.4.

Association Reaction $\mathsf{A} + \mathsf{B} \to \mathsf{C}$

In the association reaction (4.83b), we are dealing with three stochastic variables $\mathcal{X}_\mathsf{A}(t)$, $\mathcal{X}_\mathsf{B}(t)$, and $\mathcal{X}_\mathsf{C}(t)$, and two conservation relations. Following Donald McQuarrie and coworkers [384], we define the probability $P_n(t) = P\big(\mathcal{X}_\mathsf{A}(t) = n\big)$ and apply the standard initial sharp conditions $P_n(0) = \delta_{n,n_0}$ with $n_0 = a_0$, $P\big(\mathcal{X}_\mathsf{B}(0) = b\big) = \delta_{b,b_0}$, and $P\big(\mathcal{X}_\mathsf{C}(0) = c\big) = \delta_{c,0}$. From the laws of stoichiometry and mass conservation, we have

$$\mathcal{X}_\mathsf{B}(t) = \vartheta_0 + \mathcal{X}_\mathsf{A}(t) , \qquad \mathcal{X}_\mathsf{C}(t) = a_0 - \mathcal{X}_\mathsf{A}(t) ,$$

where we use $\vartheta_0 = b_0 - a_0$. Then the master equation for the association reaction is of the form

$$\frac{\partial P_n(t)}{\partial t} = k(n+1)(\vartheta_0+n+1)P_{n+1}(t) - kn(\vartheta_0+n)P_n(t) , \tag{4.85a}$$

with $n \in \mathbb{N}_\mathsf{A}^< = \{0, 1, \ldots, a_0\}$ for $a_0 < b_0$ or $n \in \mathbb{N}_\mathsf{A}^> = \{|\vartheta_0|, |\vartheta_0|+1, \ldots, a_0\}$ for $a_0 > b_0$, whence $P_n(t) = 0$, $\forall\, n \notin \mathbb{N}_\mathsf{A}^<$, or $P_n(t) = 0$, $\forall\, n \notin \mathbb{N}_\mathsf{A}^>$, respectively. It is worth noticing that, in the stochastic treatment, the distinction between $a_0 < b_0$ and $a_0 > b_0$ is made in the domains of the stochastic variables, whereas it enters directly

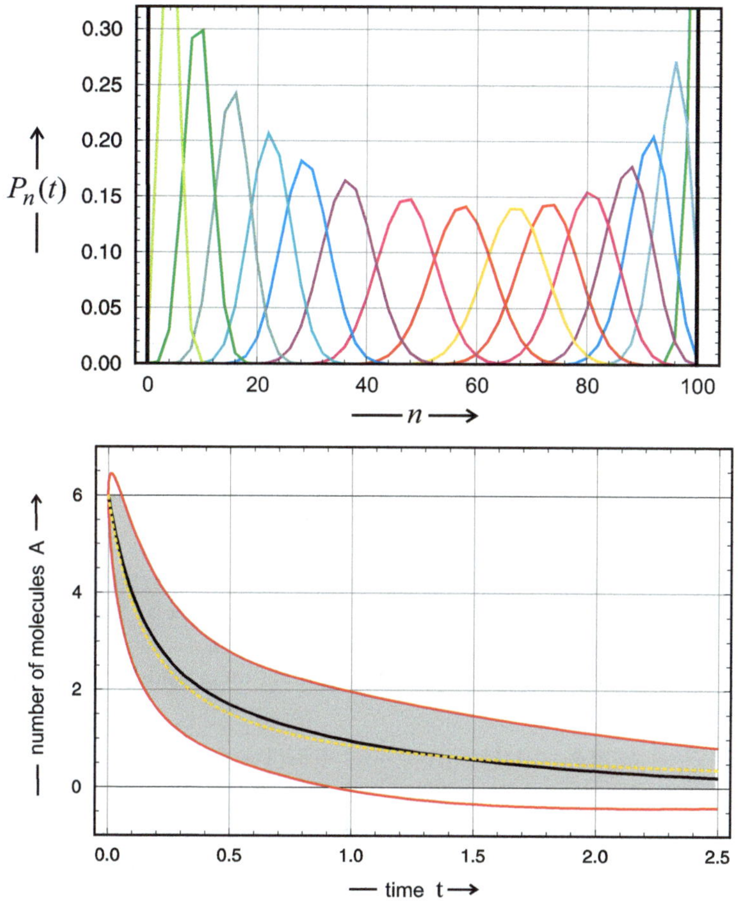

Fig. 4.26 Irreversible dimerization reaction $2A \rightarrow C$. *Upper*: Probability distribution $P_n(t) = P(\mathcal{X}_A(t) = n)$ describing the number of molecules of species A as a function of time and calculated using (4.89e). The number of molecules C is given by the distribution $P_m(t) = P(\mathcal{X}_C(t) = m)$. The initial conditions are chosen to be $\mathcal{X}_A(t) = \delta(n, a_0)$ and $\mathcal{X}_C(t) = \delta(m, 0)$, so we have $n + 2m = a_0$. With increasing time, the peak of the distribution moves from right to left. The state $n = 0$ for $a_0 = 2\lambda$ or the state $n = 1$ for $a_0 = 2\lambda + 1$ with $\lambda \in \mathbb{N}$ are absorbing states, so the long-time limit of the system is $\lim_{t\to\infty} \mathcal{X}_C(t) = \delta(m, \lfloor a_0/2 \rfloor)$ and $\lim_{t\to\infty} \mathcal{X}_A(t) = \delta(n, 0)$ or $\delta(n, 1)$, respectively. Parameter choice: $a_0 = 100$ and $k = 0.005\,[\mathrm{N}^{-1}\mathrm{t}^{-1}]$. Sampling times: $t = 0$ (*black*), 0.01 (*green*), 0.1 (*turquoise*), 0.2 (*blue*), 0.3 (*violet*), 0.5 (*magenta*), 0.75 (*red*), 1.0 (*yellow*), 1.5 (*red*), 2.25 (*magenta*), 3.5 (*violet*), 5.0 (*blue*), 7.0 (*cyan*), 11.0 (*turquoise*), 20.0 (*green*), 50.0 (*chartreuse*), and ∞ (*black*). The plots *at the bottom* and *overleaf* compare the expectation value of the stochastic solution $E(\mathcal{X}_A(t))$ (*black*) within the confidence interval $E \pm \sigma$ with the conventional deterministic solution $a(t)$ from (4.90a) (*yellow*). The curve for $E + \sigma$ becomes higher than a_0 and the curve for $E - \sigma$ goes beyond zero and thus goes outside the physically meaningful domain (*gray area* of *lower plot*).The deviations are smaller than a single particle. For comparison the deterministic solution curve, $a(t)$ is shown as well (*yellow dashed line*). Parameter choice: $k = 1\,[\mathrm{N}^{-1}\mathrm{t}^{-1}]$, $a_0 = 6$ (*lower*) and $a_0 = 40$ (*overleaf*). Continued on next page

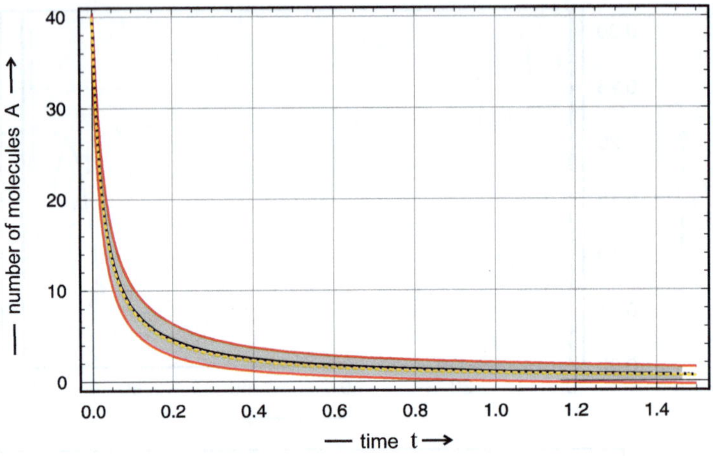

Fig. 4.26 (Cont.) Irreversible dimerization reaction 2A → C. See over for details

in the differential equations of the deterministic approach: $\mathrm{d}a/\,\mathrm{d}t = ka(a+\vartheta_0)$ for $a_0 < b_0$ and $\mathrm{d}a/\,\mathrm{d}t = ka(a-\vartheta_0)$ for $a_0 > b_0$, respectively. We remark that the step-down transition probabilities are no longer linear in n.

The corresponding PDE for the generating function is readily calculated to be

$$\frac{\partial g(s,t)}{\partial t} = k(\vartheta_0+1)(1-s)\frac{\partial g(s,t)}{\partial s} + ks(1-s)\frac{\partial^2 g(s,t)}{\partial s^2} \,. \tag{4.85b}$$

The derivation of solutions for this PDE is quite demanding, but as for the dimerization reaction, it can be achieved by separation of variables:

$$g(s,t) = \sum_{j=0}^{\infty} A_j Z_j(s) T_j(t) \,. \tag{4.85c}$$

We list here only the coefficients and functions of the solution:

$$A_j = (-1)^j \frac{(2j+\vartheta_0)\Gamma(j+\vartheta_0)\Gamma(a_0+1)\Gamma(a_0+\vartheta_0+1)}{\Gamma(j+1)\Gamma(\vartheta_0+1)\Gamma(a_0-j+1)\Gamma(a_0+\vartheta_0+j+1)} \,,$$

$$Z_j(s) = J_j(\vartheta_0, \vartheta_0+1, s) \,, \qquad T_j(t) = \exp\bigl(-j(j+\vartheta_0)\,kt\bigr) \,.$$

Here, Γ is the *gamma function* and $J_n(p,q,s)$ are the Jacobi polynomials [2, Chap. 22, pp. 773–802], named after the German mathematician Carl Jacobi. They are solutions of the differential equation

$$s(1-s)\frac{\mathrm{d}^2 J_n(p,q,s)}{\mathrm{d}s^2} + \bigl(q-(p+1)s\bigr)\frac{\mathrm{d}J_n(p,q,s)}{\mathrm{d}s} + n(n+p), J_n(p,q,s) = 0 \,.$$

These polynomials satisfy the following conditions:

$$\frac{\mathrm{d}J_n(p,q,s)}{\mathrm{d}s} = -\frac{n(n+p)}{s} J_{n-1}(p+2,q+1,s)$$

$$\int_0^1 s^{q-1}(1-s)^{p-q} J_n(p,q,s) J_\ell(p,q,s)\,\mathrm{d}s = \frac{n!\big(\Gamma(q)\big)^2 \Gamma(n+p-q+1)}{(2n+p)\Gamma(n+p)\Gamma(n+q)} \delta_{\ell,n}\,.$$

We differentiate twice at the value $s = 1$ of the dummy variable and find

$$\left.\frac{\partial g(s,t)}{\partial s}\right|_{s=1} = \sum_{j=1}^{a_0} \frac{(2j+\vartheta_0)\Gamma(a_0+1)\Gamma(a_0+\vartheta_0+1)}{\Gamma(a_0-j+1)\Gamma(a_0+\vartheta_0+j+1)} T_j(t)\,, \tag{4.85d}$$

$$\left.\frac{\partial^2 g(s,t)}{\partial s^2}\right|_{s=1} = \big\langle \mathcal{X}_A(t)\big(\mathcal{X}_A(t)-1\big)\big\rangle \tag{4.85e}$$

$$= \sum_{j=2}^{a_0} \frac{(j-1)(j+\vartheta_0+1)(2j+\vartheta_0)\Gamma(a_0+1)\Gamma(a_0+\vartheta_0+1)}{\Gamma(a_0-j+1)\Gamma(a_0-\vartheta_0+j+1)} T_j(t)\,,$$

from which we obtain the expectation value and variance as

$$E\big(\mathcal{X}_A(t)\big) = \sum_{j=1}^{a_0} \frac{(2j+\vartheta_0)\Gamma(a_0+1)\Gamma(a_0+\vartheta_0+1)}{\Gamma(a_0-j+1)\Gamma(a_0+\vartheta_0+j+1)} \exp\big(-j(j+\vartheta_0)kt\big)\,, \tag{4.85f}$$

$$\mathrm{var}\big(\mathcal{X}_A(t)\big) = \sum_{j=2}^{a_0} \frac{(j-1)(j+\vartheta_0+1)(2j+\vartheta_0)\Gamma(a_0+1)\Gamma(a_0+\vartheta_0+1)}{\Gamma(a_0-j+1)\Gamma(a_0-\vartheta_0+j+1)}$$

$$\times \exp\big(-j(j+\vartheta_0)kt\big) + \mathrm{E}\big(\mathcal{X}_A(t)\big) - \Big(\mathrm{E}\big(\mathcal{X}_A(t)\big)\Big)^2\,. \tag{4.85g}$$

As we see in the current example and we shall see again in the next section, bimolecularity substantially complicates the solution of the chemical master equations and makes the solutions quite sophisticated. Again, we postpone the calculation of closed expressions for the densities $P_n(t)$ to Sect. 4.3.4, but show a numerical example in Fig. 4.27 with different initial conditions: (i) $b_0 = a_0$ and (ii) $b_0 = 2a_0$. Because of the larger number of molecules B in the second case, the reaction rate is higher and the peak of the probability distribution moves faster.

The special case where there is a vast excess of one reaction partner, i.e., $b_0 \gg a_0$ or $|\vartheta_0| \gg a_0 > 1$, which is known as the *pseudo-first order condition* or *concentration buffering*. Then, the sums can be approximated well by the first

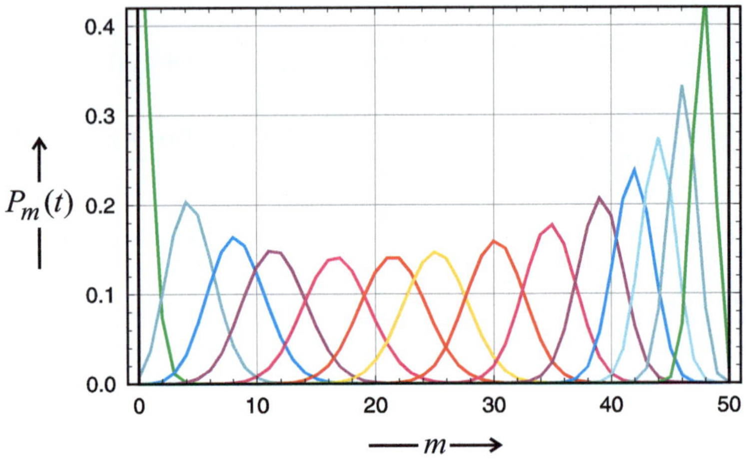

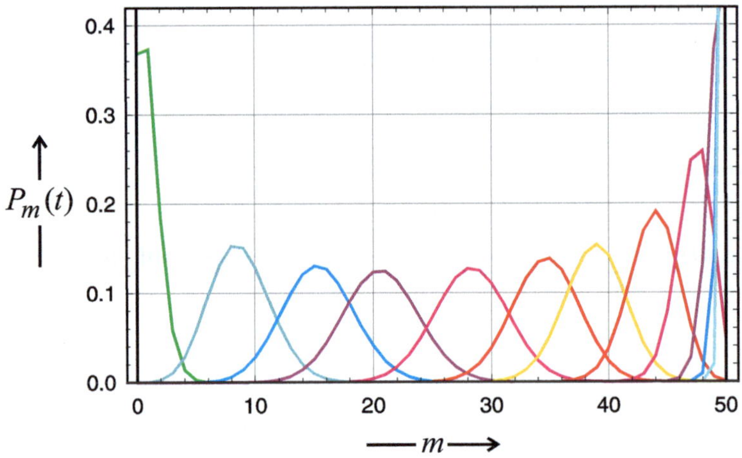

Fig. 4.27 Irreversible association reaction $\mathsf{A} + \mathsf{B} \to \mathsf{C}$. Both plots show the probability distribution $P_m(t) = P(\mathcal{X}_\mathsf{C}(t) = m)$ describing the number of molecules of species C as a function of time, as calculated using (4.88g). The initial conditions are chosen to be $\mathcal{X}_\mathsf{A}(0) = \delta(a, a_0)$, $\mathcal{X}_\mathsf{B}(0) = \delta(b, b_0)$, and $\mathcal{X}_\mathsf{C}(0) = \delta(c, 0)$. With increasing time, the peak of the distribution moves from left to right. The state $m = \min(a_0, b_0)$ is an absorbing state, so the long-time limit of the system is $\lim_{t\to\infty} \mathcal{X}_\mathsf{C}(t) = \delta\big(m, \min(a_0, b_0)\big)$. Choice of parameters: $a_0 = 50$, $b_0 = 50$ (*upper plot*) and $b_0 = 100$ (*lower plot*), $c_0 = 0$, $k = 0.02\,[\mathrm{N}^{-1}\mathrm{t}^{-1}]$. Sampling times (*upper part*): $t = 0$ (*black*), 0.01 (*green*), 0.1 (*turquoise*), 0.2 (*blue*), 0.3 (*violet*), 0.5 (*magenta*), 0.75 (*red*), 1.0 (*yellow*), 1.5 (*red*), 2.25 (*magenta*), 3.5 (*violet*), 5.0 (*blue*), 7.0 (*turquoise*), 11.0 (*seagreen*), 20.0 (*green*), and ∞ (*black*). The density $P_m(t)$ moves faster in the lower plot, reflecting higher particle numbers of B

terms and, with $k' = \vartheta_0 k$, we find

$$\left.\frac{\partial g(s,t)}{\partial s}\right|_{s=1} \approx a_0 \frac{\vartheta_0 + 2}{a_0 + \vartheta_0 + 1} \mathrm{e}^{-(\vartheta_0+1)kt} \approx a_0 \mathrm{e}^{-k't} ,$$

$$\left.\frac{\partial^2 g(s,t)}{\partial s^2}\right|_{s=1} \approx a_0(a_0 - 1)\mathrm{e}^{-2k't} .$$

Finally, we obtain

$$\begin{aligned} \mathrm{E}\big(\mathcal{X}_A(t)\big) &= a_0 \mathrm{e}^{-k't} , \\ \mathrm{var}\big(\mathcal{X}_A(t)\big) &= a_0 \mathrm{e}^{-k't}\Big(1 - \mathrm{e}^{-k't}\Big) , \end{aligned} \qquad (4.85\mathrm{h})$$

which is formally the same result as obtained for the irreversible first order reaction with $k \Longleftrightarrow k' = \vartheta_0 k$.

4.3.4 *Laplace Transform of Master Equations*

Because of its more general applicability, we consider here also the solution of chemical master equations by means of the Laplace transform [26, 271, 329], which is similar to the approach used in Sect. 3.2.4 for analyzing random walks (see also the moment generating functions in Sect. 2.2.2). The Laplace transform of a probability density $P_n(t)$ is denoted by

$$\widehat{f}_n(s) = \int_0^\infty \exp(-st) P_n(t)\, \mathrm{d}t . \qquad (4.86\mathrm{a})$$

The time derivative is calculated using integration by parts:

$$\begin{aligned} \int_0^\infty \frac{\mathrm{d}P_n(t)}{\mathrm{d}t} \mathrm{e}^{-st}\, \mathrm{d}t =& P_n(t)\mathrm{e}^{-st}\Big|_0^\infty - \int_0^\infty P_n(t)\mathrm{e}^{-st}(-s)\, \mathrm{d}t \\ =& s\int_0^\infty P_n(t)\mathrm{e}^{-st}\, \mathrm{d}t - P_n(0) = s\widehat{f}_n(s) - P_n(0) . \end{aligned} \qquad (4.86\mathrm{b})$$

We thereby obtain an algebraic equation for the Laplace transform $\widehat{f}_n(s)$, which can be solved by standard techniques. Then the probability density may be obtained through back-transformation by inverse Laplace transform. The calculations are performed for individual probabilities $P_n(t) \Longleftrightarrow \widehat{f}_n(s)$ or by linear algebra for probability vectors $\mathbf{P}(t) = \big(P_0(t), P_1(t), \ldots, P_{n_0}(t)\big)^{\mathrm{t}}$, $\mathbf{P}_n(t) \Longleftrightarrow \widehat{\mathbf{f}}_n(s)$. We give examples for both cases.

The inverse Laplace transformation by Mellin's fomula is defined by[48]

$$\mathcal{L}^{-1}\big(F(s)\big) \doteq \frac{1}{2\pi\mathrm{i}} \lim_{\vartheta\to\infty} \int_{\gamma-\mathrm{i}\vartheta}^{\gamma+\mathrm{i}\vartheta} \mathrm{e}^{st} F(s)\mathrm{d}s = \begin{cases} f(t)\ , & \text{for } t \geq 0\ , \\ 0\ , & \text{for } t < 0\ , \end{cases} \tag{4.86c}$$

where $\gamma > s_0$ is a real number chosen such that the contour of integration is the region of convergence of the Laplace transform $F(s)$ and $|f(t)| \leq C\mathrm{e}^{s_0 t}$ is satisfied for $t > \tau$ and $\int_0^\tau |f(t)|\,\mathrm{d}t < \infty$. If the integral is extended over the positive real axis and one does not invoke residue calculus, all poles of the function have to lie to the left of the imaginary axis. This somewhat complicated procedure is needed to guarantee the existence of a Laplace transform and its inverse. To illustrate this, we repeat the transformation properties of the exponential function $f(t) = \exp(-at)$:

$$\mathcal{L}\big(\mathrm{e}^{-at}\big) = \frac{1}{s+a} \iff \mathcal{L}^{-1}\Big(\frac{1}{s+a}\Big) = \mathrm{e}^{-at}\ . \tag{4.86d}$$

Here, as required, the pole is indeed situated to the left of the imaginary axis.

Considering now a birth-and-death master equation consisting of $n_0 + 1$ ODEs of the general form

$$\frac{\mathrm{d}P_n(t)}{\mathrm{d}t} = w_{n+1}^- P_{n+1}(t) + w_{n-1}^+ P_{n-1}(t) - (w_n^+ + w_n^-)P_n(t)\ , \quad n = 0, 1, \ldots, n_0\ ,$$

we obtain, after Laplace transform, an algebraic equation for the Laplace transformed probabilities $\widehat{f}_n(s)$:

$$(s + w_n^+ + w_n^-)\widehat{f}_n(s) = w_{n+1}^- \widehat{f}_{n+1}(s) + w_{n-1}^+ \widehat{f}_{n-1}(s) + P_n(0)\ , \quad n = 0, \ldots, n_0\ . \tag{4.86e}$$

For pure birth or pure death processes with an absorbing boundary at $n = n_0$ or $n = 0$, respectively, the system of equations can be solved by successive iteration. For the pure birth process we have $w_n^- = 0$ and $P_{n<0}(t) = 0$, leading to $\widehat{f}_{n<0}(s) = 0$, which results in an equation containing only $\widehat{f}_0(s)$, and from known $\widehat{f}_0$ we can calculate $\widehat{f}_0 \Rightarrow \widehat{f}_1 \Rightarrow \cdots \Rightarrow \widehat{f}_{n_0}$. Analogously, we have $w_n^+ = 0$ and $P_{n>n_0}(t) = 0$ for the pure death process, which allows direct calculation of $\widehat{f}_{n_0}$, and then we proceed via $\widehat{f}_{n_0} \Rightarrow \widehat{f}_{n_0-1} \Rightarrow \cdots \Rightarrow \widehat{f}_0$. Alternatively, solutions of the system of linear equations can be calculated by solving an eigenvalue problem. Inverse Laplace transformation yields the solutions of the master equation. Examples will be presented in the ensuing sections.

[48] Mellin's formula is named after the Finnish mathematician Hjalmar Mellin and represents one of several possible definitions of inverse Laplace transforms. Another frequently used expression is Post's inversion formula due to the Polish–American mathematician Emil Leon Post, which expands the inverse into infinite order derivatives.

The Irreversible Reaction A → B as a Tutorial

The irreversible monomolecular conversion reaction has been analyzed exhaustively already in Sect. 4.3.2. We use it here as a simple example to illustrate the application of the Laplace transform to the simple master equation

$$\frac{\mathrm{d}P_n(t)}{\mathrm{d}t} = k(n+1)P_{n+1}(t) - knP_n(t)\,, \quad n = 0, 1, \ldots, n_0\,. \tag{4.87a}$$

The Laplace transform

$$\widehat{f}_n(s) = \int_0^\infty P_n(t)\mathrm{e}^{-st}\,\mathrm{d}t\,, \qquad \int_0^\infty \frac{\mathrm{d}P_n(t)}{\mathrm{d}t}\,\mathrm{d}t = s\widehat{f}_n(s) - P_n(0)\,,$$

yields a system of linear equations

$$n = 0 \qquad s\widehat{f}_0(s) - P_0(0) = k\widehat{f}_1(s)\,, \tag{4.87b}$$

$$n = 1 \qquad s\widehat{f}_1(s) - P_1(0) = k2\widehat{f}_2(s) - k\widehat{f}_1(s)\,, \tag{4.87c}$$

$$\vdots \qquad\qquad \vdots$$

$$n = n \qquad s\widehat{f}_n(s) - P_n(0) = k(n+1)\widehat{f}_{n+1}(s) - kn\widehat{f}_n(s)\,, \tag{4.87d}$$

$$\vdots \qquad\qquad \vdots$$

$$n = n_0 \quad s\widehat{f}_{n_0}(s) - P_{n_0}(0) \;= -kn_0\widehat{f}_{n_0}(s)\,. \tag{4.87e}$$

The last equation (4.87e) allows for the calculation of $\widehat{f}_{n_0}(s)$ and, with the initial condition $P_n(0) = \delta_{n,n_0}$, we find

$$\widehat{f}_{n_0}(s) = \frac{1}{s + kn_0}\,. \tag{4.87f}$$

Successive iteration yields the solutions in Laplace space:

$$\widehat{f}_n(s) = \frac{k^{n_0-n}(n_0)_{n_0-n}}{\prod_{i=0}^{n}\bigl(s + k(n_0 - i)\bigr)}\,. \tag{4.87g}$$

The next step is partial fraction decomposition [529, pp. 569–577], and although this is a very general and useful technique to convert products into sums, we mention here the specific case required for the inverse Laplace transformation that is not readily found in most mathematics textbooks. Consider a function $f(x) = 1/Q(x)$ with $Q(x) = (x - \alpha_1)(x - \alpha_2)\cdots(x - \alpha_n)$ with n distinct real roots that is to be converted into a sum of fractions:

$$\frac{1}{Q(x)} = \frac{c_1}{x+\alpha_1} + \frac{c_2}{x+\alpha_2} + \cdots + \frac{c_n}{x+\alpha_n} \,.$$

A somewhat tedious but straightforward calculation yields

$$c_j = \frac{1}{\prod_{i=1,i\neq j}^{n}(\alpha_i - \alpha_j)} \,, \quad j = 1,2,\ldots,n \,.$$

Our special case is particularly simple because all the factors $\alpha_i - \alpha_j$ are simple multiples of k:

$$\alpha_{n_i} - \alpha_{n_j} = \big(s + k(n_0 - i)\big) - \big(s + k(n_0 - j)\big) = k(j - i) \,,$$

and, for the solutions on Laplace space that are especially prepared for the inverse transform, this yields

$$\widehat{f}_n(s) = \binom{n_0}{n_0 - n} \sum_{i=0}^{n_0-n} (-1)^i \binom{n_0 - n}{i} \frac{1}{s + k(n_0 - i)} \,. \tag{4.87h}$$

Inverse Laplace transform is now straightforward, because we have a sum of simple fractions of the type $(s + a)^{-1}$. Eventually, we find

$$\begin{aligned} P_n(t) &= \binom{n_0}{n_0 - n} \mathrm{e}^{-nkt} \left(\sum_{i=0}^{n_0-n} (-1)^i \binom{n_0 - n}{i} \mathrm{e}^{-ikt} \right) \\ &= \binom{n_0}{n} \mathrm{e}^{-nkt} \left(1 - \mathrm{e}^{-kt}\right)^{n_0-n} \,. \end{aligned} \tag{4.87i}$$

The combination of partial fraction theory and inverse Laplace transformation is often called the Heaviside method [98], after the English electrical engineer and mathematician Oliver Heaviside. Naturally, the expression coincides with (4.81c). In this particular case where the PDE for the determination of the probability generating function $g(s,t)$ was linear, the advantage in applying the Laplace transform is not obvious, but for the nonlinear PDEs handled in Sect. 4.3.3, knowledge of highly specialized functions is dispensable and the solution technique by Laplace transform is much more general, as shown in the following examples.

Irreversible Association Reaction $\mathsf{A} + \mathsf{B} \rightarrow \mathsf{C}$

We start by writing down the master equation in a slightly different form, with $\mathcal{X}_\mathsf{C}(t)$ as stochastic variable $P_m(t) = P\big(\mathcal{X}_\mathsf{C}(t) = m\big)$. The initial conditions are $P_m(0) = \delta_{m,0}$, $P\big(\mathcal{X}_\mathsf{A}(0) = a\big) = \delta_{a,a_0}$, and $P\big(\mathcal{X}_\mathsf{B}(0) = b\big) = \delta_{b,b_0}$, and $\lim_{t\to\infty} P_m(t) = m_0 =$

$\min\{a_0, b_0\}$ defines the upper limit of m. Conservation of probability yields

$$\sum_{m=0}^{m_0} P_m(t) = 1 \,, \quad m \in \mathbb{N} \,, \quad \text{and thus} \;\; P_m(t) = 0 \,, \;\; \forall\, m \notin [0, m_0] \,,$$

and the master equation now takes the form

$$\begin{aligned} \frac{\mathrm{d}P_m(t)}{\mathrm{d}t} &= k\big(a_0 - (m-1)\big)\big(b_0 - (m-1)\big)P_{m-1}(t) \\ &\quad -k(a_0 - m)(b_0 - m)P_m(t) \,, \quad 0 \le m \le m_0 \,. \end{aligned} \tag{4.88a}$$

Carrying out the Laplace transform and inserting $P_m(0) = \delta(m, 0)$ leads to

$$-1 + s\widehat{f}_0(s) = -ka_0b_0\widehat{f}_0(s) \,, \tag{4.88b}$$

$$\begin{aligned} s\widehat{f}_m(s) &= k\big(a_0 - (m-1)\big)\big(b_0 - (m-1)\big)\widehat{f}_{m-1}(s) \\ &\quad -k(a_0 - m)(b_0 - m)\widehat{f}_m(s) \,, \quad 1 \le m < m_0 \,, \end{aligned} \tag{4.88c}$$

$$s\widehat{f}_{m_0}(s) = k\big(a_0 - (m_0-1)\big)\big(b_0 - (m_0-1)\big)\widehat{f}_{m_0-1}(s) \,. \tag{4.88d}$$

The solutions in Laplace space expressed by the functions $\widehat{f}_m(s)$ are calculated by successive iteration, i.e., $\widehat{f}_0(s) \Rightarrow \widehat{f}_1(s) \Rightarrow \cdots \Rightarrow \widehat{f}_{m_0}(s)$, which yields

$$\widehat{f}_m(s) = \binom{a_0}{m}\binom{b_0}{m}(m!)^2k^n \prod_{j=0}^{n} \frac{1}{s + k(a_0 - j)(b_0 - j)} \,, \quad 0 \le m \le m_0 \,, \tag{4.88e}$$

where the product in the denominator is resolved into partial fractions

$$\begin{aligned} \prod_{j=0}^{n} \big(s + k(a_0 - j)(b_0 - j)\big)^{-1} &= \sum_{j=0}^{n} \frac{C_j}{s + k(a_0 - j)(b_0 - j)} \,, \\ C_j = (-1)^{m+j}k^{-m} &\frac{(a_0 + b_0 - 2j)(a_0 + b_0 - m - j - 1)!}{j!(m-j)!(a_0 + b_0 - j)!} \,, \end{aligned} \tag{4.88f}$$

which are suitable for inverse Laplace transform. Performing the transformation yields the final result

$$\begin{aligned} P_m(t) = (-1)^m &\binom{a_0}{m}\binom{b_0}{m} \sum_{j=0}^{m} (-1)^j \left(1 + \frac{m - j}{a_0 + b_0 - m - j}\right) \\ &\times \binom{m}{j}\binom{a_0 + b_0 - j}{m}^{-1} \mathrm{e}^{-k(a_0 - j)(b_0 - j)t} \,. \end{aligned} \tag{4.88g}$$

Illustrative examples are shown in Fig. 4.27. The difference between the two irreversible reactions, monomolecular conversion and bimolecular association (Fig. 4.23), is not spectacular, and this supports the previous statement that non-autocatalytic nonlinearities in chemical reactions make handling much more complicated without giving rise to typical nonlinear dynamical phenomena. One difference is important for calculations in practice: the dimension of the rate parameter k is different $[\mathrm{t}^{-1}]$ for the monomolecular case and $[\mathrm{M}^{-1}\mathrm{t}^{-1}]$ with $[\mathrm{M}]=[\mathrm{mol\,V}^{-1}]$ for the bimolecular reaction.

Irreversible Dimerization $2\mathsf{A} \rightarrow \mathsf{C}$

The master equation for the dimerization reaction has been solved by means of a Laplace transform [271] in full analogy to the procedure described in the previous section dealing with the association reaction. The difference according to Fig. 4.25 is that we are dealing with a random variable $\mathcal{X}_\mathsf{A}$ with $P_n(t) = P\big(\mathcal{X}_\mathsf{A}(t) = n\big)$ in two irreducible equivalence classes. As initial condition, we choose $P_n(0) = \delta(n, n_0)$, where $a_0 = n_0$ is the initial number of A molecules. Only jumps $\Delta\mathcal{X}_\mathsf{A} = 2$ are allowed, and this gives rise to two master equations, one for the even class with $n = 2\lambda$ and one for the odd class with $n = 2\lambda + 1$, where $\lambda \in \mathbb{N}$:

$$\frac{\mathrm{d}P_{2\lambda}(t)}{\mathrm{d}t} = -k(2\lambda)(2\lambda-1)P_{2\lambda}(t) + k(2\lambda+2)(2\lambda+1)P_{2\lambda+2}(t) \,, \tag{4.89a}$$

$$\frac{\mathrm{d}P_{2\lambda+1}(t)}{\mathrm{d}t} = -k(2\lambda+1)(2\lambda)P_{2\lambda+1}(t) + k(2\lambda+3)(2\lambda+2)P_{2\lambda+3}(t) \,, \tag{4.89b}$$

where

$$t \in \mathbb{R}_{\geq 0} \,, \quad P_\lambda(t) = 0 \,, \quad \forall \left\{\lambda \notin \mathbb{R} \wedge \lambda > \alpha_0 = \lfloor a_0/2 \rfloor = m_0\right\} .$$

The latter is the condition that all probabilities outside the intervals $[0, 2m_0]$ or $[1, 2m_0 + 1]$, as well as the probabilities for odd or even values of n, $P_{2\lambda+1}$ or $P_{2\lambda}$, are zero (Fig. 4.25). Here, m_0 denotes the maximal number of C molecules that can be formed from $2\alpha_0$ molecules A.

The probability distribution $P_n(t)$ is Laplace transformed, i.e.,

$$\widehat{f}_n(s) = \int_0^\infty \exp(-st)P_n(t)\,\mathrm{d}t \,,$$

yielding in the even class the set of difference equations

$$-1 + s\widehat{f}_{2\alpha_0}(s) = -k(2\alpha_0)(2\alpha_0 - 1)\widehat{f}_{2\alpha_0}(s) ,$$

$$s\widehat{f}_{2\lambda}(s) = -k(2\lambda)(2\lambda - 1)\widehat{f}_{2\lambda}(s) + k(2\lambda + 2)(2\lambda + 1)\widehat{f}_{2\lambda+2}(s) ,$$
$$1 \le \lambda \le \alpha_0 - 1 , \tag{4.89c}$$

$$s\widehat{f}_0(s) = +k \times 1 \times 2\widehat{f}_2(s) .$$

This can be solved as before by successive iteration $\widehat{f}_{2a_0} \Rightarrow \widehat{f}_{2a_0-1} \Rightarrow \ldots \Rightarrow \widehat{f}_0$ [271]:

$$\widehat{f}_{2\lambda}(s) = k^{\alpha_0-\lambda}(2\alpha_0)_{2\alpha_0-2\lambda} \prod_{j=0}^{\alpha_0-\lambda} \frac{1}{s + k\big(2(\alpha_0 - j)\big)\big(2(\alpha_0 - j) - 1\big)} , \tag{4.89d}$$

where $(2\alpha_0)_{2\alpha_0-2\lambda} = 2\alpha_0(2\alpha_0-1)\cdots(2\lambda+1)$ is the falling Pochhammer symbol. A somewhat tedious but straightforward exercise in partial fraction decomposition and calculus yields the final solution for the even class by inverse Laplace transform:

$$P_{2\lambda}(t) = (-1)^{\lambda}\frac{\alpha_0!(2\alpha_0 - 1)!!}{\lambda!(2\lambda - 1)!!} \tag{4.89e}$$
$$\times \sum_{i=\lambda}^{\alpha_0}(-1)^i \frac{(4i - 1)(2\lambda + 2i - 3)!!}{(\alpha_0 - i)!(i - \lambda)!(2\alpha_0 + 2i - 1)!!} e^{-k2i(2i-1)t} ,$$
$$\lambda = 0, 1, \ldots, \alpha_0 .$$

The solution in the odd irreducible equivalence class is derived in exactly the same way:

$$P_{2\lambda+1}(t) = (-1)^{\lambda}\frac{\alpha_0!(2\alpha_0 + 1)!!}{\lambda!(2\lambda + 1)!!} \tag{4.89f}$$
$$\times \sum_{i=\lambda}^{\alpha_0}(-1)^i \frac{(4i + 1)(2\lambda + 2i - 1)!!}{(\alpha_0 - i)!(i - \lambda)!(2\alpha_0 + 2i + 1)!!} e^{-k2i(2i+1)t} ,$$
$$\lambda = 0, 1, \ldots, \alpha_0 .$$

The easily recognizable difference between the two irreducible equivalence classes comes from the exponential function and is of a trivial nature: the probability densities for the same λ values move faster in the odd class and this is simply the result of the larger value of the exponents, i.e., $2i + 1 > 2i - 1$. The results are illustrated by means of a numerical example in Fig. 4.26.

The expectation value and variance of the dimerization reaction already shown in Fig. 4.26 deserve further attention. There is an interesting detail in the comparison of the expectation value with the deterministic solution. With $[\mathsf{A}]=\mathcal{X}_{\mathsf{A}}(t) = a(t)$ and $a(0) = a_0$, dimerization is conventionally modeled by the concentration dependence $h(a) = a^2$ in the kinetic differential equation (4.90a), for which an exact analytical solution is derived by standard integration. Picking two molecules in sequence from the reservoir leaves only $\mathcal{X}_{\mathsf{A}} - 1$ possibilities for the second choice, so a dependence $h(a) = a(a-1)$ appears more appropriate. Although the numbers of molecules in chemical reactions are commonly so large that in particle numbers a^2 is for all practical purposes indistinguishable from $a(a-1)$, it is also worth considering the corrected kinetic equation (4.90b):

$$-\frac{\mathrm{d}a}{\mathrm{d}t} = ka^2 \quad \Longrightarrow \quad a(t) = \frac{a_0}{1 + a_0kt} , \tag{4.90a}$$

$$-\frac{\mathrm{d}\widetilde{a}}{\mathrm{d}t} = k\widetilde{a}\,(\widetilde{a} - 1) \quad \Longrightarrow \quad \widetilde{a}(t) = \frac{a_0}{a_0 + (1 - a_0)\mathrm{e}^{-kt}} . \tag{4.90b}$$

In order to compare with the expectation values derived from the master equation (4.84a), $\mathrm{E}_{2\mathsf{A}}$, we compute the asymptotic tangents to the solution curves in the limit $t \to 0$ which are obtained as

$$\frac{a_0}{1 + a_0kt} \approx \widehat{a}(t) = a_0(1 - a_0kt) , \tag{4.90c}$$

$$\frac{a_0}{a_0 + (1 - a_0)\mathrm{e}^{-kt}} \approx \widehat{\widetilde{a}}(t) = a_0\big(1 - (a_0 - 1)kt\big) , \tag{4.90d}$$

for small t. Accordingly, we are dealing with two different asymptotic behaviors. The master equation for the dimerization reaction is clearly the analogue of (4.90b) and we might ask whether can find a stochastic process which comes asymptotically close to the other deterministic kinetics, i.e., $h = a^2$. A process that is a candidate for this goal is the association reaction with $a_0 = b_0$. It applies exactly the same concentration function and we shall use the expectation value of this process, viz., $\mathrm{E}_{\mathsf{A+B}}$ with $\vartheta_0 = 0$, for the purpose of comparison. Indeed, from the plots in Fig. 4.28, it follows that the expectation value $\mathrm{E}_{2\mathsf{A}}$ converges to $\widehat{\widetilde{a}}(t)$ in the limit $t \to 0$, whereas $\mathrm{E}_{\mathsf{A+B}}$ approaches $\widehat{a}(t)$. At not too long times, the expectation value of the stochastic dimerization $\mathrm{E}_{2\mathsf{A}}(t)$ lies below the deterministic solution $\widetilde{a}(t)$ and $\mathrm{E}_{\mathsf{A+B}}(t)$ below $a(t)$ (4.90b). At large t values, the conventional deterministic curve and the stochastic curve may even cross. This result is an artifact of continuous variables in the range $0 \le \mathcal{X}_{\mathsf{A}} \le 2$.

Finally, we consider also the variances of the dimerization reaction and compare with those of the association reaction with $a_0 = b_0$. At time $t = 0$ the variances are all zero because we apply sharp initial conditions. The variances increase with time, pass through a maximum, and eventually become zero again in the limit $t \to \infty$, because $\lim_{t\to\infty} \mathcal{X}_{\mathsf{C}} = \lfloor a_0/2 \rfloor$. The height of the maximum scaled by

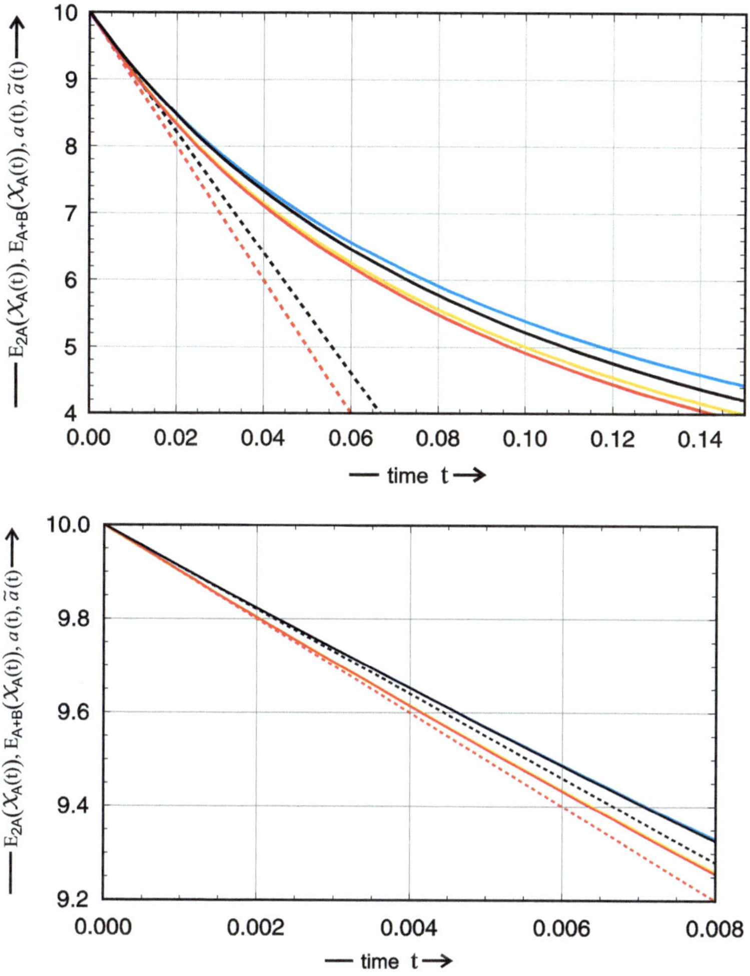

Fig. 4.28 Asymptotic behavior of expectation values in the irreversible dimerization reaction $2A \rightarrow C$. *Upper*: Comparing the stochastic and deterministic solutions. The expectation value of the dimerization reaction $E_{2A}(\mathcal{X}_A(t))$ (*black*), the expectation value of the association reaction $A + B \rightarrow C$ with $a_0 = b_0$, $E_{A+B}(\mathcal{X}_A(t))$ (*red*), the conventional deterministic solution $a(t)$ from (4.90a) (*yellow*), and the corrected deterministic solution $\widetilde{a}(t)$ from (4.90b) (*blue*). *Broken lines* represent the asymptotic tangents $\widehat{a}(t) = a_0\big(1 - k(a_0 - 1)t\big)$ (*black*) and $\widehat{a}(t) = a_0(1 - ka_0t)$ (*red*), respectively. The expectation values for the stochastic processes (*black and red curves*) lie below the deterministic solution curves in this time range. *Lower*: Enlargement of the upper plot showing the perfect convergence to the two tangents in the limit $t \rightarrow 0$. The stochastic (*black*) and the deterministic (*blue*) curve using the rate to $v \propto a(a-1)$ approach the *black dashed line*, whereas the other stochastic and deterministic curves(*red and yellow*, respectively), which apply the rate $v \propto a^2$, converge to the *broken red line*. Parameters: $k = 1\,[\mathrm{N}^{-1}\mathrm{t}^{-1}]$, $a_0 = 10$

the initial concentration a_0, i.e., $\max\{\mathrm{var}(\mathcal{X}_A(t))/a_0\}$, is surprisingly constant over a wide range of a_0 values (Fig. 4.29), and this result is another instance of the $\sqrt{N}$-law for fluctuations. Variances at sufficiently long times nicely reflect the different behaviors of the stochastic process in the even and in the odd classes (Fig. 4.30). In the time range where the expectation value comes close to $\mathrm{E}_{2A}(\mathcal{X}_A(t)) \approx 1$, we observe a pronounced difference between the variances in the even and the odd classes. The variance in the even class is substantially larger than the reference, i.e., $\mathrm{var}_{2A}^{(\mathrm{even})}(\mathcal{X}_A(t)) > \mathrm{var}_{A+B}(\mathcal{X}_A(t))$, whereas the opposite is true for the odd class, i.e., $\mathrm{var}_{2A}^{(\mathrm{odd})}(\mathcal{X}_A(t)) < \mathrm{var}_{A+B}(\mathcal{X}_A(t))$. The figure also shows the different behavior of the expectation values:

$$\lim_{t\to\infty} \mathrm{E}_{2A}^{(\mathrm{even})}(\mathcal{X}_A(t)) = \lim_{t\to\infty} \mathrm{E}_{A+B}(\mathcal{X}_A(t)) = 0\,, \quad \lim_{t\to\infty} \mathrm{E}_{2A}^{(\mathrm{odd})}(\mathcal{X}_A(t)) = 1\,.$$

These regularities are the same at larger particle numbers, but their relative importance goes down with increasing population size. Indeed the specific phenomena could be completely neglected in practice were it not for the new techniques allowing direct observation in populations with small particle numbers (Sect. 4.4).

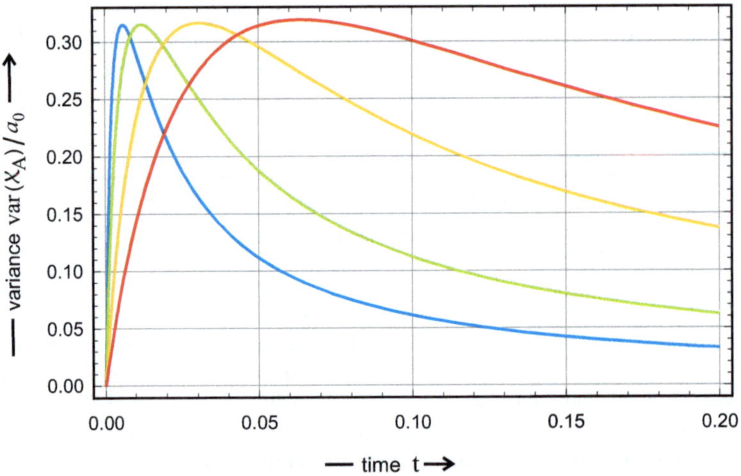

Fig. 4.29 The scaled variance of the dimerization reaction 2A → C. The scaled variance $\mathrm{var}(\mathcal{X}_A(t))/a_0$ is shown as a function of time for different initial conditions: $k = 1\,[\mathrm{N}^{-1}\mathrm{t}^{-1}]$, a_0=10 (*red*), 20 (*yellow*), 50 (*green*), and 100 (*blue*). It is remarkable that the height of the maximum is very close to $\mathrm{var}(\mathcal{X}_A(t))/a_0 = 0.32$ for all values of a_0. This finding is one more confirmation of the so-called $\sqrt{N}$ law: the variance or the square of the standard deviation is proportional to a_0 at its maximal value

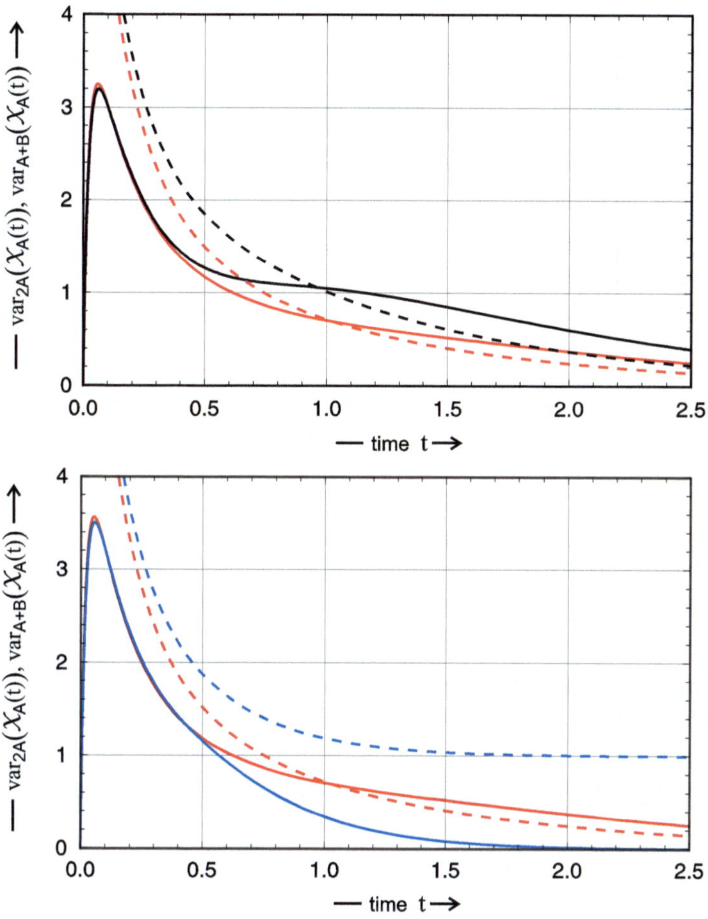

Fig. 4.30 The variance of the dimerization reaction $2A \rightarrow C$. The *upper plot* compares the variance $\mathrm{var}_{2A}^{(\mathrm{even})}\big(\mathcal{X}_A(t)\big)$ (*black*) of the dimerization reaction $2A \rightarrow C$ in the even class ($a_0 = 2\lambda$) with the variance $\mathrm{var}_{A+B}\big(\mathcal{X}_A(t)\big)$ (*red*) of the association reaction $A + B \rightarrow C$ for $a_0 = b_0$. The corresponding expectation values $\mathrm{E}_{2A}\big(\mathcal{X}_A(t)\big)$ and $\mathrm{E}_{A+B}\big(\mathcal{X}_A(t)\big)$ are shown as *dashed black* and *red curves*. The variance $\mathrm{var}_{2A}^{(\mathrm{even})}\big(\mathcal{X}_A(t)\big)$ of the dimerization reaction in the even class exhibits rather complex behavior with two inflexion points and shows a systematic deviation from that of the association reaction to higher values in the range around $\mathcal{X}_A = 1$ ($0.5\,[\mathrm{t}] < t < 1.5\,[\mathrm{t}]$). The *lower plot* shows the analogous curves for an example from the odd class ($a_0 = 2\lambda + 1$). The variance of the dimerization reaction $\mathrm{var}_{2A}^{(\mathrm{odd})}\big(\mathcal{X}_A(t)\big)$ (*blue*) shows here a systematic deviation to lower values and the *blue curve* lies below the *red curve*. As required, the expectation value in the odd class approaches $\mathrm{E}_{2A}^{(\mathrm{odd})}\big(\mathcal{X}_A(t)\big) = 1$ (*blue*). Continued on next page

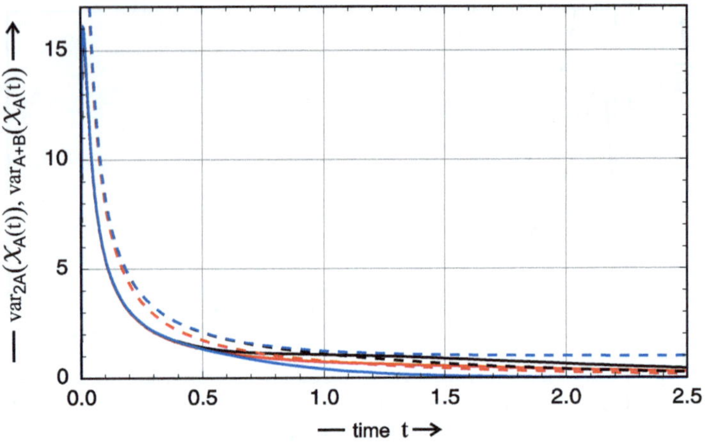

Fig. 4.30 (Cont.) The variance of the dimerization reaction $2A \rightarrow C$. The coincidence of all three variances even at moderate particle numbers. Parameter values: $k = 1\,[t^{-1}M^{-1}]$, $a_0 = 10$ and $a_0 = 11$ for the even and the odd class, as well as $a_0 = 50$ and $a_0 = 51$ for the medium sized particle numbers, respectively

Reversible Association Reaction $A + B \rightleftharpoons C$

The reversible association reaction is described by the chemical master equation

$$\frac{dP_n(t)}{dt} = k(n+1)(\vartheta_0 + n + 1)P_{n+1}(t) - kn(\vartheta_0 + n)P_n(t) + l(n_0 - n + 1)P_{n-1}(t) - l(n_0 - n)P_n(t) \,, \tag{4.91a}$$

with $P_n(t) = P\big(\mathcal{X}_A(t) = n\big)$ for $n \in \mathbb{N}$, $n \leq a_0$, $n_0 = a_0$, $\vartheta_0 = b_0 - a_0$, and the birth-and-death transition probabilities (3.96)

$$w_n^+ = \kappa(n_0 - n) \,, \quad w_n^- = n(\vartheta_0 + n) \,, \tag{4.91b}$$

where we have modified the time axis by $t \Longrightarrow \tau = k*t$, introduced the dissociation constant $\kappa = K^{-1} = l/k$, and imposed the initial condition $P_n(0) = \delta_{n,a_0}$ or $\mathcal{X}_A(0) = a_0$, $\mathcal{X}_B(0) = b_0$, $\mathcal{X}_C(0) = 0$, with the implicit assumption $b_0 > a_0$.[49]

In order to illustrate the solution of the master equation (4.91a) by application of linear algebra, Laplace transform, and its inversion [329], we define the probability vector $\mathbf{P}(t) = \big(P_n(t),\, n \in \mathbb{N}, n \leq n_0\big)^t$ and write the master equation in terms of

[49]If $a_0 > b_0$, we can simply exchange the variables $\mathcal{X}_A(t)$ and $\mathcal{X}_B(t)$ without loss of generality.

step-up and step-down transition probabilities:

$$\frac{d\mathbf{P}(t)}{dt} = \mathbf{W}\mathbf{P}(t) , \tag{4.91c}$$

where W is the general tridiagonal transition matrix

$$W = \begin{pmatrix} -(w_0^+ + w_0^-) & w_1^- & 0 & \dots & 0 & 0 \\ w_0^+ & -(w_1^+ + w_1^-) & w_2^- & \dots & 0 & 0 \\ 0 & w_1^+ & -(w_2^+ + w_2^-) & \dots & 0 & 0 \\ \vdots & \vdots & \vdots & \ddots & \vdots & \vdots \\ 0 & 0 & 0 & \dots & -(w_{n_0-1}^+ + w_{n_0-1}^-) & w_{n_0}^- \\ 0 & 0 & 0 & \dots & w_{n_0-1}^+ & -(w_{n_0}^+ + w_{n_0}^-) \end{pmatrix} .$$

Laplace transforming the probability density, viz.,

$$\widehat{f}_n(s) = \int_0^\infty \exp(-st) P_n(t)\, dt , \tag{4.86a}$$

yields a linear algebraic equation for $\widehat{\mathbf{f}}(s) = \left(\widehat{f}_n(s) : n \in \mathbb{N}, n \le a_0\right)^t$:

$$s\widehat{f}(s) = W\widehat{f}(s) + \mathbf{P}_0 , \quad \left(s\mathbb{I} - W\right)\widehat{\mathbf{f}}(s) = \mathbf{P}_0 , \tag{4.91d}$$

where $\mathbf{P}_0 = \left(P_0(0), P_1(0), \dots, P_{a_0}(0)\right)^t = (0, 0, \dots, 1)^t$. The formal solution of this linear equation is

$$\widehat{\mathbf{f}}(s) = \left(s\mathbb{I} - W\right)^{-1}\mathbf{P}_0 . \tag{4.91e}$$

The matrix inversion can be performed numerically, but then no further analytical work is possible, or in the conventional way, which can be carried out analytically in a few exceptional cases:

$$\left(s\mathbb{I} - W\right)^{-1} = \frac{1}{\left|\left(s\mathbb{I} - W\right)\right|} \operatorname{adj}\left(s\mathbb{I} - W\right) ,$$

where adj denotes the *adjugate* matrix.[50] The simple form of $\mathbf{P}_0$ makes it possible to obtain $\widehat{\mathbf{f}}$ using only the elements of the last column of the matrix $\left(s\mathbb{I} - W\right)^{-1}$. The

[50] The adjugate matrix adj A of a square matrix A is the transpose of the cofactor matrix C, which has the *minors* A_{ij} of A as elements: $C = \left(C_{ij} = (-1)^{i+j}A_{ij}\right)$. The minor A_{ij} is the determinant of the *submatrix* obtained from A by removing the ith row and the jth column. Finally, we get $\operatorname{adj} A = \left(\operatorname{adj}(A)_{ij} = C_{ji} = (-1)^{i+j}A_{ji}\right)$. For details, see any textbook on linear algebra, such as [511, pp. 231–232].

cofactors of the last row with the index $a_0 + 1$ can be calculated, and for the jth column we find

$$C_{a_0+1,j}(s) = (-1)^{a_0+1+j} D_{j-1}(s) \prod_{i=j}^{a_0} (-w_i^-) ,$$

where $D_j(s)$ is the determinant of the submatrix of $(s\mathbb{I} - \mathrm{W})$ containing the first j rows and the first j columns. The polynomials $D_n(s)$ can be constructed recursively:

$$D_n(s) - (s + w_{n-1}^+ + w_{n-1}^-) D_{n-1}(s) + w_{n-2}^+ w_{n-1}^- D_{n-2}(s) , \tag{4.91f}$$

where $D_0 = 1$ and $D_1 = s + w_0^+$. This recursion is equivalent to solving (4.88b)–(4.88d) in order to obtain the solution in Laplace space [329]:

$$\widehat{f}_n(s) = \frac{1}{(\vartheta_0 + 1)^{(n)} n!} \frac{a_0! b_0!}{\vartheta_0!} \frac{D_n(s)}{D_{a_0-1}(s)} . \tag{4.91g}$$

The last step is the inverse Laplace transform, which can be carried out by applying Mellin's formula (4.86c) and integrating using the residue theorem [21, p. 444]:

$$P_n(t) = \sum_{j=0}^{a_0} \lim_{s \to \lambda_j} \Big((s - \lambda_j) \exp(st) \widehat{f}_n(s) \Big) ,$$

where the values λ_j are the eigenvalues of the transition matrix W. Combining the two results yields the final solution:

$$P_n(t) = \frac{a_0! b_0!}{(\vartheta_0 + n)! n!} \sum_{j=0}^{a_0} \frac{D_n(\lambda_j) \exp(\lambda_j t)}{\partial D_{a_0+1}(s)/\partial s \big|_{s=\lambda_j}} . \tag{4.91h}$$

In principle, the exact probability density can be calculated from (4.91h), provided that the eigenvalues of matrix W are known. In general W is a tridiagonal matrix and the eigenvalues can be obtained only by numerical computation. Nevertheless, in special cases, analytical solutions can be obtained. We mention two examples: (i) the irreversible reaction $\mathsf{A} + \mathsf{B} \to \mathsf{C}$ and (ii) the stationary or equilibrium density $\overline{P}_n = \lim_{t\to\infty} P_n(t)$ for the reversible reaction $\mathsf{A} + \mathsf{B} \rightleftharpoons \mathsf{C}$.

For the irreversible reaction the eigenvalues are identical with the diagonal elements of the matrix W, which is upper-triangular.[51] The eigenvalues coincide with the diagonal elements in this case:

$$w_j^+ = 0 , \;\; \forall\, j \in \mathbb{N} , \;\; j \le a_0 \quad \Longrightarrow \quad \lambda_j = w_j^- = -j(\vartheta_0 + j) .$$

[51] A matrix that has no nonzero entries below the main diagonal is said to be upper-triangular. A lower-triangular matrix has only zero elements above the diagonal.

The expression for the probability density [329] then becomes:

$$P_n(t) = \frac{a_0!b_0!}{(\vartheta_0+n)!n!}\sum_{j=n}^{a_0}(-1)^{j-n}\frac{(\vartheta_0+2j)(\vartheta_0+n+j-1)!}{(a_0-j)!(j-n)!(b_0+j)!}\mathrm{e}^{-j(\vartheta_0+j)kt}\,, \tag{4.88g$'$}$$

where we have restored the original time axis $\tau \Longrightarrow \tau/k = t$. Equations (4.88g) and (4.88g′) yield exactly the same density and are mathematically equivalent, although the expressions are different, since m in (4.88g) counts the molecules C, whereas n counts the molecules A.

The equilibrium probability density may be calculated by taking advantage of an interesting relation between a function and its Laplace transform, known as the Laplace initial and final value theorem:

$$\lim_{s\to 0} s\widehat{f}_n(s) = P_n(\infty) = \overline{P}_n\,, \quad \lim_{s\to\infty} s\widehat{f}_n(s) = P_n(0)\,. \tag{4.91i}$$

In order to calculate the equilibrium density, we need to know only the limiting value of the Laplace transformed probability density:

$$\lim_{s\to 0} s\widehat{f}_n(s) = \lim_{s\to 0} s\frac{a_0!b_0!}{(\vartheta_0+n)!n!}\frac{D_n(s)}{D_{a_0+1}(s)}\,.$$

In particular, we require only the constant terms of the polynomials $D_n(s)$, which can be obtained from the recursion (4.91f):

$$D_n(0) = \prod_{j=0}^{n-1} w_j^+ = \kappa^n\frac{a_0!}{(a_0-n)!} = \frac{(a_0-n+1)^{(n)}}{K^n}\,,$$

where the rising Pochhammer symbol is applied, and $K = \kappa^{-1} = k/l$ is the association or binding constant. From the step-up transition probabilities $w_j^+ = \kappa(a_0-j)$, it follows that $D_n(0) > 0$, $\forall\, n \in [0, a_0]$ and $D_{a_0+1} = 0$. After some calculation and consideration of the normalization condition $\sum_{j=0}^{a_0} P_j(t) = 1$, we obtain the final result for the stationary distribution:

$$\overline{P}_n = \frac{(a_0-n+1)^{(n)}}{K^{-n}(\vartheta_0+1)^{(n)}n!}\frac{1}{{}_1F_1(-a_0;\vartheta_0+1;-\kappa)}\,, \tag{4.91j}$$

where ${}_1F_1(\alpha;\gamma;x)$ is the confluent hypergeometric function specified by

$${}_1F_1(\alpha;\gamma;x) = \sum_{j=0}^{\infty}\frac{(\alpha)^{(j)}}{(\gamma)^{(j)}}\frac{x^j}{j!}\,.$$

The result for the equilibrium density $\overline{P}_{\mathbf{n}}$, with $\mathbf{n} = (n_{\mathsf{A}}, n_{\mathsf{B}}, n_{\mathsf{C}})$, is more easily derived from(4.72) by eliminating the linear dependencies $n_{\mathsf{A}} = n$, $n_{\mathsf{B}} = \vartheta_0 + n$, and $n_{\mathsf{C}} = a_0 - n$:

$$\overline{p}_{\mathbf{n}} = \frac{\bar{x}_{\mathsf{A}}^{n_{\mathsf{A}}}}{n_{\mathsf{A}}!}\frac{\bar{x}_{\mathsf{B}}^{n_{\mathsf{B}}}}{n_{\mathsf{B}}!}\frac{\bar{x}_{\mathsf{C}}^{n_{\mathsf{C}}}}{n_{\mathsf{C}}!}\,, \quad \mathbf{n} = (n_{\mathsf{A}}, n_{\mathsf{B}}, n_{\mathsf{C}}) \in \mathbb{N}^3\,,$$
$$\overline{P}_{\mathbf{n}} = N\,\overline{p}_{\mathbf{n}}\,, \quad \text{with } N = \left(\sum_{i=0}^{\min\{a_0,b_0\}} \overline{p}_i\right)^{-1}, \tag{4.91k}$$

where N is the normalization factor. The equilibrium concentrations $\bar{x}_{\mathsf{A}} = \bar{x}$, $\bar{x}_{\mathsf{B}} = \vartheta_0 + \bar{x}$, and $\bar{x}_{\mathsf{C}} = a_0 - \bar{x}$ are readily obtained from the relation

$$K^{-1} = \frac{[C]}{[A][B]} = \frac{\bar{x}_{\mathsf{C}}}{\bar{x}_{\mathsf{A}}\bar{x}_{\mathsf{B}}} = K_b\,.$$

For $n_{\mathsf{A}}(0) = a_0$ and $n_{\mathsf{B}}(0) = b_0$ with $b_0 \geq a_0$ and $n_{\mathsf{C}}(0) = 0$ yields

$$\bar{x} = \frac{1}{2}\Big(a_0 - b_0 - K + \sqrt{(a_0 + b_0 + K)^2 - 4a_0b_0}\Big) \tag{4.91l}$$

for the equilibrium concentration of A. In order to generalize to arbitrary a_0 and b_0 values, we need only substitute $\vartheta_0 \leftrightarrow |b_0 - a_0|$ and $a_0 \leftrightarrow \min\{a_0, b_0\}$.

Earlier work on bimolecular chemical reactions yielded the expectation value and variance of $\mathcal{X}_{\mathsf{A}}$ by means of probability generating functions [100]:

$$\mathrm{E}(\overline{\mathcal{X}}_{\mathsf{A}}) = \mu_{\bar{a}} = K\frac{a_0}{\vartheta_0 + 1}\frac{{}_1F_1(-a_0 + 1; \vartheta_0 + 2; -K)}{{}_1F_1(-a_0; \vartheta_0 + 1; -K)}\,,$$
$$\mathrm{var}(\overline{\mathcal{X}}_{\mathsf{A}}) = -\mu_{\bar{a}}^2 - (\vartheta_0 + K)\mu_{\bar{a}} + a_0K\,. \tag{4.91m}$$

In Fig. 4.31 we consider the stochastic equilibrium in the form of the one standard deviation band $\mathrm{E}(\overline{\mathcal{X}}_{\mathsf{A}}) \pm \sigma(\overline{\mathcal{X}}_{\mathsf{A}})$ around the expectation value. As expected, the relative width of this band becomes smaller with increasing numbers of molecules in the sense of an approximate $\sqrt{N}$-law. The dependence of the probability density on the dissociation constant K for fixed values a_0 and b_0 yields a monotonic increase in the expectation value $\mathrm{E}(\overline{\mathcal{X}}_{\mathsf{A}})$ from $\lim_{K\to 0}\mathrm{E}(\overline{\mathcal{X}}_{\mathsf{A}}) = 0$ to $\lim_{K\to\infty}\mathrm{E}(\overline{\mathcal{X}}_{\mathsf{A}}) = a_0$. In contrast to the first order system A $\rightleftharpoons$ B, the expectation value $\mathrm{E}(\overline{\mathcal{X}}_{\mathsf{A}})$ does not coincide with the deterministic solution:

$$\bar{a}(a_0, b_0, K) = \frac{1}{2}\Big(a_0 - b_0 - K + \sqrt{(a_0 - b_0 - K)^2 + 4a_0K}\Big)\,. \tag{4.91l'}$$

There is a small but recognizable difference between $\mathrm{E}(\mathcal{X}_{\mathsf{A}}; K)$ and $\bar{a}(K)$ for $a_0 = b_0 = 5$, which already becomes very small at moderate particle numbers $(a_0, b_0) >$

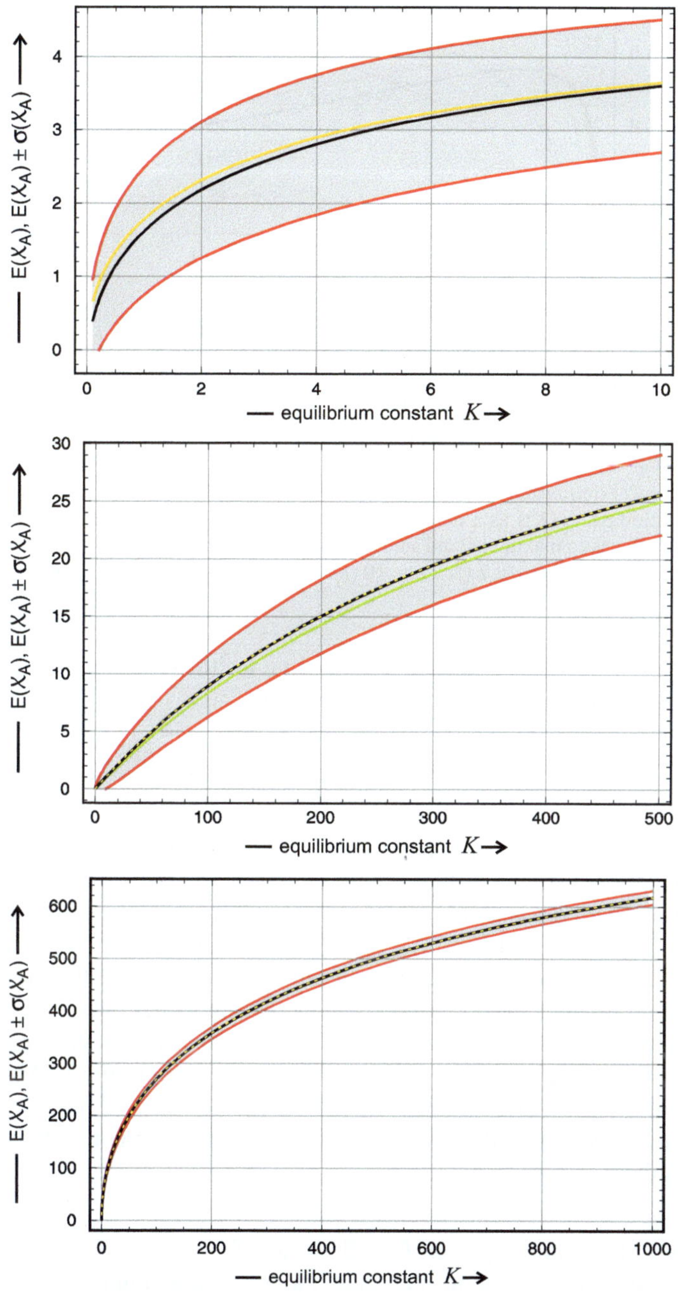

Fig. 4.31 Equilibrium of the association reaction $A + B \rightleftharpoons C$. The three figures *above* show the equilibrium expectation values $E(\overline{\mathcal{X}}_A(K))$ (*black*) embedded in the one standard deviation zone $E(\overline{\mathcal{X}}_A) \pm \sigma(\overline{\mathcal{X}}_A)$ (*gray with red borders*). The deterministic solution $\bar{a}(K)$ (*yellow*) and the pseudo-first order result $\widetilde{a}(K)$ (*green*) are also shown. Choice of parameters: $a_0 = 5, b_0 = 5$ (*upper plot*), $a_0 = 50, b_0 = 500$ (*middle plot*), and $a_0 = 1000, b_0 = 1000$ (*lower plot*). Continued on next page

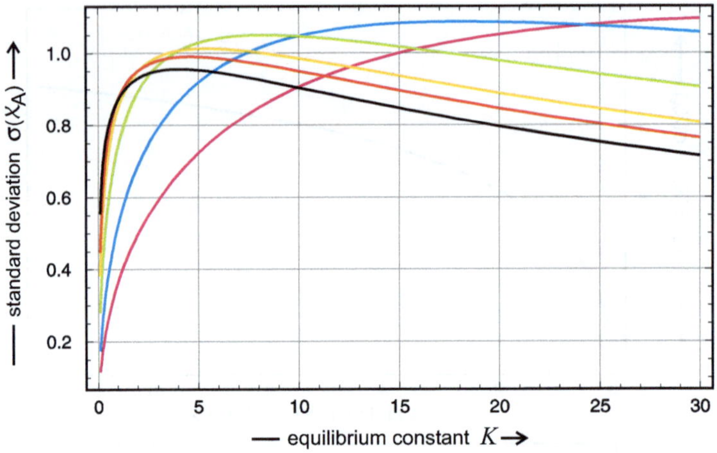

Fig. 4.31 (Cont.) Equilibrium of the association reaction $\mathsf{A}+\mathsf{B} \rightleftharpoons \mathsf{C}$. The plot *above* shows the standard deviation $\sigma\big(\overline{\mathcal{X}}_{\mathsf{A}}(K)\big)$ as a function of the equilibrium constants. These curves start from $\sigma = 0$ for $K = 0$, pass a maximum, and approach zero in the limit $K \to \infty$. Choice of parameters: $a_0 = 5$, $b_0 = 5$ (*black*), 6 (*red*), 7 (*yellow*), 10 (*green*), 20 (*blue*), and 40 (*magenta*)

10, where the two curves coincide within the line width. The limit of large b_0 values is known as the pseudo-first order condition, with

$$\bar{a} \approx \tilde{a}(a_0, b_0, K) = a_0 \frac{K}{b_0 + K}\,, \quad b_0 \gg a_0\,. \tag{4.91n}$$

A factor of $b_0 = 100a_0$ is sufficient to make all three curves $\mathrm{E}(\overline{\mathcal{X}}_{\mathsf{A}}; K)$, $\bar{a}(K)$, and $\tilde{a}(K)$ practically indistinguishable.

The variance and standard deviation of $\mathrm{E}(\overline{\mathcal{X}}_{\mathsf{A}}; K)$ are zero in both limits $\lim_{K\to 0} \mathrm{var}(\overline{\mathcal{X}}_{\mathsf{A}}) = 0$ and $\lim_{K\to\infty} \mathrm{var}(\overline{\mathcal{X}}_{\mathsf{A}}) = 0$, and pass through a maximum at some intermediate value of K. For constant a_0, the height of the maximum and the position along the K-axis increase with increasing values of b_0. We remark that the equilibrium constant K for the reaction $\mathsf{C} \rightleftharpoons \mathsf{A}+\mathsf{B}$ is not dimensionless, in fact, [K]=[number of particles×volume^{-1}] =[N] or [K]=[concentration]=[M]=[mol l^{-1}], as the equilibrium constant was in the first order scenario $\mathsf{A} \rightleftharpoons \mathsf{B}$. Analogous scenarios are thus expected to be observed for equilibrium constants scaled with particle numbers.

Reversible Conversion Reaction $\mathsf{A}+\mathsf{B} \rightleftharpoons \mathsf{C}+\mathsf{D}$

For comparison, we also consider here the reversible bimolecular conversion reaction (4.1i). As shown in Sect. 4.2.2, three conservation relations reduce the four random variables counting the molecules of each class $\mathsf{A}, \mathsf{B}, \mathsf{C}$, and D to a single one. As initial conditions we choose $\mathcal{X}_{\mathsf{A}}(0) = a_0$, $\mathcal{X}_{\mathsf{B}}(0) = b_0$, $\mathcal{X}_{\mathsf{C}}(0) = c_0$, and $\mathcal{X}_{\mathsf{D}}(0) = d_0$. We make the assumptions $c_0 = 0$, $d_0 = 0$, and introduce $b_0 - a_0 = \vartheta_0$

for the calculations reported here:

$$\mathcal{X}_{\mathsf{A}}(t) = n(t) \,, \quad \mathcal{X}_{\mathsf{B}}(t) = \vartheta_0 + n(t) \,, \quad \mathcal{X}_{\mathsf{C}}(t) = a_0 - n(t) \,, \quad \mathcal{X}_{\mathsf{D}}(t) = a_0 - n(t) \,.$$

The equilibrium probability distribution for this system was calculated using the probability generating function [100] and the results for the expectation value and variance are

$$\mathrm{E}(\overline{\mathcal{X}}_{\mathsf{A}}) = \mu_{\bar{a}} = K \frac{a_0^2}{\vartheta_0 + 1} \frac{{}_2F_1(-a_0 + 1; -a_0 + 1; \vartheta_0 + 2; K)}{{}_2F_1(-a_0; -a_0; \vartheta_0 + 1; K)} \,,$$

$$\mathrm{var}(\overline{\mathcal{X}}_{\mathsf{A}}) = \begin{cases} \mu_{\bar{a}}^2 \dfrac{b_0^2}{a_0 + b_0 - 1} \,, & \text{if } K = 1 \,, \\ -\mu_{\bar{a}}^2 - \dfrac{\vartheta_0 + 2a_0 K}{1 - K} \mu_{\bar{a}} + \dfrac{a_0^2 K}{1 - K} \,, & \text{if } K \neq 1 \,. \end{cases} \tag{4.92a}$$

For comparison we calculate the deterministic value too:

$$\bar{a}(a_0, b_0, K) = \frac{1}{2(K-1)} \left(\vartheta_0 + 2a_0 K - \sqrt{\vartheta_0^2 + 4a_0 b_0 K} \right) \,, \tag{4.92b}$$

and for $K_b = 1$ the equation simplifies to $\bar{a} = a_0^2/(a_0 + b_0)$. The illustrative example in Fig. 4.32 shows an overall picture that is very similar to the association reaction, but with two significant differences:

(i) The equilibrium constant is dimensionless and this implies that the same values of K can be used for particle numbers and other units to observe the influence on the various phenomena.
(ii) The reaction system exhibits a kind of symmetry at the equilibrium constant $K = 1$, where the expectation value and the deterministic equilibrium concentration assume the same values $\mathrm{E}(\overline{\mathcal{X}}_{\mathsf{A}}; a_0, b_0, 1) = \bar{a}(a_0, b_0, 1)$.

As in the previous example, the difference between the stochastic and the deterministic values is already very small at relatively small particle numbers $(a_0, b_0) > 10$.

4.3.5 Autocatalytic Reaction

In this section we deal with our last example of a chemical reaction for which analytical solutions are available: the bimolecular or first order autocatalytic reaction presented in (4.1g). Here, we present a solution of the master equation which makes use of the Laplace transform [26]. As we saw in our previous examples, the back-transformation into probability space, when carried out analytically, lays strong

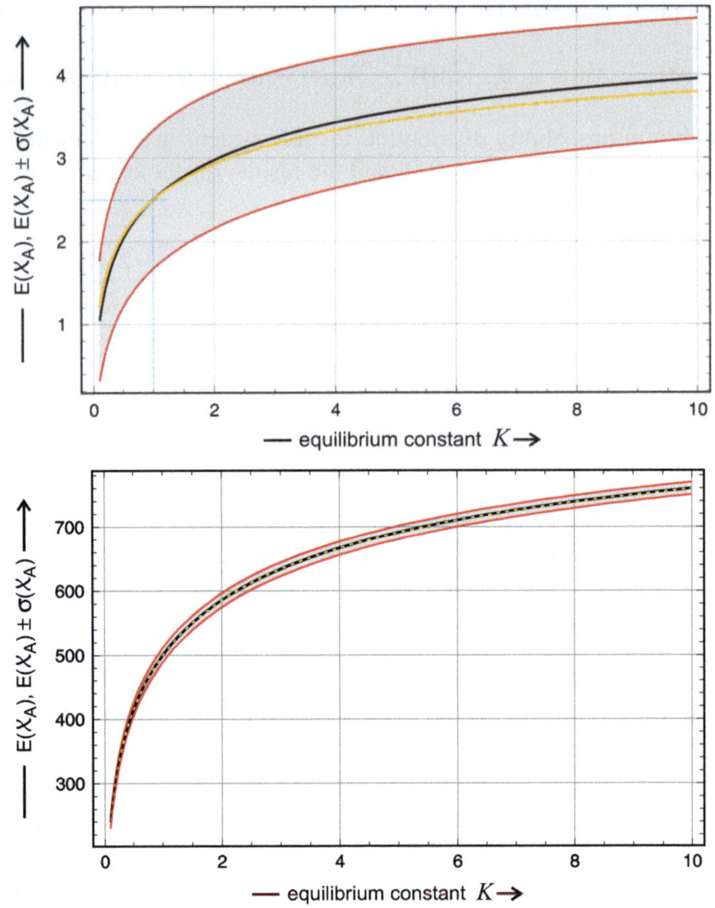

Fig. 4.32 Equilibrium of the conversion reaction $\mathsf{A} + \mathsf{B} \rightleftharpoons \mathsf{C} + \mathsf{D}$. The two plots *above* show the equilibrium expectation values $\mathrm{E}\big(\overline{\mathcal{X}}_{\mathsf{A}}(K)\big)$ (*black*) embedded in the one standard deviation zone $\mathrm{E}(\overline{\mathcal{X}}_{\mathsf{A}}) \pm \sigma(\overline{\mathcal{X}}_{\mathsf{A}})$ (*gray with red borders*). The deterministic solution $\bar{a}(K)$ (*yellow*) is also shown. Choice of parameters: $a_0 = 5, b_0 = 5$ (*upper plot*) and $a_0 = 1000, b_0 = 1000$ (*lower plot*). In the *upper plot*, we see that the curves for $\mathrm{E}\big(\overline{\mathcal{X}}_{\mathsf{A}}(K)\big)$ and $\bar{a}(K)$ cross at the point $K = 1$, as indicated by the *blue dotted lines*. Continued on next page

restrictions on the solvable cases, and the solutions are already very sophisticated even in the simplest case.

The reaction for first order autocatalysis in a closed system, i.e.,

$$\mathsf{A} + \mathsf{X} \underset{l}{\overset{k}{\rightleftharpoons}} 2\mathsf{X}\,, \tag{4.93a}$$

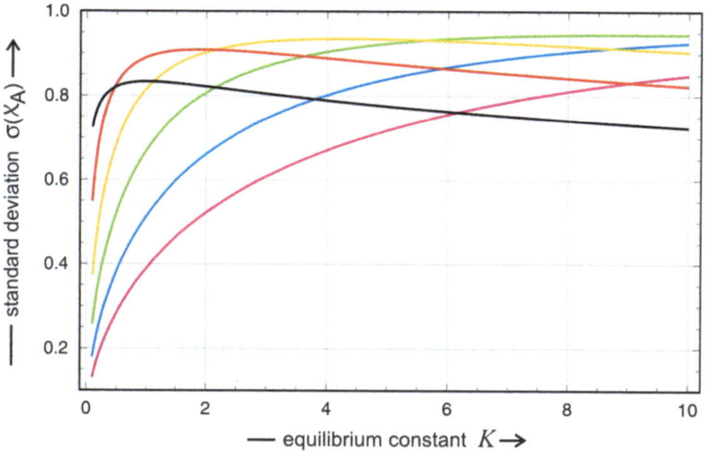

Fig. 4.32 (Cont.) Equilibrium of the conversion reaction $\mathsf{A} + \mathsf{B} \rightleftharpoons \mathsf{C} + \mathsf{D}$. Standard deviation as a function of the equilibrium constants, $\sigma\big(\mathcal{X}_{\mathsf{A}}(K)\big)$. These curves start from $\sigma = 0$ for $K = 0$, pass through a maximum and approach zero in the limit $K \to \infty$. Choice of parameters: $a_0 = 5$, $b_0 = 5$ (*black*), 10 (*red*), 20 (*yellow*), 40 (*green*), 80 (*blue*), and 150 (*magenta*)

is described by the chemical master equation

$$\frac{\mathrm{d}P_n(t)}{\mathrm{d}t} = k(n+1)(n_0-n-1)P_{n+1}(t) + l(n_0-n+1)(n_0-n)P_{n-1}(t) \\ -\Big(kn(n_0-n) + l(n_0-n)(n_0-n-1)\Big)P_n(t) \,, \tag{4.93b}$$

where $\mathcal{X}_{\mathsf{A}}(t) = n(t)$ is chosen as the single independent stochastic variable, since $\mathcal{X}_{\mathsf{X}}(t) = n_0 - n(t) = m(t)$ or $\mathcal{X}_{\mathsf{A}}(t) + \mathcal{X}_{\mathsf{X}}(t) = n_0$, with n_0 the total number of molecules. As initial conditions we choose $n(0) = a_0$, $m(0) = x_0$, $P_n(0) = \delta_{n,n(0)} = \delta_{n,a_0}$, and $P_m(0) = \delta_{m,m(0)} = \delta_{m,x_0}$, where we require $x_0 = n_0 - n(0) = n_0 - a_0 > 0$, because no reaction takes place if no autocatalyst is present. For $m(0) = x_0 = 0$ and $n(0) = a_0 = n_0$, we do indeed obtain $\mathrm{d}P_n(0)/\,\mathrm{d}t = 0$, $\forall\, n \in [0, n_0 - 1]$, $n \in \mathbb{N}$, and the probability density is constant. This follows from the master equation

$$\frac{\mathrm{d}P_n(t)}{\mathrm{d}t} = k(n+1)(n_0-n-1)P_{n+1}(t) \,,$$

where $\mathrm{d}P_n/\,\mathrm{d}t \neq 0$ if and only if $P_{n+1}(0) \neq 0$. This is satisfied exclusively for $n = a_0 - 1$ and $P_{a_0-1+1}(0) = 1$, but then $n_0 - (a_0 - 1) - 1 = 0$. The *no reaction* result is also obtained from the deterministic equation from $x(0) = 0$, which leads to $\mathrm{d}a/\,\mathrm{d}t = \mathrm{d}x/\,\mathrm{d}t = 0$. A related consequence of the autocatalytic process is the fact that the last molecule X cannot be converted into an A molecule, because two X molecules are required for the reaction, and this defines the domains $n \in [0, n_0 - 1]$, $n \in \mathbb{N}$, and $n_0 - n = m \in [1, n_0]$, $m \in \mathbb{N}$.

The master equation can be solved by Laplace transformation [26] in analogy to the procedure described for the association reaction [329], and again the transformation is facilitated by a transformation $t \Rightarrow \tau = kt$ of the time axis, by introducing the equilibrium constant $K = l/k$, which is dimensionless here, and by using step-up and step-down transition probabilities according to (3.96)[52]:

$$w_n^+ = K(n_0 - n)(n_0 - n - 1) \,, \quad w_n^- = n(n_0 - n) \,. \tag{4.93c}$$

The master equation takes the same form as shown previously:

$$\frac{\mathrm{d}\mathbf{P}(t)}{\mathrm{d}t} = \mathrm{W}\mathbf{P}(t) \,, \tag{4.91c$'$}$$

and the matrix W is identical, with the only difference that the state $n = n_0$ does not exist or, to be more precise, has probability zero, i.e., $P_{n_0}(t) = 0$. The state with $\mathcal{X}_\mathsf{X} = 0$ is an absorbing boundary, but it cannot be reached from the state $\mathcal{X}_\mathsf{X} = 1$ for the reason mentioned above.

With the previous definition $P_n(t) = 0, \forall n \notin [0, n_0 - 1]$, (4.91c′) represents a linear system of n_0 equations that may be solved by applying the Laplace transform to the components:

$$\widehat{f}_n(s) = \int_0^\infty P_n(t) \mathrm{e}^{-st} \, \mathrm{d}t \,. \tag{4.86$'$}$$

The same procedure as described in the previous section yields the solution

$$\widehat{\mathbf{f}}(s) = \left(s\mathbb{I} - \mathrm{W}\right)^{-1} \mathbf{P}_0 \,. \tag{4.91e$'$}$$

The initial condition $P_n(0) = \delta_{n,a_0}$ simplifies the calculation of the transformed probability density $\widehat{f}_n(s)$ and allows for the derivation of a closed solution:

$$\widehat{f}_n(s) = \frac{(-1)^{n+a_0} \mathrm{M}_{a_0+1,n+1}(s)}{\det(s\mathbb{I} - \mathrm{W})} \,, \tag{4.93d}$$

where $\mathrm{M}_{a_0+1,n+1}$ is a minor of the matrix $(s\mathbb{I} - \mathrm{W})$.[53]

For the irreversible reaction $\mathsf{A} + \mathsf{X} \to 2\mathsf{X}$, i.e., the case $l = K = 0$, the vectors $\widehat{\mathbf{f}}(s)$ and $\mathbf{P}_0$ have dimension $n_0 + 1$ and the matrix W has dimensions $(n_0 + 1)$

[52] The difference in the step-down transition probabilities is a result of the different stoichiometry of the association reaction and the autocatalytic reaction, as already discussed in Sect. 4.1.1 and Fig. 4.1.

[53] In general, the calculation of determinants and minors is highly nontrivial, as is the subsequent inversion of the Laplace transform, but thanks to the sharp initial conditions applied here, all steps can be performed analytically.

$\times\,(n_0+1)$. Since all step-up transitions are forbidden, i.e., $w_n^+ = 0,\ \forall\, n \in [0, n_0]$, the matrix W is upper-diagonal and the eigenvalues are

$$\lambda_j = -w_j^- \,, \quad j = 0, 1, \ldots, n_0 \,, \quad \lambda_j = \lambda_{n_0-j} \,, \quad \text{since } w_j^- = w_{n_0-j}^- \,.$$

Two cases are distinguished: (i) for $a_0 < x_0$ the eigenvalues of W are distinct, but (ii) when $a_0 \geq x_0$ degenerate pairs may occur, in particular all eigenvalues $\lambda_j, \lambda_{n_0-j}$ are degenerate for $j \in [x_0, a_0]$, except $\lambda_{n_0/2}$ if n_0 is even. In case (i), all eigenvalues of W are distinct and the inverse Laplace transformation is performed by means of the Heaviside expansion method in full analogy to the previous Sect. 4.3.4:

$$P_n(t) = \sum_{j=0}^{n_0-1} \lim_{s\to\lambda_j} (s-\lambda_j) e^{st} \widehat{f}_n(s) \,.$$

The resulting probability densities are given in Table 4.1.

In the degenerate case (ii), the straightforward Heaviside method cannot be applied. Instead, multiple eigenvalues can be handled by singular perturbation theory [26]. The degeneracy is removed by the addition of a small quantity ε and the step-down transition probabilities are replaced by

$$w_j^- = j(n_0 - j) \implies \eta_j = j(n_0 - j + \varepsilon) \,,$$

and the individual probability densities are now written as $P_n(t|\varepsilon)$ and $\widehat{f}_n(t|\varepsilon)$. Now the standard procedure is applied and the final result is obtained by taking the limit $\varepsilon \to 0$. The procedure is rather sophisticated and we dispense with the details, which can be found in the original reference [26].

Here we present only the final results for the probability density, which are available in closed expressions for four different ranges of the initial conditions $\mathcal{X}_A(0) = a_0$ and $\mathcal{X}_X(0) = x_0$. The solutions are given in terms of auxiliary functions and in the original time $t = \tau/k$:

$$\begin{aligned} P_n(t) &= \sum_j B_{j,n}^{(a_0,x_0)} T_j(t) + \sum_j C_{j,n}^{(a_0,x_0)}(t) T_j(t) \,, \\ &\text{with } T_j(t) = \exp\bigl(-kj(n_0-j)t\bigr) \,. \end{aligned} \tag{4.93e}$$

The functions $B_{j,n}^{(a_0,x_0)}$ are independent of time. Two cases are distinguished:

$$B_{j,n}^{(a_0,x_0)} = \begin{cases} \dfrac{(-1)^{j+n} a_0!(n_0-n-1)!(x_0-j-1)!(n_0-2j)}{n!(x_0-1)!(a_0-j)!(j-n)!(n_0-j-n)!} \,, & j+n \leq n_0 \,, \\[2ex] \dfrac{(-1)^{j+a_0} a_0!(j+n-n_0-1)!(n_0-n-1)!(2j-n_0)}{n!(x_0-1)!(a_0-j)!(j-n)!(j-x_0)!} \,, & j+n > n_0 \,. \end{cases} \tag{4.93f}$$

Table 4.1 Probability density of the first order irreversible autocatalytic reaction $\mathsf{A} + \mathsf{X} \rightarrow 2\mathsf{X}$. For $x_0 > a_0$, all eigenvalues of the matrix W are distinct and the probability density is obtained by a simple sum of the contributions of individual exponential decay modes. For $x_0 \le a_0$, three subcases are distinguished and $\nu = \lfloor n_0/2 \rfloor$. The expressions are taken from [26]

Case	Range of n	Probability density $P_n(t)$
$x_0 > a_0$	$[0, a_0]$	$\sum_{j=n}^{a_0} B_{j,n}^{(a_0,x_0)} T_j(t)$
$x_0 \le a_0$	$[0, x_0[$	$\sum_{j=n}^{x_0-1} B_{j,n}^{(a_0,x_0)} T_j(t) + \sum_{j=x_0}^{\nu} \frac{C_{j,n}^{(a_0,x_0)}(t) T_j(t)}{1+\delta_{j,a_0-j}}$
	$[x_0, \nu]$	$\sum_{j=n}^{\nu} \frac{C_{j,n}^{(a_0,x_0)}(t) T_j(t)}{1+\delta_{j,a_0-j}} + \sum_{j=n}^{a_0-n+1} B_{j,n}^{(a_0,x_0)} T_j(t)$
	$]\nu, a_0]$	$\sum_{j=a_0-n+1}^{a_0} B_{j,n}^{(a_0,x_0)} T_j(t)$

The second function $C_{j,n}^{(a_0,x_0)}(t)$ results from singular perturbation theory and is more involved:

$$C_{j,n}^{(a_0,x_0)}(t) = (-1)^{n-x_0} \frac{(n_0-n-1)!a_0!}{n!(x_0-1)!(a_0-j)!(j-n)!(j-x_0)!(n_0-j-n)!}$$
$$\times \left((n_0-2j)^2 \frac{k}{2} t + 2 - (n_0-2j)(h_{j-n} - h_{a_0-j} - h_{n_0-j-n} + h_{j-x_0}) \right) . \tag{4.93g}$$

The symbol h_n denotes the harmonic numbers, i.e., $h_n = \sum_{k=1}^{n} 1/k$. Table 4.1 presents the probability densities for the different cases and subcases. An example of the time dependent probability density $P_n(t)$ of the irreversible first order autocatalytic reaction is shown in Fig. 4.33.

The expectation value of the number of molecules A can also be computed by means of auxiliary functions, with some labor [26]. However, direct calculation of the mean and variance is less sophisticated:

$$\mathrm{E}\big(\mathcal{X}_{\mathsf{A}}(t)\big) = \sum_{n=0}^{a_0} n P_n(t) , \quad \mathrm{var}\big(\mathcal{X}_{\mathsf{A}}(t)\big) = \sum_{n=0}^{a_0} n^2 P_n(t) - \mathrm{E}\big(\mathcal{X}_{\mathsf{A}}(t)\big)^2 .$$

An instructive example is shown in Fig. 4.33, where we can also see a substantial difference between the deterministic solution and the expectation value. The irreversible autocatalytic reaction coupled to a simple irreversible monomolecular reaction constitutes a reaction network called the simple SIR model, which is of interest in epidemiology and will be analysed in Sect. 5.2.4.

In principle, the master equation for the reversible first-order autocatalytic reaction (4.91c) could be handled by the same procedure as the irreversible reaction. In the irreversible case the eigenvalues of the matrix W are available in analytical form. Since this does not seem to be possible for the reversible case, little is gained by the Laplace transform. We discuss $\mathsf{A} + \mathsf{X} \rightleftharpoons 2\mathsf{X}$ as an example for numerical

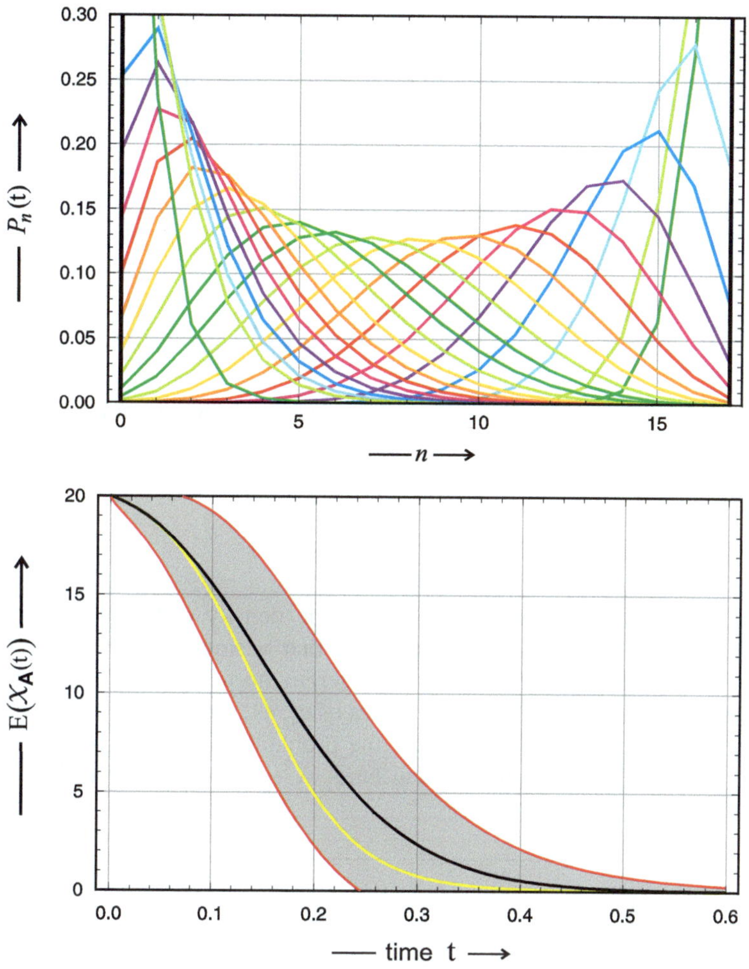

Fig. 4.33 Irreversible bimolecular autocatalytic reaction $A + X \to 2X$. The *upper plot* shows the probability distribution $P_n(t) = P\big(\mathcal{X}_A(t) = n\big)$ describing the number of molecules of species A as a function of time and calculated using the equations in Table 4.1. Parameter choice: $k = 1\,[\mathrm{N}^{-1}\mathrm{t}^{-1}]$, $a_0 = 17$, $x_0 = 5$, and sampling times: $t = 0$ (*black*), 0.005 (*chartreuse*), 0.01 (*green*), 0.02 (*turquoise*), 0.03 (*blue*), 0.04 (*violet*), 0.06 (*purple*), 0.08 (*magenta*), and ∞ (*red*). In the *lower plot*, we show the expectation value $\mathrm{E}\big(\mathcal{X}_A(t)\big)$ (*black*), together with the band $\mathrm{E} \pm \sigma$ (*red*) and the deterministic expectation value (*yellow*). The areas where the calculated values are probabilistically meaningless, viz., $\mathrm{E} + \sigma > a_0$ and $E - \sigma < 0$, are clipped. Parameter choice: $a_0 = 20$, $x_0 = 1$, and $k = 1\,[\mathrm{N}^{-1}\mathrm{t}^{-1}]$

simulations in Sect. 4.6.4. There we also discuss in detail the differences between the stochastic and deterministic solutions. The stationary solution of the reversible reaction is nevertheless readily computed by applying the results for the long-time limit:

$$\mathbf{0} = \mathbf{W}\bar{\mathbf{P}} \,, \qquad n \in [0, n_0[\,, \quad K > 0 \,, \tag{4.91c$'$}$$

which can be done for any stationary univariate master equation with the help of (3.100). By inserting the expressions from (4.93c), we find

$$\bar{P}_n^{(\mathsf{A+X \rightleftharpoons 2X})} = \binom{n_0}{n} \frac{K^n}{(1+K)^{n_0} - K^{n_0}} \,. \tag{4.93h}$$

This result is to be compared with the equilibrium of the monomolecular reaction $\mathsf{A} \rightleftharpoons \mathsf{X}$, which was calculated in Sect. 4.3.2:

$$\bar{P}_n^{(\mathsf{A \rightleftharpoons X})} = \binom{n_0}{n} \frac{K^n}{(1+K)^{n_0}} \,.$$

We recognize that the difference between the density distributions for the two reactions at equilibrium is due to the fact that a single X molecule cannot be converted into an A molecule and a different normalization is required. This deviation disappears with increasing values of n_0, as fast as $n(n-1)$ approaches n^2.

Figure 4.34 compares the width of the probability densities of the three irreversible bimolecular reactions studied here: $\mathsf{A + X \rightarrow 2X}$, $\mathsf{A + B \rightarrow C}$, and

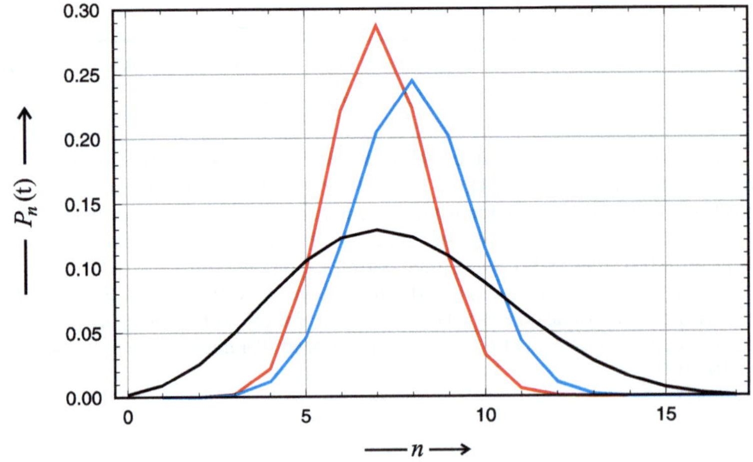

Fig. 4.34 Probability densities of bimolecular reactions. Compared are the widths of the densities $P(\mathcal{X}_\mathsf{A}(t_\mathrm{m}))$ in the middle of the state space as calculated using equations (4.93e), (4.88g), and (4.89e). Parameter choice: ($t_\mathrm{m} = 0.09$, $k = 1$) for $\mathsf{A + X \rightarrow 2X}$ (*black*), ($t_\mathrm{m} = 0.4375$, $k = 0.1$) for $\mathsf{A + B \rightarrow C}$ (*blue*), and ($t_\mathrm{m} = 0.525$, $k = 0.1$) for $\mathsf{2A \rightarrow C}$ (*red*)

$2\,\mathsf{A} \to \mathsf{C}$. All three reactions start from a sharp initial distribution $P_n(0) = \delta_{n,a0}$, and progress to a sharp distribution $\lim_{t\to\infty} P_n(t) = \delta_{n,0}$. In order to make the densities comparable, we consider them in the middle of the state space: $\mathrm{E}\big(\mathcal{X}_{\mathsf{A}}(t)\big) \approx a_0/2$. The autocatalytic process is characterized by two differences in comparison with the other two reactions: (i) the distribution is much broader and (ii) the time at which the distribution passes the middle of the state space is much shorter. Both findings are a result of the self-enhancement in autocatalysis. Fluctuations are larger and the rate of the reaction is accelerated.

4.3.6 Stochastic Enzyme Kinetics

Michaelis–Menten kinetics is more complex than the examples treated so far, since even the simple two-step mechanism, i.e., $\mathsf{S} + \mathsf{E} \rightleftharpoons \mathsf{C} \to \mathsf{E} + \mathsf{P}$ with C denoting the enzyme–substrate complex $\mathsf{C} \equiv \mathsf{S} \cdot \mathsf{E}$, cannot be reduced to a problem with a single independent variable without approximation. Instead, we have to deal with two random variables, for example, $\mathcal{X}_{\mathsf{S}}$ and $\mathcal{X}_{\mathsf{E}}$, counting substrate and enzyme molecules, respectively, and with a bivariate probability density $P_{e,n}(t)$, where e denotes the number of free enzyme molecules E and n the number of unbound substrate molecules S. The analytical model we introduce here is taken from the literature [20]. It is based on the assumption that only a single enzyme molecule is present, free or bound in the complex, E or C, respectively. The idea is to consider a volume so small that it contains only one or no enzyme molecules. Present day spectroscopic techniques make it possible to observe and study single enzyme molecules (Sect. 4.4.1) and the model presented here has found a physical realization in experimental setups with single enzyme molecules that are immobilized in compartments or on membranes.

The basic steps of irreversible substrate to product conversion, i.e., $\mathsf{S} \to \mathsf{P}$, are

$$\ldots \xrightarrow{k_2} n\mathsf{S} + \mathsf{E} \underset{l_1}{\overset{k_1 n}{\rightleftharpoons}} (n-1)\mathsf{S} + \mathsf{C} \xrightarrow{k_2} (n-1)\mathsf{S} + \mathsf{E} + \mathsf{P} \underset{l_1}{\overset{k_1(n-1)}{\rightleftharpoons}} \ldots\,,$$

where n is not the usual stoichiometric coefficient, but the number of substrate molecules that are ready for conversion. Figure 4.35 shows the entire state space for a single enzyme molecule, $e \in \{0, 1\}$ and $n \in [0, n_0]$, $n \in \mathbb{N}$. It is straightforward to write down a master equation for this scheme:

$$\frac{\mathrm{d}P_{e,n}(t)}{\mathrm{d}t} = l_1(2-e)P_{e-1,n-1}(t) + k_2(2-e)P_{e-1,n}(t) \tag{4.94a}$$

$$+k_1(e+1)(n+1)P_{e+1,n+1}(t) - \big[k_1 en + (l_1+k_2)(1-e)\big]P_{e,n}(t)\,,$$

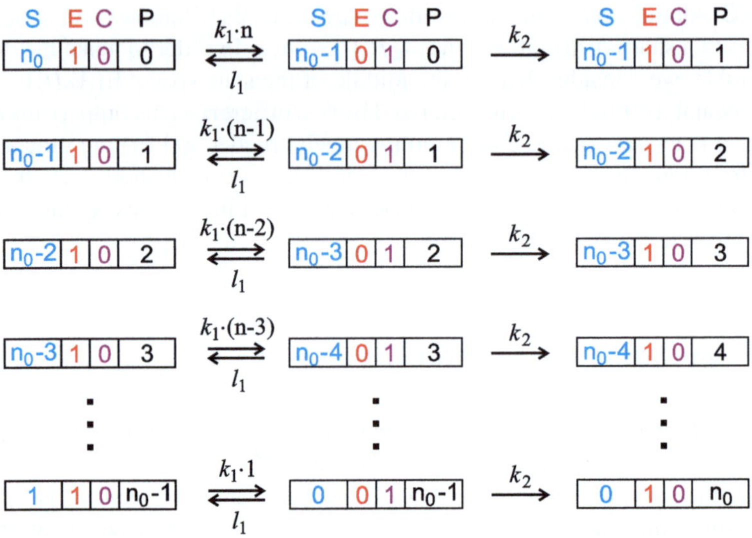

Fig. 4.35 Scheme of the Michaelis–Menten mechanism with a single enzyme molecule. We show the irreversible conversion of n substrate molecules into n product molecules that occurs in $2n$ individual reaction steps. The *boxes* contain the numbers of molecules of the four species: substrate **S** (*blue*), enzyme **E** (*red*), enzyme–substrate complex $\mathbf{C} \equiv \mathbf{S} \cdot \mathbf{E}$ (*purple*), and product **P** (*black*). All states in the third column are identical with the states of the first column in the next row, except the initial and the final state, and hence the reaction scheme consists of a single line

with the initial conditions $P_{e,n}(0) = \delta_{e,1}\delta_{n,n_0}$.[54] Since the conversion steps are irreversible, the final state is determined by the limiting density

$$\lim_{t\to\infty} P_{e,n}(t) = \delta_{e,1}\delta_{n,0} \ ,$$

all substrate molecules **S** are converted into product **P** and the enzyme is in the free state **E**.

A solution of the master equation (4.94a) can be derived by means of the marginal probability generating functions[55]:

$$g_e(s,t) = \sum_{n=0}^{n_0+e-1} s^n P_{e,n}(t) \ , \quad e \in \{0,1\} \ , \ t \geq 0 \ . \tag{4.94b}$$

[54]Here and in the following paragraphs, all probability densities and generating functions with index values outside the domains, i.e., $e \notin \{0,1\}$ and $n \notin [0,n]$, are zero.

[55]This approach is meaningful when one of the two random variables is restricted to very few values, here $\mathcal{X}_\mathsf{E} = \{0,1\}$. The use of marginal densities avoids the occurrence of second order partial derivatives, which create the difficulties encountered in solving the master equations of second order reactions.

Equations (4.94a) and (4.94b) are converted into a system of partial differential equations:

$$\frac{\partial g_e(s,t)}{\partial t} = k_1(e+1)\frac{\partial g_{e+1}(s,t)}{\partial s} - k_1 es\frac{\partial g_e(s,t)}{\partial s} - (l_1+k_2)(1-e)g_e(s,t)$$
$$+(l_1+k_2)(2-e)sg_{e-1}(s,t)\ , \quad e \in \{0,1\}\ . \tag{4.94c}$$

The solution of the PDE (4.94c) is obtained in terms of a pair of generating functions:

$$g_0(s,t) = \gamma_1 e^{-l_1(s-1)/k_2} e^{-k_2 t} + \gamma_2 \frac{l_1+k_2}{l_1 s+k_2} e^{-(k_1+k_2)t}$$
$$+\sum_{i=1}^{2}\sum_{n=0}^{\infty} \chi_i^{(n)} \left(\frac{k_2-(k_2+\lambda_i^{(n)})s}{-\lambda_i^{(n)}}\right)^{q_i^{(n)}+1} e^{\lambda_i^{(n)}t}\ , \tag{4.94d}$$

$$g_1(s,t) = \gamma_0 - \gamma_1 e^{-l_1(s-1)/k_2} e^{-k_2 t} - \gamma_2 \frac{l_1+k_2}{l_1 s+k_2} e^{-(k_1+k_2)t}$$
$$-\sum_{i=1}^{2}\sum_{n=0}^{\infty} \chi_i^{(n)} \left(\frac{k_2-(k_2+\lambda_i^{(n)})s}{-\lambda_i^{(n)}}\right)^{q_i^{(n)}+1} e^{\lambda_i^{(n)}t}\ , \tag{4.94e}$$

where the various coefficients γ_i and $\chi_i^{(k)}$ are to be determined from the conditions

$$g_0(1,t) + g_1(1,t) = 1\ , \quad g_0(s,0) = 0\ , \quad g_1(s,0) = s^{n_0}\ .$$

The eigenvalues of the $(2n_0+1)\times(2n_0+1)$ transition probability matrix corresponding to the master equation (4.94a) are obtained from n_0 quadratic equations:

$$\lambda^2 + \big(l_1 + k_1(n+1) + k_2\big)\lambda + k_1 k_2(n+1) = 0\ , \quad \forall\, n = 0,1,\ldots,n_0-1\ ,$$

including a trivial eigenvalue $\lambda = 0$:

$$\lambda_{1,2}^{(n)} = -\frac{1}{2}\left(\big(k_1(n+1)+l_1+k_2\big) \mp \sqrt{\big(k_1(n+1)+l_1+k_2\big)^2 - 4k_1k_2(n+1)}\right)\ .$$

The exponents $q_i^{(n)}$ are readily obtained from the eigenvalues:

$$q_i^{(n)} = -\frac{\big(\lambda_i^{(n)}\big)^2 + (l_1+k_1+k_2)\lambda_i^{(n)} + k_1k_2}{k_1\big(k_2+\lambda_i^{(n)}\big)}\ .$$

Although the calculations can be quite involved in practice, the coefficients γ_1 and γ_2 vanish if $l_1 \neq 0$, the summations over n contain finite numbers of terms, the $q_i^{(n)}$ values are in the range $0 \leq q_i^{(n)} \leq n_0 - 1$, and the eigenvalues $\lambda_i^{(n)}$ are distinct real numbers.

The probability densities are obtained from the generating functions in the usual way:

$$P_{e,n}(t) = \frac{1}{n!} \frac{\partial^n g_e(s,t)}{\partial s^n}\bigg|_{s=0} ,$$

As an example demonstrating the use of single molecule probabilities, the densities for a single enzyme molecule and n_0 substrate molecules constrained by the conversion of the first substrate molecule are calculated [140]:

$$P_{0,n_0-1}(t) = \frac{k_1 n_0}{\lambda_2^{(n_0-1)} - \lambda_1^{(n_0-1)}} \left(\mathrm{e}^{\lambda_2^{(n_0-1)} t} - \mathrm{e}^{\lambda_1^{(n_0-1)} t} \right) ,$$

$$P_{1,n_0}(t) = \frac{\lambda_1^{(n_0-1)} + k_1 n_0}{\lambda_2^{(n_0-1)} - \lambda_1^{(n_0-1)}} \mathrm{e}^{\lambda_2^{(n_0-1)} t} - \frac{\lambda_2^{(n_0-1)} + k_1 n_0}{\lambda_2^{(n_0-1)} - \lambda_1^{(n_0-1)}} \mathrm{e}^{\lambda_1^{(n_0-1)} t} .$$

The calculation of the time of appearance ϑ of the first product molecule is a typical first passage time problem. The probability $P_{\mathbf{P}}(t)$ of recording a product molecule is simply the probability of having neither n_0 substrate molecules **S** nor an enzyme–substrate complex **C** given by the expression $P_{\mathbf{P}}(t) = P\big(\mathcal{X}_{\mathbf{P}}(t) = 1\big) = 1 - P_{1,n_0}(t) - P_{0,n_0-1}(t)$. The expectation value of the time ϑ is then obtained from

$$\langle \vartheta \rangle = \int_0^\infty t \frac{\mathrm{d}P_{\mathbf{P}}(t)}{\mathrm{d}t} \,\mathrm{d}t = -\int_0^\infty t \left(\frac{\mathrm{d}P_{1,n_0}(t)}{\mathrm{d}t} + \frac{\mathrm{d}P_{0,n_0-1}(t)}{\mathrm{d}t} \right) \mathrm{d}t .$$

After some calculation, we obtain the final result

$$\langle \vartheta \rangle = \frac{k_1 n_0 + l_1 + k_2}{k_1 k_2 n_0} = \frac{1}{k_1 n_0} + \frac{l_1}{k_1 k_2 n_0} + \frac{1}{k_2} , \tag{4.94f}$$

which is easily interpreted: the appearance of the (first) molecule **P** takes a long time if (i) the binding rate constant k_1 multiplied by the initial number of substrate molecules is small, or if (ii) the dissociation **C** $\rightarrow$ **S** + **E** of the enzyme–substrate complex is fast as expressed by a large value of l_1, or if (iii) the rate constant k_2 of product formation is small.

For the purpose of illustration and comparison with the deterministic model, we consider the reaction of one enzyme molecule plus one substrate molecule

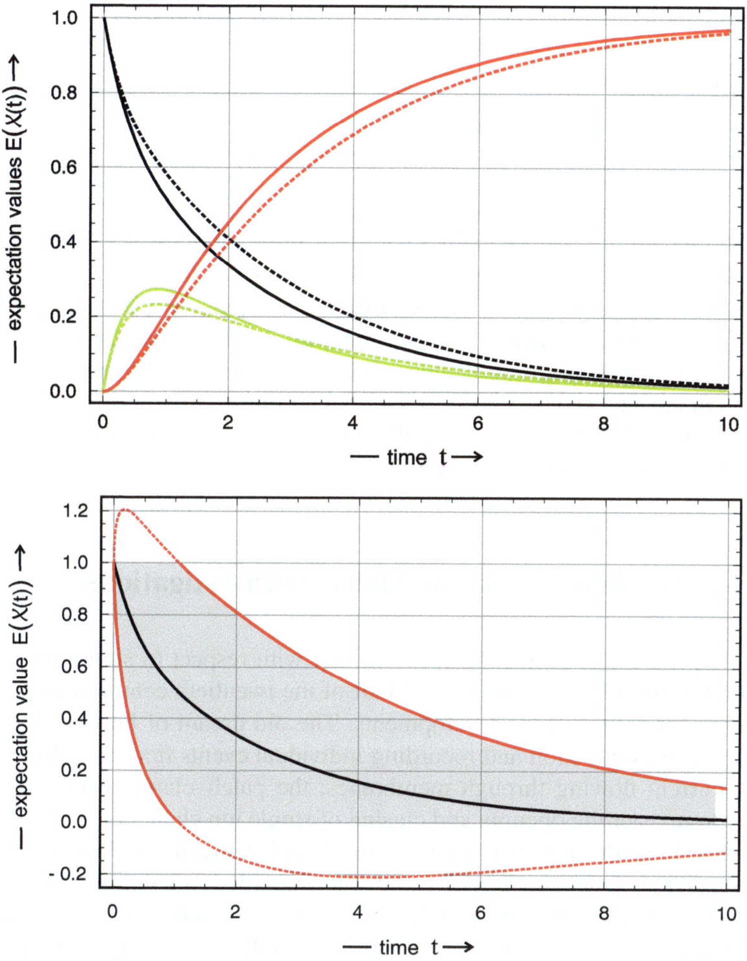

Fig. 4.36 The Michaelis–Menten reaction with single molecules. The *full curves* in the *upper plot* show the expectation values for observing the substrate molecule $\mathrm{E}(\mathcal{X}_\mathrm{S}(t))$ (*black*), the substrate–enzyme complex $\mathrm{E}(\mathcal{X}_\mathrm{C}(t))$ (*green*), and the product molecule $\mathrm{E}(\mathcal{X}_\mathrm{P}(t))$ (*red*), which are compared with the corresponding functions $s(t)$, $c(t)$, and $p(t)$ (*broken curves*) obtained by integrating the kinetic differential equations (Sect. 4.1.2) for the same conditions. The *lower plot* presents the one standard deviation error band around the expectation value of the substrate concentration, i.e., $\mathrm{E}(\mathcal{X}_\mathrm{S}(t)) \pm \sigma(\mathcal{X}_\mathrm{S}(t))$. The *gray hatched zone* is the probabilistically meaningful part of the error band. Parameter choice: $n_0 = 1$, $k_1 = 1\,[\mathrm{N}^{-1}\mathrm{t}^{-1}]$, $l_1 = k_2 = 1\,[\mathrm{t}^{-1}]$

in Fig. 4.36:

$$P_{00}(t) = \frac{k_1}{\lambda_2^{(0)} - \lambda_1^{(0)}}\left(e^{\lambda_2^{(0)} t} - e^{\lambda_1^{(0)} t}\right) ,$$

$$P_{11}(t) = \frac{\lambda_1^{(0)} + k_1}{\lambda_2^{(0)} - \lambda_1^{(0)}} e^{\lambda_2^{(0)} t} - \frac{\lambda_2^{(0)} + k_1}{\lambda_2^{(0)} - \lambda_1^{(0)}} e^{\lambda_1^{(0)} t} .$$

These probability densities are identical with the expectation values for the different states in the last line of Fig. 4.35, since for $n_0 = 1$, the random variables can assume only the values zero and one:

$$\mathrm{E}\big(\mathcal{X}_{\mathbf{S}}(t)\big) = P_{11}(t)\,, \quad \mathrm{E}\big(\mathcal{X}_{\mathbf{C}}(t)\big) = P_{00}(t)\,, \quad \mathrm{E}\big(\mathcal{X}_{\mathbf{P}}(t)\big) = 1 - P_{11}(t) - P_{00}(t)\,,$$

and from $\mathrm{var}\big(\mathcal{X}(t)\big) = \mathrm{E}\Big(\mathcal{X}(t) - \mathrm{E}\big(\mathcal{X}(t)\big)\Big)^2$ the three variances are readily obtained. In Fig. 4.36, the time dependence of the probabilities is compared with the curves obtained by integration of the kinetic differential equations (Sect. 4.1.2). We find a remarkably good agreement, despite the fact that the expectation values refer to events involving single molecules only. The variance, however, is so large that the curves $\mathrm{E}\big(\mathcal{X}(t)\big) \pm \sigma\big(\mathcal{X}(t)\big)$ also extend outside the probabilistic domain: $0 \le \mathrm{E}\big(\mathcal{X}(t)\big) \le 1$. Then, it is advisable to restrict the one standard deviation error zone to the physically meaningful domain.[56]

4.4 Fluctuations and Single Molecule Investigations

The rapid progress of molecular spectroscopy with respect to signal intensity and temporal resolution during the second half of the twentieth century has provided a foundation for entirely new developments. The old dream of being able to watch single molecules in action and recording individual events first came true with the electric current flowing through membranes: the patch–clamp technique made it possible to register the opening and closing of single ion channels [230, 420].

Another breakthrough was more general and came from fluorescence spectroscopy: signals from single molecules were recorded in solution [254, 471] and in the solid state [402, 434] (for reviews see, e.g., [321, 449, 561]), and this set the stage for the analysis of events involving single molecules. Using fluorescence, the motions of single molecules can now be traced routinely at high spatial and temporal resolution and in highly heterogeneous environments like living cells.

A third approach came from applications of scanning tunneling microscopy [52] and opened the possibility for mechanical manipulation of single molecules by means of atomic force microscopy [584]. Particularly illustrative in this context is, for example, the mechanochemistry of single nucleic acid molecules [238, 342].

[56] The same over- and undershooting of the $\mathrm{E}\big(\mathcal{X}(t)\big) \pm \sigma\big(\mathcal{X}(t)\big)$ curves has also been observed in previous cases. For most systems, however, the meaningless parts of the one standard deviation error band are negligibly small.

In essence, the single particle approaches may be grouped into three classes:

1. Methods to record the states of single molecules in solution.
2. Methods to track the motions of single particles in space.
3. Methods to manipulate single particles mechanically.

The literature on successful single molecule experiments is enormous. Here, we shall focus on two selected issues that require the stochastic methods presented in this monograph: (i) biochemical kinetics of single enzyme molecules, a new method that provides fresh insights into the mechanism of enzymatic catalysis, and (ii) fluorescence correlation spectroscopy, a general technique that allows for recording of single particles.

4.4.1 Single Molecule Enzymology

The possibility of recording signals from single protein molecules and following their time dependence is the basis for experimental single molecule enzymology. The insight into the mechanistic details gained by single molecule studies provides answers to a number of questions of the kind:

(i) Are all enzyme molecules in the same conformation and do they react with the same rate parameters or are we dealing with conformational fluctuations, which give rise to dynamical disorder?
(ii) Are the fluctuations in enzyme turnover and substrate-to-product conversion in enzymatic reactions greater than in the case of catalysis by small molecules?

In this section we shall present stochastic treatments of the two extended versions of Michaelis–Menten kinetics shown in Fig. 4.6 and begin with scheme **A** [462].

Extended Michaelis–Menten Mechanism (Scheme A)
The enzyme molecule can exist in three conformational states, namely, **E**, **S** $\cdot$ **E**, and **E** $\cdot$ **P** (Fig. 4.6a), and the random variable $\mathcal{X}(t)$ fluctuates between the three states $S_1 = \mathbf{E}$, $S_2 = \mathbf{C} \equiv \mathbf{S}\cdot\mathbf{E}$, and $S_3 = \mathbf{D} \equiv \mathbf{E}\cdot\mathbf{P}$. The probabilities for the enzyme molecule to be in one of the three states at time t are denoted by

$$P_{\mathbf{E}}(t) = P\big(\mathcal{X}(t) = \mathbf{E}\big) \;, \quad P_{\mathbf{C}}(t) = P\big(\mathcal{X}(t) = \mathbf{C}\big) \;, \quad P_{\mathbf{D}}(t) = P\big(\mathcal{X}(t) = \mathbf{D}\big) \;,$$

$$\text{with } P_{\mathbf{E}}(t) + P_{\mathbf{C}}(t) + P_{\mathbf{D}}(t) = 1 \;. \tag{4.95}$$

As in Sect. 4.1.2, we make the assumption of constant substrate and product concentration: $k_1 = k_1'[\mathbf{S}]_0 = k_1's_0$ and $l_3 = l_3'[\mathbf{P}]_0 = l_3'p_0$. Then, the time dependence of the probabilities is determined by the linear ODE (4.20). Because of the conservation relation, a superposition of two exponential functions is obtained, as follows directly from linear algebra:

$$\frac{\mathrm{d}\mathbf{P}}{\mathrm{dt}} = \mathrm{H}\mathbf{P}\,, \quad \mathbf{P} = \begin{pmatrix} P_{\mathsf{E}} \\ P_{\mathsf{C}} \\ P_{\mathsf{D}} \end{pmatrix}, \quad \mathrm{H} = \begin{pmatrix} -k_1 - l_3 & l_1 & k_3 \\ k_1 & -k_2 - l_1 & l_2 \\ l_3 & k_2 & -k_3 - l_2 \end{pmatrix}.$$

Solutions of the ODE are obtained in terms of an eigenvalue problem, where Λ is a diagonal matrix containing the eigenvalues $\lambda_{1,2}$ and $\lambda_0 = 0$:

$$\mathbf{P}(t) = \mathrm{B}\boldsymbol{\varrho}(t)\,, \quad \text{with } \mathrm{B}^{-1}\mathrm{H}B = \Lambda\,,$$

$$\frac{\mathrm{d}\boldsymbol{\varrho}}{\mathrm{dt}} = \Lambda\boldsymbol{\varrho}\,, \quad \boldsymbol{\varrho}(t) = \exp(\Lambda t)\boldsymbol{\varrho}(0)\,,$$

where the column vector $\boldsymbol{\varrho} = (\varrho_0, \varrho_1, \varrho_2)^{\mathrm{t}}$ contains the right-hand eigenvectors of the matrix H. Back-transformation to the original probabilities yields

$$\mathbf{P}(t) = \mathrm{B}\exp(\Lambda t)\big(\mathrm{B}^{-1}\mathbf{P}(0)\big)\,.$$

The eigenvalues of the matrix H have already been calculated for the deterministic system (4.21). Both eigenvalues have negative real parts and the asymptotically stable stationary state (4.19) corresponds to the macroscopic thermodynamic equilibrium. Here, we are computing normalized probabilities (for the sake of simplicity, we use numbers instead of letters as indices, viz., $1 \equiv \mathsf{S}_1 \equiv \mathsf{E}$, $2 \equiv \mathsf{S}_2 \equiv \mathsf{C}$, $3 \equiv \mathsf{S}_3 \equiv \mathsf{D}$), and find [462]

$$\begin{aligned} \bar{P}_1 &= \frac{k_2k_3 + k_3l_1 + l_1l_2}{\lambda_1\lambda_2}\,, \\ \bar{P}_2 &= \frac{k_3k_1 + k_1l_2 + l_2l_3}{\lambda_1\lambda_2}\,, \\ \bar{P}_3 &= \frac{k_1k_2 + k_2l_3 + l_3l_1}{\lambda_1\lambda_2}\,, \end{aligned} \tag{4.96}$$

The third eigenvalue $\lambda_0 = 0$ indicates a linear dependence that is given by the conservation of probabilities: $P_1 + P_2 + P_3 = 1$. Apart from normalization, (4.96) is formally identical to (4.75).

Recording single enzyme trajectories provides information on steady states and on transient kinetics, which is commonly expressed in terms of transition probabilities. Hence, $P_{ji}(t, t + \Delta t) = P\big(\mathcal{X}(t + \Delta t) = \mathsf{S}_j | \mathcal{X}(t) = \mathsf{S}_i\big)$ denotes the probability that the enzyme is in state j at time $t+\Delta t$, provided it was in state i at time t. In a stationary Markov process, the transition probabilities are independent of time t, i.e., $P_{ji}(t, t+\Delta t) = P_{ji}(\Delta t)$ (Sect. 3.1.3). The diagonal and off-diagonal stationary

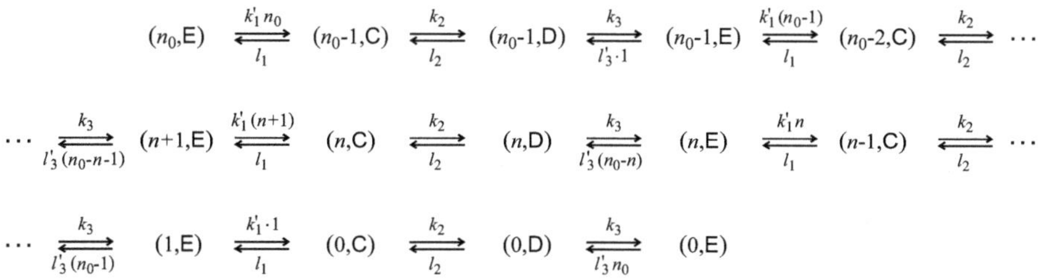

Fig. 4.37 Stochastic dynamics of substrate to product conversion. The extended Michaelis–Menten mechanism (Fig. 4.6a) is applied to model the stochasticity of the substrate to product conversion. The single enzyme molecule occurs in three conformations: (i) free enzyme (**E**), (ii) enzyme bound to substrate (**C**), and (iii) enzyme bound to product (**D**). Because of the restriction to a single enzyme molecule, the state space is a one-dimensional array (Fig. 4.36), and the stochastic model can be simplified to a (biased) continuous time random walk [462]

transition probabilities of the single molecule Michaelis–Menten mechanism are

$$\begin{aligned}
P_{jj} &= -\frac{k_j + l_{j-1} + \lambda_2(1-\bar{P}_j)}{\lambda_1 - \lambda_2}\mathrm{e}^{\lambda_1 \Delta t} + \frac{k_j + l_{j-1} + \lambda_1(1-\bar{P}_j)}{\lambda_1 - \lambda_2}\mathrm{e}^{\lambda_2 \Delta t} + \bar{P}_j \,, \\
P_{j+1,j} &= \frac{k_j + \lambda_2 \bar{P}_{j+1}}{\lambda_1 - \lambda_2}\mathrm{e}^{\lambda_1 \Delta t} + \frac{k_j + \lambda_1 \bar{P}_{j+1}}{\lambda_1 - \lambda_2}\mathrm{e}^{\lambda_2 \Delta t} + \bar{P}_{j+1} \,, \qquad (4.97) \\
P_{j-1,j} &= \frac{l_{j-1} + \lambda_2 \bar{P}_{j-1}}{\lambda_1 - \lambda_2}\mathrm{e}^{\lambda_1 \Delta t} + \frac{k_j + \lambda_1 \bar{P}_{j-1}}{\lambda_1 - \lambda_2}\mathrm{e}^{\lambda_2 \Delta t} + \bar{P}_{j-1} \,, \\
& j = 1,2,3,1,2,\ldots = (i \bmod 3) + 1 \,, \quad i = 0,1,2,3,4,\ldots \,.
\end{aligned}$$

It is straightforward to consider the transition probabilities for vanishing time intervals. As expected, we find

$$\lim_{\Delta t \to 0} P_{jj}(\Delta t) = 1 \,, \qquad \lim_{\Delta t \to 0} P_{j+1,j}(\Delta t) = \lim_{\Delta t \to 0} P_{j-1,j}(\Delta t) = 0 \,,$$

so the off-diagonal elements corresponding to proper transitions converge to zero.

In the case of single enzyme kinetics, the conversion of substrate into product can be modeled by a biased[57] continuous time random walk [462] on the finite lattice shown in Fig. 4.37. Three states are characterized by the same number of substrate molecules, i.e., (n, C), (n, D), and (n, E), and hence the number of substrate molecules is a random variable $\mathcal{X}(t)$ with the probability

$$P_n(t) = P\big(\mathcal{X}(t) = n\big) = P_n^{(\mathsf{E})}(t) + P_n^{(\mathsf{C})}(t) + P_n^{(\mathsf{D})}(t) \,,$$

[57] By 'biased' we express the fact that individual steps may have different weights.

where superscripts refer to the state of the enzyme. In other words, the three stochastic variables $\mathcal{X}_{\mathbf{E}}(t)$, $\mathcal{X}_{\mathbf{C}}(t)$, and $\mathcal{X}_{\mathbf{D}}(t)$ are lumped together in the variable $\mathcal{X}(t)$. As initial condition we assume $\mathcal{X}(0) = n_0$ substrate molecules and no product. A state of the system is fully characterized by n, the number of substrate molecules, and the state of the enzyme, viz., **E**, **C**, or **D**. Since we are dealing with a single enzyme molecule, the state space can be arranged as a one-dimensional lattice, as shown in Fig. 4.37, and the stochastic process becomes a biased random walk on this lattice. The bias is introduced by the transition probabilities, which differ for the individual transitions. Applying the same notation for the rate parameters as in the deterministic kinetic equation, we obtain for the time derivatives of the probabilities:

$$
\begin{aligned}
\frac{\mathrm{d}P_n^{(\mathbf{E})}(t)}{\mathrm{d}t} &= l_1 P_{n-1}^{(\mathbf{C})}(t) + k_3 P_n^{(\mathbf{D})}(t) - \big(k_1' n + l_3'(n_0 - n)\big) P_n^{(\mathbf{E})}(t) \,, \\
\frac{\mathrm{d}P_n^{(\mathbf{C})}(t)}{\mathrm{d}t} &= k_1'(n+1) P_n^{(\mathbf{E})}(t) + l_2 P_n^{(\mathbf{D})}(t) - (k_2 + l_1) P_n^{(\mathbf{C})}(t) \,, \qquad (4.98) \\
\frac{\mathrm{d}P_n^{(\mathbf{D})}(t)}{\mathrm{d}t} &= l_3'(n_0 - n) P_n^{(\mathbf{E})}(t) + k_2 P_n^{(\mathbf{C})}(t) - (k_3 + l_2) P_n^{(\mathbf{D})}(t) \,.
\end{aligned}
$$

The equilibrium distribution of the probabilities is readily calculated and reported in the literature [462, 526]:

$$
\begin{aligned}
\bar{P}_n^{(\mathbf{E})} &= \frac{n_0! K^{n_0-n}}{(n_0-n)!n!} \frac{1}{Q} \,, \\
\bar{P}_n^{(\mathbf{C})} &= \frac{n_0! K^{n_0-n-1}}{(n_0-n-1)!n!} K_1 \frac{1}{Q} \,, \\
\bar{P}_n^{(\mathbf{D})} &= \frac{n_0! K^{n_0-n-1}}{(n_0-n-1)!n!} K_1 K_2 \frac{1}{Q} \,, \qquad (4.99) \\
K_1 &= \frac{k_1}{l_1} \,, \quad K_2 = \frac{k_2}{l_2} \,, \quad K_3 = \frac{k_3}{l_3} \,, \quad K = K_1 K_2 K_3 \,, \\
Q &= \frac{1 + K + n_0 \kappa}{(1+K)^{n_0-1}} \,, \quad \kappa = K_1(1 + K_2) \,.
\end{aligned}
$$

The expectation value and variance of the number of substrate molecules at equilibrium can be derived from the function Q:

$$
\begin{aligned}
\mathrm{E}(n) &= n_0 - \frac{\partial \ln Q(K)}{\partial \ln K} = \frac{n_0 + K}{1 + K} - \frac{K + n_0 \kappa}{1 + K + n_0 \kappa} \,, \\
\mathrm{var}(n) &= \frac{\partial^2 \ln Q(K)}{\partial \ln K^2} = \frac{n_0 K}{(1+K)^2} + \frac{K + n_0 \kappa}{(1 + K + n_0 \kappa)^2} \,. \qquad (4.100)
\end{aligned}
$$

It is straightforward to calculate the values of the moments for large numbers of substrate molecules:

$$\mathrm{E}(n) = \frac{n_0}{1+K} \,, \quad \mathrm{var}(n) = \frac{n_0 K}{(1+K)^2} \,, \quad \text{for large } n_0 \,.$$

It is worth mentioning that precisely these expressions were obtained for the binomial distribution with the replacements $n \leftrightarrow n_0$, $p \leftrightarrow 1/(1+K)$, and $q \leftrightarrow K/(1+K)$ in (2.41).

Extended Michaelis–Menten Mechanism (Scheme B)
An alternative extension of the simple Michaelis–Menten mechanism that is suitable for handling single-molecule reactions (Fig. 4.6b) does not consider the enzyme–product complex $\mathbf{E}\cdot\mathbf{P}$ explicitly, but instead introduces a conformational change $\mathbf{E}^0 \to \mathbf{E}$ of the enzyme molecule after product release [316, 355], and this introduces a second linkage class:

$$\mathbf{S} + \mathbf{E} \underset{l_1}{\overset{k_1'}{\rightleftharpoons}} \mathbf{S}\cdot\mathbf{E} \underset{l_2'}{\overset{k_2}{\rightleftharpoons}} \mathbf{E}^0 + \mathbf{P} \,, \qquad \mathbf{E}^0 \underset{l_3}{\overset{k_3}{\rightleftharpoons}} \mathbf{E} \,. \tag{4.101}$$

Again we use primes to simplify the notation for the rate parameters under pseudo-first order conditions: $k_1 = k_1' s_0$ and $l_2 = l_2' p_0$. Empirical evidence shows that the assumption of irreversible reaction steps 2 and 3 with $l_2 \approx 0$ and $l_3 \approx 0$ fits the available data well, and therefore basic features of the original Michaelis–Menten equation (4.13d) are retained. In single molecule enzymology, it is reasonable to assume that individual turnovers do not substantially change the substrate concentration $[\mathbf{S}] = s_0 = n_0$. Then, the linear ODE describing the probabilities for the single enzyme molecule to be in one of the three states has only two degrees of freedom, because of the conservation relation $P_{\mathbf{E}} + P_{\mathbf{C}} + P_{\mathbf{E}^0} = 1$:

$$\begin{aligned}
\frac{\mathrm{d}P_{\mathbf{E}}(t)}{\mathrm{d}t} &= +k_3 P_{\mathbf{E}^0}(t) + l_1 P_{\mathbf{C}}(t) - k_1 P_{\mathbf{E}}(t) \,, \\
\frac{\mathrm{d}P_{\mathbf{C}}(t)}{\mathrm{d}t} &= +k_1 P_{\mathbf{E}}(t) - (k_2 + l_1) P_{\mathbf{C}}(t) \,, \\
\frac{\mathrm{d}P_{\mathbf{E}^0}(t)}{\mathrm{d}t} &= +k_2 P_{\mathbf{C}}(t) - k_3 P_{\mathbf{E}^0}(t) \,.
\end{aligned} \tag{4.102}$$

This linear system of ODEs can be solved exactly and the conservation relation reduces the problem to two dimensions. In particular, eigenvalues and eigenvectors can be obtained by a quadratic equation. Since the stochastic variables can assume only two values, $\mathcal{X}_S \in \{0, 1\}$ with $S \in \{\mathbf{E}, \mathbf{C}, \mathbf{E}_0\}$, the probabilities are identical with the expectation values:

$$\mathrm{E}\big(\mathcal{X}_S(t)\big) = 0 \times P_0^{(S)}(t) + 1 \times P_1^{(S)}(t) = P_1^{(S)}(t) .$$

Figure 4.38 shows the solution curves of (4.102) with and without the assumption of vanishing k_3.

Instead of dwelling further on the solutions of (4.102), we shall use it to study the first enzyme turnover cycle: with the assumption $k_3 = 0$, i.e., no recovery of the enzyme, the equations are tailored for calculating a first passage time, the time $\mathcal{T}$, which measures the time of completion of the first turnover cycle. In other words, $\mathcal{T}$ is the time until the enzyme molecule is for the first time in the conformation $\mathbf{E}^0$. This first cycle completion time $\mathcal{T}$ is a random variable with the density $f_{\mathcal{T}}(t)$:

$$f_{\mathcal{T}}(t)\,\mathrm{d}t = P(t < \mathcal{T} \le t + \mathrm{d}t) , \quad \text{with} \quad \int_0^{\infty} f_{\mathcal{T}}(t)\,\mathrm{d}t = 1 . \tag{4.103}$$

We can also interpret the density $f_{\mathcal{T}}(t)\Delta t$ as the probability that the enzyme molecule reaches the conformation $\mathbf{E}^0$ in the time interval between t and $t + \Delta t$, which can be easily calculated:

$$\Delta P_{\mathbf{E}^0}(\Delta t) = k_2 P_{\mathbf{C}} \Delta t , \qquad \lim_{\Delta t \to 0} \frac{\Delta P_{\mathbf{E}^0}}{\Delta t} = \frac{\mathrm{d}P_{\mathbf{E}^0}}{\mathrm{d}t} = k_2 P_{\mathbf{C}} = f_{\mathcal{T}}(t) .$$

The solution of (4.102) implies [355]

$$\begin{aligned} f_{\mathcal{T}}(t) &= \frac{k_1 k_2}{2\beta}\left(\mathrm{e}^{(\alpha+\beta)t} - \mathrm{e}^{(\alpha-\beta)t}\right) , \\ \alpha = -\frac{k_1 + l_1 + k_2}{2} , &\qquad \beta = \sqrt{(k_1 + l_1 + k_2)^2/4 - k_1 k_2} . \end{aligned} \tag{4.104}$$

The waiting time $f_{\mathcal{T}}(t)$ is a superposition of two exponential curves, a faster rising exponential and a slower decaying one (Fig. 4.39).

In the limit of irreversibility of the binding reaction, i.e., $\lim l_1 \to 0$, the mechanism is simply the sequence of one pseudo-first order and one first order reaction step. The waiting time distribution becomes the convolution of the waiting times for the two individual steps: $f_{\mathcal{T}}(t) = (f_1 * f_2)(t)$ or $f_{\mathcal{T}}(t) = \int_0^t \mathrm{d}\tau f_1(t-\tau) f_2(\tau)$, provided that $\mathbf{S} + \mathbf{E} \to \mathbf{S} \cdot \mathbf{E}$ and $\mathbf{S} \cdot \mathbf{E} \to \mathbf{E}^0 + \mathbf{P}$ are Poisson processes with densities $f_1 = k_1 \exp(-k_1 t)$ and $f_2 = k_2 \exp(-k_2 t)$. We obtain

$$f_{\mathcal{T}}(t) = \frac{k_1 k_2}{k_2 - k_1}\left(\mathrm{e}^{-k_1 t} - \mathrm{e}^{-k_2 t}\right) . \tag{4.104$'$}$$

The faster exponential is the rising function and the slower exponential represents the decaying function.

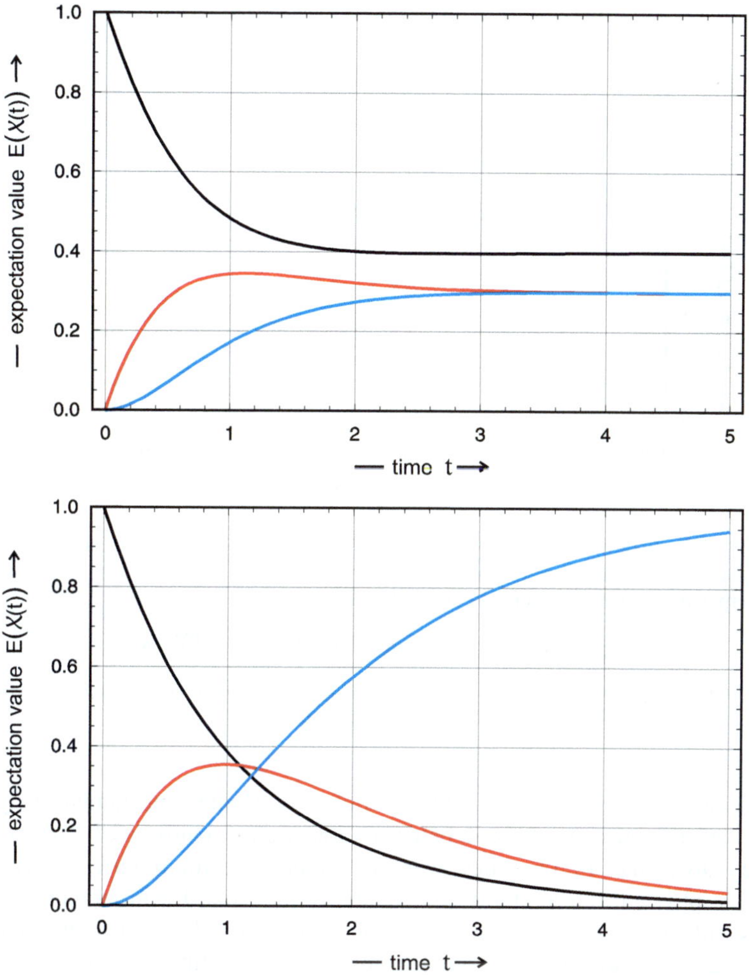

Fig. 4.38 Single enzyme turnover. The plots illustrate the single enzyme molecule mechanism (4.101). Since the stochastic variables are restricted to the values $\{0, 1\}$, the expectation value coincides with the probability of the value $\mathcal{X}_S = 1$, $\mathrm{E}(\mathcal{X}_S) = \sum_n nP_n^{(S)} = P_1^{(S)}$ and (4.102) describes the evolution of the expectation values. In the *upper plot*, we show the equilibration of the three variables in the case of multiple turnovers. The *lower plot* concerns the completion of a single turnover that is achieved by setting $k_3 = 0$. It shows the integration of (4.102) with no recovery of the enzyme being tantamount to setting $k_3 = 0$. The enzyme cycle is arrested in the state $\mathcal{X}_{\mathbf{E}_0} = 1$. Parameter choice: $k_1 = k_2 = 1$; *upper plot* $k_3 = 1$, $l_1 = 0.3$; *lower plot* $k_3 = 0$, $l_1 = 0.1$, all rate parameters in $[\mathrm{t}^{-1}]$. Color code: $[\mathbf{E}]$ *black*, $[\mathbf{C}] = [\mathbf{S} \cdot \mathbf{E}]$ *red*, and $[\mathbf{E}_0]$ *blue*

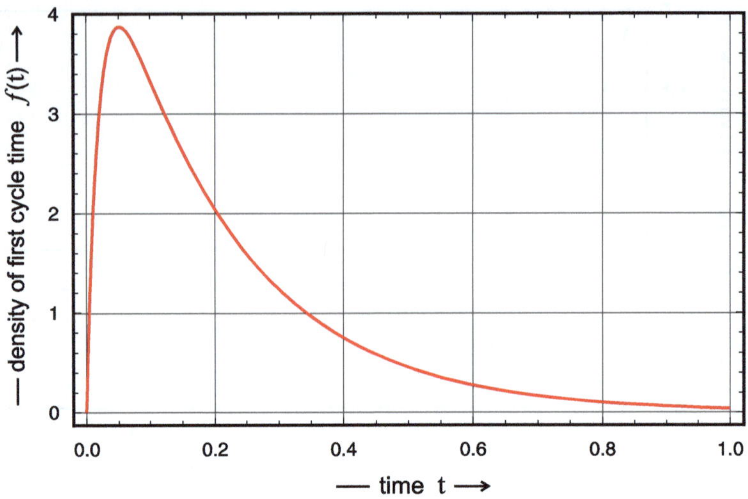

Fig. 4.39 Density of the first cycle waiting time $\mathcal{T}$. The plot shows the density of the time $\mathcal{T}$ required to complete the first turnover cycle $\mathbf{E} \rightleftharpoons \mathbf{C} \rightarrow \mathbf{E}^0 \rightarrow \mathbf{E}$, which is represented by the superposition of two exponential curves (4.104′):

$$f(t) = \frac{\alpha\beta}{\beta - \alpha}\big(\exp(-\alpha t) - \exp(-\beta t)\big), \text{ with } \alpha = k_1[\mathbf{S}] \text{ and } \beta = k_2 \, .$$

This definition requires $\alpha \neq \beta$ and implies that the fast exponential is going up, while the second one goes down, since the denominator changes sign at $\alpha = \beta$

The two extensions of the simple Michaelis–Menten mechanism (Fig. 4.56A and B) are reminiscent of the debate on allosteric mechanisms of enzyme control in the 1960s and 1970s. They were developed for multimeric proteins,[58] but the basic physical interpretation is essentially the same for monomeric enzymes. The Koshland–Némethy–Filmer mechanism of *induced fit*, postulated by Daniel Koshland [315], and worked out for cooperative binding together with George Némethy and David Filmer, assumes that the protein changes shape on ligand binding. In other words, there is only a single conformation for free protein molecules, and all complexes with ligands have their specific protein structures. The Monod–Wyman–Changeux mechanism [403], named after Jacques Monod, Jeffries Wyman, and Jean-Pierre Changeux, is based on the assumption that two or more enzyme conformations also exist in the absence of the ligand and that ligand binding shifts the equilibrium between the states of the protein. (For a more detailed but still short account on the Koshland and the Monod mechanisms, see [164, pp. 291–304].) Comparing with the two extensions of the Michaelis–Menten mechanism discussed here, we recognize Koshland's principle of induced fit in scheme **A**, and the Monod mechanism reflected by the two protein conformations $\mathbf{E}$ and $\mathbf{E}_0$ in

[58] Multimeric proteins contain several, identical or different, subunits. The protein in focus was hemoglobin, which is a tetramer.

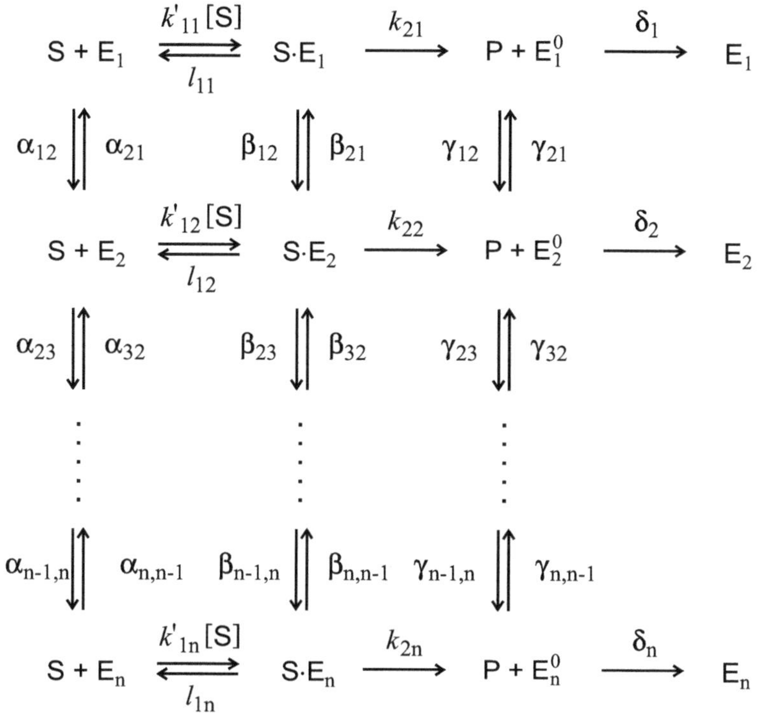

Fig. 4.40 A multistate model for enzyme reactions. The extended Michaelis–Menten mechanism (Fig. 4.6b) is augmented by the assumption that the enzyme molecule can exist in a multitude of n distinct conformations which differ in their kinetic constants. The current theory of protein folding [433] predicts the existence of a multitude of hierarchically ordered conformations and single molecule experiments are consistent with it [316]

scheme B. In contrast to binding in monomeric enzymes, cooperative binding in multimeric proteins can lead to reaction rate profiles $v([S])$ that are incompatible with the Michaelis–Menten mechanism [41, p. 267] or [42, p. 291], because they are S-shaped or sigmoid in some cases.

A multitude of distinct enzyme conformations differing in the rate parameters (Fig. 4.40) gives rise to *dynamical disorder*: both the enzyme and the enzyme–substrate complex fluctuate randomly between n different conformational states, but the general form of the Michaelis–Menten equation is nevertheless retained at the ensemble level under a variety of conditions [316]. Some conditions of general relevance are:

(i) In the limit in which the interconversion rates of the enzyme–substrate complexes $\mathbf{S} \cdot \mathbf{E}_k$ are slower than the catalytic rates, i.e., $(\beta_{ij}) < (k_{2j})$ for $i, j = 1, \ldots, n$.
(ii) In the limit in which the interconversion rates between enzyme conformations are much larger than all other rates.

(iii) In the limit in which all Michaelis constants for individual reaction channels are practically the same, i.e., $(k_{21} + l_{11})/k_{11} \approx (k_{22} + l_{12})/k_{12} \approx \ldots \approx (k_{2n} + l_{1n})/k_{1n}$.

If the first condition is satisfied and the interconversion rates between the enzyme conformations are sufficiently small, the disorder becomes quasi-stationary, and then the density of the waiting time can be approximated by a linear superposition of channel waiting times, viz.,

$$f_T(t) = \frac{1}{\sum_{i=1}^{n} w_i} \sum_{i=1}^{n} w_i \frac{k_1 k_2}{2\beta_i} \left(e^{(\alpha_i+\beta_i)t} - e^{(\alpha_i-\beta_i)t} \right) , \tag{4.105}$$

where the coefficients w_i define the weights with which the individual channels contribute to the waiting time distribution in the ensemble.

Within the last decade, dynamic disorder in enzyme reactions has been verified and analyzed in many single molecule experiments (see, e.g., [125, 126, 337, 544]), and the results thereby obtained fit well the current theory of protein folding [187, 433]. Finally, we mention that single molecule enzymology sheds new light on the mechanism of allosteric regulation of monomeric enzymes [243] and has given a clear hint that the different enzyme conformations also exist in the absence of binding partners.

4.4.2 *Fluorescence Correlation Spectroscopy*

Correlation spectroscopy aims at measuring the fluctuations of a spectroscopic signal at thermodynamic equilibrium. Consequently, interpretations require a firm understanding of the theory of stochastic processes. Fluctuations are measured in a given volume V, so by the $\sqrt{N}$-law, the smaller the number of molecules in the sample volume, the greater the relative amplitude of fluctuations. This implies that small volumes and low concentrations will facilitate such observations. In essence, fluorescence correlation spectroscopy (FCS) measures the number of molecules in a defined volume as a function of time. A few decades ago, fluctuation spectroscopy was practically hopeless since the signals were too weak. But two basic technical advances in fluorescence spectroscopy and microscopy have made it possible to observe and evaluate fluorescence fluctuations: (i) application of high-power lasers and (ii) confocal microscopy.

The spectacular improvement in laser technology has raised the signal-to-noise ratio by several orders of magnitude, and as mentioned this makes it possible to record signals from single molecules. The second breakthrough concerns the invention of the confocal microscope, which provides a way to confine the molecule to be observed to very small volumes. Recordings in volumes of about 1×10^{-15} l are now possible, corresponding to cubes with edge 1 μm. The autocorrelation function (Sect. 3.1.6) is accessible to suitable experiments, because technical devices called *autocorrelators* have been built [439] which directly record the autocorrelation by data sampling of the process under investigation.

The quantity that is commonly derived from fluctuation measurements is a characteristic time, either the relaxation time of a chemical reaction, the relaxation time of a translational or rotational diffusion process, or the residence time of a flow in the volume of observation. The theoretical basis for computating rate parameters or diffusion coefficients from fluorescence correlation data is the fluctuation–dissipation theorem: the parameters which determine the linear return to equilibrium of the system after a macroscopic perturbation are identical to the rates at which spontaneous fluctuations decay [319]. Originally, fluorescence correlation spectroscopy was used to measure relaxation times of chemical reactions of the class $\mathsf{A} + \mathsf{B} \to \mathsf{C}$, in particular the binding of a fluorescent dye to a biomolecule, e.g., *ethidium bromide* to DNA [363]. Since a chemical reaction is almost always coupled to diffusion, fluorescence correlation provides information on both binding parameters and diffusion constants.

The lower time limit for processes that can be observed by fluorescence is given by the rate of fluorescence excitation and emission of the photon. This basic photophysical process leads to the *antibunching term* in the autocorrelation function:

$$G_{\mathrm{F}}(\tau) = 1 - A_{\mathrm{F}} \exp(-\tau/\tau_{\mathrm{F}}) \ , \quad A_{\mathrm{F}} = 1 \ .$$

The excited state need not emit the fluorescence photon. It can also undergo a transition to a non-fluorescent or *dark* triplet state, and this yields another term in the autocorrelation function:

$$G_{\mathrm{T}}(\tau) = 1 + A_{\mathrm{T}} \exp(-\tau/\tau_{\mathrm{T}}) \ , \quad A_{\mathrm{T}} = \frac{\vartheta}{1-\vartheta} \ ,$$

where ϑ is the fraction of molecules *trapped* in the triplet state. Under commonly satisfied conditions, the relaxation times satisfy $\tau_{\mathrm{R,D}} \gg \tau_{\mathrm{T}} \gg \tau_{\mathrm{F}}$, the autocorrelation function can be factorized in the sense of Fig. 4.10, and we obtain

$$G(\tau) = G_{\mathrm{F}}(\tau) \times G_{\mathrm{T}}(\tau) \times G_{\mathrm{R,D}}(\tau) \ .$$

In other words, on the characteristic timescales for fluorescence spectroscopy we need only consider the contributions of chemical reactions and diffusion where the diffusion time commonly sets the upper limit for the timescale of observable processes, and accordingly we shall write $G(\tau)$ for $G_{\mathrm{R,D}}(\tau)$.

Theory of Particle Number Correlations

The theory of concentration correlations was developed in the 1960s and 1970s for the application of inelastic light scattering to the chemical kinetics of macromolecular kinetic reactions (see [55] and references therein). The theory was adapted to studies of fluorescence correlation, which in essence involves a drastic reduction of both the reaction volume and the concentration, a few years later [19, 137]. In bulk solutions the four most important processes that can be analyzed by means of the autocorrelation function are: (i) directed laminar flow through the observation volume, (ii) translational diffusion, (iii) rotational diffusion, and (iv) chemical reaction. In addition, diffusion in natural and artificial membranes has been studied and *in vivo* measurements of molecules labelled with fluorescence markers have been made in cells. Under the conditions of fluorescence correlation experiments, diffusion is strongly coupled to chemical reactions, so the fluctuation of concentrations in space and time have to be considered. The role of fluctuations is illustrated here by a brief account of the theory [318].

The sample contains M different chemical species $\mathsf{X}_1, \ldots, \mathsf{X}_M$, represented by random variables $\mathcal{X}_1, \ldots, \mathcal{X}_M$, with concentrations $x_1(\mathbf{r}, t), \ldots, x_M(\mathbf{r}, t)$, which are assumed to be functions of space and time. The sample is assumed to be at thermodynamic equilibrium, i.e., $\langle \mathcal{X}_j(\mathbf{r}, t) \rangle = x_j(\mathbf{r}, t) = \bar{x}_j(\mathbf{r}, t)$, and $\bar{x}_j$ expresses the mean-square fluctuations of $\bar{x}_j(\mathbf{r}, t)$ in a unit volume following Poisson statistics. The fluctuations understood as deviations from the equilibrium values are denoted as in Sects. 4.1.1 and 4.1.3 by

$$\eta_j(\mathbf{r}, t) = \delta x_j(\mathbf{r}, t) = x_j(\mathbf{r}, t) - \bar{x}_j \ ,$$

and used as variables to describe the linear response to displacements from equilibrium. In systems combining diffusion and chemical reactions, the fluctuations satisfy

$$\frac{\partial \eta_i(\mathbf{r}, t)}{\partial t} = D_i \nabla^2 \eta_i(\mathbf{r}, t) + \sum_{j=1}^{M} A_{ij} \eta_j(\mathbf{r}, t) \ , \quad i = 1, \ldots, M \ , \tag{4.106}$$

where D_i is the diffusion coefficient of X_i and $\mathrm{A} = (A_{ij})$ is the relaxation matrix (4.33).

Next the correlations between concentrations have to expressed in terms of the solutions $\eta_i(\mathbf{r}, t)$ of this linear PDE for the appropriate boundary conditions given by the geometry of the fluorescence experiment (Fig. 4.41) and initial conditions

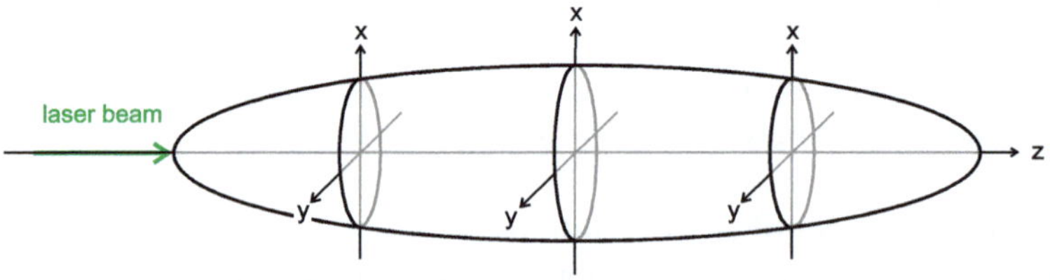

Fig. 4.41 Geometry of fluorescence measurements. A sketch of the beam waist in a fluorescence measurement with a confocal microscope. The active volume element containing the fluorescent sample is a prolate ellipsoid with a Gaussian intensity profile $I(\mathbf{r}) = I_0 \exp\big(-2(x^2+y^2)/w_{xy}^2 - 2z^2/w_z^2\big)$ with $\mathbf{r} = (x, y, z)$ and $w_z > w_{xy} = w_x = w_y$. The laser beam is oriented in the z-direction

$\eta_j(\mathbf{r},0)$ with $j = 1,\ldots,M$. The correlation function of the concentrations of two chemical species X_j and X_l at times t and $t+\tau$ and positions $\mathbf{r}_1$ and $\mathbf{r}_2$, viz.,

$$C_{jl}(\mathbf{r}_1,\mathbf{r}_2,\tau) = \big\langle \eta_j(\mathbf{r}_1,t)\eta_l(\mathbf{r}_2,t+\tau)\big\rangle \,,$$

measures the probability of finding a molecule of species X_j at position $\mathbf{r}_1$ at time t and a molecule of species X_l at position $\mathbf{r}_2$ a time interval $\Delta t = \tau$ later. Three conditions are assumed to be satisfied:

(i) microscopic reversibility,

$$C_{jl}(\mathbf{r}_1,\mathbf{r}_2,\tau) = C_{jl}(\mathbf{r}_1,\mathbf{r}_2,-\tau) = C_{jl}(\mathbf{r}_1,\mathbf{r}_2,|\tau|) \,,$$

(ii) strong or at least weak stationarity,

$$C_{jl}(\mathbf{r}_1,\mathbf{r}_2,\tau) = \big\langle \eta_j(\mathbf{r}_1,0)\,\eta_l(\mathbf{r}_2,\tau)\big\rangle \,, \tag{4.107}$$

(iii) lack of zero-time correlations between the positions of different molecules no matter whether they belong to the same or to different species,

$$\big\langle \eta_j(\mathbf{r}_1,0)\,\eta_k(\mathbf{r}_3,0)\big\rangle = \bar{x}_j\delta_{jk}\delta(\mathbf{r}_1-\mathbf{r}_3) \,. \tag{4.108}$$

The third condition is satisfied in ideal chemical solutions where there is no interaction between molecules except the collisions discussed in Sect. 4.1.4. In other words the correlation lengths are much smaller than the distances between particles.

Solutions $\eta_j(\mathbf{r},t)$ with the initial conditions $\eta_j(\mathbf{r},0)$ are derived by spatial Fourier transform. Inserting the expressions for the transformed derivatives, i.e.,

$$\mathcal{F}\left(\frac{\partial u}{\partial x}\right) = -\mathrm{i}q_x\mathcal{F}(u) \,, \quad \mathcal{F}\left(\frac{\partial^2 u}{\partial x^2}\right) = -q_x^2\mathcal{F}(u) \,,$$

into (4.106) yields a linear ODE that can be readily solved by considering an eigenvalue problem (see Sect. 4.1.3):

$$\frac{\mathrm{d}\hat{x}_i(\mathbf{q},t)}{\mathrm{d}t} = \sum_{j=1}^{M} R_{ij}\hat{x}_j(\mathbf{q},t) \,, \quad \text{with } \mathrm{R} = \big(R_{ij} = A_{ij} - D_iq^2\delta_{ij}\big) \,, \tag{4.109}$$

with the Fourier transform of the concentrations defined by

$$\hat{x}_i(\mathbf{q},t) = \frac{1}{(2\pi)^{3/2}}\int_{-\infty}^{\infty} \mathrm{d}\mathbf{r}\,\mathrm{e}^{\mathrm{i}\mathbf{q}\cdot\mathbf{r}}\eta_i(\mathbf{r},t) \,.$$

Diagonalizing the reaction–diffusion matrix, $\Lambda = \mathrm{B}^{-1}\mathrm{R}\mathrm{B}$ with $\mathrm{B} = (b_{ij})$ and $\mathrm{B}^{-1} = \mathrm{H} = (h_{ij})$, yields the solution in frequency space:

$$\hat{x}_i(\mathbf{q}, t) = \sum_{k=1}^{M} b_{ik}\beta_k(0)\mathrm{e}^{\lambda_k t}\,, \quad \text{with } \beta_k(0) = \sum_{j=1}^{M} h_{kj}\hat{x}_j(\mathbf{q}, 0)\,.$$

Inserting in (4.107) and exchanging Fourier transform and ensemble average yields

$$\begin{aligned}\big\langle \eta_j(\mathbf{r}_1, 0)\eta_l(\mathbf{r}_2, \tau)\big\rangle &= \frac{1}{(2\pi)^{3/2}}\int_{-\infty}^{\infty} \mathrm{d}\mathbf{q}\,\mathrm{e}^{-\mathrm{i}\mathbf{q}\cdot\mathbf{r}}\big\langle \eta_j(\mathbf{r}_1, 0)\hat{x}_l(\mathbf{q}, \tau)\big\rangle \\ &= \frac{1}{(2\pi)^{3/2}}\int_{-\infty}^{\infty} \mathrm{d}\mathbf{q}\,\mathrm{e}^{-\mathrm{i}\mathbf{q}\cdot\mathbf{r}} \sum_{k=1}^{M} b_{lk}\mathrm{e}^{\lambda_k\tau}\sum_{i=1}^{M} h_{ki}\big\langle \eta_j(\mathbf{r}_1, 0)\hat{x}_i(\mathbf{q}, 0)\big\rangle \\ &= \frac{1}{(2\pi)^{3}}\int_{-\infty}^{\infty} \mathrm{d}\mathbf{q}\,\mathrm{e}^{-\mathrm{i}\mathbf{q}\cdot\mathbf{r}} \sum_{k=1}^{M} b_{lk}\mathrm{e}^{\lambda_k\tau}\sum_{i=1}^{M} h_{ki}\int_{-\infty}^{\infty} \mathrm{d}\mathbf{r}\,\mathrm{e}^{\mathrm{i}\mathbf{q}\cdot\mathbf{r}}\big\langle \eta_j(\mathbf{r}_1, 0)\eta_i(\mathbf{r}_3, 0)\big\rangle\,.\end{aligned}$$

Now by (4.108), we get the final result:

$$\begin{aligned}C_{jl}(\mathbf{r}_1, \mathbf{r}_2, \tau) &= \big\langle \eta_j(\mathbf{r}_1, 0)\eta_l(\mathbf{r}_2, \tau)\big\rangle \\ &= \frac{1}{(2\pi)^3}\bar{x}_j\int_{-\infty}^{\infty} \mathrm{d}\mathbf{q}\mathrm{e}^{\mathrm{i}\mathbf{q}\cdot(\mathbf{r}_1-\mathbf{r}_2)}\sum_{k=1}^{M} b_{lk}h_{kj}\exp(\lambda_k\tau)\,.\end{aligned} \tag{4.110}$$

It is easily verified that the correlation function has the expected symmetry properties $C_{jl}(\mathbf{r}_1, \mathbf{r}_2, \tau) = C_{lj}(\mathbf{r}_1, \mathbf{r}_2, \tau)$ and $C_{jl}(\mathbf{r}_1, \mathbf{r}_2, \tau) = C_{jl}(\mathbf{r}_2, \mathbf{r}_1, \tau)$. The correlation function is proportional to the equilibrium concentration and decreases with increasing time delay $\Delta t = \tau$, since the eigenvalues $\lambda_k = -\tau_{R_k}^{-1}$ of the relaxation matrix are negative. In particular, the eigenvalues for diffusion are always negative, i.e., $\lambda = -Dq^2$, and the same is essentially true for chemical reactions, where some of the eigenvalues, but never all of them, might be zero. For vanishing delay, the autocorrelation function becomes a Dirac delta function as expected: $\lim_{\tau\to 0} C_{jj}(\mathbf{r}_1, \mathbf{r}_2, 0) = \bar{x}_j\delta(\mathbf{r}_1 - \mathbf{r}_2)$ (4.108).

Fluorescence Correlation Measurements

The quantity measured in fluorescence experiments is the number of photons $n(t)$ emitted by the sample and collected in the detector:

$$n(t) = \Delta t\int \mathrm{d}\mathbf{r}\,I(\mathbf{r})\sum_{i=1}^{M} Q_i x_i(\mathbf{r}, t)\,.$$

Here, $I(\mathbf{r})$ is the distribution of the light used to excite the sample and Q_i is the specific molecular parameter consisting of two factors: (i) the absorption cross-section and (ii) the fluorescence quantum yield of molecules X_i. Then the fluctuation in the photon count is

$$\delta n(t) = n(t) - \bar{n} = \delta t \int_{-\infty}^{\infty} \mathrm{d}\mathbf{r}\, I(\mathbf{r}) \sum_{i=1}^{M} Q_i \eta_i(\mathbf{r}, t) \,, \tag{4.111}$$

and its average or equilibrium value is obtained by Fourier transform and integration:

$$\bar{n} = \Delta t \int \mathrm{d}\mathbf{r}\, I(\mathbf{r}) \sum_{i=1}^{M} Q_i \bar{x}_i(\mathbf{r}, t) = (2\pi)^{3/2} \hat{I}(0) \Delta t \sum_{i=1}^{M} Q_i x_i(\mathbf{r}, t) \,,$$

where $\hat{I}(\mathbf{q}) = (2\pi)^{-3/2} \int \mathrm{d}\mathbf{r}\, \mathrm{e}^{-\mathrm{i}\mathbf{q}\cdot\mathbf{r}} I(\mathbf{r})$. Making use of the ergodicity of the system, we can write the fluorescence autocorrelation function as

$$\begin{aligned} G(\tau) &= \frac{1}{\bar{n}^2} \big\langle \delta n(0) \delta n(\tau) \big\rangle \\ &= \frac{(\Delta t)^2}{\bar{n}^2} \int \mathrm{d}\mathbf{r}_1\, I(\mathbf{r}_1) \int \mathrm{d}\mathbf{r}_2\, I(\mathbf{r}_2) \sum_{j,l} Q_j Q_l \big\langle \eta_j(\mathbf{r}_1, 0) \eta_l(\mathbf{r}_2) \big\rangle \\ &= \frac{(\Delta t)^2}{\bar{n}^2} \int \mathrm{d}\mathbf{q}\, |I(\mathbf{q})|^2 \sum_{j,l} Q_j Q_l \bar{x}_j \sum_{k=1}^{M} b_{lk} h_{kj} \mathrm{e}^{\lambda_k \tau} \,. \end{aligned}$$

The expression is completed by inserting a Gaussian intensity profile for the illumination of the sample (Fig. 4.41):

$$I(\mathbf{r}) = I_0 \exp\left(-\frac{2(x^2 + y^2)}{w_{xy}^2} - \frac{2z^2}{w_z^2} \right) , \tag{4.112}$$

which has the shape of a prolate ellipsoid with the shorter axes in the x- and y-direction and the longer axis in the z-direction, so $w_x = w_y = w_{xy} < w_z$, and $\omega = w_z / w_{xy} \gg 1$. Fourier transformation yields

$$I(\mathbf{q}) = \frac{I_0 w_{xy} w_z}{8} \exp\left(-\frac{w_{xy}^2}{8} (q_x^2 + q_y^2) - \frac{w_z^2}{8} z^2 \right) ,$$

and eventually we obtain the final equation for the autocorrelation function:

$$G(\tau) = \frac{1}{(2\pi)^3} \frac{1}{\left(\sum_{i=1}^{M} Q_i \bar{x}_i\right)^2} \times \int_{-\infty}^{\infty} \mathrm{d}\mathbf{q} \exp\left(-\frac{w_{xy}^2}{4}(q_x^2 + q_y^2) - \frac{w_z^2}{4} q_z^2\right) \sum_{j=1}^{M} \sum_{l=1}^{M} Q_j Q_l \bar{x}_j \sum_{k=1}^{M} b_{lk} h_{kj} \mathrm{e}^{\lambda_k \tau} . \tag{4.113}$$

We remark that according to (4.109), the eigenvalues λ_k and the eigenvectors depend on $\mathbf{q}$, and for each particular case the $\mathbf{q}$-dependence has to be calculated from the relaxation dynamics.

Examples of Fluorescence Correlations

The simplest conceivable example analyzed by fluorescence correlation is the diffusion of a single chemical species X with concentration $[\mathrm{X}] = x(\mathbf{r}, t)$ and $\eta(\mathbf{r}, t) = x(\mathbf{r}, t) - \bar{x}$. Equation (4.106) becomes a simple diffusion equation:

$$\frac{\partial \eta(\mathbf{r}, t)}{\partial t} = D\nabla^2 \eta(\mathbf{r}, t) , \qquad \hat{\eta}(\mathbf{q}, t) = \hat{\eta}(\mathbf{q}, 0) \exp(-Dq^2 t) .$$

The single eigenvalue of the matrix R is $\lambda = -Dq^2$, and the eigenvector is trivially $b = h = 1$. Inserting into (4.113) yields

$$\begin{aligned} G(\tau) &= \frac{(2\pi)^{-3}}{Q^2 \bar{x}^2} \int_{-\infty}^{\infty} \mathrm{d}\mathbf{q}\, Q^2 \bar{x} \exp\left(-\frac{w_{xy}^2}{4}(q_x^2 + q_y^2) - \frac{w_z^2}{4} q_z^2 - D(q_x^2 + q_y^2 + q_z^2) t\right) \\ &= \frac{1}{\bar{N}} \left(1 + \frac{\tau}{\tau_D}\right)^{-1} \left(1 + \frac{\tau}{\omega^2 \tau_D}\right)^{-1/2} , \qquad \bar{N} = \bar{x} V , \end{aligned} \tag{4.114}$$

where $\bar{N}$ is the number of molecules X in the effective sampling volume $V = \pi^{3/2} w_{xy}^2 w_z$, $\tau_D = w_{xy}^2 / 4D$ is the characteristic diffusion time across the illuminated ellipsoid, and $\omega^2 \tau_D = w_z^2 / 4D$ is the diffusion time along the ellipsoid. Each degree of freedom in diffusion contributes a factor $(1 - \tau/\tau_D')^{-1/2}$, where $\tau_D' = (\omega')^2 \tau_D$, and ω' is a factor depending on the geometry of the illuminated volume. For an extended prolate ellipsoid, we have $w_z \gg w_{xy}$ and then the autocorrelation function for diffusion in two dimensions, viz.,

$$G(\tau) = \frac{1}{\bar{N}} \left(1 + \frac{\tau}{\tau_D}\right)^{-1} , \tag{4.115}$$

is also a good approximation for the 3D case. The relaxation of the fluctuation of the number of molecules in the sampling volume is approximately determined by the diffusion in the smaller dimensions. Recording the autocorrelation function provides two results: (i) $G(0) = \bar{N}_{\mathsf{X}}^{-1}$, the number of particles in the beam waist, and (ii) $D = w_{xy}^2/4\tau_D$ the translational diffusion coefficient of X.

The extension to M diffusing chemical species, $\mathsf{X}_1, \ldots, \mathsf{X}_M$, is straightforward [318]:

$$G(\tau) = \frac{1}{\left(\sum_{i=1}^{M} Q_i \bar{N}_i\right)^2} \sum_{j=1}^{M} Q_j^2 \bar{N}_j \left(1 + \frac{\tau}{\tau_{D_j}}\right)^{-1} \left(1 + \frac{\tau}{\omega^2 \tau_{D_j}}\right)^{-1/2} . \tag{4.116}$$

The amplitude of the contribution of each species is weighted by its fluorescence quantum yield Q_j, $\bar{N}_j$ is the mean number of molecules X_j in the beam waist, and D_j is its diffusion coefficient.

The coupling between the translational diffusion and chemical reactions leads to a more complex expression for the autocorrelation function. An example of an excellent theoretical and experimental treatment is found in the literature [7], namely, the formation of inclusion complexes of pyronines with cyclodextrin:

$$\mathsf{G} + \mathsf{H} \underset{l}{\overset{k}{\rightleftharpoons}} \mathsf{C} , \quad \text{with } K = l/h .$$

The fluorescent guest molecule **G** binds to the non-fluorescent host **H** and forms a fluorescent inclusion complex **C**. Conditions are chosen under which the host concentration is much higher than the guest concentration, i.e., $[\mathsf{H}_0] \approx [\mathsf{H}] \gg [\mathsf{G}]$. It is useful to introduce a mean diffusion time $\bar{\tau}_D$, which is calculated from a weighted mean diffusion coefficient:

$$\bar{\tau}_D = \frac{w_{xy}^2}{4\bar{D}} , \quad \text{with } \bar{D} = x_{\mathsf{G}} D_{\mathsf{G}} + x_{\mathsf{H}} D_{\mathsf{H}} ,$$

where $x_{\mathsf{G}} = N_{\mathsf{G}}/(N_{\mathsf{G}} + N_{\mathsf{C}})$ and $x_{\mathsf{C}} = N_{\mathsf{C}}/(N_{\mathsf{G}} + N_{\mathsf{C}})$. Then the autocorrelation function is of the form

$$G_{\mathrm{R}}(\tau) = \frac{1}{N_{\mathsf{G}} + N_{\mathsf{C}}} \left(1 + \frac{\tau}{\bar{\tau}_D}\right)^{-1} \left(1 + \frac{\tau}{\omega^2 \bar{\tau}_D}\right)^{-1/2} \left(1 + A_{\mathrm{R}} e^{-\tau/\tau_{\mathrm{R}}}\right) , \tag{4.117}$$

where the relaxation amplitude and relaxation time are given by

$$A_{\mathrm{R}}\big([\mathsf{H}]\big) = \frac{N_{\mathsf{G}} N_{\mathsf{C}} (Q_{\mathsf{G}} - Q_{\mathsf{C}})^2}{(Q_{\mathsf{G}} N_{\mathsf{G}} + Q_{\mathsf{C}} N_{\mathsf{C}})^2} = \frac{K[\mathsf{H}]_0 (1 - Q)^2}{(1 + QK[\mathsf{H}]_0)^2} ,$$

$$\tau_{\mathrm{R}}\big([\mathsf{H}]\big) = \big(l(1 + K[\mathsf{H}]_0)\big)^{-1} , \quad \text{with } Q = \frac{Q_{\mathsf{C}}}{Q_{\mathsf{G}}} , \quad K[\mathsf{H}]_0 = \frac{N_{\mathsf{C}}}{N_{\mathsf{G}}} .$$

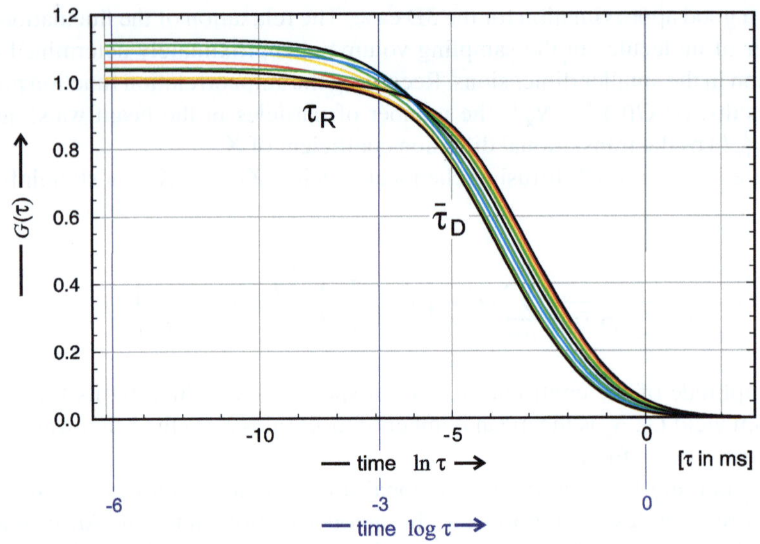

Fig. 4.42 Inclusion complexes of pyronines in cyclodextrin. The autocorrelation curves $G(\tau)$ were calculated from (4.117) with the parameters given in [7]: $N_{\mathbf{G}} + N_{\mathbf{C}} = 1$, $\tau_{\mathbf{G}} = 0.25\,\text{ms}$, $\tau_{\mathbf{C}} = 0.60\,\text{ms}$, $\omega = 5$, $K = 2\,\text{mM}^{-1}$, $Q = 0.5$, and $h = 500\,\text{ms}^{-1}$, and the cyclodextrin concentrations $[\mathbf{H}]_0$ were 12 (*black*), 6 (*red*), 3 (*yellow*), 2 (*green*), 1 (*black*), 0.5 (*green*), 0.3 (*blue*), 0.1 (*green*), 0.03 (*yellow*), 0.01 (*red*), and 0 mM (*black*) [7]

The relaxation curves were recalculated with the parameter values given in [7] and the result is shown in Fig. 4.42. The family of curves calculated for different values of the total cyclodextrin concentration $[\mathbf{H}]_0$ shows two relaxation processes, the faster one corresponding to the association reaction with a relaxation time τ_{R} and the slower process caused by diffusion of the two fluorescent species, the guest molecule **G** and the inclusion complex **C**. The amplitude of the chemical relaxation process $A_{\mathrm{R}}([\mathbf{H}])$ increases first with increasing cyclodextrin concentration, $A_{\mathrm{R}}([\mathbf{H}]) \approx K[\mathbf{H}]_0(1 - Q)^2$ for small values of $[\mathbf{H}]$, passes a maximum at $[\mathbf{H}]_0 = 1/QK$, and then decreases according to $A_{\mathrm{R}}([\mathbf{H}]) \approx (1 - Q)^2/Q^2[\mathbf{H}]_0$ for large $[\mathbf{H}]_0$ values. Coupling of the chemical reaction with the diffusion process gives results in a non-monotonic dependence of the relaxation amplitude on the host concentrations.

Provided that the parameter can be successfully estimated (see Sect. 4.1.5), fluorescence correlation spectroscopy allows for the determination of data that is otherwise hard to obtain:

(i) The local concentration in the beam waist through $G(0)$.
(ii) The local translational diffusion coefficients from diffusion relaxation times $\tau_D = w_{xy}^2/4D$.
(iii) The relaxation times of chemical reactions τ_{R}.

Rotational diffusion constants can also be derived from fluorescence correlation [128, 569] and provide direct information about the size of molecules. In particular, the formation of molecular aggregates can be detected by determining the molecular radius. Technical advances in laser techniques and microscopes have allowed for a dramatic increase in resolution, and autocorrelation data from single molecules can now be detected [345, 387, 466].

4.5 Scaling and Size Expansions

Master equations when applied to real world chemical systems encounter serious limitations with respect to both analytic solvability and numerical simulation. As we have seen, the analytical approach already becomes extremely sophisticated for simple single-step bimolecular reactions (Sect. 4.3.3), and numerical simulations cannot be carried out with reasonable resources when particle numbers become large (see Sect. 4.6). In contrast Fokker–Planck and stochastic differential equations are much easier to handle and accessible to upscaling. In the section dealing with chemical Langevin equations (Sect. 4.2.4), we discussed the approximations that allow for a transition from discrete particle numbers to continuous concentrations.

In this section we shall discuss ways to relate chemical master equations to Fokker–Planck equations. In particular, we shall solve master equations through approximation methods based on expansions in suitable parameters, as already mentioned for one case in Sect. 4.2.2, where we expanded the master equations in Taylor series with jump moments as coefficients. Truncation after the second term yields a Fokker–Planck equation. It is important to note that every diffusion process can be approximated by a jump process, but the reverse is not true. Similarly to the transition from master to Langevin equations, there are master equations for which no approximation by a Fokker–Planck equation exists. A particularly useful expansion technique based on system size has been introduced by the Dutch theoretical physicist Nico van Kampen [540, 541]. This expansion method can be used, for example, to handle and discuss fluctuations without calculating solutions with full population sizes.

4.5.1 Kramers–Moyal Expansion

The two physicists Hendrik Anthony Kramers and José Enrique Moyal proposed a general expansion of master equations, which is a kind of Taylor expansion in jump moments (Sect. 3.2.3) applied to the integral equivalent of the master equation, viz.,[59]

$$\frac{\partial P(x,t)}{\partial t} = \int \mathrm{d}z\Big(W(x|z,t)P(z,t) - W(z|x,t)P(x,t)\Big)\,, \tag{4.118}$$

The starting point is the probability of the transition from the probability density at time t to the probability density at time $t+\tau$:

$$P(x,t+\tau) = \int \mathrm{d}z W(x,t+\tau|z,t)P(z,t)\,. \tag{4.119}$$

We aim to derive an expression for the differential dP, which requires knowledge of the transition probabilities $W(x,t+\tau|z,t)$, at least for small τ, and knowledge of the jump moments $\alpha_n(z,t,\tau)$:

$$\alpha_n(z,t,\tau) = \Big\langle\big(\mathcal{X}(t+\tau)-\mathcal{X}(t)\big)^n\Big\rangle\Big|_{\mathcal{X}(t)=z} = \int \mathrm{d}x(x-z)^n W(x,t+\tau|z,t)\,. \tag{4.120}$$

Implicitly, $\mathcal{X}(t) = z$ is assumed, implying a sharp value of the random variable $\mathcal{X}(t)$ at time t. Next we introduce $z = x - \Delta x$ into the integrand in (4.118) and expand in a Taylor series in Δx around the value $x + \Delta x$:

$$\begin{aligned} W(x,t&+\tau|z,t)P(z,t) \\ &= W\big((x-\Delta x)+\Delta x, t+\tau|(x-\Delta x),t\big)P\big((x-\Delta x),t\big) \\ &= \sum_{n=0}^{\infty}\frac{(-\Delta x)^n}{n!}\frac{\partial^n}{\partial x^n}\Big(W(x+\Delta x,t+\tau|x,t)P(x,t)\Big)\,. \end{aligned}$$

Inserting in (4.119) and integrating yields

$$\begin{aligned} P(x,t+\tau) &= \int \mathrm{d}(\Delta x)\sum_{n=0}^{\infty}\frac{(-\Delta x)^n}{n!}\frac{\partial^n}{\partial x^n}\Big(W(x+\Delta x,t+\tau|x,t)P(x,t)\Big) \\ &= \sum_{n=0}^{\infty}\frac{(-1)^n}{n!}\frac{\partial^n}{\partial x^n}\int \mathrm{d}(\Delta x)(\Delta x)^n\Big(W(x+\Delta x,t+\tau|x,t)P(x,t)\Big) \\ &= \sum_{n=0}^{\infty}\frac{(-1)^n}{n!}\frac{\partial^n}{\partial x^n}\alpha_n(x,t,\tau)P(x,t)\,. \end{aligned}$$

[59] A comprehensive presentation of different ways to derive series expansions leading to the Fokker–Planck equation can be found in [468, pp. 63–76].

To derive a convenient expression, we perform also a Taylor expansion of the jump moments, i.e.,

$$\frac{\alpha_n(x,t,\tau)}{n!} = \sum_{k=0}^{\infty} \frac{\tau^k}{k!} \Theta_k^{(n)} , \quad \text{with } \Theta_k^{(n)} = \frac{1}{n!} \frac{\partial^k \alpha_n}{\partial \tau^k} ,$$

then truncate after the linear term in τ. Since $\Theta_0^{(n)}$ has to vanish, because the transition probability satisfies the initial condition $W(x,t|x-\Delta x,t) = \delta(\Delta x)$, we find

$$\frac{\alpha_n(x,t,\tau)}{n!} = \Theta_1^{(n)} \tau + \mathcal{O}(\tau^2) ,$$

where the linear term carries the only nonzero coefficient. Therefore we can drop the subscript to write $\Theta^{(n)} \equiv \Theta_1^{(n)}$, move the term with $n = 0$ to the left-hand side, and divide by τ :

$$\frac{P(x,t+\tau) - P(x,t)}{\tau} = \sum_{n=1}^{\infty} (-1)^n \frac{\partial^n}{\partial x^n} \Big(\Theta^{(n)} P(x,t) \Big) .$$

Taking the limit $\tau \to 0$ finally yields the expansion of the master equation:

$$\frac{\partial P(x,t)}{\partial t} = \sum_{n=1}^{\infty} (-1)^n \frac{\partial^n}{\partial x^n} \Big(\Theta^{(n)} P(x,t) \Big) .$$

We remark that the above derivation corresponds to a forward stochastic process, and in addition to this forward expansion, there is also a backward Kramers–Moyal expansion.

Assuming explicit time independence of the transition matrix and the jump moments, we obtain the conventional form of the Kramers–Moyal expansion:

$$\frac{\partial P(x,t)}{\partial t} = \sum_{n=1}^{\infty} \frac{(-1)^n}{n!} \frac{\partial^n}{\partial x^n} \Big(\alpha_n(x) P(x,t) \Big) ,$$
$$\text{with } \alpha_n(x) = \int_{-\infty}^{\infty} (z-x)^n W(x, z-x) \mathrm{d}z . \tag{4.121}$$

When the Kramers–Moyal expansion is terminated after the second term, the result is a Fokker–Planck equation of the form

$$\frac{\partial P(x,t)}{\partial t} = -\frac{\partial}{\partial x} \Big(\alpha_1(x) P(x,t) \Big) + \frac{1}{2} \frac{\partial^2}{\partial x^2} \Big(\alpha_2(x) P(x,t) \Big) . \tag{4.122}$$

The two jump moments represent the conventional drift and diffusion terms $\alpha_1(x) \equiv A(x)$ and $\alpha_2(x) \equiv B(x)$. We remark that we did not use the condition of a one-step birth-and-death process anywhere, and therefore (4.122) is generally valid.

4.5.2 *Small Noise Expansion*

For large particle numbers, noise satisfying a $\sqrt{N}$-law may be very small, and this can be taken advantage of by making *small noise expansions* of stochastic differential and Fokker–Planck equations. Then the SDE can be written as

$$dx = a(x)\,dt + \varepsilon b(x) dW(t) , \tag{4.123a}$$

where the solution is assumed to be of the form

$$x_\varepsilon(t) = x_0(t) + \varepsilon x_1(t) + \varepsilon^2 x_2(t) + \cdots . \tag{4.123b}$$

Solutions can be derived term by term and $x_0(t)$, for example, is the solution of the deterministic differential equation $dx = a(x)\,dt$ with initial condition $x_0(0) = c_0$.

In the small noise limit, a suitable Fokker–Planck equation is of the form

$$\frac{\partial P(x,t)}{\partial t} = -\frac{\partial}{\partial x}\Big(A(x)P(x,t)\Big) + \frac{1}{2}\varepsilon^2\frac{\partial^2}{\partial x^2}\Big(B(x)P(x,t)\Big) , \tag{4.124a}$$

where the variable x and the probability density $P(x,t)$ are scaled

$$\xi = \frac{x - x_0(t)}{\varepsilon} , \qquad P_\varepsilon(\xi,t) = \varepsilon P(x,t|c_0,0) , \tag{4.124b}$$

and the probability density is assumed to be of the form

$$P_\varepsilon(\xi,t) = P_\varepsilon^{(0)}(\xi,t) + \varepsilon P_\varepsilon^{(1)}(\xi,t) + \varepsilon^2 P_\varepsilon^{(2)}(\xi,t) + \cdots . \tag{4.124c}$$

This innocent looking approach has to face two problems:

(i) There is no guarantee that the two expansion series (4.123b) and (4.124c) will converge.
(ii) Explicit calculations based on the series expansions are commonly quite sophisticated [194, pp.169–184].

For the purpose of illustration, we consider one special example, the Ornstein–Uhlenbeck process, which is exactly solvable (see Sect. 3.2.2.3). The stochastic differential equation is of the form

$$dx = -kx\,dt + \varepsilon\,dW(t) . \tag{4.125a}$$

In the limit $\varepsilon \to 0$, the stochastic part disappears, the resulting ODE remains first order in time, and we are dealing with a non-singular limit. The exact solution of (4.125a) for the initial condition $x(0) = c_0$ is

$$x_\varepsilon(t) = c_0 \exp(-kt) + \varepsilon \int_0^t \exp\big(-k(t-\tau)\big)\mathrm{d}W(\tau) . \tag{4.125b}$$

This case is particularly simple since the partitioning according to the series expansion (4.123b) is straightforward, i.e.,

$$x_0(t) = c_0 \exp(-kt) , \qquad x_1(t) = \int_0^t \exp\big(-k(t-\tau)\big)\mathrm{d}W(\tau) ,$$

and $x_0(t)$ is indeed the solution of the ODE obtained by setting $\varepsilon = 0$ in the SDE (4.125a).

Now we consider the corresponding Fokker–Planck equation

$$\frac{\partial P(x,t)}{\partial t} = \frac{\partial}{\partial x}\Big(kxP(x,t)\Big) + \frac{1}{2}\varepsilon^2 \frac{\partial^2 P(x,t)}{\partial t^2} , \tag{4.125c}$$

where the exact solution is a Gaussian with $x_0(t)$ as expectation value, i.e.,

$$E\big(x(t)\big) = \alpha(t) = c_0 \exp(-kt) , \quad \mathrm{var}\big(x(t)\big) = \varepsilon^2\beta(t) = \varepsilon^2 \frac{1-\exp(-2kt)}{2k} , \tag{4.125d}$$

and hence,

$$P_\varepsilon(x,t|c_0,0) = \frac{1}{\varepsilon}\frac{1}{\sqrt{2\pi\beta(t)}} \exp\left(\frac{1}{\varepsilon^2}\frac{\big(x-\alpha(t)\big)^2}{2\beta(t)}\right) . \tag{4.125d$'$}$$

In the limit $\varepsilon \to 0$, we obtain once again the deterministic solution:

$$\lim_{\varepsilon\to 0} P_\varepsilon(x,t|c_0,0) = \delta\big(x-\alpha(t)\big) ,$$

which is the first order solution of the corresponding SDE and a deterministic trajectory along the path $x(t) = c_0 \exp(-kt)$. In the limit $\varepsilon \to 0$, the second order differential equation (4.125c) is reduced to a first order equation. This implies a singularity, and singular perturbation theory has to be applied. The probability density, however, cannot be expanded straightforwardly in a power series in ε, and a scaled variable must first be introduced:

$$\xi = \frac{x-\alpha(t)}{\varepsilon} , \qquad \text{or } x = \alpha(t) + \varepsilon\xi .$$

Now we can write down the probability density in ξ up to second order:

$$P_\varepsilon(\xi,t|0,0) = P_\varepsilon(x,t|c_0,0)\frac{\mathrm{d}x}{\mathrm{d}\xi} = \frac{1}{\sqrt{2\pi\beta(t)}} \exp\left(-\frac{\xi^2}{2\beta(t)}\right) .$$

Scaling has eliminated the singularity, since the probability density for ξ does not contain ε. The distribution of the scaled variable ξ is a Gaussian with mean zero and variance $\beta(t)$. The standard deviation from the deterministic trajectory $\alpha(t)$ is of order ε, as ε goes to zero. The coefficient of ε is the random variable ξ. As expected, there is no difference in interpretation between the Fokker–Planck and the stochastic differential equation.

4.5.3 Size Expansion of the Master Equation

Although quite a few representative examples and model systems can be analyzed by solving one step birth-and-death master equations exactly (Sect. 4.3), the actual applicability of this technique to specific problems of chemical kinetics is rather limited. In order to apply a chemical master equation to a problem in practice, one is commonly dealing with at least 10^{12} particles. Upscaling discloses one particular issue of size expansions, which becomes obvious in the transition from master equations to Fokker–Planck equations. The sample volume V is the best estimator of system size in condensed matter. Two classes of quantities are properly distinguished:

(i) *intensive properties* that are independent of the system size, and
(ii) *extensive properties* that grow in proportion to the system size.

Examples of intensive properties are temperature, pressure, density, or concentrations, whereas volume, particle numbers, energy, or entropy are extensive properties. In upscaling from say 1000 to 10^{12} particles extensive properties grow by a factor of 10^9, whereas intensive properties remain the same. Some *pairs* of properties, one extensive and one intensive, are of particular importance, such as particle number $\mathcal{X}$ or n and concentration $a = \mathcal{X}_\mathsf{A}/VN_L$ or mass M and (volume) density $\varrho = M/V$. The system size used for scaling will be denoted by Ω, and if not stated otherwise we shall assume $\Omega = VN_L$. Properties describing the evolution of the system are modelled by variables, and once again we distinguish extensive and intensive variables. In the case of the amount [A] of a chemical compound, we have the particle number $n(t) \propto \Omega$ as the extensive variable and the concentration $a(t) = n(t)/\Omega$ as the intensive variable, and we indicate this correspondence by $n \widehat{=} a$.[60] The system size Ω itself is, of course, also an extensive property, the special extensive property which has been chosen as reference.

[60]In order to improve clarity in the derivation of the size expansion, we shall use the lowercase letters $a, b, c, \ldots$ for intensive variables and the lowercase letters $n, m, p, \ldots$ for extensive variables. When dealing with atoms, molecules or compounds, intensive variables will be continuous and mostly concentrations, whereas the extensive variables are understood as particle numbers. In order to avoid misunderstanding, we introduce the symbol $\widehat{=}$ to express the relation between conjugate intensive and extensive variables, for example, $\varrho \widehat{=} M$.

Approximation methods have been developed which have turned out to be particularly instructive and useful in the limit of sufficiently large systems. The Dutch theoretical physicist Nico van Kampen [541, 543] expands the master equation in the inverse square root of system size Ω. A discrete random variable $\mathcal{X}_A$ with the probability density $P_n(t) = P\big(\mathcal{X}_A(t) = n(t)\big)$ is considered in the limit to macroscopic description. The limit of interest is a large value of Ω at fixed a, which is tantamount to the transition to a macroscopic system.

The transition probabilities are reformulated as

$$W(n|m) \equiv \omega\big(m; \Delta n\big) , \quad \text{with } \Delta n = n - m ,$$

and scaled according to the assumption

$$W(n|m) = \Omega\omega\Big(\frac{m}{\Omega}; \Delta n\Big) = \Omega\omega\big(a; \Delta n\big) . \tag{4.126}$$

The essential trick in the van Kampen expansion is that the size of the jump is expressed in terms of an extensive quantity Δn, whereas the intensive variable $a = n/\Omega$ is used to calculate the evolution of the system $a(t)$. The expansion is now made in a new variable z defined by

$$a = \Omega\phi(t) + \Omega^{1/2}z , \quad \text{or } z = \Omega^{-1/2}a - \Omega^{1/2}\phi(t) , \tag{4.127}$$

where the function $\phi(t)$ is still to be determined. The change of variables transforms the probability density $\Pi(a,t)$ and its derivatives according to

$$\Pi(a,t) = \Pi\big(\Omega\phi(t) + \Omega^{1/2}z, t\big) = P(z,t) , \quad \frac{\partial^n P(z,t)}{\partial z^n} = \Omega^{n/2}\frac{\partial^n \Pi(a,t)}{\partial a^n} ,$$

$$\frac{\partial P(z,t)}{\partial t} = \frac{\partial \Pi(a,t)}{\partial t} + \Omega\frac{\mathrm{d}\phi(t)}{\mathrm{d}t}\frac{\partial \Pi(a,t)}{\partial a} = \frac{\partial \Pi(a,t)}{\partial t} + \Omega^{1/2}\frac{\mathrm{d}\phi(t)}{\mathrm{d}t}\frac{\partial P(z,t)}{\partial z} .$$

The derivative moments $\alpha_n(a)$ are now proportional to the system size Ω, so we scale them accordingly: $\alpha_n(a) = \Omega\widetilde{\alpha}_n(x)$. In the next step the new variable z is introduced into the Kramers–Moyal expansion (4.121):

$$\begin{aligned}
\frac{\partial \Pi(a,t)}{\partial t} &= \frac{\partial P(z,t)}{\partial t} - \Omega^{1/2}\frac{\mathrm{d}\phi(t)}{\mathrm{d}t}\frac{\partial P(z,t)}{\partial z} \\
&= \sum_{n=1}^{\infty}(-1)^n\frac{\Omega^{1-n/2}}{n!}\frac{\partial^n}{\partial z^n}\Big(\widetilde{\alpha}_n\big(\phi(t) + \Omega^{-1/2}z\big)P(z,t)\Big) , \\
\frac{\partial P(z,t)}{\partial t} &= \Omega^{1/2}\left(\frac{\mathrm{d}\phi(t)}{\mathrm{d}t} - \widetilde{\alpha}_1\big(\phi(t)\big)\right)\frac{\partial P(z,t)}{\partial z} + \Omega^0\big(\cdots\big) \ldots .
\end{aligned}$$

For general validity of an expansion, all terms of a certain order in the expansion parameter must vanish. We make use of this property to define $\phi(t)$ in such a way that the terms of order $\Omega^{1/2}$ are eliminated by demanding

$$\frac{\mathrm{d}\phi(t)}{\mathrm{dt}} = \widetilde{\alpha}_1\big(\phi(t)\big) . \tag{4.128}$$

This equation is an ODE determining $\phi(t)$ and, of course, it is in full agreement with the deterministic equation for the expectation value of the random variable, so $\phi(t)$ is indeed the deterministic part of the solution.[61]

The next step is an expansion of $\widetilde{\alpha}_n\big(\phi(t) + \Omega^{-1/2}z\big)$ in $\Omega^{-1/2}$ and reordering of terms. This yields

$$\frac{\partial P(z,t)}{\partial t} = \sum_{m=2}^{\infty} \frac{\Omega^{-(m-2)/2}}{m!} \sum_{n=1}^{m} (-1)^n \binom{m}{n} \widetilde{\alpha}_n^{m-n}\big(\phi(t)\big) \frac{\partial^n}{\partial z^n}\Big(z^{m-n}P(z,t)\Big) .$$

In taking the limit of large system size Ω, all terms vanish except the one with $m = 2$ and we find the result

$$\frac{\partial P(z,t)}{\partial t} = -\widetilde{\alpha}_1^{(1)}\big(\phi(t)\big)\frac{\partial}{\partial z}\Big(zP(z,t)\Big) + \frac{1}{2}\widetilde{\alpha}_2\big(\phi(t)\big)\frac{\partial^2}{\partial z^2}P(z,t) , \tag{4.129}$$

where $\alpha_1^{(1)}$ stands for the linear part of the drift term. Figure 4.43 shows a specific example of partitioning a process $n(t)$ into a macroscopic part $\Omega\phi(t)$ and fluctuations $\Omega^{1/2}x(t)$ around it.

It is straightforward to compare with the result of the Kramers–Moyal expansion (4.121) truncated after two terms:

$$\frac{\partial P(x,t)}{\partial t} = -\frac{\partial}{\partial x}\Big(\alpha_1(x)P(x,t)\Big) + \frac{1}{2}\frac{\partial^2}{\partial x^2}\Big(\alpha_2(x)P(x,t)\Big) .$$

The change of variables $\xi = x/\Omega$ leads to

$$\frac{\partial P(\xi,t)}{\partial t} = -\frac{\partial}{\partial \xi}\Big(\widetilde{\alpha}_1(\xi)P(\xi,t)\Big) + \frac{1}{2\Omega}\frac{\partial^2}{\partial \xi^2}\Big(\widetilde{\alpha}_2(\xi)P(\xi,t)\Big) .$$

Applying small noise theory (Sect. 4.5.2) with $\epsilon^2 = \Omega^{-1}$ and using the substitution $\xi = \Omega^{1/2}\big(x - \phi(t)\big)$, one obtains the lowest order Fokker–Planck equation, which is exactly the same as the lowest order approximation in the van Kampen expansion. This result has an important consequence: if we are only interested in the lowest

[61] As shown in (3.94) and (3.103), this result is only true for linear first jump moments or for the linear approximation to the first jump moments (see below).

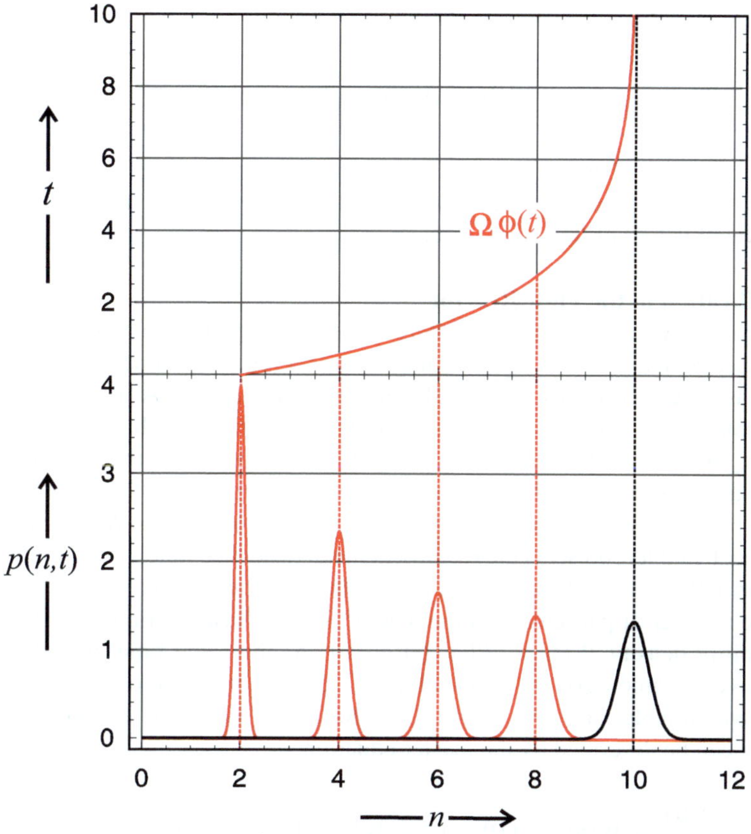

Fig. 4.43 Size expansion of a stochastic variable $\mathcal{X}(t)$. The variable n is split into a macroscopic part and the fluctuations around it, i.e., $n(t) = \Omega\phi(t) + \Omega^{1/2}x(t)$, where Ω is a size parameter, e.g., the size of the population or the volume of the system. Computations: $\Omega\phi(t) = 5n_0(1 - 0.8\mathrm{e}^{-kt})$ with $n_0 = 2$ and $k = 0.5\,[\mathrm{t}^{-1}]$ (*red*), $p(n,t) = \Omega^{1/2}x(t) = \mathrm{e}^{-(n-\Omega\phi(t))^2/2\sigma^2}/\sqrt{2\pi\sigma^2}$ with $\sigma = 0.1$, 0.17, 0.24, 0.285, 0.30 (*red*). The fluctuations at equilibrium are shown in *black*

order approximation, we may use the Kramers–Moyal equation, which is much easier to derive than the van Kampen equation.

So finally we have found a procedure for relating master equations to Fokker–Planck equations in an approximation that closes the gap between microscopic stochasticity and macroscopic behavior. It should be stressed, however, that the range of validity of a Fokker–Planck equation derived from a master equation is not independent of the kind of limiting procedure applied. If the transition is made by means of rigorous equations in a legitimate limit to continuous variables (Sect. 4.5.4), the full nonlinear dependence of $\alpha_1(x)$ and $\alpha_2(x)$ can be seriously analyzed. If on the other hand an approximately valid technique like the small noise approximation is applied, it is appropriate to consider only the linearization of the drift term, and individual solutions of these equations are represented by the trajectories of the stochastic equation:

$$dz = \widetilde{\alpha}_1^{(1)}\big(\phi(t)\big)z\;dt + \sqrt{\widetilde{\alpha}_2\big(\phi(t)\big)}dW(t)\;. \tag{4.130}$$

The choice of the best way to scale also depends on the special case to be studied, and we close this section by presenting two examples: (i) the flow reactor and (ii) the reversible first order chemical reaction.

Equilibration in the Flow Reactor

The problem we are reconsidering here is the time dependence of a single chemical substance A in a device for performing chemical reactions under controlled conditions, as described in Sect. 4.3.1. The concentration $a(t)$ of A in the reactor starts from some initial value $a_0 = a(0)$ and, after flow equilibrium has been established, $\lim_{t\to\infty} a(t) = \bar{a}$, it assumes the value $\bar{a} = \widehat{a}$, where $\widehat{a}$ is the concentration of A in the stock solution flowing into the reactor (Fig. 4.21). The flow in and out of the reactor is controlled by the flow rate r commonly measured in $[\mathrm{V\,t^{-1}}]$, e.g. $[\mathrm{cm^3/sec}]$, and it represents the reciprocal mean residence time of the solution in the reactor: $\tau_{\mathrm{v}}^{-1} = r/V$, where V is the total volume of the reactor.

The number of particles A in the reactor is a stochastic variable $\mathcal{N}_{\mathsf{A}}(t)$, with the probability density $P_n(t) = P\big(\mathcal{N}_{\mathsf{A}}(t) = n\big)$. At the same time, it is the discrete extensive variable $n_{\mathsf{A}}(t) = n(t)$ with $n \in \mathbb{N}$. The concentration is the continuous intensive variable, i.e., $a(t) = n(t)/\Omega$ with $\Omega = VN_{\mathrm{L}}$. The equilibration of the reactor can be described by the master equation

$$\begin{aligned}\frac{\partial P_n(t)}{\partial t} &= W(n|n-1)P_{n-1}(t) + W(n|n+1)P_{n+1}(t)\\ &\quad - \big(W(n-1|n) + W(n+1|n)\big)P_n(t)\;, \quad n \in \mathbb{N}\;,\end{aligned} \tag{4.79b$'$}$$

with the elements of the tridiagonal transition matrix W given by

$$W(n|m) = r\big(\delta_{n,m+1}\widehat{n} + \delta_{n,m-1}n\big)\;. \tag{4.131a}$$

The only nonzero contribution from the first term requires $n = m+1$ and describes an increase by one in the particle number in the reactor through inflow that corresponds to the step-up transition probability $w_n^+ = r\widehat{n}$. The nonzero contribution of the second term, $n = m-1$, deals with the loss of a particle A through outflow in the sense of a step-down transition with the probability $w_n^- = rn$. The equilibration of the flow reactor can thus be understood as a linear death process with *immigration* expressed by a positive constant term $r\widehat{n}$.

The reformulation of the transition matrix (4.126) in the sense of van Kampen's expansion leads to

$$W(a; \Delta n) = \Omega\big(r\widehat{a}\delta_{\Delta n,+1} + ra\delta_{\Delta n,-1}\big) ,$$

with $\Delta n = n - m$. Calculation of the first two jump moments yields

$$\alpha_1(n) = \sum_{m=0}^{\infty}(m-n)W(m|n) = r(\widehat{n} - n) = \Omega r(\widehat{a} - a) ,$$

$$\alpha_2(n) = \sum_{m=0}^{\infty}(m-n)^2 W(m|n) = r(\widehat{n} + n) = \Omega r(\widehat{a} + a) ,$$

and the deterministic equation with $\phi(t) = a(t) - n(t)/\Omega$ is of the form

$$\frac{\mathrm{d}a}{\mathrm{d}t} = r(\widehat{a} - a) , \quad a(t) = \widehat{a} + \big(a(0) - \widehat{a}\big)\mathrm{e}^{-rt} ,$$

where we recall that the equilibrium concentration of A in the reactor is equal to the influx concentration, i.e., $\bar{a} = \widehat{a}$. Following the procedure of van Kampen's expansion, we define

$$n = \Omega\phi(t) + \Omega^{1/2}z , \quad \text{or } z = \Omega^{-1/2}n - \Omega^{1/2}\phi(t) , \tag{4.127'}$$

and obtain the Fokker–Planck equation

$$\frac{\partial P(z,t)}{\partial t} = r\frac{\partial}{\partial z}\Big(zP(z,t)\Big) + \frac{r}{2}\frac{\partial^2}{\partial z^2}\Big(\big(\widehat{a} + a(t)\big)P(z,t)\Big) , \tag{4.131b}$$

which leads to the expectation value and variance in the scaled variable z:

$$E\big(z(t)\big) = z(0)\mathrm{e}^{-rt} , \quad \mathrm{var}\big(z(t)\big) = \big(\widehat{a} + a(0)\mathrm{e}^{-rt}\big)(1 - \mathrm{e}^{-rt}) .$$

Since the partition of the variable n in (4.127′) is arbitrary, we can assume $z(0) = 0$.[62] Transforming to the extensive variable, the particle number n yields

$$E\big(n(t)\big) = \widehat{n} + \big(n(0) - \widehat{n}\big)\mathrm{e}^{-rt} , \quad \mathrm{var}\big(n(t)\big) = \big(\widehat{n} + n(0)\mathrm{e}^{-rt}\big)(1 - \mathrm{e}^{-rt}) . \tag{4.131c}$$

[62]The assumption $z(0) = 0$ implies $z(t) = 0$, so the corresponding stochastic variable $\mathcal{Z}(t)$ describes the fluctuations around zero.

The stationary solution of the Fokker–Planck equation is readily calculated to be

$$\bar{P}(z) = \frac{1}{\sqrt{2\pi\widehat{a}}}\exp\left(-\frac{z^2}{2\widehat{a}}\right) ,$$

and it represents the approximation of the exact stationary Poisson density by means of a Gaussian, as mentioned in (2.52):

$$\bar{P}(n) = \frac{\widehat{n}^n}{n!}\exp(-\widehat{n}) \approx \frac{1}{\sqrt{2\pi\widehat{n}}}\exp\left(-\frac{(n-\widehat{n})^2}{2\widehat{n}}\right) .$$

A comparison of the different expansion techniques is made in the next section, where we consider the simple chemical reaction $\mathsf{A}\rightleftharpoons\mathsf{B}$ with the compound B buffered. This gives rise to a master equation that is formally identical to the one for the equilibration of the flow reactor.

The Chemical Reaction $\mathsf{A}\rightleftharpoons\mathsf{B}$

The reversible monomolecular conversion reaction is considered under a large excess of compound B: the concentration $[\mathsf{B}] = b_0 = n_{\mathsf{B}}/\Omega$ is thus constant, and we say that this compound is *buffered*. The stochastic variable counts the number of molecules A in the system, i.e., $[\mathsf{A}] = \mathcal{N}_{\mathsf{A}}(t)$, with the probability distribution $P_{n_{\mathsf{A}}}(t) = P_n(t) = P\big(\mathcal{N}_{\mathsf{A}}(t) = n\big)$ and $a(t) = n(t)/\Omega$. The elements of the transition matrix of the master equation (4.79b′) are

$$W(n|m) = \delta_{n,m+1} l n_{\mathsf{B}} + \delta_{n,m-1} k n , \tag{4.131a′}$$

where k and l are the rate parameters for the forward and backward reactions, respectively. By replacing the constant terms $ln_{\mathsf{B}} \leftrightarrow r\widehat{n}$ and $k \leftrightarrow r$, we recognize that the two problems, flow reactor and buffer reaction $\mathsf{A}\rightleftharpoons\mathsf{B}$, are formally identical. By applying van Kampen's expansion, the solutions are derived in precisely the same way as in the previous paragraph. With $n = \Omega\phi(t) + \Omega^{1/2}z$, we obtain

$$\frac{\mathrm{d}\phi(t)}{\mathrm{d}t} = lb_0 - k\phi(t) , \quad \phi(t) = \phi(0)\mathrm{e}^{-kt} + \frac{lb_0}{k}(1-\mathrm{e}^{-kt}) ,$$

$$\frac{\partial P(z)}{\partial t} = k\frac{\partial}{\partial z}\big(zP(z)\big) + \frac{1}{2}\frac{\partial^2}{\partial z^2}\Big(\big(lb_0 + k\phi(t)\big)P(z)\Big) ,$$

for the deterministic solution and the Fokker–Planck equation, respectively.

The expectation value of z is $\mathrm{E}\big(z(t)\big) = z(0)\mathrm{e}^{-kt}$. It vanishes with the usual assumption $z(0) = 0$. For the variance $\mathrm{var}\big(z(t)\big)$, we find

$$\mathrm{var}\big(z(t)\big) = \left(\frac{lb_0}{k} + \phi(0)\right)(1-\mathrm{e}^{-kt}) ,$$

and for the solutions in the variable n with $n(0) = \Omega\phi(0)$, we obtain

$$\mathrm{E}\big(n(t)\big) = \Omega\phi(t) = n(0)\mathrm{e}^{-kt} + \frac{l_1 n_\mathrm{B}}{k}(1 - \mathrm{e}^{-kt}) ,$$

$$\mathrm{var}\big(n(t)\big) = \Omega\,\mathrm{var}\big(z(t)\big) = \left(\frac{l_1 n_\mathrm{B}}{k} + n(0)\right)(1 - \mathrm{e}^{-kt}) .$$

Finally, we compare the stationary state solutions obtained from the van Kampen expansion and from the Kramers–Moyal expansion with the exact solution. The size expansion yields

$$\bar{P}(z) = \frac{1}{\sqrt{\frac{\pi\kappa}{2}}\Big(1 + \mathrm{erf}\big(\sqrt{\kappa/2}\big)\Big)} \exp\left(-\frac{(n-\kappa)^2}{2\kappa}\right) , \tag{4.132a}$$

where we have used $\kappa = ln_\mathrm{B}/k$ and replaced $z \leftrightarrow n$. The result of the truncated Kramers–Moyal expansion is calculated from the stationary solution (3.82) of a Fokker–Planck equation with $A(n) = \alpha_1(n) = ln_\mathrm{B} - kn$ and $B(n) = \alpha_2 = ln_\mathrm{B} + kn$,

$$\bar{P}(n) = N(ln_\mathrm{B} + kn)^{-1+4ln_\mathrm{B}/k}\mathrm{e}^{-2n} , \tag{4.132b}$$

where the normalization factor N is still to be determined for the special case. The exact solution is identical with the result derived for the flow reactor (4.79h), viz.,

$$\bar{P}(n) = \frac{\big(ln_\mathrm{B}/k\big)^n \exp\big(-ln_\mathrm{B}/k\big)}{n!} = \frac{\kappa^n \mathrm{e}^{-\kappa}}{n!} , \tag{4.132c}$$

which is a Poissonian. Figure 4.44 compares numerical plots. It is remarkable how well the truncated Kramers–Moyal expansion agrees with the exact probability density. It is easy to understand therefore that it is much more popular than the size expansion, which is much more sophisticated. We remark that the major difference between the van Kampen solution and the other two curves results in essence from the approximation of a Poissonian by a Gaussian (see Fig. 2.8).

4.5.4 From Master to Fokker–Planck Equations

Finally, we summarize this section by mentioning another general scaling method [194, pp. 273–274], reminiscent of the transition from continuous time random walks to diffusion, which was discussed in Sect. 3.2.4. The master equation (3.109) is converted into the partial differential equation of the Wiener process equation (3.55), a Fokker–Planck equation without drift, by taking the limit of infinitesimally small steps at infinite frequency, and which is formally identical with the

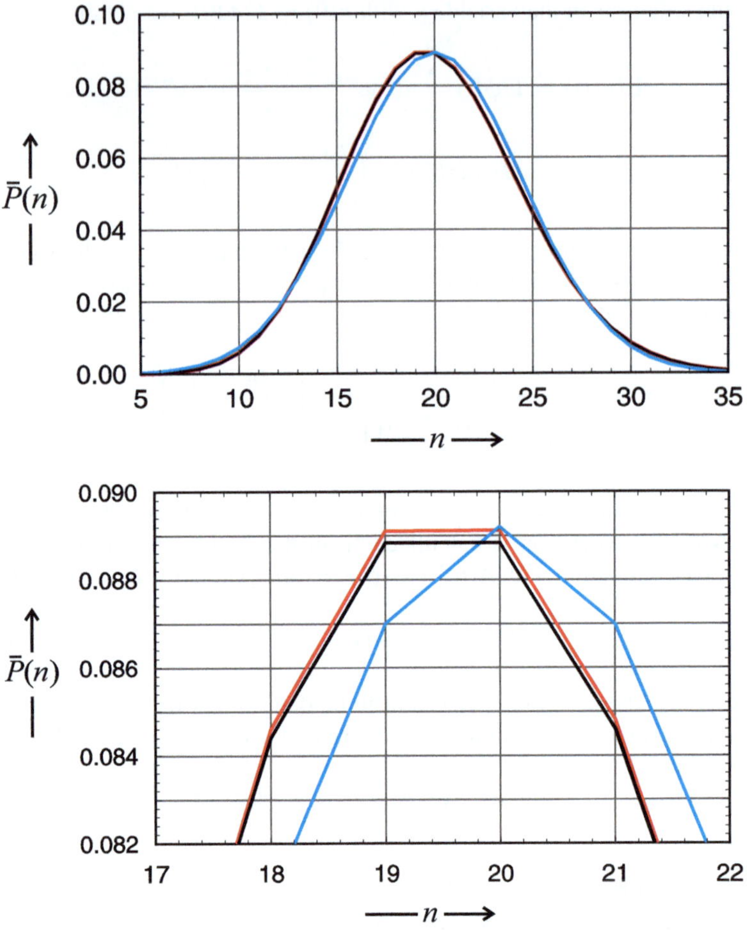

Fig. 4.44 Comparison between expansions of the master equation. The reaction A⇌B with compound B buffered, $[B] = b = b_0 = n_B/\Omega$, is chosen as an example, and the exact stationary solution (*black*) is compared with the results of the Kramers–Moyal expansion (*red*) and the van Kampen size expansion (*blue*). Parameter choice: $V = 1$, $k = 2\,[t^{-1}]$, $l = 1\,[t^{-1}]$, $n_B = 40$

1D diffusion equation. In this transition the step size was chosen to be $l = l_0\varepsilon$ and the probability of making a step was $\vartheta = \vartheta_0/\varepsilon^2$. During the transition, the jumps become simultaneously smaller and more probable, and both changes are taken care of by a *scaling assumption* based on the use of a scaling parameter ε. Hence, the average step size is proportional to ε, as is the variance of the step size,[63] and thus decreases with ε, while the jump probabilities increase as ε becomes smaller.

[63]This is automatically true when the steps follow a Poisson distribution.

Here we perform the transition from master equations to Fokker–Planck equations in a more general way, and illustrate by means of examples that a diffusion process can always be approximated by a master equation, whereas the opposite is not true. First the elements of the transition matrix are rewritten in terms of a new variable $\eta = (z - x - A(x)\varepsilon)/\sqrt{\varepsilon}$, where $A(x)$ represents the general drift term. The transition probabilities are written in the form

$$W_\varepsilon(z|x) = \varepsilon^{-3/2}\phi(\eta, x) , \tag{4.133}$$

where the function $\phi(\eta, x)$ is given by the concrete example to be studied and, in addition, satisfies the relations

$$\int \mathrm{d}\eta\, \phi(\eta, x) = I , \qquad \int \mathrm{d}\eta\, \eta\phi(\eta, x) = 0 .$$

We define consistent expressions for the first three jump moments (4.120):

$$\alpha_0(x) \doteq \int \mathrm{d}z\, W_\varepsilon(z|x) = \frac{I}{\varepsilon} , \tag{4.134a}$$

$$\alpha_1(x) \doteq \int \mathrm{d}z (z - x) W_\varepsilon(z|x) = A(x) I , \tag{4.134b}$$

$$\alpha_2(x) \doteq \int \mathrm{d}z (z - x)^2 W_\varepsilon(z|x) = \int \mathrm{d}\eta\, \eta^2 \phi(\eta, x) . \tag{4.134c}$$

These expressions are obtained from the definitions of the variable η and the two integrals of $\phi(\eta, x)$, and in the case of (4.134c), by neglecting the term of order $\mathcal{O}(\varepsilon) = A(x)^2 I\varepsilon$ in the limit $\varepsilon \to 0$. To take this limit, we shall assume further that the function $\phi(\eta, x)$ vanishes fast enough as $\eta \to \infty$ to guarantee that

$$\lim_{\varepsilon \to 0} W_\varepsilon(z|x) = \lim_{\eta \to \infty} \left(\left(\frac{x}{z - x} \right)^3 \phi(\eta, x) \right) = 0 , \quad \text{for } z \neq x .$$

Very similar to the derivation of the differential Chapman–Kolmogorov equation in Sect. 3.2 we may choose some twice differentiable function $f(z)$ and show that

$$\lim_{\varepsilon \to 0} \left\langle \frac{\partial f(z)}{\partial t} \right\rangle = \left\langle \alpha_1(z) \frac{\partial f(z)}{\partial z} + \frac{1}{2} \alpha_2(z) \frac{\partial^2 f(z)}{\partial z^2} \right\rangle .$$

Applying this result to the probability $P(x, t)$ result has the consequence that, in the limit $\varepsilon \to 0$, the master equation

$$\frac{\partial P(x, t)}{\partial t} = \int \mathrm{d}z \Big(W(x|z) P(z, t) - W(z|x) P(x, t) \Big) \tag{4.135a}$$

becomes the Fokker–Planck equation

$$\frac{\partial P(x,t)}{\partial t} = -\frac{\partial}{\partial x}\big(\alpha_1 P(x,t)\big) + \frac{1}{2}\frac{\partial^2}{\partial x^2}\big(\alpha_2 P(x,t)\big) . \tag{4.135b}$$

Accordingly, one can construct a Fokker–Planck limit for the master equation if and only if the requirements imposed by the three jump moments $\alpha_p, p = 0, 1, 2$, (4.134) can be met. If these criteria are not fulfilled, there is no approximation possible, as we shall now illustrate by means of examples.

Continuous Time Random Walk

The master equation introduced in Sect. 3.2.4, viz.,

$$\frac{\mathrm{d}P_n(t)}{\mathrm{d}t} = \vartheta\Big(P_{n+1}(t) + P_{n-1}(t) - 2P_n(t)\Big) , \quad \text{with } P_n(t_0) = \delta_{n,n_0} ,$$

as initial condition, is to be converted into a Fokker–Planck equation. First we remember that the steps were embedded in a continuous spatial coordinate $x = nl$, so the walk started at the point $x_0 = n_0 l$. The elements of the transition matrix W have the general form

$$W(z|x) = \vartheta(\delta_{z,x-l} + \delta_{z,x+l}) ,$$

and we use the three integrals over the scaled transition moments, i.e.,

$$\int_{-\infty}^{\infty} \mathrm{d}\eta\, \phi(\eta, x) = \varepsilon 2\vartheta , \quad \int_{-\infty}^{\infty} \mathrm{d}\eta\, \eta\phi(\eta, x) = (l - l)\vartheta = 0 ,$$

$$\int_{-\infty}^{\infty} \mathrm{d}\eta\, \eta^2\phi(\eta, x) = 2l^2\vartheta ,$$

where the second integral vanishes because of the intrinsic symmetry of the random walk. The first three jump moments are readily calculated from (4.134):

$$\alpha_0(x) = 2\vartheta , \quad \alpha_1(x) = 0 , \quad \alpha_2(x) = 2l^2\vartheta .$$

Introducing the variable η, we get a natural way of scaling the step size and jump probability. Assuming that we began with some discrete system (l_0, ϑ_0), reducing the step size according to $l^2 = \varepsilon l_0^2$, and raising the probability by $\vartheta = \vartheta_0/\varepsilon$, the diffusion coefficient $D = (l_0^2\varepsilon) \times (\vartheta_0/\varepsilon)$ remains constant in the scaling process. With $D = l^2\vartheta = l_0^2\vartheta_0$, we obtain a Fokker–Planck equation, the familiar stochastic diffusion equation

$$\frac{\partial P(x,t)}{\partial t} = D\frac{\partial^2 P(x,t)}{\partial x^2} . \tag{3.55$'$}$$

The final result is the same as in Sect. 3.2.4, although we used a much simpler intuitive procedure there than the transformation (4.133).

Poisson Process

The Poisson process can be viewed as a random walk restricted to one direction, hence taking place in a the (upper) half-plane with the master equation

$$\frac{\mathrm{d}P_n(t)}{\mathrm{d}t} = \vartheta\Big(P_{n+1}(t) - P_n(t)\Big) , \quad \text{with } P_n(t_0) = \delta_{n,n_0} .$$

The notation used in Sect. 3.2.2.4 is slightly modified: $\alpha \leftrightarrow \vartheta$, and with $x = nl$ we find for the transition matrix W:

$$W(x|z) = \vartheta\,\delta_{z,x+l} .$$

The calculation of the moments is exactly the same as in the previous example:

$$\alpha_0(x) = \vartheta , \qquad \alpha_1(x) = l\vartheta , \qquad \alpha_2(x) = l^2\vartheta .$$

In this case there is no way to define l and ϑ as functions of ε such that both $\alpha_1(x)$ and $\alpha_2(x)$ remain finite in the limit $l \to 0$. Applying, for example, the same model assumption as made for the one-dimensional random walk, we find $l^2 = l_0^2\varepsilon$ and $\vartheta = \vartheta_0/\varepsilon$, and hence $\lim_{\varepsilon\to 0} l^2\vartheta = D$ as before, but $\lim_{\varepsilon\to 0} l\vartheta = \lim_{\varepsilon\to 0} l_0\vartheta_0/\sqrt{\varepsilon} = \infty$. Accordingly, there is no Fokker–Planck limit for the Poisson process within the transition moment expansion scheme.

General Birth-and-Death Master Equations

Crispin Gardiner also provides a scaling analysis leading to the general Fokker–Planck equation [194]. The starting point is a master equation with the transition probability matrix

$$W_\varepsilon(z|x) = \left(\frac{A(x)}{2\varepsilon} + \frac{B(x)}{2\varepsilon^2}\right)\delta_{z,x+\varepsilon} + \left(-\frac{A(x)}{2\varepsilon} + \frac{B(x)}{2\varepsilon^2}\right)\delta_{z,x-\varepsilon} , \tag{4.136}$$

where $W_\varepsilon(z|x)$ is positive at least for sufficiently small ε: $W_\varepsilon(z|x) > 0$ if $B(x) > \varepsilon|A(x)|$. Under the assumption that this is satisfied for the entire domain of the variable x, the process takes place on an x-axis that is partitioned into integer multiples of ε.[64] In the limit $\varepsilon \to 0$, the birth-and-death master equation is converted into a Fokker–Planck equation with

$$\begin{aligned} &\alpha_0(x) = B(x)/\varepsilon^2 \,, \qquad \alpha_1(x) = A(x) \,, \qquad \alpha_2(x) = B(x) \,, \\ &\lim_{\varepsilon \to 0} W_\varepsilon(z|x) = 0 \,, \quad \text{for } z \neq x \,. \end{aligned} \tag{4.137}$$

Nevertheless, the idea of jumps converging smoothly into a continuous distribution is no longer valid, because the zeroth moment $\alpha_0(x)$ diverges as $1/\varepsilon^2$ and not as $1/\varepsilon$, as would be required by (4.134a). Notwithstanding there exists a limiting Fokker–Planck equation. This is because the limiting behavior of $\alpha_0(x)$ has no influence, since it does not show up in the final equation

$$\frac{\partial P(x,t)}{\partial t} = -\frac{\partial}{\partial x}\Big(A(x)P(x,t)\Big) + \frac{1}{2}\frac{\partial^2}{\partial x^2}\Big(B(x)P(x,t)\Big) \,. \tag{3.47$'$}$$

Equation (4.137) provides a tool for simulating a diffusion process by an approximating birth-and-death process. However, this method fails for $B(x) = 0$, for all possible ranges of x, since then $W_\varepsilon(z,x)$ cannot fulfil the criterion of being nonnegative. Otherwise, there is no restriction on the side of the Fokker–Planck equation, since (4.136) is completely general. As already mentioned, the converse is not true: there are jump processes and master equations which cannot be approximated by Fokker–Planck equations through scaling. The Poisson process discussed above may serve as an example.

Summarizing this section, we compare the size expansion described in Sect. 4.5.3 and the moment expansion presented here: in the size expansion (4.129), the system size Ω was considered as a parameter and $\lim \Omega \to \infty$ is the transition of interest that leads to the macroscopic or deterministic equations. In the moment expansion, (4.134) and (4.135b), the system size was assumed to be constant and the transition concerned the resolution of the jump size that was increased from coarse-grained to smooth or continuous variables, i.e., $\lim \varepsilon \to 0$.

4.6 Numerical Simulation of Chemical Master Equations

Historically, the basis for numerical simulation of master equations was laid down by the works of Andrey Kolmogorov and Willy Feller: Kolmogorov [310] introduced the differential equation describing Markov jump processes and Feller [156] defined the conditions under which the solutions of the Kolmogorov equations

[64] We remark that the scaling relations (4.133) and (4.136) are not the same, but both lead to a Fokker–Planck equation.

satisfied the conditions for proper probabilities. In addition, he was able to prove that the time between consecutive jumps is exponentially distributed and that the probability of the next event is proportional to the deterministic rate. In other words, he provided evidence that sampling of jump trajectories leads to a statistically correct representation of the stochastic process. Joe Doob extended Feller's derivation beyond the validity for pure jump processes [115, 116]. The implementation of a stochastic simulation algorithm for the Kolmogorov equations is due to David Kendall [292] and was applied to studies of epidemic outbreaks by Maurice Bartlett [39]. More than twenty years later, almost at the same time as the Feinberg–Horn–Jackson theory of chemical reaction networks was introduced, the American physicist and mathematical chemist Daniel Gillespie [206, 207, 209, 213] revived the formalism and introduced a popular simulation tool for stochastic chemical reactions. His algorithm became popular as a simple and powerful tool for the calculation of single trajectories. In addition, he showed that the chemical master equation and the simulation algorithm can be put together on a firm physical and mathematical basis [209]. Meanwhile the Gillespie algorithm became an essential simulation tool in chemistry and biology. Here we present the concept and the implementation of the algorithm, and demonstrate its usefulness by means of selected examples.

4.6.1 Basic Assumptions

Gillespie's general stochastic model is introduced here by means of the same definitions and notations as used in the theory of chemical reaction networks (Sect. 4.1.3). A set of M different molecular species, $\mathcal{S} = \{\mathsf{X}_1, \mathsf{X}_2, \ldots, \mathsf{X}_M\}$ in a homogeneous medium are interconverted through K elementary chemical reactions $\mathcal{R} = \{\mathbf{R}_1, \mathbf{R}_2, \ldots, \mathbf{R}_K\}$. Two conditions are assumed to be satisfied by the system:

(i) The content of a container with constant volume V is thought to be *well mixed* and spatially homogeneous (CSTR in Fig. 4.21).
(ii) The system is assumed to be in *thermal equilibrium* at constant temperature T. The primary goals of the simulation are the computation of the time courses of the stochastic variables—$\mathcal{X}_k(t)$ counting the number of molecules X_k of species K at time t—and the description of the evolution of the entire molecular population.

The computations yield *exact* trajectories of the type shown in Fig. 4.16 (Sect. 4.2.2). Within the framework of the two conditions for choosing a proper time interval for τ-leaping (Sects. 4.2.4 and 4.6.2), the trajectories provide solutions that correspond to the proper stochastic differential equations.

Variables, Reactions, and Stoichiometry

The entire population of a reaction system involving M species in K reactions is described by an M-dimensional random vector counting the numbers of molecules of individual species X_k :

$$\boldsymbol{\mathcal{X}}(t) = \big(\mathcal{X}_1(t), \mathcal{X}_2(t), \ldots, \mathcal{X}_k(t), \ldots, \mathcal{X}_M(t)\big) .$$

Molecules are discrete quantities and the random variables are discrete in the calculation of *exact* trajectories, as well as in the chemical master equation: $\mathbf{n} = \big(\mathrm{n}_1(t), \mathrm{n}_2(t), \ldots, \mathrm{n}_M(t)\big)$. Three quantities are required to fully characterize a reaction channel $\mathbf{R}_\mu$: (i) the specific probabilistic rate parameter γ_μ, (ii) the frequency functions $h_\mu(\mathbf{n})$, and (iii) the stoichiometric matrix S.

In Sect. 4.1.4, we derived the fundamental fact that a scalar rate parameter γ_μ, which is independent of dt, exists for each elementary reaction channel $\mathbf{R}_\mu$ with $\mu = 1, \ldots, K$, that is accessible to the molecules of a well mixed and thermally equilibrated system in gas phase or solution. This parameter has the property that

$$\begin{aligned} \gamma_\mu \ \mathrm{dt} = {} & \text{probability that a randomly selected combination of} \\ & \mathbf{R}_\mu \text{ reactant molecules at time } t \text{ will react within} \\ & \text{the next infinitesimal time interval } [t, t + \mathrm{dt}[\,. \end{aligned} \qquad (4.138)$$

The frequency function $h_\mu(\mathbf{n})$ is calculated from the vector $\mathbf{n}(t)$ which contains the exact numbers of all molecules at time t :

$$\begin{aligned} h_\mu(\mathbf{n}) \equiv {} & \text{the number of distinct combinations of } \mathbf{R}_\mu \text{ reactant} \\ & \text{molecules in the system when the numbers of molecules} \\ & \text{of species } \mathsf{X}_k \text{ are exactly } \mathrm{n}_k \text{ with } k = 1, \ldots, M \ . \end{aligned} \qquad (4.139)$$

The stoichiometric matrix $\mathrm{S} = (s_{k\mu};\, k = 1, \ldots, M,\, \mu = 1, \ldots, K)$ is an $M \times K$ matrix of integers, where

$$\begin{aligned} s_{k\mu} \ \equiv {} & \text{the change in the } \mathsf{X}_k \text{ molecular population caused} \\ & \text{by the occurrence of one } \mathbf{R}_\mu \text{ reaction.} \end{aligned} \qquad (4.140)$$

The functions $h_\mu(\mathbf{n})$ and the matrix S are derived from the stoichiometric equations (4.5) of the individual reaction channels, as shown in Sect. 4.1, and illustrated here by means of an example:

$$\begin{aligned} \mathbf{R}_1 &: \ \mathsf{X}_1 + \mathsf{X}_2 \longrightarrow \mathsf{X}_3 + \mathsf{X}_4 \ , \\ \mathbf{R}_2 &: \ 2\,\mathsf{X}_1 \longrightarrow \mathsf{X}_1 + \mathsf{X}_5 \ , \\ \mathbf{R}_3 &: \ \mathsf{X}_3 \longrightarrow \mathsf{X}_5 \ . \end{aligned} \qquad (4.141)$$

In particular, we find for the functions $h_\mu(\mathbf{n})$[65]

$$h_1(\mathbf{n}) = n_1 n_2 \,, \qquad h_2(\mathbf{n}) = n_1(n_1 - 1) \,, \qquad h_3(\mathbf{n}) = n_3 \,,$$

and for the stoichiometric matrix S

$$\mathrm{S} = \begin{pmatrix} -1 & -1 & 0 \\ -1 & 0 & 0 \\ +1 & 0 & -1 \\ +1 & 0 & 0 \\ 0 & +1 & +1 \end{pmatrix} ,$$

where the rows refer to molecular species $\mathbf{X} = (\mathrm{X}_1, \mathrm{X}_2, \mathrm{X}_3, \mathrm{X}_4, \mathrm{X}_5)$, and the columns to individual reactions $\mathcal{R} = (\mathbf{R}_1, \mathbf{R}_2, \mathbf{R}_3)$. The product side is considered in the stoichiometric matrix S by a positive sign of the stoichiometric coefficients, whereas reactants are accounted for by a negative sign. The column vectors corresponding to individual reactions are denoted by $\mathbf{R}_\mu : \mathbf{s}_\mu = (s_{1\mu}, \ldots, s_{M\mu})^{\mathrm{t}}$. It is worth noting that the functional form of h_μ is determined exclusively by the reactant side of $\mathbf{R}_\mu$. For mass action kinetics, there is only one difference between the deterministic and the stochastic expressions: since the particles are counted exactly in the latter approach, we have to use $n(n-1)$ instead of n^2. Only in very small systems will there be a significant difference between $n-1$ and n.

Reaction Events

The probability of occurrence of reaction events within an infinitesimal time interval dt satisfies three conditions for master equations that were formulated and discussed in Sect. 4.2.2. Here we repeat them for convenience:

Condition 1. If $\boldsymbol{\mathcal{X}}(t) = \mathbf{n}$, then the probability that *no* reaction will occur within the time interval $[t, t + \mathrm{d}t[$ is equal to

$$1 - \sum_\mu \gamma_\mu h_\mu(\mathbf{n})\,\mathrm{d}t + o(\mathrm{d}t) \,.$$

Condition 2. If $\boldsymbol{\mathcal{X}}(t) = \mathbf{n}$, then the probability that *exactly one* $\mathbf{R}_\mu$ will occur in the system within the time interval $[t, t + \mathrm{d}t[$ is equal to

$$\gamma_\mu h_\mu(\mathbf{n})\,\mathrm{d}t + o(\mathrm{d}t) \,.$$

[65] As mentioned before, there are two ways to take proper account of the combinatorics: (i) $h(\mathbf{n}) = \prod_i \binom{n_i}{\nu_i}$ and $\gamma \equiv k$ as rate parameter, or (ii) $h(\mathbf{n}) = \prod_i n_i!/(n_i - \nu_i)!$ and $\gamma \equiv k/\nu_i!$. We use here version (ii) unless stated otherwise, and indicate the factor in the denominator in the rate parameter, viz., $k_i/\nu_i!$.

Condition 3. The probability of more than one reaction occurring in the system within the time interval $[t, t + \,\mathrm{d}t[$ is of order $o(\,\mathrm{d}t)$.

The probability $P(\mathbf{n}, t + \,\mathrm{d}t|\mathbf{n}_0, t_0)$ is expressed as the sum of the probabilities for several mutually exclusive and collectively exhaustive *routes* from $\boldsymbol{\mathcal{X}}(t_0) = \mathbf{n}_0$ to $\boldsymbol{\mathcal{X}}(t + \,\mathrm{d}t) = \mathbf{n}$. These routes are distinguished from one another by the event that happened in the last time interval $[t, t + \,\mathrm{d}t[$:

$$\begin{aligned} P(\mathbf{n}, t + \,\mathrm{d}t|\mathbf{n}_0, t_0) &= P(\mathbf{n}, t|\mathbf{n}_0, t_0)\left(1 - \sum_{\mu=1}^{K} \gamma_\mu h_\mu(\mathbf{n})\,\mathrm{d}t + o(\,\mathrm{d}t)\right) \\ &\quad + \sum_{\mu=1}^{K} P(\mathbf{n} - \mathbf{s}_\mu, t|\mathbf{n}_0, t_0)\Big(\gamma_\mu h_\mu(\mathbf{n} - \mathbf{s}_\mu)\,\mathrm{d}t + o(\,\mathrm{d}t)\Big) \\ &\quad + o(\,\mathrm{d}t)\,. \end{aligned} \tag{4.142}$$

The different routes from $\boldsymbol{\mathcal{X}}(t_0) = \mathbf{n}_0$ to $\boldsymbol{\mathcal{X}}(t + \,\mathrm{d}t) = \mathbf{n}$ are obvious from the balance equation (4.142): all routes (i) and (ii) are mutually exclusive since different events are taking place within the last interval $[t, t+ \,\mathrm{d}t[$. The routes subsumed under (iii) can be neglected because they occur with probability of measure zero.

Equation (4.142) implies the multivariate *chemical master equation*, which is the reference for trajectory simulation: $P(\mathbf{n}, t|\mathbf{n}_0, t_0)$ is subtracted from both sides of (4.142), then both sides are divided by $\,\mathrm{d}t$, the limit $\,\mathrm{d}t \downarrow 0$ is taken, all $o(\,\mathrm{d}t)$ terms vanish, and finally we obtain

$$\frac{\mathrm{d}}{\mathrm{d}t} P(\mathbf{n}, t|\mathbf{n}_0, t_0) = \sum_{\mu=1}^{K} \Big(\gamma_\mu h_\mu(\mathbf{n} - \mathbf{s}_\mu) P(\mathbf{n} - \mathbf{s}_\mu|\mathbf{n}_0, t_0) - \gamma_\mu h_\mu(\mathbf{n}) P(\mathbf{n}, t|\mathbf{n}_0, t_0)\Big)\,. \tag{4.143a}$$

Initial conditions are required to calculate the time evolution of the probability $P(\mathbf{n}, t|\mathbf{n}_0, t_0)$, and we can easily express them in the form

$$P(\mathbf{n}, t_0|\mathbf{n}_0, t_0) = \begin{cases} 1\,, & \text{if } \mathbf{n} = \mathbf{n}_0\,, \\ 0\,, & \text{if } \mathbf{n} \neq \mathbf{n}_0\,, \end{cases} \tag{4.143b}$$

which is the same as the sharp initial probability distribution used implicitly in the derivation of (4.142): $P\big(n_k, t_0|n_k^{(0)}, t_0\big) = \delta_{n_k, n_k^{(0)}}$ for the molecular particle numbers at $t = t_0$.

4.6.2 *Tau-Leaping and Higher-Level Approaches*

One general problem with all stochastic simulations involving medium and large particle numbers is the enormous consumption of computer time. The necessary but prohibitive amounts of computer capacities are even required when only a single species is present at high particle numbers, and this is almost always the case even in the fairly small biological systems within cells. The clear advantage of the stochastic simulation algorithm is at the same time the ultimate cause for its failure to handle most systems in practice: considering every single event explicitly makes the simulation *exact*, but guides it directly into the *computer time requirement trap*.

Tau-Leaping

In Sect. 4.2.4 on chemical Langevin equations, τ-leaping was discussed to justify the use of stochastic differential equations in chemical kinetics. Here we mention *τ-leaping* as an attempt to accelerate the simulation algorithm, based on the same idea of lumping together all events happening with a predefined time interval $[t, t+\tau[$ [212, 213]. In contrast to the three implementations of the Monte Carlo step in the original Gillespie simulation algorithm—the direct, first-reaction, and next-reaction method, which are exact since they consider every event precisely at its time of occurrence—τ-leaping is an approximation whose degree of accuracy depends on the choice of the time interval τ. Assume, for example, that τ is chosen so small that only no reaction step or a single reaction step take place within the interval $[t, t+\tau[$. Then a calculated trajectory obtained by the exact method is indistinguishable from the results of the τ-leaping simulation, which is then also exact. Choosing a larger value of τ will introduce an error that will increase with the size of τ.

The approach is cast into a solid mathematical form by defining a function $\mathcal{P}(\kappa_1, \ldots, \kappa_K | \tau; \mathbf{n}, t)$: given $\boldsymbol{\mathcal{X}}(t) = \mathbf{n}$, $\mathcal{P}$ measures the probability that exactly κ_j reaction events will occur in the reaction channel $\mathbf{R}_j$ with $j = 1, \ldots, K$. This function $\mathcal{P}$ is the joint probability density of the integer random variables $\mathcal{K}_j(\tau; \mathbf{n}, t)$, which represent numbers counting how often the reaction channel $\mathbf{R}_j$ fires in the time interval $[t, t+\tau[$. In order to be able to calculate $\mathcal{P}(\kappa_1, \ldots, \kappa_K | \tau; \mathbf{n}, t)$ with reasonable ease, an approximation has to be made that determines an appropriate leap size:

Leap Condition. The time interval τ has to be chosen so small that none of the K propensity functions $\alpha_j(\mathbf{n}, t)$, $j = 1, \ldots, K$, will change appreciably in the interval $[t, t+\tau[$.

The word *appreciably* expresses here the relative change and excludes alterations of macroscopically non-infinitesimal size (see Sect. 4.2.4). Provided that the leap condition is satisfied and the reaction probability functions remain essentially constant within the entire time interval, i.e., $\alpha_j(\mathbf{n}) \approx$ const. $\forall j = 1, \ldots, K$, then $\alpha_j(\mathbf{n})$ dt is the probability that a reaction $\mathbf{R}_j$ will take place during any infinitesimal interval dt inside $[t, t + \tau[$, irrespective of what happens in the other reaction channels.[66] The events in the individual reaction channels are independent and the random variables follow a Poisson distribution, viz.,

$$\mathcal{K}_j(\tau; \mathbf{n}, t) = \pi_{\kappa_j}(\alpha_j) = (\alpha_j t)^{\kappa_j} \frac{e^{\alpha_j t}}{\kappa_j!} ,$$

and this yields

$$\mathcal{P}(\kappa_1, \ldots, \kappa_K | \tau; \mathbf{n}, t) = \prod_{j=1}^{K} \pi_{\kappa_j}(\chi_j t) , \tag{4.144}$$

for the probability distribution. Each event in the channel $\mathbf{R}_j$ changes the population by $\mathbf{s}_j = \mathbf{v}'_j - \mathbf{v}_j$, so we can easily express the change in the population during the entire interval $[t_i, t_i + \tau_i[$ and the whole trajectory from t_0 to t_N by

$$\boldsymbol{\lambda}_i = \sum_{j=1}^{K} \kappa_j \mathbf{s}_j , \quad \boldsymbol{\mathcal{X}}(t_N) = \boldsymbol{\mathcal{X}}(t_0) + \sum_{i=0}^{N-1} \boldsymbol{\lambda}_i . \tag{4.145}$$

The leap size τ is variable and can be adjusted to the progress of the reaction.

Tau-Leap Algorithm

A τ-leap algorithm starts from an initial set of variables $\boldsymbol{\mathcal{X}}(t_0) = \mathbf{n}(t_0) = \mathbf{n}(0)$. Then for each $j = 1, \ldots, K$, a sample value κ_j of the random variable $\mathcal{K}_j$ is drawn from the Poissonian $\pi_{\kappa_j}\big(\alpha_j(\mathbf{n}(0), t_0)\big)$ and the time and population vector are increased in increments, viz., $t_1 = t_0 + \tau$ and $\boldsymbol{\mathcal{X}}(t_1) = \mathbf{n}_1 = \mathbf{n}(0) + \boldsymbol{\lambda}_0$. Progressive iterations $t_i = t_{i-1} + \tau$ and $\mathbf{n}_i = \mathbf{n}_{i-1} + \boldsymbol{\lambda}_{i-1}$ are performed until one reaches the final time t_N. What is still missing to complete the τ-leap algorithm is a method for determining the loop sizes $\tau_i; i = 0, \ldots, N$. The obvious condition is *effective infinitesimality* of the increments $|\alpha_j(\mathbf{n} + \boldsymbol{\lambda}) - \alpha_j(\mathbf{n})|$, for all reaction channels $j = 1, \ldots, K$. Finding optimal procedures is a major part of the art of carrying out τ-leap simulations, and we refer here to the enormous literature [11, 13, 73–76, 267, 343, 464, 490, 581], where references to many other papers dealing with the choice of the best time interval can also be found.

[66] In case the τ-leap condition is fulfilled the reaction propensity $\alpha_j(\mathbf{n})$ is identical to $\chi_j(\mathbf{n})$ defined in Sect.3.4.3.

The τ-leap method is not only a valuable computational approach. It can also be seen as providing a link between the chemical master equation (CME) and the chemical Langevin equation (CLE), in the sense that a coarse-graining of time intervals of size τ is introduced (Sect. 4.2.4).

Hybrid Methods

Another class of techniques applied to speed up stochastic simulations are hybrid methods. Hybrid systems are a class of dynamical systems that integrate continuous and discrete dynamics [247, 473]. In essence, a hybrid algorithm handles fast-varying variables as continuous, whence either Langevin equations or deterministic rate equations are integrated, and restricts the discrete description to the slowly changing particle numbers. The part of the algorithm that wastes most computer time is thereby eliminated: fast variation of numerically large variables requires an enormously large number of individual jumps. Since fluctuations are relatively small by the $\sqrt{N}$-law, their neglect causes a relatively small error, and hybrid algorithms often yield highly accurate trajectories for stochastic processes, thus providing very useful tools in practice.

But despite their importance for practical purposes, hybrid methods so far lack a solid theoretical background, although many attempts at careful mathematical analysis have been made. As representative examples, we mention here two sources [78, 266].

4.6.3 *The Simulation Algorithm*

The chemical master equation (4.143) is the basis of the simulation algorithm [213], and it is important to realize how the simulation tool fits into the general theoretical framework of master equations. However, the simulation algorithm is not based on the probability function $P(\mathbf{n}, t|\mathbf{n}_0, t_0)$, but on another related probability density $p(\tau, \boldsymbol{\mu}|\mathbf{n}, t)$, which expresses the probability that, given $\boldsymbol{\mathcal{X}}(t) = \mathbf{n}$, the *next reaction* in the system will occur in the infinitesimal time interval $[t + \tau, t + \tau + \mathrm{d}\tau[$, *and* that it will be an $\mathbf{R}_\mu$ reaction.

The probability function $p(\tau, \boldsymbol{\mu}|\mathbf{n}, t)$ is the joint density of two random variables: (i) the *time to the next reaction* τ, and (ii) the *index of the next reaction* $\boldsymbol{\mu}$. The possible values of the two random variables are given by the domain of the real variable $0 \le \tau < \infty$ and the integer variable $1 \le \boldsymbol{\mu} \le K$. In order to derive an explicit formula for the probability density $p(\tau, \boldsymbol{\mu}|\mathbf{n}, t)$, we introduce the quantity

$$\alpha(\mathbf{n}) = \sum_{\mu=1}^{K} \chi_\mu(\mathbf{n}) = \sum_{\mu=1}^{K} \gamma_\mu h_\mu(\mathbf{n}) \tag{4.146}$$

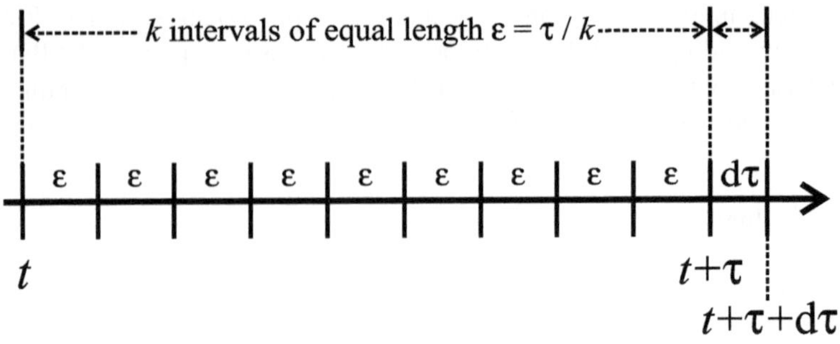

Fig. 4.45 Partitioning of the time interval $[t, t+\tau+\mathrm{d}\tau[$. The entire interval is subdivided into $(k+1)$ nonoverlapping subintervals. The first k intervals are of equal size $\varepsilon = \tau/k$ and the $(k+1)$th interval is of length $\mathrm{d}\tau$

and consider the time interval $[t, t+\tau+\mathrm{d}\tau[$ to be partitioned into $k+1$ subintervals, where $k > 1$. The first k of these intervals are chosen to be of equal length $\varepsilon = \tau/k$, and together they cover the interval $[t, t+\tau[$, leaving the interval $[t+\tau, t+\tau+\mathrm{d}\tau[$ as the remaining $(k+1)$th part (Fig. 4.45). With $\mathcal{X}(t) = \mathbf{n}$, the probability $p(\tau, \boldsymbol{\mu}|\mathbf{n}, t)$ describes the event *no reaction* occurring in each of the k ε-size subintervals and *exactly one* $\mathbf{R}_\mu$ *reaction* in the final infinitesimal $\mathrm{d}\tau$ interval. Making use of conditions 1 and 2, along with the multiplication law of probabilities, we find

$$p(\tau, \boldsymbol{\mu}|\mathbf{n}, t)\mathrm{d}\tau = \Big(1 - \alpha(\mathbf{n})\varepsilon + o(\varepsilon)\Big)^k \Big(\gamma_\mu h_\mu(\mathbf{n})\mathrm{d}\tau + o(\mathrm{d}\tau)\Big) .$$

Dividing both sides by $\mathrm{d}\tau$ and taking the limit $\mathrm{d}\tau \downarrow 0$ yields

$$p(\tau, \boldsymbol{\mu}|\mathbf{n}, t) = \Big(1 - \alpha(\mathbf{n})\varepsilon + o(\varepsilon)\Big)^k \gamma_\mu h_\mu(\mathbf{n}) .$$

This equation is valid for any integer $k > 1$, so its validity is also guaranteed for $k \to \infty$. Next we rewrite the first factor on the right-hand side of the equation as

$$\begin{aligned}\Big(1 - \alpha(\mathbf{n})\varepsilon + o(\varepsilon)\Big)^k &= \left(1 - \frac{\alpha(\mathbf{n})\, k\varepsilon + k\, o(\varepsilon)}{k}\right)^k \\ &= \left(1 - \frac{\alpha(\mathbf{n})\tau + \tau o(\varepsilon)/\varepsilon}{k}\right)^k ,\end{aligned}$$

and take the limit $k \to \infty$, whereby we make use of the simultaneously occurring convergence $o(\varepsilon)/\varepsilon \downarrow 0$:

$$\lim_{k\to\infty} \Big(1 - \alpha(\mathbf{n})\varepsilon + o(\varepsilon)\Big)^k = \lim_{k\to\infty} \left(1 - \frac{\alpha(\mathbf{n})\tau}{k}\right)^k = \mathrm{e}^{-\alpha(\mathbf{n})\tau} .$$

By substituting this result into the initial equation for the probability density of the occurrence of a reaction, we find

$$\begin{aligned} p(\tau, \boldsymbol{\mu}|\mathbf{n}, t) &= \alpha(\mathbf{n}) \frac{\gamma_\mu h_\mu(\mathbf{n})}{\alpha(\mathbf{n})} \mathrm{e}^{-\alpha(\mathbf{n})\tau} \\ &= \gamma_\mu h_\mu(\mathbf{n}) \exp\left(- \sum_{\nu=1}^{K} \gamma_\nu h_\nu(\mathbf{n}) \tau \right) . \end{aligned} \tag{4.147}$$

Equation (4.147) provides the mathematical basis for the stochastic simulation algorithm. Given $\boldsymbol{\mathcal{X}}(t) = \mathbf{n}$, the probability density consists of two independent probabilities, where the first factor describes the *time to the next reaction* and the second factor the *index of the next reaction*. These factors correspond to two statistically independent random variables η_2 and η_1.

Pseudorandom Numbers

In order to implement (4.147) for computer simulation, we consider probability densities of two unit-interval uniform random variables η_1 and η_2 in order to find the conditions to be imposed on a statistically exact sample pair $(\tau, \boldsymbol{\mu})$: η_1 has an exponential density function with decay constant $\alpha(\mathbf{n})$ from (4.146):

$$\tau = \frac{1}{\alpha(\mathbf{n})} \ln \frac{1}{\eta_1} , \tag{4.148a}$$

and m is taken to be the *smallest* integer which satisfies

$$\boldsymbol{\mu} = \inf \left\{ m \middle| \sum_{\mu=1}^{m} \gamma_\mu h_\mu(\mathbf{n}) > \alpha(\mathbf{n}) \eta_2 \right\} . \tag{4.148b}$$

After the values for τ and $\boldsymbol{\mu}$ have been determined, the action *advance the state vector $\boldsymbol{\mathcal{X}}(t)$ of the system* is carried out:

$$\boldsymbol{\mathcal{X}}(t) = \mathbf{n} \longrightarrow \boldsymbol{\mathcal{X}}(t + \tau) = \mathbf{n} + \mathbf{s}_\mu .$$

Repeated application of the advancement procedure is the essence of the stochastic simulation algorithm. It is important to realize that this advancement procedure is exact insofar as η_1 and η_2 are obtained by *fair samplings* from a unit interval uniform random number generator. In other words, the correctness of the procedure depends on the quality of the random number generator. Two further issues are important:

(i) The algorithm operates with an internal time control that corresponds to real time for the chemical process.
(ii) Contrary to the situation in differential equation solvers, the discrete time steps are not finite interval approximations of an infinitesimal time step, but instead the population vector $\boldsymbol{\mathcal{X}}(t)$ maintains the value $\boldsymbol{\mathcal{X}}(t) = \mathbf{n}$ throughout the entire finite time interval $[t, t+\mathrm{d}\tau[$ and then changes abruptly to $\boldsymbol{\mathcal{X}}(t+\tau) = \mathbf{n}+\mathbf{s}_\mu$ at the instant $t + \tau$ when the $\mathbf{R}_\mu$ reaction occurs. In other words, there is no *blind interval* during which the algorithm is unable to record changes.

Nonuniformly Distributed Random Numbers

In (4.148a), the desired distribution of the pseudorandom variable was built into the expression and the input η_1 was drawn from the uniform distribution. The general approach to deriving a continuous random variable with the (cumulative) distribution function $F_\mathcal{X}$ is called *inverse transform sampling*. If $\mathcal{X}$ has the distribution $F_\mathcal{X}$, then the random variable $\mathcal{Y} = F_\mathcal{X}(\mathcal{X})$ is uniformly distributed on the unit interval, i.e., $\mathcal{Y} : F_\mathcal{Y} = \mathcal{U} \in [0, 1]$, and this statement can be inverted to obtain $F_\mathcal{X}^{-1}(\mathcal{Y}) = F_\mathcal{X}$. The following three-step procedure can be used to calculate pseudorandom variables for an invertible distribution function F:

(i) generate a pseudorandom variable u from $\mathcal{U} \in [0, 1]$,
(ii) compute the value $x = F^{-1}(u)$ such that $u = F(x)$, and
(iii) take x as the pseudorandom variable drawn form a distribution given by F.

The procedure is used for the generation of the often required normally distributed pseudorandom numbers and is called the Box–Muller transform [62], named after the two mathematicians George Box and Mervin Muller. Generalizations to discrete variables and arbitrary invertible distribution functions can be found in the monograph [106].

Structure of the Simulation Algorithm

The time evolution of the population is described by the vector $\boldsymbol{\mathcal{X}}(t) = \mathbf{n}(t)$, which is updated after every individual reaction event. Reactions are chosen from the set $\mathcal{R} = \{\mathbf{R}_\mu, \mu = 1, \ldots, K\}$ defined by the reaction mechanism, and reaction probabilities are specified in $\boldsymbol{\alpha}(\mathbf{n}) = \big(\gamma_1 h_1(\mathbf{n}), \ldots, \gamma_K h_K(\mathbf{n})\big)^\mathrm{t}$, which is also updated after every individual reaction event. Reactions are classified according to molecularity in Table 4.2. Updating is performed by adding the stoichiometric vector $\mathbf{s}_\nu$ of the chosen reaction $\mathbf{R}_\nu$, that is, $\mathbf{n}(t + \mathrm{d}\tau) = \mathbf{n}(t) + \mathbf{s}_\nu$, where $\mathbf{s}_\nu$ represents a column of the stoichiometric matrix S.

Table 4.2 Combinatorial frequency functions $h_\mu(\mathbf{n})$ for elementary reactions. Reactions are ordered with respect to reaction order, which in the case of mass action is identical to the molecularity of the reaction. Order zero implies that no reactant molecule is involved and the products come from an external source, e.g., from the influx in a flow reactor. Orders 0, 1, 2, and 3 mean that zero, one, two, or three molecules are involved in the elementary step, respectively

Number	Reaction			Order	$h_\mu(\mathbf{n})$
1	$*$	$\longrightarrow$	products	0	1
2	A	$\longrightarrow$	products	1	n_A
3	A + B	$\longrightarrow$	products	2	$n_A n_B$
4	2A	$\longrightarrow$	products	2	$n_A(n_A - 1)$
5	A + B + C	$\longrightarrow$	products	3	$n_A n_B n_C$
6	2A + B	$\longrightarrow$	products	3	$n_A(n_A - 1)n_B$
7	3A	$\longrightarrow$	products	3	$n_A(n_A - 1)(n_A - 2)$

The Gillespie algorithm comprises five steps:

Step 0. **Initialization.** The time variable is set to $t = 0$, the initial values of all N variables $\mathcal{X}_1, \ldots, \mathcal{X}_N$ for the species, i.e., $\mathcal{X}_k$ for species X_k, are stored, the values for the K parameters of the reactions $\mathbf{R}_\mu, \gamma_1, \ldots, \gamma_K$, are stored, and the combinatorial expressions are incorporated as factors for calculating the reaction rate vector $\boldsymbol{\alpha}(\mathbf{n})$ according to Table 4.2 and the probability density $P(\tau, \mu)$. Sampling times $t_1 < t_2 < \ldots$ and the stopping time t_{stop} are specified, the first sampling time is set to t_1 and stored, and the *pseudorandom* number generator is initialized by means of *seeds* or *at random*.

Step 1. **Monte Carlo Step.** A pair of random numbers is created (τ, μ) by the random number generator according to the joint probability function $P(\tau, \mu)$. In essence, three explicit methods can be used: the *direct* method, the *first-reaction* method, and the *next-reaction* method.

Step 2. **Propagation Step.** (τ, μ) is used to advance the simulation time t and to update the population vector $\mathbf{n}$, $t \to t+\tau$ and $\mathbf{n} \to \mathbf{n}+\mathbf{s}_\mu$, then all changes are incorporated in a recalculation of the reaction rate vector $\boldsymbol{\alpha}(\mathbf{n})$.

Step 3. **Time Control.** We check whether or not the simulation time has been advanced through the next sampling time t_i, and for $t > t_i$, we send the current t and the current $\mathbf{n}(t)$ to the output storage and advance the sampling time $t_i \to t_{i+1}$. Then, if $t > t_{\text{stop}}$ or if no more reactant molecules remain leading to $h_\mu = 0, \forall\, \mu = 1, \ldots, K$, we finalize the calculation by switching to Step 4. Otherwise we continue with Step 1.

Step 4. **Termination.** We prepare for the final output by setting flags for early termination or other unforseen stops and send the final time t and the final $\mathbf{n}$ to the output storage and terminate the computation.

A caveat is needed for the integration of stiff systems, where the values of individual variables can vary by many orders of magnitude, and such a situation might trap the calculation by slowing down time progress.

The Monte Carlo Step

Pseudorandom numbers are drawn from a random number generator of sufficient quality, where the quality refers to whether or not there are very long recurrence cycles and the proximity of the distribution of the pseudorandom numbers r to the uniform distribution on the unit interval:

$$0 \leq a < b \leq 1 \quad \Longrightarrow \quad P(a \leq \eta \leq b) = b - a \,.$$

With this prerequisite, we mention three methods which use output values η of the pseudorandom number generator to generate a random pair $(\tau, \boldsymbol{\mu})$ with the prescribed probability density function $P(\tau, \boldsymbol{\mu})$.

The Direct Method

The two-variable probability density is written as the product of two one-variable density functions:

$$P(\tau, \boldsymbol{\mu}) = P_1(\tau) P_2(\boldsymbol{\mu}|\tau) \,.$$

Here, $P_1(\tau)\mathrm{d}\tau$ is the probability that the next reaction will occur between times $t + \tau$ and $t + \tau + \mathrm{d}\tau$, irrespective of which reaction it might be, and $P_2(\boldsymbol{\mu}|\tau)$ is the probability that the next reaction will be an $\mathbf{R}_\mu$, given that the next reaction occurs at time $t + \tau$.

By the probability addition theorem, $P_1(\tau)\mathrm{d}\tau$ is obtained by summing $P(\tau, \boldsymbol{\mu})\mathrm{d}\tau$ over all reactions $\mathbf{R}_\mu$:

$$P_1(\tau) = \sum_{\mu=1}^{K} P(\tau, \boldsymbol{\mu}) \,. \tag{4.149}$$

Combining the last two equations, we obtain for $P_2(\boldsymbol{\mu}|\tau)$

$$P_2(\boldsymbol{\mu}|\tau) = \frac{P(\tau, \boldsymbol{\mu})}{\sum_{\nu}^{K} P(\tau, \boldsymbol{\nu})} \,. \tag{4.150}$$

Equations (4.149) and (4.150) express the two one-variable density functions in terms of the original two-variable density function $P(\tau, \boldsymbol{\mu})$. From (4.147), we substitute into $P(\tau, \boldsymbol{\mu}) = p(\tau, \boldsymbol{\mu}|\mathbf{n}, t)$, simplifying the notation by using

$$\alpha_\mu \equiv \gamma_\mu h_\mu(\mathbf{n}) \,, \quad \alpha = \sum_{\mu=1}^{K} \alpha_\mu \equiv \sum_{\mu=1}^{K} \gamma_\mu h_\mu(\mathbf{n}) \,.$$

This leads to

$$P_1(\tau) = \alpha \exp(-\alpha\tau)\,, \quad 0 \le \tau < \infty\,,$$
$$P_2(\mu|\tau) = P_2(\mu) = \frac{\alpha_\mu}{\alpha}\,, \quad \mu = 1, \ldots, K\,. \tag{4.151}$$

As indicated, in this particular case, $P_2(\mu|\tau)$ turns out to be independent of τ. Both one-variable density functions are properly normalized over their domains of definition:

$$\int_0^\infty P_1(\tau)\mathrm{d}\tau = \int_0^\infty \alpha\, \mathrm{e}^{-\alpha\tau}\mathrm{d}\tau = 1\,, \quad \sum_{\mu=1}^{K} P_2(\mu) = \sum_{\mu=1}^{K} \frac{\alpha_\mu}{\alpha} = 1\,.$$

Thus, in the *direct* method, a random value τ is created from a random number η_1 on the unit interval, and the distribution $P_1(\tau)$ by taking

$$\tau = -\frac{\ln \eta_1}{\alpha}\,. \tag{4.152}$$

The second task is to generate a random integer $\widehat{\mu}$ according to $P_2(\mu|\tau)$ in such a way that the pair (τ, μ) will be distributed as prescribed by $P(\tau, \mu)$. For this purpose, another random number η_2 is drawn from the unit interval, and then $\widehat{\mu}$ is taken to be the integer that satisfies

$$\sum_{\nu=1}^{\mu-1} \alpha_\nu < \eta_2\alpha \le \sum_{\nu=1}^{\mu} \alpha_\nu\,. \tag{4.153}$$

The values α_1, α_2, and so on, are cumulatively added in sequence until their sum is observed to be greater than or equal to $\eta_2\alpha$, and then $\hat{\mu}$ is set equal to the index of the last α_ν term that was added. Rigorous justifications for (4.152) and (4.153) can be found in [206, pp.431–433]. If a fast and reliable uniform random number generator is available, the *direct* method can be easily programmed and rapidly executed. Thus, it represents a simple, fast, and rigorous procedure for the implementation of the *Monte Carlo* step of the simulation algorithm.

The First-Reaction Method

This alternative method for the implementation of the *Monte Carlo* step of the simulation algorithm is not quite as efficient as the *direct* method, but it is worth presenting here because it adds insight into the stochastic simulation approach. Again the notation $\alpha_\nu \equiv \gamma_\nu h_\nu(\mathbf{n})$ is adopted and then it is straightforward to derive

$$P_\nu(\tau)\mathrm{d}\tau = \alpha_\nu \exp(-\alpha_\nu\tau)\mathrm{d}\tau \tag{4.154}$$

from (4.138) and (4.139). Hence, $P_\nu(\tau)$ would indeed be the probability at time t for an $\mathbf{R}_\nu$ reaction to occur in the time interval $[t+\tau, t+\tau+d\tau[$, were it not for the fact that the number of $\mathbf{R}_\nu$ reactant combinations might have been altered between t and $t+\tau$ by the occurrence of other reactions. Taking this into account, a *tentative reaction time* τ_ν for $\mathbf{R}_\nu$ is generated according to the probability density function $P_\nu(\tau)$, and in fact, the same can be done for all reactions $\{\mathbf{R}_\mu\}$. We draw a random number η_ν from the unit interval and compute

$$\tau_\nu = -\frac{\ln \eta_\nu}{\alpha_\nu}, \qquad \boldsymbol{\nu} = 1, \ldots, K. \tag{4.155}$$

From these K tentative next reactions, the one which occurs first is chosen to be the actual next reaction:

$$\begin{aligned} \tau &= \text{smallest } \tau_\nu \text{ for all } \boldsymbol{\nu} = 1, \ldots, K, \\ \boldsymbol{\mu} &= \boldsymbol{\nu} \text{ for which } \tau_\nu \text{ is smallest}. \end{aligned} \tag{4.156}$$

Daniel Gillespie [206, pp.420–421] provides a straightforward proof that the random $(\tau, \boldsymbol{\mu})$ obtained by the first-reaction method is in full agreement with the probability density $P(\tau, \boldsymbol{\mu})$ from (4.147).

It is tempting to try to extend the first-reaction methods by letting the next reaction be the one for which τ_ν has the second smallest value. However, this conflicts with correct updating of the particle number vector $\mathbf{n}$, because the results of the first reaction are not incorporated into the combinatorial terms $h_\mu(\mathbf{n})$. Using the second earliest reaction would, for example, allow the second reaction to involve molecules already destroyed in the first reaction, but would not allow the second reaction to involve molecules created in the first reaction.

The Next-Reaction Method

In the next-reaction method, one makes use of all reaction times τ_ν that were calculated for the next reaction step as in the first reaction method [204]. Three expensive actions taking time proportional to the number of reactions K are performed in every iteration step of the first-reaction method: (i) updating all K values α_μ, (ii) generating a putative reaction time τ_ν for every $\boldsymbol{\nu}$, and (iii) identifying the shortest putative τ_μ. The next-reaction method is somewhat more involved but avoids the time-wasting calculations of future random times that are not used after the reaction event has occurred, it has been proven to be exact in the same sense as the direct and first-reaction methods.

The basic idea of the next-reaction method is to reuse the already calculated times τ_ν, wherever this is appropriate. There is, however, one important caveat: Monte Carlo simulations generally assume random numbers that are statistically independent, and therefore the reuse of random numbers and quantities derived from them is illegitimate. In the specific case of the next-reaction algorithm, however, it has been proven that all putative reaction times can be reused except the time τ_μ of the reaction which was executed. By means of a *dependency graph* that follows directly from the reaction mechanism, only the minimal number of α_μ values are updated. Storage of all τ_ν times together with the α_ν values is required and efficient implementations of the next-reaction method use special data structures [204].

Thus, the first-reaction method is just as rigorous as the direct method, and it is probably easier to implement in a computer code than the direct method. From a computational efficiency point of view, however, the direct method is preferable because, for $K \geq 3$, it requires fewer random numbers, whence the first-reaction method is wasteful. This question of economic use of computer time is not unimportant, because stochastic simulations generally tax the random number generator quite heavily. For $K \geq 3$ and in particular for large K, the direct method is probably the method of choice for the Monte Carlo step. The next-reaction method is exact too. It can be seen as a more efficient extension of the first-reaction method, which for sufficiently large M and K also beats the direct method in efficiency, because asymptotically it requires only one random number per reaction event.

Computer Codes

An early computer code of the simple version of the algorithm described (still in FORTRAN) can be found in [206]. Meanwhile many attempts were made in order to speed up computations and allow for simulation of stiff systems (see, e.g., [73]). A recent review of the simulation methods also contains a discussion of various improvements of the original code [213]. Here, we mention a few packages that are representative for many others.

Several computer codes in different languages including C++ are now available on the internet, and unless one aims at an efficient program for some special task, it would not pay to write another code except perhaps for educational purposes. The simulations reported here were performed with a *Mathematica 7,8* implementation that also runs with minor modifications under *Mathematica 9* [499]. Other equally efficient and user-friendly implementations are available for *Matlab* and other high-level user interfaces. A didactic introduction can be found in [251] and sample programs for *Matlab* are available from `personal.strath.ac.uk/d.j.higham/algfiles.html`.

A program package with the name *StochKit* has been developed. It has been designed by the group of Linda Petzold, in particular for the simulation of biochemical systems [341]. Version 2 of this software toolkit was released in 2012 [474]. A biologically motivated simulation routine for Gillespie's algorithm has been developed by Bruce Shapiro within the xCellerator software design project [499]. A slightly older simulation software called *StochSim* [168, 332] has been worked out by Carl Jason Morton-Firth, Thomas Simon Shimizu and Nicolas Le Novère. It was designed especially for modeling bacterial chemotaxis and signalling networks.

4.6.4 Examples of Simulations

In this section we shall be dealing with some selected examples of numerical simulations using the Gillespie algorithm. We begin with the reversible monomolecular reaction as a proper reference for the comparison of calculated and simulated data. Then we shall be concerned with two problems: (i) the special role of fluctuations in autocatalytic reactions and (ii) the properties of the extended Michaelis–Menten mechanism that are not accessible by the analytic approach. The application of publicly available software to stochastic simulations of chemical reactions with few molecules is reported here in order to create a feeling for the obtainable results, because most users apply the computer programs only to large networks. Rigorous mathematical analysis of stochastic simulations is quite demanding. In particular there are substantial problems arising from the slow convergence of the calculated results in the approach to long time limits. For details, we refer to one monograph out of a rich collection [27]. We remark that the results for stationary states can be readily derived analytically (see Sect. 3.2.3 and in particular (3.100)). Analytic expressions are often very complicated, but they can nevertheless be useful, because they provide exact values for comparison and they often allow for the derivation of the limits $t \to 0$ and $t \to \infty$.

Monomolecular Conversion Reaction $\mathsf{A} \rightleftharpoons \mathsf{B}$

The simple conversion or isomerization reaction is chosen here as a test example for numerical simulation that will be used to illustrate harvesting of trajectories, evaluation of samples, and convergence of the first and second moments. Monomolecular isomerization is particularly well suited for this purpose: simple analytical expressions are available since the probability density $P_n(t)$ is a binomial distribution (Sect. 2.3.2). We shall distinguish the symmetric case $k = l$, which gives rise to a symmetric probability density in the form of the binomial distribution with $p = q = 1/2$, and asymmetric cases with $p \neq q$.

In order to get a feeling for the computational efforts required to calculate the first and second moments by trajectory sampling, we consider the simple case of the equilibrium density of the monomolecular reaction. Individual trajectories are calculated and analyzed according to (3.4). The long-time density distribution $\bar{P}_n$ with $n \in \mathbb{N}$ is obtained from a sufficiently large sample of N trajectories by defining it at a sufficiently late time $t_{\max}$ after the process has reached stationarity. Ergodic theory tells us that the ensemble average equals the time average, but for most practical purposes it is much easier to compute an ensemble average than an average over long times. The expectation values and variances shown in Table 4.3 were calculated from the stationary densities. Ten samples were computed for each combination of rate parameters (k, l) and sample size N, and the values of $\bar{\mathrm{E}}(\mathcal{X}_\mathsf{A})$ and $\mathrm{var}(\mathcal{X}_\mathsf{A})$ and the standard error bands were derived by conventional statistics. The data in the table demonstrate the slow convergence of stochastic properties. An

Table 4.3 Simulation and calculation of expectation values and fluctuations for the reactions $A \rightleftharpoons B$ at equilibrium. The table shows the long-time expectation values and variances of the random variable, viz., $\bar{E}(\mathcal{X}_A)$ and $\overline{\mathrm{var}}(\mathcal{X}_A)$, the width of the one-$\sigma$ confidence interval obtained for simulations with sample sizes $N = 1000$, 10,000, and 100,000, together with the calculated values for different rate constants k and l [t^{-1}]. The results were obtained from ten individual runs with different random number seeds. Pseudorandom number generator: *Mathematica*, ExtendedCA: $s = 491, 919, 521, 877, 233, 373, 089, 773, 131$, and 631, and the standard deviations are unbiased

Parameters				Sample size	Simulation		Calculation	
a_0	b_0	k	l	N	$\bar{E}(\mathcal{X}_A)$	$\overline{\mathrm{var}}(\mathcal{X}_A)$	$\bar{E}(\mathcal{X}_A)$	$\overline{\mathrm{var}}(\mathcal{X}_A)$
5	5	5	5	1,000	4.9984 ± 0.0507	2.5155 ± 0.0908	5	2.5
				10,000	5.0015 ± 0.0208	2.4944 ± 0.0297	5	2.5
				100,000	4.9998 ± 0.0051	2.4975 ± 0.0120	5	2.5
7	3	3	7	1,000	7.0031 ± 0.0474	2.1007 ± 0.1040	7	2.1
				10,000	7.0025 ± 0.0096	2.1003 ± 0.0210	7	2.1
				100,000	6.9978 ± 0.0029	2.0997 ± 0.0087	7	2.1
9	1	1	9	1,000	9.0150 ± 0.0217	0.8662 ± 0.0581	9	0.9
				10,000	8.9985 ± 0.0101	0.8962 ± 0.0200	9	0.9
				100,000	8.9987 ± 0.0030	0.9029 ± 0.0046	9	0.9

increase in the sample size by two orders of magnitude gives rise to a reduction in the error bands by a factor of approximately 1/10, which is in agreement with a *rule of thumb* of $1/\sqrt{N}$ convergence. More details on the problems of convergence with trajectory sampling can be found in the monograph [27].

Reversible Association Reaction $A + B \rightleftharpoons 2C$

To compare the reversible autocatalytic reaction $A + X \rightleftharpoons 2X$ with a non-autocatalytic process that comes as close as possible to it from the standpoint of stoichiometry, we choose the reversible association reaction $A+B \rightleftharpoons 2C$: the single autocatalytic species X is replaced by two separate species, B and C. Recalling our results from Sect. 4.3.3, the main problem in the derivation of analytical solutions was the tridiagonal structure of the transition matrix W, which did not allow for the derivation of analytical expressions for the eigenvalues, whence we could only derive results for the equilibrium distribution.

The three random variables $\mathcal{X}_A$, $\mathcal{X}_B$, and $\mathcal{X}_C$ are reduced to a single degree of freedom by the two conservation relations: (i) $\mathcal{X}_B - \mathcal{X}_A = b_0 - a_0 = \vartheta_0$ and (ii) $\mathcal{X}_A + \mathcal{X}_B + \mathcal{X}_C = a_0 + b_0 + c_0 = n_0$ or $\mathcal{X}_C + 2\mathcal{X}_A = c_0 + 2a_0 = \gamma_0$. The deterministic kinetic rate equation can be integrated analytically, although the result looks quite complicated:

$$\widehat{a}(t) = \frac{1}{\alpha} \frac{\beta_2(\alpha a_0 + \beta_1) - \beta_1(\alpha a_0 + \beta_2)e^{-Dt}}{\alpha a_0 + \beta_2 - (\alpha a_0 + \beta_1)e^{-Dt}}, \tag{4.157}$$

where

$$D = \sqrt{4kl\gamma_0^2 + 8kl\gamma_0\vartheta_0 + k^2\vartheta_0^2}\ , \quad \alpha = 2(4l - k)\ ,$$

$$\beta_1 = -(4l\gamma_0 + k\vartheta_0 + D)\ , \quad \beta_1 = -(4l\gamma_0 + k\vartheta_0 - D)\ .$$

Nevertheless, (4.157) is useful as an alternative to numerical integration for the calculation of solution curves and for the derivation of analytical expressions for limits, e.g., $\lim_{t\to\infty} \widehat{a}(t) = \bar{a} = -\beta_2/\alpha$.

The reaction $\mathsf{A} + \mathsf{B} \rightleftharpoons 2\mathsf{C}$ has been studied here by trajectory sampling, and the results obtained for three different sample sizes are shown in Fig. 4.46. Starting from sharp initial conditions $P_n(0) = \delta_{n,n_0}$ at $t = t_0$, the one standard deviation band increases in width until it reaches the stationary value or passes through a maximum, before decreasing again towards equilibrium. The same behavior has been observed, for example, for the reversible monomolecular reaction. For small numbers of molecules, the deterministic solution $a(t)$ differs slightly but significantly from the expectation value $\mathrm{E}\big(\mathcal{X}_{\mathsf{A}}(t)\big)$, and this is in full agreement with expectations, since the first jump moment α_1 is not linear in n and contributions from higher jump moments to the expectation value vanish only in the case of linear kinetics (Sect. 3.2.3). The one-σ confidence band $\mathrm{E} \pm \sigma$ is twice as broad for $\mathcal{X}_{\mathsf{C}}$, compared to $\mathcal{X}_{\mathsf{A}}$ or $\mathcal{X}_{\mathsf{B}}$. The relative width of the fluctuation band decreases with increasing population size and for sufficiently large populations, $n_0 = 25,000$ for example (not shown in the figure), all four curves $a(t)$, $\mathrm{E}\big(n_{\mathsf{A}}(t)\big)$, $\mathrm{E}\big(n_{\mathsf{A}}(t)\big) \pm \sigma\big(n_{\mathsf{A}}(t)\big)$, coincide within the thickness of the lines, again in full agreement with expectations from the $\sqrt{n}$-law.

Another reason for the deviation between the two curves concerns the stochastic reverse reaction $2\mathsf{C} \to \mathsf{A} + \mathsf{B}$, which is determined by the rate function $\eta_{\leftarrow}^{(\mathrm{ma})}(\mathbf{n}) = \gamma h(\mathbf{n}) = -l n_{\mathsf{C}}(n_{\mathsf{C}} - 1)$ in mass action kinetics, whereas the deterministic rate function is given by $v_{\leftarrow}^{(\mathrm{ma})} = -l n_{\mathsf{C}}^2$. This difference is particularly important in the initial period of the process, when only C is available and $n_{\mathsf{A}}(0) = n_{\mathsf{B}}(0) = 0$. Then the deterministic rate is smaller in absolute value as expressed by the initial tangent of the curve $\widehat{n}(t)$, which is less steep than that of the expectation value $\mathrm{E}\big(n(t)\big)$. This difference, however, plays only a subordinate role in the approach towards equilibrium (see the reaction $\mathsf{A} \to 2\mathsf{C}$ in Fig. 4.26).

Several methods are available for calculating stationary probability densities:

(i) The stationary solution can be obtained through Laplace and inverse Laplace transformation (Sect. 4.3.4).
(ii) It can be calculated directly from the master equation through successive multiplication of step-up and step-down transition probabilities (3.100).
(iii) It can be derived from known equilibrium concentrations (4.72b).

Here, we compare the calculated values with the results of extensive sampling of long time trajectories. We use the equilibrium concentrations as initial conditions, perform the simulations over sufficiently long times, and use the last time interval to calculate the expectation value $\mathrm{E}(\mathcal{X}_{\mathsf{A}})$ and the fluctuation band $\mathrm{E} \pm \sigma$. Quantitative

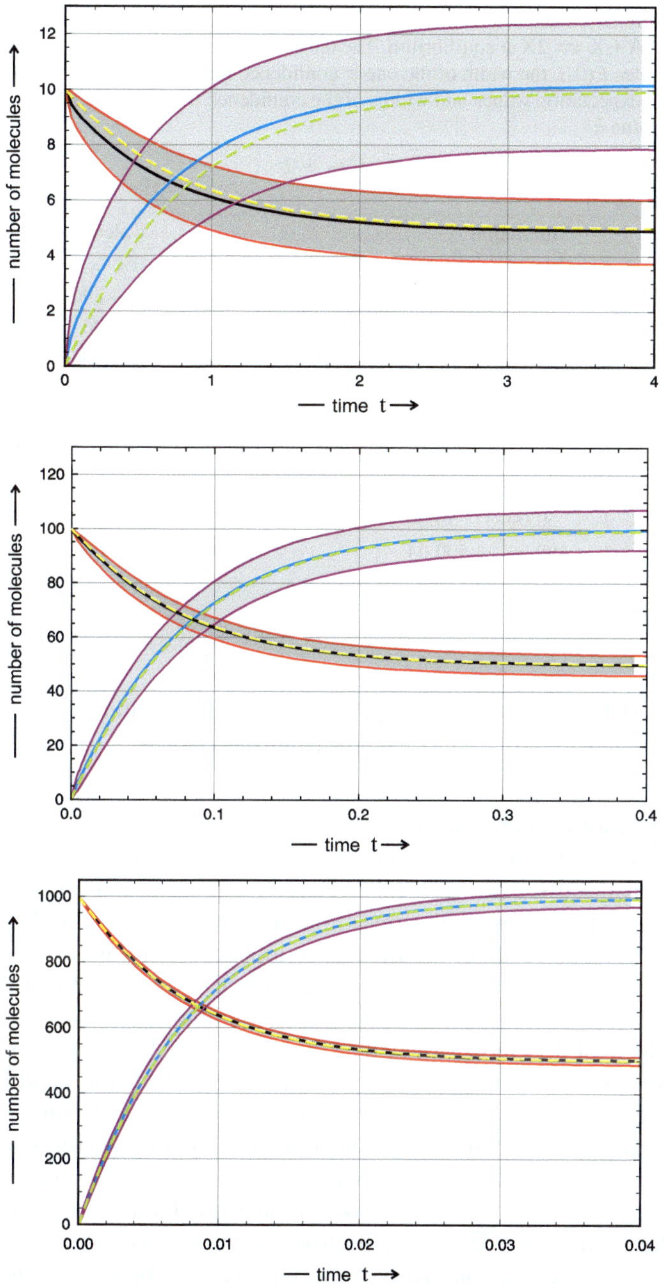

Fig. 4.46 The reversible association reaction $\mathsf{A} + \mathsf{B} \rightleftharpoons 2\mathsf{C}$. The figure shows the expectation values $\mathrm{E}\big((\mathcal{X}_{\mathsf{A}}(t)\big)$ (*black*) and $\mathrm{E}\big((\mathcal{X}_{\mathsf{C}}(t)\big)$ (*blue*), embedded between the curves $\mathrm{E}(\mathcal{X}_{\mathsf{A}}) \pm \sigma(\mathcal{X}_{\mathsf{A}})$ (*red*) and $\mathrm{E}(\mathcal{X}_{\mathsf{C}}) \pm \sigma(\mathcal{X}_{\mathsf{C}})$ (*purple*) together with the deterministic solution curves $\widehat{n}_{\mathsf{A}}(t) = a(t)$ (*yellow*) and $\widehat{n}_{\mathsf{C}} = c(t)$ (*chartreuse*). Parameter choice: $k = 0.04\,[\mathrm{M}^{-1}\mathrm{t}^{-1}]$, $l = 0.02\,[\mathrm{M}^{-1}\mathrm{t}^{-1}]$, $n_{\mathsf{C}}(0) = c_0 = 0$, and $n_{\mathsf{A}}(0) = a_0 = 10$, $n_{\mathsf{B}}(0) = b_0 = 15$ (*upper plot*), $n_{\mathsf{A}}(0) = a_0 = 100$, $n_{\mathsf{B}}(0) = b_0 = 150$ (*middle plot*), and $n_{\mathsf{A}}(0) = a_0 = 1000$, $n_{\mathsf{B}}(0) = b_0 = 1500$ (*lower plot*)

Table 4.4 Simulation and calculation of expectation values and fluctuations for the reactions $\mathsf{A}+\mathsf{B} \rightleftharpoons 2\mathsf{C}$ and $\mathsf{A}+\mathsf{X} \rightleftharpoons 2\mathsf{X}$ at equilibrium. The table shows the long-time expectation values of the random variables $\bar{E}(\mathcal{X})$, the width of the one-σ confidence interval $2\sigma(\mathcal{X})$ obtained by simulation, the calculated expectation values and widths of the confidence interval, as well as the deterministic equilibrium value $\bar{a}$

$\mathsf{A}+\mathsf{B} \rightleftharpoons 2\mathsf{C}$: $\vartheta_0 = b_0 - a_0$ $n_0 = a_0 + b_0 + 2c_0$, $l = 0.02\,[\mathrm{M}^{-1}\mathrm{t}^{-1}]$									
Parameters			Simulation				Calculation		det
ϑ_0	n_0	K	$\bar{\mathrm{E}}(\mathcal{X}_\mathsf{A})$	$\bar{\mathrm{E}}(\mathcal{X}_\mathsf{B})$	$\bar{\mathrm{E}}(\mathcal{X}_\mathsf{C})$	$2\bar{\sigma}(\mathcal{X}_\mathsf{A})$	$\bar{\mathrm{E}}(\mathcal{X}_\mathsf{A})$	$2\bar{\sigma}(\mathcal{X}_\mathsf{A})$	$\bar{a}$
5	25	2	4.885	9.885	10.229	2.301	4.896	2.414	5
50	250	2	49.89	99.89	100.22	7.527	49.90	7.566	50
500	2500	2	499.96	999.96	1000.06	23.785	499.90	23.907	500
$\mathsf{A}+\mathsf{X} \rightleftharpoons 2\mathsf{X}$: $x_0 \geq 1$, $l = 1\,[\mathrm{M}^{-1}\mathrm{t}^{-1}]$									
Parameters			Simulation				Calculation		det
a_0	x_0	K	$\bar{\mathrm{E}}(\mathcal{X}_\mathsf{A})$	$\bar{\mathrm{E}}(\mathcal{X}_\mathsf{X})$		$2\bar{\sigma}(\mathcal{X}_\mathsf{A})$	$\bar{\mathrm{E}}(\mathcal{X}_\mathsf{A})$	$2\bar{\sigma}(\mathcal{X}_\mathsf{A})$	$\bar{a}$
5	5	1	4.937	5.063		3.007	4.995	3.148	5
50	50	1	50.00	50.00		9.952	50.00	10.000	50
500	500	1	499.96	500.04		31.409	500.00	31.623	500

data for the stationary states are given in Table 4.4. The confidence band for $\mathcal{X}_\mathsf{A}$ has an approximate width of $\sqrt{\bar{\mathrm{E}}(\mathcal{X}_\mathsf{A})}$ and satisfies the $\sqrt{n}$-law.

Autocatalysis and Fluctuations

Analytical results for the irreversible autocatalytic reaction $\mathsf{A} + \mathsf{X} \rightarrow 2\mathsf{X}$ have already been discussed in Sect. 4.3.5. Here, we present and analyze the results of numerical simulations of the reversible autocatalytic process $\mathsf{A} + \mathsf{X} \rightleftharpoons 2\mathsf{X}$ in order to learn more about the difference in stochasticity between ordinary and autocatalytic processes. One feature is particularly evident and can be explained easily. When we start from the same initial conditions, the stochastic processes approach the final state more slowly than the conventional ODE solutions. Although the deterministic curve always lies within the one-σ deviation band $\mathrm{E}(t) \pm \sigma(t)$, the convergence of the stochastic solution towards the deterministic limit with increasing numbers of molecules n_0 is not obvious from the data for population sizes up to $n_0 = 10{,}000$ (Fig. 4.47). In this figure we present the simulation results for the autocatalytic process for comparison with the stoichiometrically closely related bimolecular reaction $\mathsf{A} + \mathsf{B} \rightleftharpoons 2\mathsf{C}$, discussed in the last section.

The autocatalytic reaction involves two random variables $\mathcal{X}_\mathsf{A}(t)$ and $\mathcal{X}_\mathsf{X}(t)$, and one conservation relation $\mathcal{X}_\mathsf{A} + \mathcal{X}_\mathsf{X} = a_0 + x_0 = n_0$, thus leaving a single degree of freedom. The kinetic equation $\mathrm{d}a/\,\mathrm{d}t = -kax - lx^2$ can be integrated analytically

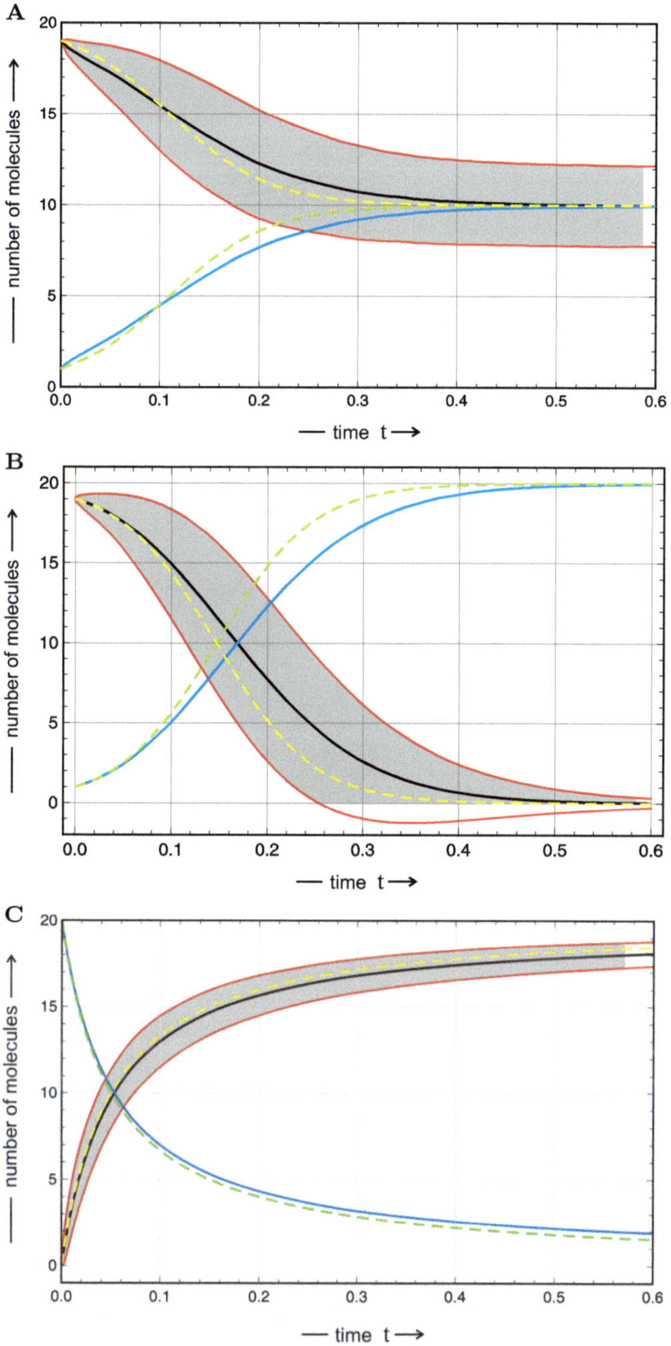

Fig. 4.47 The role of fluctuations in autocatalytic reactions. Continued on next page

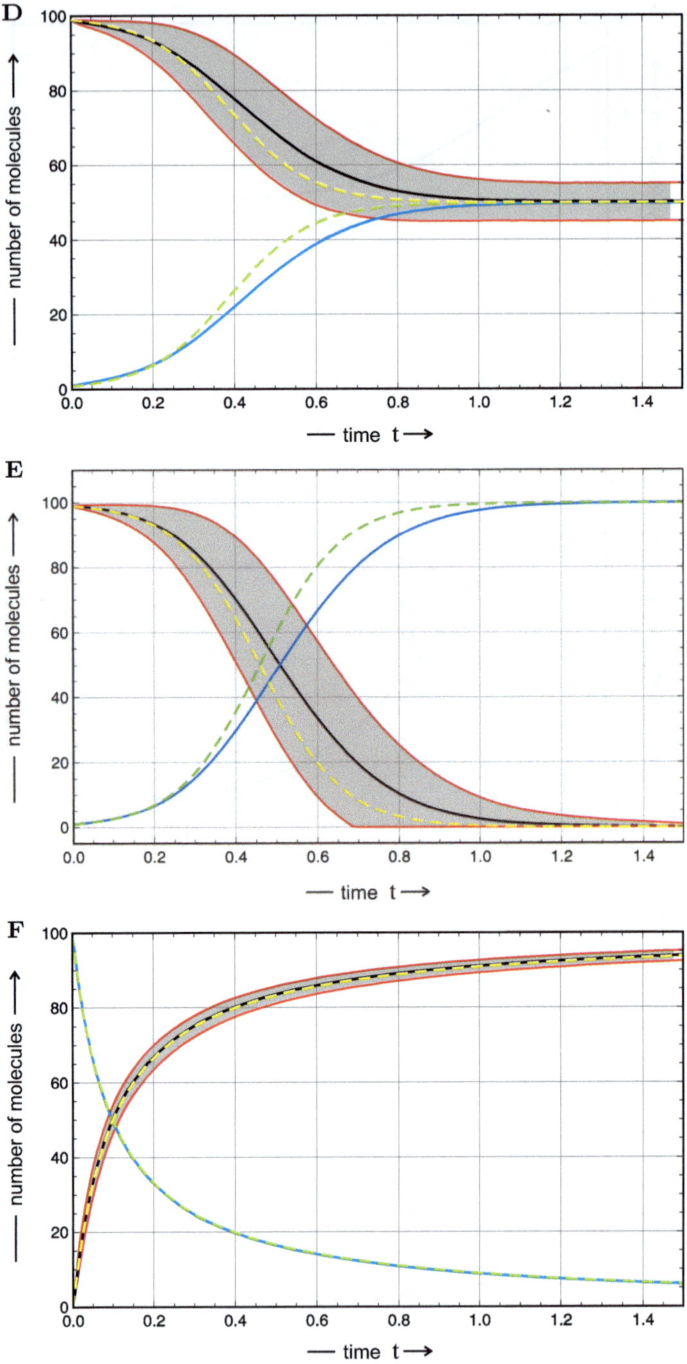

Fig. 4.47 The role of fluctuations in autocatalytic reactions. Continued on next page

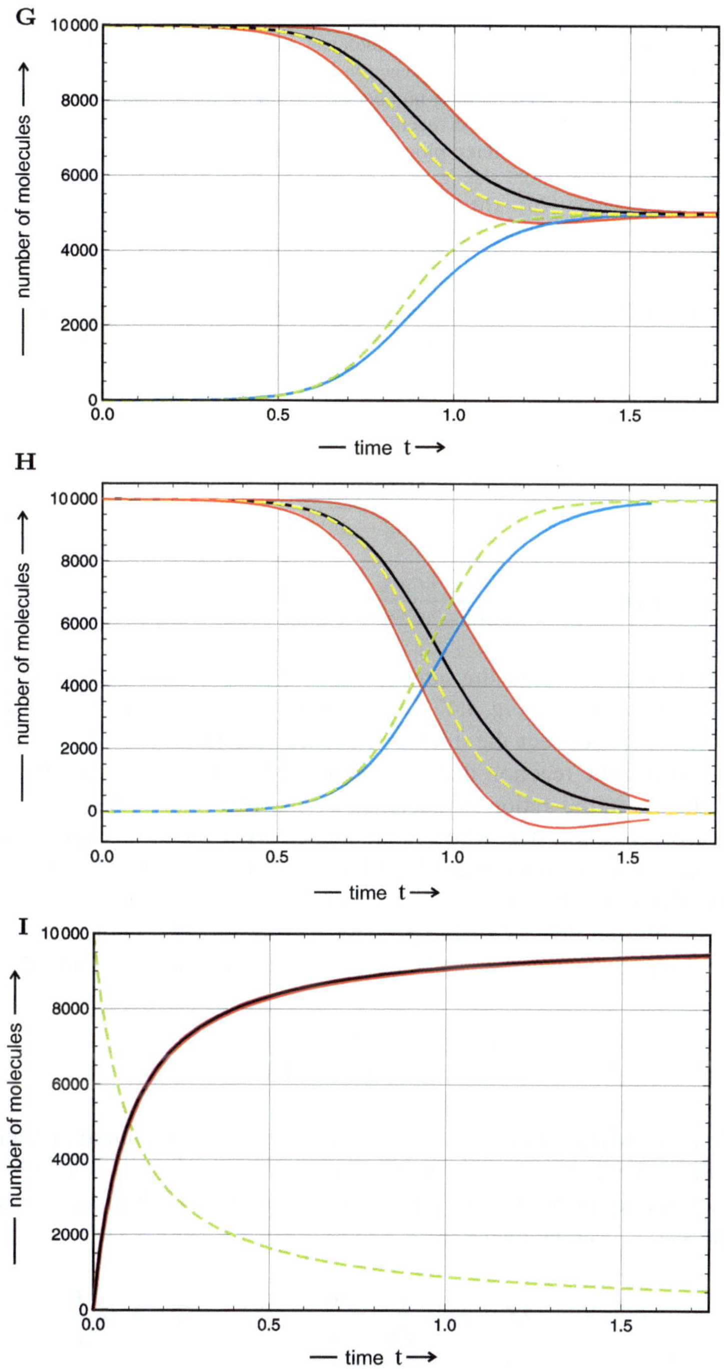

Fig. 4.47 The role of fluctuations in autocatalytic reactions. Continued on next page

Fig. 4.47 The role of fluctuations in autocatalytic reactions (see previous pages). The figure consists of six individual plots derived from the autocatalytic reaction $\mathsf{A} + \mathsf{X} \rightleftharpoons 2\mathsf{X}$. The first three plots (**a**), (**b**), and (**c**) show the reversible reaction $\mathbf{R}_{\leftrightarrow}$ together with the two irreversible reaction steps $\mathbf{R}_{\rightarrow}$ and $\mathbf{R}_{\leftarrow}$, with total particle numbers $N = n_{\mathsf{A}} + n_{\mathsf{X}} = n_0 = 20$. The next three plots (**d**), (**e**), and (**f**) show the three reactions $\mathbf{R}_{\leftrightarrow}$, $\mathbf{R}_{\rightarrow}$, and $\mathbf{R}_{\leftarrow}$ in a population of fivefold size $N = n_0 = 100$. The last three plots (**g**), (**h**), and (**i**) were computed for a population of five hundredfold or hundredfold size $N = n_0 = 10,000$, respectively. Plot (**b**) shows the results of the analytical solution [26]. The Gillespie algorithm encounters problems in the calculation of expectation values when trajectories are terminated by extinction of species. In the current example, this happens for $\mathcal{X}_{\mathsf{A}}(t) = 0$. Parameter choice: $k = l = 1$, 0.1 and 0.001 $[\mathrm{N}^{-1}\mathrm{t}^{-1}]$, for initial conditions: $(n_{\mathsf{A}}(0) = 19,\ n_{\mathsf{X}}(0) = 1)$, or $n_{\mathsf{X}}(0) = 20$, $(n_{\mathsf{A}}(0) = 99,\ n_{\mathsf{X}}(0) = 1)$ or $n_{\mathsf{X}}(0) = 100$, and $(n_{\mathsf{A}}(0) = 9999,\ n_{\mathsf{X}}(0) = 1)$ or $n_{\mathsf{X}}(0) = 10,000$, respectively. Color code: $\mathrm{E}\big(n_{\mathsf{A}}(t)\big)$ *black*, $\mathrm{E}\big(n_{\mathsf{X}}(t)\big)$ *blue*, $\mathrm{E} \pm \sigma\big(n_{\mathsf{A}}(t)\big)$ *red*, $\widehat{n_{\mathsf{A}}}(t)$ *dashed yellow*, and $\widehat{n_{\mathsf{C}}}(t)$ *dashed green*

to give

$$
\begin{aligned}
a(t) &= \frac{kn_0a_0\mathrm{e}^{-kn_0t} + ln_0(n_0 - a_0)(1 - \mathrm{e}^{-kn_0t})}{kn_0 - \big(ka_0 - l(n_0 - a_0)\big)(1 - \mathrm{e}^{-kn_0t})}\,, \\
x(t) &= \frac{kn_0x_0}{(k + l)x_0(1 - \mathrm{e}^{-kn_0t}) + kn_0\mathrm{e}^{-kn_0t}}\,.
\end{aligned}
\tag{4.158a}
$$

These solution curves show the same qualitative behavior as the expectation values of the stochastic process (Fig. 4.47) and exhibit sigmoid or S-shaped forms for $x_0 \ll n_0$. The difference between the stochastic and the deterministic curve is much greater for the autocatalytic reaction than for the uncatalyzed process $\mathsf{A} + \mathsf{B} \rightleftharpoons 2\mathsf{C}$, and varies with the initial ration of substrate A and autocatalyst X. The uncatalyzed stochastic process is faster, whereas the autocatalytic stochastic process is slower than its deterministic counterpart. In order to understand better the special features of autocatalysis, we shall analyze the reaction $\mathsf{A} + \mathsf{X} \rightleftharpoons 2\mathsf{X}$ in more detail.

The equilibrium density of the process $\mathsf{A} + \mathsf{X} \rightleftharpoons 2\mathsf{X}$ is readily obtained, for example, from (3.100), and it can be easily cast in a simple formula [26]:

$$
\bar{P}_n = \frac{1}{(1 + K)^{n_0} - 1}\binom{n_0}{n}\,. \tag{4.158b}
$$

Apart from the normalization factor, the density follows a binomial distribution and is thus the same as the density obtained for the isomerization reaction $\mathsf{A} \rightleftharpoons \mathsf{B}$. Indeed the single concentration factor $\bar{x}$ cancels in the equilibrium constant and the two expressions become identical:

$$
\mathsf{A} + \mathsf{X} \rightleftharpoons 2\mathsf{X}\,,\ K = \frac{\bar{x}^2}{\bar{a}\bar{x}} = \frac{\bar{x}}{\bar{a}}\,, \qquad \mathsf{A} \rightleftharpoons \mathsf{B}\,,\ K = \frac{\bar{b}}{\bar{a}}\,.
$$

However, stoichiometry introduces a difference. Since the last molecules of X cannot react, the state $S_{n_0} = (\mathcal{X}_{\mathsf{A}} = n_0, \mathcal{X}_{\mathsf{X}} = 0)$ has probability zero, the domains of the random variables are reduced by one, i.e., $0 \le \mathcal{X}_A \le n_0 - 1$, $\mathcal{X}_A \in \mathbb{N}$ and $1 \le \mathcal{X}_X \le n_0$, $\mathcal{X}_X \in \mathbb{N}$, and consequently the normalization factor is reduced as well.

The most striking difference between the uncatalyzed process $A + B \rightleftharpoons 2C$ and the autocatalytic reaction $A + X \rightleftharpoons 2X$ is the shape of the solution curve: in the former case we observe the usual *hyperbolic shaped* approach towards stationarity, whereas self-enhancement or autocatalysis leads to *sigmoid* or *S-shaped* curves (Fig. 4.47). The difference between the two processes becomes even more apparent when fluctuations are taken into account: for initially small numbers of autocatalysts X, we are dealing with unrestricted fluctuation enhancement, which substantially broadens the one-standard deviation envelope. From the analytic probability densities at equilibrium, we know that the fluctuation bands in the autocatalytic reaction and in the isomerization reaction differ only in the normalization factor: $\big((1+K)^{n_0} - 1\big)^{-1}$ versus $(1+K)^{-n_0}$. As a function of time, the fluctuation band increases, goes through a maximum, and then decreases to the equilibrium value (Fig. 4.48). In Fig. 4.47, the initial condition $x_0 = 1$ was chosen, and this implies that the effects of self-enhancement, broad fluctuation band and the difference between $E(\mathcal{X}_X)$ and $x(t)$, are maximal and remain large even for population sizes up in the thousands and more.

In order to learn more about the origin of the fluctuation enhancement, the reversible reaction $\mathbf{R}_{\leftrightarrow}$ has been resolved into two irreversible steps:

$$\mathbf{R}_{\rightarrow}:\ A + X \rightarrow 2X \quad \text{and} \quad \mathbf{R}_{\leftarrow}:\ 2X \rightarrow A + X\,.$$

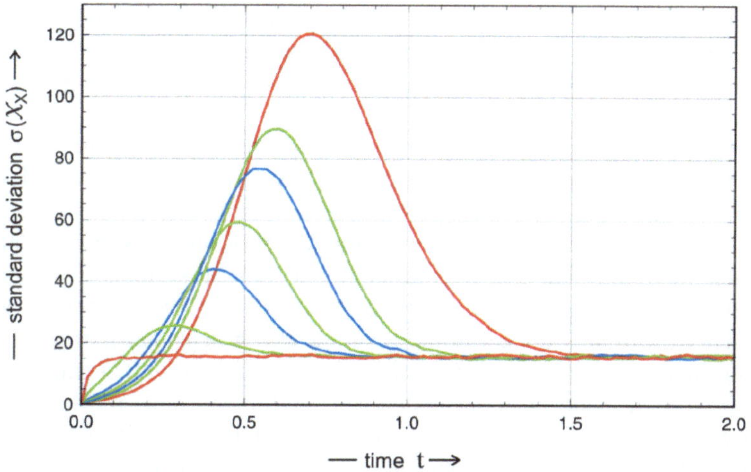

Fig. 4.48 The fluctuation band in the autocatalytic reaction $A + X \rightleftharpoons 2X$. The plot shows the standard deviation $\sigma\big(\mathcal{X}_A(t)\big) = \sigma\big(\mathcal{X}_X(t)\big)$ as a function of time for different initial conditions $\mathcal{X}_X(0) = x_0$ and $\mathcal{X}_A(0) = a_0 = n_0 - x_0$. Parameter choice: $k = l = 0.01$ [$N^{-1}\,t^{-1}$], $n_0 = 1000$, $x_0 = 1$ (*red*), $x_0 = 2$ (*chartreuse*), $x_0 = 3$ (*blue*), $x_0 = 5$ (*chartreuse*), $x_0 = 10$ (*blue*), $x_0 = 50$ (*chartreuse*), and $x_0 = 500$ (*red*)

The forward reaction step $\mathbf{R}_{\rightarrow}$ shows all the characteristic features of autocatalysis reported in the last paragraph, whereas the reverse reaction $\mathbf{R}_{\leftarrow}$ resembles a conventional non-autocatalytic reaction where the fluctuations obey the $\sqrt{n}$-law. Not unexpectedly, it is the $\mathsf{X} \rightarrow 2\mathsf{X}$ component of the process that gives rise to self-enhancement. The heuristic interpretation of the self-enhancement of fluctuations is straightforward: the initial rate of the reaction for sufficiently large values of a_0 is $\eta(\mathbf{n}) = ka_0\mathcal{X}_\mathsf{X}$ and depends only on x_0 when the factor ka_0 is absorbed into the time axis. In Fig. 4.47, this is achieved by scaling k with inverse population size, i.e., $k \propto n_0^{-1}$.

Finally, we mention that the autocatalytic reaction with buffered concentration of $[\mathsf{A}] = a_0$, corresponding to an open system

$$\mathsf{X} \xrightarrow{ka_0} 2\mathsf{X} , \tag{4.158c}$$

has already been studied by Max Delbrück [104]. The stochastic process is identical to a simple birth process with birth rate $\lambda n = ka_0$, and will be discussed in the context of other birth processes in Sect. 5.2.2. Enhancement of fluctuations to macroscopic level is observed as a characteristic for unconstrained autocatalytic growth.

Higher Order Autocatalysis: Bistability and Oscillations

So far all the chemical reactions considered have been approaching either a unique thermodynamic equilibrium or a stationary state, depending on the embedding of the system giving rise to a closed or an open state, respectively. The first order autocatalytic systems exhibited features that are otherwise uncommon in chemical kinetics, the characteristic example being self-enhancement of fluctuations. Here we consider more prominent nonlinear phenomena in chemistry, involving two or more stable steady states and oscillations [472].

An example of a simple mechanism exhibiting bistability consists of an embedding of the termolecular autocatalytic reaction step (4.1m), i.e., $\mathsf{A} + 2\mathsf{X} \leftrightharpoons 3\mathsf{X}$, in the flow reactor. In order to be consistent with chemical thermodynamics, the uncatalyzed $\mathsf{A} \leftrightharpoons \mathsf{X}$ is added. The occurrence of oscillations in concentrations is demonstrated and analyzed by means of two kinetic systems, the theoretical and highly simplified model of the Brusselator and the Oregonator model. The former mechanism was conceived as a simple theoretical model that allows for the occurrence of oscillations and spatial pattern formation in a chemical reaction network, while the latter was postulated as a simplified model mechanism of the Belousov–Zhabotinsky reaction. We use both systems here for the purpose of illustrating the influence of stochasticity on complex dynamics, in particular on bifurcations.

As mentioned in the introduction to elementary step reactions (Sect. 4.1.1), termolecular reaction steps are based on highly improbable encounters of three molecules and are therefore excluded in conventional reaction kinetics. Indeed the fully resolved multistep mechanisms of higher order autocatalytic reactions involve only mono and bimolecular steps. We mention in this context a beautiful mathematical exercise consisting of the tasks to find the smallest reaction systems exhibiting oscillations resulting from a Hopf bifurcation[67] [571] or showing bistability [570].

Bistable Reaction Networks

The termolecular autocatalytic reaction step (4.1m) together with the corresponding uncatalyzed reaction in the flow reactor give rise to the mechanism:

$$\star \xrightarrow{r a_0} \mathsf{A} \,, \tag{4.159a}$$

$$\mathsf{A} \underset{\delta l}{\overset{\delta k}{\rightleftharpoons}} \mathsf{X} \,, \tag{4.159b}$$

$$\mathsf{A} + 2\mathsf{X} \underset{l}{\overset{k}{\rightleftharpoons}} 3\mathsf{X} \,, \tag{4.159c}$$

$$\mathsf{A} \xrightarrow{r} \oslash \,, \tag{4.159d}$$

$$\mathsf{X} \xrightarrow{r} \oslash \,, \tag{4.159e}$$

which corresponds to an overall conversion of A into X. The kinetic differential equations for $[\mathsf{A}] = a$ and $[\mathsf{X}] = x$, viz.,

$$\begin{aligned} \frac{\mathrm{d}a}{\mathrm{d}t} &= -(ka - lx)(\delta + x^2) + r(a_0 - a) \,, \\ \frac{\mathrm{d}x}{\mathrm{d}t} &= +(ka - lx)(\delta + x^2) - rx \,, \end{aligned} \tag{4.160}$$

lead to

$$\frac{\mathrm{d}(a+x)}{\mathrm{d}t} = \frac{\mathrm{d}a}{\mathrm{d}t} + \frac{\mathrm{d}x}{\mathrm{d}t} = r\big(a_0 - (a+x)\big) \,,$$

[67] The Hopf or Poincaré–Andronov–Hopf bifurcation is named after Henri Poincaré, the German–US–American mathematician Eberhard Hopf, and the Russian physicist Aleksandr Andronov, and occurs, in essence, when a complex conjugate pair of eigenvalues crosses the real axis, i.e., $\lambda_{1,2} = \alpha \pm \mathrm{i}\beta$ and $\alpha < 0 \rightarrow \alpha > 0$ [496, p. 48 ff].

with the stationary solution $\bar{a} + \bar{x} = a_0$, and sustain one or three steady states $S_k(\bar{a}_k = a_0 - \bar{x}_k, \bar{x}_k)$, $k = 1, 2, 3$, which satisfy the implicit equation [451, pp. 18–27]

$$r(\bar{x}) = \frac{1}{\bar{x}}\Big(\delta k a_0 - \bar{x}\delta(k+l) + \bar{x}^2 k a_0 - \bar{x}^3(k+l)\Big) . \tag{4.161}$$

When there are three stationary states $S_1(a_0 - \bar{x}_1, \bar{x}_1)$, $S_2(a_0 - \bar{x}_2, \bar{x}_2)$, and $S_3(a_0 - \bar{x}_3, \bar{x}_3)$, S_1 and S_3 are asymptotically stable and the saddle S_2 separates the two basins of attraction, $x < \bar{x}_2$ and $x > \bar{x}_2$, respectively. The subdomains with one or three real and positive solutions for $\bar{x}$ are separated by two saddle-node bifurcations at $\bar{x}_{\mathrm{min}}$ and $\bar{x}_{\mathrm{max}}$, which are calculated straightforwardly from[68]

$$\frac{\mathrm{d}r(\bar{x})}{\mathrm{d}\bar{x}} = 0 \quad \Longrightarrow \quad \bar{x}^3_{\mathrm{crit}} 2(k+l) - \bar{x}^2_{\mathrm{crit}} k a_0 + \delta k a_0 = 0 .$$

As shown in Fig. 4.49, integration of the ODE (4.160) precisely reflects the position of S_2.

Stochasticity is readily introduced into the bistable system through Gillespie integration and sampling of trajectories. The results are shown in Fig. 4.50. For sufficiently small numbers of molecules, we observe the system switching back and forth between the two stable states S_1 and S_3, and an increase in system size changes the scenario in the sense that the system remains in essence in one stable state after it has reached it, while identical initial conditions yield either stable state, S_1 or S_3, and the dependence on initial conditions $\mathbf{C}_0 = \mathbf{n}(0) = \big(a(0), x(0)\big)$ can only be described probabilistically: $P_{S_1}(\mathbf{C}_0)$ versus $P_{S_3}(\mathbf{C}_0)$. A further increase in the system size eventually results in a situation like the one in the deterministic case. Every stable state S_k has a well defined basin of attraction $\mathcal{B}_k$, and if the initial conditions are situated within the basin, so that $\mathbf{C}_0 \in \mathcal{B}_k$, the system converges to the attractor S_k.

An elegant test for bistability consists in a series of simple experiments. The system in a stable stationary state, S_1 or S_3, is perturbed by adding increasing amounts of one compound. Then the system returns to the stable state for small perturbations, but also approaches the other stable state when the perturbation exceeds a certain critical limit.

Closely related to this phenomenon is chemical hysteresis, which is easily illustrated by means of Fig. 4.49. The formation of the stationary state is studied as a function of the flow rate r. Increasing r from $r = 0$, its value at thermodynamic equilibrium, the solution in the flow reactor state approaches state S_3, until the flow rate $r(\bar{x}_{\mathrm{max}})$ is reached. Then, further increase of r causes the system to jump to state

[68]In general, we obtain three solutions for $\bar{x}_{\mathrm{crit}}$. Two of them, $\bar{x}_{\mathrm{min}}$ and $\bar{x}_{\mathrm{max}}$, are situated on the positive $\bar{x}$ axis and correspond to horizontal tangents in the $\bar{x}, r(\bar{x})$ plot (Fig. 4.49). The corresponding vertical tangents in the $r, \bar{x}(r)$ plot separate the domains with one solution, $0 \le r(\bar{x}) \le r(\bar{x}_{\mathrm{min}})$ and $r(\bar{x}_{\mathrm{max}}) \le r(\bar{x}) < \infty$, and three solutions, $r(\bar{x}_{\mathrm{min}}) \le r(\bar{x}) \le r(\bar{x}_{\mathrm{max}})$.

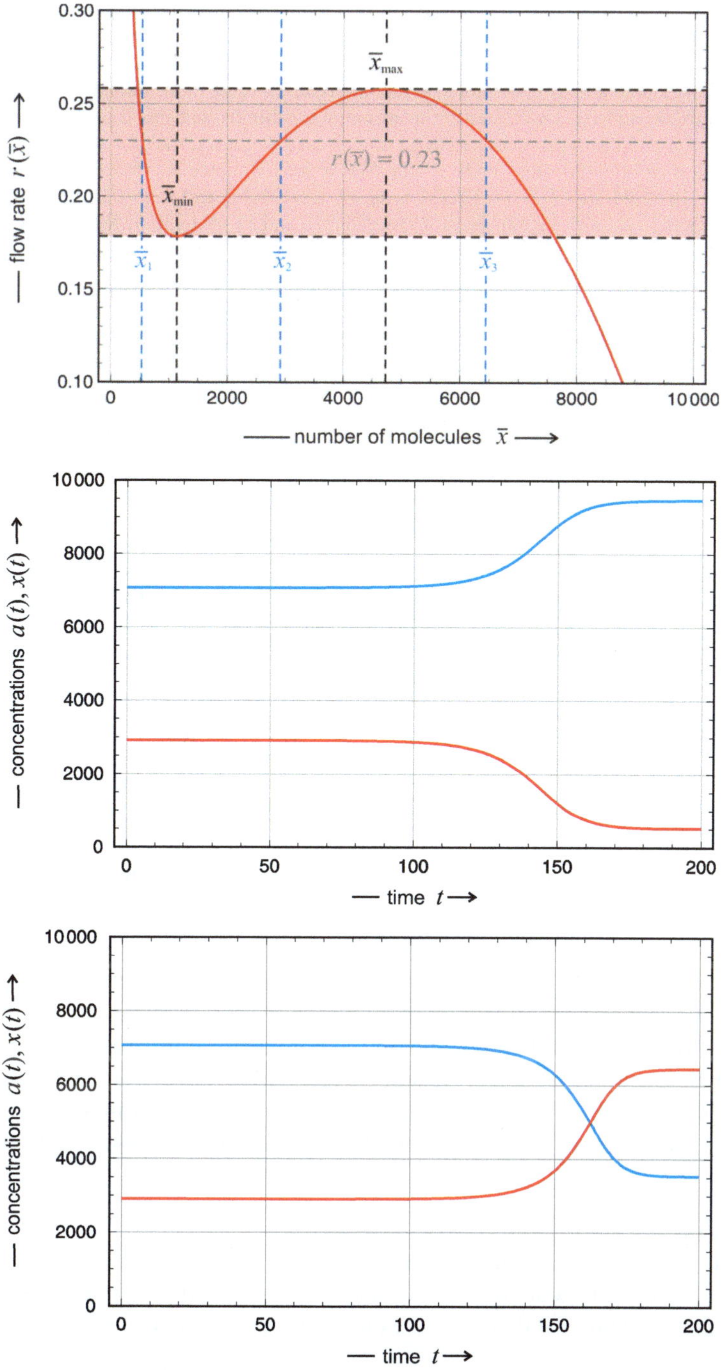

Fig. 4.49 Analysis of bistability in chemical reaction networks. Caption on next page.

Fig. 4.49 Analysis of bistability in chemical reaction networks (see previous page). The reaction mechanism (4.159) sustains three stationary states, $S_1(\bar{x} = \bar{x}_1)$, $S_2(\bar{x} = \bar{x}_2)$, and $S_3(\bar{x} = \bar{x}_3)$, in the range $r(\bar{x}_{\min}) < r(\bar{x}) < r(\bar{x}_{\max})$ with $\bar{x}_1 < \bar{x}_2 < \bar{x}_3$. The two states S_1 and S_3 are asymptotically stable and S_2 is an unstable saddle point. The plot in the *middle* shows the solution curves $a(t)$ (*blue*) and $x(t)$ (*red*), starting from initial conditions just above the unstable state, i.e., $x(0) = \bar{x}_2 - \delta$, and the system converges to state S_1. Analogously, the plot at the *bottom* starts at $x(0) = \bar{x}_2 + \delta$ and the trajectory ends at state S_1. Parameter choice: $k_1 = 1.0 \times 10^{-10}\,[\mathrm{M}^{-2}\mathrm{t}^{-1}]$, $l_1 = 1.0 \times 10^{-8}\,[\mathrm{M}^{-2}\mathrm{t}^{-1}]$, $\delta = 10^6\,[\mathrm{M}^2]$, $a_0 = 10{,}000\,[\mathrm{M}]$ and $r = 0.23\,[\mathrm{t}^{-1}]$. Steady state concentrations: $\bar{x}_1 = 525.34\,[\mathrm{M}]$, $\bar{x}_2 = 2918.978\,[\mathrm{M}]$, and $\bar{x}_3 = 6456.67\,[\mathrm{M}]$. Initial conditions: $x(0) = 2918.97\,[\mathrm{M}]$ (*middle plot*) and $x(0) = 2918.98\,[\mathrm{M}]$ (*bottom plot*)

S_1, because S_3 does not lie in the real plane any more. Alternatively, when the flow rate r is decreased from higher values where S_1 is the only stable state, S_1 remains stable until the flow rate $r(\bar{x}_{\min})$, and then the solution in the reactor jumps to S_3. Chemical hysteresis implies that the system passes through different states in the bistable region when the parameter causing bistability is raised or lowered.

An experimental reaction mechanism showing bistability is provided by a combination of the Dushman reaction [118], viz.,

$$\mathrm{IO_3^-} + 5\mathrm{I^-} + 6\mathrm{H^+} \rightleftharpoons 3\mathrm{I_2} + 3\mathrm{H_2O}\,,$$

and the Roebuck reaction [470], viz.,

$$\mathrm{I_2} + \mathrm{H_3AsO_3} + \mathrm{H_2O} \rightleftharpoons 2\mathrm{I^-} + \mathrm{H_3AsO_4} + 2\mathrm{H^+}\,,$$

leading to the overall reaction equation

$$\mathrm{IO_3^-} + 3\mathrm{H_3AsO_3} \rightleftharpoons \mathrm{I^-} + 3\mathrm{H_3AsO_4}\,,$$

the oxidation of arsenous acid by iodate. Careful studies of the Dushman–Roebuck reaction in the flow reactor revealed all the features of bistability described here by means of the simpler example [103, 237]. We mention that bistability and hysteresis have also been studied theoretically and experimentally in case of the even more complex Belousov–Zhabotinsky reaction described in the section on the Oregonator model [34, 183, 198]. In all these examples, the nonlinear phenomena originate from multistep mechanisms with only monomolecular and bimolecular reaction steps.

Brusselator

The Brusselator mechanism was invented by Ilya Prigogine and his group in Brussels [335]. The goal was to find the simplest possible hypothetical chemical system that sustains oscillations in homogeneous solution. For this purpose the

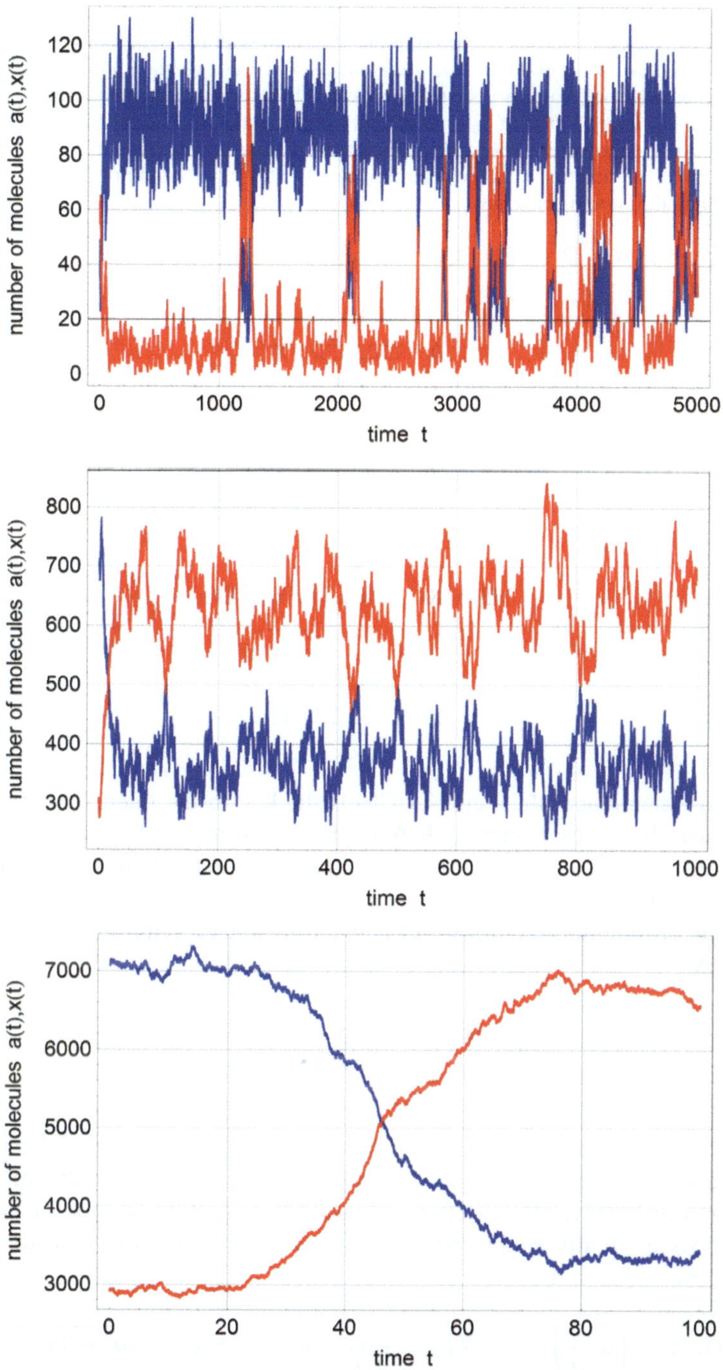

Fig. 4.50 Stochasticity in bistable reaction networks. Caption on next page.

Fig. 4.50 Stochasticity in bistable reaction networks (see previous page). The figure shows three trajectories calculated by means of the Gillespie algorithm with different numbers of molecules: $a_0 = 100$ (*top plot*), $a_0 = 1000$ (*middle plot*), and $a_0 = 10,000$ (*bottom plot*). For small system sizes, a sufficiently long trajectory switches back and forth between the two stable states S_1 and S_3 (*top plot*). For larger values of a_0 (*middle plot*), it goes either to S_1 or to S_3 with a ratio of the probabilities of approximately 0.56/0.44. At the largest population size (*bottom plot*), we encounter essentially the same situation as in the deterministic case: the initial conditions determine the state towards which the system converges

overall reaction $\mathsf{A} + \mathsf{B} \rightleftharpoons \mathsf{D} + \mathsf{E}$ is split into four reaction steps:

$$\mathsf{A} \underset{l_1}{\overset{k_1}{\rightleftharpoons}} \mathsf{X} , \tag{4.162a}$$

$$2\mathsf{X} + \mathsf{Y} \underset{l_2}{\overset{k_2}{\rightleftharpoons}} 3\mathsf{X} , \tag{4.162b}$$

$$\mathsf{B} + \mathsf{X} \underset{l_3}{\overset{k_3}{\rightleftharpoons}} \mathsf{Y} + \mathsf{D} , \tag{4.162c}$$

$$\mathsf{X} \underset{l_4}{\overset{k_4}{\rightleftharpoons}} \mathsf{E} . \tag{4.162d}$$

As already mentioned, the step (4.162b) is the key to the interesting phenomena of nonlinear dynamics. Compounds A and B are assumed to be present in buffered concentrations, $[\mathsf{A}] = a_0 = a$ and $[\mathsf{B}] = b_0 = b$, and for the sake of simplicity we consider the case of irreversible reactions, i.e., $l_1 = l_2 = l_3 = l_4 = 0$. Then the kinetic differential equations for the deterministic description of the dynamical system are

$$\begin{aligned} \frac{\mathrm{d}x}{\mathrm{d}t} &= k_1 a_0 + k_2 x^2 y - k_3 b_0 x - k_4 x , \\ \frac{\mathrm{d}y}{\mathrm{d}t} &= k_3 b_0 x - k_2 x^2 y . \end{aligned} \tag{4.163}$$

The Brusselator sustains a single steady state $S = (\bar{x}, \bar{y})$, and conventional bifurcation analysis yields the two eigenvalues $\lambda_{1,2}$. Without loss of generality, the analysis is greatly simplified by setting all rate constants equal to one, i.e., $k_1 = k_2 = k_3 = k_4 = 1$:

$$S = (\bar{x}, \bar{y}) = \left(a, \frac{b}{a}\right) , \qquad \lambda_{1,2} = \frac{1}{2}\left(b - a^2 - 1 \pm \sqrt{(b - a^2 - 1)^2 - 4a^2}\right) ,$$

The eigenvalues form a pair of complex conjugate values in the parameter range $(a-1)^2 < b < (a+1)^2$, and the real part vanishes at $b = a^2+1$. Accordingly, we are dealing with a Hopf bifurcation at $b = a^2 + 1$ with $\lambda_{1,2} = \pm 2ia$. Figure 4.51 shows computer integrations of the ODE (4.163) illustrating the analytical results. For the sake of simplicity we have chosen irreversible reactions and incorporated constant concentrations into the rate constants: $\kappa_1 = k_1 a_0$, $\kappa_2 = k_2$, $\kappa_3 = k_3 b_0$, and $\kappa_4 = k_4$.

Introducing stochasticity into the Brusselator model complicates the bifurcation scenario. At low particle numbers corresponding to a high level of parametric noise, the Hopf bifurcation disappears, leaving a scenario of more or less irregular oscillations on both sides of the deterministic position of the bifurcation. Ludwig Arnold put it as follows [23, 24]: "Parametric noise destroys the Hopf bifurcation." Increasing the system size allows for the appearance of the stable point attractor on one side of the Hopf bifurcation (Fig. 4.52).

The oscillations exhibited by the Brusselator are characteristic of so-called excitable media, in which a reservoir is filled more or less slowly with a consumable compound until a process rapidly consuming this material is ignited. In the case of the Brusselator, the consumable is the compound Y and its concentration is raised until the autocatalytic process 2X+Y→3X is triggered by an above-threshold concentration of Y. Fast consumption of Y results in a rapid increase in X that completes the wave by reducing the concentration of Y to a small value (Fig. 4.53). The easiest way to visualize an excitable medium is provided by the example of wildfires: wood grows slowly until it reaches a density that can sustain spreading fire. Once triggered by natural causes or arson, the fire consumes all the wood and thereby initiates the basis for the next refractory period. Oscillatory chemical reactions do not need an external trigger, since an internal fluctuation is sufficient to initiate the decay phase.

Finally, we mention that higher order autocatalysis is also required for the formation of spatial Turing patterns [533],[69] and there occurrence has been predicted with models where the Brusselator mechanism was coupled to diffusion [421]. The experimental verification of a standing wave pattern was achieved only 38 years later [79].

Oregonator

The prototype of an oscillatory chemical reaction and the first example that became popular in history is the Belousov–Zhabotinsky reaction, which is described by the overall process of the cerium catalyzed reaction of malonic acid by bromate ions in dilute sulfuric acid. We mention the reaction here in order to present one example showing how complicated chemical reaction networks can be in reality:

$$3\mathbf{H_2M} + 4\mathbf{BrO_3^-} \longrightarrow 4\mathbf{Br^-} + 9\mathbf{CO_2} + 6\mathbf{H_2O}\,. \qquad (4.164a)$$

[69] A Turing pattern named after the British mathematician and computer pioneer Alan Turing is a pattern in space that forms instantaneously under proper conditions.

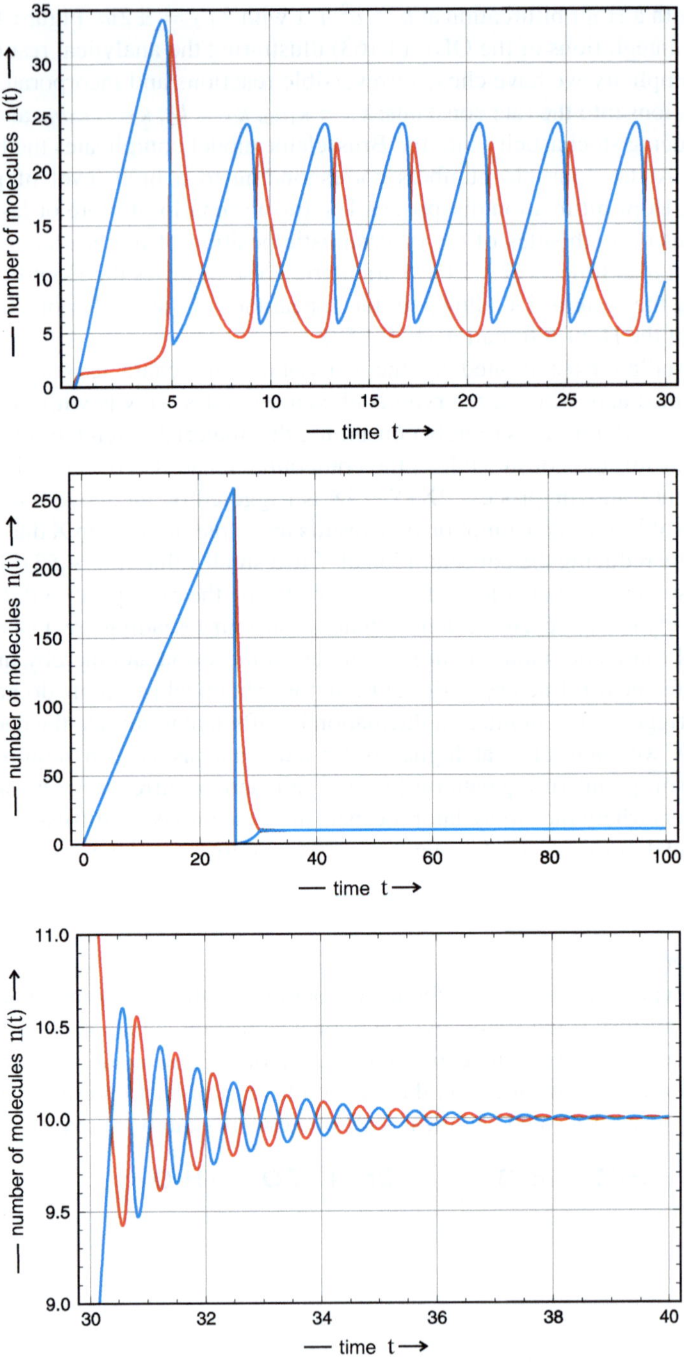

Fig. 4.51 Analysis of the Brusselator model. Caption on next page

Fig. 4.51 Analysis of the Brusselator model (see previous page). The figure presents integrations of the kinetic differential equation (4.163) in the oscillatory regime (*top plot*) and in the range of stability of the stationary point $S = (a, b/a)$ (*middle plot*). Although the integration starts at the origin (0,0), a point that lies relatively close to the stationary state (10,10), the trajectory performs a full refractory cycle before it settles down at the stable point. The plot at the *bottom* is an enlargement of the *middle plot* and illustrates the results of a complex conjugate pair of eigenvalues with negative real part: damped oscillations. Parameter choice: $\kappa_1 = k_1 a_0 = 10$, $\kappa_2 = k_2 = 0.05$, $\kappa_3 = k_3 b_0 = 6.5$, and $\kappa_4 = k_4 = 1$ (*top plot*), $\kappa_2 = k_2 = 1$ and $\kappa_3 = k_3 b_0 = 100$ (*middle plot* and *bottom plot*). Initial conditions: $x(0) = 0$ and $y(0) = 0$. Color code: $x(t)$ *red* and $y(t)$ *blue*

Malonic acid is written here as $\mathbf{CH_2(CO_2H)_2} \equiv \mathbf{H_2M}$. The process can be split into three overall reactions that follow the reaction equations

$$2\mathbf{Br}^- + \mathbf{BrO}_3^- + 3\mathbf{H}^+ + 3\mathbf{H_2M} \longrightarrow 3\mathbf{HBrM} + 3\mathbf{H_2O} \,, \tag{4.164b}$$

$$4\mathbf{Ce}^{3+} + \mathbf{BrO}_3^- + 5\mathbf{H}^+ \longrightarrow 4\mathbf{Ce}^{4+} + \mathbf{HBrO} + 2\mathbf{H_2O} \,, \tag{4.164c}$$

$$\begin{aligned} &2\mathbf{Ce}^{4+} + 2\mathbf{H_2M} + \mathbf{HBrM} + \mathbf{HBrO} + 2\mathbf{H_2O} \\ &\qquad \longrightarrow 2\mathbf{Ce}^{3+} + 2\mathbf{Br}^- + 3\mathbf{HOHM} + 4\mathbf{H}^+ \,. \end{aligned} \tag{4.164d}$$

The third reaction (4.164d) is complemented by the fragmentation of hydroximalonic acid to carbon dioxide and accompanied by further reduction of bromate to bromide. In detail the reaction mechanism is even more complicated and a network of about 20 reaction steps has been derived from the available data [121].

The reaction network of the Belousov–Zhabotinsky reaction is too complicated for a complete theoretical analysis, and therefore a simplified model, the Oregonator model, was conceived by the US physical chemists Richard Noyes and Richard Field in order to allow for a combined analytical and numerical study of a system that closely mimics the properties of the Belousov–Zhabotinsky reaction [166, 429]:

$$\mathsf{A} + \mathsf{Y} \underset{l_1}{\overset{k_1}{\rightleftharpoons}} \mathsf{X} + \mathsf{P} \,, \tag{4.165a}$$

$$\mathsf{X} + \mathsf{Y} \underset{l_2}{\overset{k_2}{\rightleftharpoons}} 2\mathsf{P} \,, \tag{4.165b}$$

$$\mathsf{A} + \mathsf{X} \underset{l_3}{\overset{k_3}{\rightleftharpoons}} 2\mathsf{X} + 2\mathsf{Z} \,, \tag{4.165c}$$

$$2\mathsf{X} \underset{l_4}{\overset{k_4}{\rightleftharpoons}} \mathsf{A} + \mathsf{B} \,, \tag{4.165d}$$

$$\mathsf{B} + \mathsf{Z} \xrightarrow{k_5} \frac{1}{2} f\mathsf{Y} \,. \tag{4.165e}$$

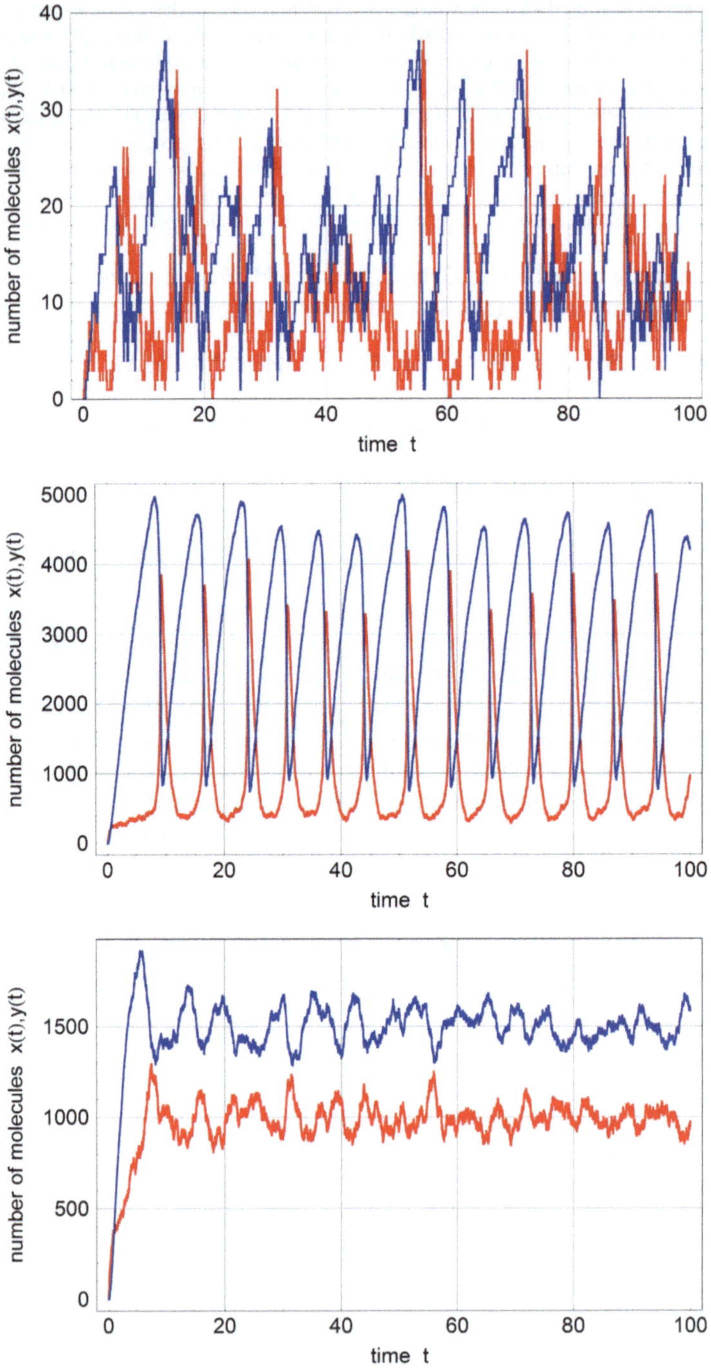

Fig. 4.52 Stochasticity in the Brusselator model. Caption on next page

Fig. 4.52 Stochasticity in the Brusselator model (see previous page). The figure shows three stochastic simulations of the Brusselator model. The *top plot* shows the Brusselator in the stable regime for low numbers of molecules ($a_0 = 10$). No settling down of the trajectories near the steady state is observed. For sufficiently high numbers of molecules ($a_0 = 1000$), the behavior of the stochastic Brusselator is close to the deterministic solutions (Fig. 4.51) in the oscillatory regime (*middle plot*), and in the range of fixed point stability, the stochastic solutions fluctuate around the stationary values (*bottom plot*). Parameter choice: $\kappa_1 = 10$, $\kappa_2 = 0.01$, $\kappa_3 = 1.5$, $\kappa_4 = 1$ (*top plot*), $\kappa_1 = 1000$, $\kappa_2 = 1 \times 10^{-6}$, $\kappa_3 = 3$, $\kappa_4 = 1$ (*middle plot*), and $\kappa_1 = 1000$, $\kappa_2 = 1 \times 10^{-6}$, $\kappa_3 = 1.5$, $\kappa_4 = 1$ (*bottom plot*). Initial conditions: $x(0) = y(0) = 0$. Color code: x *red* and y *blue*

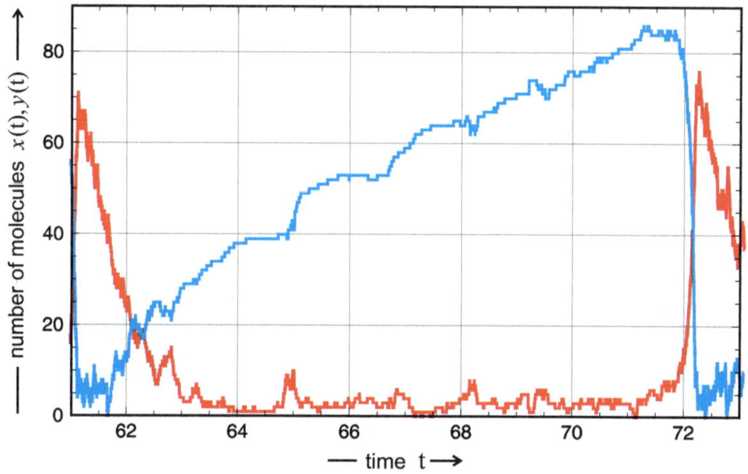

Fig. 4.53 Refractory cycle in the Brusselator model. Enlargement from a stochastic trajectory calculated with the parameters applied in the *top plot* of Fig. 4.52. It illustrates a refractory cycle consisting of filling a reservoir with a compound Y (*blue*) that is quickly emptied by conversion of Y to X after ignition triggered by a sufficiently large concentration of Y

The corresponding kinetic ODEs for irreversible reactions and buffered concentration of A and B with $[X] = x$, $[Y] = y$, and $[Z] = z$ are:

$$\frac{dx}{dt} = k_1 ay - k_2 xy + k_3 ax - 2k_4 x^2 \,,$$
$$\frac{dy}{dt} = -k_1 ay - k_2 xy + \frac{1}{2} k_5 bz \,, \tag{4.165f}$$
$$\frac{dz}{dt} = 2k_3 ax - k_5 bz \,.$$

Two features of the model are remarkable: (i) It is low-dimensional—three variables [X], [Y], and [Z] when A and B are buffered—and does not contain a termolecular step, and (ii) it makes use of a non-stoichiometric factor f. The Oregonator model has been successfully applied to reproduce experimental findings on fine details of the oscillations in the Belousov–Zhabotinsky reaction, but it fails to predict the occurrence of deterministic chaos. In later work, new models of this reaction were developed that have also been successful in this aspect [188, 227].

Figure 4.54 shows numerical integrations of (4.165f) in the open system with constant input materials and in a closed system with limited supply of A. Interestingly, the oscillations give rise to a step consumption of the resource. In his seminal paper on the simulation algorithm and its applications, Daniel Gillespie [207] provided a stochastic version of the Oregonator, which we apply here to demonstrate the stochastic simulation approach to the deterministic solution with increasing population size (Fig. 4.55).

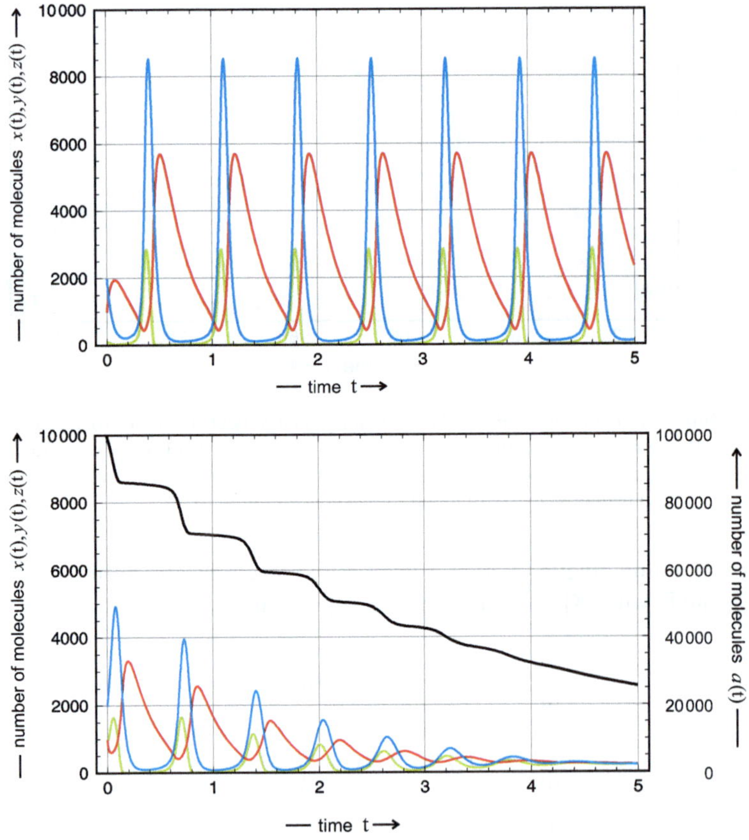

Fig. 4.54 Analysis of the Oregonator model. The kinetic ODEs (4.165f) of the Oregonator model are integrated and the undamped oscillations in the open system (with buffered concentrations of A and B) are shown in the *top plot*. The supply of A is limited in the *bottom plot*, which mimics the closed system. As A is consumed, the oscillations become smaller an eventually die out. Parameter choice: $\kappa_1 = 2$, $\kappa_2 = 0.1$, $\kappa_3 = 104$, $\kappa_4 = 0.016$, and $\kappa_5 = 26$. Initial concentrations $x(0) = 100$, $y(0) = 1000$, and $z(0) = 2000$. Color code: $x(t)$ *green*, $y(t)$ *red*, $z(t)$ *blue*, and $a(t)$ *black*

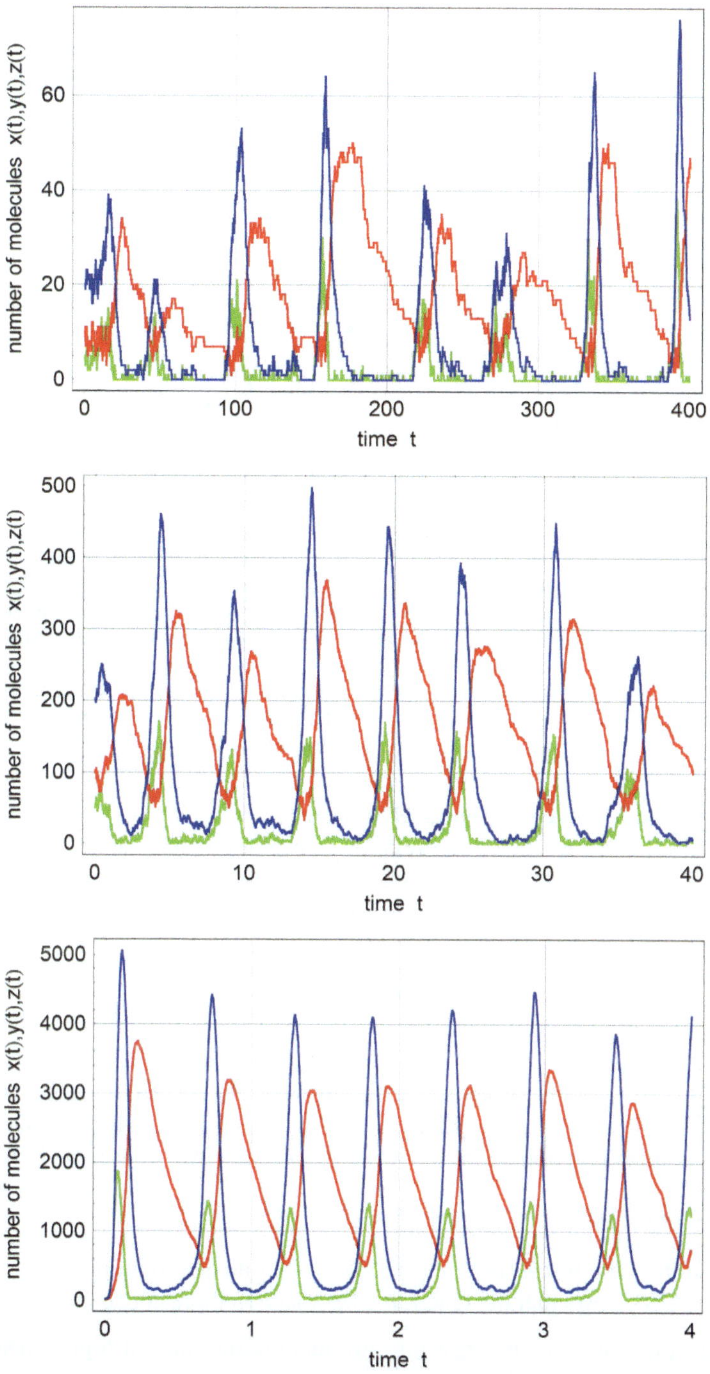

Fig. 4.55 Stochasticity in the Oregonator model. Caption on next page

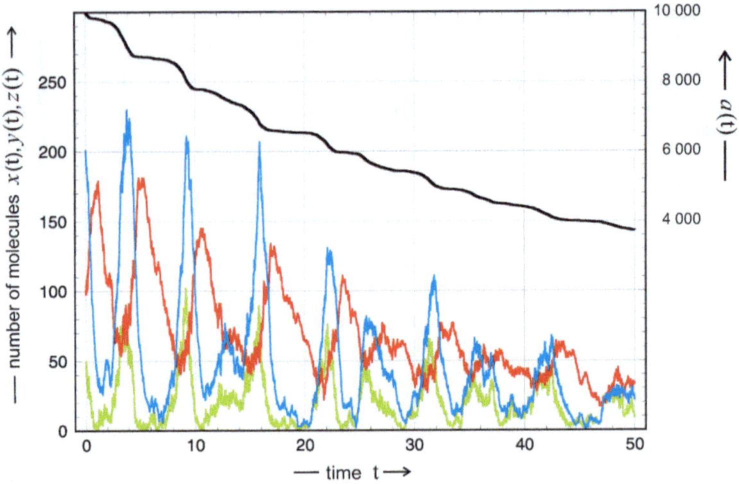

Fig. 4.55 Stochasticity in the Oregonator model (see also previous page). The figure shows stochastic simulations of the Oregonator model at different population sizes: $x(0) = 5\vartheta$, $y(0) = 10\vartheta$ (*red*), and $z(0) = 20\vartheta$ with $\vartheta = 1$ (*previous page, top plot*), $\vartheta = 10$ (*previous page, middle plot*), and $\vartheta = 100$ (*previous page, bottom plot*). A simulation of the Oregonator in a system that is closed with respect to compound A is shown *above* ($\vartheta = 10$, $a(0) = 10\,000$). The parametrization was adopted from [207]: $\bar{x} = 5$, $\bar{y} = 10$, and $\bar{z} = 20$ were used for the concentrations, and $\bar{\rho}_1 = 0.2\vartheta^2$ and $\bar{\rho}_2 = 5\vartheta^2$ for the reaction rates at the unstable stationary point. This yields for the rate parameters: $\kappa_1 = 0.02\vartheta$, $\kappa_2 = 0.1$, $\kappa_3 = 1.04\vartheta$, $\kappa_4 = 0.016\vartheta$, and $\kappa_5 = 0.26\vartheta$. Color code: $x(t)$ *green*, $y(t)$ *red*, $z(t)$ *blue*, and $a(t)$ *black*

The Extended Michaelis–Menten Reaction

In this section we investigate the relation between enzyme–substrate binding and the production of the reaction product by stochastic simulation. The extended mechanism of Michaelis–Menten type enzyme catalysis (Fig. 4.6a), i.e.,

$$\mathsf{S} + \mathsf{E} \underset{l_1}{\overset{k_1}{\rightleftharpoons}} \mathsf{SE} \underset{l_2}{\overset{k_2}{\rightleftharpoons}} \mathsf{EP} \underset{l_3}{\overset{k_3}{\rightleftharpoons}} \mathsf{E} + \mathsf{P}\,, \tag{4.166a}$$

deals with five random variables $\mathcal{X}_\mathsf{S} \equiv [\mathsf{S}] = s$, $\mathcal{X}_\mathsf{E} \equiv [\mathsf{E}] = e$, $\mathcal{X}_\mathsf{P} \equiv [\mathsf{P}] = p$, $\mathcal{X}_\mathsf{C} \equiv [\mathsf{SE}] = c$, and $\mathcal{X}_\mathsf{D} \equiv [\mathsf{EP}] = d$, which describe the five chemical species involved in three reaction steps. Accordingly, we have two conservation relations:

$$s_0 + p_0 = s + c + d + p\,, \qquad e_0 = e + c + d\,. \tag{4.166b}$$

The reaction of interest is the conversion of substrate S into product P.

Figure 4.56 shows that the reaction takes place in three concentration ranges:

(i) The conversion of substrate into product taking place at high concentrations shows relatively small scatter.

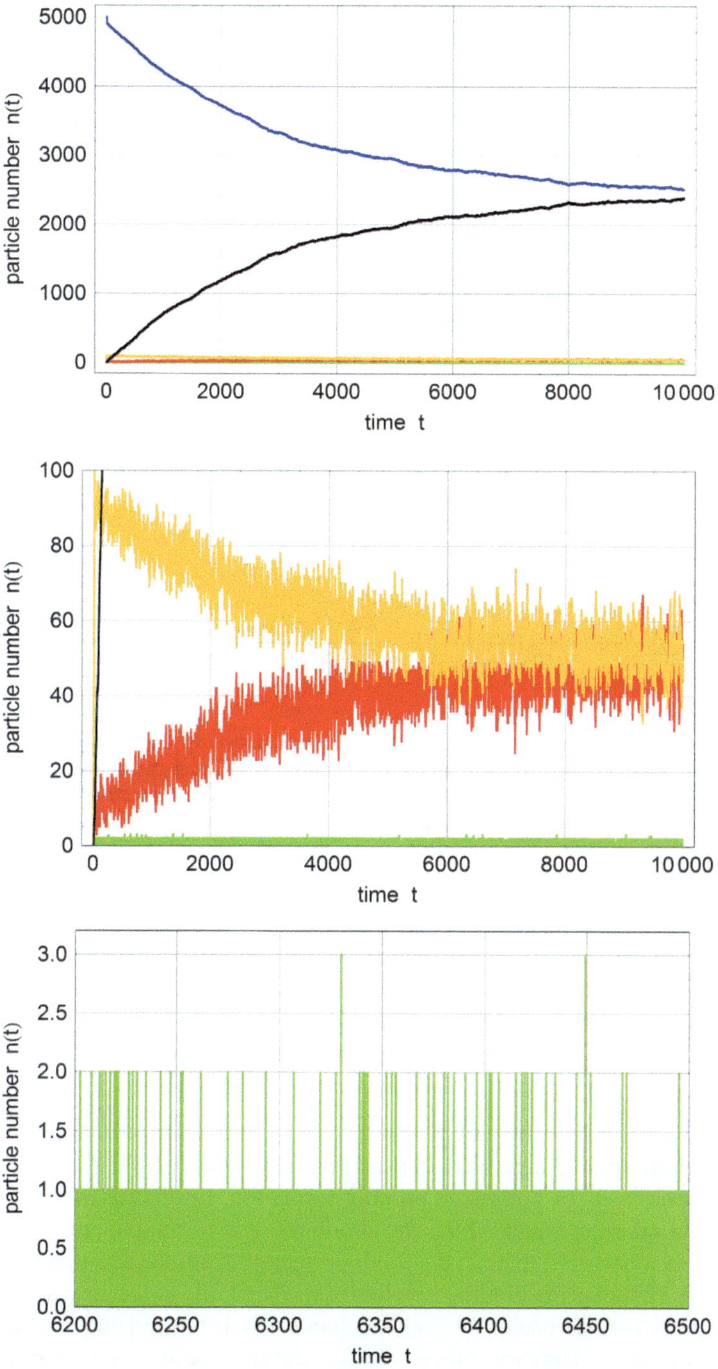

Fig. 4.56 The extended Michaelis–Menten reaction. Caption on next page

Fig. 4.56 The extended Michaelis–Menten reaction (see previous page). The fully reversible mechanism shown as version A in Fig. 4.6 is simulated in the form of a single trajectory with a large excess of substrate. The *top plot* shows the number of substrate and product molecules, $s(t)$ (*blue*) and $p(t)$ (*black*), respectively. The *middle plot* presents particle numbers for the two complexes $[\mathsf{S} \cdot \mathsf{E}] = c(t)$ (*yellow*) and $[\mathsf{E} \cdot \mathsf{P}] = d(t)$ (*red*). The *bottom plot* shows the number of free enzyme molecules $e(t)$, which almost always takes on only four different values: $e \in \{0, 1, 2, 3\}$. Initial conditions: $s_0 = 5000$, $e_0 = 100$. Parameter choice: $k_1 = l_3 = 0.1\ [\mathrm{N}^{-1}\mathrm{t}^{-1}]$, $l_2 = k_3 = 0.1\ [\mathrm{t}^{-1}]$, and $k_2 = l_2 = 0.01\ [\mathrm{t}^{-1}]$

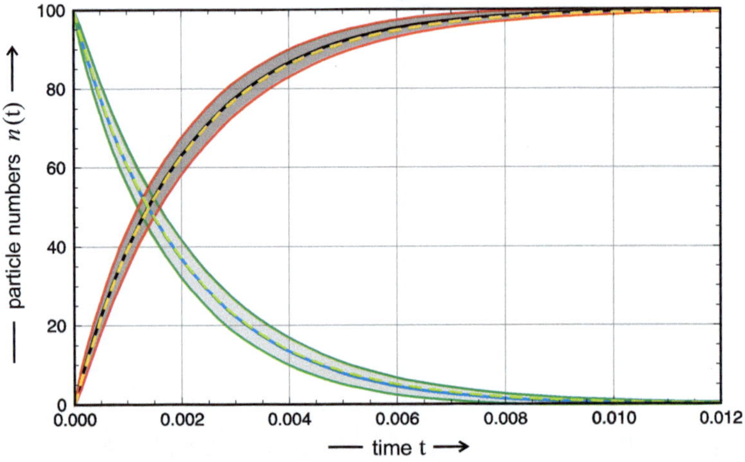

Fig. 4.57 Enzyme–substrate binding. The binding step preceding the enzymatic reaction simulated in Fig. 4.56 takes place on a much shorter time scale than the conversion of substrate into product, because (i) the rate parameters of the binding reactions are larger by one order of magnitude, and (ii) the initial substrate concentration is much larger than the total enzyme concentration, i.e., $s_0 \gg e_0$. The *two curves* show the expectation values $\mathrm{E}(\mathcal{X}_{\mathsf{SE}}(t))$ and $\mathrm{E}(\mathcal{X}_{\mathsf{E}}(t))$ within the one-σ band. Initial conditions: $s_0 = 5000$, $e_0 = 100$. Parameter choice: $k_1 = l_3 = 0.1\ [\mathrm{N}^{-1}\mathrm{t}^{-1}]$, $l_1 = k_3 = 0.1\ [\mathrm{t}^{-1}]$, and $k_2 = l_2 = 0.01\ [\mathrm{t}^{-1}]$

(ii) The concentrations of the enzyme complexes **SE** and **EP** fall into the same range as the initial enzyme concentration e_0, since the large number of substrate molecules drives almost all enzyme molecules into the bound state.

(iii) Only a few free enzyme molecules are present and the rate-determining step is product release in our example here.

The conditions applied here were chosen for the purpose of illustration and do not meet the constraints of optimized product formation. In this case one would need to choose conditions under which the product binds to the enzyme only weakly. Alternatively, one could remove the product steadily from the reaction mixture.

Figure 4.57 shows the binding kinetics of the substrate to the enzyme, $\mathsf{S} + \mathsf{E} \rightleftharpoons \mathsf{SE}$, within the full Michaelis–Menten mechanism (4.166a) and under the conditions applied in Fig. 4.56. The expectation values of the enzyme and the substrate–enzyme complex, $\mathrm{E}(\mathcal{X}_{\mathsf{E}}(t))$ and $\mathrm{E}(\mathcal{X}_{\mathsf{SE}}(t))$, coincide with the deterministic solutions, $c(t)$ and $e(t)$, almost within the line width.

Chapter 5
Applications in Biology

Nothing in biology makes sense except in the light of evolution.
Theodosius Dobzhansky 1972.

Abstract Stochastic phenomena are central to biological modeling. Small numbers of molecules regulate and control genetics, epigenetics, and cellular metabolism, and small numbers of well-adapted individuals drive evolution. As far as processes are concerned, the relation between chemistry and biology is a tight one. Reproduction, the basis of all processes in evolution, is understood as autocatalysis with an exceedingly complex mechanism in the language of chemists, and replication of the genetic molecules, DNA and RNA, builds the bridge between chemistry and biology. The earliest stochastic models in biology applied branching processes in order to give answers to genealogical questions like, for example, the fate of family names in pedigrees. Branching processes, birth-and-death processes, and related stochastic models are frequently used in biology and they are defined, analyzed, and applied to typical problems. Although the master equation is not so dominant in biology as it is in chemistry, it is a very useful tool for deriving analytical solutions, and most birth-and-death processes can be analyzed successfully by means of master equations. Kimura's neutral theory of evolution makes use of a Fokker–Planck equation and describes population dynamics in the absence of fitness differences. A section on coalescent theory demonstrates the applicability of backwards modeling to the problem of reconstruction of phylogenies. Unlike in the previous chapter we shall present and discuss numerical simulations here together with the analytical approaches. Simulations of stochastic reaction networks in systems biology are a rapidly growing field and several new monographs have come out during the last few years. Therefore only a brief account and a collection of references are given.

The population aspect is basic to biology: individuals produce mutants, but populations evolve. Accordingly, we adopt once again the notation of populations as random vectors, viz.,

$$|\Pi|\,(t) = \big(\mathcal{N}_1(t), \mathcal{N}_2(t), \ldots, \mathcal{N}_n(t)\big)\ , \quad \text{with } \mathcal{N}_k \in \mathbb{N}\ ,\ t \in \mathbb{R}_{\geq 0}\ ,$$

P. Schuster, *Stochasticity in Processes*, Springer Series in Synergetics,
DOI 10.1007/978-3-319-39502-9_5

and count the numbers of individuals of the different *subspecies* in a *biological species*[1] as time dependent random variables $\mathcal{N}_k(t)$. This definition indicates that time will be considered as a continuous variable, while the use of *counting* implies that the numbers of individuals are discrete. The basic assumptions are thus the same as in the applications of master equations to chemical reaction kinetics (Sect. 4.2.2).

There is, however, a major difference between the molecular approach based on elementary reactions on one side and macroscopic modeling as commonly used in biology on the other: the biological objects are no longer single molecules or atoms, but modules commonly consisting of a large number of atoms or individual cells or organisms. Elementary step dynamics obeys several conservation relations like conservation of mass or conservation of the numbers of atoms for every chemical element unless nuclear reactions are admitted, and the laws of thermodynamics provide additional restrictions. In macroscopic models these relations are not violated, of course, but they are hidden in complex networks of interactions, which appear in the model only after averaging on several hierarchical levels.

For example, conservation of mass and conservation of energy are encapsulated and somewhat obscured in the carrying capacity C of an ecosystem, as modeled by a differential equation that was conceived by the nineteenth century Belgium mathematician Pierre-François Verhulst and which is called the Verhulst or the logistic equation [550][2]:

$$\frac{\mathrm{d}N}{\mathrm{d}t} = fN\left(1 - \frac{N}{C}\right) , \qquad N(t) = \frac{N_0 C}{N_0 + (C - N_0)\exp(-ft)} , \tag{5.1}$$

where $N(t) = \sum_{i=1}^{n} \mathcal{N}_i(t) = \sum_{i=1}^{n} n_i(t)$ is the population size, f represents the so-called Malthus or growth parameter, and $N_0 = N(0)$ is the initial population size at time $t = 0$. The numbers of individuals in a species or subspecies may change by m without any compensation in another variable in biological models: $n_k(t) \to n_k(t + \Delta t) = n_k(t) \pm m$, where $m \in \mathbb{N}_{>0}$. The integer m is the number of individuals which are born or die in a single event. In chemistry, a single reaction event causes changes in more than one variable: products appear and reactants disappear. Exclusive changes in single variables also occur in chemistry when we are dealing with inflow into and outflow from a reactor (Sect. 4.3.1), or in the case of *buffering* in monomolecular reactions when a reservoir remains practically unchanged if a single molecule is added or withdrawn. In other cases, for example, in population genetics, the limitation of the population size is part of the specific model. Normalized relative variables are often used and the population size is not considered explicitly. As

---

[1]Throughout this monograph we use *subspecies* in the sense of *molecular species* or *variant* for the components of a population $\Pi = \{\mathsf{X}_1, \mathsf{X}_2, \ldots, \mathsf{X}_n\}$. We express its numbers by the random vector $\boldsymbol{\mathcal{N}} = |\Pi| = (\mathcal{N}_1, \mathcal{N}_2, \ldots, \mathcal{N}_n)$ and indicate by using the notion *biological species* when we mean species in biology.

[2]Verhulst himself gave no biological interpretation of the logistic equation and its parameters in terms of a carrying capacity. For a brief historical account on the origin of this equation, see [282]. The Malthus parameter is commonly denoted by r. Since r is used as the flow rate in the CSTR, we choose f here in order to avoid confusion and to stress the close relationship between the Malthus parameter and fitness.

indicated above the changes in the numbers of individuals $\pm m$ are considered for single events and this implies that the time interval considered is short enough to ensure that multiple events can be excluded. In biology, we can interpret the flow reactor (Sect. 4.3.1) as a kind of idealized ecosystem. Genuine biological processes analogous to inflow (4.1a) and outflow (4.1b) are immigration and emigration, respectively.

A stochastic process on the population level is, by the same token as in Sect. 3.1.1, a recording of successive time ordered events at times T_i:

$$T_0 < T_1 < T_2 < \ldots < T_{k-1} < T_k < T_{k+1} \ldots ,$$

along a time axis t. The application of discretized time in evolution, e.g., mimicking synchronized generations, is straightforward, and we shall discuss it in specific cases (Sect. 5.2.5). Otherwise the focus is set here on continuous time birth-and-death processes and master equations. As an example we consider a birth event or a death event at some time $t = T_r$, which creates or consumes one individual according to $\mathsf{X}_j \to 2\mathsf{X}_j$ or $\mathsf{X}_j \to \varnothing$, respectively. Then the population changes by ± 1 at time T_r:

$$|\Pi| = \begin{cases} \ldots, N_j(t) = N_j(T_{r-1}), N_k(t) = N_k(T_{r-1}), \ldots & \text{for } T_{r-1} \le t < T_r , \\ \ldots, N_j(t) = N_j(T_{r-1}) \pm 1, N_k(t) = N_k(T_{r-1}), \ldots & \text{for } T_r \le t < T_{r+1} . \end{cases}$$

This formulation of biological birth or death events reflects the convention of right-hand continuity for steps in probability theory (see Fig. 1.10).

Biology frequently deals with autocatalytic processes and stationary states far from equilibrium, where explicit consideration of fluctuations is essential. Growth commonly starts from single individuals of a new species. Then, as we have seen in Sect. 4.6.4, enhancement of fluctuations is most prominent and systems are harder to control. Simple autocatalytic processes are analyzed at the ODE level and used as deterministic references for the stochastic analysis of multiplication in biology (Sect. 5.1). Various sections discuss closed and open systems (Sects. 5.1.1 and 5.1.2), unlimited growth (Sect. 5.1.3) and growth control by the logistic equation (Sect. 5.1.4).

Section 5.2 presents an overview of various stochastic processes that are frequently used in biology: chemical master equations applied to growth processes (Sect. 5.2.1), solvable birth-and-death processes with different boundaries (Sect. 5.2.2), logistic birth-and-death processes and problems in epidemiology (Sect. 5.2.4), and branching processes (Sect. 5.2.5), while the application of Fokker–Planck equations to neutral evolution aims to describe the random drift of populations in genotype space (Sect. 5.2.3). The discussion of stochastic models in the theory of evolution (Sect. 5.3) has sections on the Wright–Fisher and Moran processes (Sect. 5.3.1), an exact solution of the master equation for the Moran process (Sect. 5.3.2), the role of mutation (Sect. 5.3.3), and the kinetic theory of molecular evolution (Sect. 5.3.3). Coalescence theory has become important in understanding evolution through the reconstruction of phylogenies (Sect. 5.4).

Examples of numerical simulations of stochastic processes in biology will be mentioned together with the analytical approaches throughout this chapter. For large scale simulations in systems biology see [400, 536, 573], for example.

5.1 Autocatalysis and Growth

Autocatalysis in its simplest form is found in the bimolecular reaction (4.1g): $\mathsf{A} + \mathsf{X} \rightarrow 2\mathsf{X}$. In Chap. 4, we analyzed bimolecular reactions, which allowed for analytical solution, although the derivation and handling of the solutions were quite sophisticated (Sect. 4.3.3). The nonlinearity in the kinetic equation became manifest in the task of finding solutions but did not change the qualitative behavior of the reaction systems, except perhaps the strong sensitivity of the autocatalytic reaction (4.1g) at small concentrations of the autocatalyst. In essence, the $\sqrt{N}$-law for the fluctuations in the stationary states retained its validity. The Brusselator (Sects. 4.1 and 4.6.4) was introduced as a theoretical model for a rare example of a higher order autocatalytic reaction in chemistry that gives rise to oscillations and spatial pattern formation. In practice, however, all known experimental autocatalytic systems and the phenomena derived from it are the result of complicated multistep mechanisms, for which the well studied BelousovZhabotinsky reaction may serve as an example [121, 583]. Even in the largely simplified Brusselator version, the reaction is too complicated for an analytic approach, which is why we studied this system by means of numerical simulation in Sect. 4.6.4.

Multiplication, the basis of biological evolution, is just a special case of autocatalysis, and we shall focus upon it in this chapter. In general, reproduction is a highly complex process and this is true even for its simplest form in the replication of nucleic acid molecules. Modeling is carried out at all levels, from molecular replicating systems to cells and whole organisms, and accordingly models are not usually resolved into elementary steps. The simplest conceivable realistic case involves two steps, reproduction and extinction, and will be presented as an example for an exactly solvable birth-and-death process in Sect. 5.2.2. Here we describe autocatalysis by simple one-step or two-step reaction mechanisms in order to derive a reference based on the deterministic approach [451, pp. 9–75] (Sects. 5.1.1 and 5.1.2). We end this section by casting a glance at the relation between autocatalysis and growth for different mechanisms (Sect. 5.1.3) and growth control in the logistic equation.

5.1.1 Autocatalysis in Closed Systems

Autocatalysis has been discussed in Sects. 4.3.5 and 4.6.4 as a chemical reaction with special properties. We derived an analytical solution to the simplest possible chemical master equation of an irreversible first order autocatalytic process, and analyzed the reversible first order autocatalytic reaction using numerical simulation and trajectory sampling. Computer studies of higher order autocatalysis gave rise to bistability and oscillations in Sect. 4.6.4. Here we shall analyze autocatalysis from a biological perspective and relate it to growth and selection in populations.

Autocatalysis in its simplest form is described by the single reaction step[3]

$$\mathsf{A} + n\mathsf{X} \underset{l}{\overset{k}{\rightleftharpoons}} (n+1)\mathsf{X}\,, \tag{5.2}$$

which is already contained in (4.1) for small n. Indeed, $n = 0$ represents the uncatalyzed isomerization reaction (4.1c) $\mathsf{A} \rightleftharpoons \mathsf{X}$, and $n = 1$ is the bimolecular reaction of first order autocatalysis (4.1g). Equation (5.2) with $n = 2$ corresponds to a termolecular reaction (4.1m),[4] which is the simplest example of second order autocatalysis and which is frequently used to model systems exhibiting nonlinear behavior such as bistability, oscillations, or deterministic chaos. Still higher autocatalytic elementary steps, i.e., $n \geq 3$, give rise to qualitative behavior that is not very different from the case $n = 2$.

Autocatalysis with mass action kinetics is modeled by the differential equation

$$\frac{\mathrm{d}x}{\mathrm{d}t} = -\frac{\mathrm{d}a}{\mathrm{d}t} = kx^n a - lx^{n+1}\,. \tag{5.3}$$

The variables are the concentrations $x(t) = [\mathsf{X}]$ and $a(t) = [\mathsf{A}]$ of molecular species with initial concentrations $x(0) = x_0$ and $a(0) = a_0$ and the conservation relation $x(t) + a(t) = c_0$.[5] Equation (5.3) can be solved by means of the integral [219, p. 106]:

$$\int \frac{\mathrm{d}x}{x^n(\alpha+\beta x)} = \sum_{i=1}^{n-1} \frac{(-1)^i \beta^{i-1}}{(n-i)\alpha^i x^{n-i}} + \frac{(-1)^n \beta^{n-1}}{\alpha^n} \ln \frac{\alpha+\beta x}{x} + C\,,$$

with $\alpha = kc_0$, $\beta = -(k+l)$, and $n \in \mathbb{N}_{>0}$. In the special case $n = 0$, we have $\int \mathrm{d}x/(\alpha+\beta x) = \ln(\alpha+\beta x)/\beta + C$. For $n \geq 2$, it is not possible to derive

[3] In this section we shall use n for the number of molecules involved in the autocatalytic reaction, as well as for the numbers of stochastic variables.

[4] Termolecular and higher reaction steps are neglected in mass action kinetics, but they are nevertheless frequently used in models and simplified kinetic mechanisms. Examples are the Schlögl model [479] and the Brusselator model [421] (Sect. 4.6.4). Thomas Wilhelm and Reinhart Heinrich provided a rigorous proof that the smallest oscillating system with only mono- and bimolecular reaction steps has to be at least three-dimensional and must contain one bimolecular term [571]. A similar proof for the smallest system showing bistability can be found in [570].

[5] This relation is a result of mass conservation in the closed system.

an explicit expression $x(t)$, but the implicit equation for $t(x)$ turns out to be quite useful too:

$$t(x) = \sum_{i=1}^{n-1} \frac{(-1)^i \beta^{i-1}}{(n-i)\alpha^i} \left(\frac{1}{x^{n-i}} - \frac{1}{x_0^{n-i}} \right) + \frac{(-1)^n \beta^{n-1}}{\alpha^n} \ln \frac{(\alpha + \beta x) x_0}{x(\alpha + \beta x_0)} . \tag{5.4}$$

For numerical calculations of the solution curves, it makes practically no difference whether one considers $x(t)$ or $t(x)$.

In Fig. 5.1, the curves $x(t)$ for first order catalysis $\mathsf{A} + \mathsf{X} \to 2\mathsf{X}$ ($n = 1$) and second order autocatalysis $\mathsf{A} + 2\mathsf{X} \to 3\mathsf{X}$ ($n = 2$) are compared with the corresponding curve for the uncatalyzed process $\mathsf{A} \to \mathsf{X}$ ($n = 0$):

$$n = 0 : \; x(t) = \frac{1}{k+l} \left(kc_0 + (lx_0 - ka_0) e^{-(k+l)t} \right) , \tag{5.5a}$$

$$n = 1 : \; x(t) = \frac{kc_0 x_0}{(k+l)x_0 \left(1 - e^{-kc_0 t}\right) + kc_0 e^{-kc_0 t}} , \tag{5.5b}$$

$$n = 2 : \; t(x) = \frac{1}{kc_0} \left(\frac{x - x_0}{x x_0} + \frac{k+l}{kc_0} \ln \frac{\big((k+l)x_0 - kc_0\big)x}{\big((k+l)x - kc_0\big)x_0} \right) . \tag{5.5c}$$

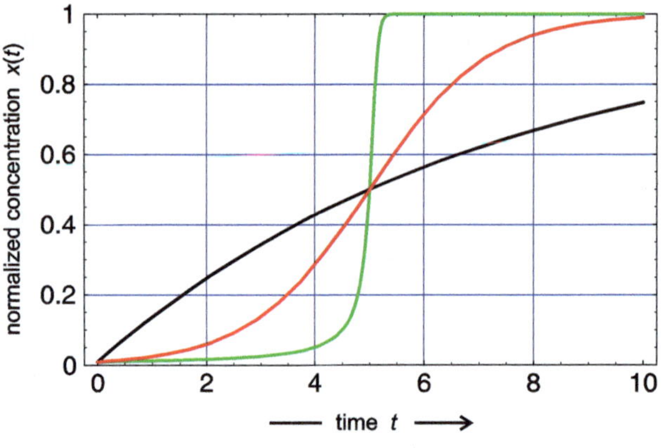

Fig. 5.1 Autocatalysis in a closed system. The concentration $x(t)$ of the substance X as a function of time, according to (5.5) is compared for the uncatalyzed first order reaction $\mathsf{A} \to \mathsf{X}$ ($n = 0$ *black curve*), for the first order autocatalytic process $\mathsf{A} + \mathsf{X} \to 2\mathsf{X}$ ($n = 1$ *red curve*), and for the second order autocatalytic process, $\mathsf{A} + 2\mathsf{X} \to 3\mathsf{X}$ ($n = 2$ *green curve*). The uncatalyzed process ($n = 0$) shows the typical hyperbolic approach towards the stationary state, whereas the two curves for the autocatalytic processes have sigmoid shape. Choice of initial conditions and rate parameters: $x_0 = 0.01$, $c_0 = a(t) + x(t) = 1$ (normalized concentrations), $l = 0$ (irreversible reaction), and $k = 0.13662\,[\mathrm{t}^{-1}]$ for $n = 0$, $0.9190\,[\mathrm{M}^{-1}\mathrm{t}^{-1}]$ for $n = 1$ and $20.519\,[\mathrm{M}^{-2}\mathrm{t}^{-1}]$ for $n = 2$, respectively. Rate parameters k are chosen such that all curves go through the point $(x, t) = (0.5, 5)$

All three curves approach the final state monotonically. This is the state of complete conversion of A into X, $\lim_{t\to\infty} x(t) = 1$, because it was assumed that $l = 0$. Both curves for autocatalysis have sigmoid shape, since they show self-enhancement at low concentrations of the autocatalyst X, pass through an inflection point, and approach the final state in the form of a relaxation curve. The difference between first and second order autocatalysis manifests itself in the steepness of the curve, i.e., the value of the tangent at the inflection point, and is remarkably large. In general, the higher the coefficient of autocatalysis, the steeper the curve at the inflection point. Inspection of (5.5) reveals three immediate results:

(i) Autocatalytic reactions require a seeding amount of X, since $x_0 = 0$ has the consequence $x(t) = 0, \forall\, t$.
(ii) For sufficiently long times, the system approaches a stationary state corresponding to thermodynamic equilibrium:

$$\lim_{t\to\infty} x(t) = \bar{x} = \frac{k}{k+l} c_0 \,, \quad \lim_{t\to\infty} a(t) = \bar{a} = \frac{l}{k+l} c_0 \,.$$

(iii) The function $x(t)$ increases or decreases monotonically for $t > 0$, depending on whether $x_0 < \bar{x}$ or $x_0 > \bar{x}$.

5.1.2 Autocatalysis in Open Systems

The continuously stirred tank reactor (CSTR) is an appropriate open system for studying chemical reactions under controlled conditions (Fig. 4.21). Material consumed during the reaction flows in solution into the reactor and the volume increase is compensated by an outflow of the reaction mixture. Here we analyze autocatalytic reactions in the CSTR. An alternative way to model autocatalysis in an open system has been proposed, for example, by Kenichi Nishiyama: species A is produced from A_0, which is kept in constant concentration by buffering. Reactions $A + X \to 2X$ and $X \to \emptyset$ are irreversible. This mechanism is closely related to (5.6) with $n = 1$, as discussed in the next section, and leads to almost the same results [422].

Autocatalysis in the Flow Reactor

The flow rate r is the reciprocal mean residence time of a volume element in the reactor: $r = \tau_V^{-1}$. Substance A flows into the reactor with concentration c_0 in the stock solution, and all substances present in the reactor flow out at the same rate r. Both parameters r and c_0 can be easily varied in real experiments. So far, species X is not present in the reactor and no reaction takes place, as discussed already in Sects. 4.3.5 and 4.6.4. The reaction is initiated by injecting a seeding amount x_0 of X. In the stochastic treatment, this corresponds to the initial condition $\mathcal{X}_X =$

1 molecule. The reaction mechanism is of the form

$$* \xrightarrow{c_0\, r} \mathsf{A}\,, \tag{5.6a}$$

$$\mathsf{A} + n\mathsf{X} \underset{l}{\overset{k}{\rightleftharpoons}} (n+1)\mathsf{X}\,, \tag{5.6b}$$

$$\mathsf{A} \xrightarrow{r} \varnothing\,, \tag{5.6c}$$

$$\mathsf{X} \xrightarrow{r} \varnothing\,. \tag{5.6d}$$

The stoichiometric factor n again distinguishes different cases: the uncatalyzed reaction with $n = 0$, first order autocatalysis with $n = 1$, and second or higher order autocatalysis with $n \geq 2$. Two kinetic differential equations are required to describe the temporal changes, because the concentrations a and x are now independent:

$$\begin{aligned} \frac{\mathrm{d}a}{\mathrm{d}t} &= -kax^n + lx^{n+1} + r(c_0 - a)\,, \\ \frac{\mathrm{d}x}{\mathrm{d}t} &= kax^n - lx^{n+1} - rx\,. \end{aligned} \tag{5.6e}$$

The sum of the concentrations, $c(t) = a(t) + x(t)$, however, converges to the concentration c_0 of A in the stock solution, since

$$\frac{\mathrm{d}c}{\mathrm{d}t} = r(c_0 - c)\,.$$

The relaxation time towards the stable steady state $c(t) = \bar{c} = c_0$ is the mean residence time, i.e., $\tau_\mathrm{V} = r^{-1}$, so different orders of autocatalysis n have no influence on the relaxation time.

Steady state analysis using $\mathrm{d}a/\mathrm{d}t = 0$ and $\mathrm{d}x/\mathrm{d}t = 0$ reveals three different scenarios sharing the limiting cases. At vanishing flow rate r the system approaches thermodynamic equilibrium with $\bar{x} = kc_0/(k+l)$, $\bar{a} = lc_0/(k+l)$ and $K = k/l$, and no reaction occurs for sufficiently large flow rates $r > r_\mathrm{cr}$, when the mean residence time is too short to sustain the occurrence of reaction events. Then we have $\bar{x} = 0$ and $\bar{a} = c_0$ for $\lim r \to \infty$. In the intermediate range, at finite flow rates $0 < r < r_\mathrm{cr}$, the steady state condition yields[6]

$$r(\bar{x}) = kc_0\bar{x}^{n-1} - (k+l)\bar{x}^n\,, \quad \bar{a} = \frac{l\bar{x}^n + r}{k\bar{x}^{n-1}}\,.$$

[6]As for the time dependence in the closed system expressed by (5.5c), we make use of the uncommon implicit function $r = f(\bar{x})$ rather than the direct relation $\bar{x} = f(r)$.

Stability analysis of the stationary states is performed by means of the Jacobian matrix. It is advantageous to split the analysis of the implicit equation for the stationary concentration $\bar{x}$ into three cases:

(i) The unique steady state for the uncatalyzed process $n = 0$, viz., $\mathsf{A} \rightleftharpoons \mathsf{X}$, satisfies

$$\bar{x} = \frac{k}{k+l+r} c_0 \,, \quad \bar{a} = \frac{l+r}{k+l+r} c_0$$

and changes monotonically from equilibrium at $r = 0$ to no reaction at $\lim r \to \infty$. Confirming stability, the two eigenvalues of the Jacobian matrix are both negative: $\lambda_1 = -r$ and $\lambda_2 = -r - (k+l)$.

(ii) In the case of first order autocatalysis $n = 1$, steady state conditions yield two solutions, S_1 and S_2:

$$\bar{x}_1 = \frac{kc_0 - r}{k+l} \,, \quad \bar{a}_1 = \frac{lc_0 + r}{k+l} \,, \quad \bar{x}_2 = 0 \,, \quad \bar{a}_2 = c_0 \,. \tag{5.6f}$$

The eigenvalues of the Jacobian matrix are $\lambda_1^{(1)} = -r$ and $\lambda_2^{(1)} = r - kc_0$ at S_1, and $\lambda_1^{(2)} = -r$ and $\lambda_2^{(2)} = -r + kc_0$ at S_2. Hence, the first solution $S_1 = (\bar{x}_1, \bar{a}_1)$ is stable in the range $0 \le r < kc_0$, whereas the second solution $S_2 = (\bar{x}_2, \bar{a}_2)$ shows stability at flow rates $r > kc_0$ above the critical value $r_{\mathrm{crit}} = kc_0$. The change from the active state S_1 to the state of extinction S_2 occurs abruptly at the transcritical bifurcation point $r = kc_0$ (see the solution for $\delta = 0$ in Fig. 5.2).[7]

(iii) Second and higher order autocatalysis ($n \ge 2$) allow for a common treatment. Points with a horizontal tangent to $r(\bar{x})$, defined by $\mathrm{d}r/\mathrm{d}\bar{x} = 0$, in an $(\bar{x}, r)$ plot are points with a vertical tangent to the function $\bar{x}(r)$, representing subcritical or other bifurcation points (Fig. 5.2). Such points correspond to maximal or minimal values of r at which branches of $\bar{x}(r)$ end, and they can be computed analytically:

$$\bar{x}(r_{\max}) = \frac{n-1}{n} \frac{kc_0}{k+l} \,, \quad \text{for } n \ge 2 \,, \qquad \bar{x}(r_{\min}) = 0 \,, \quad \text{for } n \ge 3 \,,$$

with the corresponding flow rates

$$r_{\max} = \left(\frac{n-1}{k+l} \right)^{n-1} \left(\frac{kc_0}{n} \right)^n \,, \quad r_{\min} = 0 \,.$$

[7]Bifurcation analysis is a standard topic in the theory of nonlinear systems. Monographs oriented towards practical applications are, for example, [275, 276, 496, 513].

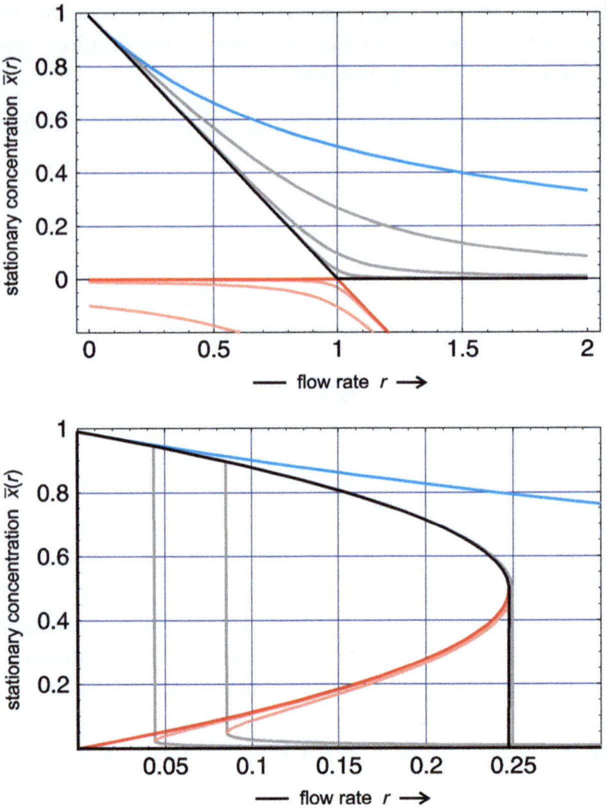

Fig. 5.2 Stationary states of autocatalysis in the flow reactor. The *upper plot* shows avoided crossing in first order autocatalysis ($n = 1$) when the uncatalyzed reaction is included. Parameter values: $k = 1\,[\mathrm{M}^{-1}\mathrm{t}^{-1}]$, $l = 0.01\,[\mathrm{M}^{-1}\mathrm{t}^{-1}]$, $c_0 = 1\,[\mathrm{M}]$, $\delta = 0$ (*black* and *red*), $\delta = 0.001, 0.01$, and 0.1 (*gray* and *pink*). The uncatalyzed reaction (*blue*) is shown for comparison. The *lower plot* refers to second order autocatalysis ($n = 2$) and shows the shrinking of the range of bistability as a function of the parameter δ. Parameter values: $k = 1\,[\mathrm{M}^{-2}\mathrm{t}^{-1}]$, $l = 0.01\,[\mathrm{M}^{-2}\mathrm{t}^{-1}]$, $c_0 = 1\,[\mathrm{M}]$, $\delta = 0$ (*black* and *red*), $\delta = 0.0005$ and 0.002 (*gray* and *pink*). Again, the uncatalyzed reaction is shown in *blue*. The upper stable branch in the bistability range is called the equilibrium branch, while the lowest branch represents the state of extinction

The state of extinction $S_3 = (\bar{x}_3, \bar{a}_3) = (0, c_0)$ is always stable since the eigenvalues of the Jacobian are $\lambda_1^{(3)} = \lambda_2^{(3)} = -r$ for $n \geq 2$. The other two stationary points are a stable point (S_1) and a saddle point (S_2), provided they exist.

Figure 5.3 compares the bifurcation patterns for second and higher order autocatalysis in the flow reactor. All four curves show a range of bistability $0 = r_{\min} < r < r_{\max}$, with one stable stationary state S_1 (black in the figure) and one unstable saddle point S_2 (red in the figure). There is a third stationary state S_3 at $\bar{x}_3 = 0$, which is always stable.

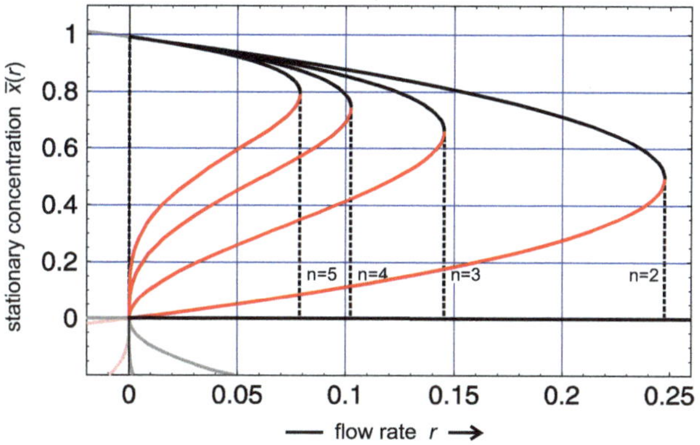

Fig. 5.3 Stationary states of higher order autocatalysis in the flow reactor. The *curves* show the range of bistability for different orders of autocatalysis ($n = 2, 3, 4$, and 5 *from right to left*) and the parameters $k = 1\,[\mathrm{M}^{-n}\mathrm{t}^{-1}]$, $l = 0.01\,[\mathrm{M}^{-n}\mathrm{t}^{-1}]$, and $c_0 = 1\,[\mathrm{M}]$. The two stable branches, the thermodynamic branch (*upper branch*) and the state of extinction ($\bar{x} = 0$), are shown in *black*, and the intermediate unstable branch is plotted in *red*. The *vertical dotted lines* indicate the critical points of the subcritical bifurcations. The analytic continuations of the curves into the non-physical ranges $r < 0$ or $\bar{x} < 0$ are shown in *light pink* or *gray*, respectively

In the case of second order autocatalysis $n = 2$, the lower limit is obtained for vanishing flow rate $r = 0$. For $n = 3$, 4, and 5 the lower limit is given by the minimum of the function $r(\bar{x})$, which coincides with $\bar{x} = 0$ (see also Sect. 4.6.4). An increase in the values of n causes the range of bistability to shrink. A vertical line corresponding to $r =$ const. intersects the curve $r(\bar{x})$ for $n = 2$, either at one or at three points corresponding to the stationary states S_1, S_2, and S_3 (Fig. 4.49).

The three cases $n = 0$, 1, and $n \geq 2$, provide an illustrative example of the role played by nonlinearity in chemical reactions. The uncatalyzed reaction shows a simple decay to the stationary state with a single negative exponential function. In closed systems, all autocatalytic processes have characteristic phases, consisting of a growth phase with a positive exponential at low autocatalyst concentrations and the (obligatory) relaxation phase with a negative exponential at concentrations sufficiently close to equilibrium (Fig. 5.1). In the flow reactor the nonlinear systems exhibit characteristic bifurcation patterns (Fig. 5.2): first order autocatalysis gives rise to a rather smooth transition in the form of a transcritical bifurcation from the equilibrium branch to the state of extinction, whereas for $n \geq 2$, the transitions are abrupt, and as is characteristic for a subcritical bifurcation, chemical hysteresis is observed.

All cases of autocatalysis in the flow reactor ($n > 0$) discussed so far contradict a fundamental theorem of thermodynamics stating the uniqueness of the equilibrium state. Only a single steady state may occur in the limit $\lim r \to 0$. The incompatibility of the model mechanism (5.6) with basic thermodynamics can be corrected by satisfying the principle that any catalyzed reaction requires the existence of an uncatalyzed process that approaches the same equilibrium state, or in other words a catalyst accelerates the forward and the backward reaction by the same factor. Accordingly, we have to add the uncatalyzed process with $n = 0$ to the reaction mechanism (5.6):

$$\mathsf{A} \underset{l\delta}{\overset{k\delta}{\rightleftarrows}} \mathsf{X}\,. \tag{5.6b'}$$

The parameter δ represents the ratio of the rate parameters of the uncatalyzed and the catalyzed reaction as applied in (4.159b). Figure 5.2 shows the effect of nonzero values of δ on the bifurcation pattern. In first order autocatalysis, the transcritical bifurcation disappears through a phenomenon known in linear algebra as avoided crossing. Two eigenvalues λ_1 and λ_2 of a 2×2 matrix A plotted as functions of a parameter p cross at some critical value: $\lambda_1(p_{\mathrm{cr}}) = \lambda_2(p_{\mathrm{cr}})$ avoid crossing when variation of a second parameter q causes an off-diagonal element of A to change from zero to some nonzero value. In the figure, the parameter p is represented by the flow rate r and parameter q by δ. The two steady states are obtained as solutions of the quadratic equation

$$\bar{x}_{1,2} = \frac{1}{2(k+l)}\left(kc_0 - \delta(k+l) - r \pm \sqrt{\left(kc_0 - \delta(k+l) - r\right)^2 + 4kc_0\delta(k+l)}\right).$$

In the limit $\delta \to 0$, we obtain the solutions (5.6f), and in the limit of vanishing flow, i.e., $\lim r \to 0$, we find $\bar{x}_1 = kc_0/(k+l)$, $\bar{x}_2 = -\delta$, and $\bar{a}_{1,2} = c_0 - \bar{x}_{1,2}$. As demanded by thermodynamics, only one solution $\bar{x}_1$, the equilibrium state $P_1 = (\bar{x}_1, \bar{a}_1)$ for $r = 0$, occurs within the physically meaningful domain of nonnegative concentrations, whereas the second steady state $P_2 = (\bar{x}_2, \bar{a}_2)$ for $r = 0$, has a negative value for the concentration of the autocatalyst.

5.1.3 *Unlimited Growth*

It is worth considering different classes of growth functions $y(t)$, where the solutions of the corresponding ODEs have qualitatively different behavior. The nature of the growth function determines general features of population dynamics and we may ask whether there exists a universal long-time behavior that is characteristic for certain classes of growth function? To answer this question, we shall study growth that is not limited by the exhaustion of resources.

The results presented below are obtained within the framework of the ODE model, i.e., neglecting stochastic phenomena. The differential equation describing unlimited growth, viz.,

$$\frac{\mathrm{d}y}{\mathrm{d}t} = fy^n\,, \tag{5.7a}$$

gives rise to two types of general solution for the initial value $y(0) = y_0$, depending on the choice of the exponent n:

$$y(t) = \left(y_0^{1-n} + (1-n)ft\right)^{1/(1-n)}, \qquad \text{for } n \neq 1, \tag{5.7b}$$

$$y(t) = y_0 e^{ft}, \qquad \text{for } n = 1. \tag{5.7c}$$

In order to make the functions comparable, we normalize them such that they satisfy $y(0) = 1$ and $dy/dt|_{t=0} = 1$. According to (5.7), this yields $y_0 = 1$ and $f = 1$. The different classes of growth functions shown in Fig. 5.4 are characterized by the following behavior:

(i) Hyperbolic growth requires $n > 1$. For $n = 2$, it yields the solution curve $y(t) = 1/(1-t)$. Characteristic here is the existence of an instability, in the sense that $y(t)$ approaches infinity at some critical time, i.e., $\lim_{t \to t_{cr}} = \infty$, which is $t_{cr} = 1$ for $n = 2$.

(ii) Exponential growth is observed for $n = 1$ and described by the solution $y(t) = e^t$. The exponential function reaches infinity only in the limit $t \to \infty$. It represents the most common growth function in biology.

(iii) Parabolic growth occurs for $0 < n < 1$, and for $n = 1/2$ has the solution curve $y(t) = (1 + t/2)^2$.

(iv) Linear growth follows from $n = 0$, and takes the form $y(t) = 1 + t$.

(v) Sublinear growth occurs for $n < 0$. In particular, for $n = -1$, it gives rise to the solution $y(t) = (1 + 2t)^{1/2} = \sqrt{1 + 2t}$.

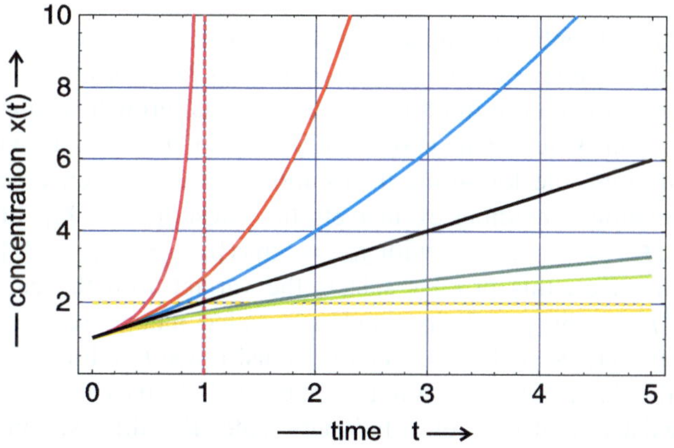

Fig. 5.4 Typical functions describing unlimited growth. All functions are normalized in order to satisfy the conditions $y(0) = 1$ and $dy/dt|_{y=0} = 1$. *Curves* show hyperbolic growth $y(t) = 1/(1-t)$ *magenta* (the *dotted line* shows the position of the instability), exponential growth $y(t) = \exp(t)$ *red*, parabolic growth $y(t) = (1 + t/2)^2$ *blue*, linear growth $y(t) = 1 + t$ *black*, sublinear growth $y(t) = \sqrt{1 + 2t}$ *turquoise*, logarithmic growth $y(t) = 1 + \log(1 + t)$ *green*, and sublogarithmic growth $y(t) = 1 + t/(1 + t)$ *yellow* (the *dotted line* indicates the maximum value y_{max}: $\lim_{t \to \infty} y(t) = y_{max}$)

In addition, we mention two other forms of weak growth that do not follow from (5.7):

(vi) Logarithmic growth, expressed by the functions $y(t) = y_0 + \ln(1 + ft)$ or $y(1) = 1 + \ln(1 + t)$ after normalization.
(vii) Sublogarithmic growth, modeled by the functions $y(t) = y_0 + ft/(1 + ft)$ or $y(t) = 1 + t/(1 + t)$ in normalized form.

In Fig. 5.4, hyperbolic growth, parabolic growth, and sublinear growth constitute families of solution curves defined by a certain parameter range, for example, a range of exponents $n_{\text{low}} < n < n_{\text{high}}$, whereas exponential growth, linear growth, and logarithmic growth represent critical curves separating zones of characteristic behavior. Logarithmic growth separates growth functions approaching infinity in the limit $t \to \infty$, $\lim_{t\to\infty} y(t) = \infty$ from those that remain finite, $\lim_{t\to\infty} y(t) = y_\infty < \infty$. Linear growth separates concave from convex growth functions, and exponential growth eventually separates growth functions that reach infinity at finite times from those that do not.

The growth functions $y(t)$ determine the population dynamics and hence also the results of long-time evolution. A useful illustration considers the internal population dynamics in a population Π of constant size N with M variants $\Pi(t) = \{\mathsf{X}_1(t), \ldots, \mathsf{X}(M)\}$, where $[\mathsf{X}_i] = x_i$, is described by the vector $\boldsymbol{\xi}(t) = \big(\xi_1(t), \ldots, \xi_M(t)\big)$ with $N = \sum_{i=1}^{M} x_i = \text{const.}$, $\xi_i = x_i/N$, and $\sum_{i=1}^{n} \xi_i = 1$. The differential equation in the internal variables

$$\frac{\mathrm{d}\xi_i}{\mathrm{d}t} = f_i\xi_i^n - \xi_i\Phi(t) , \quad i = 1, \ldots, M , \quad \text{with } \Phi(t) = \sum_{i=1}^{M} f_i\xi_i^n , \tag{5.8}$$

falls in the class of replicator equations [485] and can be solved analytically. Here we discuss only the phenomena observed at the population level.

The most important case in biology is exponential growth with $n = 1$, since it leads to Darwinian evolution in the sense of *survival of the fittest*. Provided all fitness values are different, the long-time distribution of variants converges to a homogeneous population containing exclusively the fittest species X_m: $\lim_{t\to\infty} \boldsymbol{\xi}(t) = \bar{\boldsymbol{\xi}} = (0, \ldots, \bar{\xi}_m = 1, \ldots, 0)$, with $f_m = \max\{f_i; i = 1, \ldots, M\}$. Apart from stochastic influences, this process selects the variant with the currently highest fitness value f_m in the population, and was called natural selection by Charles Darwin. Equation (5.8) with $n = 1$ can be transformed to a linear ODE by means of an integrating factor transformation [585, pp. 322–326] (see Sect. 5.1.4), and this implies the existence of only one stationary state. If a fitter variant is created in the population, for example, through mutation, the new variant will replace the previously fittest species. As already indicated, fluctuations may interfere with the selection process when the fittest species is present in only a few copies and goes extinct by accident (for details and references, see Sects. 5.2.3 and 5.2.5).

The case $n = 2$ is the best studied example of hyperbolic growth in unconstrained open systems. Populations in open systems with constant population size are described by (5.8), which is a multi-dimensional Schlögl model named after the German physicist Friedrich Schlögl [479]. They have been studied extensively [256, 485]. In contrast to Darwinian systems with $n = 1$, all uniform populations $\Pi_i : \bar{\boldsymbol{\xi}} = (0, \ldots, \bar{\xi}_i = 1, \ldots, 0)$, $\forall\, i = 1 \ldots, M$, in systems with $n = 2$ correspond to stable equilibrium points. In other words, once one particular species has been selected, it cannot be replaced by another one through a mutation creating a variant with higher f_m value. Hyperbolic growth thus leads to *once-for-ever* selection. For a visualization of hyperbolic growth, it is instructive to consider a Cartesian space of relative concentrations $\boldsymbol{\xi}$ with $\sum_{i=1}^{M} \xi_i = 1$. The physically accessible part of the space defined by $0 \leq \xi_i \leq 1$ represents a simplex S_M. In the interior of the simplex is one unstable stationary point, all corners representing the uniform populations Π_i are asymptotically stable, and the sizes of their basins of attraction correspond to the fitness parameters f_i.

Parabolic growth, which occurs in the range $0 < n < 1$, may give rise to coexistence between variants [521, 575]. This behavior has been found experimentally with autocatalytic replication of small oligonucleotides [555, 556]. Linear growth at $n = 0$ always leads to coexistence of variants, and it has been observed in replicase catalyzed replication of polynucleotides *in vitro* at high template concentrations [48].

We summarize this section by comparing growth behavior and characteristic dynamics of autocatalysis. Subexponential growth allows for coexistence, whereas hyperexponential growth gives rise to selection that depends on an initial population $\Pi(0) = \Pi_0$. Only the intermediate case of exponential growth results in population-independent selection, with the Malthus parameter or the fitness of species as selection criterion. It is not accidental therefore that, in terms of autocatalysis, exponential growth is the result of first order autocatalysis, which in discrete time corresponds to a growth and division process—$\mathbf{X} \rightarrow \mathbf{X}^* \rightarrow 2\mathbf{X}$ with $\mathbf{X}^*$ being a cell after the internal growth phase during which the genetic material has been duplicated—and which is universal for all cells in biology.

5.1.4 Logistic Equation and Selection

The logistic equation (5.1) can be used to illustrate the occurrence of selection [482] in populations with exponential growth. For this purpose the population is partitioned into n groups of individuals called subspecies or variants, $\Pi = \{\mathsf{X}_1, \ldots, \mathsf{X}_M\}$, and all individuals of one group are assumed to multiply with the same fitness factor: X_1 is multiplied by $f_1, \ldots, \mathsf{X}_M$ by f_M. Next we rewrite the logistic equation by introducing a function $\Phi(t)$ that will be interpreted later, viz., $X(t)\Phi(t) \equiv (X(t)/C)fX(t)$, to obtain

$$\frac{\mathrm{d}X}{\mathrm{d}t} = fX - \frac{X}{C}fX = fX - X\Phi(t) = X\big(f - \Phi(t)\big)\,.$$

From $\mathrm{d}X/\mathrm{d}t = 0$, we deduce that the stationary concentration equals the carrying capacity $\bar{X} = C$. The distribution of subspecies within the population,

$|\Pi(t)| = \mathbf{x}(t) = \big(x_1(t), \ldots, x_M(t)\big)$ with $X = \sum_{i=1}^{M} x_i$, is taken into account and this leads to

$$\frac{\mathrm{d}x_j}{\mathrm{d}t} = f_j x_j - x_j \Phi(t) = x_j\big(f_j - \Phi(t)\big)\,, \quad j = 1, \ldots, M\,. \tag{5.9}$$

Summing over all subspecies and taking the long-time limit yields

$$\sum_{i=1}^{M} \frac{\mathrm{d}x_i}{\mathrm{d}t} = \sum_{i=1}^{M} f_i x_i - \Phi(t) \sum_{i=1}^{M} x_i = 0 \implies \Phi = \frac{\sum_{i=1}^{M} f_i \bar{x}_i}{\sum_{i=1}^{M} \bar{x}_i} = \mathrm{E}(f)\,,$$

and we see that the function Φ can be interpreted straightforwardly as the expectation value of the fitness taken over the entire population.

The equilibration of subspecies within the population is illustrated by considering relative concentrations $\xi_i(t)$, as in (5.8), with $\sum_{i=1}^{n} \xi_i = 1$. For the time dependence of the mean fitness $\Phi(t)$, we find

$$\begin{aligned} \frac{\mathrm{d}\Phi(t)}{\mathrm{d}t} &= \sum_{i=1}^{M} f_i \frac{\mathrm{d}\xi_i}{\mathrm{d}t} = \sum_{i=1}^{M} f_i \left(f_i \xi_i - \xi_i \sum_{j=1}^{M} f_j \xi_j \right) \\ &= \sum_{i=1}^{n} f_i^2 \xi_i - \sum_{i=1}^{M} f_i \xi_i \sum_{j=1}^{M} f_j \xi_j \\ &= \mathrm{E}(f^2) - \big(\mathrm{E}(f)\big)^2 = \mathrm{var}(f) \geq 0\,. \end{aligned} \tag{5.10}$$

Since a variance is nonnegative by definition, (5.10) implies that $\Phi(t)$ is a nondecreasing function of time. The variance $\mathrm{var}(f) = 0$ defines a homogeneous population X_m with $f_m = \max\{f_j; j = 1, \ldots, M\}$ which contains only the fittest variant. Then $\Phi(t)$ is at its maximum and cannot increase any further. Accordingly, $\Phi(t)$ has been optimized during selection, as already mentioned in Sect. 5.1.3.

It is also possible to derive analytical solutions for (5.9) by a transform called integrating factors [585, pp. 322–326]:

$$\zeta_i(t) = \xi_i(t) \exp\left(\int_0^t \Phi(\tau) \mathrm{d}\tau \right)\,. \tag{5.11}$$

Inserting in (5.9) yields

$$\frac{\mathrm{d}\zeta_i}{\mathrm{d}t} = f_i \zeta_i\,, \quad \zeta_i(t) = \zeta_i(0) \exp(f_i t)\,,$$

$$\xi_i(t) = \xi_i(0) \exp(f_i t) \exp\left(- \int_0^t \Phi(\tau) \mathrm{d}\tau \right)\,,$$

$$\text{with } \exp\left(\int_0^t \Phi(\tau) \mathrm{d}\tau \right) = \sum_{j=1}^{n} \xi_j(0) \exp(f_j t)\,,$$

where we have used $\zeta_i(0) = \xi_i(0)$ and the condition $\sum_{i=1}^{M} \xi_i = 1$. The solution finally takes the form

$$\xi_i(t) = \frac{\xi_i(0)\exp(f_i t)}{\sum_{j=1}^{M} \xi_j(0)\exp(f_j t)}\,, \qquad i = 1, 2, \ldots, M\,. \tag{5.12}$$

Under the assumption that the largest fitness parameter is non-degenerate, i.e., $\max\{f_i;\ i = 1, 2, \ldots, M\} = f_m > f_i,\ \forall\, i \neq m$, every solution curve satisfying the initial condition $\xi_i(0) > 0$ approaches a homogeneous population: $\lim_{t\to\infty} \xi_m(t) = \bar{\xi}_m = 1$ and $\lim_{t\to\infty} \xi_i(t) = \bar{\xi}_i = 0,\ \forall\, i \neq m$, and the mean fitness approaches monotonically the largest fitness parameter, i.e., $\Phi(t) \to f_m$.

The process of selection is easily demonstrated by considering the sign of the differential quotient in (5.9): in the instant following time t, the concentration of all variables ξ_j with above average fitness $f_j > \Phi(t)$ will increase, whereas a decrease will be observed for all variables with $f_j < \Phi(t)$. As a consequence, the mean fitness increases $\Phi(t+\Delta t) > \Phi(t)$ and more subspecies fall under the criterion for decrease until $\Phi(t)$ assumes the maximum value and becomes constant. □

5.2 Stochastic Models in Biology

In this section we discuss models of stochastic processes that are frequently applied in biology. We present here examples of biologically relevant applications of master equations (Sect. 5.2.1) and Fokker–Planck equations (Sect. 5.2.3). Some of the approaches to biological questions are based on specific models, which provide more insights into processes than the application of general formalisms like master and Fokker–Planck equations. Such models studied here are birth-and-death processes (Sects. 5.2.2 and 5.2.4) and branching processes (Sect. 5.2.5).

5.2.1 *Master Equations and Growth Processes*

This section complements the deterministic analysis of autocatalysis in the flow reactor discussed in Sect. 5.1.2. At the same time we extend the model of the one-step autocatalytic reaction in a closed system (Sects. 4.3.5 and 4.6.4) to an embedding in the flow reactor, described by the mechanism (5.6) with $n = 1$ and $l = 0$, as an example of an open system.[8] In particular, we shall put emphasis on the distinction between true stationary states and quasi-stationarity observed in systems which exhibit a long-time distribution that is stationary for all practical purposes, but also have an absorbing boundary into which the system necessarily converges in the limit $t \to \infty$.

[8] Although confusion is highly improbable, we remark that the use of n as the exponent of the growth function in x^n and as the number of particles in P_n is ambiguous here.

The single-step autocatalytic reaction $\mathsf{A} + \mathsf{X} \to 2\mathsf{X}$ in the flow reactor, (5.6) with $n = 1$ and $l = 0$, does not satisfy the conservation relation of the closed system $\mathcal{X}_{\mathsf{A}}(t) + \mathcal{X}_{\mathsf{X}}(t) = c_0 = a_0 + x_0$, so we are dealing with two independent random variables $\mathcal{X}_{\mathsf{A}}(t)$ and $\mathcal{X}_{\mathsf{X}}(t)$ which have the joint probability density $P_{n,m}(t) = P\big(\mathcal{X}_{\mathsf{A}}(t) = n \wedge \mathcal{X}_{\mathsf{X}}(t) = m\big)$. The bivariate master equation is of the form

$$\begin{aligned}\frac{\mathrm{d}P_{n,m}(t)}{\mathrm{d}t} &= c_0 r P_{n-1,m}(t) + k(n+1)(m-1)P_{n+1,m-1}(t)\\ &+r(n+1)P_{n+1,m}(t) + r(m+1)P_{n,m+1}(t)\\ &-\big(c_0 r + knm + r(n+m)\big)P_{n,m}(t)\,,\end{aligned} \tag{5.13}$$

According to (5.6f), the deterministic system sustains two steady states:

(i) The state $\Sigma_1 = (\bar{x}_1 = c_0 - r/k, \bar{a}_1 = r/k)$.
(ii) The *state of extinction*, $\Sigma_2 = (\bar{x}_2 = 0, \bar{a}_2 = c_0)$, in which the autocatalyst vanishes.

Now the question is whether or not the two states exist in the stochastic model as well. The answer is straightforward for the state of extinction Σ_2: since the joint transition probabilities $W(n_2, m_2|n_1, m_1)$ satisfy both relations $W(n_2, 0|n_1, 1) > 0$ and $W(n_2, 1|n_1, 0) = 0$ for all values of $(n_1 > 0, n_2 > 0)$, state Σ_2 is an absorbing boundary (Sect. 3.2.3). The state Σ_2 is the only absorbing state of the system and therefore all trajectories have to end in Σ_2, either at finite times or at least in the limit $t \to \infty$. Trajectories may nevertheless also approach and reach the second state corresponding to Σ_1, and fluctuate around it for arbitrarily long times $\widehat{t}_{S_1} < \infty$. Such a state is said to be *quasi-stationary* and we indicate this property by means of a tilde, as in $\widetilde{\Sigma}_1$. Trajectories near a quasi-stationary state behave exactly like trajectories near a true stationary state, with the only exception that the system has to leave the state and converge to the absorbing state in the limit $t \to \infty$ [418, 419] (see also Sect. 5.2.4).

For initially very small numbers of autocatalysts $x_0 = 1, 2, \ldots$, the long-time expectation values lie between the quasi-stationary state $\widetilde{\Sigma}_1$ and the absorbing state Σ_2, that is, $r/k < \bar{\mathrm{E}}(\mathcal{X}_{\mathsf{A}}) < c_0$ and $c_0 - r/k > \bar{\mathrm{E}}(\mathcal{X}_{\mathsf{X}}) > 0$ with $\bar{\mathrm{E}}(\mathcal{X}_{\mathsf{A}}) + \bar{\mathrm{E}}(\mathcal{X}_{\mathsf{X}}) = c_0$. The fluctuations in the early phase of the reaction decide whether the trajectory approaches the quasi-stationary state or progresses into the absorbing boundary. Indeed, a statistical superposition of the two states $\widetilde{\Sigma}_1$ and Σ_2 is observed where the relative weights are determined by the value of x_0. As expected, the frequency $\vartheta_{\widetilde{\Sigma}_1} = n_{\widetilde{\Sigma}_1}/(n_{\widetilde{\Sigma}_1} + n_{\Sigma_2})$ at which the path towards $\widetilde{\Sigma}_1$ is chosen increases strongly with increasing numbers x_0, and this is reflected by the expectation values of extinction shown in Table 5.1 as well as the one-σ-error band in Fig. 5.5. The initial value $x_0 = 10$ is sufficient for a final deviation of less than 1 % from the deterministic limit. The frequency of becoming extinct decreases fast with increasing numbers of initially present autocatalysts, is down to about 5 % for $x_0 = 5$, and amounts to less than 1 % at $x_0 \geq 10$. Quasi-stationarity will be further discussed in Sect. 5.2.4 on logistic birth-and-death and epidemiology.

Table 5.1 Quasi-stationary state in the autocatalytic reaction $A + X \rightarrow 2X$ in the flow reactor. Long-time expectation values of the random variables $\mathcal{X}_A$ and $\mathcal{X}_X$ are given together with the band width $\bar{\Sigma} = 2\bar{\sigma}(\mathcal{X})$. In addition, we present the expectation values $E(\mathcal{N}_{\Sigma_2})$ and standard deviations $\sigma(\mathcal{N}_{\Sigma_2})$ for the numbers of extinction cases per hundred trajectories. Parameters: $a_0 = 200\,[\mathrm{N}]$, $a(0) = 0$, $r = 0.5\,[\mathrm{V\,t^{-1}}]$, and $f = 0.01\,[\mathrm{N^{-1}t^{-1}}]$. The abbreviation "det" stands for the integration of the deterministic kinetic differential equation

$\mathcal{X}_X(0)$	$\bar{E}(\mathcal{X}_A)$	$\bar{\Sigma}(\mathcal{X}_A)$	$\bar{E}(\mathcal{X}_X)$	$\bar{\Sigma}(\mathcal{X}_X)$	$E(\mathcal{N}_{\Sigma_2})$	$\sigma(\mathcal{N}_{\Sigma_2})$
1	133.9	150.4	66.1	141.1	53.7	4.5
2	101.3	143.4	98.9	143.7	32.8	6.0
3	79.1	119.2	120.7	120.9	19.4	3.6
4	66.3	93.9	133.6	96.7	12.4	1.7
5	59.5	73.6	140.4	77.8	5.4	1.8
10	51.2	26.7	149.1	37.6	0.6	0.5
det	50	–	150	–	0	–

In summary, two sources of stochasticity can be distinguished: (i) thermal fluctuations giving rise to stationary fluctuations in the long-time limit and (ii) fluctuations in the early phase of the process determining the behavior at the branching point and leading to different long-time solutions. The thermal fluctuations are described by the equilibrium density and they are analogous to the equilibrium fluctuations in every other chemical reaction (Sect. 4.6.4). In the early phase of the process, random fluctuations lead to different bifurcation behavior of trajectories progressing towards $\widetilde{\Sigma}_1$ or Σ_2, respectively.

The reversible autocatalytic reaction $A + X \rightleftharpoons 2X$ is readily embedded in the flow reactor corresponding to the mechanism (5.6b) with $n = 1$. The qualitative behavior of the reversible reaction in the open system is the same as that of the irreversible reaction. Two long-time solutions are observed: (i) a quasi-stationary state $\widetilde{\Sigma}_1$ and (ii) an absorbing state of extinction Σ_2. Initial fluctuations determine the relative weights of the two states in the final *superposition*. An expectation value between the two long-time states $\widetilde{\Sigma}_1$ and Σ_2 is observed, along with a large variance. Finally, we have to explain the difference between the long-time behavior of the autocatalytic reaction in the closed system and in the flow reactor. A single molecule of X cannot enter the reaction $2X \rightarrow A + X$. The transition $(c_0 - 1, 1) \rightarrow (c_0, 0)$ is forbidden in the closed system and this keeps the system away from the absorbing state $\Sigma_2 = \Sigma_{(c_0,0)}$. In other words, the state $\Sigma_{(c_0-1,1)}$ is a reflecting boundary. In the flow reactor, however, the last molecule of X can flow out of the reactor and then the system has reached the state of extinction, which is the only absorbing state, as in the case of the irreversible reaction.

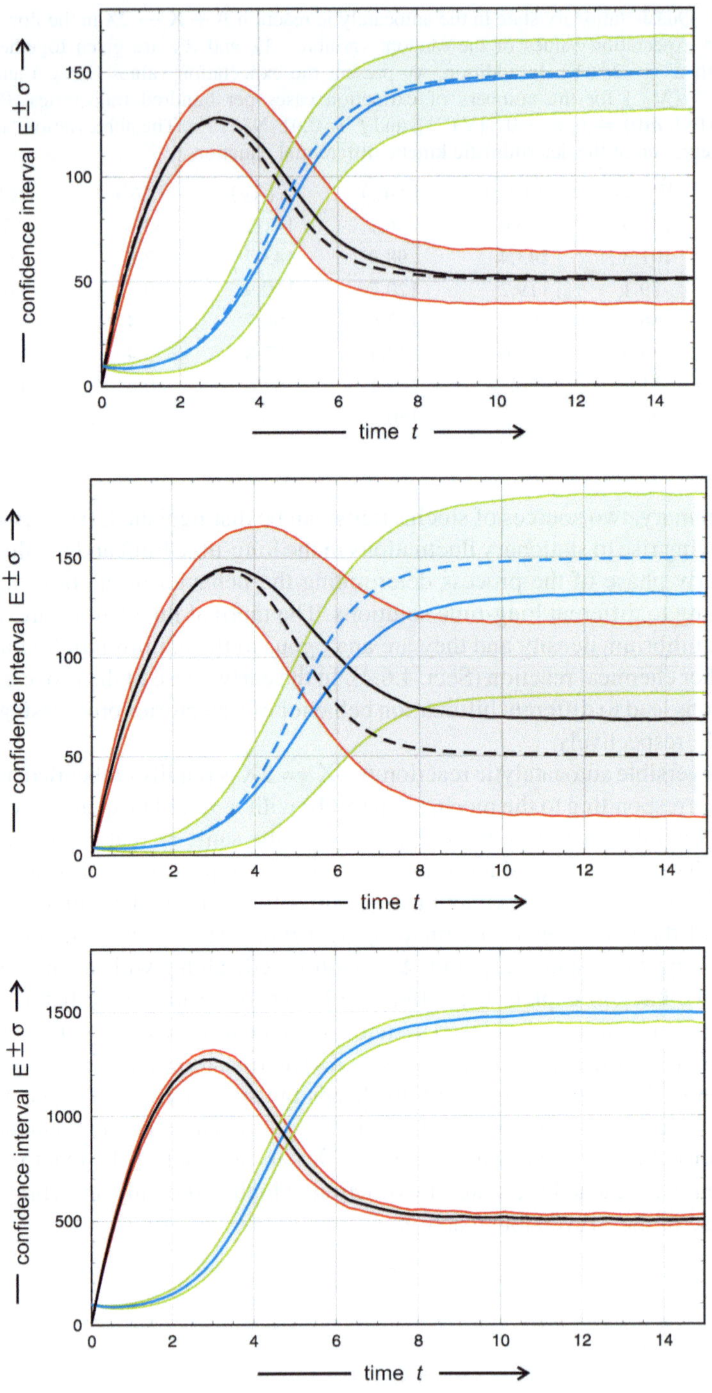

Fig. 5.5 Autocatalysis in the flow reactor. Caption on next page

Fig. 5.5 Autocatalysis in the flow reactor (see previous page). Simulations based on Gillespie's algorithm (Sect. 4.6.3) applied to the mechanism (5.6) with growth exponent $n = 1$ and rate parameters $k = f$ and $l = 0$. The figures illustrate two sources of stochasticity: (i) random thermal fluctuations as in any other chemical reaction and (ii) random fluctuations in the choice of paths leading either to the absorbing boundary at $\mathcal{X}_\mathsf{X} = 0$ or to the quasi-stationary state S_1 in the sense of a bifurcation, which is specific for autocatalytic reactions. The plots show the expectation values $\mathrm{E}(\mathcal{X}_\mathsf{A}(t))$ and $\mathrm{E}(\mathcal{X}_\mathsf{X}(t))$ of the particle numbers within the one-σ confidence intervals $\mathrm{E} \pm \sigma$, for the input material A and the autocatalyst X obtained from sampling 1000 trajectories. The expectation values are compared with the deterministic solutions (*dashed lines*) of the ODEs (5.6e) with $n = 1$. The *topmost plot* and the plot in the *middle* differ only in the initial number of autocatalyst molecules, viz., $x(0) = 10$ and $x(0) = 4$, respectively. The change in the solution curves of the deterministic ODEs concerns only the initial phase, and both curves converge to identical stationary values, but the expectation values of the stochastic process lead to much smaller long-time amounts of autocatalyst for smaller initial values $x(0)$. The conditions for the plot at the *bottom* differ from those of the plot at the top by a tenfold increase in all particle numbers. The relative widths of the one-σ bands become smaller as expected, and the deterministic solution curves coincide with the expectation values within the line width. Color code: $a(t)$ and $E(\mathcal{X}_\mathsf{A}(t))$ *black*, $E \pm \sigma(\mathcal{X}_\mathsf{A}(t))$ *red*, and $x(t)$ and $E(\mathcal{X}_\mathsf{X}(t))$ *blue*, $E \pm \sigma(\mathcal{X}_\mathsf{X}(t))$ *chartreuse*. Choice of parameters for *upper* and *middle plot*: $c_0 = 200$, $r = 0.5\,[\mathrm{V\,t^{-1}}]$ and $f = 0.001\,[\mathrm{N^{-1}t^{-1}}]$. Initial conditions: $a(0) = 1$ and $x(0) = 10$ (*top*) or $x(0) = 4$ (*middle*). Choice of parameters for *lower plot*: $c_0 = 2000$, $r = 0.5\,[\mathrm{V\,t^{-1}}]$ and $f = 0.0001\,[\mathrm{N^{-1}t^{-1}}]$. Initial conditions: $a(0) = 10$ and $x(0) = 100$

5.2.2 *Birth-and-Death Processes*

In Sect. 3.2.3, we referred to the concept of birth-and-death processes in order to introduce a well justified simplification into general master equations that turned out to be basic for applications in chemistry. Here we come back to the original biological idea of the birth-and-death process to describe births and deaths of individuals in populations. In particular, we present solutions for birth-and-death processes which are derived by the generating function method, discuss the nature and the role of boundary conditions, demonstrate the usefulness of first passage times and related concepts for finding straightforward answers to frequently asked questions, and present a collection of some analytical results that were excerpted from the monograph by Narendra Goel and Nira Richter-Dyn [216, pp. 10, 11 and 16, 17].

Birth-and-death processes like random walks are properly classified as unrestricted or restricted. A discrete process is unrestricted if its random variable covers all integers, i.e., $\mathcal{X}(t) = n \in \mathbb{Z}$, or its states Σ_n are defined on a domain that is infinite on both sides, viz., $]-\infty, +\infty[$. Such a strict definition is suitable neither for chemical reactions nor for population dynamics in biology, since particle numbers cannot become negative: $P_{n,n_0}(t) = 0,\ \forall\, n < 0$. Therefore we assume a less stringent definition and consider processes also as *unrestricted* when they are defined on the nonnegative integers $\mathcal{X}(t) = n \in \mathbb{N}$ or the semi-infinite state space $[0, +\infty[$. Processes on $\mathbb{N}$ commonly have a natural boundary at Σ_0 with $n = 0$, as exemplified by the linear birth-and-death process or the linear death process with immigration. In both cases $w_n^- = \mu_n = \mu n$ leads to $w_0^- = \mu_0 = 0$ for $n = 0$ and prevents particle numbers from becoming negative. In the first case, the state Σ_0 is an absorbing natural boundary since the birth rate $w_n^+ = \lambda_n = \lambda n$ implies $\lambda_0 = 0$ for $n = 0$. In the second example, the boundary is reflecting since $w_0^+ = \nu > 0$.

Table 5.2 compares some results for unrestricted birth-and-death processes with constant and linear rate parameters: $\lambda_n = \nu + \lambda n$ and $\mu_n = \rho + \mu n$. The processes with constant rate parameters are already well known to us: the Poisson process and the continuous time random walk. They are defined on the domains $[n_0, +\infty[$, $]-\infty, n_0]$, or $]-\infty, +\infty[$, respectively (a complete table containing also the probability generating functions can be found in [216, pp. 10, 11]). Restricted processes will be discussed in a separate paragraph. They stay finite and have either natural or designed boundaries. Such boundaries can be imagined as reflecting walls or absorbing traps, limiting, for example, the state space of a random walker.

The Unrestricted Linear Birth-and-Death Process

Reproduction of individuals is modeled by a simple duplication mechanism and death is represented by first order decay.[9] In the language of chemical kinetics, these two steps are:

$$(\mathsf{A}) + \mathsf{X} \xrightarrow{\lambda} 2\mathsf{X}\,, \tag{5.14a}$$

$$\mathsf{X} \xrightarrow{\mu} \varnothing\,. \tag{5.14b}$$

The rate parameters for reproduction and extinction are denoted by λ and μ, respectively. The material required for reproduction is tantamount to nutrition and denoted by $[\mathsf{A}] = a_0$. We assume that the pool is refilled as material is consumed, the amount of A available is *buffered*, and the constant concentration is absorbed in the birth parameter, so that $\lambda = ka_0$. The degradation products do not enter the kinetic equation, because the reaction (5.14b) is irreversible and the degraded material appears only on the product side.

The stochastic process corresponding to (5.14) with buffered A is an unrestricted linear birth-and-death process. General unrestricted birth-and-death processes, which are at most linear in n, include constant terms, and give rise to step-up and step-down transition probabilities of the form $w_n^+ = \nu + \lambda n$ and $w_n^- = \rho + \mu n$, where the individual symbols standing for birth, death, immigration,

[9]Reproduction is understood here as asexual reproduction, which under pseudo-first order conditions gives rise to linear reaction kinetics. Sexual reproduction requires two partners and gives rise to a special process of reaction order two (Table 4.2).

Table 5.2 Comparison of results for some unrestricted processes. Data are taken from [216, pp. 10, 11]. Abbreviations and notations: $\gamma \equiv \lambda/\mu$, $\eta(t) \equiv \mathrm{e}^{(\lambda-\mu)t}$, $(n, n_0) \equiv \min\{n, n_0\}$, and $I_n(x)$ is the modified Bessel function. References to literature are given in the penultimate column, and cross-references to sections in this monograph in the last column

Process	λ_n	μ_n	$P_{n,n_0}(t)$	Expectation value	Variance	Ref.	Section
Poisson	ν	0	$\frac{(\nu t)^{n-n_0}\mathrm{e}^{\nu t}}{(n-n_0)!}$, $n \ge n_0;\ n_0 > \min\{0,n\}$	$n_0 + \nu t$	νt	[91]	Sect. 3.2.2.4
Poisson	0	ρ	$\frac{(\rho t)^{n-n_0}\mathrm{e}^{\rho t}}{(n_0-n)!}$, $n \le n_0;\ n_0 < \min\{0,n\}$	$n_0 - \rho t$	ρt	[91]	Sect. 3.2.2.4
Random walk	ν	ρ	$\left(\frac{\nu}{\rho}\right)^{(n-n_0)/2} I_{n_0-n}(2t\sqrt{\nu\rho})\mathrm{e}^{-(\nu+\rho)t}$	$n_0 + (\nu - \rho)t$	$(\nu + \rho)t$	[248]	Sect. 3.2.4
Birth	λn	0	$\binom{n}{n_0}\mathrm{e}^{-n_0\lambda t}(1-\mathrm{e}^{-\lambda t})^{n-n_0}$, $n \ge n_0; n_0 > \min\{0,n\}$	$n_0\mathrm{e}^{\lambda t}$	$n_0\mathrm{e}^{\lambda t}(\mathrm{e}^{\lambda t}-1)$	[33]	Sect. 5.2.2
Death	0	μn	$\binom{n_0}{n}\mathrm{e}^{-n\mu t}(1-\mathrm{e}^{-\mu t})^{n_0-n}$, $n \le n_0; n_0 < \min\{0,n\}$	$n_0\mathrm{e}^{-\mu t}$	$n_0\mathrm{e}^{-\mu t}(1-\mathrm{e}^{-\mu t})$	[33]	Sect. 5.2.2
	ν	μn	$\exp\left(-\frac{\nu}{\mu}(1-\mathrm{e}^{-\mu t})\right) \times$ $\times \sum_{k=0}^{\min\{0,n\}} \frac{\mathrm{e}^{-\mu t k}(1-\mathrm{e}^{-\mu t})^{n+n_0-2k}}{(n-k)!}\left(\frac{\nu}{\mu}\right)^{n-k}$	$n_0\mathrm{e}^{-\mu t} +$ $+\frac{\nu(1-\mathrm{e}^{-\mu t})}{\mu}$	$\left(\frac{\nu}{\mu} + n_0\mathrm{e}^{-\mu t}\right) \times$ $\times(1-\mathrm{e}^{-\mu t})$	[91]	Sect. 5.2.2
Birth-and-death	λn	μn	$\gamma^n \sum_{k=0}^{\min\{0,n\}} (-1)^k \binom{n+n_0-k-1}{n-k}\binom{n_0}{k} \times$ $\times \left(\frac{1-\eta}{1-\gamma\eta}\right)^{n+n_0-k}\left(\frac{1-\eta/\gamma}{1-\eta}\right)^k$	$n_0\,\eta$	$\frac{n_0\eta(\gamma+1)(\eta-1)}{\gamma-1}$	[33]	Sect. 5.2.2
	λn	λn	$\left(\frac{\lambda t}{1+\lambda t}\right)^{n+n_0} \sum_{k=0}^{(n,n_0)} \binom{n_0}{k} \times$ $\times \binom{n+n_0-k-1}{n-k}\left(\frac{1-\lambda^2 t^2}{\lambda^2 t^2}\right)^k$	n_0	$2n_0\lambda\, t$		

and emigration terms are λ, μ, ν, and ρ, respectively.[10] Applying purely linear transition probabilities $w_n^+ = \lambda n$ and $w_n^- = \mu n$, we obtain a master equation of the form

$$\frac{\partial P_n(t)}{\partial t} = \lambda(n-1)P_{n-1}(t) + \mu(n+1)P_{n+1}(t) - (\lambda+\mu)nP_n(t)\,. \tag{5.14c}$$

After introducing the probability generating function $g(s,t)$, this gives rise to the PDE

$$\frac{\partial g(s,t)}{\partial t} - (s-1)(\lambda s - \mu)\frac{\partial g(s,t)}{\partial s} = 0\,. \tag{5.14d}$$

Solution of this PDE yields different expressions for equal or different replication and extinction rate coefficients, viz., $\lambda = \mu$ and $\lambda \neq \mu$, respectively.

The Case $\lambda \neq \mu$
We substitute $\gamma = \lambda/\mu$ and $\eta(t) = \exp\big((\lambda-\mu)t\big)$, and find:

$$g(s,t) = \left(\frac{\big(\eta(t)-1\big) + \big(\gamma - \eta(t)\big)s}{\big(\gamma\eta(t)-1\big) + \gamma\big(1-\eta(t)\big)s}\right)^{n_0}, \tag{5.15a}$$

$$P_n(t) = \gamma^n \sum_{m=0}^{\min(n,n_0)} (-1)^m \binom{n_0+n-m-1}{n-m}\binom{n_0}{m} \times \left(\frac{1-\eta(t)}{1-\gamma\eta(t)}\right)^{n_0+n-m}\left(\frac{\gamma-\eta(t)}{\gamma\big(1-\eta(t)\big)}\right)^m .$$

In the derivation of the expression for the probability distributions, we expanded the numerator and denominator of the expression in the generating function $g(s,t)$ by using expressions for the sums, viz.,

$$(1+s)^n = \sum_{k=0}^{n}\binom{n}{k}s^k\,, \quad (1+s)^{-n} = 1 + \sum_{k=1}^{\infty}(-1)^k\frac{n(n+1)\dots(n+k-1)}{k!}s^k\,.$$

---

[10] Here we use the symbols commonly applied in biology: $\lambda_{(n)}$ for birth, $\mu_{(n)}$ for death, ν for immigration, and ρ for emigration. Other notions and symbols are used in chemistry: $f \equiv \lambda$ for birth corresponding to the production of a molecule and $d \equiv \mu$ for death understood as decomposition or degradation through a chemical reaction. Inflow and outflow are the equivalents of immigration and emigration. Pure immigration and emigration give rise to Poisson processes and continuous time random walks, which have been discussed extensively in Chap. 3. There we denoted the parameters by γ and ϑ, instead of ν and ρ.

Multiplying, ordering terms with respect to powers of s, and comparing with the expansion $g(s,t) = \sum_{n=0}^{\infty} P_n(t)s^n$ of the generating function yields (5.15a).

In order to show some special properties of the probabilities, we write them in more detail: $P_{n,n_0}(t;\lambda,\mu) = P_n(t)$ with initial conditions $P_n(0) = \delta_{n,n_0}$. The probability of extinction P_{0,n_0} becomes

$$P_{0,n_0}(t;\lambda,\mu) = \left(\frac{\mu\eta(t)-\mu}{\lambda\eta(t)-\mu}\right)^{n_0}, \tag{5.15b}$$

which has a kind of symmetry under exchange $(\lambda,\mu) = (b,d) \leftrightarrow (d,b)$ or $(\eta \leftrightarrow \hat{\eta}^{-1}, \gamma \leftrightarrow \hat{\gamma}^{-1})$:

$$\begin{aligned} P_{0,n_0}(t;b,d) &= \left(\frac{\eta-1}{\gamma\eta-1}\right)^{n_0} = \left(\frac{1/\hat{\eta}-1}{1/\hat{\gamma}\hat{\eta}-1}\right)^{n_0} \\ &= \left(\frac{\hat{\eta}-1}{\hat{\gamma}\hat{\eta}-1}\right)^{m} \hat{\gamma}^{n_0} = \gamma^{-n_0} P_{0,n_0}(t;d,b) \; . \end{aligned}$$

By the same token, one finds for the probability of reaching the state Σ_1, viz.,

$$P_{1,n_0}(t;\lambda,\mu) = n_0(\lambda-\mu)^2 \frac{\left(\mu e^{(\lambda-\mu)t}-\mu\right)^{n_0-1}}{\left(\lambda e^{(\lambda-\mu)t}-\mu\right)^{n_0+1}} e^{(\lambda-\mu)t} ,$$

the similar relation

$$P_{1,n_0}(t;b,d) = \gamma^{1-n_0} P_{1,n_0}(t;d,b) ,$$

which yields the fully symmetric equality $P_{1,n_0}(t;b,d) = P_{1,n_0}(t;d,b)$ for the probability of remaining in state Σ_1.

Computation of the expectation value and variance is straightforward:

$$\begin{aligned} E\big(\mathcal{X}_{\mathrm{X}}(t)\big) &= n_0 e^{(\lambda-\mu)t} , \\ \mathrm{var}\big(\mathcal{X}_{\mathrm{X}}(t)\big) &= n_0 \frac{\lambda+\mu}{\lambda-\mu} e^{(\lambda-\mu)t}\left(e^{(\lambda-\mu)t}-1\right) . \end{aligned} \tag{5.15c}$$

Illustrative examples of linear birth-and-death processes with growing ($\lambda > \mu$) and decaying ($\lambda < \mu$) populations are shown in Figs. 5.6 and 5.7, respectively. For $\lambda > \mu$, the moments grow to infinity in the limit $t \to \infty$.

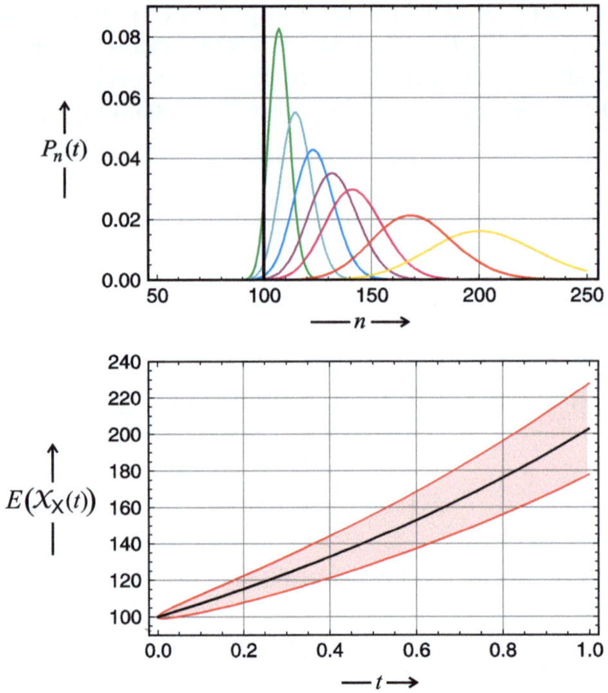

Fig. 5.6 A growing linear birth-and-death process. The two-step reaction mechanism of the process is $(\mathsf{X} \to 2\mathsf{X}, \mathsf{X} \to \varnothing)$ with rate parameters λ and μ, respectively. The growing or supercritical process is characterized by $\lambda > \mu$. *Upper*: Evolution of the probability density $P_n(t) = P(\mathcal{X}_{\mathsf{X}}(t) = n)$. The initially infinitely sharp density $P(n, 0) = \delta(n, n_0)$ becomes broader with time and flattens as the variance increases with time. *Lower*: Expectation value $E(\mathcal{X}_{\mathsf{X}}(t))$ in the confidence interval $E \pm \sigma$. Parameters used: $n_0 = 100$, $\lambda = \sqrt{2}$, and $\mu = 1/\sqrt{2}$. Sampling times (*upper*): $t =$ 0 (*black*), 0.1 (*green*), 0.2 (*turquoise*), 0.3 (*blue*), 0.4 (*violet*), 0.5 (*magenta*), 0.75 (*red*), and 1.0 (*yellow*)

The Degenerate or Neutral Case $\mu = \lambda$

Here the same procedure as used for $\lambda \neq \mu$ is applied to the PDE (5.14d) and leads to

$$g(s,t) = \left(\frac{\lambda t + (1 - \lambda t)s}{1 + \lambda t + \lambda t s} \right)^{n_0} , \tag{5.16a}$$

$$P_n(t) = \left(\frac{\lambda t}{1 + \lambda t} \right)^{n_0 + n} \sum_{m=0}^{\min(n, n_0)} \binom{n_0 + n - m - 1}{n - m} \binom{n_0}{m} \left(\frac{1 - \lambda^2 t^2}{\lambda^2 t^2} \right)^m , \tag{5.16b}$$

$$E(\mathcal{X}_{\mathsf{X}}(t)) = n_0 , \tag{5.16c}$$

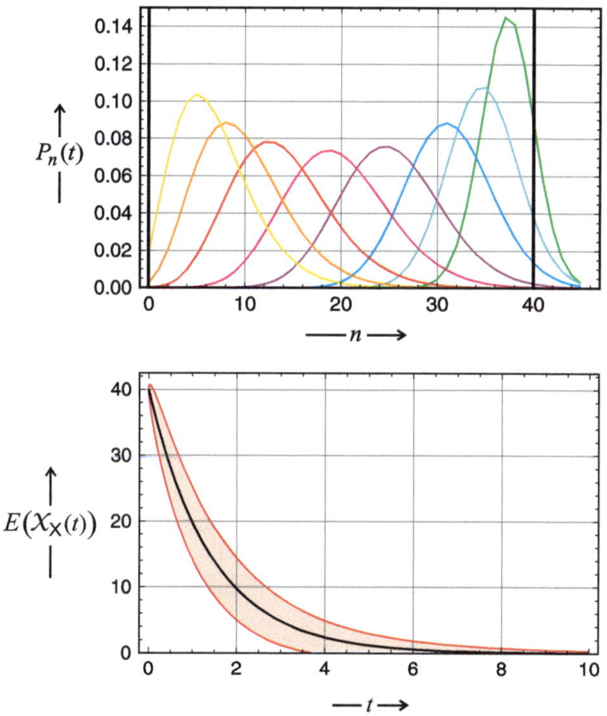

Fig. 5.7 A decaying linear birth-and-death process. The two-step reaction mechanism of the process is ($X \rightarrow 2X$, $X \rightarrow \emptyset$) with rate parameters λ and μ, respectively. The decaying or subcritical process is characterized by $\lambda < \mu$. *Upper*: Evolution of the probability density $P_n(t) = P\big(\mathcal{X}_X(t) = n\big)$. The initially infinitely sharp density $P(n, 0) = \delta(n, n_0)$ becomes broader with time and flattens as the variance increases, but then sharpens again as the process approaches the absorbing boundary at $n = 0$. *Lower*: Expectation value $E\big(\mathcal{X}_X(t)\big)$ in the confidence interval $E \pm \sigma$. Parameters used: $n_0 = 40$, $\lambda = 1/\sqrt{2}$, and $\mu = \sqrt{2}$. Sampling times (*upper*): $t = 0$ (*black*), 0.1 (*green*), 0.2 (*turquoise*), 0.35 (*blue*), 0.65 (*violet*), 1.0 (*magenta*), 1.5 (*red*), 2.0 (*orange*), 2.5 (*yellow*), and $\lim_{t\to\infty}$ (*black*)

$$\text{var}\big(\mathcal{X}_X(t)\big) = 2n_0\lambda t\,. \tag{5.16d}$$

Comparing the last two expressions reveals an inherent instability in the degenerate birth-and-death reaction system. The expectation value is constant, whereas the fluctuations increase with time. The case of steadily increasing fluctuations is in contrast to an equilibrium situation, where both expectation value and variance approach constant values. Comparing birth-and-death with the Ehrenfest urn game, we recognize an important difference. In the urn game fluctuations were negatively correlated with the deviation from equilibrium, whereas we have two uncorrelated processes, replication and extinction, in the birth-and-death system. In the latter the particle number $\mathcal{X}_X(t) = n(t)$ carries out a random walk on the natural numbers with position-dependent increments. Indeed, in the case of the random walk, we also obtained a constant expectation value $E = n_0$ and a variance that increases linearly with time, viz., $\mathrm{var}(t) = 2\vartheta(t - t_0))$ (see (3.114) in Sect. 3.2.4). The difference, however, is the existence of a trap in the form of the absorbing state Σ_0: whenever the walk reaches the trap, the walker is caught in it and the process ends. An example of the degenerate birth-and-death process is illustrated in Fig. 5.8.

First Passage Times and Extinction

The first passage time $\mathcal{T}_{w,n_0}$ is defined as the time when a process starting from state Σ_{n_0} reaches state Σ_w for the first time. The random variable $\mathcal{T}_{w,n_0}$ has probability density $f_{w,n_0}(t)$ and raw moments $\hat{\mu}_i^{(\mathcal{T})}$, with $i \in \mathbb{N}_{>0}$. The relation between the first passage times and the probabilities $P_{n,n_0}(t)$ is given by the convolution

$$P_{n,n_0}(t) = \int_0^t \mathrm{d}\tau f_{w,n_0}(t-\tau) P_{n,w}(\tau) \;, \quad n_0 < w < n \quad \text{or} \quad n < w < n_0 \;, \tag{5.17}$$

where the state Σ_w is intermediate between Σ_{n_0} and Σ_n. The interpretation is straightforward: in order to reach Σ_{n_0} from Σ_m, one has to pass through Σ_w before one arrives at the target. The standard procedure for solving convolution integrals is Laplace transform, which converts it into a product (3.28)[11]:

$$\mathcal{L}\left(\int_0^t f(t-\tau)g(\tau)\mathrm{d}\tau\right) = \mathcal{L}\big(f(t)\big)\mathcal{L}\big(g(t)\big) = F(\hat{s})\,G(\hat{s}) \;,$$

and the Laplace transforms of the probability density of the particle number $\Pi_{n,n_0}(\hat{s}) = \mathcal{L}\big(P_{n,n_0}(t)\big)$ and the first passage time $\phi_{w,n_0}(\hat{s}) = \mathcal{L}\big(f_{w,n_0}(t)\big)$ satisfy

$$\Pi_{n,n_0}(\hat{s}) = \Pi_{n,w}(\hat{s})\phi_{w,n_0}(\hat{s}) \quad \text{or} \quad \phi_{w,n_0}(\hat{s}) = \frac{\Pi_{n,n_0}(\hat{s})}{\Pi_{n,w}(\hat{s})} \,.$$

Since the probability densities are known, the calculation of the first passage time densities via Laplace transform and inverse Laplace transform is standard, but it may nevertheless be quite complicated. We dispense with the details, but consider one particularly important case, the extinction problem.

The application of first passage times to analyze birth-and-death processes can be nicely demonstrated by the population extinction problem. The time of extinction is tantamount to the first passage time $\mathcal{T}_{0,n_0}$, where n_0 stands for the initial population size: $P_n(0) = \delta_{n,n_0}$. Here we shall consider the simple linear birth-and-death process with $\lambda_n = \lambda n$ and $\mu_n = \mu n$. The probability density of the first passage time $\mathcal{T}_{0,n_0}$ is $f_{0,n_0}(t)$ and can be described by the backward master equation

$$\frac{\mathrm{d}f_{0,n_0}(t)}{\mathrm{d}t} = \lambda n_0\big(f_{0,n_0+1}(t) - f_{0,n_0}(t)\big) + \mu n_0\big(f_{0,n_0-1}(t) - f_{0,n_0}(t)\big) \,, \quad n_0 \in \mathbb{N}_{>0} \,. \tag{5.18}$$

[11] Here the conjugated Laplace variable is denoted by $\hat{s}$ in order to avoid confusion with the dummy variable s in the generating function.

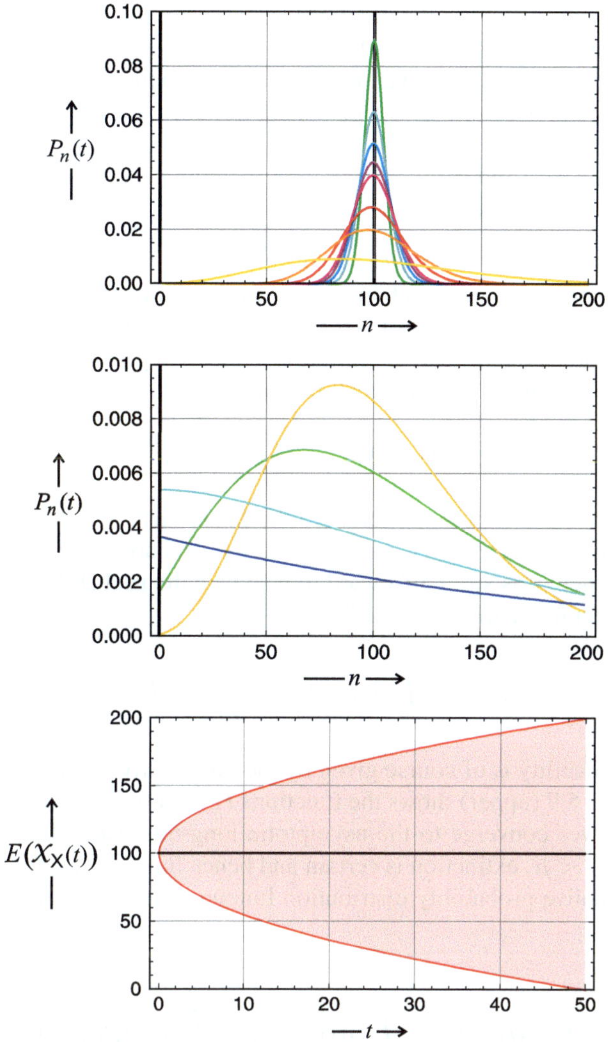

Fig. 5.8 Probability density of a linear birth-and-death process with equal birth and death rate. The two-step reaction mechanism of the critical process is (X → 2X, X → ∅) with rate parameters $\lambda = \mu$. The *upper* and the *middle plots* show the evolution of the probability density, $P_n(t) = P(\mathcal{X}_X(t) = n)$. The initially infinitely sharp density $P(n, 0) = \delta(n, n_0)$ becomes broader with time and flattens as the variance increases, but then sharpens again as the process approaches the absorbing boundary at $n = 0$. In the *lower plot*, we show the expectation value $E(\mathcal{X}_X(t))$ in the confidence interval $E \pm \sigma$. The variance increases linearly with time, and at $t = n_0/2\lambda = 50$, the standard deviation is as large as the expectation value. Parameters used: $n_0 = 100$, $\lambda = 1$. Sampling times for *upper plot*: $t = 0$ (*black*), 0.1 (*green*), 0.2 (*turquoise*), 0.3 (*blue*), 0.4 (*violet*), 0.49999 (*magenta*), 0.99999 (*red*), 2.0 (*orange*), 10 (*yellow*). Sampling times for *middle plot*: $t =$ 10 (*yellow*), 20 (*green*), 50 (*cyan*), 100 (*blue*), and $\lim_{t\to\infty}$ (*black*)

The state Σ_0 with $n = 0$ is a natural absorbing state, the birth rate and the death rate in Σ_0 vanish, $\lambda_0 = \mu_0 = 0$ for $\lambda_n = \lambda n$ and $\mu_n = \mu n$, and therefore we have $f_{0,0} = 0$, which has the trivial meaning that an extinct species cannot become extinct. The equation for the next higher state Σ_1 takes the form

$$\frac{\mathrm{d}f_{0,1}(t)}{\mathrm{d}t} = \lambda_1 f_{0,2}(t) - (\lambda_1 + \mu_1) f_{0,1}(t) \,,$$

which follows from (5.18) and $f_{0,0} = 0$. In order to find solutions of the master equation, we consider a relation between the probabilities $P_{1,n_0}(t)$ of being in Σ_1 at time t and the first passage times [216, p. 18]: in order to reach Σ_0 for the first time at $t + \Delta t$, the process has to be in Σ_1 at time t and then go to Σ_0 within the interval $[t, t + \Delta t]$. For an infinitesimally small interval, we find

$$f_{0,n_0}(t)\,\mathrm{d}t = P_{1,n_0}(t)\mu\,\mathrm{d}t \,, \tag{5.19}$$

where the right-hand expression refers to the linear birth-and-death process. From the probability density $P_{n,n_0}(t)$, we calculate the probability of reaching the state Σ_1, and from here it is straightforward to calculate the probability that the process becomes extinct in the time interval $0 \leq \tau \leq t$:

$$F_{0,n_0}(t) = \int_0^t \mathrm{d}\tau f_{0,n_0}(\tau) = \int_0^t \mathrm{d}\tau \mu P_{1,n_0}(\tau) \,. \tag{5.20}$$

The same probability is of course given by the probability of extinction: $P_{0,n_0}(t) = F_{0,n_0}(t)$. Figure 5.9 (upper) shows the functions $F_{0,n_0}(t)$ for $n_0 = 1, 2, 3$, and $\lambda > \mu$, where the curves converge to the asymptotic long-time value $\lim_{t\to\infty} F_{0,n_0}(t) = (\mu/\lambda)^{n_0}$. For $\lambda < \mu$, extinction is certain and hence $\lim_{t\to\infty} F_{0,n_0}(t) = 1$. As in the case of cumulative probability distribution functions, the integral can be split into intervals

$$F_{0,n_0}(t_2 - t_1) = \int_{t_1}^{t_2} \mathrm{d}\tau f_{0,n_0}(\tau) = \int_{t_1}^{t_2} \mathrm{d}\tau \mu P_{1,n_0}(\tau) \,,$$

which yields the probability that the population dies out between t_1 and t_2. Table 5.3 shows a partitioning of samples from numerical calculations of extinction times. Not surprisingly, the random scatter is large, but there is no doubt that the Gillespie algorithm reproduces very well the values predicted by the analytical approach. It is a straightforward matter to compute the expectation value of the extinction time:

$$\mathrm{E}(\mathcal{T}_{0,n_0}) = \frac{\int_0^\infty \mathrm{d}t\, t f_{0,n_0}(t)}{\int_0^\infty \mathrm{d}t f_{0,n_0}(t)} = \gamma^{n_0} \int_0^\infty \mathrm{d}t\, t f_{0,n_0}(t) \,. \tag{5.21}$$

In contrast to the conventional probability distributions, the integral over the entire time range has to be normalized, because extinction does not occur with probability

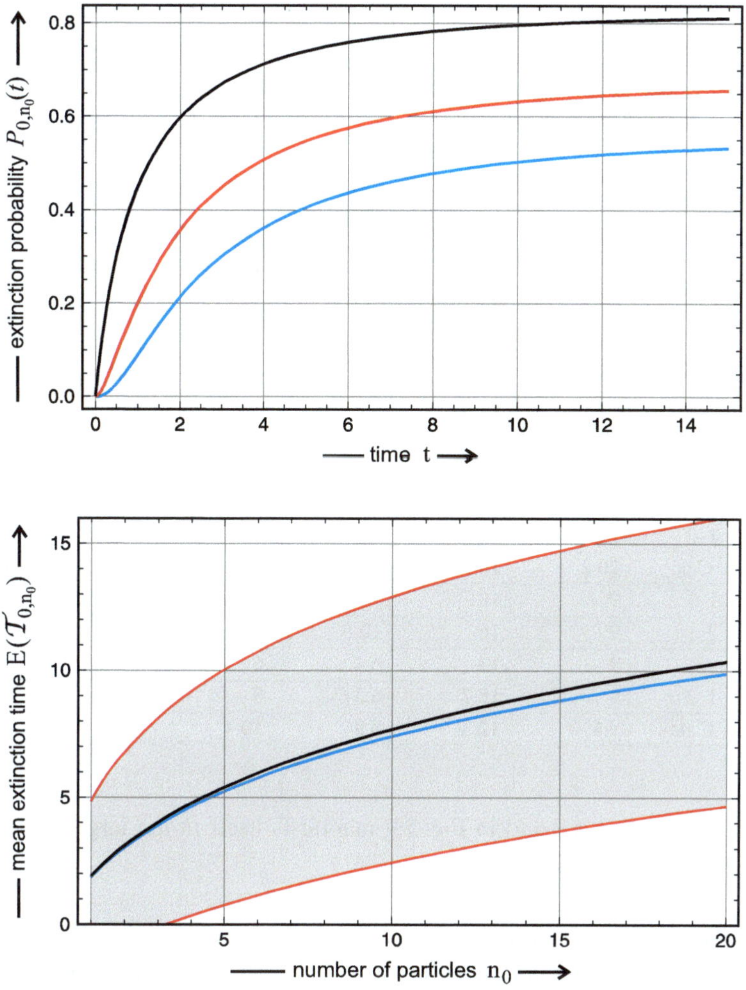

Fig. 5.9 Probability of extinction and extinction time. We consider here the case $\lambda \neq \mu$. *Upper*: Probability of extinction P_{0,n_0} as a function of time t for $n_0 = 1$ (*black*), $n_0 = 2$ (*red*), and $n_0 = 3$ (*blue*). The asymptotic limits are given by $\lim_{t\to\infty} P_{0,n_0} = \gamma^{-n_0}$. *Lower*: Expected time of extinction $E(\mathcal{T}_{0,n_0})$ as a function of m (*black*) together with the one standard deviation band $E \pm \sigma$ (*red curves*). The *blue curve* shows $\hat{E}(\mathcal{T}_{0,n_0}, t_{\max})$, the result of taking the expectation value from a finite time interval $[0, t_{\max}]$. Choice of parameters: $\lambda = 1.1$, $\mu = 0.9$, and $t_{\max} = 25$

one for $\lambda > \mu$. The variance and standard deviation are obtained from the second raw moment

$$\hat{\mu}_2(\mathcal{T}_{0,n_0}) = \frac{\int_0^\infty \mathrm{d}t\, t^2 f_{0,n_0}(t)}{\int_0^\infty \mathrm{d}t f_{0,n_0}(t)}\,,$$
$$\mathrm{var}(\mathcal{T}_{0,n_0}) = \hat{\mu}_2(\mathcal{T}_{0,n_0}) - \big(E(\mathcal{T}_{0,n_0})\big)^2\,, \quad \sigma(\mathcal{T}_{0,n_0}) = \sqrt{\mathrm{var}(\mathcal{T}_{0,n_0})}\,. \tag{5.22}$$

Table 5.3 Statistics of extinction times. Comparison of the probability distribution $P_{0,n_0}(t)$ with the extinction times obtained from numerical simulations of the linear birth-and-death mechanism $\mathbf{X} \rightarrow 2\mathbf{X}$ and $\mathbf{X} \rightarrow \oslash$, with rate parameters $\lambda = 1.1$ and $\mu = 0.9$, respectively. Three initial values were chosen: $n_0 = 1$, $n_0 = 2$, and $n_0 = 3$. The values for the six slots, viz., $0 \leq \mathcal{T}_{0,n_0} < 1$, $1 \leq \mathcal{T}_{0,n_0} < 2$, $2 \leq \mathcal{T}_{0,n_0} < 3$, $3 \leq \mathcal{T}_{0,n_0} < 5$, $5 \leq \mathcal{T}_{0,n_0} < 10$, and $\mathcal{T}_{0,n_0} > 10$ were sampled from one hundred extinction times for each run. *Bold values* refer to all 300 simulations for each value of n_0

		Extinction time interval					
n_0	Run	$0 \rightarrow 1$	$1 \rightarrow 2$	$2 \rightarrow 3$	$3 \rightarrow 5$	$5 \rightarrow 10$	> 10
1	1	53	13	7	5	6	16
	2	47	12	6	6	8	21
	3	46	14	11	4	5	20
	1–3	**48.7**	**13**	**8**	**4**	**6.3**	**19**
	Calc	**44.9**	**14.8**	**7.3**	**7.0**	**5.6**	**20.5**
2	1	19	11	8	9	1	52
	2	25	13	11	11	11	29
	3	20	21	10	4	7	38
	1–3	**21.3**	**15**	**9.7**	**8**	**6.3**	**39.7**
	Calc	**20.2**	**15.5**	**9.2**	**9.9**	**8.6**	**36.7**
3	1	9	18	3	9	7	54
	2	9	16	10	9	7	49
	3	9	13	6	6	8	58
	1–3	**9**	**15.7**	**6.3**	**8**	**7.3**	**53.7**
	Calc	**9.1**	**12.3**	**8.8**	**10.4**	**9.8**	**49.7**

The very broad $\mathrm{E} \pm \sigma$ band in Fig. 5.9 manifests itself in the large scatter of the counts in Table 5.3.

Sequential Extinction Times

In the degenerate birth-and-death process with $\lambda = \mu$, a constant expectation value is accompanied by a variance that increases with time, and this has an easy-to-visualize consequence (Fig. 5.8): there is a critical time $t_{\mathrm{cr}} = n_0/2\lambda$ above which the standard deviation exceeds the expectation value. From this instant on, predictions about the evolution of the system based on the expectation value become obsolete, and we have to rely on individual probabilities or other quantities. The probability of extinction of the entire population is useful in this context, and it is readily computed:

$$P_0(t) = \left(\frac{\lambda t}{1 + \lambda t}\right)^{n_0} . \tag{5.23}$$

Provided we wait long enough, the system will die out with probability one, since we have $\lim_{t \to \infty} P_0(t) = 1$. This seems to be in contradiction with the constant expectation value. As a matter of fact it is not: in almost all individual runs, the system will go extinct, but there are very few cases of probability measure zero, where the particle number grows to infinity for $t \to \infty$. These rare cases are responsible for the finite expectation value.

Equation (5.23) can be used to derive a simple model for *random selection* [486]. We assume a population of n different species

$$(\mathsf{A}) + \mathsf{X}_j \xrightarrow{\lambda_j} 2\mathsf{X}_j\ , \quad j = 1, \ldots, n\ , \tag{5.14a$'$}$$

$$\mathsf{X}_j \xrightarrow{\mu_j} \mathsf{B}\ , \quad j = 1, \ldots, n\ . \tag{5.14b$'$}$$

The joint probability distribution of the entire population is described by

$$P_{x_1 \ldots x_n} = P\big(\mathcal{X}_1(t) = x_1, \ldots, \mathcal{X}_n(t) = x_n\big) = P^{(1)}_{x_1} \cdots P^{(n)}_{x_n}\ , \tag{5.24}$$

where all probability distributions for individual species are given by (5.16b). The independence of all individual birth events and death events allows for the simple product expression. In the spirit of Motoo Kimura's neutral theory of evolution [304], all birth and all death parameters are assumed to be equal, i.e., $\lambda_j = \lambda$ and $\mu_j = \mu$ for all $j = 1, \ldots, n$, and $\lambda = \mu$. For convenience, we assume that every species is initially present in a single copy: $P_{n_j}(0) = \delta_{n_j,1}$. We introduce a new random variable $\mathcal{T}_k$ that has the nature of a first passage time. It is the time up to the extinction of $n - k$ species, and we characterize it as *sequential extinction time*. Accordingly, n species are present in the population between $\mathcal{T}_n$, which satisfies $\mathcal{T}_n \equiv 0$ by definition, and $\mathcal{T}_{n-1}$, $n-1$ species between $\mathcal{T}_{n-1}$ and $\mathcal{T}_{n-2}$, and eventually a single species between $\mathcal{T}_1$ and $\mathcal{T}_0$, which is the moment of extinction for the entire population. After $\mathcal{T}_0$, no further individual exists.

Next we consider the probability distribution of the sequential extinction times

$$H_k(t) = P(\mathcal{T}_k < t)\ . \tag{5.25}$$

The probability of extinction of the population is readily calculated. Since individual reproduction and extinction events are independent, we find

$$H_0 = P_{0,\ldots,0} = P^{(1)}_0 \cdots P^{(n)}_0 = \left(\frac{\lambda t}{1 + \lambda t}\right)^n .$$

The event $\mathcal{T}_1 < t$ can happen in several ways. Either X_1 is present and all other species have become extinct already, or only X_2 is present, or only X_3, and so on, but $\mathcal{T}_1 < t$ is also satisfied if the whole population has died out:

$$H_1 = P_{x_1 \neq 0,0,\ldots,0} + P_{0,x_2 \neq 0,\ldots,0} + P_{0,0,\ldots,x_n \neq 0} + H_0\ .$$

The probability that a given species has not yet disappeared is obtained by exclusion, since existence and nonexistence are complementary:

$$P_{x\neq 0} = 1 - P_0 = 1 - \frac{\lambda t}{1+\lambda t} = \frac{1}{1+\lambda t} ,$$

which yields the expression for the presence of a single species

$$H_1(t) = (n+\lambda t)\frac{(\lambda t)^{n-1}}{(1+\lambda t)^n} .$$

By similar arguments a recursion formula is found for the extinction probabilities with higher indices:

$$H_k(t) = \binom{n}{k}\frac{(\lambda t)^{n-k}}{(1+\lambda t)^n} + H_{k-1}(t) ,$$

and this eventually leads to the expression

$$H_k(t) = \sum_{j=0}^{k}\binom{n}{j}\frac{(\lambda t)^{n-j}}{(1+\lambda t)^n} .$$

The moments of the sequential extinction times are computed straightforwardly by means of a handy trick: H_k is partitioned into terms for the individual powers of λt, $H_k(t) = \sum_{j=0}^{k} h_j(t)$, and then differentiated with respect to time t:

$$h_j(t) = \binom{n}{j}\frac{(\lambda t)^{n-k}}{(1+\lambda t)^n} ,$$

$$\frac{\mathrm{d}h_j(t)}{\mathrm{dt}} = h_j' = \frac{\lambda}{(1+\lambda t)^{n+1}}\left(\binom{n}{j}(n-j)(\lambda t)^{n-j-1} - \binom{n}{j}j(\lambda t)^{n-j}\right) .$$

The summation of the derivatives is simple because $h_k'+h_{k-1}'+\cdots+h_0'$ is a *telescopic sum* and we find

$$\frac{\mathrm{d}H_k(t)}{\mathrm{dt}} = \binom{n}{k}(n-k)\,\lambda^{n-k}\frac{t^{n-k-1}}{(1+\lambda t)^{n+1}} .$$

Making use of the definite integral [219, p. 338]

$$\int_0^\infty \frac{t^{n-k}}{(1+\lambda t)^{n+1}}\,\mathrm{dt} = \frac{\lambda^{-(n-k+1)}}{\binom{n}{k}k} ,$$

we finally obtain

$$E(\mathcal{T}_k) = \int_0^\infty \frac{\mathrm{d}H_k(t)}{\mathrm{d}t} t \,\mathrm{d}t = \frac{n-k}{k}\frac{1}{\lambda}, \quad n \geq k \geq 1, \tag{5.26}$$

and $E(\mathcal{T}_0) = \infty$ for the expectation values of the sequential extinction times (Fig. 5.10). It is worth recognizing here another paradox of probability theory: although extinction is certain, the expectation value for the time to extinction diverges. In a similar way to the expectation values, we calculate the variances of the sequential extinction times:

$$\mathrm{var}(\mathcal{T}_k) = \frac{n(n-k)}{k^2(k-1)}\frac{1}{\lambda^2}, \quad n \geq k \geq 2, \tag{5.27}$$

from which we see that the variances diverge for $k = 0$ and $k = 1$.

For distinct birth parameters $\lambda_1, \ldots, \lambda_n$, and different initial particle numbers, $x_1(0), \ldots, x_n(0)$, the expressions for the expectation values become considerably more complicated, but the main conclusion remains unaffected: $E(\mathcal{T}_1)$ is finite, whereas $E(\mathcal{T}_0)$ diverges.

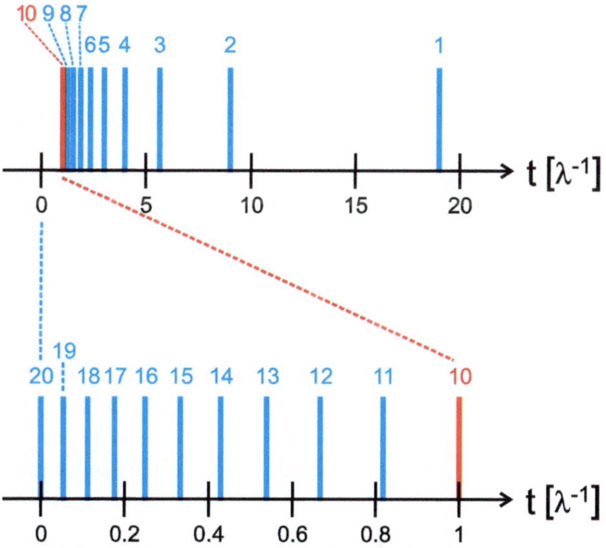

Fig. 5.10 The distribution of sequential extinction times $\mathcal{T}_k$. Expectation values $E(\mathcal{T}_k)$ for $n = 20$ according to (5.26). Since $E(\mathcal{T}_0)$ diverges, $\mathcal{T}_1$ is the extinction that appears on the average at a finite value. A single species is present above $\mathcal{T}_1$ and random selection has occurred in the population

Restricted Birth-and-Death Processes and Boundaries

Unrestricted processes, whether or not confined on one side by a natural boundary, are an idealized approximation. A requirement imposed by physics in a finite world demands that all changes in state space should be finite, so all probabilities of reaching infinity must vanish: $\lim_{n\to\pm\infty} P_{n,n_0} = 0$. Reversible chemical reactions approaching thermodynamic equilibrium provide excellent examples of stochastic processes that are restricted by two reflecting natural boundaries. Irreversible autocatalytic processes commonly have one absorbing boundary. In addition, artificial boundaries, either absorbing or reflecting, are easily introduced into birth-and-death processes. Here we shall give a brief account of restricted processes, referring the reader to the chapter in [216, pp. 12–31] for a wealth of useful details.

First we mention again the influence of boundaries on stochastic processes. As in Fig. 3.25, we define an interval $[l, u]$ representing the domain of the stochastic variable: $l \leq \mathcal{X}(t) \leq u$. Two classes of boundaries are distinguished, characterized as *absorbing* and *reflecting*. In the case of an absorbing boundary, a particle that crossed the boundary is not allowed to return, whereas a reflecting boundary implies that it is forbidden to exit from the interval. Boundary conditions are easily implemented by *ad hoc* definitions of transition probabilities:

	reflecting	absorbing
lower boundary at l	$w_l^- = 0$	$w_{l-1}^+ = 0$
upper boundary at u	$w_u^+ = 0$	$w_{u+1}^- = 0$

The reversible chemical reaction $\mathsf{A} \rightleftharpoons \mathsf{B}$ (Sect. 4.3.2) with $w_n^- = kn$ and $w_n^+ = l(n_0 - n)$, for example, had two reflecting boundaries at $l = 0$ with $w_0^- = 0$ and at $u = n_0$ with $w_{n_0}^+ = 0$. Among the examples of birth-and-death processes we have discussed so far, we were dealing with an absorbing boundary in the replication–extinction process at $\mathcal{X} = 0$ that is tantamount to the lower boundary at $l = 1$ satisfying $w_0^+ = 0$: the absorbing state Σ_0 with $n = 0$ is the end point or ω-limit of all trajectories reaching it. A particularly interesting case was the reaction $2\mathsf{X} \to \mathsf{A} + \mathsf{X}$ (Sect. 4.6.4). The last molecule of X is unable to react, we observe an absorbing barrier at $\mathcal{X}_\mathsf{X} = n_0 - n = 1$, with $n_0 = \mathcal{X}_\mathsf{A} + \mathcal{X}_\mathsf{X}$, and the domains of the two dependent random variables are $1 \leq \mathcal{X}_\mathsf{X} \leq n_0$ and $n_0 - 1 \leq \mathcal{X}_\mathsf{A} \leq 0$, respectively.

Compared, for example, to unrestricted random walks, which are defined on positive and negative integers $n \in \mathbb{Z}$, a chemical reaction or a biological process has to be restricted to positive integers, so $n \in \mathbb{N}$, since negative particle numbers are not allowed. In general, the one step birth-and-death master equation (3.97), viz.,

$$\frac{\mathrm{d}P_n(t)}{\mathrm{d}t} = w_{n-1}^+ P_{n-1}(t) + w_{n+1}^- P_{n+1}(t) - \left(w_n^+ + w_n^-\right) P_n(t) ,$$

is not restricted to $n \in \mathbb{N}$ and thus does not automatically satisfy the proper boundary conditions to model a chemical reaction unless we have $w_0^- = 0$. A modification of the equation at $n = 0$ is required, thereby introducing a proper boundary:

$$\frac{\mathrm{d}P_0(t)}{\mathrm{d}t} = w_1^- P_1(t) - w_0^+ P_0(t) . \tag{3.97'}$$

This occurs *naturally* if w_n^- vanishes for $n = 0$, which is always the case for birth-and-death processes with $w_n^- = \nu + \mu n$, when the constant term referring to immigration vanishes, that is, $\nu = 0$. With $w_0^- = 0$, we only need to make sure that $P_{-1}(t) = 0$ and obtain (3.97′). $P_{-1}(t) = 0$ will always be satisfied for proper initial conditions, for example, $P_n(0) = 0,\ \forall\, n < 0$, and it is certainly true for the conventional initial condition $P_n(0) = \delta_{n,n_0}$ with $n_0 \geq 0$. By the same token, we prove that the upper reflecting boundary for chemical reactions, viz., $u = n_0$, satisfies the condition of being *natural* too, like most other boundaries we have encountered so far. Equipped with natural boundary conditions, the stochastic process can be solved for the entire integer range $n \in \mathbb{Z}$, and this is often much easier than with *artificial* boundaries.

Goel and Richter-Dyn present a comprehensive table of analytical solutions for restricted birth-and-death processes [216, pp. 16, 17]. A few selected examples are given in Table 5.4. Previously analyzed processes are readily classified according to restriction as well as natural or artificial boundaries. The backward Poisson process (Sect. 3.3.3, for example, is a restricted process with an unnatural boundary at $n = 0$, because from the point of view of the stochastic process it could be extended to negative numbers of events, while this makes no sense if we think about phone calls, mail arrivals, or other countable events. The conventional Poisson process and the random walk, on the other hand, are unrestricted processes in one or both directions since, in principle, they can be continued to infinite time and reach $n = \infty$ or $n = \pm\infty$, respectively. The chemical reactions of Chap. 4 are restricted processes with natural boundaries defined by stoichiometry and mass conservation.

5.2.3 *Fokker–Planck Equation and Neutral Evolution*

Mere reproduction without mutation gives rise to selection, and even in the absence of fitness differences, a kind of random selection is observed, as was pointed out by the Japanese geneticist Motoo Kimura [302–304]. He investigated the evolution of the distribution of alleles[12] at a given gene locus, and solved the problem by means of drift and diffusion processes in an abstract allele frequency space that he assumed to be continuous. Like the numbers of molecules in chemistry, the numbers

[12]The notion of *allele* was invented in genetics as a short form of *allelomorph*, which means *other form*, for the variants of a gene.

Table 5.4 Comparison of results for some restricted processes. Data from [216, pp.16,17]. Abbreviation and notations: $\gamma \equiv \lambda/\mu$, $\eta \equiv e^{(\lambda-\mu)t}$, $\alpha \equiv (\nu/\rho)^{(n-n_0)/2} e^{(\nu+\rho)t}$, and $I_n = I_{-n} \equiv I_n\big(2(\nu\rho)^{1/2}t\big)$, where $I_n(x)$ is a modified Bessel function. $G_n \equiv G_n(\xi_j, \gamma)$, where G_n is a Gottlieb polynomial $\hat{G}_n \equiv G_n(\hat{\xi}_j, \gamma)$, $G_n(x, \gamma) \equiv \gamma^n \sum_{k=0}^{n} (1-\gamma^{-1})^k \binom{n}{k}\binom{x-k+1}{k} = \gamma^n {}_2F_1(-n, -x; 1; 1-\gamma^{-1})$, where ${}_2F_1$ is a hypergeometric function, ξ_j and $\hat{\xi}_j$ are the roots of $G_{u-l}(\xi_j, \gamma) = 0$, $j = 0, \ldots, u-l-1$ and $G_{u-l+1}(\hat{\xi}_j, \gamma) = \gamma G_{u-l}(\hat{\xi}_j, \gamma)$, $j = 0, \ldots, u-l$, respectively. $H_n \equiv H_n(\zeta_j, \gamma)$, $\hat{H}_n \equiv H_n(\hat{\zeta}_j, \gamma)$, $H_n(x, \gamma) = G_n(x, \gamma^{-1})$, $H_{u-l}(\zeta_j, \gamma) = 0$, $j = 0, \ldots, u-l-1$, and finally, $H_{u+l-1}(\hat{\zeta}_j, \gamma) = H_{u-l}(\hat{\zeta}_j, \gamma)/\gamma$, respectively

λ_n	μ_n	Boundaries	$P_{n,n_0}(t)$	Ref.
ν	ρ	u : abs; l : $-\infty$	$\alpha\big(I_{n-n_0} - I_{2u-n-n_0}\big)$	[91, 404]
ν	ρ	u : $+\infty$; l : abs	$\alpha\big(I_{n-n_0} - I_{n+n_0-2l}\big)$	[91, 404]
ν	ρ	u : refl; l : $-\infty$	$\alpha\left(I_{n-n_0} + \left(\frac{\nu}{\rho}\right)^{1/2} I_{2u+l-n-n_0} + \left(1-\frac{\rho}{\nu}\right)\right)\sum_{j=2}^{\infty}\left(\frac{\nu}{\rho}\right)^{j/2} I_{2u-n-n_0+j}\Big)$	[91, 404]
ν	ρ	u : $+\infty$; l : refl	$\alpha\left(I_{n-n_0} + \left(\frac{\nu}{\rho}\right)^{1/2} I_{n+n_0+l-2u} + \left(1-\frac{\rho}{\nu}\right)\right)\sum_{j=2}^{\infty}\left(\frac{\nu}{\rho}\right)^{j/2} I_{n+n_0-2l+j}\Big)$	[91, 404]
ν	ρ	u : abs; l : abs	$\alpha\Big(\sum_{k=-\infty}^{\infty} I_{n-n_0+2k(u-l)} - \sum_{k=0}^{\infty}\big(I_{n+n_0-2l+2k(u-l)} + I_{2l-n-n_0+2k(u-l)}\big)\Big)$	[91, 404]
$\lambda(n-l+1)$	$\mu(n-l)$	u : abs; l : refl	$\gamma^{l-n}\sum_{k=0}^{u-l-1} G_{n_0-l}G_{n-l}\,\eta^{\xi_k}\left(\sum_{j=0}^{u-l-1}\frac{G_j}{\gamma^j}\right)^{-1}$	[406, 502]
$\lambda(n-l+1)$	$\mu(n-l)$	u : refl; l : refl	$\gamma^{l-n}\sum_{k=0}^{u-l} \hat{G}_{n_0-l}\hat{G}_{n-l}\,\eta^{\hat{\xi}_k}\left(\sum_{j=0}^{u-l}\frac{\hat{G}_j}{\gamma^j}\right)^{-1}$	[406, 502]
$\lambda(u-n)$	$\mu(u-n+1)$	u : refl; l : abs	$\gamma^{u-n}\sum_{k=0}^{u-l-1} H_{u-n_0}H_{u-n}\eta^{-\zeta_k}\left(\sum_{j=0}^{u-l-1} H_j\gamma^j\right)^{-1}$	[406, 502]
$\lambda(u-n)$	$\mu(u-n+1)$	u : refl; l : refl	$\gamma^{u-n}\sum_{k=0}^{u-l} \hat{H}_{u-n_0}\hat{H}_{u-n}\eta^{-\hat{\zeta}_k}\left(\sum_{j=0}^{u-l}\hat{H}_j\gamma^j\right)^{-1}$	[406, 502]

of frequencies or alleles are discrete. If N is the size of a diploid population,[13] the total number of copies of a given gene in this population is $2N$. The numbers of copies of allele A at this locus may take the values $n_{\mathsf{A}} = 0, 1, 2, \ldots, 2N$, and the allele frequencies $x_{\mathsf{A}} = n_{\mathsf{A}}/2N$ are $x_{\mathsf{A}} = 0/2N, 1/2N, 2/2N, \ldots, 1$.

The probability of observing a certain frequency x_{A} of the allele A at time t is denoted $f_{\mathsf{A}}(x_{\mathsf{A}}, t)$. In population genetics, time t is commonly counted in numbers of generations and is, in principle, a discrete variable in the case of non-overlapping generations. The problem to be solved is the determination of the evolution of the allele distribution in the population, i.e., $f(x, t)$ given an initial condition $f(x_0, 0)$. A problem of primary importance is fixation of alleles described by the probability $f(1, t)$. If an allele has become fixed, all other alleles at this gene locus have died out. In the limit $\lim N \to \infty$, the variables x become continuous, the domain of allowed allele frequencies is the closed interval $x \in [0, 1]$, and the probabilities are described by densities $f(x, t) = p(x, t|x_0, t_0)\mathrm{d}x$ under the condition of a sharp initial value $p(x, 0) = \delta(x - x_0)$ at $t_0 = 0$. Kimura modeled the evolution of the allele distribution by a Fokker–Planck equation on this domain [96, pp. 367–432] and we begin by sketching this approach here.

For simplicity we consider a gene locus with two alleles A and B, and denote their frequencies by $x_{\mathsf{A}} = n_{\mathsf{A}}/2N = x$ and $x_{\mathsf{B}} = n_{\mathsf{B}}/2N = 1 - x$. The points $x = 0$ and $x = 1$ are absorbing boundaries and correspond to fixation of allele A or allele B, respectively. The change in allele frequency per generation is denoted by δx and its first and second moment due to selection and random sampling are

$$\mathrm{E}_{\delta x}(x, t) = x(1 - x)\varrho(x, t) , \quad \mathrm{var}_{\delta x}(x) = \frac{x(1 - x)}{2N} , \tag{5.28a}$$

where $\varrho(x, t)$ is the selection coefficient of the allele. The coefficient is related to the relative fitness of an allele A through the relation $f_{\mathsf{A}} = f(1 + \varrho_{\mathsf{A}})$, where f is the reference fitness that is commonly assigned to the *wild type*.[14] The moments are introduced into a conventional Fokker–Planck equation (3.47) and we obtain

$$\begin{aligned} \frac{\partial p(x, t)}{\partial t} &= -\frac{\partial}{\partial x}\big(\mathrm{E}_{\delta x}(x)p(x, t)\big) + \frac{1}{2}\frac{\partial^2}{\partial x^2}\big(\mathrm{var}_{\delta x}(x)p(x, t)\big) \\ &= -\varrho\frac{\partial}{\partial x}\big(x(1 - x)p(x, t)\big) + \frac{1}{4N}\frac{\partial^2}{\partial x^2}\big(x(1 - x)p(x, t)\big) \end{aligned} \tag{5.28b}$$

---

[13]Here we use $2N$ for the number of alleles in a population of size N, which refers to diploid organisms. For haploid organisms, $2N$ has to be replaced by N. In real populations, the population size is corrected for various other factors and taken to be $2N_e$ or N_e, respectively.

[14]The selection coefficient is denoted here by ϱ instead of s in order to avoid confusion with the auxiliary variable of the probability generation function. The definition here is the same as used by Kimura [96, 304]: $\varrho > 0$ implies greater fitness than the reference and an advantageous allele, $\varrho < 0$ reduced fitness and a deleterious allele. We remark that the conventional definition in population genetics uses the opposite sign: $s = 1$ means fitness zero, no progeny, and a lethal variant. In either case, selective neutrality occurs for $s = 0$ or $\varrho = 0$ (see also Sect. 5.3.3).

by assuming a constant selection coefficient ϱ. Restriction to neutral evolution $\varrho = 0$ yields the PDE

$$\frac{\partial p(x,t)}{\partial t} = \frac{1}{2N}\frac{\partial^2}{\partial x^2}\bigl(x(1-x)p(x,t)\bigr)\,, \tag{5.28c}$$

which has singularities at the boundaries $x = 0$ and $x = 1$. These two points correspond to fixation of allele B or A, respectively, and have to be treated separately. Equation (5.28c) has been solved by Kimura [302, 303] (for more recent work on the problem see, e.g., [1, 283]). The form of the PDE (5.28c) suggests applying a solution based on separation of variables: $p(x,t) = \Xi(x)\Phi(t)$. Dividing both sides by $\Xi(x)\Phi(t)$ yields

$$\frac{1}{\Phi(t)}\frac{\partial\Phi(t)}{\partial t} = \lambda = \frac{1}{4N}\frac{1}{\Xi(x)}\frac{\partial^2}{\partial x^2}\bigl(x(1-x)\Xi(x)\bigr)\,,$$

where λ depends neither on time t nor on gene frequency x. Care is needed when there are singularities, and here we shall apply special handling of the points $x = 0$ and $x = 1$ that correspond to fixation of allele A or B, respectively. The PDE is transformed into two ODEs

$$\frac{\mathrm{d}\Phi(t)}{\mathrm{d}t} = \lambda\Phi(t)\,, \qquad \frac{\mathrm{d}^2}{\mathrm{d}x^2}\Bigl(x(1-x)\Xi(x)\Bigr) = 4N\lambda\Xi(x)\,.$$

The time-dependent equation is readily solved and yields

$$\Phi(t) = \theta\exp(\lambda t)\,,$$

where θ is a constant factor that remains to be determined. The solution of the second ODE is facilitated by a transformation of the independent variable:

$$z = 1-2x\,, \quad x = \frac{1}{2}(1-z)\,, \quad \text{and } z_0 = 1-2x_0\,, \quad x_0 = \frac{1}{2}(1-z_0)\,.$$

This introduces symmetry with respect to the origin into the open interval: $]0,1[\ \to\ ]-1,+1[$. The resulting eigenvalue equation is known as the Gegenbauer differential equation:

$$\frac{\mathrm{d}^2}{\mathrm{d}z^2}\Bigl((z^2-1)\,\Xi(z)\Bigr) = \lambda\,\Xi(z)\,, \quad |z| < 1\,.$$

The solutions of this equation are the Gegenbauer polynomials,[15] corresponding to the eigenvalues $\lambda_k = (k+1)(k+2)$ [412, pp. 782–783]:

$$\Xi_k(z) = T_k^{(1)}(z) = \frac{(k+1)(k+2)}{2}\,{}_2F_1\Big(k+3, -k, 2, \frac{1-z}{2}\Big)\,,$$

where ${}_2F_1$ is the conventional hypergeometric function. The general solution is obtained as a linear combination of the eigenfunctions $\Xi_k(z)$ for $k = 0, 1, \ldots, \infty$, where the coefficients are determined by the initial conditions. Back-transformation into the original gene frequencies yields the desired solution:

$$\begin{aligned} p(x,t|x_0,0) = x_0(1-x_0)\sum_{i=1}^{\infty} i(i+1)(2i+1)\,{}_2F_1(1-i, i+2, 2, x_0) \\ \times\,{}_2F_1(1-i, i+2, 2, x)\mathrm{e}^{-i(i+1)t/4N}\,, \quad x \in]0,1[\,. \end{aligned} \tag{5.28d}$$

Figure 5.11 shows an example. The initially sharp density $p(x,0) = \delta(x-x_0)$ broadens with the height of the peak becoming smaller until an almost uniform distribution is reached on $]0,1[$. Then, the height of the quasi-uniform density decreases and becomes zero in the limit $t \to \infty$. The process has a lot in common with 1D diffusion of a substance in a leaky vessel.

Finally, we derive expressions for the calculation of the gene frequencies at the singular points $x = 0$ and $x = 1$, $f(0,t)$ and $f(1,t)$, respectively. For this purpose, we recall the probability current defined in Sect. 3.2.3 for master equations and generalize to the continuous case:

$$\begin{aligned} \varphi(x,t) = -\int \mathrm{d}x\,\frac{\partial p(x,t)}{\partial t} &= -\frac{1}{2}\frac{\partial}{\partial x}\big(\mathrm{var}_{\delta x}p(x,t)\big) + \mathrm{E}_{\delta x}p(x,t)\,, \\ \frac{\partial p(x,t)}{\partial t} &= -\frac{\partial \varphi(x,t)}{\partial x}\,. \end{aligned} \tag{5.28e}$$

At the lower boundary $x = 0$, we find

$$-\varphi(0,t) = \lim_{x\to 0}\Big(\frac{1}{4N}\frac{\partial}{\partial x}\big(x(1-x)p(x,t)\big) - x(1-x)\varrho(x,t)p(x,t)\Big) = \frac{1}{4N}p(0,t)\,.$$

In the absence of selection, i.e., $\varrho = 0$, we calculate

$$\frac{\mathrm{d}f(0,t)}{\mathrm{d}t} = \frac{1}{4N}p(0,t)\,, \qquad \frac{\mathrm{d}f(1,t)}{\mathrm{d}t} = \frac{1}{4N}p(1,t)\,,$$

[15]The definition of Gegenbauer polynomials here is slightly different from the one given in Sect. 4.3.3: $T_n^{(\beta)}(z) = (2\beta-1)!!C_n^{(\beta+1/2)}(z)$.

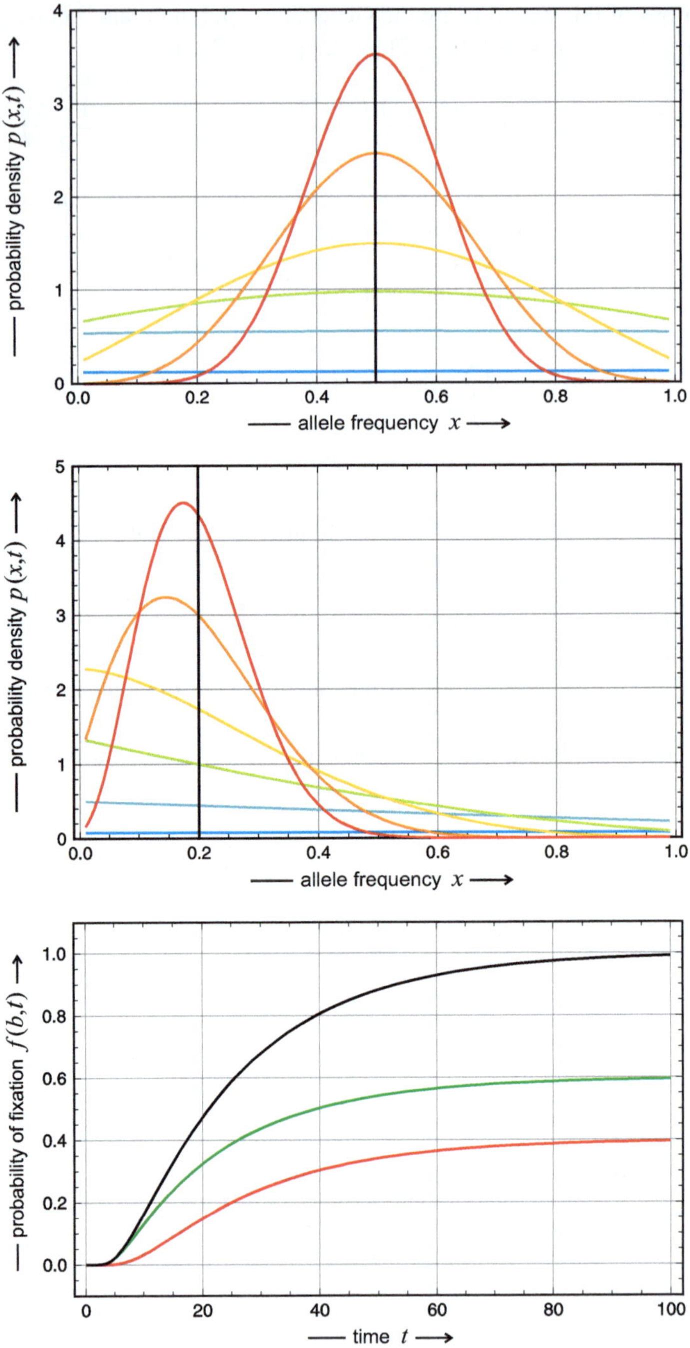

Fig. 5.11 Random selection as a diffusion process. Caption on next page

Fig. 5.11 Random selection as a diffusion process (see previous page). *Upper*: Spreading of allele A for symmetric initial conditions, $p(x,0) = \delta(x - x_0)$ with $x_0 = 1/2$. Three phases of the process can be recognized: (i) broadening of the peak within the interval $]0,1[$, with zero probability of extinction and fixation $f(0,t) = f(1,t) = 0$, (ii) the broadening distribution has reached the absorbing boundaries and the probability of fixation has begun to rise, and (iii) the distribution has become flat inside the interval, the almost uniform distribution decreases further, and the probability of fixation approaches $\lim_{t\to\infty} f(0,t) = \lim_{t\to\infty} f(1,t) = 1/2$. *Middle*: The same process for an asymmetric initial condition $x_0 = 2/10$. Choice of parameters: $N = 20$, $t = 0$ (*black*),1 (*red*), 2 (*orange*), 5 (*yellow*), 10 (*chartreuse*), 20 (*seagreen*), and 50 (*blue*). *Lower*: Probability of fixation as a function of time $f(1,t)$ (fixation of A *red*), $f(0,t)$ (fixation of B *green*), and $f(1,t) + f(0,t)$ (*black*) for $N = 20$ and $x_0 = 2/5$. In the limit $t \to \infty$ fixation of A or B is certain (recalculated after [302])

for the differential increase in the probability of fixation of alleles B and A at the two absorbing boundaries. A somewhat lengthy calculation yields a convenient expression for the probability of fixation as an infinite sum of Legendre polynomials $P_k(z_0)$ ($z_0 = 1 - 2x_0$):

$$\begin{aligned} f(1,t) &= x_0 + \sum_{k=1}^{\infty} \frac{(-1)^k}{2} \Big(P_{k-1}(z_0) - P_{k+1}(z_0) \Big) \mathrm{e}^{-k(k+1)t/4N} \,, \\ f(0,t) &= 1 - x_0 - \sum_{k=1}^{\infty} \frac{1}{2} \Big(P_{k-1}(z_0) - P_{k+1}(z_0) \Big) \mathrm{e}^{-k(k+1)t/4N} \,. \end{aligned} \tag{5.28f}$$

The sum of fixed alleles is readily calculated and yields the expression

$$f(1,t) + f(0,t) = 1 - \sum_{k=0}^{\infty} \Big(P_{2k}(z_0) - P_{2k+1}(z_0) \Big) \mathrm{e}^{-k(k+1)t/4N} \,,$$

which becomes zero for $t = 0$ and approaches one in the limit $t \to \infty$. The mathematical analysis thus provides a proof that random drift leads to fixation of alleles, and one might characterize this phenomenon therefore as *random selection*. A concrete numerical example is shown in Fig. 5.11.

It is worth mentioning that the simple sequential extinction model described in Sect. 5.2.2 gave the same qualitative result when interpreted properly: the numbers of copies refer to allele B and extinction of this allele is tantamount to fixation of allele A. In Sect. 5.3.2, we shall come back to the problem of random drift and model it by means of a master equation.

5.2.4 Logistic Birth-and-Death and Epidemiology

In order to introduce finite resources into birth-and-death processes, model considerations can be used in full analogy with the approach by Pierre-François Verhulst, described in Sect. 5.1.4. The logistic birth-and-death process, although interesting in its own rights, has found important applications in theoretical epidemiology. The basis for the restriction is again the logistic equation (5.1): λ_n and μ_n are modeled as functions of n, in which λ_n decreases with n and μ_n increases. A typical restricted birth-and-death process has a lower absorbing boundary at Σ_0 with $n = 0$, since $w_0^+ = \lambda_0 = 0$, and a reflecting upper boundary at some value $n = N$, whence all trajectories have to end in Σ_0. For certain parameter functions λ_n and μ_n, however, the systems may stay at, or more precisely very close to, quasi-stationary states $\widetilde{\Sigma}$ for arbitrarily long but finite times. As we have seen in the case of autocatalysis (Sect. 5.2.1), the analysis of quasi-stationarity requires some special treatment.

Quasi-Stationarity

We consider a restricted birth-and-death process on the states Σ_n with $n = 0, 1, 2, \ldots, N$, with an absorbing boundary at Σ_0 with $n = 0$ and a reflecting boundary at Σ_N with $n = N$. The boundaries result from the step-up transition probabilities $w_0^+ = \lambda_0 = 0$ and $w_N^+ = \lambda_N = 0$. For $\lambda_k > \mu_k$ ($k = 1, 2, \ldots$ and $k < N$), the process may fluctuate for long time around a state $\widetilde{\Sigma}$, which corresponds to a stable stationary state of the deterministic approach, before it eventually ends up in the only absorbing state Σ_0. A *quasi-stationary state* $\widetilde{\Sigma}$ is characterized by the same long term behavior as a stationary state in the sense that the corresponding probability density is approached by almost all processes from the environment of the state. The final drift into the absorbing state occurs only after extremely long times. Here we calculate the probability density of the quasi-stationary state and the time to extinction using an approach suggested in [418, 419].

The master equation of the birth-and-death process is applied in matrix form (4.91c):

$$\frac{\mathrm{d}\mathbf{P}}{\mathrm{d}t} = \mathrm{W}\mathbf{P}\,, \quad \text{with } \mathrm{W} = \begin{pmatrix} -\kappa_0 & \mu_1 & 0 & \ldots & 0 \\ \lambda_0 & -\kappa_1 & \mu_2 & \ldots & 0 \\ 0 & \lambda_1 & -\kappa_2 & \ldots & 0 \\ \vdots & \vdots & \vdots & \ddots & \vdots \\ 0 & 0 & 0 & \ldots & -\kappa_N \end{pmatrix}, \quad \mathbf{P} = \begin{pmatrix} P_0 \\ P_1 \\ P_2 \\ \ldots \\ P_N \end{pmatrix},$$

where the diagonal elements of the transition matrix are $-\kappa_n = -(\lambda_n + \mu_n)$ and $P_n(t)$, $n = 0, 1, \ldots, N$, with $\sum_{n=0}^{N} P_n(t) = 1$ are the probability densities.

The quasi-stationary distribution is a conditional stationary distribution defined with the condition that the process has not yet become extinct at time t, i.e., $\mathcal{X}(t) > 0$. The probabilities of the individual states Σ_n are contained in the column vector $\mathbf{Q}(t) = \big(Q_1(t), Q_2(t), \ldots, Q_N(t)\big)^{\mathrm{t}}$ that depends on the initial distribution $\mathbf{Q}(0)$. In order to derive the time dependence of $\mathbf{Q}(t)$, we introduce a truncated vector $\mathbf{P}(t)$, without P_0, defined by $\widehat{\mathbf{P}}(t) = \big(P_1(t), P_2(t), \ldots, P_N(t)\big)^{\mathrm{t}}$, and assume a positive initial state Σ_{n_0} with $n_0 > 0$. Using $\mathrm{d}P_0/\,\mathrm{d}t = \mu_1 P_1$, we obtain from the normalized vector $\mathbf{Q}(t) = \widehat{\mathbf{P}}(t)/\big(1 - P_0(t)\big)$:

$$\frac{\mathrm{d}\mathbf{Q}}{\mathrm{d}t} = \frac{\mathrm{d}}{\mathrm{d}t}\left(\frac{\widehat{\mathbf{P}}}{1-P_0}\right) = \frac{\mathrm{d}\widehat{\mathbf{P}}/\mathrm{d}t + \mu_1 Q_1 \widehat{\mathbf{P}}}{1-P_0} = \widehat{\mathrm{W}}\mathbf{Q} + \mu_1 Q_1 \mathbf{Q}\,, \tag{5.29}$$

where the truncated matrix $\widehat{\mathrm{W}}$ represents the $N \times N$ square matrix obtained from W by elimination of the first column and the first row:

$$\widehat{\mathrm{W}} = \begin{pmatrix} -\kappa_1 & \mu_2 & 0 & \dots & 0 \\ \lambda_1 & -\kappa_2 & \mu_3 & \dots & 0 \\ 0 & \lambda_2 & -\kappa_3 & \dots & 0 \\ \vdots & \vdots & \vdots & \ddots & \vdots \\ 0 & 0 & 0 & \dots & -\kappa_N \end{pmatrix}.$$

The stationary solution of (5.29) satisfies the eigenvalue equation

$$\frac{\mathrm{d}\mathbf{Q}}{\mathrm{d}t} = 0 \quad \Longrightarrow \quad \widehat{\mathrm{W}}\widetilde{\mathbf{Q}} = -\mu_1 \tilde{Q}_1 \widetilde{\mathbf{Q}}\,. \tag{5.30}$$

Here $-\mu\tilde{Q}_1$ turns out to be the largest (nonzero) eigenvalue of $\widehat{\mathrm{W}}$ and the dominant right-hand eigenvector of $\widehat{\mathrm{W}}$, viz., $\widetilde{\mathbf{Q}}$, represents the quasi-stationary probability density [452]. Equation (5.29) is suitable for numerical computation of the density and will be applied in the next section to calculate the quasi-stationary density in the logistic model.

The analysis of the computed densities and the calculation of extinction times $\mathcal{T}$ is facilitated by the definition of two auxiliary birth-and-death processes [419], i.e., $\{\mathcal{X}^{(0)}(t)\}$ and $\{\mathcal{X}^{(1)}(t)\}$, which are illustrated in Fig. 5.12. The process $\{\mathcal{X}^{(0)}(t)\}$ is derived from the original process $\{\mathcal{X}(t)\}$ by setting the death rate $\mu_1^{(0)} = 0$ and keeping all other birth and death rates unchanged. Thereby the state of extinction is simply removed. The process $\{\mathcal{X}^{(1)}(t)\}$ differs from $\{\mathcal{X}(t)\}$ by assuming the existence of one *immortal* individual. This is achieved by setting $\mu_n^{(1)} = \mu_{n-1}$ or shifting all death rates to the next higher state and leaving the birth rates unchanged.

The stationary distributions for the two auxiliary processes $\bar{\mathbf{P}}^{(0)}$ and $\bar{\mathbf{P}}^{(1)}$ can be calculated from the general expressions derived for birth-and-death processes in Sect. 3.2.3. For the auxiliary process $\{\mathcal{X}^{(1)}(t)\}$, we obtain:

$$\varrho_1 = 1\,, \quad \varrho_n = \frac{\lambda_1 \lambda_2 \cdots \lambda_{n-1}}{\mu_1 \mu_1 \cdots \mu_{n-1}}\,, \qquad n = 2, 3, \dots, N\,, \tag{5.31a}$$

$$\bar{P}_n^{(1)} = \varrho_n \bar{P}_1^{(1)}\,, \quad \bar{P}_1^{(1)} = \frac{1}{\sum_{n=1}^{N} \varrho_n}\,. \tag{5.31b}$$

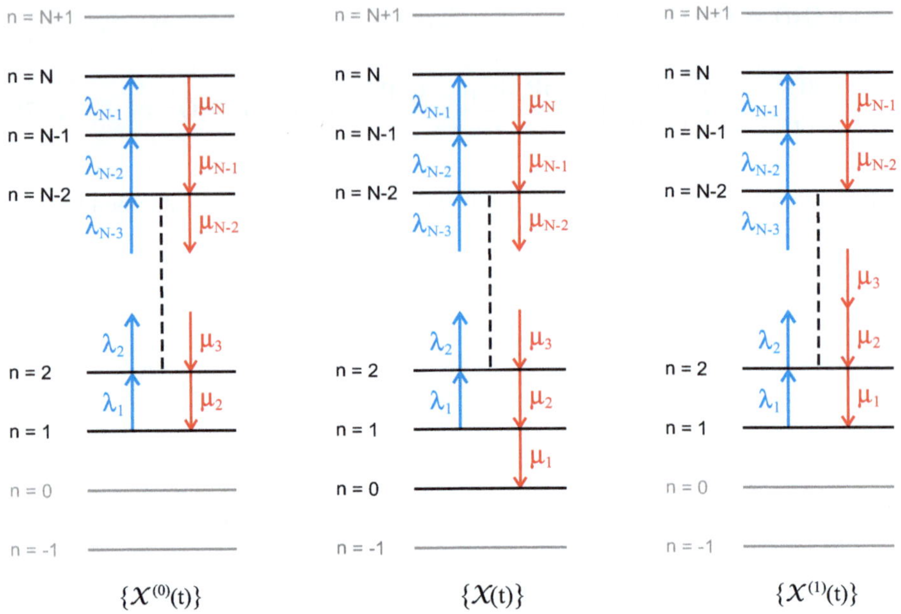

Fig. 5.12 A birth-and-death process between an absorbing lower and a reflecting upper boundary and two modified processes with the lower boundary reflecting. The process $\{\mathcal{X}(t)\}$ shown in the middle is accompanied by two auxiliary processes $\{\mathcal{X}^{(0)}(t)\}$ and $\{\mathcal{X}^{(1)}(t)\}$, in which the lower absorbing boundary has been replaced by a reflecting boundary. This is achieved either by setting $\mu_1^{(0)} = 0$ for $\{\mathcal{X}^{(0)}(t)\}$ or by shifting the death rates $\mu_n^{(1)} = \mu_{n-1}$ for $\{\mathcal{X}^{(1)}(t)\}$. States outside the domains of the random variable are shown in *gray*

For the process $\{\mathcal{X}^{(0)}(t)\}$, we have to take into account the difference in the birth rates and find:

$$\pi_n = \frac{\mu_1}{\mu_n}\varrho_n\,, \quad n = 1, 2, \ldots, N\,, \tag{5.31c}$$

$$\bar{P}_n^{(0)} = \pi_n \bar{P}_1^{(0)}\,, \quad \bar{P}_1^{(0)} = \frac{1}{\sum_{n=1}^{N} \pi_n}\,. \tag{5.31d}$$

Both stationary distributions $\bar{\mathbf{P}}^{(0)}$ and $\bar{\mathbf{P}}^{(1)}$ represent approximations of the quasi-stationary distribution $\widetilde{\mathbf{Q}}$ for different ranges of parameter values and will be discussed for the special case of logistic birth-and-death processes in the next section. The expressions for ϱ_n and π_n are also used in the calculations of extinction times (see below) and in iteration methods for the determination of the quasi-stationary distribution [418, 419].

Finally, we consider the times to extinction for the process $\{\mathcal{X}(t)\}$ and distinguish two different initial conditions: (i) the extinction time $\widetilde{\mathcal{T}}$ from the quasi-stationary distribution, and (ii) the extinction time from a defined initial state Σ_n, denoted by $\mathcal{T}_n$. As shown in [539], the first case is relevant, because every process approaches the quasi-stationary state $\widetilde{\Sigma}$ provided it has not been previously absorbed in Σ_0. Since the eigenvalue of the transition matrix $\widehat{W}$ is known, the determination of $\widetilde{\mathcal{T}}$ is straightforward: the time to extinction has an exponential distribution

$$P(\widetilde{\mathcal{T}} \leq t) = 1 - \exp(-\mu_1 \widetilde{Q}_1 t) ,$$

and the expectation value is given by

$$E(\widetilde{\mathcal{T}}) = \frac{1}{\mu_1 \widetilde{Q}_1} . \tag{5.32}$$

We remark that the distribution of the extinction time $\widetilde{\mathcal{T}}$ is completely determined by the probability $\widetilde{Q}_1$.

The expectation value of the time to extinction from a pure state Σ_n has already been calculated for the unrestricted birth-and-death process in Sect. 3.2.3. In contrast to the unrestricted process, the probability of extinction of the restricted process $\{\mathcal{X}(t)\}$ is one under all conditions: $Q_i = 1, \forall\, i = 1, 2, \ldots, N$. In order to obtain the time to extinction for the restricted process with the reflecting barrier at Σ_N, we need only truncate the infinite sums at $n = N$[16]:

$$\vartheta_n(N) = E(\mathcal{T}_n) = \frac{1}{\mu_1} \sum_{i=1}^{n} \frac{\sum_{j=i}^{N} \pi_j}{\varrho_i} . \tag{5.33}$$

For extinction from the state Σ_1, we need only put $n = 1$ and find

$$\vartheta_1(N) = E(\mathcal{T}_1) = \frac{1}{\mu_1} \sum_{j=i}^{N} \pi_j = \frac{1}{\mu_1 P_1^{(0)}} , \tag{5.34}$$

which illustrates the importance of the auxiliary processes. We can use (5.34) to rewrite the expression for the mean extinction time:

$$E(\mathcal{T}_n) = E(\mathcal{T}_1) \sum_{i=1}^{n} \frac{1 - \sum_{j=1}^{k-1} P_j^{(0)}}{\varrho_i} . \tag{5.33'}$$

Logistic Birth-and-Death Processes

Here we make birth-and-death models more realistic and adjust the state space to the carrying capacity of the ecosystem. For a proper modification of the unrestricted process, the birth rates and/or the death rates could by adapted to a limitation of resources as done in the logistic equation. In principle, the birth rate has to

[16]The definitions of the product terms of the ratios π_k and ϱ_k differ from those used in Sect. 3.2.3.

be successively decreased or the death rate has to be increased with growing population density, or both. The two transition rates λ_n and μ_n can be modeled by the expressions [419]

$$\lambda_n = \begin{cases} \lambda\left(1 - \alpha_1 \frac{n}{N}\right) n\,, & \text{for } n = 0, 1, \ldots, N-1\,, \\ 0\,, & \text{for } n = N\,, \end{cases} \tag{5.35}$$

$$\mu_n = \mu\left(1 + \alpha_2 \frac{n}{N}\right) n\,, \qquad \text{for } n = 0, 1, \ldots, N\,.$$

The birth rate decreases with increasing n and for $\alpha_1 = 1$ becomes zero at $n = N$, and the death rate increases monotonically with increasing n for $\alpha_2 > 0$. Accordingly, $\mu_1 = 1 + \alpha_2/N > 0$ and $\lambda_0 = 0$ and the state Σ_0 with $n = 0$ is an absorbing barrier independently of the choice of parameters, so the system will inevitably become extinct. Nevertheless, as we shall show in the following the process sustains a quasi-stationary state $\widetilde{\Sigma}$ for $\lambda > \mu$.

The deterministic equation for the process (5.35) can be derived, for example, by the law of large numbers applied to the birth-and-death process [385, 386], and takes the form

$$\frac{\mathrm{d}x}{\mathrm{d}t} = \lambda x\left(1 - \alpha_1 \frac{x}{N} - \frac{\mu}{\lambda}\left(1 - \alpha_2 \frac{x}{N}\right)\right) = ax(1 - bx) = kx - lx^2\,. \tag{5.36}$$

Equation (5.36) can be solved by standard integration and has the solution

$$x(t) = x(0)\frac{1}{bx(0) + \big(1 - bx(0)\big)\mathrm{e}^{at}}\,. \tag{5.37}$$

The rates of the birth-and-death process were combined to parameters in two different ways:

(i) $a = \lambda - \mu$ and $b = (\lambda\alpha_1 + \mu\alpha_2)/(\lambda - \mu)N$, or
(ii) $k = \lambda - \mu$ and $l = (\lambda\alpha_1 + \mu\alpha_2)/N$.

The first choice is particularly useful for calculating the solution curve and for qualitative analysis, which yields two steady states:

$$\bar{x}_{(1)} = b^{-1} = \frac{\lambda - \mu}{\lambda\alpha_1 + \mu\alpha_2} N\,, \quad \text{and } \bar{\lambda}_{(1)} = -a = -(\lambda - \mu)\,, \tag{5.38}$$

$$\bar{x}_{(2)} = 0\,, \quad \text{and } \bar{\lambda}_{(2)} = a = \lambda - \mu\,.$$

State $\bar{\Sigma}_{(1)}$ is situated on the positive x-axis and stable, i.e., $\bar{x}_{(1)} > 0$, $\bar{\lambda}_{(1)} < 0$, for $\lambda > \mu$. For $\lambda < \mu$, the state $\bar{\Sigma}_{(1)}$ is unstable and lies outside the physically meaningful range. The second stationary state $\bar{\Sigma}_{(2)}$ is the state of extinction and the condition for its asymptotic stability is $\bar{\lambda}_{(2)} < 0$ or $\mu > \lambda$, while it is unstable for $\lambda > \mu$.

Using the parameters k and l suggests a comparison of the birth-and-death process (5.14) with the reversible autocatalytic reaction

$$(\mathsf{A}) + \mathsf{X} \underset{l}{\overset{k}{\rightleftarrows}} 2\mathsf{X}\,. \tag{4.93a$'$}$$

Since both processes are described by the same ODEs, they have identical solutions. The only difference lies in the physically acceptable range of parameters. The rate parameter of a chemical reaction has to be positive, so we have $k \in \mathbb{R}_{>0}$ but $(\lambda - \mu) \in \mathbb{R}$ and the birth-and-death process may sustain a stable extinction state $\bar{\Sigma}_{(2)}$, whereas Σ_1 is reflecting for the autocatalytic reaction in the closed system (Sect. 5.2.1).

Finally, we would like to stress that the logistic equation contains only two independent parameters [181]

$$k = a = \lambda - \mu \quad \text{and} \quad l = ab = \frac{\lambda\alpha_1 + \mu\alpha_2}{N}\,.$$

The additional parameters may facilitate illustration and interpretation, but they are overparameterizations and give rise to mathematical redundancies. Provided we allow for a linear scaling of the time axis only the ratio called the basic reproduction ratio $\gamma = \lambda/\mu$ is relevant. The range of parameter values $\gamma > 1$ implies long times to extinction, whereas the extinction time is short for $\gamma < 1$ [419].

The Russian microbiologist Gregorii Frantsevich Gause [196] favors a different model in which the birth rate is linear, i.e., $\lambda_n = \lambda n$, and the population size effect is included in a death rate that accounts for the constraint on population growth: $\mu_n = \mu n^2/C$. The actual decision, which model is best suited has to be determined empirically by measuring growth and death rates in competition studies [195]. A comparison of various equations modeling constrained growth can be found in [282].

In contrast to the deterministic approach, the stochastic model (5.35) is not overparameterized since different sets of the five parameter values $(\lambda, \mu, \alpha_1, \alpha_2, N)$ give rise to different probability densities for the same initial condition $n(0) = n_0$ [426]. As already mentioned, the steady state $\bar{\Sigma}_{(2)} \equiv \Sigma_0$ is the only absorbing boundary, so the population becomes extinct with probability one. The second steady state $\bar{\Sigma}_{(1)}$ is described by the quasi-stationary distribution of the stochastic process, viz., $\tilde{\Sigma} \equiv \bar{\Sigma}_{(1)}$, as outlined in general terms in the last section. Here we shall introduce the special features of the logistic birth-and-death process. A function of the four parameters N, γ, α_1, and α_2,

$$\rho = \frac{\gamma - 1}{\sqrt{\alpha_1 + \alpha_2}}\sqrt{N}\,, \tag{5.39}$$

is useful for separating into three ranges where the quasi-stationary distribution has different behavior:

(i) the region $\rho \geq \rho_{cr}$ of long times to extinction,
(ii) an intermediate region $0 < \rho < \rho_{cr}$ with moderately long times to extinction, and
(iii) the region $\rho \ll 0$ of short times to extinction.

Ingemar Nåsell suggests applying $\rho_{cr} = 3$ [419]. The characteristic behavior for negative ρ values is found for $|\rho| \approx 1$. Figure 5.13 compares the quasi-stationary density $\widetilde{\mathbf{Q}}$ of the stochastic logistic process with the two auxiliary distributions and $\bar{\mathbf{P}}^{(1)}$. In the first region with $\rho > 3$, the auxiliary density $\bar{\mathbf{P}}^{(0)}$ is an excellent approximation to the quasi-stationary density $\widetilde{\mathbf{Q}}$—for the example shown in Fig. 5.13c, the two densities differ by less than the line width—and the auxiliary density $\bar{\mathbf{P}}^{(1)}$ is close to $\widetilde{\mathbf{Q}}$. In the core of the densities, i.e., in the range around the maximum, all three distributions are very well represented by a normal density. Major deviations from the Gaussian curve are found at the left-hand end of the densities, but even there $\widetilde{\mathbf{Q}}$ and $\bar{\mathbf{P}}^{(0)}$ are very close for $n > 20$. In the intermediate region, the density $\bar{\mathbf{P}}^{(1)}$ becomes a better and better approximation to $\widetilde{\mathbf{Q}}$ (Fig. 5.13a and b) the smaller the value of ρ. Finally, at distinct negative values in the third region, the two densities coincide (Fig. 5.13d).

Mean times to extinction were calculated using (5.32) and (5.33) and the results are shown in Fig. 5.14. There is a characteristic change in the dependence of the mean times to extinction on the initial population size n at the value $\gamma = 1$: for $\lambda > \mu$, the approximate time to extinction grows fast with increasing n and soon reaches a value that is kept constant for the entire domain until $n = N$, whereas an exponential approach over the full range $1 \leq n \leq N$ is observed with $\lambda < \mu$. The long plateau exhibited by the values of $\mathrm{E}(\mathcal{T}_n)$ is accompanied by the observation that $\mathrm{E}(\widetilde{\mathcal{T}}) \approx \mathrm{E}(\mathcal{T}_N)$ (Table 5.5), which has a straightforward explanation. Almost all states contributing with significant frequency to the quasi-stationary density are situated on the plateau and have approximately the same mean time to extinction, i.e., $\mathrm{E}(\mathcal{T}_n) \approx \mathrm{E}(\mathcal{T}_N)$. As expected, $\mathrm{E}(\widetilde{\mathcal{T}})$ is smaller than $\mathrm{E}(\mathcal{T}_N)$, because $\mathrm{E}(\mathcal{T}_n) < \mathrm{E}(\mathcal{T}_N)$, $\forall\, n < N$. In addition, the difference $45 < \mathrm{E}(\mathcal{T}_N) - \mathrm{E}(\widetilde{\mathcal{T}}) < 47$ is remarkably constant over the whole range of calculated values: $958 \leq \mathrm{E}(\mathcal{T}_N) \leq 51{,}467$. No such regularity is observable in the range of short extinction times.

Applications in Theoretical Epidemiology

The application of mathematical models for a better understanding of epidemics is not new. The famous Swiss mathematician Daniel Bernoulli conceived a model of smallpox epidemics and tried to convince his contemporaries of the advantage of inoculation by means of mathematical analysis [44]. Bernoulli's seminal paper saw a revival about ten years ago when a debate on the importance of vaccination was launched by medically uninformed people [45, 110]. Three hundred years after Bernoulli, theoretical epidemiology has become a discipline in its own right and it is impossible to provide a full account of models on the population dynamics

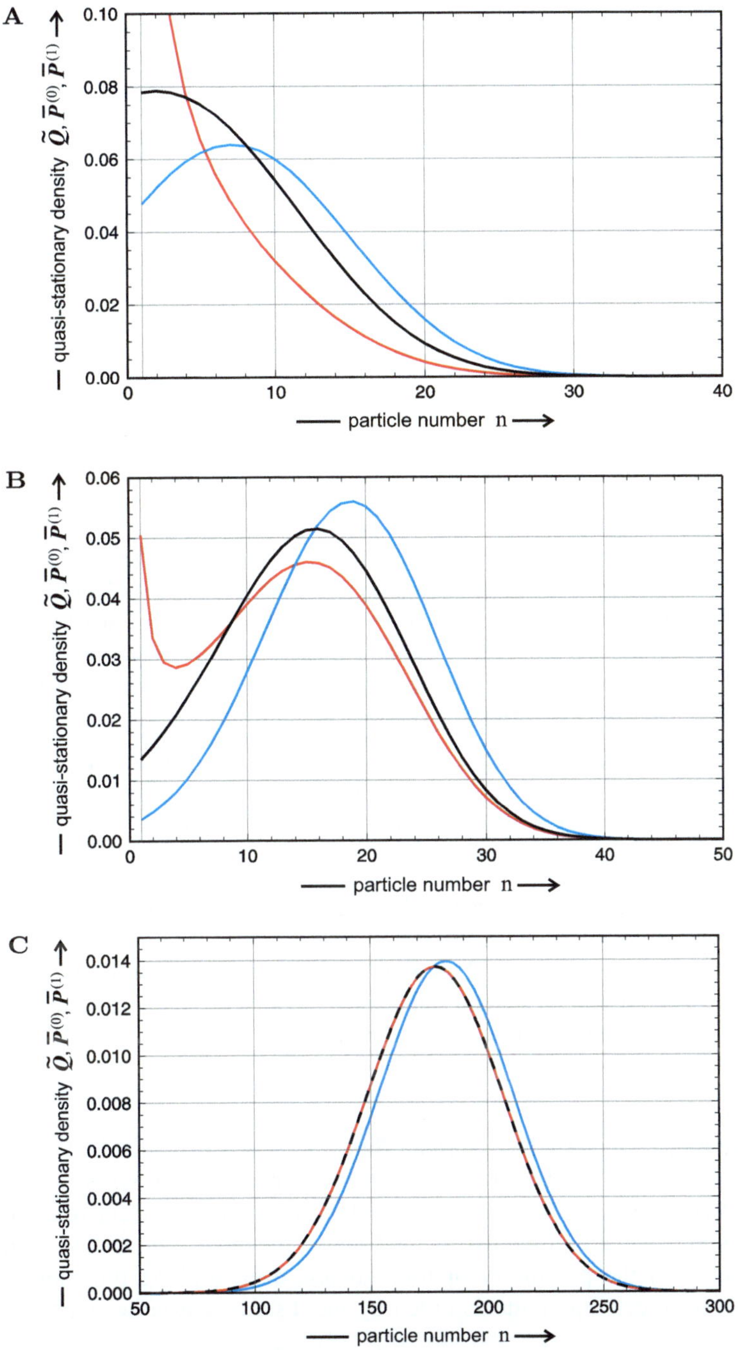

Fig. 5.13 Quasi-stationary density of logistic birth-and-death processes. Caption on next page

D

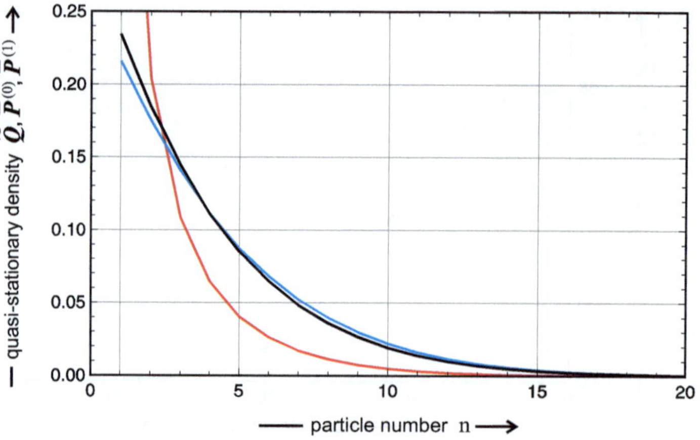

Fig. 5.13 Quasi-stationary density of logistic birth-and-death processes (see previous page). Plots of the quasi-stationary density $\widetilde{Q}_n$ (*black*) and the stationary densities of the two auxiliary processes $\bar{P}_n^{(0)}$ (*red*) and $\bar{P}_n^{(1)}$ (*blue*) in different regions defined by the parameter ρ. (**a**) A typical intermediate case with a positive ρ-value below the critical value, i.e., $0 < \rho < \rho_{cr}$, where the quasi-stationary density is weakly described by the auxiliary processes, although $\bar{P}_n^{(1)}$ does a little better then $\bar{P}_n^{(0)}$. The ρ-value in (**b**) is chosen almost exactly at the critical value $\rho = \rho_{cr} = 3$ and the density $\bar{P}_n^{(0)}$ is the better approximation. (**c**) At a value $\rho \gg \rho_{cr}$, the function $\bar{P}_n^{(0)}$ coincides with the exact density $\widetilde{Q}_n$ and $\bar{P}_n^{(1)}$ represents an acceptable approximation. (**d**) Example with a negative value of ρ, where $\bar{P}_n^{(1)}$ is close to $\widetilde{Q}_n$. Choice of parameters and calculated moments: $\alpha_1 = 1$, $\alpha_2 = 0$, and (**a**) $\lambda = 1.05, \mu = 0.95, N = 70, \rho = 0.88, \widetilde{\mu \pm \sigma} = (8.03 \pm 5.38, 5.31 \pm 4.83, 9.70 \pm 5.82)$, (**b**) $\lambda = 1.15$, $\mu = 0.85, N = 70, \rho = 2.95, \widetilde{\mu \pm \sigma} = (15.58 \pm 7.26, 14.08 \pm 7.84, 18.45 \pm 6.95)$, (**c**) $\lambda = 1.1$, $\mu = 0.9, N = 1000, \rho = 7.03, \widetilde{\mu \pm \sigma} = (177.0 \pm 29.1, 177.0 \pm 29.1, 181.8 \pm 28.6, 9.70 \pm 5.82)$, and (**d**) $\lambda = 0.9, \mu = 1.0, N = 100, \rho = -1.82, \widetilde{\mu \pm \sigma} = (3.95 \pm 3.13, 2.33 \pm 2.09, 4.21 \pm 3.32)$

of epidemics here. (For a reviews of the beginnings and the early development of modeling epidemics in the twentieth century see, e.g., [15, 16, 109]. More recent monographs are [17, 107, 108].) In addition, we mention a humoristic but nevertheless detailed deterministic analysis of models dealing with zombie infection of human society that is well worth reading [417].

SIS Model

A simple model suggested by Norman Bailey [32] was extended by George Weiss and Menachem Dishon [566] to produce the susceptive–infectious–susceptive (SIS) model of epidemiology: uninfected individuals denoted as susceptive are infected, become cured, and are susceptive again. Clearly, the model ignores two well known phenomena: (i) the possibility of long-lasting, even lifelong immunity, and (ii) the death of infected individuals killed by the disease. Nevertheless, the model is

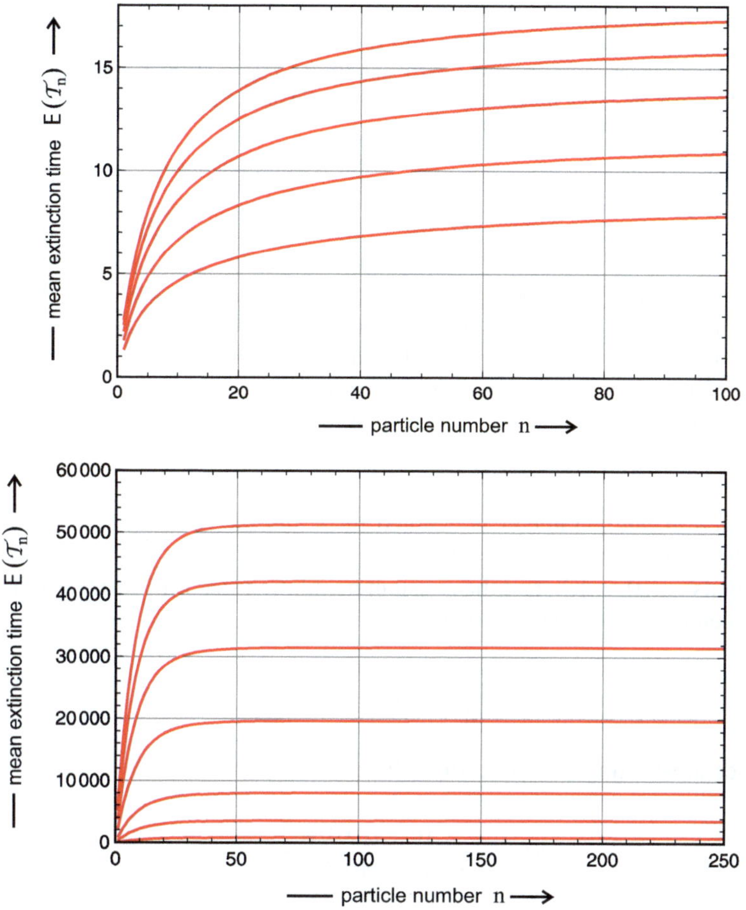

Fig. 5.14 Mean extinction times of logistic birth-and-death processes. Mean extinction times $E(\mathcal{T}_n)$ as functions of the initial number n of individuals in the population. *Upper*: Characteristic examples of short mean extinction times corresponding to region (iii, $\rho \ll 0$), where $E(\mathcal{T}_n)$ increases gradually up to $n = N$. *Lower*: Typical long extinction times in region (i, $\rho > 3$), where the extinction times become practically independent of n at values $n \ll N$. Choice of parameters: $N = 100$, $\alpha_1 = 1.0$, $\alpha_2 = 0.0$ with $(\lambda, \mu) = (0.99, 1.01), (0.975, 1.025), (0.95, 1.05), (0.9, 1.1)$, and $(0.8, 1.2)$ (*upper plot*, curves from top to bottom), and $N = 1000$, $\alpha_1 = 1.0$, $\alpha_2 = 0.0$ with $(\lambda, \mu) = (1.15, 1.00), (1.148, 1.00), (1.145, 1.00), (1.14, 1.00), (1.13, 1.00), (1.12, 1.00)$, and $(1.10, 1.00)$ (*lower plot*, curves from top to bottom)

interesting in its own right, because it is mathematically close to the Verhulst model. Susceptive and infectious individuals are denoted by S and I, respectively:

$$\begin{aligned} \mathsf{S} + \mathrm{I} &\xrightarrow{\lambda} 2\mathrm{I}\,, \\ \mathrm{I} &\xrightarrow{\mu} \mathsf{S}\,. \end{aligned} \tag{5.40a}$$

Table 5.5 Extinction times of logistic birth-and-death processes. Mean times to extinction $\mathrm{E}(\widetilde{\mathcal{T}})$ from the quasi-stationary distribution, and $\mathrm{E}(\mathcal{T}_N)$ from the highest state, for two ranges: (i) the region of long extinction times $\rho > 3$ (*left*), and (ii) the region of short extinction times $\rho < 0$ (*right*). All times are given in arbitrary time units [t]. Choice of parameters: $\alpha_1 = 1.0$, $\alpha_2 = 0.0$ and $N = 1000$ (*left*) and $N = 100$ (*right*)

Long extinction times					Short extinction times				
λ	μ	ρ	$\mathrm{E}(\widetilde{\mathcal{T}})$	$\mathrm{E}(\mathcal{T}_{1000})$	λ	μ	ρ	$\mathrm{E}(\widetilde{\mathcal{T}})$	$\mathrm{E}(\mathcal{T}_{100})$
1.150	1.00	4.74	51, 422	51, 467	0.999	1.001	−0.020	9.060	18.450
1.148	1.00	4.68	42, 228	42, 273	0.990	1.010	−0.198	8.200	17.325
1.145	1.00	4.59	31, 590	31, 635	0.975	1.025	−0.488	7.032	15.739
1.140	1.00	4.43	19, 752	19, 797	0.950	1.050	−0.952	5.606	13.684
1.130	1.00	4.11	8, 156	8, 203	0.900	1.100	−1.818	3.879	10.913
1.120	1.00	3.80	3, 633	3, 680	0.800	1.200	−3.333	2.304	7.868
1.100	1.00	3.16	911	958	0.700	1.300	−4.616	1.609	6.206

The infection rate parameter is denoted by λ and the recovery rate parameter by μ. If the number of infected individuals is $x(t) = [\mathrm{I}]$, the number of susceptive individuals is $s(t) = [\mathsf{S}]$, and the constant population size is $c = x(t) + s(t)$, we find for the deterministic kinetic equation

$$\frac{\mathrm{d}x}{\mathrm{d}t} = \lambda x(c - x) - \mu x\,, \tag{5.40b}$$

and after substituting $\varrho = \mu/\lambda$, the solution is

$$x(t) = \frac{(c - \varrho)x(0)\mathrm{e}^{\lambda(c-\varrho)t}}{c - \varrho + x(0)(\mathrm{e}^{\lambda(c-\varrho)t} - 1)}\,. \tag{5.40c}$$

The ODE (5.40b) sustains two stationary states:

1. $\bar{\Sigma}_{(1)}$ with $\bar{x}_{(1)} = c - \varrho$ and the eigenvalue $\bar{\lambda}_{(1)} = -(c - \varrho)$.
2. $\bar{\Sigma}_{(2)}$ with $\bar{x}_{(2)} = 0$ and the eigenvalue $\bar{\lambda}_{(2)} = c - \varrho$.

Stability analysis reveals that $\bar{\Sigma}_{(1)}$ is asymptotically stable for $c > \varrho$ and $\bar{\Sigma}_{(2)}$ for $c \leq \varrho$.

Strictly speaking, the deterministic result contradicts the stochastic expectation. Since the state $x = 0$ is absorbing, the number of I individuals must be zero in the limit of infinite time. In other words within the framework of the SIS model, every epidemic has to disappear some day. As we have seen in case of the stochastic logistic process, however, there exist quasi-stationary states, and the expectation value of the extinction time from these states $\mathrm{E}(\widetilde{\mathcal{T}})$ may be very long for $\varrho \ll 1$. Figure 5.16 illustrates the ultimate reason for the enormous scatter using numerical simulations: out of four trajectories, two die out and two survive within the time span considered. In the case of one trajectory, the enlargement shows that the process passed through the bottleneck of a single infected individual three times.

SIR Model

A somewhat more elaborate model considers susceptible (S), infectious (I), and refractory or immunized individuals (R), which cannot develop the disease for some period of time or even during their whole lifetime. An example of such an SIR model is illustrated in Fig. 5.15. In the original version [297], which is also the simplest, we are dealing with three species in two reactions:

$$\begin{aligned} \mathsf{S} + \mathrm{I} &\xrightarrow{\quad \lambda \quad} 2\mathrm{I}\,, \\ \mathrm{I} &\xrightarrow{\quad \mu \quad} \mathsf{R}\,. \end{aligned} \tag{5.41a}$$

In the language of chemical kinetics the SIR model considers two consecutive irreversible reactions, and the analysis would be rather trivial were the first reaction not autocatalytic. The concentrations are $[\mathsf{S}] = s$, $[\mathrm{I}] = x$, and $[\mathsf{R}] = r$, with the conservation relation $[\mathsf{S}] + [\mathrm{I}] + [\mathsf{R}] = c$. They satisfy the kinetic differential equations

$$\begin{aligned} \frac{\mathrm{d}s}{\mathrm{d}t} &= -\lambda x s\,, \\ \frac{\mathrm{d}x}{\mathrm{d}t} &= \lambda x s - \mu x\,. \end{aligned} \tag{5.41b}$$

Although naïvely one would not expect something special from a simple consecutive reaction network of one irreversible bimolecular and one irreversible monomolecular elementary step, the fact that the first reaction is autocatalytic in (5.41a) provides

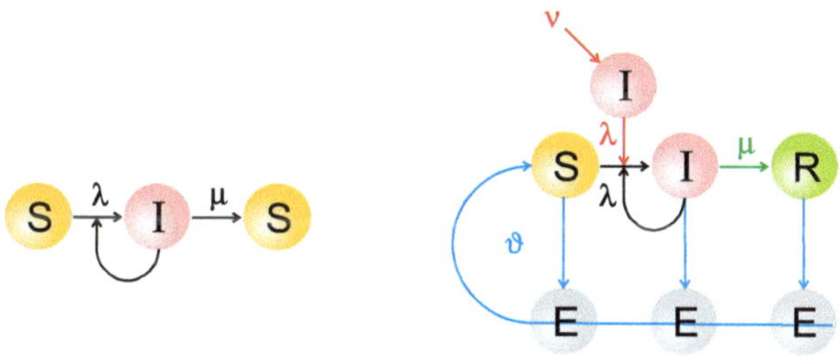

Fig. 5.15 Infection models in epidemiology. Theoretical epidemiology uses several models for the spreading of disease in populations. The SIS model shown on the *left* is about the simplest conceivable model. It distinguishes between susceptive (S) and infectious individuals (I), and considers neither immunity nor infections from outside the population. Infectious individuals become cured and are susceptible to infection again. The model on the *right* is abbreviated to the SIR model. It is more elaborate, considers recovered individuals (R), and corrects for both flaws of the SIS model mentioned above: (i) cured or recovered individuals have acquired lifelong immunity and cannot be infected again, and (ii) infection from outside the population is admitted. In the variant of SIR shown here all three classes of individuals are mortal and die with the same rate, giving rise to empty sites (E), which are instantaneously filled by susceptive individuals

the basis for a surprise (Fig. 5.17). The only stationary state of the deterministic system satisfies

$$\bar{x} = 0 \,, \qquad \bar{s} + \bar{r} = c \,. \tag{5.41c}$$

As a matter of fact this is not a single state but a whole 1D manifold of marginally stable states. In other words, all combinations of acceptable concentrations $\bar{s} \geq 0$,

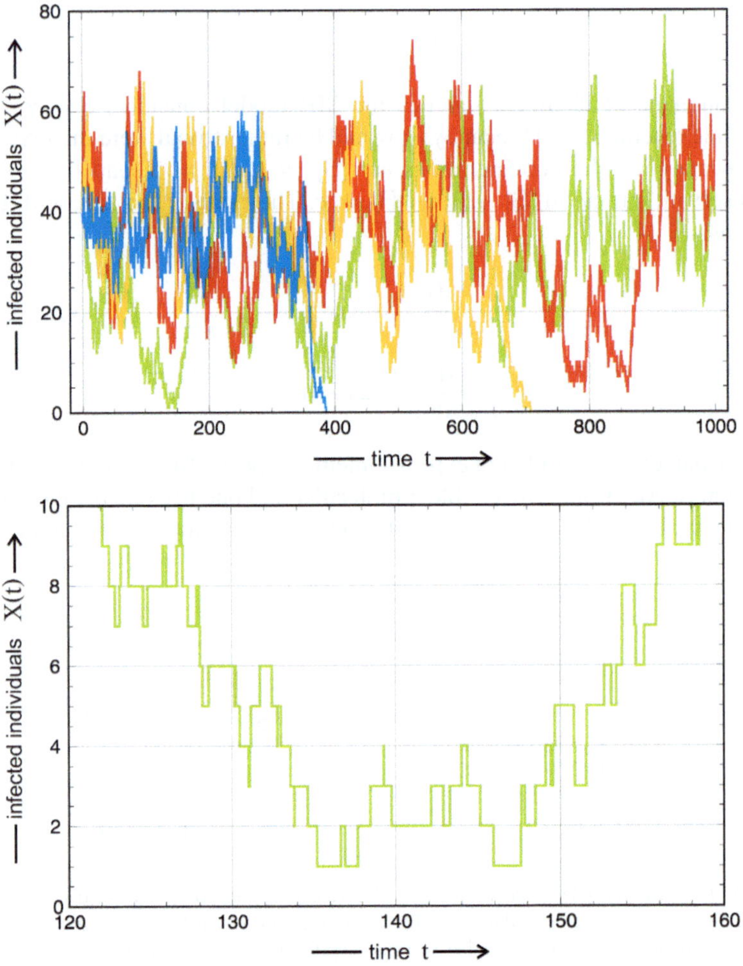

Fig. 5.16 The susceptible–infectious–susceptible (SIS) model in epidemics. The *upper plot* shows the number of infected individuals $\mathcal{X}(t)$, for four individual runs (*red*, *green*, *yellow*, *blue*). The values of $\mathcal{X}(t)$ fluctuate around the quasi-stationary state $\widetilde{\Sigma} : \bar{x} = c - \mu/\lambda = 40$. Whenever a trajectory reaches the absorbing state Σ_0, it stays there for ever, since no reaction can take place if $\bar{x} = 0$. The *lower plot* shows an enlargement of the green trajectory, a case where $\mathcal{X}(t)$ assumes the value $x = 1$ several times without being caught by extinction. Parameter choice: $\lambda = 0.00125$, $\mu = 0.2$, $\mathcal{X}(0) = 40$, and $\mathcal{S}(0) = 160$

$\bar{r} \geq 0$, and $\bar{s} + \bar{r} = c$ are solutions of (5.41b) with $\mathrm{d}s/\mathrm{d}t = 0$ and $\mathrm{d}x/\mathrm{d}t = 0$. The Jacobian matrix at the positions of the steady manifold has the two eigenvalues $\lambda_1 = \lambda\bar{s} - \mu$ and $\lambda_2 = 0$. The eigenvalue λ_1 is negative in the range $\bar{s} < \mu/\lambda$, implying stability with respect to fluctuations in x. Any fluctuation $\Delta x > 0$ returns to the steady state $\bar{x} = 0$, accompanied by some conversion of S into R. For $\bar{s} > \mu/\lambda$, the state $\bar{x} = 0$ is unstable in the sense that a fluctuation $\Delta x > 0$ will first increase, whereupon the trajectory progresses into the interior of the space $s + x + r = c$. Since there the system has no other steady state except the manifold $\bar{x} = 0$, the trajectory will go through a maximum value of x and eventually return to $\bar{x} = 0$. During the loop excursion of the trajectory (Fig. 5.17), S is converted into R.

In order to study the stochastic analogue of the marginally stable manifold, we performed simulations of the SIR model and recorded the number of S individuals, $\lim_{t\to\infty} \mathcal{S}(t) = \widehat{\mathcal{S}}$, which were never infected and therefore not converted into R because the infection died out before they were infected (Fig. 5.17). Table 5.6 shows the results recorded for several samples of one hundred trajectories each with the initial condition $\big(S(0) = 90, X(0) = 10, C = 100\big)$ and different values for the parameters λ and μ. The first three samples refer to identical parameter values $\lambda = 0.03$ and $\mu = 1.0$, and exhibit substantial scatter as should be expected. Interestingly, the standard deviation in the numbers of uninfected individuals is more than twice the standard deviation in the time of extinction. The following rows deal with the dependence of $\widehat{\mathcal{S}}$ on λ and μ: increasing the rate parameter of recovery μ leads to a pronounced increase in the number of uninfected individuals $\widehat{\mathcal{S}}$, whereas a growing infection parameter λ has the opposite effect, i.e., $\mathrm{E}(\widehat{\mathcal{S}})$ becomes smaller. The interpretation is straightforward: the greater the number of infected individuals $\mathcal{I}$, the more susceptive individuals are infected and ultimately turned into $\mathcal{R}$. The amount of $\mathcal{I}$ is the intermediate of two consecutive irreversible reactions and so grows larger when λ is increased and μ is reduced: more $\mathcal{I}$ implies that more $\mathcal{S}$ is converted into $\mathcal{R}$ and the fraction of uninfected individuals $\widehat{\mathcal{S}}$ becomes smaller. The dependence of the time to extinction on the intensity of infection $\mathcal{T}_0$ is less pronounced but clearly detectable: it decreases with increasing λ and with increasing μ, reflecting the fact that faster reactions lead to earlier disappearance of the infected fraction of the population.

Extended SIR Model

The last example is the extended SIR model shown in Fig. 5.15, which includes infection from an external source as well as direct recovery from the infected and

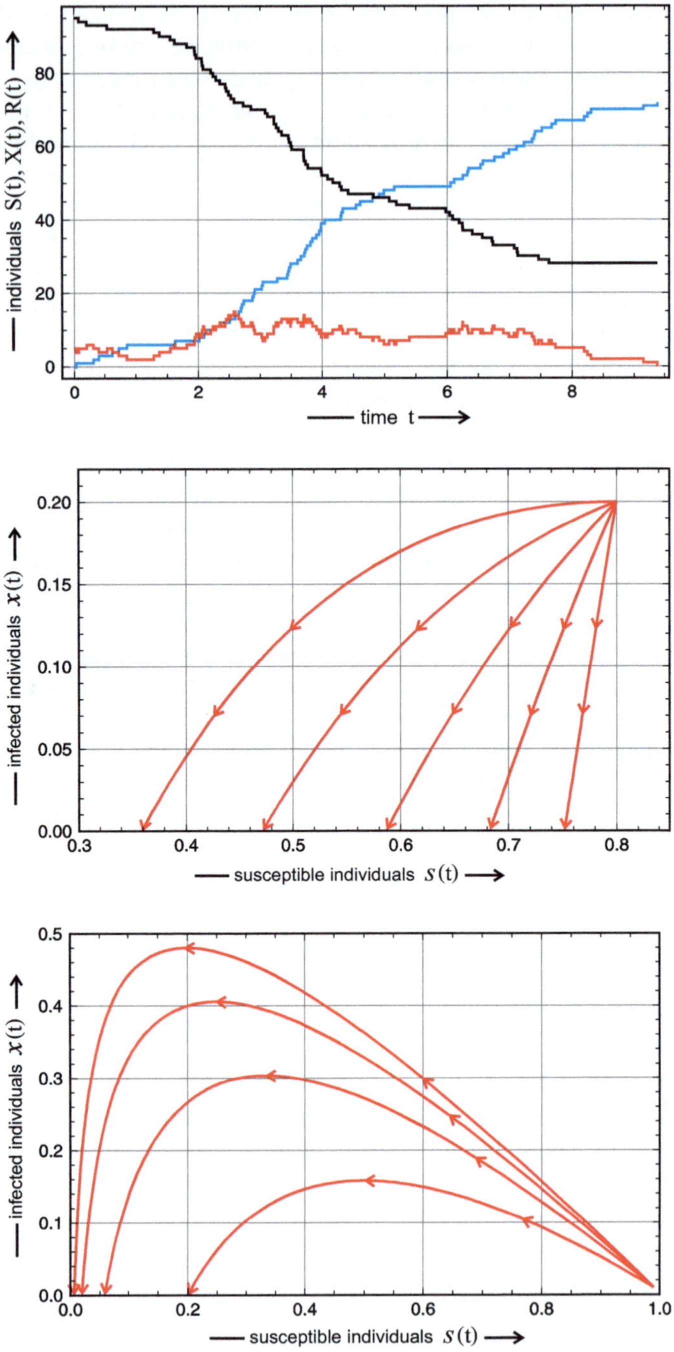

Fig. 5.17 The simple susceptible–infectious–recovered (SIR) model in epidemiology. Caption on next page

Fig. 5.17 The simple susceptible–infectious–recovered (SIR) model in epidemiology (see previous page). *Upper*: Typical trajectory of the simple SIR model. The stochastic variables denote the number of susceptible individuals $\mathcal{S}(t)$ (*black*), the number of infected individuals $\mathcal{X}(t)$ (*red*), and the number of recovered individuals $\mathcal{R}(t)$ (*blue*). The process ends at an absorbing boundary at time $t = t_{\max}$ when $\mathcal{X}(t)$ reaches the state Σ_0, no matter what the values of $\mathcal{S}(t_{\max})$ and $\mathcal{R}(t_{\max})$. In the case shown, we have $\mathcal{S}(t_{\max}) = 28$ and $\mathcal{R}(t_{\max}) = 72$. The *two other plots* show the stationary manifold at $\bar{x} = 0$ of the deterministic system. In the range $0 < \bar{s} < \mu/\lambda$, the state at the manifold is stable and any fluctuation $|\Delta x| > 0$ will be instantaneously compensated by the force driving the population towards the manifold $\bar{x} = 0$ (*middle plot*). The *lower plot* describes the more complicated situation in the range $\mu/\lambda < \bar{s} < c$. In the presence of a sufficiently high concentration of S, a fluctuation $|\Delta x| > 0$ is instantaneously amplified because of the autocatalytic process $\mathsf{S} + \mathsf{X} \rightarrow 2\mathsf{X}$. Since the only stationary state requires $\bar{x} = 0$, the trajectories progress in a loop and eventually end up on the manifold $\bar{x} = 0$, in the stable range $\bar{s} < \mu/\lambda$. Choice of parameters: (i) *upper plot*: $\lambda = 0.02$, $\mu = 1.0$, $\mathcal{X}(0) = 5$, $\mathcal{S}(0) = 95$,and $C = 100$, colors: $\mathcal{S}(t)$ *black*, $\mathcal{X}(t)$ *red*, and $\mathcal{R}(t)$ *blue*, (ii) *middle plot*: $\lambda = 0.25, 0.50, 0.75, 1.00$, and 1.25, $\mu = 1.0$, $x_0 = 0.2$, $s_0 = 0.8$, and $c = 1$, and (iii) *lower plot*: $\lambda = 2.0, 3.0, 4.0$, and 5.0, $\mu = 1.0$, $x_0 = 0.01$, $s_0 = 0.99$, and $c = 1$

the immunized state, in addition to infection and immunization [9]:

$$\begin{aligned}
\mathsf{S} + \mathrm{I} &\xrightarrow{\ \lambda\ } 2\mathrm{I}\,, \\
\mathsf{S} &\xrightarrow{\ \nu\ } \mathrm{I}\,, \\
\mathrm{I} &\xrightarrow{\ \mu\ } \mathsf{R}\,, \\
\mathrm{I} &\xrightarrow{\ \vartheta\ } \mathsf{S}\,, \\
\mathsf{R} &\xrightarrow{\ \vartheta\ } \mathsf{S}\,.
\end{aligned} \tag{5.42a}$$

The concentrations are $[\mathsf{S}] = s$, $[\mathrm{I}] = x$, and $[\mathsf{R}] = r$, and they satisfy the corresponding kinetic differential equations:

$$\begin{aligned}
\frac{\mathrm{d}s}{\mathrm{d}t} &= -(\lambda x + \nu)s + \vartheta(x + r)\,, \\
\frac{\mathrm{d}x}{\mathrm{d}t} &= (\lambda x + \nu)s - (\mu + \vartheta)x\,, \\
\frac{\mathrm{d}r}{\mathrm{d}t} &= \mu x - \vartheta r\,.
\end{aligned}$$

Table 5.6 Uninfected individuals and extinction times in the simple SIR model. The table presents mean numbers of uninfected individuals $E(\widehat{\mathcal{S}})$ and mean times to extinction $E(\mathcal{T}_0)$, together with their standard deviations $\sigma(\widehat{\mathcal{S}})$ and $\sigma(\mathcal{T}_0)$, respectively, for different values of the parameters λ and μ. Each sample consists of 100 independent recordings, and (2.115) and (2.118) were used to compute sample means and variances. Initial conditions: $S(0) = 90$, $X(0) = 10$, $C = 100$

Sample	Parameters		Uninfected individuals		Times to extinction	
	λ	μ	$E(\widehat{\mathcal{S}})$	$\sigma(\widehat{\mathcal{S}})$	$E(\mathcal{T}_0)$	$\sigma(\mathcal{T}_0)$
1	0.03	1.00	6.07	3.63	7.01	1.69
2	0.03	1.00	6.29	3.63	6.78	1.47
3	0.03	1.00	6.37	4.36	7.11	1.83
1–3	0.03	1.00	6.24	3.88	6.97	1.67
4	0.03	1.25	12.42	6.68	6.09	1.54
5	0.03	1.50	20.88	10.55	5.39	1.45
6	0.04	1.50	8.79	4.53	4.78	1.14
7	0.05	1.50	4.17	3.16	4.46	0.69

The ODE satisfies a conservation relation $[\mathsf{S}] + [\mathsf{I}] + [\mathsf{R}] = s + x + r = c$, which allows for a reduction of the variables from three to two:

$$\begin{aligned} \frac{\mathrm{d}s}{\mathrm{d}t} &= -(\lambda x + \nu)s + \vartheta(c - s)\,, \\ \frac{\mathrm{d}x}{\mathrm{d}t} &= (\lambda x + \nu)s - (\mu + \vartheta)x\,. \end{aligned} \tag{5.42b}$$

The dynamical system sustains two stationary states $\bar{\Sigma}_1$ and $\bar{\Sigma}_2$ with the concentrations

$$\begin{aligned} \bar{s}_{1,2} &= \frac{1}{2\lambda\vartheta}\Bigg(\vartheta(\lambda c + \mu + \nu + \vartheta) + \mu\nu \\ &\qquad \mp \sqrt{\big(\vartheta(\lambda c + \mu + \nu + \vartheta) + \mu\nu\big)^2 - 4\vartheta^2\lambda c(\mu + \vartheta)}\Bigg)\,, \\ \bar{x}_{1,2} &= \frac{1}{2\lambda(\mu + \vartheta)}\Bigg(\vartheta(\lambda c - \mu - \nu - \vartheta) - \mu\nu \\ &\qquad \pm \sqrt{\big(\vartheta(\lambda c - \mu - \nu - \vartheta) - \mu\nu\big)^2 - 4\vartheta^2\lambda c(\mu + \vartheta)}\Bigg)\,. \end{aligned} \tag{5.42c}$$

The state $\bar{\Sigma}_1 = (\bar{s}_1, \bar{x}_1)$ is the only physically acceptable stationary state since we find $\bar{s}_2 > c$ and $\bar{x}_2 < 0$ (Fig. 5.18).

The calculation of the 2×2 Jacobian matrix is straightforward, giving

$$J = \begin{pmatrix} -(\lambda\bar{x} + \nu + \vartheta) & -\lambda\bar{s} \\ \lambda\bar{x} + \nu & \lambda\bar{s} - (\mu + \vartheta) \end{pmatrix}\,,$$

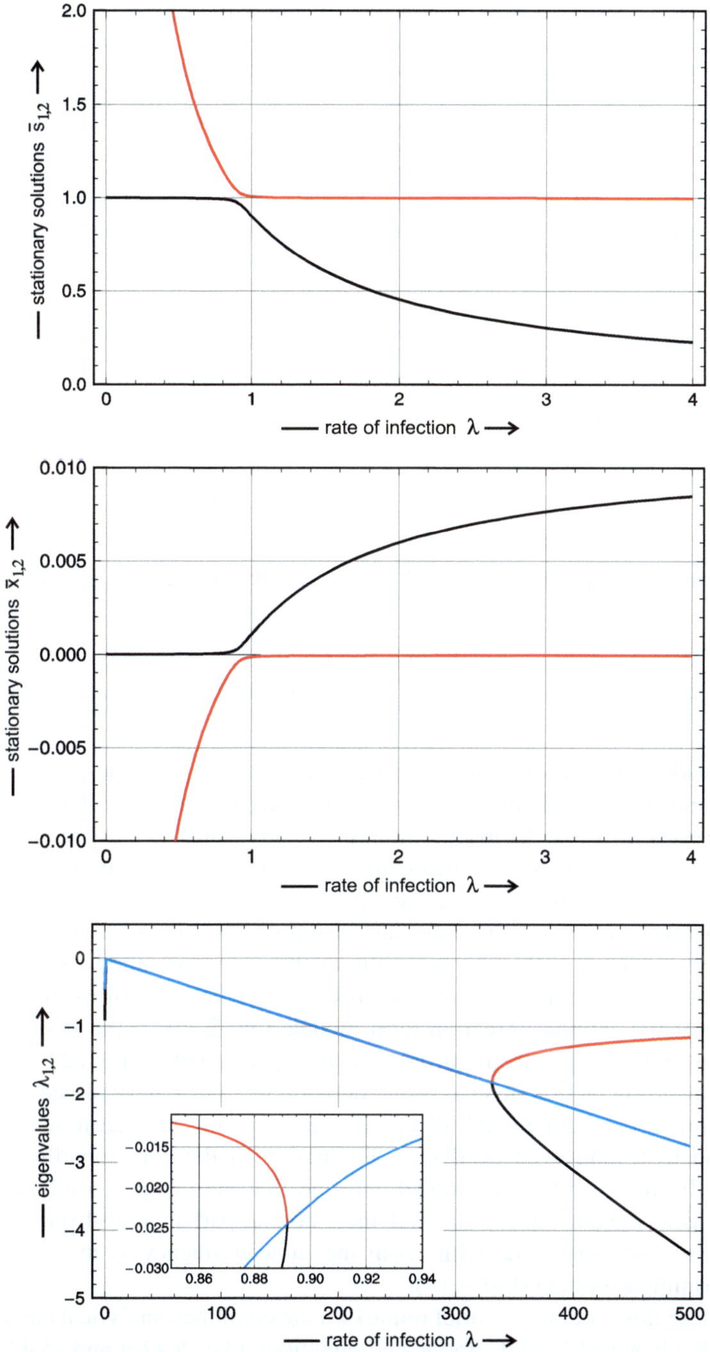

Fig. 5.18 The extended susceptible–infectious–recovered (SIR) model in epidemics. Caption on next page

Fig. 5.18 The extended susceptible–infectious–recovered (SIR) model in epidemics (see previous page). *Upper*: Concentration of susceptible individuals as a function of the infection rate parameter λ. The solution $\bar{s}_1$ (*black*) is the physically acceptable solution since $\bar{s}_2 > c$ (*red*), and the two solutions show avoided crossing near the parameter value $\lambda \approx 0.9$. *Middle*: Analogous plot for the solution $\bar{x}_{1,2}$. Once again $\bar{x}_1$ (*black*) is the acceptable solution since $\bar{x}_2 < 0$ (*red*). *Lower*: The two eigenvalues of the Jacobian matrix $\lambda_{1,2}$ (5.42d) as a function of the infection parameter λ. The *three curves* are λ_1 (*red*), λ_2 (*black*), and $\Re(\lambda_{1,2}) = (\lambda_1 + \lambda_2)/2$ (*black*). In the entire range $0.89 < \lambda < 330$, the two eigenvalues form a complex conjugate pair with negative real part. The *insert* shows an enlargement of the left-hand bifurcation. Parameter choice: $\mu = 0.9$, $\nu = 1 \times 10^{-5}$, $\vartheta = 0.01$ and $c = 1.0$

and two eigenvalues are obtained in the conventional manner:

$$\lambda_{1,2} = \frac{1}{2}\Bigg(-\big(2\vartheta + \mu + \nu - \lambda(\bar{s} - \bar{x})\big) \pm \sqrt{(\mu - \nu)^2 + \big(\lambda(\bar{s} - \bar{x})\big)^2 - 2\lambda\bar{s}(\mu + \nu) - 2\lambda\bar{x}(\mu - \nu)}\Bigg) . \tag{5.42d}$$

Inspection of the plots in Fig. 5.18 shows the basic features of the dynamical model:

(i) The state $\bar{\Sigma}_1 = (\bar{s}_1, \bar{x}_1)$ is the only physically acceptable state.
(ii) It is asymptotically stable since the eigenvalues have a negative real part, i.e., $\Re(\lambda_{1,2}) < 0$.

Over a wide range of values of the parameter λ, the eigenvalues form a complex conjugate pair and we therefore expect an approach to the steady state with damped oscillations (Fig. 5.19). A closer look at the steady states as a function of the birth parameter λ reveals an interesting detail: the eigenvalues of the two states approach each other very closely at some critical parameter value $\lambda_{\mathrm{cr}} \approx 0.9$ in Fig. 5.18, and then separate again, whereupon the curves appear to exchange their global shapes. This phenomenon is well known in quantum physics and is called *avoided crossing*.

The stochastic simulation of the SIR model is straightforward (Fig. 5.19). Because of the external infection term modeled by $\mathsf{S} \to \mathrm{I}$, the infection cannot die out as long as $[\mathsf{S}] > 0$, and instantaneous replacement expressed by $\mathrm{I} \to \mathsf{S}$ and $\mathsf{R} \to \mathsf{S}$ avoids depletion of susceptible individuals. As we have already seen in the case of the Brusselator model, damped oscillations in the deterministic approach may lead to a long-lived oscillating fluctuation in the corresponding stochastic process. In the extended SIR model, we find that the damped oscillations of the deterministic dynamical system find their counterpart in long-lived fluctuations around the stationary state, with about the same frequency as the oscillations in the deterministic system (Fig. 5.19).

We close this section by a brief remark on the combined analytical and numerical approach advocated here. Complicated equations like (5.42c) and (5.42d) can be derived by computer-based symbolic computation. Of course, in most cases they are not useful for further analytical work, but they are exact and provide a useful

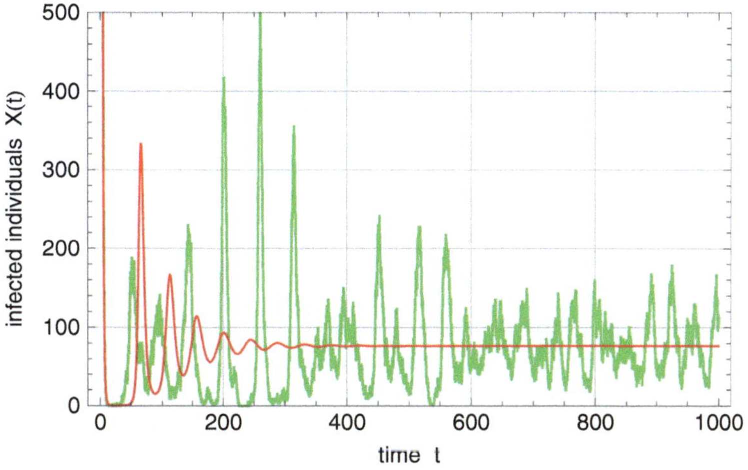

Fig. 5.19 The extended susceptible–infectious–recovered (SIR) model in epidemics. The plot shows a stochastic trajectory $\mathcal{X}(t)$ (*green*) that fluctuates around the deterministic stationary state at $\bar{X} = 77$, and the corresponding deterministic solution curve $X(t)$ (*red*), which shows damped oscillations. Interestingly, the frequency of these oscillations is very close to the mean frequency of the stochastic fluctuations. Parameter choice: $\mu = 0.9$, $\nu = 1 \times 10^{-5}$, $\vartheta = 0.01$. For the stochastic plot $\lambda = 3 \times 10^{-5}$, $\mathcal{X}(0) = 1000$, $\mathcal{S}(0) = 9000$, and $C = 10,000$. For the analogue deterministic solution $\lambda = 3$, $x(0) = 0.1$, $s(0) = 0.9$, and $c = 1.0$ [9]

basis for a continuation by numerical computation or test cases for approximations, as we have shown here. Finally, we would like to mention that logistic birth-and-death as well as other epidemic models were used by Nico van Kampen as examples showing the applicability of his expansion of master equations [541, pp. 251, 265].

5.2.5 Branching Processes

According to David Kendall's historical accounts of the beginnings of stochastic thinking in population mathematics [293, 294], the term *branching process* was coined rather late on by Kolmogorov and Dmitriev in their 1947 paper [313]. However, the interest in the stochasticity of evolving populations is much older and focussed on the genealogy of human males, which is reflected by the development of family names or *surnames* in the population. The stock of family names is typically *eroded* in the sense that there is a steady disappearance of families, particularly in small communities. The problem was clearly stated 1873 in a book by Alphonse de Candolle [102] and was brought up by Sir Francis Galton after he had read de Candolle's book.

The first rigorous mathematical analysis of a problem by means of a branching process is commonly attributed to Galton and the Reverend Henry William Watson

[562], and the Galton–Watson process named after them has become a standard problem in the theory of branching processes. Apparently, Galton and Watson were not aware of previous work on this topic [250], carried out and published almost thirty years earlier by Jules Bienaymé [49]. Most remarkably, Bienaymé already discussed the criticality theorem, which expresses the different behavior of the Galton–Watson process for $m < 1$, $m = 1$, and $m > 1$, where m denotes the expected or mean number of sons per father. The three cases were called *subcritical*, *critical*, and *supercritical*, respectively, by Kolmogorov [312]. Watson's original work contained a serious error in the analysis of the supercritical case and this was not detected or reported for more than fifty years until Johan Steffensen published his work on this topic [505].

In the years following 1940, the Galton–Watson model received plenty of attention because of the analogies between genealogies and nuclear chain reactions. In addition, mathematicians became generally more interested in probability theory and stochasticity. The pioneering work on nuclear chain reactions and criticality of nuclear reactors was carried out by Stan Ulam at the Los Alamos National Laboratory [143–146, 246]. Many other applications to biology and physics were found, and branching processes have since been intensively studied. By now, it seems, we have a clear picture of the Galton–Watson process and its history [294].

The Galton–Watson Process

A *Galton–Watson process* [562] counts objects which are derived from objects of the same kind by reproduction. These objects may be neutrons, bacteria, higher organisms, or men as in the family name genealogy problem. The Galton–Watson process is the simplest possible description of consecutive reproduction and falls into the class of branching processes. We consider a population of $\mathcal{Z}_n$ individuals in generation n that reproduce asexually and independently. Only the population sizes of successive generations are recorded, thus forming a sequence of random variables $\mathcal{Z}_0, \mathcal{Z}_1, \mathcal{Z}_2, \ldots$, with $P(\mathcal{Z}_i = k) = p_k$ for $k \in \mathbb{N}$. A question of interest, for example, is the extinction of a population in generation n, and this simply means $\mathcal{Z}_n = 0$, from which it follows that all random variables in future generations are zero: $\mathcal{Z}_{n+1} = 0$ if $\mathcal{Z}_n = 0$. Indeed, the extinction or disappearance of aristocratic family names was the problem that Galton wanted to model by means of a stochastic process. The following presentation and analysis are adapted from two books [29, 240].

The Galton–Watson process describes an evolving population of particles or individuals, and it may sometimes be useful, although not always necessary, to define a time axis. The process starts with $\mathcal{Z}_0$ particles at time $t = 0$, each of which produces a random number of offspring at time $t = 1$, independently of the others, according to the probability mass function (pmf) $f(k) = p_k$ with $k \in \mathbb{N}$, $p_k \geq 0$ and $\sum_{k=0}^{\infty} p_k = 1$. The total number $\mathcal{Z}_1$ of particles in the first generation is the sum of all random variables counting the offspring of the $\mathcal{Z}_0$ individuals of generation $\mathcal{Z}_0$, where each number was drawn according to the probability mass function $f(p_k)$. The first generation produces $\mathcal{Z}_2$ particles at time $t = 2$ by the same rules, the second generation gives rise to the third with $\mathcal{Z}_3$ particles at time $t = 3$, and so on.

Since discrete times t_n are equivalent to the numbers of generations n, we shall refer only to generations in the following.

In mathematical terms the Galton–Watson process is a Markov chain on the nonnegative integers, $\mathcal{Z}_n$ with $n \in \mathbb{N}$, where the Markov property implies that knowing $\mathcal{Z}_i$ provides full information on all future generations $\mathcal{Z}_j$ with $j > i$. The random variable $\mathcal{Z}_i$ in generation i is characterized by its probability mass function $f_{\mathcal{Z}_i}(k)$. The transition probabilities for consecutive generations satisfy

$$W(j|i) = P(\mathcal{Z}_{n+1} = j|\mathcal{Z}_n = i) = \begin{cases} p_j^{*i}, & \text{if } i \geq 1\,,\ j \geq 0\,, \\ \delta_{0,j}\,, & \text{if } i = 0\,,\ j \geq 0\,, \end{cases} \tag{5.43a}$$

where δ_{ij} is the *Kronecker delta*, p_j^{*i} with $i, j \in \mathbb{N}$ is the i-fold convolution of p_j (see Sect. 3.1.6), and i is the number of individuals in generation n. These transition probabilities constitute the Markov property since the full future development is given by (5.43a). Accordingly, the probability mass function $f_{\mathcal{Z}}(k) = p_k$ is the only datum of the process. The use of the convolution of the probability distribution is an elegant mathematical trick for rigorous analysis of the problem. Convolutions are quite difficult to handle explicitly, as we shall see in the case of the generating function. Nowadays, one can use computer-assisted symbolic computation, but during Galton's lifetime in the second half of the nineteenth century, handling higher convolutions was quite hopeless.

The number of offspring in the nth generation produced by a single parent is a random variable $\mathcal{Z}_n^{(1)}$, where the superscript indicates $\mathcal{Z}_0 = 1$. In general, we shall write for the branching process $(\mathcal{Z}_n^{(i)}\,;\ n, i \in \mathbb{N})$ whenever the process starts with i particles in generation zero. Since $i = 1$ is by far the most common case, we write simply $\mathcal{Z}_n$ instead of $\mathcal{Z}_n^{(1)}$. Equation (5.43a) says that $\mathcal{Z}_{n+k} = 0$, $\forall\, k \geq 0$, if $\mathcal{Z}_n = 0$. Accordingly, the state $\mathcal{Z} = 0$ is absorbing and reaching $\mathcal{Z} = 0$ is tantamount to becoming extinct.

In order to analyze the process, we shall make use of the probability generating function

$$g(s) = \sum_{k=0}^{\infty} p_k s^k\,, \quad |s| \leq 1\,, \tag{5.43b}$$

where s is complex in general, but we shall assume here $s \in \mathbb{R}^1$. In addition, we define the iterates of the generating function:

$$g_0(s) = s\,, \quad g_1(s) = g(s)\,, \quad g_{n+1}(s) = g\big(g_n(s)\big)\,, \quad n = 1, 2, \ldots\,. \tag{5.43c}$$

Expressed in terms of transition probabilities, the generating function is of the form

$$\sum_{j=0}^{\infty} W(j|1)s^j = g(s)\,, \quad \sum_{j=0}^{\infty} W(j|i)s^j = \big(g(s)\big)^i\,, \quad i \geq 1\,. \tag{5.43d}$$

If we denote the n-step transition probability by $W_n(j|i)$ and make use of the Chapman–Kolmogorov equation, we obtain

$$\begin{aligned}\sum_{j=0}^{\infty} W_{n+1}(j|1)s^j &= \sum_{j=0}^{\infty}\sum_{k=0}^{\infty} W_n(k|1)W(j|k)s^j \\ &= \sum_{k=0}^{\infty} W_n(k|1)\sum_{j=0}^{\infty} W(j|k)s^j \\ &= \sum_{k=0}^{\infty} W_n(k|1)\big(g(s)\big)^k .\end{aligned}$$

Writing $g_{(n)} = \sum_j W_n(j|1)s^j$, the last equation shows that

$$g_{(n+1)}(s) = g_{(n)}\big(g(s)\big) ,$$

which yields the fundamental relation

$$g_{(n)}(s) = g_n(s) , \tag{5.43e}$$

and by making use of (5.43d), we find

$$\sum_{j=0}^{\infty} W_n(j|i)s^j = \big(g_n(s)\big)^i . \tag{5.43f}$$

Equation (5.43e) can be expressed in words by saying that the generating function of $\mathcal{Z}_n$ is the n-iterate $g_n(s)$. It provides a powerful tool for calculating the generating function. As stated in (5.43a), the probability distribution of $\mathcal{Z}_n$ is obtained as the n th convolution or iterate of $g(s)$. The explicit form of an n th convolution is hard to compute, and the true value of (5.43e) lies in the calculation of the moments of $\mathcal{Z}_n$ and in the possibility of deriving asymptotic laws for large n.

For the purpose of illustration, we present the first iterates of the simplest useful generating function, namely,

$$g(s) = p_0 + p_1 s + p_2 s^2 .$$

The first convolution $g_2(s) = g\big(g(s)\big)$ already contains ten terms:

$$\begin{aligned}g_2(s) &= p_0 + p_1(p_0 + p_1 s + p_2 s^2) + p_2(p_0 + p_1 s + p_2 s^2)^2 \\ &= p_0 + p_0 p_1 + p_0^2 p_2 + (p_1^2 + 2p_0 p_1 p_2)s \\ &\quad +(p_1 p_2 + p_1^2 p_2 + 2p_0 p_2^2)s^2 + 2p_1 p_2^2 s^3 + p_2^3 s^4 .\end{aligned}$$

The next convolution $g_3(s)$ already contains nine constant terms that contribute to the probability of extinction $g_n(0)$, and $g_4(s)$ already 29 terms (for a numerical calculation see Fig. 5.20). It is nevertheless straightforward to compute the moments of the probability distributions from the generating function:

$$\frac{\partial g(s)}{\partial s} = \sum_{k=0}^{\infty} k p_k s^{k-1} , \qquad \left.\frac{\partial g(s)}{\partial s}\right|_{s=1} = E(\mathcal{Z}_1) = m , \tag{5.43g}$$

$$\frac{\partial^2 g(s)}{\partial s^2} = \sum_{k=0}^{\infty} k(k-1) p_k s^{k-2} , \qquad \left.\frac{\partial^2 g(s)}{\partial s^2}\right|_{s=1} = E(\mathcal{Z}_1^2) - m ,$$

$$\mathrm{var}(\mathcal{Z}_1) = \left.\frac{\partial^2 g(s)}{\partial s^2}\right|_{s=1} + m - m^2 = \sigma^2 . \tag{5.43h}$$

Next we calculate the moments of the distribution in higher generations and differentiate the last expression in (5.43c) at $|s| = 1$:

$$\begin{aligned}\left.\frac{\partial g_{n+1}(s)}{\partial s}\right|_{s=1} &= \frac{\partial g(s)}{\partial s}\Big(g_n(s)\big|_{s=1}\Big)\left.\frac{\partial g_n(s)}{\partial s}\right|_{s=1} \\ &= \left.\frac{\partial g(s)}{\partial s}\right|_{s=1}\left.\frac{\partial g_n(s)}{\partial s}\right|_{s=1} ,\end{aligned}$$

$$E(\mathcal{Z}_{n+1}) = E(\mathcal{Z})E(\mathcal{Z}_n) , \quad \text{whence } E(\mathcal{Z}_n) = m^n , \tag{5.43i}$$

by induction. Provided the second derivative of the generating function at $|s| = 1$ is finite, (5.43c) can be differentiated twice:

$$\left.\frac{\partial^2 g_{n+1}(s)}{\partial s^2}\right|_{s=1} = \left.\frac{\partial g(s)}{\partial s}\right|_{s=1}\left.\frac{\partial^2 g_n(s)}{\partial s^2}\right|_{s=1} + \left.\frac{\partial^2 g(s)}{\partial s^2}\right|_{s=1}\left(\left.\frac{\partial g_n(s)}{\partial s}\right|_{s=1}\right)^2 .$$

Then $\partial^2 g(s)/\partial s^2\,\big|_{s=1}$ is obtained by repeated application and the result is

$$\mathrm{var}(\mathcal{Z}_n) = E(\mathcal{Z}_n^2) - E(\mathcal{Z}_n)^2 = \begin{cases} \dfrac{\sigma^2 m^n (m^n - 1)}{m(m-1)} , & \text{if } m \neq 1 , \\ n\sigma^2 , \quad \text{if } m = 1 . \end{cases} \tag{5.43j}$$

Thus, we have derived $E(\mathcal{Z}_n) = m^n$ and, provided that $\mathrm{var}(\mathcal{Z}_1) < \infty$, we have also derived the variances in the different generations, as given by (5.43j).

Two more assumptions are made to simplify the analysis:

(i) Neither the probabilities p_0 and p_1 nor their sum are equal to one, i.e., $p_0 < 1$, $p_1 < 1$, and $p_0 + p_1 < 1$, and this implies that $g(s)$ is strictly convex on the unit interval $0 \leq s \leq 1$.

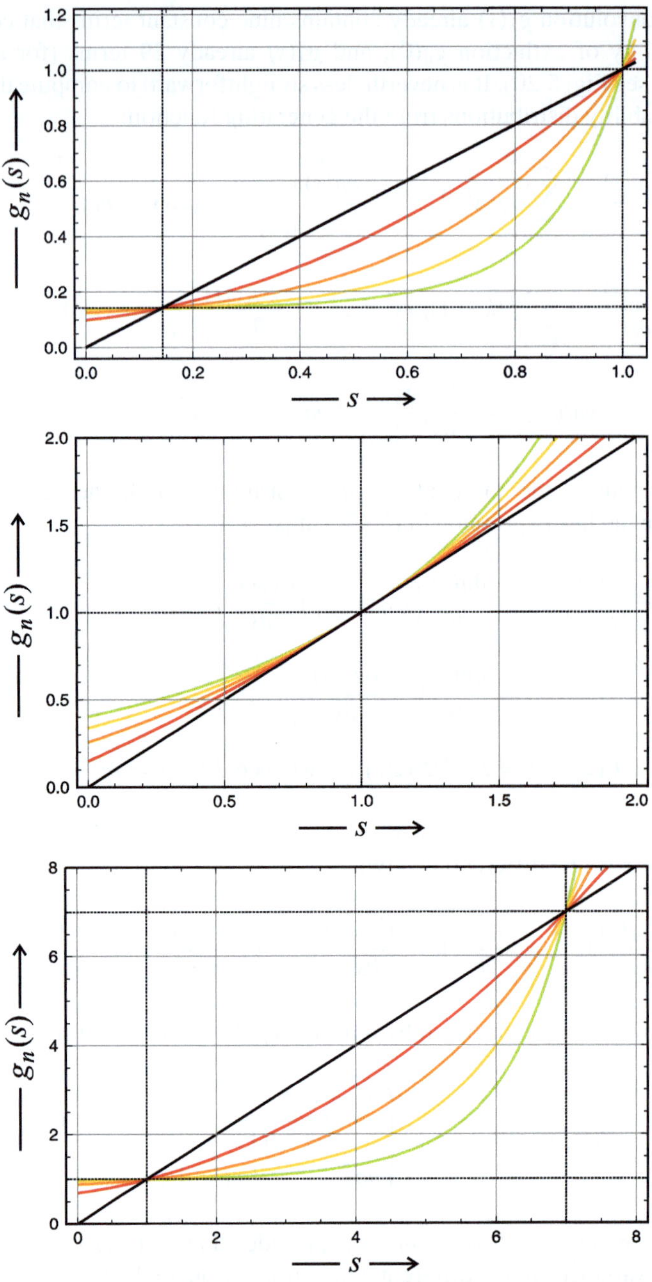

Fig. 5.20 Calculation of extinction probabilities for the Galton–Watson process. *Individual curves* show the iterated generating functions of the Galton–Watson process: $g_0(s) = s$ (*black*), $g_1(s) = g(s) = p_0 + p_1 s + p_2 s^2$ (*red*), $g_2(s)$ (*orange*), $g_3(s)$ (*yellow*), and $g_4(s)$ (*green*), for different probability densities $\mathbf{p} = (p_0, p_1, p_2)$. Choice of parameters: supercritical case (*upper*) $\mathbf{p} = (0.1, 0.2, 0.7)$, $m = 1.6$; critical case (*middle*) $\mathbf{p} = (0.15, 0.7, 0.15)$, $m = 1$; subcritical case (*lower*) $\mathbf{p} = (0.7, 0.2, 0.1)$, $m = 0.4$

(ii) The expectation value $E(\mathcal{Z}_1) = \sum_{k=0}^{\infty} kp_k$ is finite, and from the finiteness of the expectation value, it follows that $\partial g/\partial s|_{s=1}$ is also finite since $|s| \leq 1$.

Finally we can now consider Galton's problem of the extinction of family names. The straightforward definition of extinction is given in terms of a random sequence $(\mathcal{Z}_n;\, n = 0, 1, 2, \ldots, \infty)$, which consists of zeros except for a finite number of positive integer values at the beginning of the series. The random variable $\mathcal{Z}_n$ is integer valued, so extinction is tantamount to the event $\mathcal{Z}_n \to 0$. The relation $W(\mathcal{Z}_{n+1} = 0|\mathcal{Z}_n = 0) = 1$ implies the equality

$$\begin{aligned} P(\mathcal{Z}_n \to 0) &= P(\mathcal{Z}_n = 0\,,\ \text{for some}\ n) \\ &= P\big((\mathcal{Z}_1 = 0) \cup (\mathcal{Z}_2 = 0) \cup \ldots \cup (\mathcal{Z}_n = 0)\big) \\ &= \lim_{n\to\infty} P\big((\mathcal{Z}_1 = 0) \cup (\mathcal{Z}_2 = 0) \cup \ldots \cup (\mathcal{Z}_n = 0)\big) \\ &= \lim_{n\to\infty} P(\mathcal{Z}_n = 0) = \lim_{n\to\infty} g_n(0)\,, \end{aligned} \tag{5.43k}$$

and the fact that $g_n(0)$ is a non-decreasing function of n (see also Fig. 5.21).

We define a probability of extinction $q = P(\mathcal{Z}_n \to 0) = \lim g_n(0)$ and show that, for$m = E(\mathcal{Z}_1) \leq 1$, the probability of extinction satisfies $q = 1$, and the family names disappear in finite time. For $m > 1$, however, an extinction probability $q < 1$

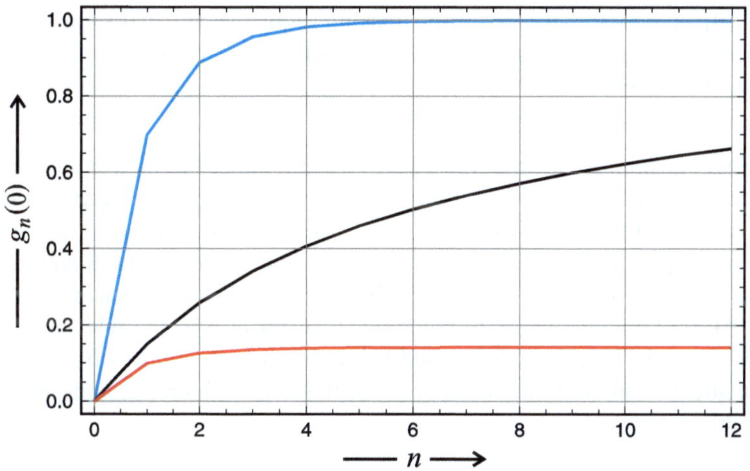

Fig. 5.21 Extinction probabilities in the Galton–Watson process. Extinction probabilities for the three Galton–Watson processes discussed in Fig. 5.20. The supercritical process ($\mathbf{p} = (0.1, 0.2, 0.7)$, $m = 1.6$ *red*) is characterized by a probability of extinction of $q = \lim g_n < 1$, leaving room for a certain probability of survival, whereas both the critical ($\mathbf{p} = (0.15, 0.7, 0.15)$, $m = 1$ *black*) and the subcritical process ($\mathbf{p} = (0.7, 0.2, 0.1)$, $m = 0.4$, *blue*) lead to certain extinction, i.e., $q = \lim g_n = 1$. In the critical case, we observe much slower convergence than in the super- or subcritical cases, representing a nice example of *critical slowing down*

is the unique solution of the equation

$$s = g(s) \,, \quad \text{for } 0 \le s < 1 \,. \tag{5.43l}$$

It is straightforward to show by induction that $g_n(0) < 1$, $n = 0, 1, \ldots$. From (5.43k), we know that

$$0 = g_n(0) \le g_1(0) \le g_2(0) \le \ldots \le q = \lim g_n(0) \,.$$

Making use of $g_{n+1}(0) = g\big(g_n(0)\big)$ and $\lim g_n(0) = \lim g_{n+1}(0) = q$, we deduce that $q = g(q)$ for $0 \le q \le 1$ is trivially fulfilled for $q = 1$ since $g(1) = 1$:

(i) If $m \le 1$, then $\partial g(s)/\partial s < 1$ for $0 \le s < 1$. We use the *law of the mean*,[17] express $g(s)$ in terms of $g(1)$, and find $g(s) > s$ for $m \le 1$ in the entire range $0 \le s < 1$. Hence, there is only the trivial solution $q = g(q)$ in the physically acceptable domain with $q = 1$, and extinction is certain. □

(ii) If $m > 1$, then $g(s) < s$ for s slightly less than one, because we have $\partial g/\partial s|_{s=1} = m > 1$, whereas for $s = 0$, we have $g(0) > 0$, so we have at least one solution $s = g(s)$ in the half-open interval $[0, 1[$. Assuming there are two solutions, for example, s_1 and s_2 with $0 \le s_1 < s_2 < 1$, then Rolle's theorem named after the French mathematician Michel Rolle would require the existence of ξ and η with $s_1 < \xi < s_2 < \eta < 1$ such that $\partial g(s)/\partial s|_{s=\xi} = \partial g(s)/\partial s|_{s=\eta} = 1$, but this contradicts the fact that $g(s)$ is strictly convex. In addition, $\lim g_n(0)$ cannot be one, because $(g_n(0); n = 0, 1, \ldots)$ is a non-decreasing sequence. If $g_n(0)$ were slightly less than one, then $g_{n+1}(0) = g\big(g_n(0)\big)$ would be less than $g_n(0)$ and the series would be decreasing. Accordingly, $q < 1$ is the unique solution of (5.43l) in $[0, 1[$. □

The answer is simple and straightforward: when a father has on average one son or less, the family name is doomed to disappear, but when he has more than one son, there is a finite probability of survival $0 < 1-q < 1$, which of course increases with increasing expectation value m, the average number of sons. The Reverend Henry William Watson correctly deduced that the extinction probability is given by a root of (5.43l). He failed, however, to recognize that, for $m > 1$, the relevant root is the one with $q < 1$ [192, 562]. It is remarkable that it took almost fifty years for the mathematical community to detect an error that has such drastic consequences for the result.

[17]The law of the mean expresses the difference in the values of a function $f(x)$ in terms of the derivative at one particular point $x = x_1$ and the difference in the arguments

$$f(b) - f(a) = (b-a) \left. \frac{\partial f}{\partial x} \right|_{x=x_1} \,, \quad a < x < b \,.$$

The law of the mean is satisfied for at least one point x_1 on the arc between a and b.

Reproduction and Mutation as Multitype Branching Process

The kinetics of reproduction and mutation has been studied in great detail. *In vitro* evolution provides an additional access to population dynamics that is easily traced down to the molecular level, where correct replication and mutation are understood as parallel chemical reactions [48, 130–132]. Evolution of RNA molecules in cell-free replication assays and reproduction of viroids and RNA viruses is presently understood at the same mechanistic resolution as other chemical reactions. We present here a stochastic treatment of the problem as a branching process and follow the derivation and analysis described in [105]. In particular, we shall consider the development of the population in discrete time steps corresponding to non-overlapping generations, and move on later to continuous time. Finally, we compare the discrete and continuous stochastic models with their deterministic analogues (for another model based on master equations, see Sect. 5.3.3).

Discrete Time Branching Process

In the focus of these models is the evolution of a population of N individuals chosen from m distinct classes or subspecies $\mathsf{X}_k \in \{\mathsf{X}_1, \mathsf{X}_2, \ldots, \mathsf{X}_m\}$:

$$\Pi(n) = \mathcal{Z}(n) = \big(\mathcal{Z}_1(n), \mathcal{Z}_2(n), \ldots, \mathcal{Z}_m(n)\big) , \quad \text{with } \mathcal{Z}_k \in \mathbb{N} , \; n \in \mathbb{N} .$$

The random variables $\mathcal{Z}_k(n)$, $k = 1, \ldots, m$, count the numbers of individuals X_k in generation n, $n \in \mathbb{N}$. For a description of the evolution in discrete time, we introduce multitype branching (Fig. 5.22). A simple initial condition is used for the purpose of illustration:

$$\mathcal{Z}(0) = \mathbf{e}_i , \quad \text{with } \mathbf{e}_i = (0, \ldots, 1, \ldots, 0) ,$$

where $\mathbf{e}_i$ is the unit vector pointing in the direction of type X_i. In other words, the initial condition is one individual X_i at generation $n = 0$. Now we define the probability of obtaining a certain distribution of species through replication and mutation in the first generation, viz.,

$$P_i^{(1)}(z_1, \ldots, z_m) = P\Big(\mathcal{Z}_1(1) = z_1, \ldots, \mathcal{Z}_m(1) = z_m\Big) ,$$

and analogously in the n th generation:

$$P_i^{(n)}(z_1, \ldots, z_m) = P\Big(\mathcal{Z}_1(n) = z_1, \ldots, \mathcal{Z}_m(n) = z_m\Big) . \tag{5.44a}$$

Next we introduce the generating functions $g_i(\mathbf{s})$, where $\mathbf{s} = (s_1, \ldots, s_m)$ is the vector of auxiliary variables, with the subscript referring to the subspecies of the individual in generation zero. Accordingly, we obtain for the generating function of

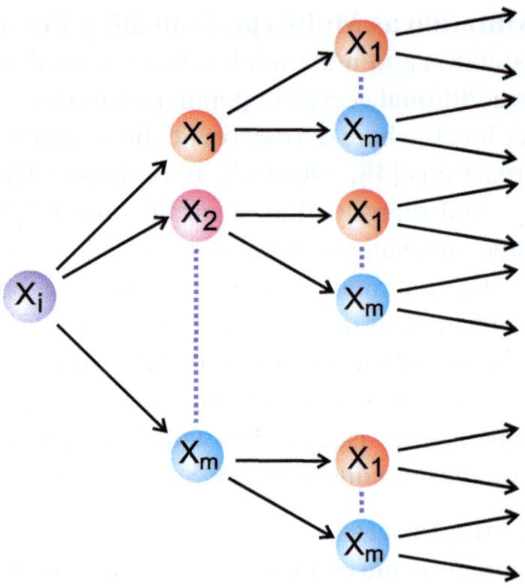

Fig. 5.22 Reproduction as a discrete multitype branching process. An individual X_i has progeny $\mathsf{X}_k \in \{\mathsf{X}_1, \mathsf{X}_2, \ldots, \mathsf{X}_i, \ldots, \mathsf{X}_m\}$, which consists of correct copies X_i or mutations X_j, $j \neq i$. Reproduction is assumed to be homogeneous in time and to occur independently of the other individuals present in the population, and in discrete generations. The probabilities for an individual of type X_i to produce γ_1 offspring of type X_1, γ_2 offspring of type X_2, ..., and γ_m offspring of type X_m, are given by $P_i(\gamma_1^{(i)}, \gamma_2^{(i)}, \ldots, \gamma_i^{(i)}, \ldots, \gamma_m^{(i)})$. They are independent of the generation, but of course depend on the subspecies X_i to which the individual belongs

the first generation $\mathcal{Z}(1)$:

$$G^{(1)}(\mathbf{s}) = g_i(\mathbf{s}) = \sum_{z_1,\ldots,z_m \geq 0} P^{(1)}(z_1, \ldots, z_m) s_1^{z_1} \cdots s_m^{z_m} . \tag{5.44b}$$

It is easy to generalize to further generations. If $\mathcal{Z}(n) = (z_1, \ldots, z_m)$ represents the distribution of individuals at generation n, then $\mathcal{Z}(n+1)$ is the sum of $z_1 + \cdots + z_m$ random vectors, out of which z_1 have the generating function $g_1(\mathbf{s})$, z_2 the generating function $g_2(\mathbf{s})$, and so on. As is typical for convolutions (see Sect. 5.2.5), the explicit formula is rather lengthy and provides little additional insight, and we shall not show it here.

Instead, we compute a *mean matrix*, which contains the expectation value for X_j to be obtained through replication of X_i :

$$\mathrm{M} = \Big(m_{ji} = E\big(\mathcal{Z}_j(1) | \mathcal{Z}(0) = \mathbf{e}_i\big) , \quad \forall\, i,j = 1, \ldots, m \Big) . \tag{5.44c}$$

For obvious reasons, we take it for granted that first moments exist for all i and j.[18] The matrix element m_{ji} is the mean number of X_j individuals derived from one X_i individual within one generation and this number is readily obtained from the generating function:

$$m_{ji} = \left(\frac{\partial g_i}{\partial s_j}\right)_{s_1=\ldots=s_m=1} , \quad i,j = 1,\ldots,m . \tag{5.44d}$$

In general, we are dealing with nonnegative first moments $m_{ji} \geq 0$, and unless stated otherwise, we shall assume that the matrix $\mathrm{M} = (m_{ji})$ is positively regular, i.e., there exists an $n > 0$ such that M^n has strictly positive elements, and M is irreducible, which implies that each type X_j can be derived from each type X_i through a finite chain of mutations.[19]

The Perron–Frobenius theorem [492] applies to irreducible matrices M and states that the mean matrix admits a unique simple largest eigenvalue λ_0, which is dominant in the sense that $|\lambda_k| < \lambda_0$ is satisfied for every other eigenvalue λ_k of M with $k \neq 0$. Since λ_0 is non-degenerate, a unique strictly positive right eigenvector $\mathbf{u} = (u_1,\ldots,u_m)$ with $u_i > 0$, $\forall\, i = 1,\ldots,m$, and a unique strictly positive left eigenvector $\mathbf{v} = (v_1,\ldots,v_m)$ with $v_i > 0$, $\forall\, i = 1,\ldots,m$, such that

$$\mathrm{M}\mathbf{u}^{\mathrm{t}} = \lambda_0 \mathbf{u}^{\mathrm{t}} , \qquad \mathbf{v}\mathrm{M} = \lambda_0 \mathbf{v} . \tag{5.44e}$$

No other eigenvalue λ_k with $k \neq 0$ has a left or right eigenvector with only strictly positive components. The left eigenvector is normalized according to an L^1-norm and for the right eigenvector we use a peculiar *scalar product normalization*:

$$\sum_{i=1}^{m} v_i = 1 , \qquad (\mathbf{v},\mathbf{u}) = \mathbf{v}\cdot\mathbf{u}^{\mathrm{t}} = 1 .$$

The use of the L^1-norm rather than the more familiar L^2-norm is a direct consequence of the existence of conservation laws based on addition of particle numbers or concentrations. The somewhat strange normalization has the consequence that the matrix $\mathrm{T} = \mathbf{u}^{\mathrm{t}}\cdot\mathbf{v} = (t_{ij} = u_i v_j)$ is idempotent or a *projection operator*:

$$\mathrm{T}\cdot\mathrm{T} = \mathbf{u}^{\mathrm{t}}\cdot\mathbf{v}\cdot\mathbf{u}^{\mathrm{t}}\cdot\mathbf{v} = \mathbf{u}^{\mathrm{t}}\cdot 1\cdot\mathbf{v} = \mathrm{T} , \quad \text{whence } \mathrm{T} = \mathrm{T}^2 = \ldots = \mathrm{T}^n ,$$

[18] In real systems, we are always dealing with finite populations in finite time, and then expectation values do not diverge (but see, for example, the unrestricted birth-and-death process in Sect. 5.2.2).

[19] Situations may exist where it is for all practical purposes impossible to reach one population from another one through a chain of mutations in any reasonable time span. Then M is not irreducible in reality, and we are dealing with two independently mutating populations. In particular, when more involved mutation mechanisms comprising point mutations, deletions, and insertions are considered, it may be advantageous to deal with disjoint sets of subspecies.

and hence we have in addition

$$\mathrm{T} \cdot \mathrm{M} = \mathrm{M} \cdot \mathrm{T} = \lambda \mathrm{T} , \quad \text{and} \quad \lim_{n \to \infty} \lambda^{-n} \mathrm{M}^n = \mathrm{T} , \tag{5.44f}$$

despite the fact that λ^n goes to zero, diverges, or stays at $\lambda^n = 1$, a situation of probability measure zero, depending on whether $\lambda < 1$, $\lambda > 1$, or $\lambda = 1$, respectively.

A population is said to become extinct if $\mathcal{Z}(n) = 0$ for some $n > 0$. We denote the probability of extinction for the initial condition $\mathcal{Z}(n) = \mathbf{e}_i$ by q_i and define

$$q_i = P\big(\exists\, n \text{ such that } \mathcal{Z}(n) = 0 | \mathcal{Z}(0) = \mathbf{e}_i\big) . \tag{5.44g}$$

The vector $\mathbf{q} = (q_1, \ldots, q_m)$ is given by the smallest nonnegative solution of the equation

$$\mathbf{g}(\mathbf{q}) = \mathbf{q} \ \text{ or } \ \mathbf{g}(\mathbf{q}) - \mathbf{q} = 0 , \tag{5.44h}$$

where $\mathbf{g}(\mathbf{s}) = \big(g_1(\mathbf{s}), \ldots, g_m(\mathbf{s})\big)$, with the functions $g_i(\mathbf{s})$ defined by (5.44b). The conditions for extinction can be expressed in terms of the dominant eigenvector λ of the mean matrix M:

(i) If $\lambda \leq 1$ then $q_i = 1$, $\forall\, i$ and extinction is certain.
(ii) If $\lambda > 1$ then $q_i < 1$, $\forall\, i$ and there is a positive probability of survival to infinite time.

In the second case, it is of interest to compute *asymptotic frequencies*, where frequency stands for a normalized random variable

$$\mathcal{X}_i(n) = \frac{\mathcal{Z}_i(n)}{\sum_{k=1}^m \mathcal{Z}_k(n)} , \quad \text{with } \mathcal{Z}_i(n) > 0 , \ \forall\, i . \tag{5.44i}$$

If $\lambda > 1$, then there exists a random vector $\boldsymbol{\omega} = (\omega_1, \ldots, \omega_m)$ and a scalar random variable w such that we have

$$\lim_{n \to \infty} \lambda^{-n} \mathcal{Z}(n) = \boldsymbol{\omega} , \qquad \boldsymbol{\omega} = w\mathbf{u} , \tag{5.44j}$$

with probability one, where $\mathbf{u}$ is the right eigenvector of M given by (5.44e). It then follows that

$$\lim_{n \to \infty} \mathcal{X}_i(n) = \frac{u_i}{\sum_{k=1}^m u_k} \tag{5.44k}$$

holds almost always, provided that the population does not become extinct. Equation (5.44k) states that the random variable for the frequency of type X_i, $\mathcal{X}_i(n)$, converges almost certainly to a constant value (provided $w \neq 0$). The asymptotic behavior of the random vector $\mathcal{X}(n)$ contrasts sharply with the behavior of the total population size $|\mathcal{Z}(n)| = \sum_{k=1}^{m} \mathcal{Z}_k(n)$ and that of the population distribution $\mathcal{Z}(n)$, which may both undergo large fluctuations accumulating in later generations, because of the autocatalytic nature of the replication process. In late generations, the system has either become extinct or grown to very large size where fluctuations in relative frequencies become small by the law of large numbers.

The behavior of the random variable w can be completely described by means of the results given in [298]. We have either

(i) $w = 0$ with probability one, which is always the case if $\lambda \leq 1$, or
(ii) $\mathrm{E}\big(w|\mathcal{Z}(0) = \mathbf{e}_i\big) = v_i$,

where v_i is the ith component of the left eigenvector $\mathbf{v}$ of matrix M. A necessary and sufficient condition for the validity of the second condition here is

$$\mathrm{E}\big(\mathcal{Z}_j(1)\log \mathcal{Z}_j(1)|\mathcal{Z}(0) = \mathbf{e}_i\big) < \infty\,, \quad \text{for } 1 \leq i,j \leq m\,,$$

which is a condition of finite population size that is always satisfied in realistic systems.

Continuous Time Branching Process

For intermixing generations, in particular for *in vitro* evolution [288] or in the absence of generation synchronizing pacemakers, the assumption of discrete generations is not justified, because any initially given synchronization is lost within a few reproduction cycles. Continuous time multitype branching processes offer an appropriate but technically more complicated description for such cases. Since the results are similar to the discrete case, we will sketch them only briefly here.

For the continuous time model we suppose that an individual of type X_i, independently of other individuals present in the population, persists for an exponentially distributed time with mean α^{-1} (see also (4.147) in Sect. 4.6.1) and then generates a copy by reproduction and mutation according to a distribution whose generating function is $g_i(\mathbf{s})$. As discussed and implemented in the case of chemical master equations (Sect. 4.2.2), we assume that, in a properly chosen time interval of length Δt, exactly one of the following three alternatives occurs up to probability $o(\Delta t)$:

(i) no change,
(ii) extinction, or
(iii) survival and production of a copy of subspecies X_j ($j = 1,\ldots,m$).

The probabilities for the events (ii) and (iii) are homogeneous in time and up to some $o(\Delta t)$ proportional to Δt. As before, $\mathcal{Z}_i(t)$ denotes the number of individuals of subspecies X_i at time t, and $\mathcal{Z}(t)$ the distribution of subspecies. Once again, we define a mean matrix

$$\mathrm{M} = \Big(m_{ij} = \mathrm{E}\big(\mathcal{Z}_i(t)|\mathcal{Z}(0) = \mathbf{e}_j\big)\Big)\,, \tag{5.44c$'$}$$

where we assume that all first moments are finite for all $t \geq 0$. The mean matrix satisfies the semigroup and continuity properties

$$\mathrm{M}(t+u) = \mathrm{M}(t)\mathrm{M}(u)\,, \qquad \lim_{t \to +0} M(t) = \mathbb{I}\,. \tag{5.45a}$$

Condition (5.45a) implies the existence of a matrix A, called the infinitesimal generator, which satisfies for all $t > 0$:

$$\mathrm{M}(t) = \mathrm{e}^{\mathrm{A}t}\,, \quad \text{with } \mathrm{A} = \big(a_{ij} = \alpha_i (b_{ij} - \delta_{ij})\big)\,, \quad b_{ij} = \left(\frac{\partial g_i}{\partial s_j}\right)_{s_1 = \ldots = s_m = 1}\,. \tag{5.45b}$$

Again we assume that each type can produce every other type. As in the discrete time case, we have $m_{ij}(t) > 0$ for $t > 0$, A is strictly positive, and the Perron–Frobenius theorem holds. The matrix A has a unique dominant real eigenvalue λ with strictly positive right and left eigenvectors **u** and **v**, respectively. The dominant eigenvalue of $\mathrm{M}(t)$ is $\mathrm{e}^{\lambda t}$, again we normalize $\sum_{i=1}^m v_i = \sum_{i=1}^m u_i v_i$, and with $\mathrm{T} = \mathbf{u}^{\mathrm{t}} \cdot \mathbf{v}$, we have

$$\lim_{t \to \infty} \mathrm{e}^{-\lambda t} \mathrm{M}(t) = \mathrm{T}\,, \tag{5.44f'}$$

which guarantees the existence of finite solutions in relative particle numbers. As in the discrete case the extinction conditions are determined by λ : if $\mathbf{q} = (q_1, \ldots, q_m)$ denotes the extinction probabilities, then $\mathbf{q}$ is the unique solution of $\mathbf{g}(\mathbf{q}) - \mathbf{q} = 0$, where $\mathbf{g}(\mathbf{s}) = (s_1, \ldots, s_m)$ as before, and accordingly (i) $q_i = 1, \forall\, i$ if $\lambda \leq 0$ and (ii) $q_i < 1, \forall\, i$ if $\lambda > 0$. Again we obtain

$$\lim_{t \to \infty} \mathcal{X}_i(t) = \frac{u_i}{u_1 + \cdots + u_m}\,, \tag{5.44k'}$$

whenever the process does not lead to extinction.

The Deterministic Reference

We shortly sketch here the solutions of the deterministic problem [130–132], which is described by the differential equation

$$\frac{\mathrm{d}x_i}{\mathrm{d}t} = \sum_{j=1}^m w_{ij} x_j - x_i \left(\sum_{r=1}^m \sum_{s=1}^m w_{rs} x_s \right)\,, \tag{5.46a}$$

or in vector notation,

$$\frac{\mathrm{d}\mathbf{x}^{\mathrm{t}}}{\mathrm{d}t} = \mathrm{W}\mathbf{x}^{\mathrm{t}} - (\mathbf{1} \cdot \mathrm{W}\,\mathbf{x}^{\mathrm{t}})\mathbf{x}^{\mathrm{t}} \tag{5.46a'}$$

with $W = (w_{ij};\ i,j = 1,\ldots,m)$, $\mathbf{x} = (x_1 \ldots, x_m)$ restricted to the unit simplex

$$\mathbb{S}_m = \left\{\mathbf{x} \in \mathbb{R}^m : x_i \geq 0\,,\ \sum\nolimits_{i=1}^m x_i = 1\right\}\,, \tag{5.46b}$$

and $\mathbf{1} = (1,\ldots,1)$. The matrix W is characterized as a *value matrix* and commonly written as product of a *fitness matrix* F and a *mutation matrix* Q[20]:

$$W = \begin{pmatrix} Q_{11}f_1 & Q_{12}f_2 & \ldots & Q_{1m}f_m \\ Q_{21}f_1 & Q_{22}f_2 & \ldots & Q_{2m}f_m \\ \vdots & \vdots & \ddots & \vdots \\ Q_{m1}f_1 & Q_{m2}f_2 & \ldots & Q_{mm}f_m \end{pmatrix} = \mathrm{QF}\,. \tag{5.46c}$$

The fitness matrix is a diagonal matrix whose elements are the fitness values of the individual species: $\mathrm{F} = (f_{ij} = f_i\delta_{ij})$. The mutation matrix corresponds to the branching diagram in Fig. 5.22: $\mathrm{Q} = (Q_{ij})$, where Q_{ij} is the frequency with which subspecies X_i is obtained through copying of subspecies X_j. Since every copying event results either in a correct copy or a mutant, we have $\sum_{i=1}^m Q_{ij} = 1$ and Q is a *stochastic matrix*. Some model assumptions, for example the *uniform error rate model* [520], lead to symmetric Q-matrices, which are then *bistochastic matrices*.[21] It is worth considering the second term on the right-hand side of (5.46a) in the explicit formulation

$$\mathbf{1}\cdot\mathrm{W}\mathbf{x}^{\mathrm{t}} = \sum_{r=1}^{m}\sum_{s=1}^{m} w_{rs}x_s = \sum_{r=1}^{m}\sum_{s=1}^{m} Q_{rs}f_s x_s = \sum_{s=1}^{m} f_s x_s \sum_{r=1}^{m} Q_{rs} = \bar{f} = \phi\,,$$

where the different notation indicates three different interpretations. The term $\mathbf{1}\cdot\mathrm{W}\mathbf{x}^{\mathrm{t}}$ is the mean excess productivity of the population, which has to be compensated in order to avoid net growth. In mathematical terms, $\phi(t)$ maintains the population normalized, and in an experimental setup, $\phi(t)$ is an externally controllable *dilution flow* that is suggestive of a flow reactor (Fig. 4.21). It is straightforward to check that $\mathbb{S}_m$ is invariant under (5.46a): if $\mathbf{x}(0) \in \mathbb{S}_m$, then $\mathbf{x}(t) \in \mathbb{S}_m$ for all $t > 0$. Equation (5.46a) was introduced as a phenomenological equation describing the kinetics of *in vitro* evolution under the constraint of constant population size. Here the aim is to relate deterministic replication–mutation kinetics to multitype branching processes.

---

[20] In the case of the mathematically equivalent *Crow–Kimura mutation–selection equation* [96, p. 265, Eq. 6.4.1], additivity of the fitness and mutation matrix is assumed rather than factorizability (see, e.g., Sect. 5.3.3 and [484]).

[21] The selection–mutation equation (5.46a) in the original formulation [130, 132] also contains a degradation term $d_j x_j$, and the corresponding definition of the value matrix reads $\mathrm{W} = (w_{ij} = Q_{ij}f_j - d_j)$. If all individuals follow the same death law, i.e., $d_j = d$, $\forall\, j$, the parameter d can be absorbed into the population size conservation relation and need not be considered separately.

Some preliminary remarks set the stage for the comparison:

1. The linear differential equation

$$\frac{\mathrm{d}\mathbf{y}^{\mathrm{t}}}{\mathrm{d}t} = \mathrm{W}\mathbf{y}^{\mathrm{t}} \quad \text{and} \quad \mathbf{x}(t) = \frac{1}{\sum_{j=1}^{m} y_j(t)}\mathbf{y}(t) , \tag{5.46d}$$

with a positive or nonnegative irreducible matrix W and the initial condition $\mathbf{y}(0) \in \mathbb{R}^m_{>0}$ satisfies (i) $\mathbf{y}(t) \in \mathbb{R}^m_{>0}$ and (ii) $\mathbf{x}(t) \in \mathbb{S}_m$, $\forall\, t \geq 0$, and it is a solution of (5.46a).

2. As noted in the references [286, 530], (5.46d) can be obtained from (5.46a) through the transformation

$$\psi(t) = \int_0^t \phi(\tau)\mathrm{d}\tau \quad \text{and} \quad \mathbf{y}(t) = \mathbf{x}(t)\mathrm{e}^{\psi(t)} .$$

3. Accordingly, the nonlinear equation (5.46a) is easy to solve and any equilibrium of this equation must satisfy

$$\mathrm{W}\,\boldsymbol{\xi}^{\mathrm{t}} = \varepsilon\boldsymbol{\xi}^{\mathrm{t}} ,$$

and therefore be a right-hand eigenvector of W. By Perron–Frobenius, there exists such a unique right eigenvector in $\mathbb{S}_m$, which we denoted by $\boldsymbol{\xi}$. It represents the stationary solution called *quasispecies*, and the corresponding eigenvalue ε is just the dominant eigenvalue of W. From the correspondence between (5.46a) and (5.46d), it follows that all orbits of equations (5.46a) converge to $\boldsymbol{\xi}$: $\lim_{t\to\infty} \mathbf{x}(t) = \boldsymbol{\xi}$.

4. Difference and differential equations are associated in the canonical way:

$$\mathbf{v}_{n+1} = \mathbf{F}(\mathbf{v}_n) \iff \frac{\mathrm{d}\mathbf{v}^{\mathrm{t}}}{\mathrm{d}t} = \mathbf{F}(\mathbf{v})^{\mathrm{t}} - \mathbf{v}^{\mathrm{t}} . \tag{5.46e}$$

An unreflected *passage to continuous time* is not always justifiable, but for a generation length of one, the difference equation $\mathbf{v}_{n+1} - \mathbf{v}_n = \mathbf{F}(\mathbf{v}_n) - \mathbf{v}_n$ can be written as

$$\mathbf{v}(1) - \mathbf{v}(0) = \mathbf{F}\big(\mathbf{v}(0)\big) - \mathbf{v}(0) .$$

Provided we assume blending generations, the change $\mathbf{v}(1/n) - \mathbf{v}(0)$ during the time interval $1/n$ can be approximated by $\big(\mathbf{F}\big(\mathbf{v}(0)\big) - \mathbf{v}(0)\big)/n$, or

$$\frac{\mathbf{v}(\Delta t) - \mathbf{v}(0)}{\Delta t} = \mathbf{F}\big(\mathbf{v}(0)\big) - \mathbf{v}(0) ,$$

which in the limit $\Delta t \to 0$ yields the differential equation (5.46e).

The relationship between branching processes and the mutation–selection equation (5.46a) is illustrated in Fig. 5.23. Starting out from the discrete multitype branching process $\mathcal{Z}(n)$, the expectation value $\mathrm{E}(\mathcal{Z}(n)) = \mathbf{Y}(n)$ satisfies $\mathbf{Y}(n)^{\mathrm{t}} = \mathrm{M}^n\mathbf{Y}(0)^{\mathrm{t}}$, where M is the mean matrix (5.44c). Hence, $\mathbf{Y}(n)$ is obtained by iteration from the difference equation $\mathbf{y}_{n+1}^{\mathrm{t}} = M\mathbf{y}_n^{\mathrm{t}}$. From here, one can reach the mutation–selection equation in two ways: (i) by passing first to continuous time as expressed by the differential equation $\mathrm{d}\mathbf{y}^{\mathrm{t}}/\mathrm{d}t = \mathrm{V}\mathbf{y}^{\mathrm{t}}$ with $\mathrm{V} = \mathrm{M} - \mathbb{I}$ followed by normalization, which yields

$$\frac{\mathrm{d}\mathbf{x}^{\mathrm{t}}}{\mathrm{d}t} = \mathrm{V}\,\mathbf{x}^{\mathrm{t}} - \mathbf{x}^{\mathrm{t}}(\mathbf{1}\cdot\mathrm{V}\mathbf{x}^{\mathrm{t}})\,, \tag{5.46f}$$

or (ii) by following the opposite sequence, so first normalizing the difference equation

$$\mathbf{x}_{n+1}^{\mathrm{t}} = \frac{1}{\mathbf{1}\cdot\mathrm{M}\mathbf{x}_n^{\mathrm{t}}}\mathrm{M}\mathbf{x}_n^{\mathrm{t}}$$

on $\mathbb{S}_m$, and then passing to continuous time, which yields

$$\frac{\mathrm{d}\mathbf{x}^{\mathrm{t}}}{\mathrm{d}t} = \big(\mathrm{M}\mathbf{x}^{\mathrm{t}} - \mathbf{x}^{\mathrm{t}}(\mathbf{1}\cdot\mathrm{M}\mathbf{x}^{\mathrm{t}})\big)\frac{1}{\mathbf{1}\cdot\mathrm{M}\mathbf{x}^{\mathrm{t}}}\,. \tag{5.46g}$$

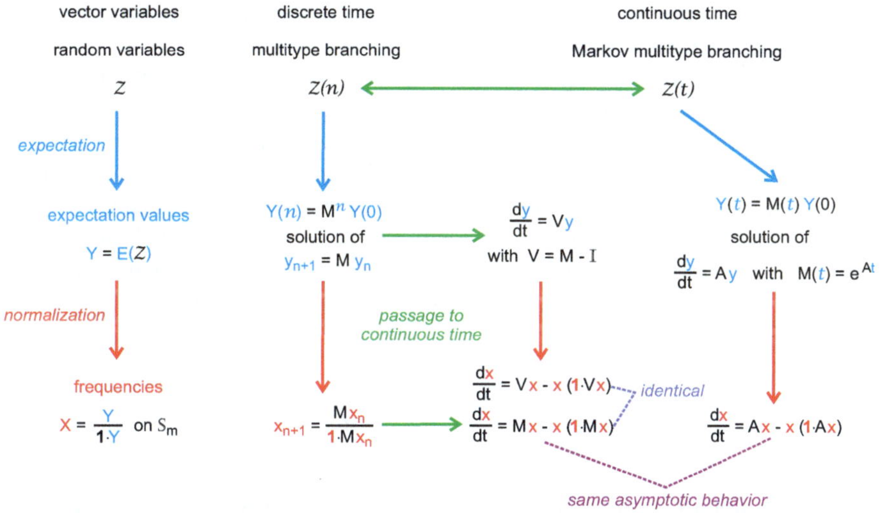

Fig. 5.23 Comparison of mutation–selection dynamics and branching processes. The sketch summarizes the different transformations discussed in the text. The distinct classes of transformation are color coded: forming expectation values in *blue*, normalization in *red*, and transformation between discrete and continuous variables in *green* (for details see the text and [105])

Multiplication by the factor $\mathbf{1} \cdot \mathrm{M}\mathbf{x}^{\mathrm{t}}$, which is independent of i and always strictly positive on $\mathbb{S}_m$, merely results in a transformation of the time axis that is tantamount to a change in the velocity, and the orbits of (5.46g) are the same as those of

$$\frac{\mathrm{d}\mathbf{x}^{\mathrm{t}}}{\mathrm{dt}} = \mathrm{M}\mathbf{x}^{\mathrm{t}} - \mathbf{x}^{\mathrm{t}}(\mathbf{1} \cdot \mathrm{M}\mathbf{x}^{\mathrm{t}}) \ . \tag{5.46g$'$}$$

Since $\mathrm{V} = \mathrm{M} - \mathbb{I}$, the two equations (5.46a) and (5.46g') are identical on $\mathbb{S}_m$.

Alternatively, we can begin with a continuous Markovian multitype branching process $\mathcal{Z}(t)$ for $t \geq 0$ and either reduce it by discretization to the discrete branching process $\mathcal{Z}(n)$, or else obtain $\mathbf{Y}(t)^{\mathrm{t}} = \mathrm{M}(t)\mathbf{Y}(0)^{\mathrm{t}}$ for the expectation values $\mathrm{E}\big(\mathcal{Z}(t)\big) = \mathbf{Y}(t)$, where $\mathrm{M}(t)$ is again the mean matrix with $\mathrm{M}(1) = \mathrm{M}$. The expectation value $\mathbf{Y}(t)$ is then the solution of the linear differential equation

$$\frac{\mathrm{d}\mathbf{y}^{\mathrm{t}}}{\mathrm{dt}} = \mathrm{A}\mathbf{y}^{\mathrm{t}} \ , \quad \text{with } \mathrm{A} = \lim_{t \to +0} \frac{\mathrm{M}(t) - \mathbb{I}}{t} \ , \tag{5.46h}$$

as infinitesimal generator of the semigroup $\mathrm{M}(t)$, and $\mathrm{M}(t) = \mathrm{e}^{\mathrm{A}t}$. Normalization leads to

$$\frac{\mathrm{d}\mathbf{x}^{\mathrm{t}}}{\mathrm{dt}} = \mathrm{A}\mathbf{x}^{\mathrm{t}} - \mathbf{x}^{\mathrm{t}}(\mathbf{1} \cdot \mathrm{A}\mathbf{x}^{\mathrm{t}}) \ , \quad \text{on } \mathbb{S}_m \ . \tag{5.46i}$$

This equation generally has different dynamics to (5.46g$'$), but the asymptotic behavior is the same, because A and $\mathrm{M} = \mathrm{e}^{\mathrm{A}}$ have the same eigenvectors, so $\mathbf{u}$ is the global attractor for both equations (5.46g$'$) and (5.46i).

Three simple paths lead from branching processes to an essentially unique version of the mutation–selection equation (5.46a), and the question is whether or not such a reduction from a stochastic to a deterministic system is relevant. A superficial analysis may suggest that it is not. Passing from the random variables $\mathcal{Z}_i(n)$ $(i = 1, \ldots, m)$ to the expectation values $\mathrm{E}\big(\mathcal{Z}_i(n)\big)$ may be misleading, because the variances grow too fast, as can be easily verified for single-type branching. If $\mu = \mathrm{E}\big(\mathcal{Z}(1)\big)$ and $\sigma^2 = \mathrm{var}\big(\mathcal{Z}(1)\big)$ are the mean and the variance of a single individual in the first generation, then the mean and variance of the n th generation grow in the supercritical case as

$$m^n \quad \text{and} \quad \sigma^2 \frac{m^n(m^n - 1)}{m(m-1)} = \sigma^2 \sum_{k=n-1}^{2n-2} m^k \ ,$$

respectively, and the ratio from the standard deviation and the mean converges to a positive constant

$$\frac{\sqrt{\mathrm{var}\big(\mathcal{Z}(n)\big)}}{\mathrm{E}\big(\mathcal{Z}(n)\big)} = \sqrt{\sum_{k=n-3}^{2n-4} m^k} \ .$$

Accordingly, the *window* of probable values of the random variable $\mathcal{Z}(n)$ is rather large. For a critical process, the situation is still worse: the mean remains constant, whereas the variance grows to infinity (see Fig. 5.8). For multitype branching, the situation is similar, but the expressions for the variance and correlations get rather complicated, and again the second moments grow so fast that the averages tell us precious little about the process (see [239] for the discrete process and [29] for the continuous process).

However, normalization changes the situation. The transition from expectation values to relative frequencies cancels the fluctuations, or more precisely, if the process does not go to extinction, the relative frequencies of the random variables, viz.,

$$\mathcal{X}_i = \frac{\mathcal{Z}_i}{\mathcal{Z}_1 + \cdots + \mathcal{Z}_m} ,$$

converge almost certainly to the value u_i $(i - 1, \ldots, m)$, which are at the same time the limits of the relative frequencies of the expectation values:

$$x_i = \frac{y_i}{y_1 + \cdots + y_m} .$$

In this sense, the deterministic mutation–selection equation (5.46a) yields a description of the stochastic evolution of the internal structure of the population which is much more reliable than the dynamics of the unnormalized means. The qualitative features of the selection process condense the *variance free* part of the deterministic approach.

Finally, we shall come back to the numerical analysis of the stochastic replication–mutation problem in Sect. 5.3.3, and mention here a few other attempts to find stochastic solutions [249, 270, 287, 338, 379].

5.3 Stochastic Models of Evolution

In this section we shall compare two specific models from population biology, the Wright–Fisher process named after the US American population geneticist Sewall Wright and the English statistician Ronald Fisher, and the Moran process that got its name from the Australian statistician Patrick Moran, with the replication–mutation model from biochemical kinetics, which we used in the previous Sect. 5.2.5 as an example for the application of multitype branching processes.

In Sect. 5.2.2, we used master equations to find solutions for simple birth-and-death processes. Here we consider more general models and start out from the standard Markov chain

$$\begin{aligned} P(n, t+1) &= \sum_m p_{nm} P(m, t) \ , \\ P(n, t+1) - P(n, t) &= \sum_m p_{nm} P(m, t) - \sum_m p_{mn} P(n, t) \ , \end{aligned} \tag{5.47}$$

where we used the relation $\sum_m p_{mn} = 1$ in the last term on the right-hand side. The two terms with $m = n$ can be omitted due to cancelation, while t, which could be considered as an integer label for generations, is now interpreted as time. Then the intervals Δt have to be taken small enough to ensure that at most one sampling event occurs between t and $t + \Delta t$. Division by Δt yields

$$\frac{P(n, t + \Delta t) - P(n, t)}{\Delta t} = \sum_m \left(\frac{p_{nm}}{\Delta t} \right) P(m, t) - \sum_m \left(\frac{p_{mn}}{\Delta t} \right) P(n, t) \ .$$

Instead of assuming that exactly one sampling event happens per generation, including $n \to n$ where no actual transition occurs, we now consider sampling events at *unit rate*, and one event per generation will take place *on average*. If t is sufficiently large, then by far the most likely number of events that will have occurred is equal to t and we can expect continuous time and discrete time processes to be barely distinguishable in the long run.

The transition probability is replaced by the transition rate per unit time

$$p_{nm} = W(n|m)\Delta t + \mathcal{O}(\Delta t)^2 \ , \quad \text{for } n \neq m \ , \tag{5.48}$$

where the terms of order $(\Delta t)^2$ and higher express the probabilities that two or more events take place during the time interval Δt. Taking the limit $\Delta t \to 0$ yields the familiar master equation

$$\frac{\partial P(n, t)}{\partial t} = \sum_{m \neq n} \Big(W(n|m) P(m, t) - W(m|n) P(n, t) \Big) \ . \tag{4.66}$$

The only difference with the general form of the master equation is the assumption that the transition rates per unit time are rate parameters, which are independent of time. Accordingly, we can replace the conditional probabilities by the elements of a square matrix $W = \big(W_{nm} = W(n|m) \big)$.

For the purpose of illustration we derive solutions for the Moran model by means of a master equation. The solution also allows one to handle the neutral case and provides an alternative approach to random selection that has already been discussed in Sect. 5.2.3 as Motoo Kimura's model for neutral evolution, based on a Fokker–Planck equation, which represents an approximation to the discrete process.

5.3.1 *The Wright–Fisher and the Moran Process*

Here we shall introduce two common stochastic models in population biology, the *Wright–Fisher model*, named after Sewall Wright and Ronald Fisher, and the *Moran model*, named after the Australian statistician Pat Moran. The Wright–Fisher model and the Moran model are stochastic models for the evolution of allele distributions in populations with constant population size [56]. The first model [174, 579], also referred to as *beanbag* population genetics, probably the simplest process for illustrating genetic drift and definitely the most popular one [96, 147, 241, 372], deals with strictly separated generations, whereas the Moran process [410, 411], based on continuous time and overlapping generations, is generally more appealing to statistical physicists. Both processes are introduced here for the simplest scenarios: haploid organisms, two alleles of the gene under consideration, and no mutation. Extension to more complicated cases is straightforward. The primary question addressed by the two models is the evolution of populations in the case of neutrality for selection.

The Wright–Fisher Process

The Wright–Fisher process is illustrated in Fig. 5.24. A single reproduction event is modeled by a sequence of four steps: (i) A gene is randomly chosen from the gene pool of generation T containing exactly N genes distributed over M alleles, (ii) it is replicated, (iii) the original is put back into the gene pool T, and (iv) the copy

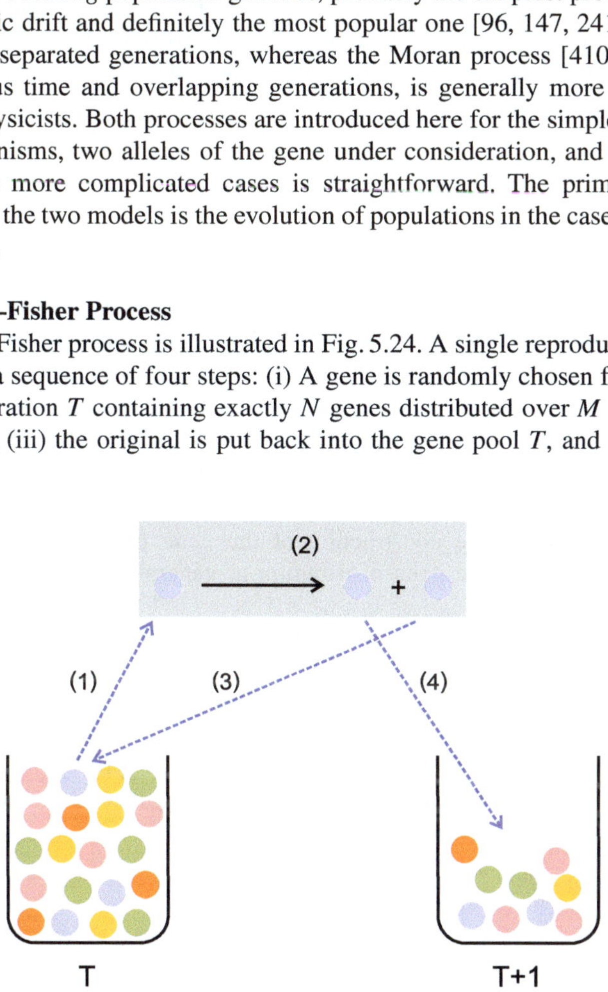

Fig. 5.24 The Wright–Fisher model of *beanbag* genetics. The gene pool of generation T contains N gene copies chosen from m alleles. Generation $T + 1$ is built from generation T by ordered cyclic repetition of a four-step event: (1) random selection of one gene from the gene pool T, (2) error-free copying of the gene, (3) return of the original into the gene pool T, and (4) insertion of the copy into the gene pool of the next generation $T + 1$. The procedure is repeated until the gene pool $T + 1$ contains exactly N genes. No mixing of generations is allowed

is put into the gene pool of the next generation $T+1$. The process is terminated when the next generation gene pool has exactly N genes. Since filling the gene pool of generation $T+1$ depends exclusively on the distribution of genes in the pool of generation T and earlier gene distributions have no influence on the process, the Wright–Fisher model is Markovian.

In order to simplify the analysis, we assume two alleles A and B, which are present in a_T and b_T copies in the gene pool at generation T. Since the total number of genes is constant, $a_T + b_T = N$ and $b_T = N - a_T$, we are dealing with a single discrete random variable a_T with $T \in \mathbb{N}$. A new generation $T+1$ is produced from the gene pool at generation T by picking a gene at random N times and replacing it. The probability of obtaining $n = a_{T+1}$ alleles A in the new gene pool is given by the binomial distribution:

$$P(a_{T+1} = n) = \binom{N}{n} p_{\mathsf{A}}^n p_{\mathsf{B}}^{N-n} .$$

The probabilities of picking A or B from the gene pool in generation T are $p_{\mathsf{A}} = a_T/N$ and $p_{\mathsf{B}} = b_T/N = (N-a_T)/N$, respectively, with $p_{\mathsf{A}}+p_{\mathsf{B}} = 1$. The transition probability from m alleles A at time T to n alleles A at time $T+1$ is simply given by[22,23]

$$W_{nm} = \binom{N}{n} \left(\frac{m}{N}\right)^n \left(1 - \frac{m}{N}\right)^{N-n} . \tag{5.49a}$$

As already mentioned the construction of the gene pool at generation $T+1$ is fully determined by the gene distribution at generation T, and the process is Markovian.

In order to study the evolution of populations, an initial state has to be specified. We assume that the number of alleles A was n_0 at generation $T = 0$, whence we are calculating the probability $P(n, T|n_0, 0)$, which we denote by $p_n(T)$. Since the Wright–Fisher model does not contain any interactions between alleles or mutual dependencies between processes involving alleles, the process is best modeled by means of linear algebra. We define a probability vector $\mathbf{p}$ and a transition

[22]The notation applied here is the conventional way of writing transitions in physics: W_{nm} is the probability of the transition $n \leftarrow m$, whereas many mathematicians would write W_{mn}, indicating $m \rightarrow n$.

[23]When doing actual calculations, one has to use the convention $0^0 = 1$ used in probability theory and combinatorics, but not usually in analysis, where 0^0 is an indefinite expression.

matrix W :

$$\mathbf{p}(T) = \big(p_0(T), p_1(T), \ldots, p_N(T)\big) ,$$

$$\mathrm{W} = \begin{pmatrix} W_{00} & W_{01} & \cdots & W_{0N} \\ W_{10} & W_{11} & \cdots & W_{1N} \\ \vdots & \vdots & \ddots & \vdots \\ W_{N0} & W_{N1} & \cdots & W_{NN} \end{pmatrix} = \begin{pmatrix} 1 & W_{01} & \cdots & 0 \\ 0 & W_{11} & \cdots & 0 \\ \vdots & \vdots & \ddots & \vdots \\ 0 & W_{N1} & \cdots & 1 \end{pmatrix} . \tag{5.49b}$$

Conservation of probability provides two conditions: (i) The probability vector has be normalized to a L^1-norm, i.e., $\sum_n p_n(T) = 1$, and (ii) it has to remain normalized in future generations, i.e., $\sum_n W_{nm} = 1$.[24] The evolution is now simply described by the matrix equation

$$\mathbf{p}(T+1)^{\mathrm{t}} = \mathrm{W}\mathbf{p}(T)^{\mathrm{t}} , \quad \text{or } \mathbf{p}(T)^{\mathrm{t}} = \mathrm{W}^T\mathbf{p}(0)^{\mathrm{t}} . \tag{5.49c}$$

Equation (5.49c) is identical with the matrix formulation of linear *difference equations*, i.e., $\mathbf{p}_{k+1}^{\mathrm{t}} = \mathrm{W}\mathbf{p}_k^{\mathrm{t}}$, which were used in Sect. 5.2.5 to discuss multitype branching, and which are presented and analyzed, for example, in the monograph [97, pp. 179–216].

Solutions of (5.49c) are known in the from of an analytical expression for the eigenvalues of the transition matrix W [159]:

$$\lambda_k = \binom{N}{k}\frac{k!}{N^k} , \quad k = 0, 1, 2, \ldots . \tag{5.49d}$$

Although we do not have analytical expressions for the eigenvectors of the transition matrix W, the stationary state of the Wright–Fisher process can be deduced from the properties of a Markov chain by asking what the system would look like in the limit of an infinite number of generations when the probability density might assume a stationary distribution $\bar{\mathbf{p}}$. If such a stationary state exists the density must satisfy $\mathrm{W}\bar{\mathbf{p}} = \bar{\mathbf{p}}$, or in other words, $\bar{\mathbf{p}}$ will be a right eigenvector of W with the eigenvalue $\lambda = 1$.

By intuition we guess that a final absorbing state of the system must be either all B, corresponding to $\bar{n} = 0$ and fixation of allele B, or all A with $\bar{n} = N$ and fixation of allele A. Both states are absorbing and the general solution will be a mixture of the two states. The probability density of such a mixed steady state is

$$\bar{\mathbf{p}}^{\mathrm{t}} = (1-\vartheta, 0, \ldots, 0, \vartheta) . \tag{5.49e}$$

It satisfies $\mathrm{W}\bar{\mathbf{p}} = \bar{\mathbf{p}}$, as is easily confirmed by inserting W from (5.49b).

[24] A matrix W with this property is called a stochastic matrix.

Next we compute the expected number of alleles A as a function of the generation number:

$$\begin{aligned}\big\langle n(T+1)\big\rangle &= \sum_{n=0}^{N} n p_n(T+1) = \sum_{n=0}^{N} n \sum_{m=0}^{N} W_{nm} p_m(T) \\ &= \sum_{m=0}^{N} p_m(T) \sum_{n=0}^{N} n W_{nm} = \sum_{m=0}^{N} m p_m(T) = \big\langle n(T)\big\rangle\,,\end{aligned} \tag{5.49f}$$

where we have used the expectation value of the binomial distribution (2.41a) in the last line:

$$\sum_{n=0}^{N} n W_{nm} = \sum_{n=0}^{N} n \binom{N}{n} \left(\frac{m}{N}\right)^n \left(1-\frac{m}{N}\right)^{N-n} = N\frac{m}{N} = m\,.$$

The expectation values of the numbers of alleles is independent of the generation T and this implies $\langle n(T)\rangle = \langle n(0)\rangle = n_0$. This result enables us to determine the probability ϑ for the fixation of allele A. From (5.49e), we deduce two possible states in the limit $T \to \infty$, viz., $n = N$ with probability ϑ and $n = 0$ with probability $1 - \vartheta$. We thus have

$$\lim_{T\to\infty} \big\langle n(T)\big\rangle = \vartheta N + (1-\vartheta)0 \implies n_0 = \vartheta N\,, \quad \text{and } \vartheta = \frac{n_0}{N}\,. \tag{5.49g}$$

So finally, we have found the complete expression for the stationary state of the Wright–Fisher process and the probability of fixation of allele A, which amounts to $\vartheta = n_0/N$.

The Moran Process

The Moran process introduced by Pat Moran [410] is a continuous time process and deals with transitions that are defined for single events. As in the Wright–Fisher model, we consider an example with two alleles A and B in a haploid population of size N, and the probabilities for choosing A or B are p_{A} and p_{B}, respectively. Unlike the Wright–Fisher model, there is no well defined previous generation from which a next generation is formed by sampling N genes, so overlapping generations make it difficult, if not impossible, to define generations unambiguously. The *event* in the Moran process is a *combined birth-and-death step*: two genes are picked, one is copied, both template and copy are put back into the urn, and the second one is deleted (see Fig. 5.25). The probabilities are calculated form the state of the urn just before the event $p_{\mathsf{A}} = m(t)/N$ and $p_{\mathsf{B}} = \big(N - m(t)\big)/N$, where $m(t)$ is the number of alleles A, $N - m(t)$ the number of alleles B, and N the constant total number of genes. After the event, we have exactly n alleles of type A and $N - n$ alleles of type

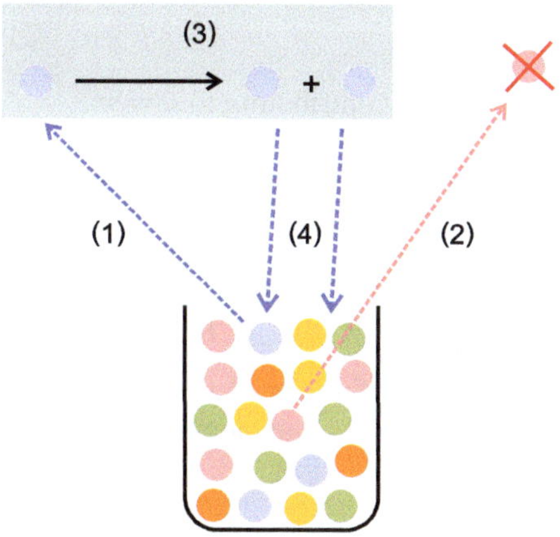

Fig. 5.25 The Moran process. The Moran process is a continuous time model for the same problem as the one handled by the Wright–Fisher model (Fig. 5.24). The gene pool of a population of N genes chosen from m alleles is represented by the urn in the figure. Evolution proceeds via successive repetition of a four-step process: (1) One gene is chosen from the gene pool at random, (2) a second gene is randomly chosen and deleted, (3) the first gene is copied, and (4) both genes, original and copy, are put back into the urn. The Moran process has overlapping generations and, in particular, the notion of generation is not well defined

B, with $\Delta n = n - m = \pm 1, 0$, depending on the nature of the process. In particular, two different ways of picking two genes are commonly used in the literature:

(i) In the more intelligible first counting, one explicitly considers the numerical reduction by one as a consequence of the first pick [442].
(ii) In the second procedure, the changes introduced in the urn by picking the first gene are ignored in the second draw (see, e.g., [56]).[25]

We shall present the (almost identical) results of the two picking procedures here, starting with the first, which is perhaps easier to motivate count.

Before the combined birth-and-death event, we have m genes with allele A out of N genes in total. Because of the first pick, the total number of genes and the number of genes with allele A are reduced by one for the coupled second pick, viz., $N \rightarrow N - 1$ and $m \rightarrow m - 1$, respectively. If the first pick chose a B allele, the changes in the numbers of genes would be $N \rightarrow N - 1$ and $N - m \rightarrow N - m - 1$. After the event, the numbers will have changed to n and $N - n$, respectively, and

[25]The second procedure can be visualized by a somewhat strange but nevertheless precise model assumption: after the replication event, the parent but not the offspring is put back into the pool from which the individual, which is doomed to die, is chosen in the second draw.

$n - m = 0, \pm 1$. Now we compute the probabilities for the four possible sequential draws and find:

(i) A + A: $p_{\mathsf{A+A}} = \dfrac{m}{N}\dfrac{m-1}{N-1}$ contributing to $n = m$.

(ii) A + B: $p_{\mathsf{A+B}} = \dfrac{m}{N}\left(1 - \dfrac{m-1}{N-1}\right)$ contributing to $n = m + 1$.

(iii) B + A: $p_{\mathsf{B+A}} = \left(1 - \dfrac{m}{N}\right)\dfrac{m}{N-1}$ contributing to $n = m - 1$.

(iv) B + B: $p_{\mathsf{B+B}} = \left(1 - \dfrac{m}{N}\right)\left(1 - \dfrac{m}{N-1}\right)$ contributing to $n = m$.

It is readily verified that the probabilities of the four possible events sum to one: $p_{\mathsf{A+A}} + p_{\mathsf{A+B}} + p_{\mathsf{B+A}} + p_{\mathsf{B+B}} = 1$. The elements of the transition matrix can be written as

$$W_{nm} = \begin{cases} \dfrac{m}{N}\dfrac{m-N}{N-1} \,, & \text{if } n = m+1 \,, \\ \dfrac{m(m-1) + (N-m)(N-m-1)}{N(N-1)} \,, & \text{if } n = m \,, \\ \dfrac{m}{N}\dfrac{m-N}{N-1} \,, & \text{if } n = m-1 \,. \end{cases} \tag{5.50}$$

We check easily that W is a stochastic matrix, i.e., $\sum_n W_{nm} = 1$. The transition matrix W of the Moran model is tridiagonal since only the changes $\Delta n = 0, \pm 1$ can occur.

In the slightly modified version of the model—the second procedure above—we assume that the replicated individual, but not the offspring, is returned to the pool from which the dying individual is chosen after the replication event. Then the elements of the transition matrix are:

$$W_{nm} = \begin{cases} \dfrac{m(m-N)}{N^2} \,, & \text{if } n = m+1 \,, \\ \dfrac{m^2 + (N-m)^2}{N^2} \,, & \text{if } n = m \,, \\ \dfrac{m(m-N)}{N^2} \,, & \text{if } n = m-1 \,. \end{cases} \tag{5.50$'$}$$

Clearly, $\sum_n W_{nm} = 1$ is satisfied, as in the first procedure.

The transition matrix $\mathrm{W} = (W_{nm})$ has tridiagonal form and the eigenvalues and eigenvectors are readily calculated [147, 410, 411]. The results for the different picking procedures are almost the same. For the first procedure, we find

$$\lambda_k = 1 - \frac{k(k-1)}{N(N-1)} \,, \quad k = 0, 1, 2, \ldots \,, \tag{5.51}$$

and for the second, we get

$$\lambda_k = 1 - \frac{k(k-1)}{N^2} \,, \quad k = 0, 1, 2, \ldots \,. \tag{5.51$'$}$$

For the Moran model, the eigenvectors are the same for both procedures, and they are available in analytical form [411]. The first two eigenvectors belong to the doubly degenerate largest eigenvalue $\lambda_0 = \lambda_1 = 1$:

$$\zeta_0 = (1,0,0,0,0,\ldots,0)^{\mathrm{t}} , \qquad \zeta_1 = (0,0,0,0,0,\ldots,1)^{\mathrm{t}} , \tag{5.52}$$

and they describe the long-time behavior of the Moran process, since stationarity does indeed imply $\mathbf{p}(T+1)^{\mathrm{t}} = \mathbf{p}(T)^{\mathrm{t}} = \bar{\mathbf{p}}$, or $\mathrm{W}\bar{\mathbf{p}}^{\mathrm{t}} = \bar{\mathbf{p}}^{\mathrm{t}}$, and hence $\lambda = 1$. As in the Wright–Fisher model, we are dealing here with twofold degeneracy, and we recall that, in such a case, any properly normalized linear combination of the eigenvectors is a legitimate solution of the eigenvalue problem. Here we have to apply the L^1-norm and obtain

$$\eta = \alpha\zeta_0 + \beta\,\zeta_1 , \qquad \alpha + \beta = 1 ,$$

whence for the general solution of the stationary state, we find

$$\eta = (1-\vartheta,0,0,0,0,\ldots,\vartheta)^{\mathrm{t}} . \tag{5.53}$$

The interpretation of the stationary state, which is identical with the result for the Wright–Fisher process, is straightforward: the allele A goes into fixation in the population with probability ϑ, and it is lost with probability $1-\vartheta$. The Moran model, like the Wright–Fisher model, provides a simple explanation for gene fixation by random drift. The calculation of the value for ϑ, which depends on the initial conditions,[26] again assumed to be $n(0) = n_0$, follows the same reasoning as for the Wright–Fisher model in (5.49f) and (5.49g). From the generation-independent expectation value $\langle n(T)\rangle = n_0$, we obtain

$$\lim_{T\to\infty}\langle n(T)\rangle = N\vartheta = n_0 , \qquad \vartheta = \frac{n_0}{N} , \tag{5.49g$'$}$$

and finally, the probability of fixation of A is n_0/N. From the value of ϑ, it follows immediately that $\alpha = 1-\vartheta = (N-n_0)/N$ and $\beta = \vartheta = n_0/N$.

The third eigenvector belonging to the eigenvalue $\lambda_2 = 1 - 2/N(N-1)$ can be used to calculate the evolution towards fixation [56]:

$$\mathbf{p}(t) \approx \begin{pmatrix} 1-n_0/N \\ 0 \\ \vdots \\ 0 \\ n_0/N \end{pmatrix} + \frac{6n_0(N-n_0)}{N(N^2-1)} \begin{pmatrix} (N-1)/2 \\ -1 \\ \vdots \\ -1 \\ (N-1)/2 \end{pmatrix} \left(1-\frac{2}{N^2}\right)^T .$$

[26] In the non-degenerate case, stationary states do not depend on initial conditions, but this is no longer true for linear combinations of degenerate eigenvectors: α and β, and ϑ are functions of the initial state.

After a sufficiently long time, the probability density function becomes completely flat, except at the two boundaries $n = 0$ and $n = N$. We encountered the same form of the density for continuous time in the solution of the Fokker–Planck equation (Sect. 5.2.3), and we shall encounter it again with the solutions of the master equation (Sect. 5.3.2).

5.3.2 Master Equation of the Moran Process

Revisiting the two-allele Moran model (Sect. 5.3.1 and Fig. 5.25), we construct a master equation for the continuous time process and then make the approximations for large population sizes in the spirit of a Fokker–Planck equation. We recall the probabilities for the different combinations of choosing genes from the pool, and adopt the second procedure, which is simpler to calculate (Sect. 5.3.1). Again we have a gene pool of N genes, exactly m alleles of type A, and $N-m$ alleles of type B before the picking event. After the event the numbers have changed to n and $N-n$, respectively:

(i) $\mathsf{A}+\mathsf{A}$: $p_{\mathsf{A}+\mathsf{A}} = \dfrac{m^2}{N^2}$ contributing to $n = m$.

(ii) $\mathsf{A}+\mathsf{B}$: $p_{\mathsf{A}+\mathsf{B}} = \dfrac{m(N-m)}{N^2}$ contributing to $n = m+1$.

(iii) $\mathsf{B}+\mathsf{A}$: $p_{\mathsf{B}+\mathsf{A}} = \dfrac{(N-m)m}{N^2}$ contributing to $n = m-1$.

(iv) $\mathsf{B}+\mathsf{B}$: $p_{\mathsf{B}+\mathsf{B}} = \dfrac{(N-m)^2}{N^2}$ contributing to $n = m$.

These probabilities give rise to the same transition rates as before:

$$\begin{aligned} W(n+1|n) &= \kappa\frac{m(N-m)}{N^2}\,, \\ W(n|n) &= \kappa\frac{m^2+(N-m)^2}{N^2}\,, \\ W(n-1|n) &= \kappa\frac{(N-m)m}{N^2}\,, \end{aligned} \tag{5.54a}$$

where κ is a rate parameter. Apart from the two choices that do not change the composition of the urn, we have only two allowed transitions, as in the single-step birth-and-death process: (i) $n \to n+1$ with w_n^+ as transition probability and (ii) $n \to n-1$ with w_n^- as transition probability (see Sect. 3.2.3), and moreover the analytical expressions are the same for both. Therefore we are dealing with a *symmetric single-step process*:

$$w_n^+ = w_n^- = \kappa\frac{n(n-N)}{N^2}\,. \tag{5.54b}$$

It is of advantage to handle the neutral case and the natural selection simultaneously. We therefore introduce a selective advantage for allele A in the form of a factor $(1+\varrho)$. Then, for the reproduction of the fitter variant A, we have[27]

$$w_n^+ = \kappa \frac{n(n-N)}{N^2}(1+\varrho) \; . \tag{5.54b$'$}$$

The process is no longer symmetric, but we can return to the neutral case by putting $\varrho = 0$. The constant factor κ/N^2 can be absorbed in the time, which is measured in units $[N^2/\kappa]$. Then the master equation is of the form

$$\begin{aligned} \frac{\partial P_n(t)}{\partial t} &= w_{n+1}^- P_{n+1}(t) + w_{n-1}^+ P_{n-1}(t) - (w_n^- + w_n^+)P_n(t) \\ &= (n+1)(N-n+1)(1+\varrho)P_{n+1}(t) \\ &\quad +(n-1)(N-n-1)P_{n-1}(t) - n(N-n)(2+\varrho)P_n(t) \; . \end{aligned} \tag{5.54c}$$

An exact solution of the master equation (5.54c) has been derived by Bahram Houchmandzadeh and Marcel Vallade [264] for the neutral ($\varrho = 0$) and the natural selection case ($\varrho \neq 0$). It provides an exact reference and also gives unambiguous answers to a number of open questions. The approach to analytical solution of (5.54c) is the conventional one based on generating functions and partial differential equations, as used to solve the chemical master equations (Sect. 4.3). We repeat the somewhat technical procedure here, because it has general applicability, and one more example is quite instructive.

First we introduce the usual probability generating function (2.27),

$$g(s,t) = \sum_{n=0}^{N} s^n P_n(t) \; , \tag{2.27$'$}$$

and the following PDE is obtained in the way shown before:

$$\frac{\partial g(s,t)}{\partial t} = (1-s)\big(1-(1+\varrho)s\big)\frac{\partial}{\partial s}\left(Ng(s,t) - s\frac{\partial g(s,t)}{\partial s}\right) \; . \tag{5.54d}$$

Equation (5.54d) must now be solved for a given initial condition, for example, exactly n_0 alleles of type A at time $t = 0$:

$$P_n(0) = \delta_{n,n_0} \; , \quad \text{or } g(s,0) = s^{n_0} \; . \tag{5.54e}$$

The definition of the probability generating function implies the boundary conditions

$$g(1,t) = 1 \; . \tag{5.54f}$$

[27] In population genetics, the fitness parameter is conventionally denoted by s, but here we use ϱ in order to avoid confusion with the auxiliary variable s.

For $\varrho = 0$, we have the symmetric transition probabilities $w_n^+ = w_n^-$, whence the expectation value is

$$\mathrm{E}\big(n(t)\big) = \left.\frac{\partial g(s,t)}{\partial s}\right|_{s=1} = n_0 \,. \tag{5.54g}$$

For $\varrho \neq 0$, the point $s = 1/(1+\varrho) = \gamma$, where $\gamma = f^{-1}$, the reciprocal of the fitness, represents a fixed point of the PDE (5.54d) since $\big(1-(1+\varrho)s\big) = 0$:

$$\left.\frac{\partial g(s,t)}{\partial s}\right|_{s=\gamma} = 0 \,, \qquad g(\gamma,t) = g(\gamma,0) = \gamma^{n_0} \,. \tag{5.54h}$$

The beauty of this approach [264] is that the PDE (5.54d) with the initial condition (5.54e) and the boundary conditions (5.54fh) constitute a well defined problem, in contrast to the stochastic diffusion equation used in population genetics, which requires separate *ad hoc* assumptions for the limiting gene frequencies $x = 0$ and $x = 1$ (see Sect. 5.2.3 or [96, pp. 379–380]).

The Neutral Case $\varrho = 0$

The master equation for the Moran process (5.54c) is a system of $N + 1$ first order linear differential equations with the constraint $\sum_{n=0}^{N} P_n(t) = 1$, leaving N independent probability functions $P_n(t)$. Houchmandzadeh and Vallade [264] derive an explicit expression for these functions and we shall sketch their procedure here.

By separation of variables as shown for the Fokker–Planck equation (5.28c), the PDE (5.54d) is transformed into two ODEs:

$$\frac{\mathrm{d}\Phi(t)}{\mathrm{d}t} = \lambda\Phi(t) \,, \qquad (1-s)^2\frac{\mathrm{d}}{\mathrm{d}s}\left(N\psi(s) - s\frac{\mathrm{d}\psi(s)}{\mathrm{d}s}\right) = \lambda_n\psi(s) \,.$$

The time dependent equation is readily solved and yields

$$\Phi(t) = \theta\exp(\lambda t) \,,$$

where θ is a constant factor that has still to be determined. The solutions of the second ODE are obtained as eigenvalues λ_n and eigenfunctions $\psi_n(s)$ of the eigenvalue problem

$$\begin{aligned}\lambda_n\psi_n(s) &= (1-s)^2\frac{\mathrm{d}}{\mathrm{d}s}\left(N\psi_n(s) - s\frac{\mathrm{d}\psi_n(s)}{\mathrm{d}s}\right)\\ &= \mathcal{M}\psi_n(s) \,, \quad \text{with } \mathcal{M} = (1-s)^2\frac{\mathrm{d}}{\mathrm{d}s}\left(N - s\frac{\mathrm{d}}{\mathrm{d}s}\right) .\end{aligned} \tag{5.55a}$$

So far the strategy is completely analogous to the separation of variables procedure applied to the Fokker–Planck equation for the diffusion approximation to random selection in Sect. 5.2.3. Indeed, the solutions of (5.55a) can be given in terms of hypergeometric functions ${}_2F_1$ here too.

Houchmandzadeh and Vallade [264] present a direct derivation which makes use of the polynomial character of the solutions. For $\lambda_0 = 0$, we obtain the stationary solution since this is the only time independent term in the expansion. By straightforward integration of (5.55a), we find

$$N\bar{g}(s) - s\frac{\mathrm{d}\bar{g}(s)}{\mathrm{d}s} = K = \text{const.}$$

Second integration and determination of the two integration constants by the two boundary conditions (5.54f) for $\varrho = 0$ yields for ψ_0

$$\bar{g}(s) = \vartheta_N s^N + \vartheta_0 \,, \quad \text{with } \vartheta_N = \frac{n_0}{N} \,, \quad \vartheta_0 = \frac{N - n_0}{N} \,,$$

$$\lambda_0 = 0: \quad \psi_0(s) = \vartheta_0 + \vartheta_N s^N = \bar{g}(s) \,. \tag{5.55b}$$

For $\lambda \neq 0$, we try a polynomial expansion of $\psi(s)$. Equation (5.55a) implies that $s = 1$ is a double root of $\psi(s) = 0$, so we try polynomials in $(1 - s)$:

$$\psi(s) = \sum\nolimits_{k=0}^{N-1} a_k (1-s)^{k+1} \,. \tag{5.55c}$$

The first coefficient has to be zero, i.e., $a_0 = 0$, since the lowest term in the polynomial is $a_1(1 - s)^2$. The other coefficients are determined by expanding the expressions for $\mathrm{d}\psi/\mathrm{d}s$ and $\mathrm{d}^2\psi/\mathrm{d}s^2$ and collecting the terms of the same power in $(1 - s)$. One thereby obtains the recursion:

$$\big(\lambda + k(k+1)\big)a_k = k(k-N)a_{k-1} \,, \quad k = 1, \ldots, N-1 \,.$$

The relation for the first coefficient, i.e., $a_0 = 0$, implies that nontrivial solutions exist only for $\lambda = -n(n+1)$, for an integer n that is also used to label the eigenvalues λ_n and the eigenfunctions $\psi_n(s)$:

$$\lambda_n = -n(n+1): \ \psi_n(s) = \sum_{k=n}^{N-1} a_k^{(n)} (1-s)^{k+1} \,, \quad n = 1, \ldots, N-1 \,,$$

$$a_n^{(n)} = 1 \,, \quad a_k^{(n)} = \frac{k(N-k)}{n(n+1) - k(k+1)} a_{k-1}^{(n)} \,, \tag{5.55d}$$

$$k = n+1, \ldots, N-1 \,, \quad n, k \in \mathbb{N} \,.$$

Making use of the stationary solution ψ_0, we can express the probability generating function in terms of the eigenfunctions

$$g(s,t) = \vartheta_0 + \vartheta_N s^N + \sum_{n=1}^{N-1} C_n \psi_n(s) e^{\lambda_n t} , \tag{5.55e}$$

where the coefficients C_n are to be determined from the initial conditions, in particular, from $g(s,0) = s^{n_0}$. The probabilities $P_n(t)$ then follow in the conventional way from the expansion of the probability generating function $g(s,t)$ in powers of the variable s and identification of coefficients:

$$\begin{aligned} P_n(t) &= \vartheta_0 \delta_{n,0} + \vartheta_N \delta_{n,N} + (-1)^n \sum_{k=n}^{N} \binom{n}{k} \alpha_{k-1}(t) , \\ \text{where } \alpha_{k-1}(t) &= \sum_{n=1}^{N-1} C_n a_k^{(n)} e^{\lambda_n t} , \quad k = 1, \ldots, N-1 , \end{aligned} \tag{5.55f}$$

with $\alpha_{-1}(t) = \alpha_0(t) = 0$.

The recurrence relation (5.55d) also allows for direct computation of the coefficients:

$$a_k^{(n)} = \binom{k}{n} \frac{(1-N+n)^{(k-n)}}{(2n+2)^{(k-n)}} , \tag{5.55d$'$}$$

where the uncommon binomial coefficients are defined by $\binom{k}{n} = 0, \forall\, k < n$, and $x^{(n)} = \Gamma(x+n)/\Gamma(x)$ is the rising Pochhammer symbol. The coefficients $a_k^{(n)}$ are zero, except in the range $k \ge n$, and hence the relevant values fill an upper triangular $(N-1) \times (N-1)$ matrix $\mathrm{A} = (a_{nk} = a_k^{(n)})$ with all diagonal elements equal to unity, i.e., $a_n^{(n)} = 1$. In order to determine the coefficients C_n, we apply the initial condition $g(s,0) = s^{n_0}$, whence (5.55d) and (5.55e) imply, for $t = 0$,

$$s^{n_0} = \vartheta_0 + \vartheta_N s^N + \sum_{n=1}^{N-1} C_n \sum_{k=1}^{N-1} a_k^{(n)} (1-s)^{k+1} ,$$

$$\sum_{n=1}^{N-1} \sum_{k=1}^{N-1} C_n a_k^{(n)} (1-s)^{k+1} = s^{n_0} - \vartheta_0 - \vartheta_N s^N .$$

The C_k can be calculated directly by means of a hypergeometric function[28]:

$$C_n = (-1)^{n+1} n_0 \frac{(1-N)_n}{(n+1)_n} \, {}_3F_2(1-n_0, -n, n+1; 2, 1-N; 1) . \tag{5.55d$''$}$$

[28]The function ${}_3F_2$ belongs to the class of extended hypergeometric functions, referred to in *Mathematica* as HypergeometricPFQ.

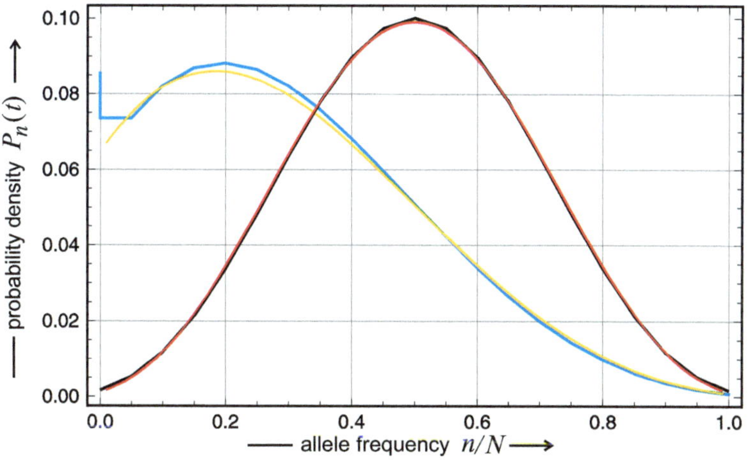

Fig. 5.26 Solution of the master equation of the Moran process. The figure compares exact solutions of the master equation for the Moran process [264] with the diffusion approximation of Motoo Kimura. The solution curves of the master equation computed from (5.55f) (*black* and *blue*) are compared with the results of the Fokker–Planck equation (5.28d) (*red* and *yellow*). The master equation provides results for the entire domain $n/N = x \in [0, 1]$ (*blue curve*) whereas the Fokker–Planck equation does not cover the margins $x \in]0, 1[$ (*yellow curve*). Choice of parameters for (i) symmetric initial conditions: $N = 20$, $n_0 = 10$, $t = 0.075$ [t] (*black* and *red*) and (ii) asymmetric initial conditions: $N = 20$, $n_0 = 6$, $t = 0.12$ [t] (*blue* and *yellow*)

This completes the exact solution of the neutral Moran master equation. Figure 5.26 compares the probability density computed by means of (5.55f) with the corresponding solutions (5.28d) of the Fokker–Planck equation for diffusion in genotype space. The agreement is excellent, apart from the values at the margins $x = 0$ and $x = 1$, which are perfectly reproduced by the solution of the master equation, but are not accessible by the diffusion approximation.

The Selection Case $\varrho \neq 0$
In the presence of selection, we have $\varrho \neq 0$ and the eigenvalue equation changes to

$$\lambda_n \gamma \psi_n(s) = (1-s)(\gamma - s)\frac{\mathrm{d}}{\mathrm{d}s}\left(N\psi_n(s) - s\frac{\mathrm{d}\psi_n(s)}{\mathrm{d}s}\right) , \tag{5.56a}$$

with $\gamma = 1/(1+\varrho)$. This ODE is known as Heun's equation [25]. The Heun polynomials and their eigenvalues have not yet been investigated in detail, in contrast to the hypergeometric functions, and there are no explicit formulas for Heun's polynomials [264]. Nevertheless, knowledge of the results for the small ϱ limit is often sufficient and then solutions of (5.56a) can be obtained by perturbation theory on powers of ϱ. First order results can be obtained by proper scaling from the solution of pure genetic drift ($\varrho = 0$). A change in the auxiliary variable, viz., $s \rightarrow y = 1 - s/\sqrt{\gamma}$, is appropriate and leads to[29]

$$-\sqrt{\gamma}\lambda\psi = \left(y^2 - \epsilon(y-1)\right)\frac{\mathrm{d}}{\mathrm{d}y}\left(N\psi + (1-y)\frac{\mathrm{d}\psi}{\mathrm{d}s}\right) ,$$
$$\text{where } \epsilon = \gamma + \frac{1}{\sqrt{\gamma}} - 2 = \frac{\varrho^2}{4} + \mathcal{O}(\varrho^3) . \qquad (5.56\text{b})$$

Since the first non-vanishing term in the perturbation expansion of ϵ is $\propto \varrho^2$, it is neglected in the first order perturbation calculation. After neglecting the residual term, (5.56b) has the same formal structure as (5.55a) for pure genetic drift, and this has already been solved. The probability generating function is now of the form

$$g(s,t) = \vartheta_0 + \vartheta_N s^N \sum_{n=1}^{N-1} C_n^{(1)} \psi_n^{(1)} \mathrm{e}^{-n(n+1)t/\sqrt{\gamma}} + \mathcal{O}(r^2) ,$$
$$\text{with } \psi_n^{(1)} = \sum_{k=1}^{N-1} a_k^{(n)} \left(1 - \frac{s}{\sqrt{\gamma}}\right)^{k+1} , \quad \lambda_n^{(1)} = -\frac{n(n+1)}{\sqrt{\gamma}} . \qquad (5.56\text{c})$$

The coefficients $a_k^{(n)}$ are the same as before, as given in (5.55d′), and the amplitudes $C_n^{(1)}$ are obtained again for the initial condition $g(s,0) = s^{n_0}$. Second and higher order perturbation theory can be used to extend the range of validity of the approach, but this gives rise to quite sophisticated expressions.

Another approximation is valid for large values of Ns, based on the fact that the term $s\,\partial g(s,t)/\partial s$ is then comparable in size to $Ng(s)$ only in the immediate neighborhood of $s = 1$ and can thus be neglected in the range $s \in [0, \gamma]$. The remaining approximate equation

$$\sigma \frac{\partial g}{\partial t} = N(1-s)(\sigma - s)\frac{\partial g}{\partial s} \qquad (5.56\text{d})$$

can be solved exactly and yields

$$g(s,t) = \left(\frac{(\gamma - s)\mathrm{e}^{-Nst} - \gamma(1-s)}{(\gamma - s)\mathrm{e}^{-Nst} - (1-s)}\right)^{n_0} . \qquad (5.56\text{e})$$

[29]The result for ε is easily obtained by making use of the infinite series

$$\sqrt{1+x} = 1 + \frac{1}{2}x - \frac{1}{8}x^2 + \frac{1}{16}x^3 + \cdots , \quad 1/\sqrt{1+x} = 1 - \frac{1}{2}x + \frac{3}{8}x^2 - \frac{5}{16}x^3 + \cdots ,$$

for small x.

This equation was found to be a good approximation for the probability generating function for $Ns \geq 2$ on the interval $[0, \gamma]$, but (5.56e) is not polynomial for $g(s, t)$ and the determination of the probabilities $P_n(t)$ is numerically ill-conditioned, except for small n. In particular, the expression for the probability of the loss of the allele A is very accurate:

$$P_0(t) = \left(\frac{1 - e^{-Nrt}}{1 + r - e^{-Nrt}} \right)^{n_0} . \tag{5.56f}$$

Finally, we consider the stationary solution $\lim t \to \infty$ in the natural selection case ($\varrho \neq 0$). Then the boundary condition (5.54f) has to be replaced by (5.54f′) and we obtain

$$\vartheta_N = \frac{1 - \gamma^{n_0}}{1 - \gamma^N} , \qquad \vartheta_0 = \frac{\gamma^{n_0} - \gamma^N}{1 - \gamma^N} , \tag{5.55b′}$$

for the two constants, where $\gamma = 1/(1 + \varrho)$ as before. The stationary probability can be calculated by comparing coefficients:

$$\lim_{t \to \infty} P_n(t) = \bar{P}_n = \vartheta_N \delta_{n,N} + \vartheta_0 \delta_{n,0} , \tag{5.56g}$$

where we can now identify ϑ_N and ϑ_0 as the total probability of fixation and the total probability for the loss of allele A, respectively.

5.3.3 *Models of Mutation*

Mutation is readily introduced into the Wright–Fisher model and the Moran model for the two allele case [56]: A mutates to B with probability u, while the back mutation B into A occurs with probability v. These parameters are probabilities per generation and they differ in the Wright–Fisher model by a factor N from those in the Moran model. In the two-allele case, we need only minor changes to calculate the solutions. The mutational event is introduced before we put the copy back into the urn: the offspring is mutated with probability u for A $\to$ B and v for B $\to$ A, or chosen to be identical with the parent with probabilities $(1 - u)$ or $(1 - v)$, respectively. Now the probabilities of the two alleles just after the event are

$$\begin{aligned} p_{\mathsf{A}}(n) &= (1 - u)\frac{n}{N} + v\left(1 - \frac{n}{N}\right) , \\ p_{\mathsf{B}}(n) &= u\frac{n}{N} + (1 - v)\left(1 - \frac{n}{N}\right) , \end{aligned} \tag{5.57a}$$

and we have to remember that, in the Wright–Fisher model, the new generation is created by sampling N times.

Mutation in the Wright–Fisher Process

In the Wright–Fisher model with mutation, the transition probability from m alleles A at generation T to n alleles A in the next generation $T+1$ is given by

$$W_{nm} = \binom{N}{n} p_{\mathsf{A}}(m)^n p_{\mathsf{B}}(m)^{N-n} \,,$$
$$\sum_{n=0}^{N} n W_{nm} = N p_{\mathsf{A}} = (1-u)m + v(N-m) \,. \tag{5.57b}$$

The calculation of the expectation value is straightforward and yields

$$\langle n_{T+1} \rangle = (1-u)\,\langle n_T \rangle + v\big(N - \langle n_T \rangle\big) \,. \tag{5.57c}$$

In (5.57c), the expectation value satisfies exactly the same difference equation as the deterministic variables in the equation for mutational change:

$$a_{T+1} = (1-u)a_T + v b_T \,,$$
$$\text{with solution } a_T = \frac{v}{u+v}\big(1 - (1-u-v)^T\big) + a_0(1-u-v)^T \,. \tag{5.57d}$$

Since $1-u-v$ is inevitably smaller than one, (5.57c) converges to the unique stable stationary state

$$\bar{a} = \frac{v}{u+v} \,, \qquad \bar{b} = \frac{u}{u+v} \,, \tag{5.57e}$$

and for non-vanishing mutation rates, no allele will die out, in contrast to the mutation-free case. Calculation of the probability density is more involved, but the eigenvalues of the transition matrix are readily obtained in analytical form:

$$\lambda_k = (1-u-v)^k \binom{N}{k} \frac{k!}{N^k} \,, \qquad k = 0, 1, \ldots, N \,. \tag{5.57f}$$

The largest eigenvector in the Wright–Fisher model is $\lambda_0 = 1$, and it corresponds to the long-time solution of the process.

Mutation in the Moran Process

For the Moran process with mutation, we use the following procedure: one allele from the current population is chosen to die and another one is drawn at random for replication, the mutational event is introduced, and we have once again

$\Delta n = n - m = \pm 1, 0$. This leads to the following expressions for the elements of the tridiagonal transition matrix in the Moran model:

$$W_{nm} = \begin{cases} (1-u)\left(1-\frac{n}{N}\right)\frac{n}{N} + v\left(1-\frac{n}{N}\right)^2, & \text{if } n = m+1\,, \\ (1-u)\left(\frac{n}{N}\right)^2 + (u+v)\frac{n}{N}\left(1-\frac{n}{N}\right) + (1-v)\left(1-\frac{n}{N}\right)^2, & \text{if } n = m\,, \\ u\left(1-\frac{n}{N}\right)^2 + (1-v)\left(1-\frac{n}{N}\right)\frac{n}{N}\,, & \text{if } n = m-1\,. \end{cases} \tag{5.57g}$$

Because of the mutation terms, the expectation value of the fraction of A alleles is no longer constant, and by calculating $\sum_n nW_{nm}$, we obtain

$$\langle n(t+\mathrm{d}t)\rangle = \langle n(t)\rangle + v\left(1 - \frac{\langle n(t)\rangle}{N}\right)\mathrm{d}t - u\frac{\langle n(t)\rangle}{N}\mathrm{d}t\,.$$

As in the Wright–Fisher case, the expectation value $\langle n(t)\rangle$ coincides with the deterministic frequencies of the allele A, viz., $a(t) = \langle n(t)\rangle/N$, and allele B, viz., $b(t) = 1 - a(t)$. We obtain the differential equation

$$N\frac{\mathrm{d}a(t)}{\mathrm{d}t} = v\big(1 - a(t)\big) - ua(t)\,. \tag{5.57h}$$

The factor N can be absorbed in the time axis, i.e., $\mathrm{d}t \to \mathrm{d}t/N$ or, as mentioned before, the mutation rates for the comparable Wright–Fisher process are a factor N larger than those of the Moran process. The solution of (5.57h) is obtained by integration:

$$a(t) = \frac{1}{u+v}\Big(v - \big(v - (u+v)a(0)\mathrm{e}^{-(u+v)t}\big)\Big)\,. \tag{5.57i}$$

The solution curve satisfies the following two limits: $\lim_{t\to 0} a(t) = a(0)$ and $\lim_{t\to\infty} a(t) = \bar{a} = v/(u+v)$, as in the case of the Wright–Fisher model. Nonzero mutation rates imply that neither of the two alleles can become fixed or die out, and this also implies that the temporal behaviour of the model with mutation is more complicated than the one for the mutation-free case. Nevertheless, solutions can be obtained [289]. We finish by giving the eigenvalues of the transition matrix:

$$\lambda_k = 1 - (u+v)\frac{k}{N} - (1-u-v)\frac{k(k-1)}{N^2}\,, \quad \varrho = 0, 1, \ldots, N\,. \tag{5.57j}$$

Mutation at the Molecular Level

Molecular genetics has provided insights into the mechanisms of mutation that provide a basis for a more adequate formulation of the mutation process, and the results are in agreement with replication kinetics [48]. As shown in Fig. 5.27, a carrier of genetic information, a DNA or RNA template, is bound to a replicating

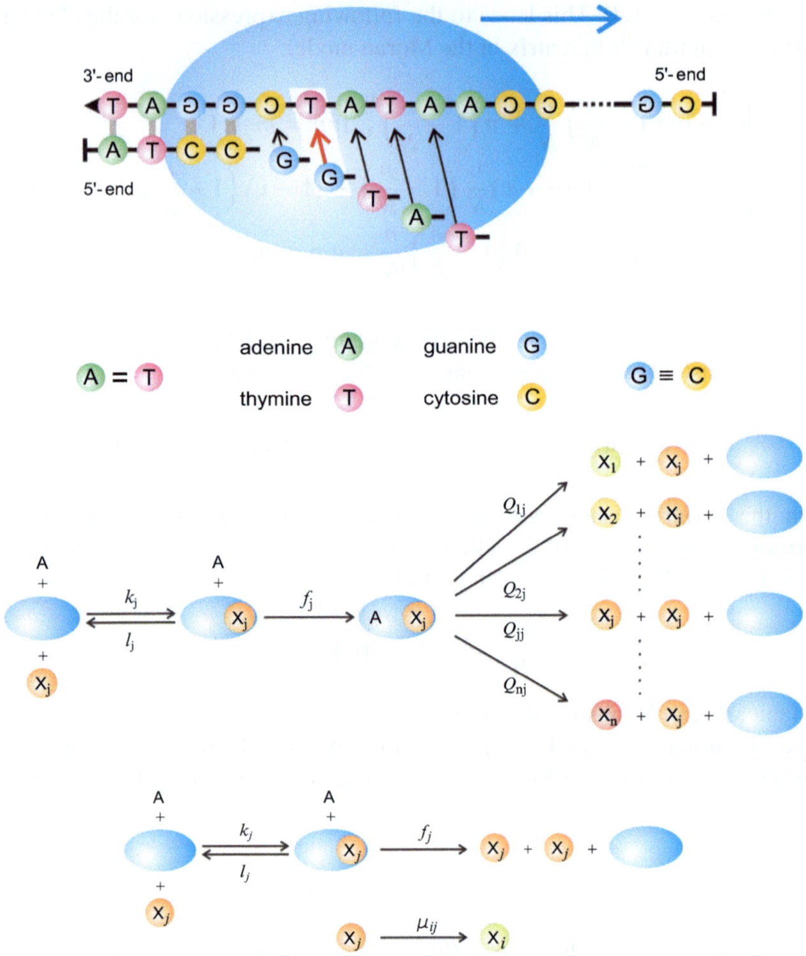

Fig. 5.27 Mechanisms of replication and mutation. *Upper*: Molecular principle of replication: a single stranded polynucleotide is completed to a double helix by making use of the base-pairing principle A = T and G ≡ C. Mutations are the result of mismatch pairings (as indicated in the *white rhomboid*). An example of this replication mechanism is the polymerase chain reaction (PCR), which constitutes the standard laboratory protocol for multiplying genetic information [141]. The replicating enzyme is a heat stable DNA polymerase isolated from the bacterium *Thermus aquaticus*. Cellular DNA replication is a much more complicated reaction network that involves some twenty different enzymes. The *other two pictures* show two different mutation mechanisms. *Middle*: Mechanism proposed by Manfred Eigen [130] and verified *in vitro* with RNA replicated by a phage-specific RNA replicase [48]. The template is bound to the enzyme and replicated digit by digit, as shown in the *top plot*. The reactions leading to correct copies and mutants represent parallel channels of the polymerisation reaction. The reaction parameters k_j and l_j describe binding and release of the template I_j, f_j measures the fitness of I_j, and Q_{ij} gives the frequency with which I_i is produced by copying the template I_j. The mechanism *at the bottom* interprets the reproduction–mutation model proposed by Crow and Kimura [96, p. 265, Eq. 6.4.1]: reproduction and mutation are completely independent processes, f_j is the fitness parameter, and μ_{ij} is the rate parameter for the mutation $I_j \rightarrow I_i$

enzyme and then copied digit by digit from the 3′-end to the 5′-end.[30] Correct copies require the complementary digit at every position, and mutations arise from mismatches in base pairs. Replication and mutation are parallel reaction channels in the model proposed by Manfred Eigen [130].

A simple but very useful and fairly accurate model is the uniform error-rate model, which makes the assumption that the accuracy of digit incorporation is independent of the nature of the digit, **A**, **T**, **G**, or **C**, and the position on the polynucleotide string. Then, all mutations for strings of length l can be expressed in terms of just two parameters, namely, the single-digit mutation rate per generation p, which is the probability of making a copying error, and the Hamming distance[31] $d_{\mathrm{H}}(i,j)$ between the template I_j and the mutant I_i. The probability for correct reproduction of a digit is $1-p$, and

$$Q_{ij} = (1-p)^l \varepsilon^{d_{\mathrm{H}}(i,j)} , \quad \text{with } \varepsilon = \frac{p}{1-p} , \tag{5.58a}$$

since $l - d_{\mathrm{H}}(i,j)$ digits have to be copied correctly, while $d_{\mathrm{H}}(i,j)$ digits mutate.

The mutation frequencies are subsumed in the mutation matrix Q. The molecular replication–mutation mechanism (Fig. 5.27 middle) requires that each copy is either correct or a mutant. Accordingly, Q is a stochastic $m \times m$ matrix:

$$\mathrm{Q} = \left(Q_{ij};\ i,j = 1,\ldots,m,\ \sum_{i=1}^{m} Q_{ij} = 1 \right) . \tag{5.58b}$$

Some simplifying assumptions like the uniform error-rate model lead to a symmetric matrix $Q_{ij} = Q_{ji}$, which is then a *bistochastic* matrix. The value matrix W is the product of a diagonal fitness matrix $\mathrm{F} = (f_{ij} = f_i \delta_{ij})$, $\mathrm{W} = \mathrm{QF}$, according to the mechanism shown in Fig. 5.27.

Crow and Kimura present a model [96] that leads to a formally identical mathematical problem as far as deterministic reaction kinetics is concerned (Sect. 5.2.5 and [484]):

$$\mathrm{W} = \mathrm{F} + \boldsymbol{\mu} , \tag{5.58c}$$

where $\boldsymbol{\mu} = (\mu_{ij})$ is the rate parameter for a mutation $\mathrm{I}_i \rightarrow \mathrm{I}_j$. Despite the formal identity, the interpretation of the Eigen and the Crow–Kimura model is different. As

[30] Nucleic acids are linear polymers and have two different ends with the hydroxy group in the 5′ position or in the 3′ position, respectively.

[31] The Hamming distance between two end-to-end aligned strings is the number of digits in which the two strings differ [235, 236].

shown in Fig. 5.27 (bottom), replication and mutation are two completely different processes and the Crow–Kimura approach refers exclusively to mutations of the genetic material occurring during the lifetime of individual, independently of the reproduction process, whereas the Eigen model considers only replication errors. Regarding the probabilistic details of reproduction–mutation kinetics, we refer to the molecular model presented in the next section.

Simulation of Molecular Evolution

Replication–mutation dynamics was studied in Sect. 5.2.5, where we were especially interested in the relation between continuous and discrete time models, as well as their asymptotic behavior for large particle numbers. Here we shall consider the role of fluctuations in the replication–mutation network of reactions. Since the number of subspecies or polynucleotide sequences increases exponentially with chain lengths, i.e., 2^l for binary and 4^l for four-letter alphabets, we can investigate only the smallest possible example with $l = 2$.

The deterministic reaction kinetics of the replication–mutation system has been extensively studied with the constraint of constant total concentrations of all subspecies [130–132]. Direct implementation of the mechanism in a master equation, however, leads to an instability. The expectation value for the total concentration is constant, but the variance diverges [287]. In order to study stochastic phenomena in replication–mutation dynamics and to avoid the instability, the dynamical system has to be embedded in some real physical device. Here the mechanism of replication and mutation was implemented in the flow reactor:

$$* \xrightarrow{a_0 r} \mathsf{A} \,, \tag{5.59a}$$

$$\mathsf{A} + \mathrm{I}_j \xrightarrow{w_{ij}=Q_{ij}f_j} \mathrm{I}_i + \mathrm{I}_j \,, \quad i,j = 1,\ldots,m \,, \tag{5.59b}$$

$$\mathsf{A} \xrightarrow{r} \varnothing \,, \tag{5.59c}$$

$$\mathrm{I}_j \xrightarrow{r} \varnothing \,, \quad j = 1,\ldots,m \,. \tag{5.59d}$$

Considering strings of digits to be replicated, for example polynucleotide sequences, m grows exponentially with the chain length l, and in the smallest possible case with $l = 2$, we are dealing with five independent variables. Although it is not difficult to write down a master equation for this mechanism, there is practically no chance to obtain analytical solutions.

However, simulation of the replication–mutation system by means of Gillespie's algorithm is straightforward. For $l = 2$, the 16 entries of the value matrix W are readily computed from the mutation rate p per site and generation, and the fitness values f_j for the four subspecies:

$$w_{ij} = (1-p)^l \varepsilon^{l-d_{\rm H}(i,j)} f_j \,, \quad i,j = 1,\ldots,m \,, \quad m = \kappa^l \,, \tag{5.59e}$$

where κ denotes the size of the alphabet from which the strings are built. The system can be further simplified by assuming a single fitter subspecies and assigning the same fitness value to all other variants: $f_1 = f_0$ and $f_2 = f_3 = \ldots = f_{\kappa^l} = f_n$. The notion of *single-peak fitness landscape* is common for this assignment. The replication–mutation system sustains a unique stationary state which has been called a *quasispecies* and which is characterized by a dominant subspecies, the *master sequence* I_0 with concentration x_0, surrounded by a cloud of less frequent mutants I_j with concentration x_j:

$$\bar{x}_0 = \frac{Q-\sigma_0^{-1}}{1-\sigma_0^{-1}} \,, \quad \bar{x}_j = \varepsilon^{l-d_{\rm H}(i,j)} \frac{Q-\sigma_0^{-1}}{(1-\sigma_0^{-1})^2} \,, \quad j = 1,\ldots,m-1 \,, \tag{5.59f}$$

with $Q = (1-p)^l$ and $\sigma_0 = f_0 / \big(\sum_{i=1}^{m-1} f_i x_i / \sum_{i=1}^{m-1} x_i \big)$. Equation (5.59f) is an approximation that already gives excellent results at small chain lengths l and becomes even better with increasing l. One prominent result is the existence of an *error threshold*: the existence of a well defined and unique stationary distribution of subspecies requires a replication accuracy above an error threshold. At mutation rates higher than the critical threshold value, no stationary distribution exists, and random replication is observed in the sense of diffusion in sequence space (for more detail on quasispecies theory, see, for example, [113, 131, 132, 484]).

Two examples of trajectory sampling for the replication–mutation system with $l = 2$ and different population sizes are shown in Fig. 5.28. Starting far away from equilibrium concentrations, the system passes through an initial phase where the expectation value $\mathrm{E}\big(\mathcal{X}_j(t)\big)$ of the stochastic model and the solution of the deterministic system are rather close. For long times, the expectation value converges to the stationary value of the deterministic system. However, the convergence is slow in the intermediate time range and substantial differences are observed, in particular, for small mutation rates and small particle numbers (Fig. 5.28 upper). A second question that can be addressed with stochastic simulations is the following: is the most frequent subspecies of the deterministic system also the most likely subspecies in the stochastic population? In other words, are the one standard deviation bands of the individual variants well separated or not? Figure 5.28 shows two scenarios. The bands are separated for sufficiently large populations, but they will overlap in smaller populations and then there is no guarantee that the variant which is found by isolating the most frequent subspecies is also the one with highest fitness. The metaphor of Darwinian selection as a hill-climbing process in genotype space [580] is only useful in sufficiently large populations.

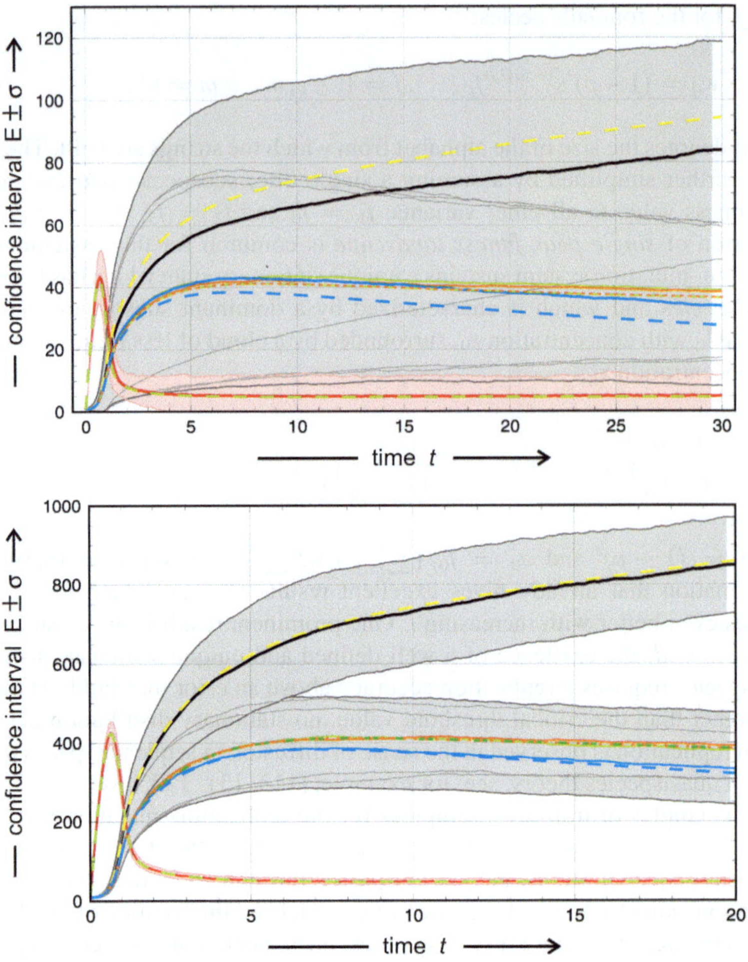

Fig. 5.28 Stochastic replication–mutation dynamics. Expectation values and one standard deviation error bands for quasispecies dynamics in the flow reactor, computed by stochastic simulation with the Gillespie algorithm. Individual curves show the numbers of building material molecules A (*red*) and the numbers of different subspecies: I_1 (*black*), I_2 (*orange*), I_3 (*chartreuse*), and I_4 (*blue*). The error bands are shown in *pink* for A and in *gray* for I_k, $k = 1, 2, 3, 4$. The *upper* and *lower plots* refer to concentrations $a_0 = 100$ and $a_0 = 1000$ molecules/V, respectively. In addition, the deterministic solution curves are shown as *dashed lines* for A (*chartreuse*), I_1 (*yellow*), I_2 (*orange*), I_3 (*chartreuse*), and I_4 (*blue*). Choice of other parameters: $r = 0.5\,[\mathrm{Vt}^{-1}]$, $l = 2$, $p = 0.025$, and $f_1 = 0.11\,[\mathrm{N}^{-1}\mathrm{t}^{-1}]$, and $f_2 = f_3 = f_4 = 0.10\,[\mathrm{N}^{-1}\mathrm{t}^{-1}]$ or $f_1 = 0.011\,[\mathrm{N}^{-1}\mathrm{t}^{-1}]$ and $f_2 = f_3 = f_4 = 0.010\,[\mathrm{N}^{-1}\mathrm{t}^{-1}]$ for the *upper* and *lower plots*, respectively

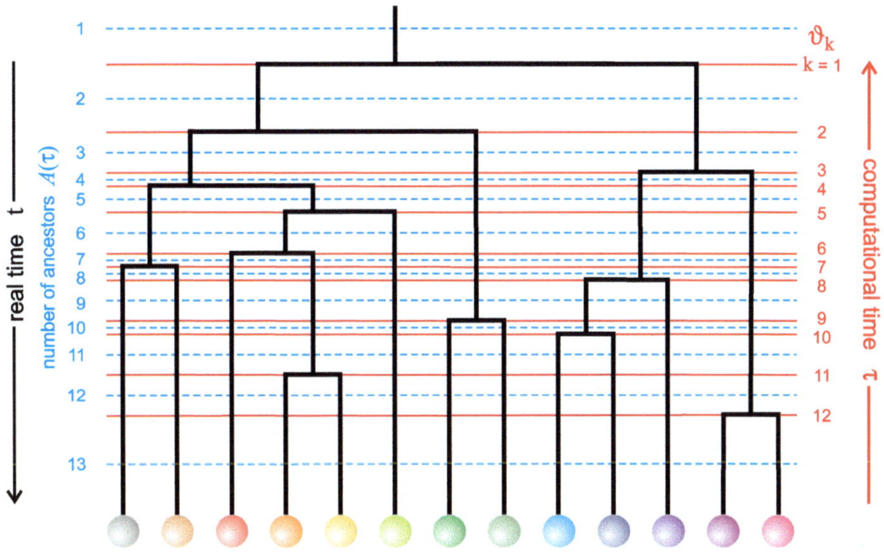

Fig. 5.29 Coalescence in ancestry. Reconstruction of the ancestry of a present day population containing 13 different alleles in the form of a phylogenetic tree that traces back to the most recent common ancestor (MRCA). We distinguish real time t (*black*) and computational time τ (*red*). Coalescence events ϑ_n are characterized by the number of ancestors $A(\tau)$ present at times before the event: $A(\tau) \leq n, \forall \tau \geq \vartheta_n$. Accordingly, we have exactly n ancestors in the time interval $\vartheta_n \leq \tau < \vartheta_{n-1}$

5.4 Coalescent Theory and Phylogenetic Reconstruction

The typical problem to be solved by coalescent theory is encountered in phylogenetic reconstruction. Given the current distribution of variants with known DNA sequences, the problem is to reconstruct the lines of descent (Fig. 5.29). Historically, the theory of phylogenetic coalescence was developed in the 1970s by the British mathematician John Kingman [307], and became popular through the reconstruction of a genealogical tree of human mitochondrial DNA, and the idea of a *mitochondrial Eve* [72]: the common ancestral mitochondrial genome was found to date back about $\tau_{\mathrm{MRCA}} = 200,000$ years (MRCA=most recent common ancestor).[32] The reconstruction of genealogies by sequence comparison from a sample of present day alleles, in principle a rather simple matter, is enormously complicated by the fact of variable and finite population sizes, recombination events, migration, and incomplete mixing in mating populations [114]. Mitochondrial DNA is haploid and mitochondria are inherited exclusively from the mother. The reconstruction of their genealogies is therefore simpler than that of autosomal genes. The maternal coalescent derived in [72] has been misinterpreted in the popular literature as saying

[32]Time is running backwards from the present $\tau = -t$, with today as the origin (Fig. 5.29).

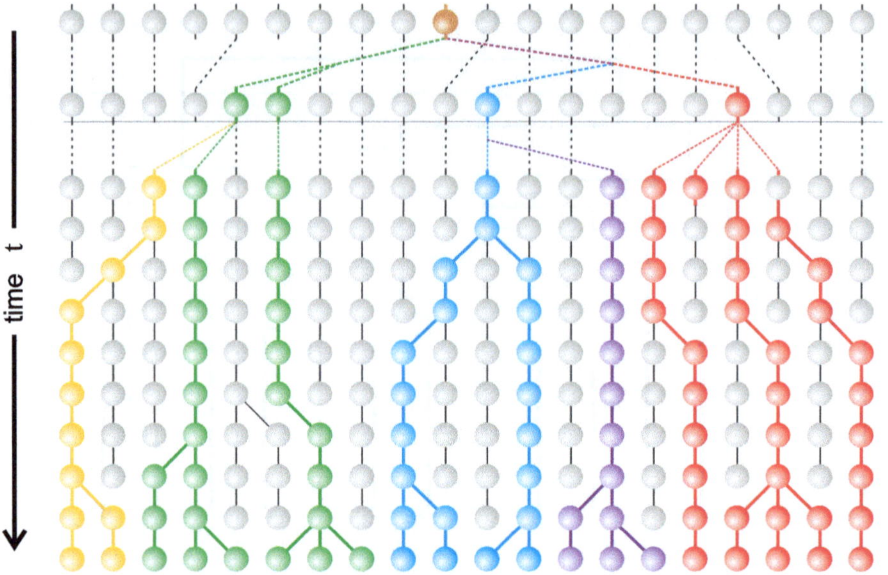

Fig. 5.30 Ancestral populations. The coalescence of all present day alleles in a population of constant size. Whenever a coalescence event happens as we go backwards in time, exactly one other branch that does not reach the current population has to die out. Coalescence events become rarer and rarer, the further we progress into the past. The last three generations shown are separated by many generations as indicated by the *broken lines*

that the entire present day human population descended from a *single woman* (see Fig. 5.30).

About eight years later the paternal counterpart was published in the form of the *Y-chromosomal Adam* [234].[33] Like the mitochondrial Eve, the Y-chromosomal Adam strongly supported the 'out-of-Africa' theory of modern humans, but the timing of the coalescent provided a kind of mystery: the mitochondrial Eve lived about 84,000 years earlier than the Y-chromosomal Adam. A very careful and technically elaborate evaluation of the available data confirmed this discrepancy [522]. Only very recently, and using new sets of data, has the somewhat disturbing issue been resolved [71, 185, 457]: the timing of the coalescent is $120,000 < t_{\mathrm{MRCA}} < 156,000$ for the Y-chromosomal Adam and $99,000 < t_{\mathrm{MRCA}} < 148,000$ for the mitochondrial Eve, and this time coincides roughly with the data from palaeoanthropology for the 'out-of-Africa' migration of *Homo sapiens*, dated between 125,000 and 60,000 years ago [393].

In order to illustrate coalescent theory [306, 525], we consider a haploid population with discrete nonoverlapping generations Γ_n $(n = 0, 1, \ldots)$ which evolves according to the Wright–Fisher model (Sect. 5.3.1). For the sake of convenience,

[33]The Y-chromosome in males is haploid and non-recombining.

the generation label is taken to run backwards in time: Γ_{n-1} is the generation of the immediate descendants of Γ_n and Γ_0 is the present day generation. Haploidy of genes is tantamount to the absence of recombination events that would substantially complicate the calculations. We are dealing with constant population size, so in each generation Γ_n the population contains exactly N alleles $X_i(n)$ of a gene that we assume to be labelled by indices $1, 2, \ldots, N$. Each member of generation Γ_{n-1} is the descendant of exactly one member of generation Γ_n, but the number of progeny of $X_i(n)$ is a random variable $\mathcal{N}_i$ subject to the constraint $\sum_{i=1}^{N} \mathcal{N}_i = N$. If all alleles are present in the population, the copy number has to be $\mathcal{N}_i = 1, \forall\, i = 1, \ldots, N$, or in other words, each allele occurs in a single copy only (see, for example, the generation Γ_0 in Fig. 5.29). The numbers $\mathcal{N}_i$ are assumed to have a symmetric multinomial distribution in the neutral Wright–Fisher model[34]:

$$P_{\nu_1,\ldots\nu_N} = P\big(\mathcal{N}_i = \nu_i\,,\ \ \forall\, i = 1, \ldots, N\big) = \frac{1}{N^N}\,\frac{N!}{\nu_1!\nu_2!\cdots\nu_N!}\,. \tag{2.43'}$$

For specific finite values of N and for arbitrary distributions of the ν values, the calculation of the probabilities is generally quite involved, but under the assumption of the validity of (2.43′), the process has a rather simple *backwards* structure and becomes tractable: the assumption of (2.43) implies equivalence with a process in which each member of generation Γ_{n-1} chooses its parent at random, independently and uniformly from the N individuals of generation Γ_n.

Let $A(\tau)$ be the number of ancestors of the present day population Γ_0 in generation Γ_n. Because of unique branching in the forward direction (branching backwards would violate the condition of unique ancestors), extrapolation backwards leads to fewer and fewer ancestors of the alleles from the present day populations, and $A(\tau)$ is a non-increasing step function in the direction of computational time, itself tantamount to the generation number $\tau = n : A(n+1) \le A(n)$. Coalescence events ϑ_k are characterized by their time of occurrence and the number of ancestors A which are present earlier than the event. Accordingly, we have k ancestors present in the interval $\tau(\vartheta_k) \le \tau < \tau(\vartheta_{k-1})$ (Fig. 5.29), and the last coalescence event corresponds to the time of the most recent common ancestor, i.e., $\vartheta_1 = \tau_{\mathrm{MRCA}}$. John Kingman provided a simple and straightforward estimate for the ϑ values. We consider two particular members $X_i(n)$ and $X_j(n)$ of generation Γ_n. They have the same parent in generation Γ_{n+1} with probability[35] N^{-1} and different parents with probability $1 - N^{-1}$. Accordingly, the probability that $X_i(n)$ and $X_j(n)$ have distinct parents but the same grandparent in generation Γ_{n+2} is simply $(1 - N^{-1})N^{-1}$, and the probability that they have no common ancestor in generation Γ_{n+s} is $(1 - N^{-1})^s$.

[34]This means that reproduction lies in the domain of neutral evolution, i.e., all fitness values are assumed to be the same, or in other words no effects of selection are observable and the numbers of descendants of the individual alleles $\mathcal{N}_1, \mathcal{N}_2, \ldots, \mathcal{N}_N$, are entirely determined by random events.

[35]Assume that $X_i(n)$ has the ancestor $X_i(n+1)$. The probability that $X_j(n)$ has the same ancestor is simply one out of N, i.e., $1/N$.

What we want to know, however, is the probability $\pi(N,s)$ that the entire generation Γ_n has a single common ancestor in the generation Γ_{n+s}. Again it is easier to calculate an estimate for the complement, which is the probability that all pairs of different alleles $X_i(n)$ and $X_j(n)$ with $i<j$ have distinct ancestors in Γ_{n+s}, that is, $1-\pi(N,s)$, for which upper and lower bounds can be obtained by simply summing the probabilities $(1-N^{-1})^s$ for one and for all possible pairs:

$$(1-N^{-1})^s \le 1-\pi(N,s) \le \frac{N(N-1)}{2}(1-N^{-1})^s \,.$$

As it turns out, the upper bound is very crude and it can be improved by replacing the number of all pairs by $\gamma(N-1)/(N+1)$, where γ is a constant for which the best choice is $\gamma = 3$ [305]. Then for large N, we obtain

$$sN^{-1} \le \pi(N,s) \le 3sN^{-1} - 2 \,.$$

For $s = N$, the two boundaries coincide at the value $\pi = 1$. In other words, the probability that all alleles in Γ_n have a common ancestor approaches one as s approaches N, and we have $\tau_{\mathrm{MRCA}} \approx N$ generations ago.

An alternative straightforward derivation yielding essentially the same result for the time of the most recent common ancestor [56, 261] assumes m different alleles in the initial generation Γ_0. Since we have $m(m-1)/2$ pairs of alleles in the generation immediately before, and $1/N$ is the probability of a coalescence event between two arbitrarily chosen alleles, we find $m(m-1)/2N$ for the probability that some pair of alleles coalesces. From this probability, it follows that

$$P_{m-1}(n) = P\big(\tau(\vartheta_{m-1}) = n\big) = \left(1-\frac{m(m-1)}{2N}\right)^{n-1}\frac{m(m-1)}{2N}\,, \tag{5.60}$$

for the time ϑ_{m-1} of the first coalescence event (Fig. 5.29), and

$$\mathrm{E}\big(\tau(\vartheta_{m-1})\big) = \sum_{k=0}^{\infty} k\left(1-\frac{m(m-1)}{2N}\right)^{k-1}\frac{m(m-1)}{2N} = \frac{2N}{m(m-1)}\,, \tag{5.61}$$

for the mean time back to this event, where we have used the expression

$$\sum_{k=0}^{\infty} ka(1-a)^{k-1} = \frac{1}{a}$$

for the infinite sum. The problem can now be solved by means of a nice recursion argument due to John Kingman. Since we now know the mean time until the first coalescence event, we can start with $m-1$ alleles and calculate the mean time span until the next event ϑ_{m-2}, and continue the series until we reach the last interval

$\tau(\vartheta_1) - \tau(\vartheta_2)$. To evaluate the finite sum, we may use the relation

$$\sum_{k=m+1}^{n} \frac{1}{k(k-1)} = \frac{1}{m} - \frac{1}{n} \,,$$

and obtain for $m = 1$ $\big(\tau(\vartheta_1) = \tau_{\mathrm{MRCA}}\big)$:

$$\begin{aligned} \langle \tau_{\mathrm{MRCA}} \rangle &= \sum_{k=2}^{n} \mathrm{E}\big(\tau(\vartheta_k)\big) = \sum_{k=2}^{n} \frac{2N}{k(k-1)} \\ &= 2N\left(1 - \frac{1}{n}\right) \approx 2N \,. \end{aligned} \tag{5.62}$$

Since coalescence events in this model are uncorrelated, we may generalize to intervals between arbitrarily chosen coalescence events:

$$\langle \tau(\vartheta_n) - \tau(\vartheta_m) \rangle = \sum_{k=m+1}^{n} \mathrm{E}\big(\tau(\vartheta_k)\big) = 2N\left(\frac{1}{m} - \frac{1}{n}\right) \,. \tag{5.63}$$

We point at a striking similarity between (5.63) and the intervals between sequential extinction times: the further we progress into the past, the longer the time spans between individual events. For example, the time τ_{MRCA} is about twice as long as the time to the last but one coalescence event ϑ_2.

Finally, we mention that realistic populations differ in many aspects from the idealized model introduced here. Structured populations may deviate from random choice, spatial effects and migration introduce deviations from the idealized concept, adaptive selection and recombination events complicate the neutral evolution scenario, and this is far from being a complete list of the relevant phenomena. In short, coalescent theory is a complex subject. Some features, but not all, are captured by the introduction of an effective population size N_{eff}. Sequence comparison as used for the reconstruction of phylogenies for data is an art in its own right (see, e.g., [414]). Maximum likelihood methods are frequently used in phylogeny [162]. For an application of Bayesian methods in the reconstruction of phylogenetic trees, see [259].

Notation

Mathematical Symbols

Symbol	Usage	Interpretation
$\{\ \}$	$\{A, B, \ldots\}$	A set consisting of elements $A, B, \ldots$
$\emptyset$		Empty set
Ω		Entire sample space, universe
$[\ ]$	$[a, b]$	An interval usually on the real line $a, b \in \mathbb{R}$
$\mid$	$\{A \mid C(A)\}$	Elements of A, which satisfy condition $C(A)$
$:$	$\{A : C(A)\}$	Elements of A, which satisfy condition $C(A)$
$\doteq$	$a \doteq b$	Definition
$\circ$	$T_2 \circ T_1(\cdot)$	Composition, sequential operation on $(\cdot)$
$*$	$f(t) * g(t)$	Convolution $\big(f * g\big)(t) \doteq \int_{-\infty}^{\infty} f(\tau)\, g(t-\tau)\mathrm{d}\tau$
$\star$	$f(t) \star g(t)$	Cross-correlation $\big(f \star g\big)(t) \doteq \int_{-\infty}^{\infty} f^*(\tau)\, g(t+\tau)\mathrm{d}\tau$
$\xrightarrow{\mathrm{d}}$	$\lim_{n\to\infty} \langle f(\mathcal{X}_n)\rangle \xrightarrow{\mathrm{d}} \langle f(\mathcal{X})\rangle$	Convergence in distribution
$\times$	$\mathbf{x} \times \mathbf{y}$	Cross product used for vectors in 3D space
$\times$	$\mathbf{e}_1 \times \ldots \times \mathbf{e}_n$	Cartesian product used for n-dimensional space
$\otimes$	$\mathcal{B}_1 \otimes \mathcal{B}_2$	Kronecker product of two Borel algebras
$\bigotimes_{k=1}^{n}$	$\mathcal{B}_1 \otimes \ldots \otimes \mathcal{B}_n$	Kronecker product used for n Borel algebras
log		Logarithm in general and logarithm to base 10
ln		Natural logarithm or logarithm to base e
ld		Logarithm to base 2

P. Schuster, *Stochasticity in Processes*, Springer Series in Synergetics,
DOI 10.1007/978-3-319-39502-9

Intervals	Symbols	Definitions	
Closed intervals	$[a,b]$	$[a,b]$	$= \{x \in \mathbb{R} : a \leq x \leq b\}$
Open intervals	$]a,b[$	$]a,b[$	$= \{x \in \mathbb{R} : a < x < b\}$
Left-hand half-open intervals	$]a,b]$	$]a,b]$	$= \{x \in \mathbb{R} : a < x \leq b\}$
Right-hand half-open intervals	$[a,b[$	$[a,b[$	$= \{x \in \mathbb{R} : a \leq x < b\}$

Number Systems and Function Spaces

Natural numbers	$\mathbb{N}$	$\{0, 1, 2, 3, \ldots\}$
	$\mathbb{N}_{>0}$	$\{1, 2, 3, \ldots\}$
Integers	$\mathbb{Z}$	$\{\ldots, -3, -2, -1, 0, 1, 2, 3, \cdots\}$
	$\mathbb{Z}_{>0} = \mathbb{N}_{>0}$	$\{1, 2, 3, \ldots\}$
	$\mathbb{Z}_{\geq 0} = \mathbb{N}$	$\{0, 1, 2, 3, \ldots\}$
	$\mathbb{Z}_{<0}$	$\{-1, -2, -3, \ldots\}$
	$\mathbb{Z}_{\leq 0}$	$\{0, -1, -2, -3, \ldots\}$
Rational numbers	$\mathbb{Q}$	$\left\{\frac{m}{n} \,\vert\, (m,n) \in \mathbb{Z} \wedge n \neq 0\right\}$
Real numbers	$\mathbb{R}$	$\left\{x \,\vert\, x \text{ is rational or irrational}\right\}$
Complex numbers	$\mathbb{C}$	$\left\{z = a + b\mathrm{i} \,\vert\, (a,b) \in \mathbb{R}\,,\ \mathrm{i} = \sqrt{-1}\right\}$

Functions	$\mathcal{C}(a,b)$	Continuous on the interval $[a,b]$
	$\mathcal{C}^{-1}$	Curves including discontinuities
	$\mathcal{C}^{0}$	Continuous curves
	$\mathcal{C}^{1}$	Continuous first derivatives
	$\mathcal{C}^{n}$	Continuous first through n th derivatives

Variables	Symbols	Functions
Discrete variables	$i, j, k, (l), n, m, \ldots, T$	$f_k\,,\ P_k(t)\,,\ \ldots$
Continuous variables	$x, y, z, \ldots, r, s, t$	$f(x)\,,\ P(x,t)\,,\ \ldots$
Random variables	$\mathcal{A}, \mathcal{B}, \ldots, \mathcal{X}, \mathcal{Y}, \mathcal{Z}$	
discrete	$(\mathcal{X}, \mathcal{Y}, \mathcal{Z}, \ldots) \in \mathbb{N}$	$f(\mathcal{N}) = f_n\,, \ldots$
continuous	$(\mathcal{X}, \mathcal{Y}, \mathcal{Z}, \ldots) \in \mathbb{R}$	$f(\mathcal{X}) = f(x)\,, \ldots$

Vectors and Matrices

$\mathbf{a}$	(a_x, a_y, a_z)	Vector in 3D space
$\mathbf{a}$	$(a_1, a_2, \ldots, a_n)$	Vector in nD space
$\mathbf{e}_k$	$(0, \ldots, 1, \ldots, 0)$	Unit vector in the direction of the kth coordinate
A	$(a_{ij};\ i = 1, \ldots, n;$ $j = 1, \ldots, m)$	$n \times m$ rectangular matrix with n rows and m columns
A	$(a_{ij};\ i, j = 1, \ldots, n)$	$n \times n$ square matrix

Some Common Distributions

Symbol	Notation	Name
π	$\pi(\alpha)$	Poisson distribution
B	$B(n, p)$	Binomial distribution
$\mathcal{N}$	$\mathcal{N}(\mu, \sigma^2)$	Normal distribution
φ	$\mathcal{N}(0, 1)$	Standardized normal distribution
$\mathcal{U}$	$\mathcal{U}(\Omega)$	Uniform distribution over sample space Ω

Some Special Functions	Symbols	Expressions
Gamma function	$\Gamma(x)$	$(x-1)!$
Beta function	$B(x, y)$	$\int_0^1 t^{x-1}(1-t)^{y-1}\,\mathrm{d}t$
Pochhammer symbols		
Rising	$x^{(n)}$	$x(x+1)\cdots(x+n-1) = \frac{\Gamma(x+n)}{\Gamma(x)}$
Falling	$(x)_n$	$x(x-1)\cdots(x-n+1) = \frac{\Gamma(x+1)}{\Gamma(x-n+1)}$

Frequently Used Transforms

Fourier transform: $\mathcal{F}\big(f(x)\big)(k) = \tilde{f}(k) = \frac{1}{\sqrt{2\pi}} \int\limits_{-\infty}^{\infty} f(x)\, e^{\imath\, \mathrm{k}\, \mathrm{x}}\, \mathrm{d}x$

Laplace transform: $\mathcal{L}\big(f(t)\big)(s) = \hat{f}(s) = \int\limits_{0}^{\infty} e^{-st} f(t)\ \mathrm{d}t$

Fourier-Laplace transform:

$$\mathcal{LF}\big(f(x,t)\big)(k,s) = \hat{\tilde{f}}(k,s) = \frac{1}{\sqrt{2\pi}} \int\limits_{0}^{\infty} \int\limits_{-\infty}^{\infty} e^{-st}\, e^{\imath\, \mathrm{k}\, \mathrm{x}} f(x,t)\, \mathrm{d}x\ \mathrm{d}t$$

Dimensions	Symbols	Arbitrary Unit	Common unit
Temperature	T	[T]	Degree Kelvin: 1 °K
Time, cont.	t	[t]	Second: 1 s
discrete	T	[T]	Generation: 1 generation
Length	l	[l]	meter: 1 m = 100 cm = 1000 mm
Volume	V	[V]	liter: $1\,\mathrm{l} = 1\,\mathrm{dm}^3 = 0.001\,\mathrm{m}^3$
Mass	m	[m]	gram: $1\,\mathrm{g} = 1000\,\mathrm{mg} = 1\times10^{-6}\,\mu\mathrm{g}$
Number density	N/V	[N]	1 particle per liter
Mole	mol	[mol]	mole: 1 mol = N_{L} particles
Concentration	mol/V	[M]	1 molar = 1 mol per liter

Chemical Species	Variables	Random variables
Specific entities: **A**, **B**, ... ,	$a = [\mathbf{A}]$, $b = [\mathbf{B}]$, ... ,	$\mathcal{X}_{\mathbf{A}}$, $\mathcal{X}_{\mathbf{B}}$, ... ,
Unspecific entities: A, B, ...,	$a = [\mathsf{A}]$, $b = [\mathsf{B}]$, ...,	$\mathcal{X}_{\mathsf{A}}$, $\mathcal{X}_{\mathsf{B}}$, ... ,
$\mathsf{X}_1, \mathsf{X}_2$, ...,	$x_1 = [\mathsf{X}_1]$, $x_2 = [\mathsf{X}_2]$, ...,	$\mathcal{X}_{\mathsf{X}_1}$, $\mathcal{X}_{\mathsf{X}_2}$, ... ,
Autocatalysts: X, Y, ...,	$x = [\mathsf{X}]$, $y = [\mathsf{Y}]$, ...,	$\mathcal{X}_{\mathsf{X}}$, $\mathcal{X}_{\mathsf{Y}}$, ... ,
Individual molecules: A, B, ...		

Kinetics	Symbol	Usage
Rate function:	$v_\mu(\mathbf{x}) = \gamma_\mu h_\mu(\mathbf{x})$	Deterministic reaction $\mathbf{R}_\mu$
	$\chi_\mu(\mathbf{n}) = \gamma_\mu h_\mu(\mathbf{n})$	Stochastic reaction $\mathbf{R}_\mu$
Rate parameter:	γ_μ	Reaction $\mathbf{R}_\mu$
	k_μ, l_μ	Deterministic reaction $\mathbf{R}_\mu$
	κ_μ, λ_μ	Stochastic reaction $\mathbf{R}_\mu$
Stoichiometric factor:	$h_\mu(\mathbf{x}) = \prod_{j=1}^{M} x_j^{\nu_{j\mu}}$	Deterministic reaction $\mathbf{R}_\mu$
	$h_\mu(\mathbf{n}) = \prod_{j=1}^{M} (n_j)_{\nu_{j\mu}}$	Stochastic reaction $\mathbf{R}_\mu$
Extent of reaction:	$\boldsymbol{\xi}_\mu(\mathbf{x}) = \frac{\mathbf{x}-\bar{\mathbf{x}}}{\nu_\mu' - \nu_\mu}$	Deterministic reaction $\mathbf{R}_\mu$

References

1. Aase, K.: A note on a singular diffusion equation in population genetics. J. Appl. Probab. **13**, 1–8 (1976)
2. Abramowitz, M., Segun, I.A. (eds.): Handbook of Mathematical Functions with Formulas, Graphs, and Mathematical Tables. Dover Publications, New York (1965)
3. Abramson, M., Moser, W.O.J.: More birthday surprises. Am. Math. Monthly **77**, 856–858 (1970)
4. Acton, F.S.: Numerical Methods That Work. Harper & Row, New York (1970)
5. Acton, F.S.: Numerical Methods That (Usually) Work, fourth printing edn. Mathematical Association of America, Washington, DC (1990)
6. Adams, W.J.: The Life and Times of the Central Limit Theorem, *History of Mathematics*, vol. 35, 2nd edn. American Mathematical Society and London Mathematical Society, Providence, RI (2009). Articles by A. M. Lyapunov translated from the Russian by Hal McFaden.
7. Al-Soufi, W., Reija, B., Novo, M., Kelekyan, S., Kühnemuth, R., Seidel, C.A.M.: Fluorescence correlation sprctroscopy, a tool to inverstigate supramolecular dynamics: Inclusion complexes of pyronines with cyclodextrin. J. Am. Chem. Soc. **127**, 8775–8784 (2005)
8. Aldrich, J.: R. A. Fisher and the making of the maximum likelihood 1912–1922. Stat. Sci. **12**, 162–176 (1997)
9. Alonso, D., McKane, A.J., Pascual, M.: Stochastic amplifications in epidemics. J. Roy. Soc. Interface **4**, 575–582 (2007)
10. Anderson, B.D.O.: Reverse-time diffusion equation models. Stoch. Process. Appl. **12**, 313–326 (1982)
11. Anderson, D.F.: Incorporating postleap checks in tau-leaping. J. Chem. Phys. **128**, e 054103 (2008)
12. Anderson, D.F., Craciun, G., Kurtz, T.G.: Product-form stationary distributions for deficiency zero chemical reaction networks. Bull. Math. Biol. **72**, 1947–1970 (2010)
13. Anderson, D.F., Ganguly, A., Kurtz, T.G.: Error analysis of tau-leap simulation methods. Ann. Appl. Probab. **6**, 2226–2262 (2011)
14. Anderson, P.W.: More is different. Broken symmetry and the nature of the hierarchical stucture of science. Science **177**, 393–396 (1972)
15. Anderson, R.M., May, R.M.: Population biology of infectious diseases: Part I. Nature **280**, 361–367 (1979)
16. Anderson, R.M., May, R.M.: Population biology of infectious diseases: Part II. Nature **280**, 455–461 (1979)

P. Schuster, *Stochasticity in Processes*, Springer Series in Synergetics,
DOI 10.1007/978-3-319-39502-9

17. Anderson, R.M., May, R.M.: Infectious Diseases of Humans: Dynamics and Control. Oxford University Press, New York (1991)
18. Applebaum, D.: Lévy processes – From probability to finance and quantum groups. Not. Am. Math. Soc. **51**, 1336–1347 (2004)
19. Aragón, S.R., Pecora, R.: Fluorescence correlation spectroscopy and Brownian rotational diffusion. Biopolymers **14**, 119–138 (1975)
20. Arányi, P., Tóth, J.: A full stochastic description of the Michaelis-Menten reaction for small systems. Acta Biochim. et Biophys. Acad. Sci. Hung. **12**, 375–388 (1977)
21. Arfken, G.B., Weber, H.J.: Mathematical Methods for Physicists, fifth edn. Harcourt Academic Press, San Diego (2001)
22. Arnold, L.: Stochastic Differential Equations. Theory and Applications. Wiley, New York (1974)
23. Arnold, L.: Random Dynamical Systems. Springer, Berlin (1998). Second corrected printing 2003
24. Arnold, L., Bleckert, G., Schenk-Hoppé, K.R.: The stochastic brusselator: Parametric noise destroys hopf bifurcation. In: Crauel, H., Gundlach, M. (eds.) Stochastic Dynamics, chap. 4, pp. 71–92. Springer, New York (1999)
25. Arscott, F.M.: Heun's equation. In: Ronveau, A. (ed.) Heun's Differential Equations, pp. 3–86. Oxford University Press, New York (1955)
26. Arslan, E., Laurenzi, I.J.: Kinetics of autocatalysis in small systems. J. Chem. Phys. **128**, e 015101 (2008)
27. Asmussen, S., Glynn, P.W.: Stochastic Simulation: Algortihms and Analysis. Springer, New York (2007)
28. Aster, R.C., Borchers, B., Thurber, C.H.: Parameter Estimation and Inverse Problems, 2nd edn. Academic Press, Elsevier, Singapore (2013)
29. Athreya, K.B., Ney, P.E.: Branching Processes. Springer, Heidelberg, DE (1972)
30. Atkins, P.W., Friedman, R.S. (eds.): Molecular Quantum Mechanics, fifth edn. Oxford University Press, Oxford (2010)
31. Bachelier, L.: Théorie de la spéculation. Annales scientifiques de l'É.N.S. 3e série **17**, 21–86 (1900)
32. Bailey, N.T.J.: A simple stochastic epidemic. Biometrika **37**, 193–202 (1950)
33. Bailey, N.T.J.: The Elements of Stochastic Processes with Application in the Natural Sciences. Wiley, New York (1964)
34. Bar-Eli, K., Noyes, R.M.: Detailed calculations of multiple steady states during oxidation of cerous ion by bromate in a stirred flow reactor. J. Phys. Chem. **82**, 1352–1359 (1978)
35. Bartholomay, A.F.: On the linear birth and death processes of biology as Markoff chains. Bull. Math. Biophys. **20**, 97–118 (1958)
36. Bartholomay, A.F.: Stochastic models for chemical reactions: I. Theory of the unimolecular reaction process. Bull. Math. Biophys. **20**, 175–190 (1958)
37. Bartholomay, A.F.: Stochastic models for chemical reactions: II. The unimolecular rate constant. Bull. Math. Biophys. **21**, 363–373 (1959)
38. Bartholomay, A.F.: A stochastic approach to statistical kinetics with applications to enzyme kinetics. Biochemistry **1**, 223–230 (1962)
39. Bartlett, M.S.: Stochastic processes or the statistics of change. J. R. Stat. Soc. C **2**, 44–64 (1953)
40. Bazley, N.W., Montroll, E.W., Rubin, R.J., Shuler, K.E.: Studies in nonequilibrium rate processes: III. The vibrational relaxation of a system of anharmonic oscillators. J. Chem. Phys. **28**, 700–704 (1958). Erratum: *J.Chem.Phys.*, 29:1185–1186
41. Berg, J.M., Tymoczko, J.L., Stryer, L.: Biochemistry, fifth edn. W. H. Freeman and Company, New York (2002)
42. Berg, J.M., Tymoczko, J.L., Stryer, L.: Biochemistry, seventh edn. W. H. Freeman and Company, New York (2012)
43. Bergström, H.: On some expansions of stable distribution functions. Ark. Math. **2**, 375–378 (1952)

44. Bernoulli, D.: Essai d'une nouvelle analyse de la mortaltié causée par la petite vérole et des avantages de l'inoculation pour la prévenir. Mém. Math. Phys. Acad. Roy. Sci.,Paris **T5**, 1–45 (1766). English translation: 'An Attempt at a New Analysis of the Mortality Caused by Smallpox and of the Advantages of Inoculation to Prevent It.' In: L. Bradley, Smallpox Inoculation: An Eighteenth Century Mathematical Controversy. Adult Education Department: Nottingham 1971, p. 21
45. Bernoulli, D., Blower, S.: An attempt at a new analysis of the mortality caused by smallpox and of the advantages of inoculation to prevent it. Rev. Med. Virol. **14**, 275–288 (2004)
46. Berry, R.S., Rice, S.A., Ross, J.: Physical Chemistry, 2nd edn. Oxford University Press, New York (2000)
47. Berry, R.S., Rice, S.A., Ross, J.: Physical and Chemical Kinetics, 2nd edn. Oxford University Press, New York (2002)
48. Biebricher, C.K., Eigen, M., William C. Gardiner, J.: Kinetics of RNA replication. Biochemistry **22**, 2544–2559 (1983)
49. Bienaymé, I.J.: Da la loi de Multiplication et de la durée des familles. Soc. Philomath. Paris Extraits **Ser. 5**, 37–39 (1845)
50. Billingsley, P.: Probability and Measure, 3rd edn. Wiley-Interscience, New York (1995)
51. Billingsley, P.: Probability and Measure, Anniversary edn. Wiley-Interscience, Hoboken (2012)
52. Binnig, G., Quate, C.F., Gerber, C.: Atomic force microscopy. Phys. Rev. Lett. **56**, 930–933 (1986)
53. Birkhoff, G.D.: Proof of the ergodic theorem. Proc. Natl. Acad. sci. USA **17**, 656–660 (1931)
54. Björck, Å.: Numerical Methods for Least Square Problems. Other Titles in Applied Mathematics. SIAM Society for Industrial & Applied Mathematics, Philadelphia (1996)
55. Bloomfield, V.A., Benbasat, J.A.: Inelastic light-scattering study of macromolecular reaction kinetics. I: The reactions $A \rightleftharpoons B$ and $2A \rightleftharpoons A_2$. Macromolecules **4**, 609–613 (1971)
56. Blythe, R.A., McKane, A.J.: Stochastic models of evolution in genetics, ecology and linguistics. J. Stat. Mech. Theor. Exp. (2007). P07018
57. Boas, M.L.: Mathematical Methods in the Physical Sciences, 3rd edn. Wiley, Hoboken (2006)
58. Boole, G.: An Investigation of the Laws of Thought on which Are Founded the Mathematical Theories of Logic and Probabilities. MacMillan, London (1854). Reprinted by Dover Publ. Co., New York, 1958
59. Born, M., Oppenheimer, R.: Zur Quantentheorie der Moleküle. Annalen der Physik **84**, 457–484 (1927). In German
60. Börsch, A., Simon, P. (eds.): Carl Friedrich Gauß: Abhandlungen zur Methode der kleinsten Quadrate. P. Stankiewicz, Berlin (1887). In German
61. Bouchaud, J.P., Georges, A.: Anomalous diffusion in disordered madia: Statistical mechanisms, models and physical applications. Phys. Rep. **195**, 127–293 (1990)
62. Box, G.E.P., Muller, M.E.: A note on the generation of random normal deviates. Ann. Math. Stat. **29**, 610–611 (1958)
63. Brenner, S.: Theoretical biology in the third millenium. Philos. Trans. R. Soc. Lond. B **354**, 1963–1965 (1999)
64. Brenner, S.: Hunters and gatherers. Scientist **16**(4), 14 (2002)
65. Briggs, G.E., Haldane, J.B.S.: A note on the kinetics of enzyme action. Biochem. J. **19**, 338–339 (1925)
66. Brockmann, D., Hufnagel, L., Geisel, T.: The scaling laws of human travel. Nature **439**, 462–465 (2006)
67. Brockwell, P.J., Davis, R.A.: Introduction to Time Series and Forecasting. Springer, New York (1996)
68. Brockwell, P.J., Davis, R.A., Yang, Y.: Continuous-time Gaussian autoregression. Stat. Sin. **17**, 63–80 (2007)
69. Brown, R.: A brief description of microscopical observations made in the months of June, July and August 1827, on the particles contained in the pollen of plants, and on the general existence of active molecules in organic and inorganic bodies. Phil. Mag. Ser. 2 **4**, 161–173

(1828). First Publication: The Edinburgh New Philosophical Journal. July-September 1828, pp. 358–371
70. Calaprice, A. (ed.): The Ultimate Quotable Einstein. Princeton University Press, Princeton (2010)
71. Cann, R.L.: Y weigh in again on modern humans. Science **341**, 465–467 (2013)
72. Cann, R.L., Stoneking, M., Wilson, A.C.: Mitochondrial DNA and human evolution. Nature **325**, 31–36 (1987)
73. Cao, Y., Gillespie, D.T., Petzold, L.R.: Efficient step size selection for the tau-leaping simulation method. J. Chem. Phys. **124**, 044,109 (2004)
74. Cao, Y., Gillespie, D.T., Petzold, L.R.: Avoiding negative populations in explicit Poisson tau-leaping. J. Chem. Phys. **123**, e054,104 (2005)
75. Cao, Y., Gillespie, D.T., Petzold, L.R.: Efficient step size selection for the tau-leaping simulation method. J. Chem. Phys. **124**, e044,109 (2006)
76. Cao, Y., Gillespie, D.T., Petzold, L.R.: Adaptive explicit-implicit tau-leaping method with automatic tau selection. J. Chem. Phys. **126**, e224,101 (2007)
77. Carter, M., van Brunt, B.: The Lebesgue-Stieltjes Integral. A Practical Introduction. Springer, Berlin (2007)
78. Cassandras, C.G., Lygeros, J. (eds.): Stochastic Hybrid Systems. Control of Engineering Series. CRC Press, Taylor & Francis Group, Boca Raton (2007)
79. Castets, V., Dulos, E., Boissonade, J., De Kepper, P.: Exprimental evidence of a sustained standing Turing-type nonequilibrium xhemical pattern. Phys. Rev. Lett. **64**, 2953–2956 (1990)
80. Chang, C., Gzyl, H.: Parameter estimation in superposition of decaying exponentials. Appl. Math. Comput. **96**, 101–116 (1998)
81. Chechkin, A.V., Metzler, R., Klafter, J., Gonchar, V.Y.: Introduction to the theory of Lévy flights. In: R. Klages, G. Radons, I.M. Sokolov (eds.) Anomalous Transport: Foundations and Applications, chap. 5, pp. 129–162. Wiley-VCH Verlag GmbH, Weinheim, DE (2008)
82. Child, M.S.: Molecular Collision Theory. Dover Publications, Mineola (1996). Originally publisher: Academic Press, London (1974)
83. Chung, K.L.: A Course in Probability Theory, *Probability and Mathematical Statistics*, vol. 21, 2nd edn. Academic Press, New York (1974)
84. Chung, K.L.: Elementary Probability Theory with Stochastic Processes, 3rd edn. Springer, New York (1979)
85. Cochran, W.G.: The distribution of quadratic forms in normal systems, with applications to the analysis of covariance. Math. Proc. Camb. Philos. Soc. **30**, 178–191 (1934)
86. Conrad, K.: Probability distributions and maximum entropy. Expository paper, University of Connecticut, Storrs, CT (2005)
87. Cook, M., Soloveichik, D., Winfree, E., Bruck, J.: Programmability of chemical reaction networks. In: Condon, A., Harel, D., Kok, J.N., Salomaa, A., Winfree, E. (eds.) Algorithimc Bioprocesses, *Natural Computing Series*, vol. XX, pp. 543–584. Springer, Berlin (2009)
88. Cooper, B.E.: Statistics for Experimentalists. Pergamon Press, Oxford (1969)
89. Cortina Borja, M., Haigh, J.: The birthday problem. Significance **4**, 124–127 (2007)
90. Cover, T.M., Thomas, J.A.: Elements of Information Theory, 2nd edn. Wiley, Hoboken (2006)
91. Cox, D.R., Miller, H.D.: The Theory of Stochastic Processes. Methuen, London (1965)
92. Cox, R.T.: The Algebra of Probable Inference. The John Hopkins Press, Baltimore (1961)
93. Craciun, G., Tang, Y., Feinberg, M.: Understanding bistability in complex enzyme-driven reaction networks. Proc. Natl. Acad. Sci. USA **103**, 8697–8702 (2006)
94. Cramér, H.: Mathematical Methods of Statistics. Princeton Univ. Press, Priceton (1946)
95. Crank, J.: The Mathematics of Diffusion. Clarendon Press, Oxford (1956)
96. Crow, J.F., Kimura, M.: An Introduction to Population Genetics Theory. Sinauer Associates, Sunderland (1970). Reprinted at **The Blackburn Press**, Caldwell (2009)
97. Cull, P., Flahive, M., Robson, R.: Difference Equations. From Rabbits to Chaos. Undergraduate Texts in Mathematics. Springer, New York (2005)

98. Dalla Valle, J.M.: Note on the Heaviside expansion formula. Proc. Natl. Acad. Sci. USA **17**, 678–684 (1931)
99. Darvey, I.G., Ninham, B.W.: Stochastic models for second-order chemical reaction kinetics. Time course of reactions. J. Chem. Phys. **46**, 1626–1645 (1967)
100. Darvey, I.G., Ninham, B.W., Staff, P.J.: Stochastic models for second-order chemical reaction kinetics. The equilibirum state. J. Chem. Phys. **45**, 2145–2155 (1966)
101. Darvey, I.G., Staff, P.J.: Stochastic approach to first-order chemical reaction kinetics. J. Chem. Phys. **44**, 990–997 (1966)
102. De Candolle, A.: Zur Geschichte der Wissenschaften und Gelehrten seit zwei Jahrhunderten nebst anderen Studien über wissenschaftliche Gegenstände insbesondere über Vererbung und Selektion beim Menschen. Akademische Verlagsgesellschaft, Leipzig, DE (1921). Deutsche Übersetzung der Originalausgabe **"Histoire des sciences et des savants depuis deux siècle"**, Geneve 1873, durch Wilhelm Ostwald.
103. DeKepper, P., Epstein, I.R., Kustin, K.: Bistability in the oxidatiion of arsenite by iodate in a stirred flow reactor. J. Am. Chem. Soc. **103**, 6121–6127 (1981)
104. Delbrück, M.: Statistical fluctuations in autocatalytic reactions. J. Chem. Phys. **8**, 120–124 (1940)
105. Demetrius, L., Schuster, P., Sigmund, K.: Polynucleotide evolution and branching processes. Bull. Math. Biol. **47**, 239–262 (1985)
106. Devroye, L.: Non-Uniform Random Variate Generation. Springer, New York (1986)
107. Diekmann, O., Heesterbeek, J.A.P.: Mathematical Epidemiology of Infectious Diseases: Model Building, Analysis and Interpretation. Wiley Series in Mathematical and Computational Biology. Princeton University Press, Hoboken (2000)
108. Diekmann, O., Heesterbeek, J.A.P., Britton, T.: Mathematical Tools for Understanding Infectious Disease Dynamics. Princeton Series in Theoretical and Computational Biology. Princeton University Press, Princeton (2012)
109. Dietz, K.: Epidemics and rumors: A survey. J. R. Stat. Soc. A **130**, 505–528 (1967)
110. Dietz, K., Heesterbeeck, J.A.P.: Daniel Bernoulli's epidemiological model revisited. Math. Biosci. **180**, 1–21 (2002)
111. Djermoune, E.H., Tomczak, M.: Statistical analysis of the Kumaresan-Tufts and matrix pencil methods in estimating damped sinusoids. In: Hlawatsch, F., Matz, G., Rupp, M., Wistawel, B. (eds.) Proceedings of the XII. European Signal Processing Conference, vol. II, pp. 1261–1264. Technische Universität Wien, Wien (2004)
112. Domingo, E., Parrish, C.R., John J, H. (eds.): Origin and Evolution of Viruses, 2nd edn. Elsevier, Academic Press, Amsterdam, NL (2008)
113. Domingo, E., Schuster, P. (eds.): Quasispecies: From Theory to Experimental Systems, *Current Topics in Microbiology and Immunology*, vol. 392. Springer, Berlin (2016)
114. Donnelly, P.J., Tavaré, S.: Coalescents and genealogical structure under neutrality. Annu. Rev. Genet. **29**, 401–421 (1995)
115. Doob, J.L.: Topics in the theory of Markoff chains. Trans. Am. Math. Soc. **52**, 37–64 (1942)
116. Doob, J.L.: Markoff chains – Denumerable case. Trans. Am. Math. Soc. **58**, 455–473 (1945)
117. Dudley, R.M.: Real Analysis and Probability. Wadsworth and Brooks, Pacific Grove (1989)
118. Dushman, S.: The reaction between iodic and hydroiodic acid. J. Phys. Chem. **8**, 453–482 (1903)
119. Dyson, F.: A meeting with Enrico Fermi. How one intuitive physicist rescued a team from fruitless research. Nature **427**, 297 (2004)
120. Eddy, S.R.: What is Bayesian statistics? Nat. Biotechnol. **22**, 1177–1178 (2004)
121. Edelson, D., Field, R.J., Noyes, R.M.: Mechanistic details of the Belousov-Zhabotinskii oscillations. Int. J. Chem. Kinet. **7**, 417–423 (1975)
122. Edgeworth, F.Y.: On the probable errors of frequence-constants. J. R. Stat. Soc. **71**, 381–397 (1908)
123. Edgeworth, F.Y.: On the probable errors of frequence-constants (contd.). J. R. Stat. Soc. **71**, 499–512 (1908)

124. Edgeworth, F.Y.: On the probable errors of frequence-constants (contd.). J. R. Stat. Soc. **71**, 651–678 (1908)
125. Edman, L., Földes-Papp, Z., Wennmalm, S., Rigler, R.: The fluctuating enzyme: A single moleculae approach. Chem. Phys. **247**, 11–22 (1999)
126. Edman, L., Rigler, R.: Memory landscapes of single-enzyme molecules. Proc. Natl. Acad. Sci. USA **97**, 8266–8271 (2000)
127. Edwards, A.W.F.: Are Mendel's resulta really too close. Biol. Rev. **61**, 295–312 (1986)
128. Ehrenberg, M., Rigler, R.: Rotational Brownian motion and fluorescence intensity fluctuations. Chem. Phys. **4**, 390–401 (1974)
129. Ehrenfest, P., Ehrenfest, T.: Über zwei bekannte Einwände gegen das Boltzmannsche H-Theorem. Z. Phys. **8**, 311–314 (1907)
130. Eigen, M.: Selforganization of matter and the evolution of biological macromolecules. Naturwissenschaften **58**, 465–523 (1971)
131. Eigen, M., McCaskill, J., Schuster, P.: The molecular quasispecies. Adv. Chem. Phys. **75**, 149–263 (1989)
132. Eigen, M., Schuster, P.: The hypercycle. A principle of natural self-organization. Part A: Emergence of the hypercycle. Naturwissenschaften **64**, 541–565 (1977)
133. Einstein, A.: Über die von der molekular-kinetischen Theorie der Wärme geforderte Bewegung von in ruhenden Flüssigkeiten suspendierten Teilchen. Annal. Phys. (Leipzig) **17**, 549–560 (1905)
134. Einstein, A.: Investigations on the Theory of the Brownian Movement. Dover Publications, New York (1956). Five original publications by Albert Einstein edited with notes by R. Fürth
135. Elliot, R.J., Anderson, B.D.O.: Reverse-time diffusions. Stoch. Process. Appl. **19**, 327–339 (1985)
136. Elliot, R.J., Kopp, A.E.: Mathematics of Financial Markets, 2nd edn. Springer, New York (2005)
137. Elson, E., Magde, D.: Fluorescence correlation spectroscopy. I. Conceptual basis and theory. Biopolymers **13**, 1–27 (1974)
138. Engl, H.W., Flamm, C., Kügler, P., Lu, J., Müller, S., Schuster, P.: Inverse problems in systems biology. Inverse Prob. **25**, 123,014 (2009)
139. Engl, H.W., Hanke, M., Neubauuer, A.: Regularization of Inverse Problems. Kluwer Academic, Boston (1996)
140. Érdi, P., Lente, G.: Stochastic Chemical Kinetics. Theory and (Mostly) Systems Biological Applications. Understanding Complex Systems. Springer, Berlin (2014)
141. Erlich, H.A. (ed.): PCR Technology. Principles and Applications for DNA Amplification. Stockton Press, New York (1989)
142. Evans, M., Hastings, N.A.J., Peacock, J.B.: Statistical Distributions, 3rd edn. Wiley, New York (2000)
143. Everett, C.J., Ulam, S.: Multiplicative systems I. Proc. Natl. Acad. Sci. USA **34**, 403–405 (1948)
144. Everett, C.J., Ulam, S.M.: Multiplicative systems in several variables I. Tech. Rep. LA-683, Los Alamos Scientific Laboratory (1948)
145. Everett, C.J., Ulam, S.M.: Multiplicative systems in several variables II. Tech. Rep. LA-690, Los Alamos Scientific Laboratory (1948)
146. Everett, C.J., Ulam, S.M.: Multiplicative systems in several variables III. Tech. Rep. LA-707, Los Alamos Scientific Laboratory (1948)
147. Ewens, W.J.: Mathematical Population Genetics. I. Theoretical Introduction, 2nd edn. Interdisciplinary Applied Mathematics. Springer, Berlin (2004)
148. Eyring, H.: The activated complex in chemical reactions. J. Chem. Phys. **3**, 107–115 (1935)
149. Farlow, S.J.: Partial Differential Equations for Scientists and Engineers. Dover Publications, New York (1982)
150. Feigenbaum, M.J.: Universal behavior in nonlinear systems. Physica D **7**, 16–39 (1983)
151. Feinberg, M.: Complex balancing in general kinetic systems. Arch. Ration. Mech. Anal. **49**, 187–194 (1972)

152. Feinberg, M.: Mathematical aspects of mass action kinetics. In: Lapidus, L., Amundson, N.R. (eds.) Chemical Reactor Theory – A Review, pp. 1–78. Prentice Hall, Englewood Cliffs (1977)
153. Feinberg, M.: Lectures on Chemical Reaction Networks. Chemical Engineering & Mathematics. The Ohio State University, Columbus (1979)
154. Feinberg, M.: Chemical oscillations, multiple equilibria, and reaction network structure. In: Stewart, W.E., Ray, W.H., Conley, C.C. (eds.) Dynamics and Modelling of Reactive Systems, pp. 59–130. Academic Press, New York (1980)
155. Feinberg, M.: Chemical reaction network structure and the stability of complex isothermal reactors – II. Multiple steady states for networks of deficiency one. Chem. Eng. Sci. **43**, 1–25 (1988)
156. Feller, W.: On the integro-differential equations of purely discontinuous Markoff processes. Trans. Am. Math. Soc. **48**, 488–515 (1940)
157. Feller, W.: The general form of the so-called law of the iterated logarithm. Trans. Am. Math. Soc. **54**, 373–402 (1943)
158. Feller, W.: On the theory of stochastic processes, with particular reference to applications. In: The Regents of the University of California (ed.) Proceedings of the Berkeley Symposium on Mathematical Statistics and Probability, pp. 403–432. University of California Press, Berkeley (1949)
159. Feller, W.: Diffusion processes in genetics. In: Neyman, J. (ed.) Proc. 2nd Berkeley Symp. on Mathematical Statistics and Probability. University of Caifornia Press, Berkeley (1951)
160. Feller, W.: An Introduction to Probability Theory and Its Application, vol. I, 3rd edn. Wiley, New York (1968)
161. Feller, W.: An Introduction to Probability Theory and Its Application, vol. II, 2nd edn. Wiley, New York (1971)
162. Felsenstein, J.: Inferring Phylogenies. Sinauer Associates, Sunderland (2004)
163. Fernández-Ramos, A., Miller, J.A., Klippenstein, S.J., Truhlar, D.G.: Modeling the kinetics of bimolecular reactions. Chem. Rev. **106**, 4518–4584 (2006)
164. Fersht, A.: Structure and Mechanism in Protein Science: A Guide to Enzyme Catalysis and Protein Folding. W. H. Fremman and Company, New York (1999)
165. Fick, A.: Über diffusion. Annalen der Physik und Chemie **170**(4. Reihe 94), 59–86 (1855)
166. Field, R.J., Körös, E., Noyes, R.M.: Oscillations in chemical systems. II. Thorough analysis of temporal oscillations in the bromate-cerium-malonic acid system. J. Am. Chem. Soc. **94**, 8649–8664 (1972)
167. Field, R.J., Noyes, R.M.: Oscillations in chemical systems. IV. Limit cycle behavior in a model of a real chemical reaction. J. Chem. Phys. **60**, 1877–1884 (1974)
168. Firth, C.J.M., Bray, D.: Stochastic simulation of cell signalling pathways. In: Bower, J.M., Bolouri, H. (eds.) Computational Modeling of Genetic and Biochemical Networks, pp. 263–286. MIT Press, Cambridge (2000)
169. Fisher, R.A.: On an absolute criterion for fitting frequency curves. Messeng. Math. **41**, 155–160 (1912)
170. Fisher, R.A.: On the mathematical foundations of theoretical statistics. Philos. Trans. R. Soc. Lond. A **222**, 309–368 (1922)
171. Fisher, R.A.: Applications of "Student's" distribution. Metron **5**, 90–104 (1925)
172. Fisher, R.A.: Theory of statistical estimation. Proc. Camb. Philos. Soc. **22**, 700–725 (1925)
173. Fisher, R.A.: Moments and product moments of sampling distributions. Proc. Lond. Math. Soc. **Ser.2, 30**, 199–238 (1928)
174. Fisher, R.A.: The Genetical Theory of Natural Selection. Oxford University Press, Oxford (1930)
175. Fisher, R.A.: The logic of inductive inference. J. R. Stat. Soc. **98**, 39–54 (1935)
176. Fisher, R.A.: Has Mendel's work been rediscovered? Ann. Sci., 115–137 (1936)
177. Fisher, R.A.: The Design of Experiments, 8th edn. Hafner Publishing Company, Edinburgh (1966)
178. Fisk, D.L.: Quasi-martingales. Trans. Am. Math. Soc. **120**, 369–389 (1965)

179. Fisz, M.: Probability Theory and Mathematical Statistics, 3rd edn. Wiley, New York (1963)
180. Fisz, M.: Wahrscheinlichkeitsrechnung und mathematische Statistik. VEB Deutscher Verlag der Wissenschaft, Berlin (1989). In German
181. Fletcher, R.I.: The quadratic law of damped exponential growth. Biometrics **20**, 111–124 (1974)
182. Fofack, H., Nolan, J.P.: Tail behavior, modes and other characteristics of stable distributions. Extremes **2**, 39–58 (1999)
183. Föllner, H.H., Geiseler, W.: A model of bistability in an open homogeneous chemical reaction system. Naturwissenschaften **64**, 384 (1977)
184. Foster, D.P.: Law of the iterated logarithm. Wikipedia entry, University of Pennsylvania, Philadelphia, PA (2009). Retrieved April 07, 2009 from en.wikipedia.org/wiki/Law_of_the_iterated_logarithm
185. Francalacci, P., Morelli, L., Angius, A., Berutti, R., Reinier, F., Atzeni, R., Pilu, R., Busonero, F., Maschino, A., Zara, I., Sanna, D., Useli, A., Urru, M.F., Marcelli, M., Cusano, R., Oppo, M., Zoledziewska, M., Pitzalis, M., Deidda, F., Porcu, E., Poddie, F., Kang, H.M., Lyons, R., Tarrier, B., Gresham, J.B., Li, B., Tofanelli, S., Alonso, S., Dei, M., Lai, S., Mulas, A., Whalen, M.B., Uzzau, S., Jones, C., Schlessinger, D., Abecasis, G.R., Sanna, S., Sidore, C., Cucca, F.: Low-pass DNA sequencing of 1200 Sardinians reconstructs European Y-chrmosome phylogeny. Science **341**, 565–569 (2013)
186. Franklin, A., Edwards, A.W.F., Fairbanks, D.J., Hartl, D.L., Seidenfeld, T.: Ending the Mendel-Fisher Controversy. University of Pittburgh Press, Pittsburgh (2008)
187. Frauenfelder, H., Sligar, S.G., Wolynes, P.G.: The eenergy landscape and motions of proteins. Science **254**, 1598–1603 (1991)
188. Freire, J.G., Field, R.J., Gallas, J.A.C.: Relative abundance and structure of chaotic behavior: The nonpolynomial belousov-zhabotinsky reaction kinetics. J. Chem. Phys. **131**, e044,105 (2009)
189. Fubini, G.: Sugli integrali multipli. Rom. Acc. L. Rend. V **16**, 608–614 (1907). Reprinted in Fubini, G. Opere scelte 2, Cremonese pp. 243–249, 1958
190. Gadgil, C., Lee, C.H., Othmer, H.G.: A stochastic analysis of first-order reaction networks. Bull. Math. Biol. **67**, 901–946 (2005)
191. Galton, F.: The geometric mean in vital and social statistics. Proc. Roy. Soc. Lond. **29**, 365–367 (1879)
192. Galton, F.: Natural Inheritance, second american edn. Macmillan, London (1889). App. F, pp. 241–248
193. Gardiner, C.W.: Handbook of Stochastic Methods, 1st edn. Springer, Berlin (1983)
194. Gardiner, C.W.: Stochastic Methods. A Handbook for the Natural Sciences and Social Sciences, fourth edn. Springer Series in Synergetics. Springer, Berlin (2009)
195. Gause, G.F.: Experimental studies on the struggle for existence. J. Exp. Biol. **9**, 389–402 (1932)
196. Gause, G.F.: The Struggle for Existence. Willans & Wilkins, Baltimore (1934). Also published by Hafner, New York (1964) and Dover, Mineola (1971 and 2003)
197. Gauß, C.F.: Theoria motus corporum coelestium in sectionibus conicis solem ambientium. Perthes et Besser, Hamburg (1809). English translation: Theory of the Motion of the Heavenly Bodies Moving about the Sun in Conic Sections. Little, Brown. Boston, MA. 1857. Reprinted by Dover, New York (1963)
198. Geisler, W., Föllner, H.H.: Three steady state situation in an open chemical reaction system. I. Biophys. Chem. **6**, 107–115 (1977)
199. Gelman, A., Carlin, J.B., Stern, H.S., Rubin, D.B.: Baysian Data Analysis, 2nd edn. Texts in Statistical Science. Chapman & Hall / CRC, Boca Raton (2004)
200. George, G.: Testing for the independence of three events. Math. Gaz. **88**, 568 (2004)
201. Georgii, H.: Stochastik. Einführung in die Wahrscheinlichkeitstheorie und Statistik, 3rd edn. Walter de Gruyter GmbH & Co., Berlin (2007). In German. English translation: *Stochastics. Introduction to Probability and Statistics.* Walter de Gruyter GmbH & Co. Berlin (2008).

202. Gibbs, J.W.: Elementary Principles in Statistical Mechanics. Charles Scribner's Sons, New York (1902). Reprinted 1981 by Ox Bow Press, Woodbridge, CT
203. Gibbs, J.W.: The Scientific Papers of J. Willard Gibbs, vol.I, Thermodynamics. Dover Publications, New York (1961)
204. Gibson, M.A., Bruck, J.: Efficient exact stochastic simulation of chemical systems with many species and many channels. J. Phys. Chem. A **104**, 1876–1889 (2000)
205. Gihman, I.F., Skorohod, A.V.: The Theory of Stochastic Processes. Vol. I, II, and III. Springer, Berlin (1975)
206. Gillespie, D.T.: A general method for numerically simulating the stochastic time evolution of coupled chemical reactions. J. Comp. Phys. **22**, 403–434 (1976)
207. Gillespie, D.T.: Exact stochastic simulation of coupled chemical reactions. J. Phys. Chem. **81**, 2340–2361 (1977)
208. Gillespie, D.T.: Markov Processes: An Introduction for Physical Scientists. Academic Press, San Diego (1992)
209. Gillespie, D.T.: A rigorous derivation of the chemical master equation. Physica A **188**, 404–425 (1992)
210. Gillespie, D.T.: Exact numerical simulation of the Ornstein-Uhlenbeck process and its integral. Phys. Rev. E **54**, 2084–2091 (1996)
211. Gillespie, D.T.: The chemical Langevin equation. J. Chem. Phys. **113**, 297–306 (2000)
212. Gillespie, D.T.: Approximate accelerated stochastic simulation of chemically reacting systems. J. Chem. Phys. **115**(4), 1716–1733 (2001)
213. Gillespie, D.T.: Stochastic simulation of chemical kinetics. Annu. Rev. Phys. Chem. **58**, 35–55 (2007)
214. Gillespie, D.T., Seitaridou, E.: Simple Brownian Diffusion. An Introduction to the Standard Theoretical Models. Oxford University Press, Oxford (2013)
215. Gillies, D.: Varieties of propensity. Br. J. Philos. Sci. **51**, 807–853 (2000)
216. Goel, N.S., Richter-Dyn, N.: Stochastic Models in Biology. Academic Press, New York (1974)
217. Goutsias, J., Jenkinson, G.: Markovian dynamics on complex reaction networks. Phys. Rep. **529**, 199–264 (2013)
218. Goychuk, I.: Viscoelastic subdiffusion: Generalized langevin equation approach. Adv. Chem. Phys. **150**, 187–253 (2012)
219. Gradstein, I.S., Ryshik, I.M.: Tables of Series, Products, and Integrals, vol. 1. Verlag Harri Deutsch, Thun, DE (1981). In German and English. Translated from Russian by Ludwig Boll, Berlin
220. Gray, R.M.: Entropy and Information Theory, 2nd edn. Springer, New York (2011)
221. Griffiths, A.J.F., Wessler, S.R., Caroll, J.B., Doebley, J.: An Introduction to Genetic Analysis, 10th edn. W. H. Freeman, New York (2012)
222. Grimmett, G., Stirzaker, D.: Probability and Random Processes, 3rd edn. Oxford University Press, Oxford (2001)
223. Grünbaum, B.: Venn diagrams and independent falilies of sets. Math. Mag. **48**, 12–23 (1975)
224. Grünbaum, B.: The construction of Venn diagrams. Coll. Math. J. **15**, 238–247 (1984)
225. Guckenheimer, J., Holmes, P.: Nonlinear Oscillations, Dynamical Systems, and Bifurcations of Vector Fields, *Applied Mathematical Sciences*, vol. 42. Springer, New York (1983)
226. Gunawardena, J.: Chemical reaction network theory for *in-silico* biologists. Tech. rep., Bauer Center for Genomics Research at Harvard University, Cambridge, MA (2003)
227. Györgyi, L., Field, R.J.: A three-variable model of deterministic chaos in the belousov-zhabotinsky reaction. Nature **355**, 808–810 (1992)
228. Hájek, A.: Interpretations of probability. In: Zalta, E.N. (ed.) The Stanford Encyclopedia of Philosophy, Winter 2012 edn. The Metaphysics Research Lab, Center for the Study of Language and Information, Stanford University, Stanford Universiy, Stanford, CA. World Wide Web URL: http://plato.stanford.edu/entries/probability-interpret/ (2013). Retrieved January 23, 2013

229. Hajek, B.: An exploration of random processes for engineers. Lecture Notes ECE 534, University of Illinois at Urbana-Champaign, Urbana-Champaign, IL (2014). Retrieved March 16, 2014 from www.ifp.illinois.edu/~hajek/Papers/randomprocesses.html
230. Hamill, O.P., Marty, A., Neher, E., Sakmann, B., Sigworth, F.J.: Improved patch-clamp techniques for high-resolution current recording from cels and cell-free mambrane patches. Pflügers Archiv. Eur. J. Physiol. **391**, 85–100 (1981)
231. Hamilton, J.D.: Time Series Analysis. Princeton University Press, Princeton (1994)
232. Hamilton, W.R.: On a general method in dynamics. Philos. Trans. R. Soc. Lond. **II** for 1834, 247–308 (1834)
233. Hamilton, W.R.: Second essay on a general method in dynamics. Philos. Trans. R. Soc. London **I** for 1835, 95–144 (1835)
234. Hammer, M.F.: A recent common ancestry for human Y chromosomes. Nature **378**, 376–378 (1995)
235. Hamming, R.W.: Error detecting and error correcting codes. Bell Syst. Tech. J. **29**, 147–160 (1950)
236. Hamming, R.W.: Coding and Information Theory, 2nd edn. Prentice-Hall, Englewood Cliffs (1986)
237. Hanna, A., Saul, A., Showalter, K.: Detailed studies of propagating frints in the iodate oxidation of arsenous acid. J. Am. Chem. Soc. **104**, 3838–3844 (1982)
238. Hansma, H.G., Kasuya, K., Oroudjev, E.: Atomic force microscopy imaging and pulling of nucleic acids. Curr. Op. Struct. Biol. **14**, 380–385 (2004)
239. Harris, T.E.: Branching Processes. Springer, Berlin (1963)
240. Harris, T.E.: The Theory of Branching Processes. Dover Publications, New York (1989)
241. Hartl, D.L., Clark, A.G.: Principles of Population Genetics, 3rd edn. Sinauer Associates, Sunderland (1997)
242. Hartman, P., Wintner, A.: On the law of the iterated logarithm. Am. J. Math. **63**, 169–173 (1941)
243. Hatzakis, N.S., Wei, L., Jorgensen, S.K., Kunding, A.H., Bolinger, P.Y., Ehrlich, N., Makarov, I., Skjot, M., Svendsen, A., Hedegård, P., Stamou, D.: Single enzyme studies reveal the existence of discrete states for monomeric enzymes and how they are "selected" upon allosteric regulation. J. Am. Chem. Soc. **134**, 9296–9302 (2012)
244. Haubold, H.J., Mathai, M.A., Saxena, R.K.: Mittag-Leffler functions and their applications. J. Appl. Math. **2011**, e298,628 (2011). Hindawi Publ. Corp.
245. Haussmann, U.G., Pardoux, E.: Time reversal of diffusions. Ann. Probab. **14**, 1188–1205 (1986)
246. Hawkins, D., Ulam, S.: Theory of multiplicative processes I. Tech. Rep. LADC-265, Los Alamos Scientific Laboratory (1944)
247. Hazeltine, E.L., Rawlings, J.B.: Approximate simulation of coupled fast and slow reactions for stochastic chemical kinetics. J. Chem. Phys. **117**, 6959–6969 (2002)
248. Heathcote, C.R., Moyal, J.E.: The random walk (in continuous time) and its application to the theory of queues. Biometrika **46**, 400–411 (1959)
249. Heinrich, R., Sonntag, I.: Analysis of the selection equation for a multivariable population model. Deterministic anad stochastic solutios and discussion of the approach for populations of self-reproducing biochemical networks. J. Theor. Biol. **93**, 325–361 (1981)
250. Heyde, C.C., Seneta, E.: Studies in the history of probability and statistics. xxxi. the simple branching porcess, a turning point test and a fundmanetal inequality: A historical note on I. J. Bienaymé. Biometrika **59**, 680–683 (1972)
251. Higham, D.J.: Modeling and somulationg chemical reactions. SIAM Rev. **50**, 347–368 (2008)
252. Hinshelwood, C.N.: On the theory of unimolecular reactions. Proc. R. Soc. Lond. A **113**, 230–233 (1926)
253. Hirsch, M.W., Smale, S.: Differential Equations, Dynamical Systems, and an Introduction to Chaos, 2nd edn. Elsevier, Amsterdam (2004)
254. Hirschfeld, T.: Optical microscopic observation of small molecules. Appl. Opt. **15**, 2965–2966 (1976)

255. Hocking, R.L., Schwertman, N.C.: An extension of the birthday problem to exactly *k* matches. Coll. Math. J. **17**, 315–321 (1986)
256. Hofbauer, J., Schuster, P., Sigmund, K., Wolff, R.: Dynamical systems und constant organization II: Homogenoeous growth functions of degree $p = 2$. SIAM J. Appl. Math. **38**, 282–304 (1980)
257. Hogg, R.V., McKean, J.W., Craig, A.T.: Introduction to Mathematical Statistics, 7th edn. Pearson Education, Upper Saddle River (2012)
258. Hogg, R.V., Tanis, E.A.: Probability and Statistical Inference, 8th edn. Pearson – Prentice Hall, Upper Saddle River (2010)
259. Holder, M., Lewis, P.O.: Phylogeny estimation: Traditional and Bayesian approaches. Nat. Rev. Genet. **4**, 275–284 (2003)
260. Holdren, J.P., Lander, E., Varmus, H.: Designing a Digital Future: Federally Funded Research and Development in Networking and Information Technology. President's Council of Advisors on Science and Technology, Washington, DC (2010)
261. Holsinger, K.E.: Lecture Notes in Population Genetics. University of Connecticut, Dept. of Ecology and Evolutionary Biology, Storrs, CT (2012). Licensed under the Creative Commons Attribution-ShareAlike License: http://creativecommons.org/licenses/by-sa/3.0/
262. Horn, F.: Necessary and sufficient conditions for complex balancing in chemical kinetics. Arch. Ration. Mech. Anal. **49**, 172–186 (1972)
263. Horn, F., Jackson, R.: General mass action kinetics. Arch. Ration. Mech. Anal. **47**, 81–116 (1972)
264. Houchmandzadeh, B., Vallade, M.: An alternative to the diffusion equation in population genetics. Phys. Rev. E **83**, e051,913 (2010)
265. Houston, P.L.: Chemical Kinetics and Reaction Dynamics. The McGraw-Hill Companies, New York (2001)
266. Hu, J., Lygeros, J., Sastry, S.: Towards a theory of stochastic hybrid systems. In: Lynch, N., Krogh, B. (eds.) Hybrid Systems: Computation and Control, *Lecture Notes in Computer Science*, vol. 1790, pp. 160–173. Springer, Berlin (2000)
267. Hu, Y., Li, T.: Highly accurate tau-leaping methiods with random corrections. J. Chem. Phys. **130**, e124,109 (2009)
268. Hua, Y., Sarkar, T.K.: Matrix pencil method for estimating parameters of exponentially damped/undamped sinusoids in noise. IEEE Trans. Acoust. Speech Signal Process. **38**, 814–824 (1990)
269. Humphries, N.E., Queiroz, N., Dyer, J.R.M., Pade, N.G., Musyl, M.K., Schaefer, K.M., Fuller, D.W., Brunnschweiler, J.M., Doyle, T.K., Houghton, J.D.R., Hays, G.C., Jones, C.S., Noble, L.R., Wearmouth, V.J., Southall, E.J., Sims, D.W.: Environmental context explains Lévy and Brwonian movement patterns of marine predators. Nature **465**, 1066–1069 (2010)
270. Inagaki, H.: Selection under random mutations in stochastic Eigen model. Bull. Math. Biol. **44**, 17–28 (1982)
271. Ishida, K.: Stochastic model for bimolecular reaction. J. Chem. Phys. **41**, 2472–2478 (1964)
272. Itō, K.: Stochastic integral. Proc. Imp. Acad. Tokyo **20**, 519–524 (1944)
273. Itō, K.: On stochastic differential equations. Mem. Am. Math. Soc. **4**, 1–51 (1951)
274. Jachimowski, C.J., McQuarrie, D.A., Russell, M.E.: A stochastic approach to enzyme-substrate reactions. Biochemistry **3**, 1732–1736 (1964)
275. Jackson, E.A.: Perspectives of Nonlinear Dynamics, vol. 1. Cambridge University Press, Cambridge (1989)
276. Jackson, E.A.: Perspectives of Nonlinear Dynamics, vol. 2. Cambridge University Press, Cambridge (1989)
277. Jacobs, K.: Stochastic processes for Physicists. Understanding Noisy Systems. Cambridge University Press, Cambridge (2010)
278. Jahnke, T., Huisinga, W.: Solving the chemical master equation for monomolecular reaction systems analytically. J. Math. Biol. **54**, 1–26 (2007)
279. Jaynes, E.T.: Information theory and statistical mechanics. Phys. Rev. **106**, 620–630 (1957)
280. Jaynes, E.T.: Information theory and statistical mechanics. II. Phys. Rev. **108**, 171–190 (1957)

281. Jaynes, E.T.: Probability Theory. The Logic of Science. Cambridge University Press, Cambridge (2003)
282. Jensen, A.L.: Comparison of logistic equations for population growth. Biometrics **31**, 853–862 (1975)
283. Jensen, L.: Solving a singular diffusion equation occurring in population genetics. J. Appl. Probab. **11**, 1–15 (1974)
284. Johnson, N.L., Kotz, S., Balakrishnan, N.: Continuous Univariate Distributions, *Probability and Mathematical Statistics. Applied Probability and Statistics*, vol. 1, 2nd edn. Wiley, New York (1994)
285. Johnson, N.L., Kotz, S., Balakrishnan, N.: Continuous Univariate Distributions, *Probability and Mathematical Statistics. Applied Probability and Statistics*, vol. 2, 2nd edn. Wiley, New York (1995)
286. Jones, B.L., Enns, R.H., Rangnekar, S.S.: On the theory of selection of coupled macromolecular systems. Bull. Math. Biol. **38**, 15–28 (1976)
287. Jones, B.L., Leung, H.K.: Stochastic analysis of a non-linear model for selection of biological macromolecules. Bull. Math. Biol. **43**, 665–680 (1981)
288. Joyce, G.F.: Forty years of *in vitro* evolution. Angew. Chem. Internat. Ed. **46**, 6420–6436 (2007)
289. Karlin, S., McGregor, J.: On a genetics model of moran. Math. Proc. Camb. Philos. Soc. **58**, 299–311 (1962)
290. Karlin, S., Taylor, H.M.: A First Course in Stochastic Processes, 2nd edn. Academic Press, New York (1975)
291. Kassel, L.S.: Studies in homogeneous gas reactions I. J. Phys. Chem. **32**, 225–242 (1928)
292. Kendall, D.G.: An artificial realization of a simple "birth-and-death" process. J. R. Stat. Soc. B **12**, 116–119 (1950)
293. Kendall, D.G.: Branching processes since 1873. J. Lond. Math. Soc. **41**, 386–406 (1966)
294. Kendall, D.G.: The genalogy of genealogy: Branching processes before (an after) 1873. Bull. Lond. Math. Soc. **7**, 225–253 (1975)
295. Kenney, J.F., Keeping, E.S.: Mathematics of Statistics, 2nd edn. Van Nostrand, Princeton (1951)
296. Kenney, J.F., Keeping, E.S.: The k-Statistics. In Mathematics of Statistics. Part I, §7.9, 3rd edn. Van Nostrand, Princeton (1962)
297. Kermack, W.O., McKendrick, A.G.: A contribution to the mathematical theory of epidemics. Proc. R. Soc. Lond. A **115**, 700–721 (1927)
298. Kesten, H., Stigum, B.P.: A limit theorem for multidimensional Galton-Watson processes. Ann. Math. Stat. **37**, 1211–1223 (1966)
299. Keynes, J.M.: A Treatise on Probability. MacMillan, London (1921)
300. Khinchin, A.Y.: Über einen Satz der Wahrscheinlichkeitsrechnung. Fundam. Math. **6**, 9–20 (1924). In German
301. Kim, S.K.: Mean first passage time for a random walker and its application to chemical knietics. J. Chem. Phys. **28**, 1057–1067 (1958)
302. Kimura, M.: Solution of a process of random genetic drift with a continuous model. Proc. Natl. Acad. Sci. USA **41**, 144–150 (1955)
303. Kimura, M.: Diffusion models in population genetics. J. Appl. Probab. **1**, 177–232 (1964)
304. Kimura, M.: The Neutral Theory of Molecular Evolution. Cambridge University Press, Cambridge (1983)
305. Kingman, J.F.C.: Mathematics of Genetic Diversity. Society for Industrial and Applied Mathematics, Washigton, DC (1980)
306. Kingman, J.F.C.: The genealogy of large populations. J. Appl. Probab. **19**(Essays in Statistical Science), 27–43 (1982)
307. Kingman, J.F.C.: Origins of the coalescent: 1974 – 1982. Genetics **156**, 1461–1463 (2000)
308. Knuth, D.E.: Two notes on notation. Am. Math. Monthly **99**, 403–422 (1992)
309. Kolmogorov, A.N.: Über das Gesetz es interierten Logarithmus. Math. Ann. **101**, 126–135 (1929). In German

310. Kolmogorov, A.N.: Über die analytischen Methoden in der Wahrscheinlichkeitsrechnung. Math. Ann. **104**, 415–458 (1931)
311. Kolmogorov, A.N.: Grundbegriffe der Wahrscheinlichkeitsrechnung. Ergebnisse der Mathematik und ihrer Grenzgebiete. Springer, Berlin (1933). English translation: *Foundations of Probability*. Chelsea Publ. Co. New York (1950)
312. Kolmogorov, A.N., Dmitriev, N.A.: "Zur Lösung einer biologischen Aufgabe". Isvestiya Nauchno-Issledovatel'skogo Instituta Matematiki i Mekhaniki pri Tomskom Gosudarstvennom Universitete **2**, 1–12 (1938)
313. Kolmogorov, A.N., Dmitriev, N.A.: Branching stochastic processes. Doklady Akad. Nauk U.S.S.R. **56**, 5–8 (1947)
314. Koroborov, V.I., Ochkov, V.F.: Chemical Kinetics with *Mathcad* and *Maple*. Springer, Wien (2011)
315. Koshland, D.E.: Application of a theory of enzyme specificity to protein synthesis. Proc. Natl. Acad. Sci. USA **44**, 98–104 (1958)
316. Kou, S.C., Cherayil, B.J., Min, W., English, B.P., Xie, X.S.: Single-molecule Michaelis-Menten equations. J. Phys. Chem. B **109**, 19,068–19,081 (2005)
317. Kowalski, C.J.: Non-normal bivariate distributions with normal marginals. Am. Statistician **27**, 103–106 (1973)
318. Krichevsky, O., Bonnet, G.: Fluorescence correlation spectroscopy: The technique and its applications. Rep. Prog. Phys. **65**, 251–297 (2002)
319. Kubo, R.: The fluctuation-dissipation theorem. Rep. Prog. Phys. **29**, 255–284 (1966)
320. Kügler, P., Gaubitzer, E., Müller, S.: Perameter identification for chemical reaction systems using sparsity enforcing regularization A case study for the chlorite–iodide reaction. J. Phys. Chem. A **113**, 2775–2785 (2009)
321. Kulzer, F., Orrit, M.: Single-molecule optics. Annu. Rev. Phys. Chem. **55**, 585–611 (2004)
322. Kumaresan, R., Tufts, D.W.: Estimating the parameters of exponentially damped sinusoids and pole-zero modeiling in noise. IEEE Trans. Acoust. Speech Signal Process. **30**, 833–840 (1982)
323. Laidler, K.J.: Chemical Kinetics, 3rd edn. Addison Wesley, Boston (1987)
324. Laidler, K.J., King, M.C.: The development of transition-state theory. J. Phys. Chem. **87**, 2657–2664 (1983)
325. Langevin, P.: Sur la théorie du mouvement Brownien. Comptes Rendues hebdomadaires des Séances de L'Academié des Sceinces **146**, 530–533 (1908)
326. Laplace, P.S.: Mémoirs sur la probabilité des causes par les évènemens. Mémoires de Mathématique et de Physique, Presentés à l'Académie Royale des Sciences, par divers Savans & lûs dans ses Assemblées **6**, 621–656 (1774). Reprinted in Laplace's Ouevres complète **8**, 27–65. English translation: Stat. Sci. **1**, 364–378 (1986)
327. Laplace, P.S.: Théorie analytique des probabililtés. Courcies Imprimeur, Paris (1812)
328. Laplace, P.S.: Essai philosophique les probabililtés. Courcies Imprimeur, Paris (1814). English edition: A Philosophical Eaasay on Probabilties. Dover Publications, New York (1951)
329. Laurenzi, I.J.: An analytical solution of the stochastic master equation for reversible bimolecular reaction kinetics. J. Chem. Phys. **113**, 3315–3322 (2000)
330. Lauritzen, S.L.: Time series analysis in 1880: A discussion of contributions made by t. n, thiele. Int. Stat. Rev. **49**, 319–331 (1981)
331. Le Cam, L.: Maximum likelihood: An introduction. Int. Stat. Rev. **58**, 153–171 (1990)
332. Le Novère, N., Shimizu, T.S.: StochSim: Modeling of stochastic biomolecular processes. Bioinformatics **17**, 575–576 (2001)
333. Lee, P.M.: Bayesian Statistics, 3rd edn. Hodder Arnold, London (2004)
334. Leemis, L.: Poisson to normal. College of William & Mary, Department of Mathematics, Williamsburg, VA (2012). URL: www.math.wm.edu/~leemis/chart/UDR/PDFs/PoissonNormal.pdf
335. Lefever, R., Nicolis, G., Borckmans, P.: The Brusselator: It does oscillate all the same. J. Chem. Soc. Faraday Trans. **1**, 1013–1023 (1988)

336. Legendre, A.M.: Nouvelles méthodes pour la détermination des orbites des comètes. F. Didot, Paris (1805). In French
337. Lerch, H.P., Rigler, R., Mikhailov, A.S.: Functional conformational motions in the turnover cycle of cholesterol oxidase. Proc. Natl. Acad. Sci. USA **102**, 10,807–10,812 (2005)
338. Leung, K.: Expansion of the master equation for a biomolecular selection model. Bull. Math. Biol. **47**, 231–238 (1985)
339. Lévy, P.: Calcul de probabilités. Geuthier-Villars, Paris (1925). In French
340. Lewis, W.C.M.: Studies in catalysis. Part IX. The calculation in absolute measure of velocity constants and equilibrium constants in gaseous systems. J. Chem. Soc. Trans. **113**, 471–492 (1918)
341. Li, H., Cao, Y., Petzold, L.R., Gillespie, D.T.: Algortihms and software for stochastic simulation of biochemical reacting systems. Biotechnol. Prog. **24**, 56–61 (2008)
342. Li, P.T.X., Bustamante, C., Tinoco, Jr., I.: Real-time control of the energy landscape by force directs the folding of RNA molecules. Proc. Natl. Acad. Sci. USA **104**, 7039–7044 (2007)
343. Li, T.: Analysis of explicit tau-leaping schemes for simulating chemically reacting systems. Multiscale Model. Simul. **6**, 417–436 (2007)
344. Li, T., Kheifets, S., Medellin, D., Raizen, M.G.: Measurement of the instantaneous velocity of a Brownian particle. Science **328**, 1673–1675 (2010)
345. Liao, D., Galajda, P., Riehn, R., Ilic, R., Puchalla, J.L., Yu, H.G., Craighead, H.G., Austin, R.H.: Single molecule correlation spectroscopy ib continuous flow mixers with zero-mode waveguides. Opt. Express **16**, 10,077–10,090 (2008)
346. Limpert, E., Stahel, W.A., Abbt, M.: Log-normal distributions across the sciences: Keys and clues. BioScience **51**, 341–352 (2001)
347. Lin, H., Truhlar, D.G.: QM/MM: What have we learned, where are we, and where do we go from here? Theor. Chem. Acc. **117**, 185–199 (2007)
348. Lin, S.H., Lau, K.H., Richardson, W., Volk, L., Eyring, H.: Stochastic model of unimolecular reactions and the RRKM theory. Proc. Natl. Acad. Sci. USA **69**, 2778–2782 (1972)
349. Lindeberg, J.W.: Über das Exponentialgesetzes in der Wahrscheinlichkeitsrechnung. Ann. Acad. Sci. Fenn. **16**, 1–23 (1920). In German.
350. Lindeberg, J.W.: Eine neue Herleitung des Exponentialgesetzes in der Wahrscheinlichkeitsrechnung. Math. Z. **15**, 211–225 (1922). In German
351. Lindemann, F.A.: Discussion on the radiation theory on chemical action. Trans. Farad. Soc. **17**, 598–606 (1922)
352. Liouville, J.: Note sur la théorie de la variation des constantes arbitraires. Journal de Mathématiques pure et appliquées **3**, 342–349 (1838). In French.
353. Liouville, J.: Mémoire sur l'intégration des équations différentielles du mouvement quelconque de points matériels. Journal de Mathématiques pure et appliquées **14**, 257–299 (1849). In French.
354. Lorenz, E.N.: Deterministic nonperiodic flow. J. Atmos. Sci. **20**, 130–141 (1963)
355. Lu, H.P., Xun, L., Xie, X.S.: Single-molecule enzyme dynamics. Science **282**, 1877–1882 (1998)
356. Lu, J., Engl, H.W., Rainer Machné, Schuster, P.: Inverse bifurcation analysis of a model for the mammalian G1/S regulatory module. Lect. Notes Comput. Sci. **4414**, 168Ű184 (2007)
357. Lu, J., Engl, H.W., Schuster, P.: Inverse bifurcation analysis: Application to simple gene systems. ABM – Algorithms Mol. Biol. **1**, e11 (2006)
358. Lu, Z., Wang, Y.: An introduction to dissipative particle dynamics. In: Monticelli, L., Salonen, E. (eds.) Biomolecular Simulations: Methods and Protocols, *Methods in Molecular Biology*, vol. 924, chap. 24, pp. 617–233. Springer, New York (2013)
359. Lukacs, E.: Characteristic Functions. Hafner Publ. Co., New York (1970)
360. Lukacs, E.: A survey of the theory of characteristic functions. Adv. Appl. Probab. **4**, 1–38 (1972)
361. Lyapunov, A.M.: Sur une proposition de la théorie des probabilités. Bull. Acad. Imp. Sci. St. Pétersbourg **13**, 359–386 (1900)

362. Lyapunov, A.M.: Nouvelle forme du théorème sur la limite des probabilités. Mem. Acad. Imp. Sci. St. Pétersbourg, Classe Phys. Math. **12**, 1–24 (1901)
363. Magde, D., Elson, E., Webb, W.W.: Thermodynamic fluctuations in a reating system – Measurement by fluorescence correlation spectroscopy. Phys. Rev. Lett. **29**, 705–708 (1972)
364. Mahnke, R., Kaupužs, J., Lubashevsky, I.: Physics of Stochastic Processes. How Randomness Acts in Time. Wiley-VCh Verlag, Weinheim (Bergstraße), DE (2009)
365. Mallows, C.: Anothre comment on O'Cinneide. Am. Statistician **45**, 257 (1991)
366. Mandelbrot, B.B.: The Fractal Geometry of Nature, updated edn. W. H. Freeman Company, New York (1983)
367. Mansuy, R.: The origins of the word "martingale". Electron. J. Hist. Probab. Stat. **5**(1), 1–10 (2009). Translated by Ronald Sverdlove from the French *Histoire des martigales*. Mathématiques Sciences Humaines **43**(169), 105–113 (2005)
368. Marcus, R.A.: Unimolecular dissociations and free radical recombination reactions. J. Chem. Phys. **20**, 359–364 (1952)
369. Marcus, R.A.: Vibrational nonadiabaticity and tunneling effects in thranition state theory. J. Chem. Phys. **83**, 204–207 (1979)
370. Marcus, R.A.: Unimolecular reactions, rates and quantum state distributions of products. Philos. Trans. R. Soc. Lond. A **332**, 283–296 (1990)
371. Marcus, R.A., Rice, O.K.: The kinetics of the recombination of methyl radical and iodine atoms. J. Phys. Colloid Chem. **55**, 894–908 (1951)
372. Maruyama, T.: Stochastic Problems in Population Genetics. Springer, Berlin (1977)
373. Marx, D., Jürg Hutter: *Ab initio* Molecular Dynamics. Basic Theory and Advanced Methods. Cambridge University Press, Cambridge (2009)
374. Mathai, A.M., Saxena, R.K., Haubold, H.J.: A certain class of Laplace transforms with applications to reaction and reaction-diffusion equations. Astrophys. Space Sci. **305**, 283–288 (2006)
375. Maxwell, J.C.: Illustartions of the dynamical theory of gases. Part I. on the motions and collisions of perfectly elastic spheres. Philos. Mag. 4th Ser. **19**, 19–32 (1860)
376. Maxwell, J.C.: Illustartions of the dynamical theory of gases. Part II. on the process of diffusion of two or more kinds of particles among one another. Philos. Mag. 4th Ser. **20**, 21–37 (1860)
377. Maxwell, J.C.: On the dynamical theory of gases. Philos. Trans. R. Soc. Lond. **157**, 49–88 (1867)
378. McAlister, D.: The law of the geometric mean. Proc. R. Soc. Lond. **29**, 367–376 (1879)
379. McCaskill, J.S.: A stochastic theory of macromolecular evolution. Biol. Cybern. **50**, 63–73 (1984)
380. McKean, Jr., H.P.: Stochastic Integrals. Wiley, New York (1969)
381. McQuarrie, D.A.: Kinetics of small systems. I. J. Chem. Phys. **38**, 433–436 (1962)
382. McQuarrie, D.A.: Stochastic approach to chemical kinetics. J. Appl. Probab. **4**, 413–478 (1967)
383. McQuarrie, D.A.: Mathematical Methods for Scientists and Engineers. University Science Books, Sausalito (2003)
384. McQuarrie, D.A., Jachimowski, C.J., Russell, M.E.: Kinetics of small systems. II. J. Chem. Phys. **40**, 2914–2921 (1964)
385. McVinish, R., Pollett, P.K.: A central limit theorem for a discrete time SIS model with individual variation. J. Appl. Probab. **49**, 521–530 (2012)
386. McVinish, R., Pollett, P.K.: The deterministic limit of a stochastic logistic model with individual variation. J. Appl. Probab. **241**, 109–114 (2013)
387. Medina, M.Ángel., Schwille, P.: Fluorescence correlation spectroscopy for the detection and study of single molecules in biology. BioEssays **24**, 758–764 (2002)
388. Medvegyev, P.: Stochastic Integration Theory. Oxford University Press, New York (2007)
389. Meinhardt, H.: Models of Biological Pattern Formation. Academic Press, London (1982)
390. Meintrup, D., Schäffler, S.: Stochastik. Theorie und Anwendungen. Springer, Berlin (2005). In German

391. Melnick, E.L., Tenenbein, A.: Misspecifications of the normal distribution. Am. Statistician **36**, 372–373 (1982)
392. Mendel, G.: Versuche über Pflanzen-Hybriden. Verhandlungen des naturforschenden Vereins in Brünn **IV**, 3–47 (1866). In German
393. Meredith, M.: Born in Africa: The Quest for the Origins of Human Life. Public Affairs, New York (2011)
394. Merkle, M.: Jensen's inequality for medians. Stat. Probab. Lett. **71**, 277–281 (2005)
395. Messiah, A.: Quantum Mechanics, vol. II. North-Holland Publishing, Amsterdam (1970). Translated from the French by J. Potter
396. Metzler, R., Klafter, J.: The random walk's guide to anomalous diffusion: A fractional dynamics approach. Phys. Rep. **339**, 1–77 (2000)
397. Michaelis, L., Menten, M.L.: The kinetics of the inversion effect. Biochem. Z. **49**, 333–369 (1913)
398. Miller, R.W.: Propensity: Popper or Peirce? Br. J. Philos. Sci. **26**, 123–132 (1975)
399. Mittag-Leffler, M.G.: Sur la nouvelle fonction $E_\alpha(x)$. C. R. Acad. Sci. Paris Ser. II **137**, 554–558 (1903)
400. Mode, C.J., Sleeman, C.K.: Stochastic Processes in Genetics and Evolution. Computer Experiments in the Quantification of Mutation and Selection. World Scientific Publishing, Singapore (2012)
401. Moeendarbarry, E., Ng, T.Y., M.Zangeneh: Dissipative particle dynamics: Introduction, methodology and complex fluid applications – A review. Int. J. Appl. Mech. **1**, 737–763 (2009)
402. Moerner, W.E., Kador, L.: Optical detection and spectroscopy of single molecules in a solid. Phys. Rev. Lett. **62**, 2535–2538 (1989)
403. Monod, J., Wyman, J., Changeaux, J.P.: On the natur of allosteric transitions: A plausible model. J. Mol. Biol. **12**, 88–118 (1965)
404. Montroll, E.W.: Stochastic processes and chemical kinetics. In: Muller, W.M. (ed.) Energetics in Metallurgical Phenomenon, vol. 3, pp. 123–187. Gordon & Breach, New York (1967)
405. Montroll, E.W., Shuler, K.E.: Studies in nonequilibrium rate processes: I. The relaxation of a system of harmonic oscillators. J. Chem. Phys. **26**, 454–464 (1956)
406. Montroll, E.W., Shuler, K.E.: The application of the theory of stochastic processes to chemical kinetics. Adv. Chem. Phys. **1**, 361–399 (1958)
407. Montroll, E.W., Weiss, G.H.: Random walks on lattices. II. J. Math. Phys. **6**, 167–181 (1965)
408. Moore, C.C.: Ergodic theorem, ergodic theory and statistical mechanics. Proc. Natl. Acad. Sci. USA **112**, 1907–1911 (2015)
409. Moore, G.E.: Cramming more components onto intergrated circuits. Electronics **38**(8), 4–7 (1965)
410. Moran, P.A.P.: Random processes in genetics. Proc. Camb. Philos. Soc. **54**, 60–71 (1958)
411. Moran, P.A.P.: The Statistical Processes of Evolutionary Theroy. Clarendon Press, Oxford (1962)
412. Morse, P.M., Feshbach, H.: Methods of Theoretical Physics, vol. I. McGraw-Hill, Boston (1953)
413. Motulsky, H.J., Christopoulos, A.: Fitting Models to Biological Data Using Linear and Nonlinear Regression. A Practical Guide to Curve Fitting. GraphPad Software Inc., San Diego (2003)
414. Mount, D.W.: Bioinformatics. Sequence and Genome Analysis, 2nd edn. Cold Spring Harbor Laboratory Press, Cold Spring Harbor (2004)
415. Moyal, J.E.: Stochastic processes and statistical physics. J. R. Stat. Soc. B **11**, 150–210 (1949)
416. Müller, S., Regensburger, G.: Generalized mass action systems: Complex balanding equilibria and sign vectors of the stoichiometric and kinetic-order subspaces. SIAM J. Appl. Math. **72**, 1926–1947 (2012)

417. Munz, P., Hudea, I., Imad, J., Smith, R.J.: When zombies attack: Mathematical modelling of an outbreak of zombie infection. In: Tchuenche, J.M., Chiyaka, C. (eds.) Infectious Disease Modelling Research Progress, chap. 4, pp. 133–156. Nova Science Publishers, Hauppauge (2009)
418. Nåsell, I.: On the quasi-stationary distribution of the stochastic logistic epidemic. Math. Biosci. **156**, 21–40 (1999)
419. Nåsell, I.: Extiction and quasi-stationarity in the Verhulst logistic model. J. Theor. Biol. **211**, 11–27 (2001)
420. Neher, E., Sakmann, B.: Single-cheannel currents recorded from membrane of denervated frog muscle fibres. Nature **260**, 799–802 (1976)
421. Nicolis, G., Prigogine, I.: Self-Organization in Nonequilibrium Systems. Wiley, New York (1977)
422. Nishiyama, K.: Stochastic approach to nonlinear chemical reactions having multiple steatdy states. J. Phys. Soc. Jpn. **37**, 44–49 (1974)
423. Nolan, J.P.: Stable Distributions: Models for Heavy-Tailed Data. Birkhäuser, Boston (2013). Unfinished manuscript. Online at academic2.american.edu/~jpnolan
424. Norden, R.H.: A survey of maximum likelihood estimation I. Int. Stat. Rev. **40**, 329–354 (1972)
425. Norden, R.H.: A survey of maximum likelihood estimation II. Int. Stat. Rev. **41**, 39–58 (1973)
426. Norden, R.H.: On the distribution of the time to extinction in the stochastic logistic population model. Adv. Appl. Probab. **14**, 687–708 (1982)
427. Novitski, C.E.: On Fisher's criticism of Mendel's results with the garden pea. Genetics **166**, 1133–1136 (2004)
428. Novitski, C.E.: Revision of Fisher's analysis of Mendel's garden pea experiments. Genetics **166**, 1139–1140 (2004)
429. Noyes, R.M., Field, R.J., Körös, E.: Oscillations in chemical systems. I. Detailed mechanism in a system showing temporal oscillations. J. Am. Chem. Soc. **94**, 1394–1395 (1972)
430. Nyman, J.E.: Another generalization of the birthday problem. Math. Mag. **48**, 46–47 (1975)
431. Øksendal, B.K.: Stochastic Differential Equations. An Introduction with Applications, 6th edn. Springer, Berlin (2003)
432. Olbregts, J.: Termolecular reaction of nitrogen monoxide and oxygen. A still unsolved problem. Int. J. Chem. Kinetics **17**, 835–848 (1985)
433. Onuchic, J.N., Luthey-Schulten, Z., Wolynes, P.G.: Theory of protein folding: The energy landscape perspective. Annu. Rev. Phys. Chem. **48**, 545–600 (1997)
434. Orrit, M., Bernard, J.: Single pentacene molecules detected by fluorescence exitation in a p-terphenyl crystal. Phys. Rev. Lett. **65**, 2716–2719 (1990)
435. Oster, G.F., Perelson, A.S.: Chemical reaction dynamics. Part I: Geometrical structure. Arch. Ration. Mech. Anal. **55**, 230–274 (1974)
436. Papapantoleon, A.: An Introduction to Lévy Processes with Applications in Finance. arXiv, Princeton, NJ (2008). ArXiv:0804.0482v2 retrieved July 27, 2015
437. Papoulis, A., Pillai, S.U.: Probability, Random Variables and Stochastic Processes, 4th edn. McGraw-Hill, New York (2002)
438. Park, S.Y., Bera, A.K.: Maximum entropy autoregressive conditional heteroskedasticy model. J. Econ. **150**, 219–230 (2009)
439. Paschotta, R.: Field Guide to Laser Puls Generation. SPIE Press, Bellingham (2008)
440. Patrick, R., Golden, D.M.: Third-order rate constants of atmospheric importance. Int. J. Chem. Kinetics **15**, 1189–1227 (1983)
441. Pearson, E.S., Wishart, J.: "Student's" Collected Papers. Cambridge University Press, Cambridge (1942). Cambridge University Press for the Biometrika Trustees
442. Pearson, J.A.: Advanced Statistical Physics. University of Manchester, Manchester, UK (2009). URL: http://www.joffline.com/
443. Pearson, K.: Contributions to the mathematical theory of evolution. II. Skew variation in homogeneous material. Philos. Trans. R. Soc. Lond. A **186**, 343–414 (1895)

444. Pearson, K.: On the criterion that a given system of deviations form the probable in the case of a correlated system of variables is such that it can be reasonably supposed to have arisen from random sampling. Philos. Mag. Ser. 5 **50**(302), 157–175 (1900)
445. Pearson, K.: The problem of the random walk. Nature **72**, 294 (1905)
446. Pearson, K.: Notes on the history of correlation. Biometrika **13**, 25–45 (1920)
447. Pearson, K., Filon, L.N.G.: Contributions to the mathematical theory of evolution. IV. On the probable errors of frequency constants and on the influence of random selection on variation and correlation. Philos. Trans. R. Soc. Lond. A **191**, 229–311 (1898)
448. Peirce, C.S.: Vol.7: Science and philosophy and Vol.8: Reviews, correspondence, and bibliography. In: Burks, A.W. (ed.) The Collected Papers of Charles Sanders Peirce, vol. 7–8. Belknap Press of Harvard University Press, Cambridge (1958)
449. Peterman, E.J.G., Sosa, H., Moerner, W.E.: Single-molecule flourescence spectrocopy and microscopy of biomolecular motors. Annu. Rev. Phys. Chem. **55**, 79–96 (2004)
450. Philibert, J.: One and a half century of diffusion: Fick, Einstein, before and beyond. Diffusion Fundamentals **4**, 6.1–6.19 (2006)
451. Phillipson, P.E., Schuster, P.: Modeling by Nonlinear Differential Equations. Dissipative and Conservative Processes, *World Scientific Series on Nonlinear Science A*, vol. 69. World Scientific, Singapore (2009)
452. Picard, P.: Sur les Modèles stochastiques logistiques en Démographie. Ann. Inst. H. Poincaré B **II**, 151–172 (1965)
453. Plass, W.R., Cooks, R.G.: A model for energy transfer in inelasitc molecular collisions applicable at steady state and non-steady state and for an arbitrary distribution of collision energies. J. Am. Soc. Mass Spectrom. **14**, 1348–1359 (2003)
454. Pollard, H.: The representatioin of $e^{-x^{\lambda}}$ as a Laplace intgeral. Bull. Am. Math. Soc. **52**, 908–910 (1946)
455. Popper, K.: The propensity interpretation of the calculus of probability and of the quantum theory. In: S. Körner, M.H.L. Price (eds.) Observation and Interpretation in the Philosophy of Physics: Proceedings of the Ninth Symposium of the Colston Research Society. Butterworth Scientific Publications, London (1957)
456. Popper, K.: The propensity theory of probability. Br. J. Philos. Sci. **10**, 25–62 (1960)
457. Poznik, G.D., Henn, B.M., Yee, M.C., Sliwerska, E., Lin, A.A., Snyder, M., Quintana-Murci, L., Kidd, J.M., Underhill, P.A., Bustamante, C.D.: Sequencing Y chromosomes resolves discrepancy in time to common ancestor of males versus females. Science **341**, 562–565 (2013)
458. Press, W.H., Flannery, B.P., Teukolsky, S.A., Vetterling, W.T.: Numerical Recipes. The Art of Scientific Computing. Cambridge University Press, Cambridge (1986)
459. Price, R.: LII. an essay towards soliving a problem in the doctrine of chances. By the late Ref. Mr. Bayes, communicated by Mr. Price, in a letter to John Canton, M.A. and F.R.S. Philos. Trans. R. Soc. Lond. **53**, 370–418 (1763)
460. Protter, P.E.: Stochastic Intergration and Differential Equations, *Applications of Mathematics*, vol. 21, 2nd edn. Springer, Berlin (2004)
461. Provencher, S.W., Dovi, V.G.: Direct analysis of continuous relaxation spectra. J. Biophys. Biochem. Methods **1**, 313–318 (1979)
462. Qian, H., Elson, E.L.: Single-molecule enzymology: Stochastic Michaelis-Menten kinetics. Biophys. Chem. **101–102**, 565–576 (2002)
463. Rao, C.R.: Information and the acuracy attainable in the estimation of statistical parameters. Bull. Calcutta Math. Soc. **37**, 81–89 (1945)
464. Rathinam, M., Petzold, L.R., Cao, Y., Gillespie, D.T.: Stiffness in stochastic chemically reacting systems: The implicit τ-leaping method. J. Chem. Phys. **119**, 12,784–12,794 (2003)
465. Rice, O.K., Ramsperger, H.C.: Theories of unimolecular gas reactions at low pressures. J. Am. Chem. Soc. **49**, 1617–1629 (1927)
466. Rigler, R., Mets, U., Widengren, J., Kask, P.: Fluorescence correlation spectroscopy with high count rate and low-background-analysis of translational diffusion. Eur. Biophys. J. **22**, 169–175 (1993)

467. Riley, K.F., Hobson, M.P., Bence, S.J.: Mathematical Methods for Physics and Engineering, 2nd edn. Cambridge University Press, Cambridge (2002)
468. Risken, H.: TheFokker-Planck Equation. Methods of Solution and Applications, 2nd edn. Springer, Berlin (1989)
469. Robinett, R.W.: Quantum Mechanics. Classical Results, Modern Systems, and Visualized Examples. Oxford University Press, New York (1997)
470. Roebuck, J.R.: The rate of the reaction between arsenious acid and iodine in acid solution, the rate of the reverse reaction, and the equilibrium between them. J. Phys. Chem. **6**, 365–398 (1901)
471. Rotman, B.: Measurement of activity of single molecules of β-d-galactosidase. Proc. Natl. Acad. Sci. USA **47**, 1981–1991 (1961)
472. Sagués, F., Epstein, I.R.: Nonlinear chemical dynamics. J. Chem. Soc. Dalton Trans. **2003**, 1201–1217 (2003)
473. Salis, H., Kaznessis, Y.: Accurate hybrid stochastic simulation of a system of coupled chemicel or biochemical reactions. J. Chem. Phys. **122**, e054,103 (2005)
474. Sanft, K.R., Wu, S., Roh, M., Fu, J., Lim, R.K., Petzold, L.R.: StochKit2: Software for discrete stochastic simulation of biochemical systems with events. Bioinformatics **27**, 2457–2458 (2011)
475. Sato, K.: Lévy Processes and Infinitely Divisible Distributions, 2nd edn. Cambridge University Press, Cambridge (2013)
476. Scatchard, G.: The attractions of proteins for smal molecules and ions. Ann. New York Acad. Sci. **51**, 660–672 (1949)
477. Scher, H., Shlesinger, M.F., Bendler, J.T.: Time scale invariance in transport and relaxation. Phys. Today **44**(1), 26–34 (1991)
478. Schilling, M.F., Watkins, A.E., Watkins, W.: Is human height bimodal? Am. Statistician **56**, 223–229 (2002)
479. Schlögl, F.: Chemical reaction models for non-equilibrium phase transitions. Z. Physik **253**, 147–161 (1972)
480. Schoutens, W.: Lévy Processes in Finance. Wiley Series in Probability and Statistics. Wiley, Chichester (2003)
481. Schubert, M., Weber, G.: Quantentheorie. Grundlagen und Anwendungen. Spektrum Akademischer Verlag, Heidelberg, DE (1993). In German
482. Schuster, P.: Mathematical modeling of evolution. Solved and open problems. Theory Biosci. **130**, 71–89 (2011)
483. Schuster, P.: Are computer scientists the sutlers of modern biology? Bioinformatics is indispesible for progress in molecular life sciences but does not get credit for its contributions. Complexity **19**(4), 10–14 (2014)
484. Schuster, P.: Quasispecies on fitness landscapes. In: Domingo, E., Schuster, P. (eds.) Quasispecies: From Theory to Experimental Systems, *Current Topics in Microbiology and Immunology*, vol. 392, chap. 4, pp. ppp–ppp. Springer, Berlin (2016). DOI 10.10007/82_2015_469
485. Schuster, P., Sigmund, K.: Replicator dynamics. J. Theor. Biol. **100**, 533–538 (1983)
486. Schuster, P., Sigmund, K.: Random selection - A simple model based on linear birth and death processes. Bull. Math. Biol. **46**, 11–17 (1984)
487. Schwabl, F.: Quantum Mechanics, 4th edn. Springer, Berlin (2007)
488. Schwarz, G.: Kinetic analysis by chemical relaxation methods. Rev. Mod. Phys. **40**, 206–218 (1968)
489. Seber, G.A., Lee, A.J.: Linear Regression Analysis. Wile Series in Probabiity and Statistics. Wiley-Intersceince, Hoboken (2003)
490. Sehl, M., Alekseyenko, A.V., Lange, K.L.: Accurate stochastic simulation via the step anticipation τ-leaping (SAL) algorithm. J. Comp.,Biol. **16**, 1195–1208 (2009)
491. Selmeczi, D., Tolić-Nørrelykke, S., Schäffer, E., Hagedorn, P.H., Mosler, S., Berg-Sørensen, K., Larsen, N.B., Flyvbjerg, H.: Brownian motion after Einstein: Some new applications and new experiments. Lect. Notes Phys. **711**, 181–199 (2007)
492. Seneta, E.: Non-negative Matrices and Markov Chains, 2nd edn. Springer, New York (1981)

493. Seneta, E.: The central limit problem and lienear least squares in pre-revolutionary Russia: The background. Math. Scientist **9**, 37–77 (1984)
494. Senn, H.M., Thiel, W.: QM/MM Methods for biological systems. Top. Curr. Chem. **268**, 173–290 (2007)
495. Senn, H.M., Thiel, W.: QM/MM Methods for biomolecular systems. Angew. Chem. Int. Ed. **48**, 1198–1229 (2009)
496. Seydel, R.: Practical Bifurcation and Stability Analysis. From Equilibrium to Chaos, *Interdisciplinary Applied Mathematics*, vol. 5, 2nd edn. Springer, New York (1994)
497. Shannon, C.E.: A mathematical theory of communication. Bell Syst. Tech. J. **27**, 379–423 (1948)
498. Shannon, C.E., Weaver, W.: The Mathematical Theory of Communication. University of Illinois Press, Urbana (1949)
499. Shapiro, B.E., Levchenko, A., World, E.M.M.B.J., Mjolsness, E.D.: Cellerator: Extending a computer algebra system to include biochemical arrows for signal transduction simulations. Bioinformatics **19**, 677–678 (2003)
500. Sharpe, M.J.: Transformations of diffusion by time reversal. Ann. Probab. **8**, 1157–1162 (1980)
501. Shuler, K.E.: Studies in nonequilibrium rate processes: II. The relaxation of vibrational nonequilibrium distributions in chemical reactions and shock waves. J. Phys. Chem. **61**, 849–856 (1957)
502. Shuler, K.E., Weiss, G.H., Anderson, K.: Studies in nonequilibrium rate processes. V. The relaxation of moments derived from a master equation. J. Math. Phys. **3**, 550–556 (1962)
503. Sotiropoulos, V., Kaznessis, Y.N.: Analytical derivation of moment equations in stochastic chemical kinetics. Chem. Eng. Sci. **66**, 268–277 (2011)
504. Stauffer, P.H.: Flux flummoxed: A proposal for consistent usage. Ground Water **44**, 125–128 (2006)
505. Steffensen, J.F.: "deux problème du calcul des probabilités". Ann. Inst. Henri Poincaré **3**, 319–344 (1933)
506. Stepanow, S., Schütz, G.M.: The distribution function os a semiflexible polymer and random walks with constraints. Europhys. Lett. **60**, 546–551 (2002)
507. Stevens, J.W.: What is Bayesian Statistics? What is ... ? Hayward Medical Communications, a division of Hayward Group Ltd., London (2009)
508. Stigler, S.M.: Laplace's 1774 memoir on inverse probability. Stat. Sci. **1**, 359–378 (1986)
509. Stigler, S.M.: The epic story of maximum likelihood. Stat. Sci. **22**, 598–620 (2007)
510. Stone, J.V.: Bayes' Rule. A Tutorial to Bayesian Analysis. Sebtel Press, England (2013)
511. Strang, G.: Linear Algebra and its Applications, 3rd edn. Brooks Cole Publishing Co, Salt Lake City (1988)
512. Stratonovich, R.L.: Introduction to the Theory of Random Noise. Gordon and Breach, New York (1963)
513. Strogatz, S.H.: Nonlinear Dynamics and Chaos. With Applications to Physics, Biology, Chemistry, and Engineering. Westview Press at Perseus Books, Cambridge (1994)
514. Stuart, A., Ord, J.K.: Kendall's Advanced Theory of Statistics. Volume 1: Distribution Theory, 5th edn. Charles Griffin & Co., London (1987)
515. Stuart, A., Ord, J.K.: Kendall's Advanced Theory of Statistics. Volume 2: Classical Inference and Relationship, 5th edn. Edward Arnold, London (1991)
516. Student: The probable error of a mean. Biometrika **6**, 1–25 (1908)
517. Suber, P.: A crash course in the mathematics of infinite sets. St.John's Rev. **XLIV**(2), 35–59 (1998)
518. Suppes, P.: Axiomatic Set Theory. Dover Publications, New York (1972)
519. Swamee, P.K.: Near lognormal distribution. J. Hydrol. Eng. **7**, 441–444 (2007)
520. Swetina, J., Schuster, P.: Self-replication with errors - A model for polynucleotide replication. Biophys. Chem. **16**, 329–345 (1982)
521. Szathmáry, E., Gladkih, I.: Sub-exponential growth and coexistence of non-enzymatically replicating templates. J. Theor. Biol. **138**, 55–58 (1989)

522. Tang, H., Siegmund, D.O., Shen, P., Oefner, P.J., Feldman, M.W.: Frequentist estimation of coalescence times from nucleotide sequence data using a tree-based partition. Genetics **161**, 448–459 (2002)
523. Tao, T.: An Introduction to Measure Theory, *Graduate Studies in Mathematics*, vol. 126. American Mathematical Society, Providence (2011)
524. Tarantola, A.: Inverse Problem Theory and Methods for Model Parameter Estimation. Society for Industrial and Applied Mathematics, Philadelphia (2005)
525. Tavaré, S.: Line-of-descent and genealogical processes, and their application in population genetics models. Theor. Popul. Biol. **26**, 119–164 (1984)
526. Taylor, H.M., Karlin, S.: An Introduction to Stochastic Modeling, 3rd edn. Academic press, San Diego (1998)
527. Taylor, M.E.: Measure Theory and Intergration, *Graduate Studies in Mathematics*, vol. 76. American Mathematical Society, Providence (2006)
528. Thiele, T.N.: Om Anvendelse af midste Kvadraters Methode i nogle Tilfælde, hvor en Komplikation af visse Slags uensartede tilfædige Feijlkilder giver Feijlene en 'systenatisk' Karakter. Vidensk. Selsk. Skr. 5. rk., naturvid. og mat. Afd. **12**, 381–408 (1880). In Danish
529. Thomas, G.B., Finney, R.L.: Calculus and Analytic Geometry, 9th edn. Addison-Wesley, Reading (1996)
530. Thompson, C.J., McBride, J.L.: On Eigen's theory of the self-organization of matter and the evolution of biological macromolecules. Math. Biosci. **21**, 127–142 (1974)
531. Tolman, R.C.: The Principle of Statistical Mechanics. Oxford University Press, Oxford (1938)
532. Tsukahara, H., Ishida, T., Mayumi, M.: Gas-phase oxidation of nitric oxide: Chemical kinetics and rate constant. Nitric Oxide Biol. Chem. **3**, 191–198 (1999)
533. Turing, A.M.: The chemical basis of morphogenesis. Philos. Trans. R. Soc. Lond. B **237**(641), 37–72 (1952)
534. Uhlenbeck, G.E., Ornstein, L.S.: On the theory of the Brownian motion. Phys. Rev. **36**, 823–841 (1930)
535. Ullah, M., Wolkenhauer, O.: Family tree of Markov models in systems biology. IET Syst. Biol. **1**, 247–254 (2007)
536. Ullah, M., Wolkenhauer, O.: Stochastic Approaches for Systems Biology. Springer, New York (2011)
537. van den Berg, T.: Calibrating the Ornstein-Uhlenbeck-Vasicek model. Sitmo – Custom Financial Research and Development Services, www.sitmo.com/article/calibrating-the-ornstein-uhlenbeck-model/ (2011). Retrieved April 20, 2014
538. van den Bos, A.: Parameter Estimation for Scientists and Engineers. Wiley, Hoboken (2007)
539. Van Doorn, E.A.: Quasi-stationary distribution and convergence to quasi-stationarity of birth-death processes. Adv. Appl. Probab. **23**, 683–700 (1991)
540. van Kampen, N.G.: A power series expansion of the master equation. Can. J. Phys. **39**, 551–567 (1961)
541. van Kampen, N.G.: The expansion of the master equation. Adv. Chem. Phys. **34**, 245–309 (1976)
542. van Kampen, N.G.: Remarks on non-markov processes. Braz. J. Phys. **28**, 90–96 (1998)
543. van Kampen, N.G.: Stochastic Processes in Physics and Chemistry, 3rd edn. Elsevier, Amsterdam (2007)
544. van Oijen, A.M., Blainey, P.C., Crampton, D.J., Richardson, C.C., Ellenberger, T., Xie, X.S.: Single-moleucles kinetics of λ exconuclease reveal base dependence and dynamic disorder. Science **301**, 1235–1238 (2003)
545. van Slyke, D.D., Cullen, G.E.: The mode of action of urease and of enzymes in general. J. Biol. Chem. **19**, 141–180 (1914)
546. Vasicek, O.: An equlibrium characterization of the term structure. J. Financ. Econ. **5**, 177–188 (1977)
547. Venn, J.: On the diagrammatic and mechanical representation of propositions and reasonings. Lond. Edinb. Diblin Philos. Mag. J. Sci. **9**, 1–18 (1880)

548. Venn, J.: Sybolic Logic. MacMillan, London (1881). Second edition, 1984. Reprinted by Lenox Hill Pub. & Dist. Co., 1971
549. Venn, J.: The Logic of Chance. An Essay on the Foundations and Province of the Theory of Probability, with Especial Reference to its Logical Bearings and its Application to Moral and Social Science, and to Statistics, 3rd edn. MacMillan, London (1888)
550. Verhulst, P.: Notice sur la loi que la population pursuit dans son accroisement. Corresp. Math. Phys. **10**, 113–121 (1838)
551. Viswanathan, G.M., Raposo, E.P., da Luz, M.G.E.: Lévy flights and superdiffusion in the context of biological encounters and random searches. Phys. Life Rev. **5**, 133–150 (2008)
552. Vitali, G.: Sul problema della misura dei gruppi di pinti di una retta. Gamberini E. Parmeggiani, Bologna (1905)
553. Vitali, G.: Sui gruppi di punti e sulle funzioni di variabili reali. Atti dell'Accademia delle Science di Torino **43**, 75–92 (1908)
554. Volkenshtein, M.V.: Entropy and Information, *Progress in Mathematical Physics*, vol. 57. Birkhäuser Verlag, Basel, CH (2009). German version: W. Ebeling, Ed. Entropie und Information. Wissenschaftliche Taschenbücher, Band 306, Akademie-Verlag, Berlin (1990). Russian Edition: Nauka Publ., Moscow (1986)
555. von Kiedrowski, G.: A self-replicating hexanucleotide. Angew. Chem. Internat. Ed. **25**, 932–935 (1986)
556. von Kiedrowski, G., Wlotzka, B., Helbig, J., Matzen, M., Jordan, S.: Parabolic growth of a self-replicating hexanucleotide bearing a 3'-5'-phosphoamidate linkage. Angew. Chem. Int. Ed. **30**, 423–426 (1991)
557. von Mises, R.: Über Aufteilungs- und Besetzungswahrscheinlichkeiten. Revue de la Faculté des Sciences de l'Université d'Istanbul, N.S. **4**, 145–163 (1938–1939). In German. Reprinted in *Selected Papers of Richard von Mises*, vol.2, American Mathematical Society, 1964, pp. 313–334
558. von Neumann, J.: Proof of the quasi-ergodic hypothesis. Proc. Natl. Acad. Sci. USA **4**, 70–82 (1932)
559. von Smoluchowski, M.: Zur kinetischen Theorie der Brownschen Molekularbewegung und der Suspensionen. Ann. Phys. (Leipzig) **21**, 756–780 (1906)
560. Waage, P., Guldberg, C.M.: Studies concerning affinity. J. Chem. Educ. **63**, 1044–1047 (1986). English translation by Henry I. Abrash
561. Walter, N.G.: Single molecule detection, analysis, and manipulation. In: Meyers, R.A. (ed.) Encyclopedia of Analytical Chemistry, pp. 1–10. Wiley, Hoboken (2008)
562. Watson, H.W., Galton, F.: On the probability of the extinction of families. J. Anthropol. Inst. G. Br. Irel. **4**, 138–144 (1875)
563. Weber, N.A.: Dimorphism of the African *oecophylla* worker and an anomaly (*hymenoptera formicidae*). Ann. Entomol. Soc. Am. **39**, 7–10 (1946)
564. Wegscheider, R.: Über simultane Gleichgewichte und die Beziehungen zwischen Thermodynamik und Reaktionskinetik homogener Systeme. Mh. Chem. **32**, 849–906 (1911). In German
565. Wei, W.W.S.: Time Series Analysis. Univariate and Multivariate Methods. Addison-Wesley Publishing, Redwood City (1990)
566. Weiss, G.H., Dishon, M.: On the asympotitic behavior of the stochastic and deterministic models of an epidemic. Math. Biosci. **11**, 261–265 (1971)
567. Weisstein, E.W.: Cross-Correlation. MathWorld - A Wolfram Web Resource. The Wolfram Centre, Long Hanborough, UK. http://www.Mathworld.wolfram.com/Cross-Correlation.html, retrieved July 17, 2015
568. Weisstein, E.W.: Fourier Transform. MathWorld - A Wolfram Web Resource. The Wolfram Centre, Long Hanborough, UK. http://www.Mathworld.wolfram.com/FourierTransform.html, retrieved July 17, 2015
569. Widengren, J., Mets, Ülo., Rigler, R.: Photodynamic properties of green fluorescent proteins investigated by fluoresecence correlation spectroscopy. Chem. Phys. **250**, 171–186 (1999)

570. Wilheim, T.: The smallest chemical rwaction system with bistability. BMC Syst. Biol. **3**, e90 (2009)
571. Wilheim, T., Heinrich, R.: Smallest chemical rwaction system with Hopf bifurcation. J. Math. Chem. **17**, 1–14 (1995)
572. Wilkinson, D.J.: Stochastic modeling for quatitative description of heterogeneous biological systems. Nat. Rev. Genet. **10**, 122–133 (2009)
573. Wilkinson, D.J.: Stochastic Modelling for Systems Biology, 2nd edn. Chapman & Hall/CRC Press – Taylor and Francis Group, Boca Raton (2012)
574. Williams, D.: Diffusions, Markov Processes and Martingales. Volume 1: Foundations. Wiley, Chichester (1979)
575. Wills, P.R., Kauffman, S.A., Stadler, B.M.R., Stadler, P.F.: Selection dynamics in autocatalytic systems: Templates replicating through binary ligation. Bull. Math. Biol. **60**, 1073–1098 (1998)
576. Winzor, D.J., Jackson, C.M.: Interpretation of the temperature dependence of equilibrium and rate contants. J. Mol. Recognit. **19**, 389–407 (2006)
577. Wolberg, J.: Data Analysis Using the Method of Least Squares. Extracting the Most Information from Experiments. Springer, Berlin (2006)
578. Wold, H.: A Study in the Analysis of Time Series, second revised edn. Almqvist and Wiksell Book Co., Uppsala, SE (1954). With an appendix on *Recent Developments in Time Series Analysis* by Peter Whittle
579. Wright, S.: Evolution in Mendelian populations. Genetics **16**, 97–159 (1931)
580. Wright, S.: The roles of mutation, inbreeding, crossbreeding and selection in evolution. In: Jones, D.F. (ed.) Int. Proceedings of the Sixth International Congress on Genetics, vol. 1, pp. 356–366. Brooklyn Botanic Garden, Ithaca (1932)
581. Yang, Y., Rathinam, M.: Tau leaping of stiff stochastical chemical systems via local central limit approximation. J. Comp. Phys. **242**, 581–606 (2013)
582. Yashonath, S.: Relaxation time of chemical reactions from network thermodynamics. J. Phys. Chem. **85**, 1808–1810 (1981)
583. Zhabotinsky, A.M.: A history of chemical oscillations and waves. Chaos **1**, 379–386 (1991)
584. Zhang, W.K., Zhang, X.: Single molecule mechanochemistry of macromolecules. Prog. Polym. Sci. **28**, 1271–1295 (2003)
585. Zwillinger, D.: Handbook of Differential Equations, 3rd edn. Academic Press, San Diego (1998)

Author Index

Acton, F.S., 411
Andronov, A., 553
Arányi, P., 431
Arnold, L., 241, 559
Arrhenius, S., 389
Avogadro, A., 3

Bachelier, L., 4, 319
Bailey, N., 620
Bartholomay, A., 350
Bartlett, M.S., 527
Bayes, T., 14, 190
Belousov, B.P., 552, 572
Bernoulli, D., 182, 618
Bernoulli, J., 11, 112, 210
Bernstein, S.N., 41
Bessel, F.W., 170
Bienaymé, I.J., 171, 632
Bjerhammar, A., 410
Boltzmann, L., 5, 100, 393
Boole, G., 12
Borel, E., 45
Born, M., 406
Box, G.E.P., 536
Brenner, S., ix
Briggs, G.E., 363
Brown, R., 3, 218

Cantor, G., 16, 19, 22, 51
Cardano, G., 6
Cauchy, A.L., 58, 88, 156
Changeux, J.-P., 498
Chapman, S., 201, 225
Chebyshev, P.L., 134
Chung, K.L., x
Cochran, W.G., 144
Cramér, H., 182
Cullen, G.E., 363

Darboux, G., 60
Darwin, C., 582
de Candolle, A., 631
de Fermat, P., 6
de Moivre, A., 126, 131
De Morgan, A., 24
Dedekind, R., 16
Delbrück, M., 552
Dirac, P., 34, 250
Dirichlet, G.L., 66
Dishon, M., 620
Dmitriev, N.A., 631
Doob, J.L., 210, 527

Edgeworth, F.Y., 184
Ehrenfest, P., 447, 595
Ehrenfest-Afanassjewa, T., 595
Eigen, M., ix, 645, 668
Einstein, A., x, 4, 203, 212, 214, 319
Erlang, A.K., 259, 313
Euler, L., 68, 195
Eyring, H., 389

Feinberg, M., 372
Feller, W., 14, 136, 166, 526
Fick, A.E., 4, 239

P. Schuster, *Stochasticity in Processes*, Springer Series in Synergetics,
DOI 10.1007/978-3-319-39502-9

Field, R.J., 353, 561
Filmer, D., 498
Filon, L.N.G., 182
Fisher, R.A., 10, 12, 15, 143, 172, 174, 182, 649, 651
Fisk, D.L., 329
Fisz, M., 168
Fokker, A.D., 201
Frobenius, F.G., 641
Fubini, G., 76

Galton, F., 11
Galton, Sir Francis, 137, 631
Gardiner, C., x, 201, 228, 320
Gause, G.F., 617
Gauss, C.F., 73, 115, 170, 182, 407
Gegenbauer, L., 453, 608
Gibbs, J.W., 235, 355
Gillespie, D., 419, 527, 564
Goel, N.S., 589, 605
Gosset, W.S., 143
Grötschel, M., ix
Gram, J.P., 409
Guinness, A., 143
Guldberg, C.M., 353

Haldane, J.B.S., 363
Hamilton, W.R., 237
Heaviside, O., 30, 462, 481
Heinrich, R., 573
Heisenberg, W., 391
Heun, K., 663
Hinshelwood, C., 402
Hooke, R., 356
Hopf, E.F.F., 553
Horn, F., 372
Houchmandzadeh, B., 268, 659
Huisinga, W., 449
Hurst, H.E., 284

Itō, K., 70, 325

Jackson, R., 372
Jacobi, C., 456
Jahnke, T., 449
Jaynes, E.T., 99

Kassel, L.S., 403
Kendall, D.G., 527, 631
Kendall, M., 168
Keynes, J.M., 25
Khinchin, A.Y., 135, 223, 286
Kimura, M., 601, 605
Kingman, J.F.C., 673
Kleene, S.C., 23
Kolmogorov, A.N., 21, 135, 201, 225, 526, 631
Koshland, Jr., D.E., 498
Kramers, H.A., 509
Kronecker, L., 254

Lévy, P.P., 131, 161, 210, 284
Lagrange, J.L., 182
Langevin, P., 4, 231, 319
Laplace, P.-S., 12, 15, 115, 126, 131, 150
Laurent, P.A., 291
Le Novère, N., 541
Lebesgue, H.L., 35, 45, 61
Legendre, A.-M., 407
Leibniz, G.W., 225
Lindeberg, J.W., 131
Lindemann, F., 402
Liouville, J., 234, 235
Lipschitz, R., 338
Lorentz, H.A., 156
Lorenz, E., 5
Loschmidt, J., 3
Lyapunov, A.M., 131

MacLaurin, C., 107
Mandelbrot, B., 295
Marcus, R.A., 403
Markov, A., 203, 214
Maxwell, J.C., 5, 11, 393
McAlister, D., 137
McKean, H.P., 136
McQuarrie, D., 454
Mellin, H., 460
Mendel, G., 9, 174, 178
Menten, M., 358
Michaelis, L., 358
Mittag-Leffler, M.G., 298
Monod, J., 498
Montroll, E.W., 282, 297
Moore, E.H., 410
Moore, G., ix
Moran, P.A.P., 649, 651
Morton-Firth, C.J., 541
Moyal, J.E., 509
Muller, M.E., 536

Némethy, G., 498
Nåsell, I., 618
Newton, I., 225
Neyman, J., 12
Nicolis, G., 353
Nishiyama, K., 575
Noyes, R.M., 353, 561

Oppenheimer, J.R., 406
Ornstein, L.S., 200, 249
Ostwald, W., 5

Pareto, V., 115, 151
Pascal, B., 6
Pearson, E.S., 12
Pearson, K., 26, 115, 143, 169, 174, 182, 273
Peirce, C.S., 14
Penrose, R., 410
Perron, O., 641
Petzold, L.R., 541
Planck, M., 100, 201
Pochhammer, L.A., 94, 662
Poincaré, H., 5
Poisson, S.D., 109, 253
Popper, K., 14
Post, E.L., 460
Prigogine, I., 353, 556

Ramsperger, H.C., 402
Rao, C.R., 182
Reichenbach, H., 12
Rice, O.K., 402
Richter-Dyn, N., 589, 605
Riemann, B., 55, 61
Rolle, M., 638

Scatchard, G., 412
Schlögl, F., 573, 583
Schrödinger, E., 391
Schwarz, H.A., 88
Shannon, C.E., 95
Shapiro, B.E., 541
Shimizu, T.S., 541
Steffensen, J.F., 632
Stieltjes, T.J., 61
Stigler, S.M., 182
Stirling, J., 94, 100, 126, 342
Stratonovich, R.L., 329
Student, *see* Gosset, W.S.

Tóth, J., 431
Taylor, B., 107
Thiele, T.N., 4
Tolman, R.C., 267
Turing, A.M., 559

Uhlenbeck, G.E., 200, 249
Ulam, S.M., 632
Ullah, M., 228

Vallade, M., 659
van Kampen, N.G., 264, 509, 515, 631
van Slyke, D.D., 363
Venn, J., 12, 18
Verhulst, P.-F., 570
Verhulst, P.-F., 611
Vitali, G., 46
Volkenshtein, M.V., 101
von Mises, R., 8, 12
von Smoluchowski, M., 4, 203, 212, 214, 319

Waage, P., 353
Watson, H.W., 11, 631
Weiss, G.H., 282, 297, 620
Wiener, N., 218, 223, 234, 238
Wigner, E., 92
Wilhelm, T., 573
Wold, H., 248
Wolkenhauer, O., 228
Wright, S., 649, 651
Wyman, J., 498

Zhabotinsky, A.M., 552, 572

Index

additivity
  σ, 21, 46
algebra
  σ, 24, 49, 50, 70
  Borel-σ, 50
allele, 605
antibunching term, 501
anticipation, *see* process, nonanticipating
approximation
  Poisson-normal, 119, 341
  pre-equilibrium, 363
  steady state, 363, 365
  Stirling's, 100, 126, 342
arrival time, *see* time, arrival
assumption
  scaling, 277
asymptotic frequencies, 642
autocatalysis, 352
Avogadro's constant, 3, 5

balancing
  complex, 377
  detailed, 267, 377
barrier, *see* boundary
Bernoulli trials, 210
bifurcation
  Hopf, 553, 559
  saddle-node, 554
  subcritical, 579
  transcritical, 579
bijection, 48
bit, 95
Borel algebra, *see* algebra, Borel-σ
Borel field, *see* algebra, Borel-σ
boundary
  absorbing, 268, 313, 442, 604
  artificial, 605
  natural, 316, 425, 442, 445, 604
  noflux, 315
  reflecting, 268, 313, 445, 604
Brownian motion, 3, 218, 241
  fractal, 298
Brusselator, 353, 552
buffer, 457, 520, 570, 590

cardinality (set theory), 17
catalysis, 352
chain reaction, nuclear, 632
characteristic manifold, 439
characteristics, method of, 438
closure, 23
coalescent theory, 673
coefficient
  binding, 413
collisions
  classical, 393
  elastic, 394
  inelastic, 394
  molecular, 348
  nonreactive, 397
  reactive, 394, 397
compatibility class
  stoichiometric, 374
complement (set theory), 17
condition
  final, 217, 304
  growth, 338
  initial, 200, 217, 304

P. Schuster, *Stochasticity in Processes*, Springer Series in Synergetics,
DOI 10.1007/978-3-319-39502-9

Lindeberg's, 132
Lipschitz, 338
Lyapunov's, 131
confidence interval, 117, 440
contingency table, 180
convergence
pointwise, 57, 65
radius of, 102
uniform, 57
convolution, 220, 281, 496, 633
cooperative binding, 499
coordinates
labor, 394
correction
Bessel, 170
correlation
coefficient, 88
countable additivity, *see* σ-additivity
covariance, 88
sample, 172
crossing, avoided, 630
cumulant, 91, 104
curvature, 185

decomposition
partial fraction, 461
Wold's, 248
deficiency, 385
one, 387
zero, 386
density
joint, 85, 208
spectral, 222
density matrix
classical, 237
deterministic chaos, 5
diagram
Venn, 18
difference (set theory), 19
difference equation, 653
diffusion
anomalous, 298
normal, 280
diffusion coefficient, 4, 239, 278
disjoint sets (set theory), 19
disorder, dynamical, 499
displacement
mean square, 302
distance
Hamming, 208, 669
distribution
Bernoulli, 112
bimodal, 89
binomial, 112
chi-squared, 140
Erlang, 259, 313
exponential, 147
geometric, 150
heavy-tailed, 156
joint, 43, 120
log-normal, 137
logistic, 154
marginal, 43, 44, 77
Maxwell–Boltzmann, 396
Maxwell-Boltzmann, 395
multinomial, 114, 427, 449, 675
normal, 72, 91, 115, 241
Poisson, 109, 148, 449
ratio, 158
stable, 115, 159, 161
strictly stable, 162
Student's, 143
symmetrically stable, 162
uniform, 25, 47
distribution, Laplace, 150
distribution, Poisson, 188
divisibility, infinite, 161, 284
double factorial, 118
dynamics
complex, 5

energy
activation, 390
ensemble average, 224
entropy
information, 95
thermodynamic, 95
epidemiology
theoretical, 618
equation
Arrhenius, 390
backward, 305, 306
birth-and-death master, 266
Chapman–Kolmogorov, 225, 322
chemical Langevin, 533
chemical master, 421, 530
differential C.K., 228
diffusion, 234, 239
Fokker–Planck, 233, 322, 350
forward, 305
Langevin, 231, 303, 320, 350
Liouville, 235
logistic, 570, 615
master, 234, 350, 418
reaction–diffusion, 206
Schrödinger, 390

stationary Fokker–Planck, 251
stoichiometric, 351
equations
normal, 409
equilibrium
binding, 412
constant, 355, 442
thermal, 395
equivalence class
irreducible, 427, 451, 464
ergodicity, *see* theory ergodic
error
mean squared, 171
error function, 74
error rate, uniform, 669
error threshold, 671
estimator, 169
efficient, 182
event, 7, 23
space, 49
system, 49
excitable medium, 559
exit problem, 305
expectation value, 28, 62, 76, 84
extinction
state of, 586
extinction time, *see* time, extinction

factor
geometric, 394
factorial
falling, 94
fixation, 607
flow
dilution, 645
mass, 437
volume, 437
flow reactor, 436, 518, 571, 575, 645
fluctuations
natural, 3, 5
flux, 437
formula
Lévy-Khinchin, 288
fractal, 283, 292
frequentism
finite, 12
hypothetical, 12
function
association, 371
autocorrelation, 220, 500
autocovariance, 216, 222
characteristic, 101, 105
cumulant generating, 101
cumulative distribution, 27, 35, 38, 90
density, 71, 80, 238
Dirac delta, 34
distribution, 72
frequency, 419, 528, 537
Gamma, 141
generating, 93
Heaviside, 30
hyperbolic shaped, 551
hypergeometric, 453
indicator, 33, 64
logistic, 154
measurable, 64
Mittag-Leffler, 298
moment generating, 101, 103
nonanticipating, 327, 337
objective, 183
probability generating, 101
probability mass, 26, 27, 34
rate, 355, 371, 423
score, 184
sigmoid shaped, 551
signum, 33
simple, 63
tent, 35
transition, 286

generator
infinitesimal, 644
random number, 250, 535
set theory, 51
genetics
Mendelian, 9

half-life, 148
harmonic number, 482
heavy tail, 156, 291, 299
heteroscedasticity, 410
homogeneity, spatial, 395
homoscedasticity, 410
hysteresis, chemical, 554

immigration, 518
independence
stochastic, 40, 77, 121
index
Pareto, 152
tail, 152
induced fit, 498
inequality
Cauchy–Schwarz, 88

median–mean, 89
median-mean, 149
infinite divisibility, 294
information
content, 95
Fisher, 184
observed, 186
inhibition
product, 359
integral
Cauchy-Euler, 330
improper, 63, 67
Itō, 70
Lebesgue, 35, 61
Lebesgue-Stieltjes, 32
Riemann, 55, 61
Riemann–Stieltjes, 35
Stieltjes, 61, 322, 323
stochastic, 323
Stratonovich, 329
integrand, 61
integration, *see* integral
integrator, 61
intensity function, *see* reaction probability, differential
intersection (set theory), 18
irreducibility, 427
isotherm, 414

jump length, 279

kinematics, 393
kinetics
higher level, 358
mass action, 351
Michaelis–Menten, 358
kinetics, fractional, 285
Kleene star, 23
Kronecker delta, 254
kurtosis, 91
excess, 91

Lévy flights, 299
law
$\sqrt{N}$, 2, 111, 125, 212, 257, 440, 447, 468, 500, 512, 533, 544, 552, 572
deterministic, 8
Hook's, 356
large numbers, 133
power, 295
statistical, 9
least squares, *see* regression, least squares
limit
almost certain, 56
in distribution, 56, 130
in probability, 56
mean square, 56, 325
stochastic, 56
linear span, 374
linkage class, 378
location parameter, 163, 294
logarithm
law of iterated, 135
Loschmidt's constant, 3

macroscopic infinitesimal, 344
manifold
one-dimensional, 624
Markov chain, *see* process, Markov
martingale, 210, 274
local, 325
mass action, 353, 529
master sequence, 671
matrix
adjugate, 471
bistochastic, 645, 669
complex, 373
design, 408
diffusion, 231
fitness, 645
Gramian, 409
idempotent, 641
mean, 640
mutation, 645
pseudoinverse, 410
stochastic, 645, 669
stoichiometric, 374, 376
tridiagonal, 472
upper-triangular, 472
value, 645
maximum likelihood, 182, 188, 192, 193
mean
displacement, 5
sample, 169
value, 10
mean absolute deviation, 172
measure
Borel, 45
complete, 50
Lebesgue, 45, 53
product, 52
mechanics
quantum, 348
statistical, 348

mechanism
    Koshland–Némethy–Filmer, 498
    Monod–Wyman–Changeux, 498
median, 88
memory effect, 203
memorylessness, 150
method
    direct, 538
    first reaction, 539
    Heaviside expansion, 462, 481
    next reaction, 540
Michaelis constant, 363
mitochondrial Eve, 673
mode, 89
model
    moving average, 248
molecularity, *see* reaction, molecularity of
moment
    centered, 86
    factorial, 111
    jump, 263, 269, 424, 509
    raw, 86, 117
    sample, unbiased, 170
moments
    factorial, 94
motion
    Brownian, 320
mutation, *see* point mutation

neutral evolution, *see* selection, random
noise
    additive, 321
    colored, 224
    multiplicative, 337
    real, 337
    small, 512
    white, 223, 238, 323, 337
nonanticipation, *see* process, nonanticipating
normality, asymptotic, 184
notation
    multi-index, 355
null hypothesis, 8, 25
number density, 3
numbers
    irrational, 45
    natural, 19
    rational, 19, 45, 54
    real, 19

operator
    linear, 84
    projection, 641

Oregonator, 353, 552

p-value, 177
parameter
    rate, 148
    survival, 148
peakedness, 91
physical time, *see* time real
pivotal quantity, 144
Pochhammer symbol, 94, 465, 473, 662
point mutation, 208
polynomial
    Gegenbauer, 453
    Jacobi, 453, 456
power law, 151
powerset, 22, 24, 45, 49
preimage, 48, 52, 64
principle
    of indifference, 12, 25
principle of
    detailed balance, 214, 377
    indifference, 96
    maximum entropy, 99
probability
    classical, 11
    conditional, 38
    current, 609
    density, 25, 62, 71
    distribution, 62, 72
    elementary, 75
    evidential, 12
    extinction, 270
    frequency, 12
    inverse, 192, 194
    joint, 42, 77
    measure, 21
    physical, 12
    posterior, 192, 194
    prior, 192, 194
    propensity, 14
    transition, 261
    triple, 29, 70
problem
    forward, 349
    inverse, 349
    parameter identification, 349, 407
process
    Lévy, 284
    adapted, 213, 327
    ambivalent, 299
    AR(n), 248
    AR(1), 251
    autoregressive, 248

Bernoulli, 22, 112, 209
birth-and-death, 264, 351
linear, 590
restricted, 589
unrestricted, 589
càdlàg, 32, 245
compensated Poisson, 289
compound Poisson, 289
counting, 260, 424
death-and-birth, 264
diffusion, 335
elementary, 350
Galton–Watson, 632
Gaussian, 247, 251
independent, 210
Liouville, 234
Markov, 203, 210, 213, 321, 350
Markov homogeneous, 215, 286
Markov stationary, 215
nonanticipating, 213, 327, 337
Pareto, 290
Poisson, 109, 147, 254, 288
recurrent, 273
separable, 208
stationary, 215, 251
transient, 273
unit Poisson, 260, 424
Wiener, 218, 223, 234, 238, 303, 326, 330
process ambivalent, 299
product, reaction, 351, 375
propensity function, *see* reaction probability, differential
property
extensive, 100, 514
intensive, 100, 514
pseudorandom number, 13
uniform, 535
pseudorandom numbers
nonuniform, 536
pseudoreaction, 437

quantile, 90
quantum mechanical calculations, 390
quasi-stationary state, *see* stationarity
quasidiffusion, 299
quasispecies, 646, 671

radioactive decay, 366
random drift, 3
random number, *see* pseudorandom number
random walk
continuous time, 274
one dimension, 274
one-sided, 254
rate constant, *see* rate parameter
rate function, *see* function, rate
rate parameter, 349, 354, 373, 376, 381, 389, 401, 407, 423, 430, 442, 453, 464, 491, 501, 528
reactant, 351, 375
reaction
$2X \rightarrow A + X$, 604
$2A \rightarrow C$, 451, 464
$A + 2X \rightarrow 3X$, 353, 451
$A + B \rightleftharpoons 2C$, 543
$A + B \rightleftharpoons C$, 470
$A + B \rightleftharpoons C + D$, 380, 476
$A + B \rightarrow C$, 379, 454, 462
$A + X \rightleftharpoons 2X$, 546
$A + X \rightarrow 2X$, 478, 546
$A \rightleftharpoons B$, 427, 445, 520, 542, 550
$A \rightarrow B$, 442, 460
$A + BC \rightarrow AB + C$, 406
$A + X \rightleftharpoons 2X$, 587
$A + X \rightarrow 2X$, 586, 587
$A \rightleftharpoons B$, 604
bimolecular, 352, 435
complex, 373
conversion, *see* reaction, isomerization
coordinate, 392
cross-section, 399
Dushman–Roebuck, 556
extent of, 356
irreversible, 353
isomerization, 401, 441, 460, 550
Michaelis–Menten, 566
molecularity of, 350, 435
monomolecular, 352, 401, 435, 441, 461, 542
order, 435
propensity, 341
pseudo-first order, 435, 457, 476, 495
reversible, 354
termolecular, 352, 403
vector, 385
zero-molecular, 352, 435
reaction probability, differential, 419, 425
reaction rate, deterministic, *see* function, rate
reaction rate, stochastic, *see* reaction probability, differential
reaction system, 376
real time, *see* time real
regression
analysis, 407
generalized least squares, 411
least squares, 409

linear, 408
ordinary least squares, 410
relaxation
chemical, 356, 368, 383
vibrational, 350
residual, 408
reversibility
strong, 378
weak, 378, 385
Riemann sum, *see* sum, Darboux

sample
point, 49
space, 49
sample path, *see* trajectory
sample point, 16
sampling
inverse transform, 536
scale parameter, 163, 294
scaling, 522
Scatchard plot, 412
scattering
reactive, 390
selection
natural, 582, 659
random, 601, 605, 611, 661
self-information, 95
semiclassical calculations, 390
semimartingale, 33, 325
sequence
random, 13
set
Borel, 46, 51
Cantor, 51, 55
countable, 19
dense, 45
disjoint, 19
empty, 17
null, 50
uncountable, 19
Vitali, 49, 55
shape parameter, 163
singleton, 37
skewness, 91
skewness parameter, 163, 294
slowing down
critical, 637
space
concentration, 372
genotype, 206
measure, 50
phase, 204, 235
probabiity, 30
sample, 16
state, 17, 261
spectrum, 220
stability
shape, 161
strict, 161
stability parameter, 163, 290, 294
standard deviation, 87
sample, 169
stationarity
quasi-stationarity, 363, 500, 586, 612, 620
second order, 216
strong, 215
weak, 215
statistic
sufficient, 186
statistics
Bayesian, 14, 190
inferential, 140
step
elementary, 350
rate determining, 368
string
empty, 23
subdiffusion, 297, 302
submartingale, 213
subset, 17
subspace
stoichiometric, 374
subspecies, 570, 583
sum
Darboux, 61
superdiffusion, 302
supermartingale, 213
symmetric difference (set theory), 19
system
closed, 387, 442, 445
isolated, 100, 442
open, 352, 387, 436, 575

tail, heavy, *see* heavy tail
tau-leaping, 434, 531
telescopic sum, 58, 602
test statistic, 175
theorem
Cantor, 22
central limit, 36, 72, 115, 131, 159, 163, 297
compound probabilities, 39
convolution, 220
de Moivre-Laplace, 126
deficiency one, 387
deficiency zero, 386, 426

factorization, 186
final value, 473
fluctuation–dissipation, 501
initial value, 473
mutliplication, 85
Perron–Frobenius, 641
Wiener–Khinchin, 223
theory
ergodic, 224
large sample, 133, 136
Maxwell-Boltzmann, 399
transition state, 389, 391
time
arrival, 188, 258
computational, 303
extinction, 270, 311, 600
first passage, 305, 488, 596, 601
mean waiting, 258, 283
real, 217, 303
sequential extinction, 601
waiting, 258, 277, 279
time homogeneity, 262
time series, 220
trajectory, 204
trajectory sampling, 205, 416, 542
transform
Fourier, 106, 220, 245, 282
inverse Laplace, 105, 460, 463
Laplace, 105, 221, 282, 459, 477, 596
transition
step-down, 266, 422
step-up, 266, 422
transition probability, 232
transition state, 391
translation, 54
trimolecular, *see* termolecular
triple
measure, 50
Turing pattern, 559

uncertainty
deterministic, 6
quantum mechanical, 5
uncorrelatedness, 121
unimolecular, *see* monomolecular
union (set theory), 18
universality exponent, 283
universality exponents, 282

variable
continuous, 71
discrete, 71
random, 30
stochastic, 27
variables
explanatory, 409
extensive, 514
intensive, 514
separation of, 608
variables, separation of, 660
variance, 28, 76, 87
sample, 169
variant, *see* subspecies
vector
drift, 231
random, 120
Verhulst equation, *see* equation, logistic
volume
generalized, 53

waiting time, *see* time, waiting

Y-chromosomal Adam, 674

MIX
Papier aus verantwortungsvollen Quellen
Paper from responsible sources
FSC® C105338

If you have any concerns about our products,
you can contact us on
ProductSafety@springernature.com

In case Publisher is established outside the EU,
the EU authorized representative is:
Springer Nature Customer Service Center GmbH
Europaplatz 3, 69115 Heidelberg, Germany

Printed by Libri Plureos GmbH
in Hamburg, Germany